T0190169

Preface

CCIS 1205 comprises the post proceedings of the 11th International Symposium on Intelligence Computation and Applications (ISICA 2019) held in Guangzhou, China, November 16–17, 2019. This volume features the most up-to-date research in evolutionary algorithms, parallel and quantum computing, evolutionary multi-objective and dynamic optimization, intelligent multimedia systems, virtualization and AI applications, smart scheduling, intelligent control, big data and cloud computing, deep learning, and hybrid machine learning systems.

CCIS 1205 is dedicated in memory of Lishan Kang on the 10th anniversary of his death. Prof. Kang was the founder of ISICA, who organized the first ISICA in 2005. Besides his research book on evolutionary computation, *Non-Numerical Algorithms: (II) Genetic Algorithms* published by China Science Press in 1995, Prof. Kang gave hundreds of public talks and lectures on both domain decomposition methods and evolutionary computation at many universities in China starting in the 1980s. In the late 1980s, Prof. Kang foresaw that evolutionary computation was the foundation of computational intelligence while computational intelligence was the future of computational science. Nowadays thousands of students and researchers in China are following in his footsteps. Evolutionary computation will bring us to creative evolution beyond deep learning from the available big data and powerful hardware.

On behalf of the Organizing Committee, we would like to warmly thank the sponsors: South China Agricultural University, Jiangxi University of Science and Technology, Intelligent Simulation Optimization and Scheduling Committee of China Simulation Federation, and Computing Intelligence of Guangdong Computer Academy, who helped in one way or another to achieve our goals for the conference. We wish to express our appreciation to Springer for publishing the proceedings of ISICA 2019. We also wish to acknowledge the dedication and commitment of both the staff at the Springer Beijing Office and the CCIS editorial staff. We would like to thank the authors for submitting their work, as well as the Program Committee members and reviewers for their enthusiasm, time, and expertise. The invaluable help of active members from the Organizing Committee, including Lixia Zhang, Lei Yang, Yan Chen, Hui Wang, Zhiping Tan, Ying Feng, Dunmin Chen, Yaohua Liu, Wenbiao Chen, Xiangzheng Fu, Qiong Liu, Daisy Kansal, Jalil Hassan, and Nwokedi Kingsley Obumneme, in setting up and maintaining the online submission systems by EasyChair, assigning the papers to the reviewers, and preparing the camera-ready version of the proceedings is highly appreciated. We would like to thank them personally for their help in making ISICA 2019 a success.

March 2020

Kangshun Li
Wei Li
Hui Wang
Yong Liu

Organization

Honorary Chairs

Kay Chen Tan	City University of Hong Kong, China
Qingfu Zhang	City University of Hong Kong, China
Ling Wang	Tsinghua University, China

General Chairs

Kangshun Li	South China Agricultural University, China
Zhangxing Chen	University of Calgary, Canada
Zhijian Wu	Wuhan University, China

Program Chairs

Yiu-ming Cheung	Hong Kong Baptist University, China
Jing Liu	Xidian University, China
Hailin Liu	Guangdong University of Technology, China
Yong Liu	University of Aizu, Japan

Local Arrangement Chair

Zhiping Tan	South China Agricultural University, China

Publicity Chairs

Lixia Zhang	South China Agricultural University, China
Yan Chen	South China Agricultural University, China
Lei Yang	South China Agricultural University, China

Program Committee

Ehsan Aliabadian	University of Calgary, Canada
Rafael Almeida	University of Calgary, Canada
Ehsan Amirian	University of Calgary, Canada
Zhangxing Chen	University of Calgary, Canada
Iyogun Christopher	University of Calgary, Canada
Lixin Ding	Wuhan University, China
Xin Du	Fujian Normal University, China
Zhun Fan	Shantou University, China
Zhaolu Guo	Jiangxi University of Science and Technology, China
Guoliang He	Wuhan University, China

Contents

Virtualization and AI Applications

Intelligent Control

Big Data and Cloud Computing

New Frontier in Evolutionary Algorithms

Citrus Disease and Pest Recognition Algorithm Based on Migration Learning

Kangshun Li[1(✉)], Miaopeng Chen[1], Juchuang Lin[1], and Shanni Li[2]

[1] College of Mathematics and Informatics, South China Agricultural University, Guangzhou 510642, China
likangshun@sina.com

[2] Digital Grid Research Institute, China Southern Power Grid, Guangzhou 510660, China

Abstract. Citrus is the largest fruit production in the world. Owing to the damage by various pest diseases, the production of citrus is reduced and the quality is getting worse and worse every year. The recognition and control of the citrus diseases are very important. By now the main measures we take to control them is sowing pesticides, which is not good for the environment and do harm to the soil greatly. The technology of image identification can recognize what kind of citrus disease they have with high efficiency and low cost, which is also environmentally friendly and is not limited by time and space. It is our top priority to apply it to recognize and prevent the disease from citrus. In order to detect citrus pest disease and control them automatically, we studied the pests and traits of citrus leaves and their multi-fractal characteristics and methods for figuring pests and diseases, and created a model for detecting leaf images of citrus. We use Keras and Tensorflow to build the model. To reduce recognition loss and improve accuracy, we put the citrus photos into the model and train it persistently. After examining, the recognition accuracy of citrus greening disease of 120 images can reach 96%. The experimental result shows that the model can recognize citrus diseases with high accuracy and robustness.

Keywords: Image recognition · Machine learning · Deep learning · Convolutional neural network · Migration learning · VGG16 model

1 Introduction

There are many precedents of plant disease recognition at home and abroad. In 2017, Zhao et al. [1] adapted the Otus threshold segmentation algorithm to extract 4 kinds of diseased potato leaves images and extract underlying visual feature vectors. She used the SVMc classifier, and the recognition rate is 92%. In 2018, Shi JiHong [2] tried to combine traditional database with service of WeChat public platform, focused on agricultural disease and pest recognition and realize image database construction based on WeChat public account which provided the users a convenient query, identify and disease prediction platform. Sharada et al. [3] trained a deep convolutional network to detect 26 kinds of diseases of 14 kinds of plants, its classification accuracy reached 99.35% in 54306 training disease photos. And it highlights the importance of deep learning and convolutional network.

K. Li et al. (Eds.): ISICA 2019, CCIS 1205, pp. 3–20, 2020.
https://doi.org/10.1007/978-981-15-5577-0_1

Rastogi A et al. [4] proposed a universal system for leaf disease identification, the first stage is based on feature extracting and artificial neural network recognition, the second stage is based on Kemans segmentation and ANN disease classification.

All the above researches show the hot topic of plant disease recognition based on computer vision combined with the popular interconnecting devices. But the recognition with high accuracy and low response time is the guarantee for the promotion of the identification technology. This paper researches on the recognition algorithms based on migration learning, and finds its great advantage in the recognition of citrus disease and pest.

2 Deep Learning and Migration Algorithm

2.1 Machine Learning

Machine learning specializes in how computer simulates or realizes human learning behavior to acquire new knowledge or skills, and how to reorganize knowledge structures and continuously develop its performance. It is central to artificial intelligence and the base to make computers intelligent. Machine learning mainly refers to that computer acquires knowledge from experience(data) and we could deem it as figuring out patterns and then learned from it. And machine learning is also called pattern recognition [5].

Automatically learning from data instead of following certain rules is a data analysis method or technique, and experience-based learning is the focus of machine learning. Machine learning is not programmed to perform a task, but programmed to learn to perform a task [6]. Machine learning can be divided into supervised learning, semi-supervised learning, and unsupervised learning in the light of to what extend it is manually intervened. The division basis of supervised learning and unsupervised one is whether their input data needs labels. The algorithms of supervised learning mainly contain classification and regression, while unsupervised learning's is clustering. Artificial neural network abstracts the neural networks in human brain into certain models according to different linking mode from an informatic processing standpoint. And it is a computing paradigm which is composed of a large number of nodes, also known as artificial neural units, connecting to each other. The model may contain several layers, which could be grouped into input layer, hidden layer and output layer. Each layer could have lots of neural units in it. The neural unit is depicted in Fig. 1.

The regular present of a neural unit is shown in Fig. 1. Each neural unit in the network has its input and output. The unit gathers its input from other neural units in the preceding layer to form the weighted output.

Z is the weighed input of neural unit i (n is the total number of the unit's inputs; w_n represents the bias of the unit, and its weight is 1):

$$Z = \sum_{i=1}^{n} x_i \cdot w_i \tag{1}$$

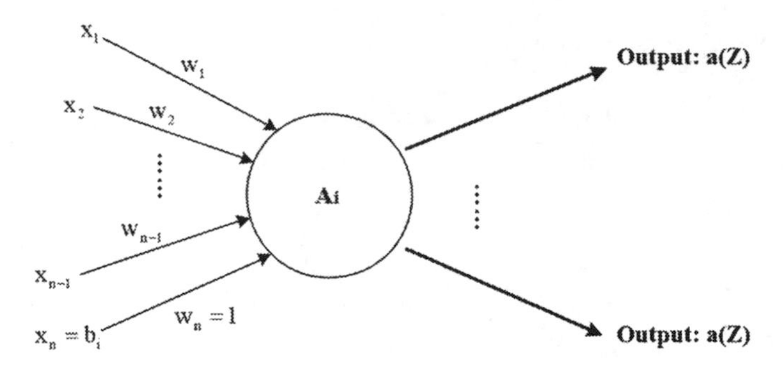

Fig. 1. Artificial neural unit

Then uses the weighted input Z as the input of its activation function, and finally outputs the result to other units.

A_i is the output of the neural unit i (a is the activation function of unit i):

$$A_i = a(Z) \tag{2}$$

2.2 Deep Learning

2.2.1 Deep Learning Network Structure

Machine learning could be divided into swallow learning and deep learning. Different from swallow learning which has only one hidden layer, deep learning has a lot of hidden layers. The overall structure of neural network is shown in Fig. 2.

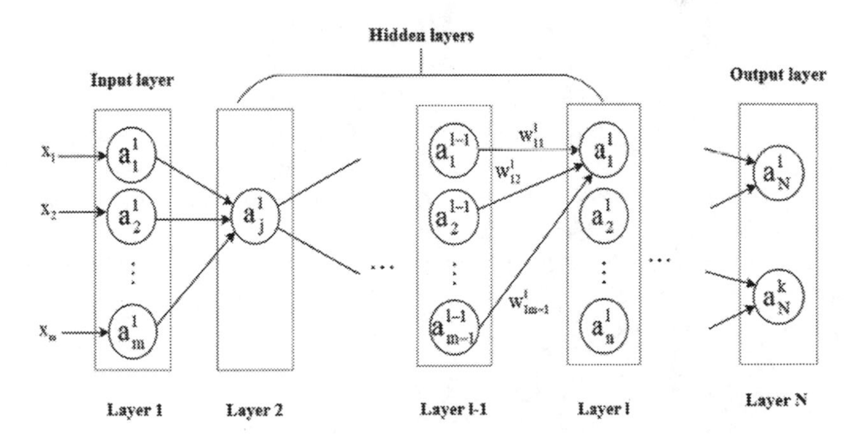

Fig. 2. Sequential neural network structure

The symbols definition of the sequential neural network is shown in Table 1.

Table 1. Sequential neural network symbols

Neural unit symbols	Symbol meanings
a_i^l	The output of unit i in layer l, if it represents the output of input layer, it can also be replaced by x_i
w_{ij}^l	The weight of the input from unit j in layer l-1 to unit i in layer l
b_i^l	The bias of the weighted input of unit i in layer l
z_i^l	The weighted input of unit i in layer l

2.2.2 The Forward Propagation Process

The units in the input layer get the inputs and use them as their output, the activation function of the units in the input layer can be described as a(x) = x, and the output of the first layer in the model(input layer, l = 1) can be depicted as, M is the total number of units in input layer:

$$a_i^1 = x_i (i = 1, 2, \ldots, M) \tag{3}$$

Each unit in the hidden layers receives other units' outputs in the preceding layer as its input if it's fully connected. And it uses the weighed input as the input of its activation function and gets the output.

The output of unit i in layer l, l > 1(m − 1 is the total number of units in layer l-1; a() is the activation function of the unit):

$$z_i^l = \left(\sum\nolimits_{j=1}^m a_j^{l-1} \cdot w_{ij}^l\right) + b_i^l \tag{4}$$

$$a_i^l = a\left(z_i^l\right) \tag{5}$$

Finally we get the outputs $a_1^N, a_2^N, \ldots, a_k^N$, assume there are k units in the output layer, we gather them as model's result. And the predict result is usually evaluated by the loss functions, some of which are listed below. $\hat{y}$ represents the output of the model, y represents the input data's label, n is the total number of the input data. Common loss functions are shown as below:

squared error loss function

$$L(\hat{y}, y) = \frac{1}{2}\sum\nolimits_{i=1}^n (y - \hat{y})^2 \tag{6}$$

cross entropy loss function

$$L(\hat{y}, y) = \sum\nolimits_{i=1}^n -y \cdot \log\hat{y} - (1 - y) \cdot \log(1 - \hat{y}) \tag{7}$$

Since the weighted input procedure presents the linear function and activation functions are usually nonlinear. Combined with linear and nonlinear function, the deep learning network could stimulate a large amount of transform in the world theoretically. The network itself is usually the approach of some kind of algorithm or function in nature, or some expression of strategy [7].

2.2.3 The Back Propagation Process

The model relies on the gradient decline algorithm, shown in formula (8) below, which is the mathematical basis of supervised learning model. The gradient descent method is a typical method which calculate the minimum values of the target function by slowly moving the point in the define domain to explore rather than finding the solution of the equation with partial derivatives equal to 0 [8]. w and b are weights and bias of the model, η is a small positive number which we call learning rate, Loss presents the loss function of the model, later we use L to represent it instead. w^* *and* b^* are the updated values of w and b.

$$(\Delta w, \Delta b) = -\eta\left(\frac{\partial Loss}{\partial w}, \frac{\partial Loss}{\partial b}\right)$$

$$(w^*, b^*) = (w + \Delta w, b + \Delta b) \tag{8}$$

We define neural unit error of unit i in layer l (l > 1) as below:

$$\delta_i^l = \frac{\partial L}{\partial z_i^l} \tag{9}$$

Neural unit error has relations with the gradient of weights and bias below. As long as we get the value of the neural unit errors of the layers from hidden layer to output layer, we could get the gradients of each parameter of the model easily.

$$\frac{\partial L}{\partial w_{ij}^l} = \delta_j^l \cdot a_i^{l-1}\left(\frac{\partial C}{\partial z_j^l} = \delta_j^l, \frac{\partial z_j^l}{\partial w_{ji}^l} = a_i^{l-1}\right) \tag{10}$$

$$\frac{\partial C}{\partial b_j^l} = \delta_j^l\left(\frac{\partial C}{\partial z_j^l} = \delta_j^l, \frac{\partial z_j^l}{\partial b_j^l} = 1\right) \tag{11}$$

The neural unit error in output layer, shown as below (N represents the output layer, a() is the activation function of unit i in the output layer, there are k units in the output layer):

$$\delta_i^N = \frac{\partial L}{\partial z_i^N} = \frac{\partial L}{\partial a_i^N} \cdot a'\left(z_i^N\right)(i = 1, 2, \ldots, k) \tag{12}$$

And the neural unit errors in layer l and in layer l-1 has relations, shown as follows (m, n are the total number of units in layer l-1 and layer l; a() is the activation function of unit j in layer l-1):

$$\delta_j^{l-1} = \sum\nolimits_{i=1}^{n} \frac{\partial C}{\partial z_i^l} \cdot \frac{\partial z_i^l}{\partial a_j^{l-1}} \cdot \frac{\partial a_j^{l-1}}{\partial z_j^{l-1}} = \sum\nolimits_{i=1}^{n} \delta_i^l \cdot w_{ij}^l \cdot a'\left(z_j^{l-1}\right)(j = 1, 2, \ldots, m) \quad (13)$$

The back propagation algorithm of sequential network model:

1) Get the output of the model, count the neural unit error in the output layer according to formula (12).
2) If the preceding layer is not the input layer, use formula (10), (11) to count the gradients of the weights and bias of this layer, turn to step 3; else turn to step 4.
3) Use formula (13) to count the neural unit errors in the preceding layer, turn to step 2.
4) Get all the gradients of the parameters in the model, then update the parameters according to formula (8).

2.3 Migration Learning

Migration learning is a sort of machine learning which refers to adapting a pretrained model to another recognition task. The migration learning refers to the migration from the original task and data to the target task and data, using the weight parameters in the original data domain to improve the predictive function of the target task [9]. It can efficiently reduce the over-fitting degree of the normal convolution neural network.

3 Citrus Pest and Diseases Identification Based on Deep Learning and Migration Learning

3.1 Problem Description

At present, the fruit plantation area reaches 1130 thousand hectares, among which citrus's occupy 266 thousand, in Guangdong province, China. And citrus is the main type of fruit in Guangdong. Due to the numerous citrus diseases, the planting area of citrus in Guangdong reduces by more than 30% and the diseases causes the direct economic loss of about 4 billion yuan each year. Therefore the recognition of citrus diseases is of great importance. In the past, people used the convolutional neural network (CNN) model to distinguish the disease citrus from healthy one. The convolution network can better solve the problem that it's hard to find an appropriate feature to train because of the citrus leaves' great similarity, and we don't have to choose features manually for the training, CNN is capable of learning from the original 2D photo. And It can extract new features of the input photos as well as renewing the features it learned persistently. But CNN also has drawbacks like overfitting.

This paper focus on the sick recognition of citrus, including citrus greening disease, Citrus canker disease and Citrus Anthracnose. We try to build a recognition model with the CNN and later we build the model based on migration learning methods.

3.2 The Structure of Convolutional Neural Network

Convolutional neural network is a kind of feed forward neural network. CNN's network structure which is local-sensitive and weight-sharing makes it more like a biological neural network, and it decreases the number of neural units' weights, and largely reduces the complexity of the model.

CNN can be divided into convolutional layer, sampling layer, flatten layer, and fully connected layer. The structure is shown in Fig. 6. The first three layers form the basic unit for CNN.

3.2.1 Convolutional Layer

Convolutional layer is used to extract the 2-dimension feature of the photo, using different kernels to detect different edge of the interest region. C is short for convolutional layer, which is used to extract features from the picture. Different from other neural network, CNN uses a matrix to store the values of the weights, which is defined as the convolutional kernel. When the layer gets the input photo data, it will use the kernel to scan the photo from the up-left side to the right-bottom side, and each step of the convolution will generate a new pixel in the output feature map according to the linear transformation between the kernel and the region of photo it cover. The schematic diagram of convolution is shown in Fig. 3.

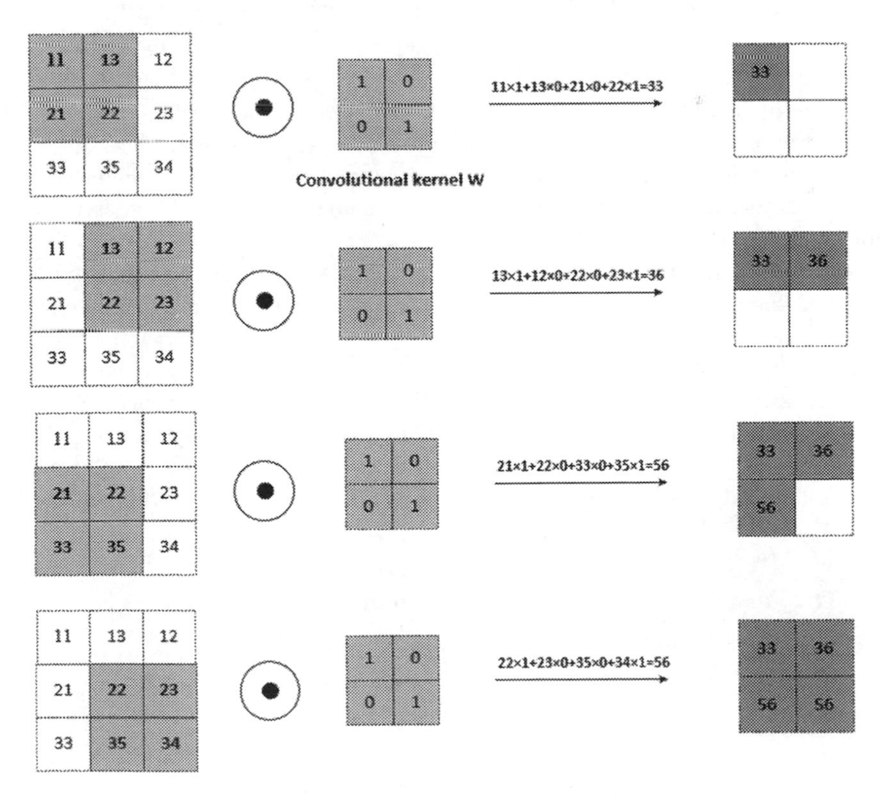

Fig. 3. Schematic diagram of convolution with kernel size (2 × 2), input size (3 × 3)

3.2.2 Sampling Layer

Sampling layer mainly aims at reducing the size of the feature map and extract the primary features. Through the process of sampling, the number of parameters drops. Sampling layer execute the pooling function which select the max, min, average value of the region it covers as its output. The sampling filter does not store any weights like the convolutional kernel. The schematic diagram of max-pooling is depicted in Fig. 4.

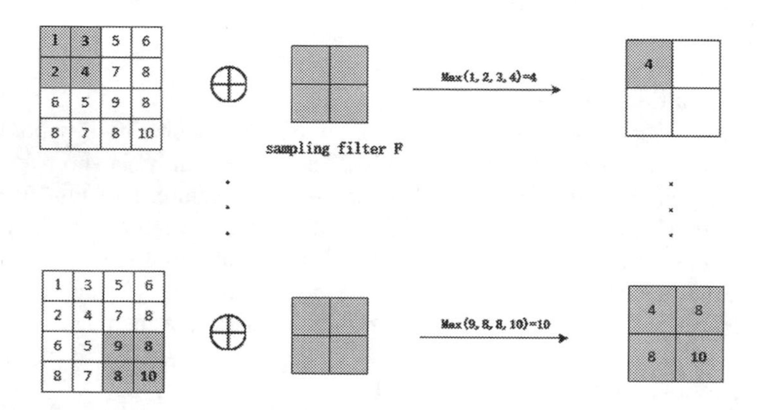

Fig. 4. Schematic diagram of max-pooling with filter size (2 × 2), input size (4 × 4)

3.2.3 Dropout Layer and Flatten Layer

The usage of dropout layer is to disable some neural units randomly, preventing the model from overfitting and gradient vanishing. CNN use the dropout function to pick neural units in the hidden layer randomly and disable them during the training process. Because the disabled units could not transmit the signal forward, it averts the overfitting problem effectively [10].

Flatten layer is designed to flatten the output result of sampling layer, it's usually connected between the parts of feature extraction and pattern recognition in the model.

3.2.4 Fully Connected Layer

Fully connected means each neural unit in the layer are fully connected to the preceding layers' units. Convolutional layer, activation layer and sampling layer are the layers to extract features, and the duty of fully connected layer is to integrate the features and ready for classification and identification procedure [11]. It is just like the neural network in Fig. 2.

3.3 The Forward Propagation of Convolution Neural Network

3.3.1 Symbols Definition of the Convolutional Neural Network

CNN's symbols definition is shown in Table 2.

Table 2. Convolutional neural network symbols

CNN layers	Layer's symbol	Symbol meaning
Input layer	x_{ij}	It represents the input in row i, column j
	a_{ij}^{I}	It represents the output of the input layer in row i, column j
Filter of CNN	w_{ij}^{Fk}	It represents the weight in row i, column j in the filter in CNN's sublayer k
Convolutional layer	z_{ij}^{Fk}	It represents the weighted input of the neural unit in row i, column j in CNN's sublayer k
	b^{Fk}	It represents the bias of the neural unit in CNN's sublayer k
	a_{ij}^{Fk}	It represents the output of the neural unit in row i, column j in CNN's sublayer k
Sampling layer	z_{ij}^{Pk}	It represents the input of the sampling layer in row i, column j in CNN's sublayer k
	a_{ij}^{Pk}	It represents the output of the sampling layer in row i, column j in CNN's sublayer k
Flatten layer	a_{i}^{F}	It represents the output of the flatten layer of unit i
Output layer	w_{ij}^{O}	It represents the weight in output layer which is from neural unit j in the flatten layer to unit i in the output layer
	z_{n}^{O}	It represents the weighted input of the nth neural unit in the output layer
	b_{n}^{O}	It represents the bias of the unit n in the output layer
	a_{n}^{O}	It represents the output of the unit n in the output layer

3.3.2 The Input Process of the CNN

The input process of CNN is shown in Fig. 5, and the relation of input layer is depicted below, in formula (15) (we use the input size of (5, 5) as an example):

$$a_{ij}^{I} = x_{ij} \tag{15}$$

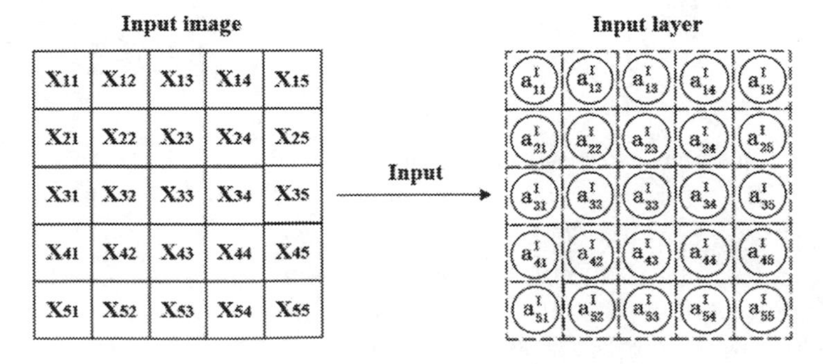

Fig. 5. Input process of the convolutional neural network

3.3.3 The Convolution Process of CNN

The convolution process of CNN is shown in Fig. 6, and the relations in convolutional layer are depicted below, assuming that the CNN has 3 sublayers (W_{in}, H_{in} are the input size of the convolutional layer, depicted as W_i, H_i below; W_{out}, H_{out} are the output size of the convolutional layer, depicted as W_o, H_o below; W_f, H_f are the kernel's size, P_w, P_h are the padding number of the output feature map, $stride_w, stride_h$ are the strides of the kernel in the horizontal direction and vertical direction; $a_f()$ represents the activation of the convolutional layer):

$$W_{out} = \frac{W_{in} - F_w + 2 * P_w}{Stride_w} + 1 \tag{16}$$

$$H_{out} = \frac{H_{in} - H_w + 2 * P_h}{Stride_h} + 1 \tag{17}$$

$$c_{ij}^{Fk} = \sum_{k=1}^{W_f} \sum_{l=1}^{H_f} w_{kl}^{Fk} * a_{i+k-1j+l-1}^{I} (i = 1, 2, \ldots, W_o; j = 1, 2, \ldots, H_o) \tag{18}$$

$$z_{ij}^{Fk} = c_{ij}^{Fk} + b^{Fk} (i = 1, 2, \ldots, W_o; j = 1, 2, \ldots, H_o) \tag{19}$$

$$a_{ij}^{Fk} = a_f\left(z_{ij}^{Fk}\right) \tag{20}$$

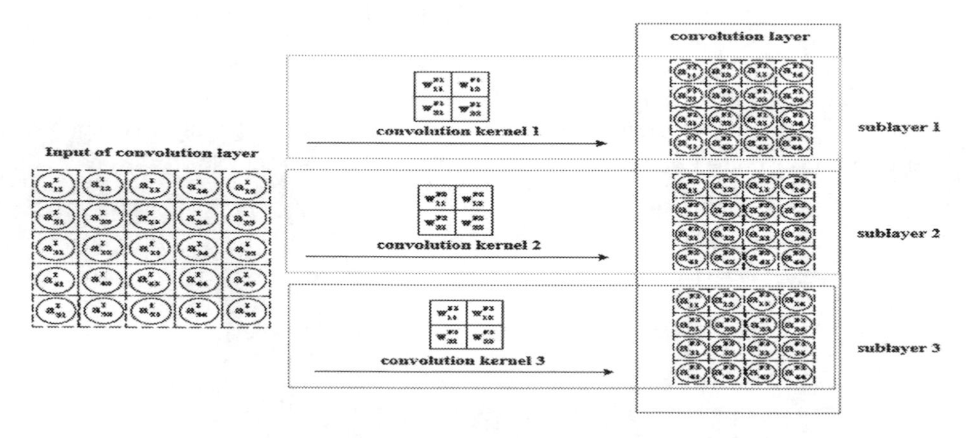

Fig. 6. Convolution process of the convolutional neural network

3.3.4 The Sampling Process of CNN

The max-pooling process of CNN is shown in Fig. 7, and the relations in sampling layer are depicted below (W_{in}, H_{in} are the input size of the sampling layer, depicted as W_i, H_i below; W_p, H_p are the size of the sampling filter; W_{out}, H_{out} are the output size of the sampling layer, depicted as W_o, H_o below):

$$W_{out} = \frac{W_{in}}{W_p}, H_{out} = \frac{W_{out}}{H_p} \tag{21}$$

$$z_{ij}^{Pk} = Max\left(a_{kl}^{Fk}\right)(i = 1, 2, \ldots W_o; j = 1, 2, \ldots, H_o; k = 1, 2, \ldots, W_p; l = 1, 2, \ldots, H_p) \tag{22}$$

$$a_{ij}^{Pk} = z_{ij}^{Pk} \tag{23}$$

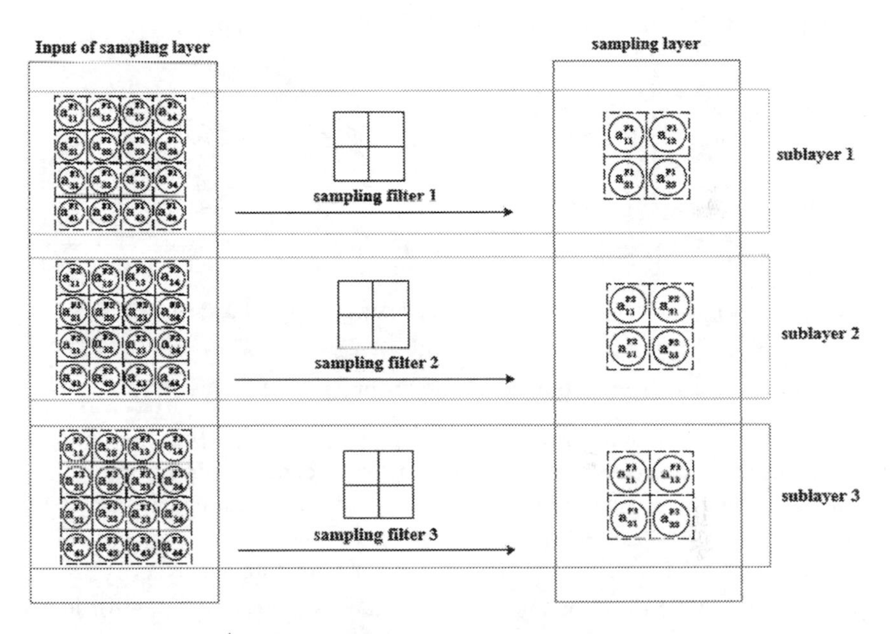

Fig. 7. Sampling process of the convolutional neural network

3.3.5 The Flatten Process and Output Process of CNN

The flatten and output process of CNN is shown in Fig. 8, and the relations in flatten layer and output layer are depicted below (W_{in}, H_{in} are the input size of the flatten layer, assume that M = $W_{in} \cdot H_{in}$ and there are K sublayers in the model, and there are N = M $\cdot$ K units in the flatten layer; there are O units in the output layer; a_o is the activation function of the output layer):

$$a_i^F = a_{jk}^{Pl}\left(i = 1, 2, \ldots, N; l = \frac{i}{M} + 1; j = \frac{i\%M}{W_{in}} + 1; k = i\%W_{in}\right) \tag{24}$$

$$z_i^O = \left(\sum\nolimits_{j=1}^{n} w_{ij}^O \cdot a_j^F\right) + b_i^O (i = 1, 2, \ldots, O) \tag{25}$$

$$a_i^O = a_o\left(z_i^O\right) \tag{26}$$

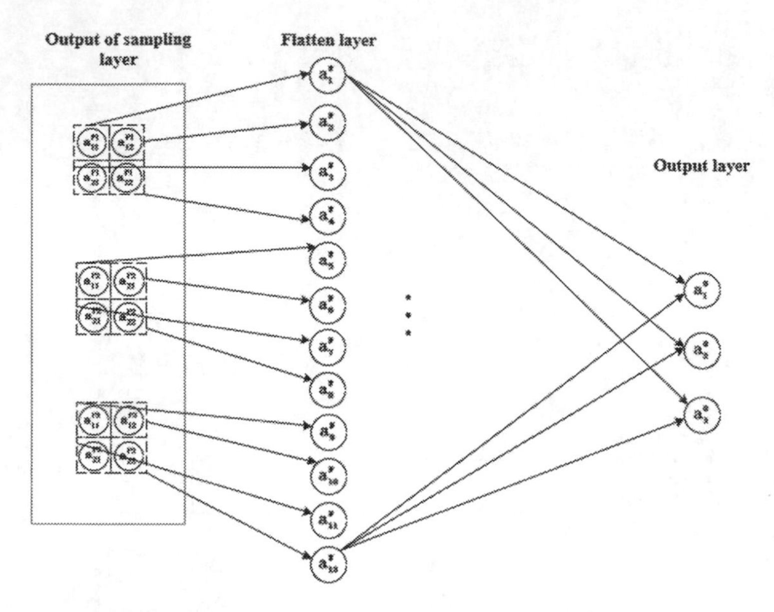

Fig. 8. Output process of the convolutional neural network

3.4 The Backward Propagation of Convolution Neural Network

The backward propagation algorithm of CNN is similar to other deep learning model, but the major difference is that convolutional layer's weights are in the convolutional kernels. The flatten layer and the subsequent layers backward propagation process is the same as deep learning model.

Define neural unit errors in the output layers and convolutional layers as below (δ_n^O means the neural unit error of output unit n; δ_{ij}^{Fk} means the neural unit error in row i, column j of convolutional layer in sublayer k):

$$\delta_n^O = \frac{\partial L}{\partial z_n^O} \tag{27}$$

$$\delta_{ij}^{Fk} = \frac{\partial L}{\partial z_{ij}^{Fk}} \tag{28}$$

Through the neural unit errors, it is easy to get the gradient of each parameter of the model. The relation between gradients and neural unit errors are shown below (W_f, H_f are the kernel's size; W_o, H_o are the output size of the convolutional layer):

$$\frac{\partial L}{\partial w_{ij}^O} = \delta_n^O \cdot a_j^F, \frac{\partial L}{\partial b_i^O} = \delta_n^O \tag{29}$$

$$\frac{\partial L}{\partial w_{ij}^{Fk}} = \sum_{k=1}^{w_o} \sum_{l=1}^{h_o} \delta_{kl}^{Fk} \cdot a_{i+k-1j+l-1}^{Fk} \left(i = 1, 2, \ldots, w_f; j = 1, 2, \ldots, h_f\right) \quad (30)$$

$$\frac{\partial C}{\partial b^{Fk}} = \sum_{k=1}^{w_o} \sum_{l=1}^{h_o} \delta_{kl}^{Fk} \quad (31)$$

Then we try to figure out the expressions of neural unit errors in output layer and convolutional layers:

$$\delta_n^O = \frac{\partial L}{\partial a_n^O} \cdot \frac{\partial a_n^O}{\partial z_n^O} = \frac{\partial L}{\partial a_n^O} \cdot a_o^{'}\left(z_n^O\right) \quad (32)$$

$$\delta_{ij}^{Fk} = \left\{\sum_{i=1}^{O} \delta_i^O \cdot w_{ik}^O\right\} \cdot (v) \cdot a_{Fk}^{'}\left(z_{ij}^{Fk}\right) \quad (33)$$

v values 1 or 0 depend on whether a_{ij}^{Fk} is the largest number among the region of the filter of the convolutional layer F_k. Then we get all the neural unit errors and we could calculate all the gradients of the parameters in the convolutional network and update the parameters according to gradient decline algorithm.

3.5 Data Preprocessing

In order to reduce overfitting phenomenon, we extend the dataset by flipping the original photos horizontally and vertically and scaling randomly [12], the training photos are shown in Fig. 9. Then we set the plant images to (224, 224), and we divide the dataset according to the ratio of 5:1 into the training set and test set.

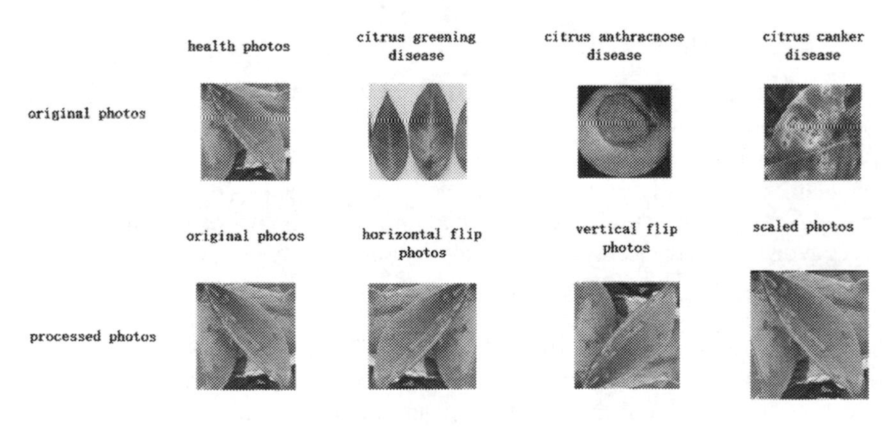

Fig. 9. Data preprocessing

3.6 Convolutional Network Model

The preliminary experiment is to build a convolutional network with 4 convolutional layers. Its structure is depicted in Table 3.

Table 3. Convolutional neural network structure

Layer (type)	Output shape	Parameter count
Convolution_1 (convolution)	(224, 224, 16)	448
Max_pooling_1 (Max pooling)	(112, 112, 16)	0
Convolution_2 (convolution)	(112, 112, 32)	4640
Max_pooling_2 (Max pooling)	(56, 56, 32)	0
Convolution_3 (convolution)	(56, 56, 64)	18496
Max_pooling_3 (Max pooling)	(28, 28, 64)	0
Convolution_4 (convolution)	(28, 28, 128)	73856
Max_pooling_4 (Max pooling)	(14, 14, 128)	0
Flatten_1 (Flatten)	(25088)	0
Dense_1 (Dense)	(512)	12845568
Dropout_1 (Dropout)	(512)	0
Dense_2 (Dense)	(11)	5643
Dropout_2 (Dropout)	(11)	0
Dense_3 (Dense)	(2)	24

We test the model in different situations, such as different convolutional kernel size and different learning rate. And the test results are shown in title 4.

3.7 Migration Learning Model

The preliminary experiment shows that the CNN has high accuracy rate on test dataset, but it can easily become overfitting. In order to solve this problem, we decide to alter our model based on the migration learning algorithm. We choose the VGG-16 model as the base of our recognition model. VGG model is a typical CNN with high classification and recognition rate. It increases the depth of the network steadily by adding more convolutional layers. And very small convolutional filters (3 * 3) make it work successfully [13].

3.7.1 The Structure of VGG16

The structure of VGG16 is shown in Fig. 10 below. There are totally 16 weighted layers in the model, 13 convolutional layers and the last 3 fully connected layers. Its basic unit is a convolutional layer followed by a sampling layer. The kernel size of convolutional layers are all 3×3, using relu as its activation function to train the model quickly. And the kernel of sampling layer is 2×2, adopting max polling.

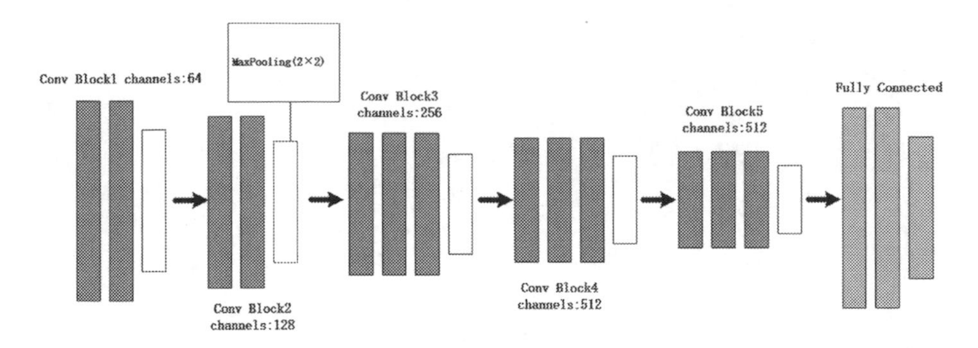

Fig. 10. Structure of VGG16

3.7.2 Fine-Tuning Method

We choose the fine-tuning method to construct our model, shown in Fig. 11. Fine-tuning means adjusting the model which trained by others to train our data. It can be seen as using the front layers of the original models to extract the features of the photos, and the newly add layers to classify [14]. The increase of network layers will not lead to the explosion of the parameters, because the parameters are mainly concentrated in the last three fully connected layers [15]. As in Fig. 16, the layers before the fully connected layers can not be trained, only the last three layers can be trained and its parameters could be updated. Since the parameters in convolutional layers are untrainable, the time for training the model is less and it's with high efficiency.

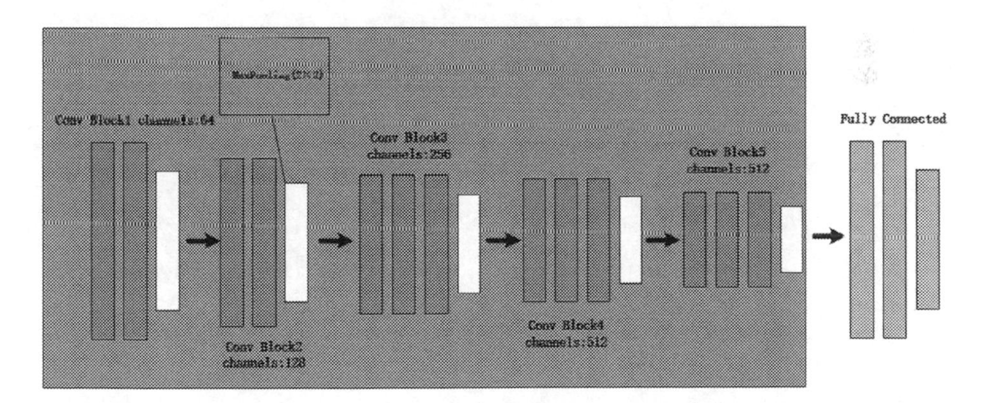

Fig. 11. Structure of VGG16 with fine-tuning

3.7.3 Migration Model Construction

Our migration model is shown in Table 4, we remove the top three FC layers and add a flatten, dropout and 2 dense layers for training.

Table 4. Migration model design

Layer (type)	Output shape	Parameter count
Vgg16 (Model)	(7, 7, 512)	14714688
Flatten_1 (Flatten)	(25088)	0
Dense_1 (Dense)	(1024)	25691136
Dropout_1 (Dropout)	(1024)	0
Dense_2 (Dense)	(2)	2050

4 Experimental Simulation and Analysis

The three citrus diseases are cgd (short for citrus greening disease), cad (short for citrus anthracnose disease) and ccd (short for citrus canker disease), and the experiment result are listed as below.

The test result of the simple convolutional network is shown in Table 5.

Table 5. Test result of simple convolutional network

Dataset	Train number	Train accuracy	Test number	Test accuracy
cgd	250	79%	50	67%
cad	200	78%	40	71%
ccd	180	82%	35	65%

It seems that a simple CNN could have train accuracy of 80% on average and test accuracy of 67%. But the problem of overfitting is still serious.

Then we change the kernel size to find out its influence on the accuracy, using dataset of citrus greening disease as training data. The result shows in Table 6.

Table 6. Test result of simple CNN with different kernel size

Dataset	Train number	Train accuracy	Test number	Test accuracy	Kernel size
cgd	250	83%	50	76%	(3, 3)
cgd	200	79%	40	75%	(5, 5)
cgd	180	78%	35	72%	(7, 7)

The test result shows that convolutional kernel size values (3, 3) gets optimize results, the main reason is that a smaller kernel can extract more local features, and develop the accuracy of citrus disease classification.

Then we alter the learning rate of the model to find out its influence on the accuracy, using dataset of citrus greening disease as dataset. The result is shown in Table 7.

Table 7. Test result of simple CNN with different learning rate

Dataset	Train number	Train accuracy	Test number	Test accuracy	Learning rate
cgd	250	76%	50	62%	0.01
cgd	200	77%	40	63%	0.001
cgd	180	83%	35	64%	0.0001

Through the result we find that the learning rate 0.0001 gets the best result, learning rate's value is important for the construction of the model. If it is too small, the convergence time would be long. If it is too big, it may cause oscillating around the minimal value.

At last we test the accuracy of these diseases on the migration model based on the fine-tuning VGG16. The result is shown in Table 3, it's clear that the recognition model based on migration learning with fine-tuning method has higher accuracy both on the test dataset and the validation dataset than the simple CNN. The test result of citrus disease recognition is shown in Table 8.

Table 8. Test result of fine-tuning convolutional network

Dataset	Train number	Train accuracy	Test number	Test accuracy
cgd	250	97%	50	93%
cad	200	98%	40	95%
ccd	180	98%	35	92%

5 Conclusion

Image recognition is a technology that dependents on deep learning. It can be done through extracting different level of features, from the insignificant characteristics next to the input end to the abstract features like the semantic of the photo near the output end. In our paper, in order to find the better disease recognition model of citrus, we study the algorithm of convolutional network and the model based on migration learning, the result of the experiment shows that a transfer model based on migration learning can ease overfitting, increase recognition accuracy and has a wider range of applications.

Acknowledgements. This work is supported by National Key R&D Program of China with the Grant No. 2018YFC0831100, the National Natural Science Foundation Youth Fund Project of China under Grant No. 61703170, the National Natural Science Foundation of China with the Grant No. 61773296, Foreign Science and Technology Cooperation Program of Guangzhou with the grant No. 201907010021, the Open Foundation of Key Lab of Data Analysis and Processing of Guangdong Province in Sun Yat-sen University with No. 201901, the Major Science and Technology Project in Dongguan with No. 2018215121005. Key R&D Program of Guangdong Province with No. 2019B020219003, Foreign Science and Technology Cooperation Program of Huangpu District of Guangzhou with No. 2018GH09 Technology Cooperation Program of Huangpu District of Guangzhou.

References

1. Zhao, J., Xue, X., Li, Q.: Potato disease recognition system based on machine vision. J. Jiangsu Agric. Sci. **45**(02), 198–202 (2017)
2. Shi, J.: Database design of agricultural pest and disease identification system based on WeChat public account. Hebei Agric. Mach. **11**, 15–16 (2018)
3. Mohanty, S.P., Hughes, D., Salath, M.: Using deep learning for image-based plant disease detection. (3), 42–48 (2015). Digital Epidemiology Lab, EPFL, Switzerland
4. Rastogi, A., Arora, R., Sharma, S.: Leaf disease detection and grading using computer vision technology & fuzzy logic. In: International Conference on Signal Processing and Integrated Networks, pp. 500–505. IEEE (2015)
5. Peng, L.: Deep Learning Practice Computer Vision. Tsinghua University Press, Beijing (2019)
6. Lin, H., Liu, T.T., Li, H., et al.: Prospects for machine learning research and its application in agriculture. Inf. Inst. Chin. Acad. Agric. Sci. **31**(10), 13 (2019)
7. Lin, H., Liu, T.T., Li, H., et al.: Prospects for machine learning research and its application in agriculture. Inf. Inst. Chin. Acad. Agric. Sci. **31**(10), 15 (2019)
8. Wakui, Y., Wakui, S.: Mathematics for Deep Learning. Yang Y.L., trans. Posts & Telecom Press (2019). 84
9. Liu, H.: Study on the automatic detection of tomato diseases and pests based on leaf images. Dalian JiaoTong University, Dalian (2017). 37
10. Qiu, N., Wang, X., Wang, P., et al.: Research on convolutional neural network algorithm combined with transfer learning model. Comput. Eng. Appl. **56**(5), 45 (2020)
11. Ran, D.: Research and implementation of pest detection algorithm and mobile client based on convolutional neural network. Anhui University, Anhui (2018). 10
12. Zuo, Yu., Tao Qian, W., Lian, W.Y.: Research of plant classification methods based on convolutional neural network. Internet of Things Technol. **10**(03), 72–75 (2020)
13. Zeng, G.: Fruit and vegetables classification system using image saliency and convolutional neural network. IEEE Beijing Section, Global Union Academy of Science and Technology, Chongqing Global Union Academy of Science and Technology, Chongqing Geeks Education Technology Co., Ltd. In: Proceedings of 2017 IEEE 3rd Information Technology and Mechatronics Engineering Conference (ITOEC 2017). IEEE Beijing Section, Global Union Academy of Science and Technology, Chongqing Global Union Academy of Science and Technology, Chongqing Geeks Education Technology Co., Ltd. IEEE BEIJING SECTION (2017). 615
14. Liu, H.: Study on the automatic detection of tomato diseases and pests based on leaf images. Dalian JiaoTong University, Dalian (2017). 39
15. Duan, C.: Image classification of fashion-MNIST data set based on VGG network. In: Proceedings of 2019 2nd International Conference on Information Science and Electronic Technology (ISET 2019). International Informatization and Engineering Associations: Computer Science and Electronic Technology International Society (2019). 19

Artificial Bee Colony Based on Adaptive Selection Probability

Songyi Xiao[1,2], Hui Wang[1,2](✉), Minyang Xu[1,2], and Wenjun Wang[3]

[1] Jiangxi Province Key Laboratory of Water Information Cooperative Sensing and Intelligent Processing, Nanchang Institute of Technology, Nanchang 330099, China
huiwang@whu.edu.cn

[2] School of Information Engineering, Nanchang Institute of Technology, Nanchang 330099, China

[3] School of Business Administration, Nanchang Institute of Technology, Nanchang 330099, China

Abstract. Because of the powerful searching ability of artificial bee colony algorithm, it has applications in various fields. However, it still has a drawback on local search ability. Therefore, an adaptive selection probability ABC algorithm (called PABC) is proposed to improve its local search ability. In the multi-strategy search solutions, a probability is assigned to each strategy and the probability is adaptive adjusted to control the choice of strategy. Meanwhile, a modified mean center is introduced to replace the global best solution to guide search. The proposed PABC is proved to have better optimization ability than some other improved ABCs by testing classical 12 functions.

Keywords: Artificial bee colony · Swarm intelligence · Multi-strategy · Adaptive selection probability

1 Introduction

Traditional optimization methods cannot effectively solve more and more complex practical problems. Swarm intelligence algorithm can effectively solve such problems. Genetic algorithm [1], differential evolution algorithm (DE) [2], particle swarm optimization (PSO) [3], and artificial bee colony algorithm (ABC) [4] are some representative swarm intelligence algorithms.

Artificial bee colony algorithm (ABC) was proposed by Karaboga in 2005, which simulated the foraging behaviors of bees. Although ABC has strong searching ability, it still has the inherent defect of swarm intelligence algorithm, that is, it is easy to fall into local optimization. Some scholars have made efforts to solve this problem. To improve the exploitation ability of ABC, [5] integrated the information of the global best solution into the search equation and proposed GABC. In order to allocate more space to better food sources, [6] used the DFS framework. Experiments show that the DFS framework is better than other ABC algorithms in solving quality when applied to ABC.

K. Li et al. (Eds.): ISICA 2019, CCIS 1205, pp. 21–30, 2020.
https://doi.org/10.1007/978-981-15-5577-0_2

The selection of strategy is very important in multi-search strategy integration, so this paper proposes an adaptive selection probability ABC (called PABC) for multi-strategy integration. First, an adaptive selection probability was proposed to control the selection of multiple search strategies. Then, a modified mean center is introduced to replace the current global best solution.

The structure of the work is as follows. The second part introduces the definition of basic ABC. The third part introduces PABC in detail. The fourth part compares PABC with the other five ABCs. Finally, the work is summarized.

2 Artificial Bee Colony

In the algorithm, populations are divided into two groups. The first part is called the employed bee. Employed bee mainly performs global searches. The second part is called the onlooker bee Onlooker bee to gain the employed bee's search experience first, and then search more carefully based on the experience gained. When an employed bee and onlooker bee cannot find a better food source for a while, they are abandoned. And then a scout bee will come up with a new solution to replace it. Next, we will introduce the ABC algorithm in details.

Initially, N solutions are randomly generated in the search space as follows [4]:

$$X_{ij} = low_j + \mu_{ij} * (up_j - low_j) \tag{1}$$

where i = 1, …, N, j = 1, 2, …, D, and D is the dimension size. low_j represents the lower limit of the search space, up_j represents the upper limit of the search space, μ_{ij} is a random number in the range [0, 1].

For each solution X_i in the swarm, a different solution to X_i, X_k, is selected so that the difference between X_i and X_k is searched as a step in the employed bee phase. Finally, a new candidate solution V_i is generated as follows [4]:

$$V_{ij} = X_{ij} + \varphi_{ij} * (X_{ij} - X_{kj}) \tag{2}$$

where $\varphi_{ij} \in [-1, 1]$ is a random number.

The onlooker bees will select some excellent food sources in a probabilistic manner based on the employed bees for the search. The probability is calculated as follows [4]:

$$p_i = \frac{fit_i}{\sum\limits_{i=1}^{N} fit_i} \tag{3}$$

where fit_i is the fitness value of X_i and it is calculated by [4]:

$$fit_i = \begin{cases} 1/(1+f_i) \\ 1+abs(f_i) \end{cases} \tag{4}$$

where f_i is the function value of X_i. It can be seen from Eqs. (3) and (4) that a better solution has a higher probability of selection. So the onlooker bees have a higher probability of choosing a better solution.

For each the employed and onlooker bee search, a candidate solution V_i is created. When the candidate solution V_i is better than the solution X_i, X_i is replaced by the candidate solution V_i.

When a solution cannot be changed in a long time, it is considered to be in trouble. The scout's job is to help the solution out of its current predicament by creating an unconditional alternative to a new solution.

3 Proposed Approach

3.1 Adaptive Selection Probability

Due to the defects of the search strategy, it easily affects the performance of an algorithm. Multiple search strategies can solve this problem to a certain extent. Currently, there are two kinds of multiple search strategies. The one is fixed and the other is random. In [7], five search strategies are fixed, and they are selected in rotation. In [8], a strategy pool is constructed. The switching of search strategies is random and passive. So, the generated candidate solutions are relatively random and cannot accelerate the convergence. Though the fixed form of rotation selection can make up the disadvantages of single strategy, it is easy to lose the advantages of the search strategies in rotation.

Therefore, we propose a probabilistic selection method for multiple search strategies. Kiran et al. [9] also constantly adjust the selection probability of each search strategy. In this paper, the rotation and random selection are directly replaced by the probability selection.

Search strategy 1 (S_1)	p_1
Search strategy 2 (S_2)	p_2

Fig. 1. The search strategy pool of PABC.

Assume that S_1 and S_2 represent two different search solutions. p_1 and p_2 are the corresponding selection probability of S_1 and S_2. Then construct the new strategy pool as shown in Fig. 1. GABC is used as the first search strategy [10]:

$$x_{ij} = x_{ij} + \phi_{ij}(x_{ij} - x_{kj}) + \varphi_{ij}(Gbest_j - x_{ij}) \tag{5}$$

where ϕ_{ij} is chosen at random in [0, 1], and φ_{ij} is the random number between [0, 1.5].

ABC/best/1 as the second search strategy [11]:

$$x_{ij} = Gbest_j + \phi_{ij}(x_{rj} - x_{kj}) \tag{6}$$

where $X_r \neq X_k$ are two different solutions randomly selected.

The selection probability p is related to a search strategy. When a candidate solution generated by a search strategy is better than its parent, the probability corresponding to the search strategy will be increased. The adjustment of the probability is defined by Eq. (7). If the search is performed by S_1 and the new solution outperforms the old one, the corresponding probability p_1 is increased by Δ (an incremental unit). When the probability has been updated, it will be re-balanced as shown in Eq. (8). For multiple search strategies, each strategy has different convergence characteristics. At the beginning, the initial probability is set to 0.5, i.e. $p_1(0) = p_2(0) = 0.5$.

$$\begin{cases} p_1(t+1) = p_1(t) + \Delta \\ p_2(t+1) = p_2(t) + \Delta \end{cases} \tag{7}$$

$$\begin{cases} p_1(t+1) = \dfrac{p_1(t+1)}{p_1(t+1) + p_2(t+1)} \\ p_2(t+1) = \dfrac{p_2(t+1)}{p_1(t+1) + p_2(t+1)} \end{cases} \tag{8}$$

3.2 Modified Mean Center

The global best solution is used in many ABCs to guide searches because of the variety of information it contains. Then, all solutions in the population will follow the search direction of the super solution. This will make the evolution direction of the population deviate from the true global optimal. In [12], a concept of mean center was proposed, in which the mean of all solutions in the population is calculated as the mean center. The global best solution is replaced by the mean center to guide the search. When the population converges slowly, the mean center can guide the population to converge to better solutions. Even if there are super solutions, the evolutionary direction and location distribution of most solutions in the population will not be changed. The mean center is defined as follows [12]:

$$C_j = \frac{1}{SN}(\sum_{i=1}^{SN} x_{ij}) \tag{9}$$

where $i = 1, 2 \ldots SN, j = 1, 2, \ldots, D$, SN is the population size. C_j is the arithmetic mean of all individuals in the j-th dimension.

When some solutions fall into local minima, the mean center cannot guide the search. In [12], the partial mean center IC is constructed to solve this problem. The partial mean center is defined as the mean position of some good solutions. Firstly, the mean fitness values $\bar{f}$ of all solutions is calculated based on their fitness values. Then, the solutions which is better than $\bar{f}$ are selected to form a high-quality group. Finally, the IC is the mean position of the high-quality group.

$$\bar{f} = \frac{1}{SN}\sum_{i=1}^{SN} f_i \tag{10}$$

where f_i is objective function of the i-th solution. $n(1 \leq n \leq SN)$ as the number of solutions for the high quality group. Then, the partial mean IC can be defined by:

$$IC_j = \frac{1}{n}\sum_{i=1}^{n} x_{ij} \tag{11}$$

where n is the amount of superior particles. IC_j represents the j dimension value of IC.

The mean centers C and IC have their own advantages in different situations. The improved center of mean (MC) is obtained by combining the advantages of C and IC [12]:

$$MC = better(C, IC) \tag{12}$$

Then, the MC is used to replace the $Gbest$ as follows.

$$x_{ij} = x_{ij} + \phi_{i,j}(x_{ij} - x_{kj}) + \varphi_{ij}(MC_j - x_{ij}) \tag{13}$$

$$x_{ij} = MC_j + \phi_{ij}(x_{rj} - x_{kj}) \tag{14}$$

The pseudo code of the adaptive selection probability is listed in Algorithm 1.

Algorithm 1: Adaptive selection probability

```
Begin
If rand(0,1)<p1 then
   Generate Vi by Eq. (13);
   If f(Vi) < f(Xi)
      Replaced Xi by Vi;
      p1=p1+Δ;
      p1=p1/(p1+p2);
      p2=1-p1;
      triali = 0;
   End if
   Else
      triali = triali + 1;
   End Else
End if
Else
   Generate Vi by Eq. (14);
   If f(Vi) < f(Xi)
      replaced Xi by Vi;
      p2=p2+Δ;
      p2=p2/(p1+p2);
      p1=1-p2;
      triali = 0;
   End if
   Else
      triali = triali + 1;
   End Else
End Else
End
```

4 Experiments on Benchmark Functions

In the following experiments, we used 12 functions to test some performance of PABC in 30 dimensions. Table 1 shows some descriptions of the 12 basic test functions used to test the optimization [13–15].

The results of PABC and several other ABCs were compared on the $D = 30$ basis. The involved ABCs are as follows:

- Standard ABC;
- ABC with variable search strategy (ABCVSS) [9];
- *Gbest* guided ABC (GABC) [10];

Table 1. Benchmark functions.

Functions	Search range	Global optium
Sphere (f_1)	[−100, 100]	0
Schwefel 2.22 (f_2)	[−10, 10]	0
Schwefel 1.2 (f_3)	[−100, 100]	0
Schwefel 2.21 (f_4)	[−100, 100]	0
Rosenbrock (f_5)	[−30, 30]	0
Step (f_6)	[−100, 100]	0
Quartic with noise (f_7)	[−1.28, 1.28]	0
Schwefel 2.26 (f_8)	[−500, 500]	−12569.5
Rastrigin (f_9)	[−5.12, 5.12]	0
Ackley (f_{10})	[−32, 32]	0
Griewank (f_{11})	[−600, 600]	0
Penalized (f_{12})	[−50, 50]	0

- Modified ABC (MABC) [11];
- CABC [16]
- Our approach, PABC.

To get a fair comparison across all comparison algorithms, we used the same Settings for some common parameters. The population size is 50, the limit is 100, and the MaxFEs is 1.5E+05, each problem is run 30 times. The respective parameters are as follows: the number of search equations used in ABCVSS is 5 [9]. GABC adopts constant value C = 1.5 [10]. MABC adopts the parameter p = 0.7 [11]. In PABC, Δ is set to 0.0005.

Table 2 shows the comparison between PABC and other 5 ABCs, where Mean represents the "*Mean*" function value and "*Std*" represents the standard deviation. The term "*w/t/l*" is used to describe the winner of 12 comparison problems between PABC and the five improved algorithms. The symbol w represents the number of problems where PABC outperforms the comparison algorithm on 12 problems. The symbol l represents the number of PABC minus the comparison algorithm. The final symbols t, PABC and the comparison algorithm get the same result on t problems. As seen in Table 2, among all the comparison algorithms, PABC obtained the best values on 11 problems. Compared with GABC, the algorithm has better results on 8 functions. PABC outperforms ABC on all functions except f_5 and f_6. GABC obtains the best solution among all algorithms on f_5, while PABC achieves the worst value. The results of PABC and MABC on f_4 and f_{10} are similar, while PABC performs better than MABC on other functions. Similarly, PABC is not as good as MABC on f_5. Compared with ABCVSS and CABC, PABC has obvious advantages on f_1, f_2, f_3, f_7 and f_{11}, and they achieve the same results on f_6, f_8, f_9 and f_{12}. The best results for each problem are shown in bold.

The convergence curves of PABC and the other 5 ABCs are shown in Fig. 2. On f_1, f_2, PABC does not have a prominent advantage at the beginning of the iteration, and at the middle and later iterations, PABC converges faster than the other five ABCs. On f_3,

Table 2. Results of PABC and five other ABC algorithms for D = 30.

Fun	ABC		GABC		MABC		ABCVSS		CABC		PABC	
	Mean	Std	Mean	Std	Mean	Std	Mean	Std	Mean	Std	Mean	Std
f_1	2.60E−17	3.58E−16	4.55E−16	2.76E−16	9.63E−42	6.67E−41	1.10E−36	3.92E−36	2.47E−50	3.53E−49	**6.73E−56**	**2.45E−54**
f_2	1.49E−10	2.35E−10	1.49E−15	3.53E−15	1.50E−21	6.64E−22	8.39E−20	1.60E−19	1.17E−26	3.04E−26	**1.89E−29**	**9.79E−29**
f_3	1.05E+04	3.38E+03	4.22E+03	2.15E+03	1.48E+04	1.44E+04	9.94E+03	9.37E+03	1.63E+04	1.16E+04	**2.68E+01**	**4.94E+01**
f_4	4.07E+01	1.70E+01	1.18E+01	6.36E+00	4.69E+00	4.08E+00	2.58E+01	2.70E+01	**2.31E+01**	**1.29E+01**	**1.72E+00**	**1.06E+00**
f_5	1.28E+00	1.12E+00	**2.30E−01**	**3.41E−01**	1.10E+00	3.45E+00	1.20E+00	1.03E+01	1.81E+00	6.37E+00	2.44E+01	6.66E+00
f_6	**0.00E+00**	**0.00E+00**	**0.00E+00**	**0.00E+00**	**0.00E+00**	**0.00E+00**	**0.00E+00**	**0.00E+00**	**0.00E+00**	**0.00E+00**	**0.00E+00**	**0.00E+00**
f_7	1.54E−01	2.93E−01	5.63E−02	3.66E−02	2.77E−02	6.36E−03	4.13E−02	4.72E−02	5.72E−02	8.39E−02	**5.28E−04**	**3.49E−03**
f_8	−12490.5	5.87E+01	**−12569.5**	**3.25E−10**	**−12569.5**	**1.97E−13**	**−12569.5**	**1.94E−11**	**−12569.5**	**1.09E−11**	**−12569.5**	**1.41E−11**
f_9	7.11E−15	2.28E−15	**0.00E+00**	**0.00E+00**	**0.00E+00**	**0.00E+00**	**0.00E+00**	**0.00E+00**	**0.00E+00**	**0.00E+00**	**0.00E+00**	**0.00E+00**
f_{10}	1.60E−09	4.32E−09	3.97E−14	2.83E−14	7.07E−14	2.36E−14	3.02E−14	2.04E−14	2.95E−14	2.66E−14	**8.55E−15**	**2.12E−14**
f_{11}	1.04E−13	3.58E−13	1.54E−16	2.26E−16	**0.00E+00**	**0.00E+00**	1.85E−17	3.87E−16	1.11E−10	2.76E−09	**0.00E+00**	**0.00E+00**
f_{12}	5.46E−16	3.45E−16	4.03E−16	2.39E−16	**1.57E−32**	**4.50E−47**	**1.57E−32**	**4.50E−47**	**1.57E−32**	**4.50E−47**	**1.57E−32**	**4.50E−47**
$w/t/l$	**10/1/1**		**8/3/1**		**6/5/1**		**7/4/1**		**7/4/1**		–	

PABC is much better than other algorithms. With respect to f_{10}, PABC is beaten by all compared algorithms in convergence accuracy.

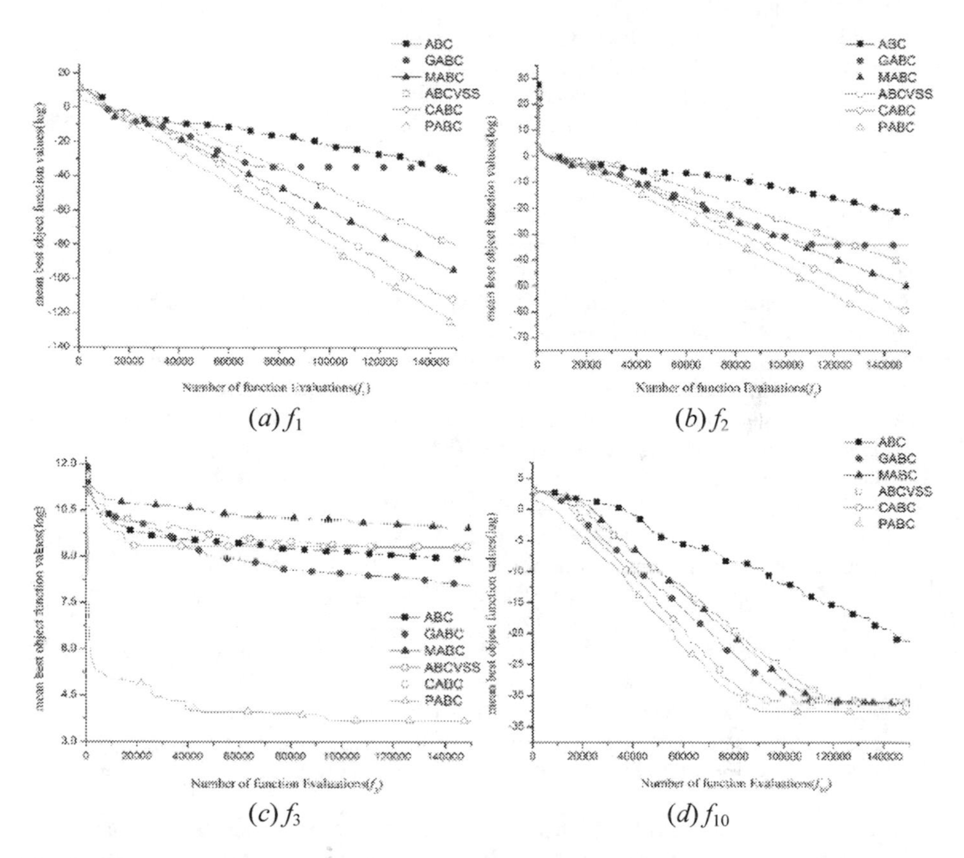

Fig. 2. The convergence curves of PABC, CABC, ABCVSS, MABC, GABC and ABC on selected functions.

5 Conclusions

This paper presents an improved ABC with probabilistic adaptive selection and modified mean center (PABC). The first strategy is applied to multi-strategy search and the selection probability of the corresponding search equation is changed adaptively when the solution quality changes. In addition, the modified mean center is introduced to replace the current optimal value to guide the current search. Comparing PABC with 5 improved ABCs out of 12 classical benchmark functions, the results show that PABC performs better than these improved ABCs. There will be more attempts at adaptive probabilistic selection and more experiments with complex problems.

Acknowledgement. This work was supported by the National Natural Science Foundation of China (No. 61663028).

References

1. Patel, N., Padhiyar, N.: Modified genetic algorithm using box complex method. Application to optimal 533 control problems. J. Process Control **26**, 35–50 (2015)
2. Meang, Z., Pan, J.S., Kong, L.P.: Parameters with adaptive learning mechanism (PALM) for the enhancement of differential evolution. Knowl.-Based Syst. **141**, 92–112 (2018)
3. Liang, J.J., Qin, A.K., Suganthan, P.N., Baskar, S.: Comprehensive learning particle swarm optimizer for global optimization of multimodal functions. IEEE Trans. Evol. Comput. **10** (3), 281–295 (2006)
4. Karaboga, D.: An idea based on honey bee swarm for numerical optimization. Technical report-TR06, Erciyes University, Engineering Faculty, Computer Engineering Department (2005)
5. Zhu, G., Kwong, S.: Gbest-guided artificial bee colony algorithm for numerical function optimization. Appl. Math. Comput. **217**, 3166–3173 (2010)
6. Cui, L.Z., et al.: A novel artificial bee colony algorithm with depth-first search framework and elite-guided search equation. Inf. Sci. **367**(368), 1012–1044 (2016)
7. Wang, Z.G., Shang, X.D., Xia, H.M., Ding, H.: Artificial bee colony algorithm with multi-search strategy cooperative evolutionary. Control Decis. **33**(02), 235–241 (2018)
8. Wang, H., Wu, Z.J., Rahnamayan, S., Sun, H., Liu, Y., Pan, J.S.: Multi-strategy ensemble artificial bee colony algorithm. Inf. Sci. **279**, 587–603 (2014)
9. Kiran, M.S., Hakli, H., Guanduz, M., Uguz, H.: Artificial bee colony algorithm with variable search strategy for continuous optimization. Inf. Sci. **300**, 140–157 (2015)
10. Zhu, G.P., Kwong, S.: Gbest-guided artificial bee colony algorithm for numerical function optimization. Appl. Math. Comput. **217**(7), 3166–3173 (2010)
11. Gao, W.F., Liu, S.Y.: A modified artificial bee colony algorithm. Comput. Oper. Res. **39**(3), 687–697 (2012)
12. Sun, H., Deng, Z.C., Zhao, J., Wang, H., Xie, H.H.: Mixed mean center reverse learning particle swarm optimization algorithm. Electron. J. **47**(09), 1809–1818 (2019)
13. Wang, H., et al.: Firefly algorithm with neighborhood attraction. Inf. Sci. **382**, 374–387 (2017)
14. Wang, H., Cui, Z.H., Sun, H., Rahnamayan, S., Yang, X.S.: Randomly attracted firefly algorithm with neighborhood search and dynamic parameter adjustment mechanism. Soft. Comput. **21**, 5325–5339 (2017)
15. Wang, H., Sun, H., Li, C.H., Rahnamayan, S., Pan, J.S.: Diversity enhanced particle swarm optimization with neighborhood search. Inf. Sci. **223**, 119–135 (2013)
16. Gao, W.F., Liu, S.Y., Huang, L.L.: A novel artificial bee colony algorithm based on modified search equation and orthogonal learning. IEEE Trans. Cybern. **43**(3), 1011–1024 (2013)

Average Convergence Rate of Evolutionary Algorithms II: Continuous Optimisation

Yu Chen[1(✉)] and Jun He[2]

[1] School of Science, Wuhan University of Technology, Wuhan 430070, China
ychen@whut.edu.cn

[2] School of Science and Technology, Nottingham Trent University, Nottingham, UK

Abstract. Previously a theoretical study of the average convergence rate was conducted for discrete optimisation. This paper extends it to a further analysis for continuous optimisation. First, the strategies of generating new solutions are classified into two categories: landscape-invariant and landscape-adaptive. Then, it is proven that the average convergence rate of evolutionary algorithms using positive-adaptive generators is asymptotically positive, but that of algorithms using landscape-invariant generators and zero-adaptive generators asymptotically converges to zero. A case study is made to validate the applicability of theoretical results. Besides the theoretical study, numerical simulations are presented to show the feasibility of the average convergence rate in practical applications. In case of unknown optimum, an alternative definition of the average convergence rate is also considered.

Keywords: Evolutionary algorithms · Continuous optimisation · Convergence rate · Markov chains · Approximation error · Search strategies

1 Introduction

In the theoretical study of evolutionary algorithms (EAs), a fundamental question is how fast an EA converges to an optimal solution of an optimisation problem. In discrete optimisation, this can be measured by the number of generations (hitting time) or the number of fitness evaluations (running time) when an EA finds an optimal solution [13,16]. However, in continuous optimisation, computation time is normally infinite because the optimal solution set of a continuous optimisation problem is usually a zero-measure set. So, computation time is preferred to being evaluated when EAs reach an ε-neighbour of the optimal solution set [1,10,19] that forms a positive-measure set.

The first author was supported by the National Science Foundation of China (NSFC) under Grant No. 61303028; The second author was supported by EPSRC under Grant No. EP/I009809/1.

K. Li et al. (Eds.): ISICA 2019, CCIS 1205, pp. 31–45, 2020.
https://doi.org/10.1007/978-981-15-5577-0_3

Alternatively, the convergence rate is another popular way to quantifies how fast population X_t converges to the optimal solution set $\mathcal{S}^*$ in the continuous decision space. For multi-modal problems it is also rational to investigate how fast the fitness $f(X_t)$ of population X_t converges to the optimal fitness f^*. Numerous theoretical work discussed the convergence rate from different perspectives [3,6,14,15,17,18,22,23].

Denote $f_t = \mathbb{E}[f(X_t)]$ and $e_t = |f_t - f^*|$. By investigating the one-generation error ratio e_t/e_{t-1}, it is straightforward to derive the geometric convergence $e_t \leq e_0 c^t$ under the condition $e_t/e_{t-1} \leq c < 1$ [23]. But unfortunately, randomness in EAs results in too much noise on e_t/e_{t-1}, which in turn hinders its practical application in numerical experiments. This motivates us to utilise the average of the error ratio e_t/e_{t-1}. He and Lin [15] considered its geometric average over consecutive t generations and proposed the average convergence rate (ACR), by which it is straightforward to draw an exact expression of the approximation error: $e_t = (1-R_t)^t e_0$. More importantly, R_t is more stable than e_t/e_{t-1} in computer simulation, which makes it practical in numerical experiments to evaluate convergence rate of EAs.

For discrete optimisation, it has been proven that under random initialisation, R_t converges to a positive; and under particular initialisation, R_t always equals to this positive [15]. The current paper extends the analysis of the ACR to continuous optimisation. The paper is organized as follows: Sect. 2 introduces the related work. Section 3 presents some preliminaries for the study. In Sect. 4, theoretical analysis of the ACR is performed, and Sect. 5 provides a case study to confirm applicability of the theoretical results. Section 6 discusses a method of designing landscape-adaptive generators, and finally, Sect. 7 concludes the paper.

2 Related Work

The convergence rate of EAs has been investigated from different perspectives and in varied terms. He, Kang and Ding [11,14] studied the convergence in distribution $\parallel \mu_t - \pi \parallel$ where μ_t is the probability distribution of X_t and π a stationary probability distribution. Based on the Doeblin condition, they obtained bounds on $\parallel \mu_t - \pi \parallel \leq (1-\delta)^{t-1}$ for some $\delta \in (0,1)$. He and Yu [17] also derived lower and upper bounds on $1 - \mu_t(X_\delta^*)$ where $\mu_t(\mathcal{S}_\delta^*)$ denotes the probability of X_t entering in a δ-neighbour of $\mathcal{S}^*$.

Rudolph [23] proved under the condition $e_t/e_{t-1} \leq c < 1$, the sequence e_t converges in mean geometrically fast to 0, that is, $q^t e_t = o(1)$ for some $q > 1$. For a superset of the class of quadratic functions, sharp bounds on the convergence rate is obtained. Rudolph [22] also compared Gaussian and Cauchy mutation on minimising the sphere function in terms of the rate of local convergence, $\mathbb{E}[\min\{\parallel X_{t+1} \parallel^2 / \parallel X_t \parallel^2, 1\} \mid X_t]$, where $\parallel \cdot \parallel$ denotes the Euclidean norm. He proved the rate is identical for Gaussian and spherical Cauchy distributions, whereas nonspherical Cauchy mutations lead to slower local convergence.

Beyer [7] developed a systematic theory of evolutionary strategies (ES) based on the progress rate and quality gain. The progress rate measures the distance change to the optimal solution in one generation, $\mathbb{E}[\| X_t - \mathcal{S}^* \| - \| X_{t-1} - \mathcal{S}^* \|]$. The quality gain is the fitness change in one generation, $\mathbb{E}[\bar{f}(X_t) - \bar{f}(X_{t-1})]$, where $\bar{f}(X)$ is the fitness mean of individuals in population X. Recently Beyer et al. [8,9] analyzed dynamics of ES with cumulative step size adaption and ES with self-adaption and multi-recombination on the ellipsoid model and derived the quadratic progress rate. Akimoto et al. [2] investigated evolution strategies with weighted recombination on general convex quadratic functions and derived the asymptotic quality gain. However, Auger and Hansen [4] argued the limits of the predictions based on the progress rate.

Auger and Hansen [5] developed the theory of ES from a new perspective using stability of Markov chains. Auger [3] investigated the $(1,\lambda)$-SA-EA on the sphere function and proved the convergence of $(\ln \| X_t \|)/t$ based on Foster-Lyapunov drift conditions. Jebalia et al. [20] investigated convergence rate of the scale-invariant (1+1)-ES in minimizing the noisy sphere function and proved a log-linear convergence rate in the sense that: $(\ln \| X_t \|)/t \to \gamma$ for some γ as $t \to +\infty$. Auger and Hansen [6] further investigated the comparison-based step-size adaptive randomized search on scaling-invariant objective functions and proved as $t \to +\infty$, $\ln(\| X_t \| / \| X_0 \|_t)/t \to -CR$ for some positive CR. This log-linear convergence is an extension of the average rate of convergence in deterministic iterative methods [24].

3 Preliminaries

3.1 Convergence and Average Convergence Rate

A continuous minimisation problem is given as follows:

$$\min f(\mathbf{x}), \quad \mathbf{x} = (x_1, \cdots, x_d) \in \mathcal{D} \subset \mathbb{R}^d, \tag{1}$$

where $f(\mathbf{x})$ is a continuous function defined on a closed set $\mathcal{D}$. $f^* = \min f(\mathbf{x})$. An individual $x \in \mathcal{D}$ is a single point and a population X is a collection of individuals. For many optimisation problems (for example, most benchmark problems), we may assume that their optimal solution sets have Lebesgue measure 0. A general framework of EAs for solving optimisation problems is described in Algorithm 1, where two types of genetic operators are employed. One is the generation operator such as mutation or crossover, and the other is the selection operator.

The fitness of population X is defined as $f(X) = \min\{f(\mathbf{x}) \mid \mathbf{x} \in X\}$, and its quality is evaluated by the approximation error $e(X) = |f(X) - f^*|$. If $e(X_{t+1}) \leq e(X_t)$ for any t, then the EA is called elitist. Since population Y_t in Algorithm 1 only depends on X_t and then X_{t+1} only depends on X_t, the population sequence $\{X_t; t = 0, 1, \cdots\}$ is a Markov chain [14,17].

Definition 1. *The sequence* $\{e(X_0), e(X_1), \cdots\}$ *is called* ***convergent in mean*** *if* $\lim_{t\to+\infty} \mathbb{E}[e(X_t)] = 0$*; it is called* ***convergent almost surely*** *if* $\Pr(\lim_{t\to+\infty} e(X_t) = 0) = 1$.

Algorithm 1. Evolutionary Algorithms (EAs)

1: counter $t \leftarrow 0$;
2: $X_0 \leftarrow$ initialise a population of individuals subject to a probability distribution $\Pr(X_0)$ over the definition domain $\mathcal{D}$;
3: **while** the stopping criterion is not satisfied **do**
4: $\quad Y_t \leftarrow$ generate a population of individuals from X_t subject to a conditional transition probability $\Pr(Y_t \mid X_t)$;
5: $\quad X_{t+1} \leftarrow$ select a population of individuals from $X_t \cup Y_t$ subject to a conditional transition probability $\Pr(X_{t+1} \mid X_t, Y_t)$;
6: $\quad$ counter $t \leftarrow t+1$;
7: **end while**

If $\mathbb{E}[e(X_t)] \leq e(X_{t-1})$ (e.g., elitist EAs), then the sequence $\{e(X_t); t = 0, 1, \cdots\}$ is a supermartingale [12]. According to Doob's convergence theorem [12], convergence in mean implies almost sure convergence [23]. Non-convergent EAs are not considered in the current paper, so, we always assume that $\{e(X_t); t = 0, 1, \cdots\}$ converges in mean.

Let $f_t = \mathbb{E}[f(X_t)]$ and $e_t = \mathbb{E}[e(X_t)]$. The convergence rate[1] of the sequence $\{e_t; t = 0, 1, \cdots\}$ for one generation is e_t/e_{t-1}. But calculating the above ratio is unstable in computer simulation. Thus the geometric average convergence rate (ACR) is employed to evaluate the average convergence speed of EAs for consecutive t generations [15].

Definition 2. *The geometric average convergence rate (ACR) of an EA for t generations is for $e_0 \neq 0$,*

$$R_t = 1 - \left(\frac{e_t}{e_0}\right)^{1/t}. \tag{2}$$

In (2), the term $(e_t/e_0)^{1/t}$ represents a geometric average of the reduction factor over t generations. $1 - (e_t/e_0)^{1/t}$ normalizes the average in the interval $(-\infty, 1]$. For elitist EAs, it always holds $e_t \leq e_0$. Then, we always have $0 \leq R_t \leq 1$.

The above definition is applicable to benchmark functions in which the optimal fitness f^* is known in advance. In case of unknown f^*, an alternative definition is given as follows [15].

Definition 3. *An alternative average convergence rate of an EA is*

$$R_t^{\dagger} := 1 - \left|\frac{f_{t+k} - f_t}{f_t - f_{t-k}}\right|^{1/k}, \tag{3}$$

where k is an appropriate time interval. Its value relies on an EA and a problem.

In numerical calculation, it is important to set k to a large value. Otherwise there might exist too much noise on $R_t^{\dagger}$.

[1] We don't add a converge order because for discrete optimisation, e_t converges linearly to 0 according to [15, Theorem 1]. For continuous optimisation, a conjecture is that e_t also converges linearly unless gradient information is used in the search.

3.2 Links Between Two Average Convergence Rates

There exists some links between the two rates R_t and $R_t^\dagger$. First, we consider the scenario: $\forall t$, $e_t/e_{t-1} = \gamma$. Then $e_t = e_0\gamma^t$ and $f_t = e_0\gamma^t - f^*$. We have

$$R_t = 1 - \left(\frac{e_t}{e_0}\right)^{1/t} = 1 - \gamma, \qquad R_t^\dagger = \left|\frac{f_t - f_{t+k}}{f_{t-k} - f_t}\right|^{1/k} = 1 - \gamma. \tag{4}$$

Secondly, we consider the scenario: $\lim_{t\to+\infty} e_t/e_{t-1} = \gamma$. Then $\lim_{t\to+\infty} e_t/e_{t-k} = \gamma^k$. We prove that the limit of the ACR R_t is $1-\gamma$.

$$\lim_{t\to+\infty} R_t = 1 - \lim_{t\to+\infty}\left(\frac{e_t}{e_0}\right)^{1/t} = 1 - \lim_{t\to+\infty}\left(\prod_{k=1}^{t}\frac{e_k}{e_{k-1}}\right)^{1/t} = 1 - \gamma.$$

We also prove that $\forall\ k \in \mathbb{Z}^+$, the limit of the alternative ACR $R_t^\dagger$ is $1-\gamma$.

$$\begin{aligned}\lim_{t\to+\infty} R_t^\dagger &= 1 - \lim_{t\to+\infty}\left|\frac{f_t - f_{t+k}}{f_{t-k} - f_t}\right|^{1/k} = 1 - \lim_{t\to+\infty}\left|\frac{e_t - e_{t+k}}{e_{t-k} - e_t}\right|^{1/k}\\ &= 1 - \lim_{t\to+\infty}\left|\frac{e_t}{e_{t-k}} \times (1 - \frac{e_{t+k}}{e_t}) \div (1 - \frac{e_t}{e_{t-k}})\right|^{1/k} = 1 - \gamma.\end{aligned}$$

So, for sufficiently large t, we can approximate R_t by $R_t^\dagger$. In this paper, we focus on analysing R_t. The analysis of $R_t^\dagger$ will be left for future research.

3.3 Numerical Calculation of Average Convergence Rate

We show the application of ACR through a popular example. Consider the problem of minimising the 2-dimensional sphere function

$$\min f_S(\mathbf{x}) = x_1^2 + x_2^2, \quad \mathbf{x} \in \mathbb{R}^2.$$

The minimal point to this function is $\mathbf{x}^* = (0,0)$ with $f(\mathbf{x}^*) = 0$. The solver is the (1+1) elitist EA described in Algorithm 2.

Algorithm 2. (1+1) elitist EA

1: counter $t \leftarrow 0$;
2: individual $\mathbf{x}_0 \leftarrow$ initialise a solution;
3: **while** the stopping criterion is not satisfied **do**
4: candidate solution $\mathbf{y}_t \leftarrow$ generate a new solution from $\mathbf{x}_t$ by Gaussian mutation;
5: individual $\mathbf{x}_{t+1} \leftarrow$ select the best of individuals in $\mathbf{y}_t$ and $\mathbf{x}_t$;
6: $t \leftarrow t+1$;
7: **end while**

The candidate solution $\mathbf{y}$ is generated by Gaussian mutation $\mathbf{y} = \mathbf{x} + \mathbf{z}$, where $\mathbf{x}$ is the parent, $\mathbf{y}$ the child and $\mathbf{z} = (z_1, \cdots, z_d)$ a Gaussian random vector obeying the probability distribution $z_i \sim \mathcal{N}(0, \sigma_i)$. Two variants of Gaussian mutation with a covariance matrix $\mathbf{\Sigma^2} = diag(\sigma_1^2, \sigma_2^2)$ are used where one adopts invariant-σ and the other adopts adaptive-σ. In **invariant-σ mutation**, σ is set to a constant for all $\mathbf{x}$. In computer simulation, set $\sigma_1 = \sigma_2 = 1$. In **adaptive-σ mutation**, σ takes varied values on different $\mathbf{x}$. In computer simulation, set $\sigma_1 = \sigma_2 = \parallel \mathbf{x} \parallel_2$. For the sake of terms, the EA using invariant-σ mutation is called an invariant EA and the EA using adaptive-σ mutation is called an adaptive EA.

In the experiment, both invariant and adaptive EA are tested with an initial solution $\mathbf{x}_0$ randomly selected from $[-100, 100]^2$, the number of running an EA equal to 10,000, the maximum number of generations equal to 300, and the time interval in the $R_t^\dagger$ formula k equal to 50. Because $k = 50$, the alternative ACR $R_t^\dagger$ has no value for $t < 50$ and $t > 250$ according to formula (3).

The trend plots of the approximation error e_t and error ratio e_{t+1}/e_t are illustrated in Fig. 1. Figure 1(a) shows the adaptive EA converges faster than the invariant one. Figure 1(b) displays the error ratio e_{t+1}/e_t with noise caused by randomness in EAs, so, e_{t+1}/e_t is not appropriate as the measurement of the convergent rate.

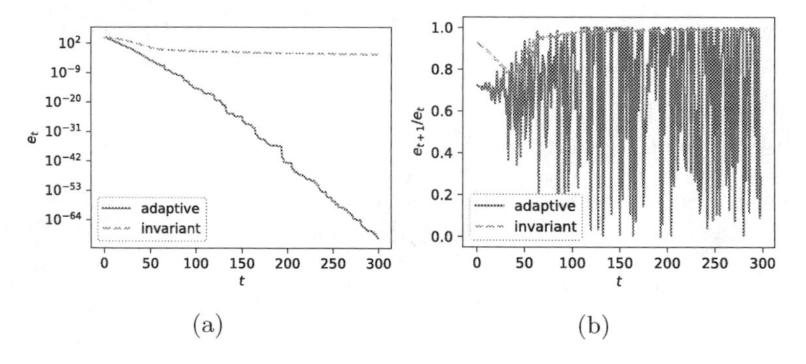

Fig. 1. Approximation error e_t (left figure) and error ratio e_{t+1}/e_t (right figure) of the invariant EA and adaptive EA

The trend plots of the ACR R_t and alternative ACR $R_t^\dagger$ are illustrated in Fig. 2, which indicates that the ACR of the adaptive EA tends to a positive value, while that of the invariant EA tends to zero. The shapes of R_t and $R_t^\dagger$ are similar but $R_t^\dagger$ suffers bigger noise than R_t.

4 General Analyses

Consider the sequence $\{R_t, t = 1, \dots\}$. It is crucial to investigate the limit property of R_t when $t \to +\infty$. If $\lim_{t \to +\infty} R_t = 0$, the average convergence rate asymptotically reduces to zero and then the approximation error e_t converges slowly to 0;

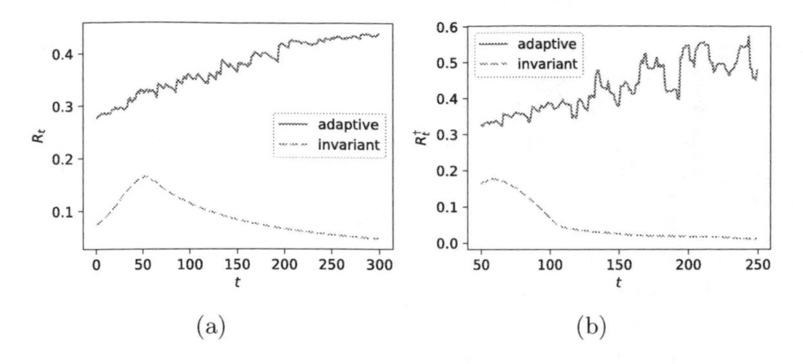

Fig. 2. The average convergence rate R_t (left figure) and alternative rate $R_t^\dagger$ (right figure) of the invariant EA and adaptive EA

otherwise, the average convergence rate is always positive so that the approximation error e_t converges fast to 0.

4.1 Landscape-Invariant and -Adaptive Generators

Operations of an EA, including generation and election, can be represented by transition probabilities. Let $\mathcal{S}$ denote the set consisting of all populations. A population is represented by a capital letter such as X, and the tth generation population is represented by X_t. A population X satisfying $f(X) = f^*$ is called an optimal population, and the collection of all optimal populations is denoted as $\mathcal{S}^*$. Given a contraction factor $\rho \in (0, 1]$ and a population X, the set $\mathcal{S}$ can be divided into two disjoint subsets:

$$\mathcal{S}(X, \rho) = \{Y \in \mathcal{S} | e(Y) < \rho e(X)\}, \tag{5}$$

$$\overline{\mathcal{S}}(X, \rho) = \{Y \in \mathcal{S} | e(Y) \geq \rho e(X)\}. \tag{6}$$

The set $\mathcal{S}(X, \rho)$ is called a ρ-promising region. When $\rho = 1$, the set $\mathcal{S}(X, 1)$ is called a promising region.

The generation of Y_t via X_t, denoted as $X_t \Rightarrow Y_t$, can be characterised by a probability transition. Given a population $X \in \mathcal{S}$ and a population set $\mathcal{A} \subset \mathcal{S}$, the transition probability kernel $P_g(X; \mathcal{A})$ is defined as

$$P_g(X; \mathcal{A}) = \int_{\mathcal{A}} p_g(X; Y) dY,$$

where $p_g(X; Y)$ is a transition probability density function [21]. Generally, the operators of generating new individuals may be classified into two categories.

Definition 4. *Let $p_g(X; Y)$ be the probability density function depicting the generation transition from X to Y. In this paper, we assume it is continuous and bounded.*

1. ***Landscape-invariant:*** *a generator $X \Rightarrow Y$ is called* ***landscape-invariant*** *if $Y = X + Z$ and Z is a multivariate random variable whose joint probability density function $p_g(0; Z)$ is independent on X. Here $X + Z$ represents $(\mathbf{x}_1 + \mathbf{z}_1, \cdots, \mathbf{x}_N + \mathbf{z}_N)$, and for simplicity, denote $p_g(0; Z)$ as $p_z(Z)$.*
2. ***Landscape-adaptive:*** *otherwise, a generator $X \Rightarrow Y$ is called* ***landscape-adaptive****.*

A landscape-invariant generator generates candidate solutions subject to the same probability distribution no matter where a parent population locates. In Algorithm 2, invariant-σ mutation is a landscape-invariant generator but adaptive-σ mutation is a landscape-adaptive generator.

4.2 Analysis of Landscape-Invariant Generators

For the landscape-invariant generator, we first demonstrate that the infinum of the transition probability to the promising region is zero. Then, an elitist EAs using landscape-invariant generators have an ACR tending to 0.

Lemma 1. *If the number of optimal solutions is finite and the generator is landscape-invariant, then the transition probability to the promising region satisfies*

$$\inf\{P_g(X, \mathcal{S}(X,1)); X \notin \mathcal{S}^*\} = 0, \tag{7}$$

where inf *is the abbreviation of mathematical infimum.*

Proof. In order to prove (7), it is sufficient to prove $\lim_{e(X)\to 0} P_g(X, \mathcal{S}(X,1)) = 0$. That is, $\forall \varepsilon > 0, \exists \delta > 0$, $\forall X \in \mathcal{A}(\mathcal{S}^*, \delta) \setminus \mathcal{S}^*$ (where the set $\mathcal{A}(\mathcal{S}^*, \delta) = \{X; e(X) \leq \delta\}$), it holds

$$P_g(X, \mathcal{S}(X,1)) < \varepsilon. \tag{8}$$

Let $m(\mathcal{A})$ be the Lebesgue measure of a Lebesgue-measurable set $\mathcal{A} \subset \mathcal{S}$. Because $p_z(Z)$ is continuous and bounded, the probability of $X + Z$ falling in a small area is small for a fixed X. That is, $\forall \varepsilon > 0$, $\exists \delta' > 0$ (set $\delta' = \varepsilon / \sup p_z(Z)$), it holds

$$\Pr(X + Z \in \mathcal{A}) = \int_{Z: X+Z \in \mathcal{A}} p_z(X + Z) dZ < \varepsilon, \tag{9}$$

$\forall\$\mathcal{A} \subset \mathcal{S} : m(\mathcal{A}) \leq \delta'$ and $\forall X \in \mathcal{S}$. Because the number of optimal solutions is finite (then $m(\mathcal{S}^*) = 0$) and f is continuous, for the set $\mathcal{A}(\mathcal{S}^*, \delta)$, we can choose δ sufficiently small so that $m(\mathcal{A}(\mathcal{S}^*, \delta)) \leq \delta'$.

Because f is continuous, we may choose δ sufficiently small so that $\forall X \in \mathcal{A}(\mathcal{S}^*, \delta)$ and $Y \notin \mathcal{A}(\mathcal{S}^*, \delta)$: $f(X) < f(Y)$. This implies that $\mathcal{S}(X,1) \subset \mathcal{A}(\mathcal{S}^*, \delta)$. According to (9) and $m(\mathcal{A}(\mathcal{S}^*, \delta)) \leq \delta'$, $\forall X \in \mathcal{A}(\mathcal{S}^*, \delta) \setminus \mathcal{S}^*$, we have

$$\Pr(X + Z \in \mathcal{A}(\mathcal{S}^*, \delta)) < \varepsilon.$$

Because $\mathcal{S}(X,1) \subset \mathcal{A}(\mathcal{S}^*, \delta)$, we have

$$P_g(X, \mathcal{S}(X,1)) \leq \Pr(X + Z \in \mathcal{A}(\mathcal{S}^*, \delta)) < \varepsilon.$$

Then we get (8), and the proof is completed. □

Theorem 1. *For Problem (1) and Algorithm 1, if*

1. *the number of optimal solutions is finite;*
2. *the sequence* $\{e_t; t = 0, 1, \cdots\}$ *converges to 0;*
3. *the generator is landscape-invariant;*

then $\lim_{t\to+\infty} R_t = 0$.

Proof. Proof of $\lim_{t\to+\infty} R_t = 0$ can be achieved by demonstrating that $\lim_{t\to+\infty} e_t/e_{t-1} = 1$, equivalently, $\lim_{t\to+\infty} (e_{t-1} - e_t)/e_{t-1} = 0$. According to the definition of limit, it is sufficient to prove that $\forall \varepsilon > 0, \exists t_0 > 0$, $\forall t \geq t_0$,

$$e_{t-1} - e_t < \varepsilon e_{t-1}. \tag{10}$$

From (8) in Lemma 1, we know $\forall \varepsilon > 0, \exists \delta > 0$, let $\mathcal{A}(\mathcal{S}^*, \delta) = \{X; e(X) \leq \delta\}$, then $\forall X \in \mathcal{A}(\mathcal{S}^*, \delta) \setminus \mathcal{S}^*$, it holds

$$P_y(X, \mathcal{S}(X, 1)) < \varepsilon. \tag{11}$$

Since the sequence $\{e(X_t); t = 0, 1, \cdots\}$ converges in mean, then it converges almost surely to 0, that is, $\Pr(\lim_{t\to+\infty} e(X_t) = 0) = 1$. Denote

$$\mathcal{S}_1 = \{\omega \in \mathcal{S} | \lim_{t\to+\infty} e(X_t(\omega)) = 0\}, \ \mathcal{S}_2 = \{\omega \in \mathcal{S} | \lim_{t\to+\infty} e(X_t(\omega)) \neq 0\}.$$

It is obvious that

$$\Pr(\omega \in \mathcal{S}_2) = 0, \tag{12}$$

and for the given $\delta > 0$, $\exists t_0 > 0$, then $\forall t > t_0$, it holds

$$e(X_{t-1}(\omega)) < \delta, \quad \forall \omega \in \mathcal{S}_1.$$

From (11) we know

$$P_g(X, \mathcal{S}(X_{t-1}(\omega), 1)) \leq \varepsilon, \quad \omega \in \mathcal{S}_1.$$

Then we obtain

$$\mathbb{E}[e(X_{t-1}(\omega)) - e(X_t(\omega)) \mid X_{t-1}(\omega)] \leq \varepsilon e(X_{t-1}(\omega)), \quad \forall \omega \in \mathcal{S}_1. \tag{13}$$

While $\forall \omega \in \mathcal{S}_2$, we know there exists a positive B:

$$\mathbb{E}[e(X_{t-1}(\omega)) - e(X_t(\omega)) \mid X_{t-1}(\omega)] \leq B. \tag{14}$$

Combining (12), (13) and (14) together, we get

$$\begin{aligned}
e_{t-1} - e_t =& \int_{\mathcal{S}_1} \mathbb{E}[e(X_{t-1}(\omega)) - e(X_t(\omega)) \mid X_{t-1}(\omega)] \Pr(d\omega) \\
&+ \int_{\mathcal{S}_2} \mathbb{E}[e(X_{t-1}(\omega)) - e(X_t(\omega)) \mid X_{t-1}(\omega)] \Pr(d\omega) \\
\leq& \varepsilon \int_{\mathcal{S}_1} e(X_{t-1}(\omega)) \Pr(d\omega) + B \cdot 0 \leq \varepsilon e_{t-1}.
\end{aligned}$$

So (10) is true, and we complete the proof. □

Note that theorem 1 may not hold if the Lebesgue measure of $\mathcal{S}^*$ is positive. However, for many continuous optimisation problems, $\mathcal{S}^*$ is a zero-measure set. Thus, Theorem 1 states that for EAs using landscape-invariant generators, the limit of their ACR is 0 as $t \to +\infty$. This implies that landscape-invariant generators are not good for solving most of continuous optimisation problems.

4.3 Analysis of Landscape-Adaptive Generators

A natural way to improve performance of EAs is to employ landscape-adaptive generators. According to the probability of locating promising regions, they can be split into two types:

1. **zero-adaptive:** a landscape-adaptive generator $X \Rightarrow Y$ is called **zero-adaptive** if the transition probability to the promising region satisfies

$$\inf\{P_g(X;\mathcal{S}(X,1)); X \notin \mathcal{S}^*\} = 0. \tag{15}$$

2. **positive-adaptive:** a landscape-adaptive generator $X \Rightarrow Y$ is called **positive-adaptive** if $\exists \rho \in (0,1)$, the transition probability to the ρ-promising region satisfies

$$C_\rho = \inf\{P_g(X;\mathcal{S}(X,\rho)); X \notin \mathcal{S}^*\} > 0. \tag{16}$$

Theorem 2. *For Problem (1) and Algorithm 1, if*

1. *the number of optimal solutions is finite;*
2. *the sequence $\{e_t; t = 0, 1, \cdots\}$ converges to 0;*
3. *the generation operator is zero-adaptive,*

then $\exists$ a sequence $\{X_t; t = 0, 1, \cdots\}$ such that $\lim_{t\to+\infty} R_t = 0$.

Proof. For the zero-adaptive generators, (15) implies that there exists a sequence $\{X_t; t = 0, 1, \cdots\}$ such that $\lim\limits_{k\to+\infty} P_g(X_k, S(X_k, 1)) = 0$. Similar to proof of Theorem 1, we know that $\lim_{t\to+\infty} R_t = 0$.

Thus, we can conclude that the zero-adaptive generator is not efficient because the ACR tends to 0. However, a positive-adaptive generator is always efficient because it results in a positive lower bound of the ACR.

Theorem 3. *For Problem (1) and Algorithm 1, if*

1. *the sequence $\{e_t; t = 0, 1, \cdots\}$ converges to 0;*
2. *the generation operator is positive-adaptive with a contraction factor $\rho \in (0,1)$,*

then $\exists C > 0$ such that $R_t \geq C$.

Proof. From (5), we know that for any $k-1 \geq 0$,

$$\mathcal{S}(X_{k-1}, \rho) = \{Y \in \mathcal{S} \mid e(Y) \leq \rho e(X_{k-1})\}.$$

It follows that $\mathcal{S}(X_{k-1}, \rho) \subset \mathcal{S}(X_{k-1}, 1)$, and for any $Y \in \mathcal{S}(X_{k-1}, \rho)$,

$$f(X_{k-1}) - f(Y) \geq (1-\rho)(f(X_{k-1}) - f^*). \tag{17}$$

So we get

$$\begin{aligned}
&\mathbb{E}\left[f(X_{k-1}) - f(X_k) | X_{k-1}\right] = \int_{\mathcal{S}(X_{k-1},1)} (f(X_{k-1}) - f(Y)) p_g(X_{k-1}; Y) dY \\
&\geq \int_{\mathcal{S}(X_{k-1},\rho)} (f(X_{k-1}) - f(Y)) p_g(X_{k-1}; Y) dY \\
&\geq \int_{\mathcal{S}(X_{k-1},\rho)} (1-\rho)\left(f(X_{k-1}) - f^*\right) p_g(X_{k-1}; Y) dY \quad \text{(from (17))} \\
&\geq (1-\rho) C_\rho \left(f(X_{k-1}) - f^*\right). \quad \text{(from (16))}
\end{aligned} \tag{18}$$

Then

$$\begin{aligned}
\frac{e_k}{e_{k-1}} &= 1 - \frac{f_{k-1} - f_k}{f_{k-1} - f^*} \\
&= 1 - \frac{\mathbb{E}[\mathbb{E}\left[f(X_{k-1}) - f(X_k) | X_{k-1}\right]]}{f_{k-1} - f^*} \\
&\leq 1 - \frac{\mathbb{E}\left[(1-\rho) C_\rho \left(f(X_{k-1}) - f^*\right)\right]}{f_{k-1} - f^*} \leq 1 - (1-\rho) C_\rho.
\end{aligned} \tag{19}$$

Let $C = (1-\rho)C_\rho$. It holds that

$$R_t = 1 - \left(\frac{e_t}{e_0}\right)^{1/t} = 1 - \left(\prod_{k=1}^{t} \frac{e_k}{e_{k-1}}\right)^{1/t} \geq (1-\rho) C_\rho = C. \qquad \square$$

5 Case Study

Theorems 1–3 indicate an EA with a landscape-invariant or zero-adaptive generator has a bad ACR (towards 0) but an EA with a positive-adaptive generator has a good ACR (greater than a positive). For case study this section investigates maximisation of a linear function

$$\max L(\mathbf{x}) = \sum_{i=1}^{d} c_i |x_i|, \quad \mathbf{x} = (x_1, \ldots, x_d) \in \mathbb{R}^d, \tag{20}$$

This definition is a natural extension of linear functions in pseudo-Boolean optimisation [13,16] in which x_i takes Boolean values. It is solved by the random rotation search described in Algorithm 3, which can be regarded as the parallel version of (1+1) EA using one-bit mutation for continuous optimisation.

Algorithm 3. Random rotation search

1: counter $t \leftarrow 0$;
2: individual $\mathbf{x}_0 \leftarrow$ initialise a solution;
3: **while** the stopping criterion is not satisfied **do**
4: candidate solution $\mathbf{y}_t \leftarrow$ generate a new solution from $\mathbf{x}_t$ by 1-D Gaussian mutation performed on a randomly chosen dimension;
5: individual $\mathbf{x}_{t+1} \leftarrow$ select the best of individuals in $\mathbf{y}_t$ and $\mathbf{x}_t$;
6: $t \leftarrow t+1$;
7: **end while**

Denote $\mathbf{x} = (x_1, \ldots, x_d)$ be the present parent. The offspring $\mathbf{y}$ is generated on a randomly selected dimension j with probability $1/d$. That is,

$$\mathbf{y}_j = (x_1, \ldots, x_{j-1}, y_j, x_{j+1}, \ldots, x_d),$$

where $y_j = x_j + \mathcal{N}(0, \sigma_j)$. Thanks to elitist selection, $\mathbf{y}_j$ is accepted if and only it satisfies $|y_j| < |x_j|$. To simplify notation, we assume $x_j > 0$ without loss of generality. We first find a positive-adaptive strategy for the studied case, and then, investigate the connection between ACR and the dimension d of the decision space.

5.1 Positive-Adaptive Generator

It is trivial to validate that the probability of hitting the promising region is

$$P_g(\mathbf{x}; \mathcal{S}(\mathbf{x}, 1)) = \sum_{j=1}^{d} \frac{1}{\sqrt{2\pi}\sigma_j d} \int_{-x_j}^{x_j} e^{-\frac{(y-x_j)^2}{2\sigma_j^2}} dy = \frac{1}{d} \sum_{j=1}^{d} \left(\frac{1}{2} - \Phi\left(-\frac{2x_j}{\sigma_j}\right)\right) \tag{21}$$

If σ_j is a constant for any j, then the mutation is landscape-invariant. When the (1+1) EA converges to the optimal solution, x_j converges to 0. As a result, the value of (21) also converges to 0 since σ is a constant. This means that Lemma 1 holds. According to Theorem 1, R_t converges to 0 when $t \to +\infty$.

A positive ACR can be obtained when the generator is positive-adaptive, that is, $\exists C > 0$, $\rho \in (0, 1)$, $\forall \mathbf{x} \notin \mathcal{S}^*$,

$$P_g(\mathbf{x}; \mathcal{S}(\mathbf{x}, \rho)) \geq C.$$

Set $x_j/\sigma_j = C_0$ for all j where $C_0 > 0$. Then we have

$$P_g(\mathbf{x}; \mathcal{S}(\mathbf{x}, 1)) = \frac{1}{2} - \Phi(-2C_0).$$

Take $P_g(\mathbf{x}, \mathcal{S}(\mathbf{x}, \rho))$ as a function of ρ defined in the interval $(0, 1]$. Obviously $P_g(\mathbf{x}, \mathcal{S}(\mathbf{x}, \rho))$ is continuous, that is, $\forall\, \varepsilon > 0$, $\exists \rho(\varepsilon) \in (0, 1)$ such that

$$P_g(\mathbf{x}, \mathcal{S}(\mathbf{x}, \rho(\varepsilon))) > P_g(\mathbf{x}, \mathcal{S}(\mathbf{x}, 1)) - \varepsilon. \tag{22}$$

Set $\rho = \rho(\varepsilon)$ and $C_\rho = \frac{1}{2} - \Phi(-2C_0) - \varepsilon$ for any given $\frac{1}{2} - \Phi(-2C_0) > \varepsilon > 0$. (22) and (16) confirm that the generator with $x_j/\sigma_j = C_0$ is positive-adaptive. Then, Theorem 3 claims that the ACR is not less than the positive

$$C = (1 - \rho(\varepsilon))(\frac{1}{2} - \Phi(-2C_0) - \varepsilon). \tag{23}$$

5.2 Connection Between ACR and Problem Dimension

The lower bound presented by (23) is general, and does not show how the ACR is connected to the dimension d. In the following, we demonstrate a connection between R_t and d.

Performing the positive-adaptive mutation with $x_j/\sigma_j = C_0, \forall j$, the improvement induced by the jth mutation is

$$\begin{aligned}\mathbb{E}[f(\mathbf{x}) - f(\mathbf{y}_j)|\mathbf{x}] &= \frac{1}{\sqrt{2\pi}\sigma_j}\int_{-x_j}^{x_j} c_j(x_j - |y|)\exp\left\{-\frac{(y-x_j)^2}{2\sigma_j^2}\right\}dy \\ &= 2c_j x_j\left(\Phi\left(2C_0\right) - \Phi\left(C_0\right)\right) + \frac{c_j x_j}{\sqrt{2\pi}C_0}\left(1 - 2e^{-\frac{C_0^2}{2}} + e^{-2C_0^2}\right).\end{aligned} \tag{24}$$

So,

$$\begin{aligned}e_{t-1} - e_t &= \sum_{j=1}^{d}\frac{1}{d}\mathbb{E}[f(\mathbf{x}) - f(\mathbf{y}_j)|\mathbf{x}] \\ &= \frac{1}{d}\sum_{j=1}^{d} 2c_j x_j\left(\Phi\left(2C_0\right) - \Phi\left(C_0\right)\right) + \frac{1}{d}\sum_{j=1}^{d}\frac{c_j x_j}{\sqrt{2\pi}C_0}\left(1 - 2e^{-\frac{C_0^2}{2}} + e^{-2C_0^2}\right) \\ &= \frac{1}{d}\sum_{j=1}^{d} c_j x_j\left(2\left(\Phi\left(2C_0\right) - \Phi\left(C_0\right)\right) + \frac{1}{\sqrt{2\pi}C_0}\left(1 - 2e^{-\frac{C_0^2}{2}} + e^{-2C_0^2}\right)\right),\end{aligned}$$

and we know that

$$\frac{e_{t-1} - e_t}{e_{t-1}} = \frac{1}{d}\left(2\left(\Phi\left(2C_0\right) - \Phi\left(C_0\right)\right) + \frac{1}{\sqrt{2\pi}C_0}\left(1 - 2e^{-\frac{C_0^2}{2}} + e^{-2C_0^2}\right)\right).$$

Then,

$$\begin{aligned}R_t &= 1 - \left(\prod_{k=1}^{t}\frac{e_k}{e_{k-1}}\right)^{1/t} \\ &= \frac{1}{d}\left(2\left(\Phi\left(2C_0\right) - \Phi\left(C_0\right)\right) + \frac{1}{\sqrt{2\pi}C_0}\left(1 - 2e^{-\frac{C_0^2}{2}} + e^{-2C_0^2}\right)\right),\end{aligned} \tag{25}$$

which demonstrates that R_t is inversely proportional to the dimension d.

6 Discussion of Positive-Adaptive Generators

Foregoing analysis on positive-adaptive generators requires knowledge of the optimum. The assumption of known optimum can make the analysis more intuitive, and the related computation can be easily implemented, but f^* usually is not available in real applications. If we do not know f^*, how can we design landscape-adaptive generator? For example, in the previous case study, how can we adapt the variance to make the Gaussian mutation positive-adaptive?

Consider the studied case presented in Sect. 5. Denote the individual at the tth generation to be $\mathbf{x}(t) = (x_1(t), \ldots, x_d(t))$. The positive-adaptive generator is obtained by setting $x_j(t)/\sigma_j(t) = C_0$ for all $j \in \{1, \ldots, d\}$, which leads to a theoretical setting of $\sigma_j(t) = x_j(t)/C_0$, $\forall\, j \in \{1, \ldots, d\}$. With the prerequisite that $\lim_{t\to+\infty} e_t = 0$, we know $\lim_{t\to+\infty} \mathbf{x}_t = \mathbf{x}^*$ *a.s.*, and consequently, $\lim_{t\to+\infty} \frac{1}{t}\sum_{k=0}^{t} \mathbf{x}_k = \mathbf{x}^*$ *a.s.* If the optimal solution $\mathbf{x}^*$ is unknown, the positive-adaptive Gaussian mutation can be obtained by setting $\sigma_t = \|\mathbf{x}_t - \frac{1}{t}\sum_{k=0}^{t} \mathbf{x}_k\|$.

7 Conclusions

The work in this paper extends the analysis of the average convergence rate (ACR) [15] from discrete optimisation to continuous optimisation. In terms of the ACR, this paper proves the necessity of using adaptive generators for solving continuous optimisation problems. Theorems 1 and 2 states that for EAs using landscape-invariant or zero-adaptive generators, the limit of their ACR is 0. These results don't exist in discrete optimisation. Theorem 3 indicates that for EAs using positive-adaptive generators, it converges to the optimal set with a positive ACR. A case study on (1+1) EA for minimising linear functions validates the applicability of general theoretical results. A simple way to design positive-adaptive generators is also proposed. Our future work is to develop methods for estimating lower and upper bounds on the ACR in both continuous and discrete optimisation.

References

1. Agapie, A., Agapie, M., Rudolph, G., Zbaganu, G.: Convergence of evolutionary algorithms on the n-dimensional continuous space. IEEE Trans. Cybern. **43**(5), 1462–1472 (2013)
2. Akimoto, Y., Auger, A., Hansen, N.: Quality gain analysis of the weighted recombination evolution strategy on general convex quadratic functions. In: Proceedings of the 14th ACM/SIGEVO Conference on Foundations of Genetic Algorithms, pp. 111–126. ACM (2017)
3. Auger, A.: Convergence results for the $(1, \lambda)$-sa-es using the theory of ϕ-irreducible markov chains. Theoret. Comput. Sci. **334**(1–3), 35–69 (2005)
4. Auger, A., Hansen, N.: Reconsidering the progress rate theory for evolution strategies in finite dimensions. In: Proceedings of the 8th Annual Conference on Genetic and Evolutionary Computation, pp. 445–452. ACM (2006)

5. Auger, A., Hansen, N.: Theory of evolution strategies: a new perspective. In: Theory of Randomized Search Heuristics: Foundations and Recent Developments, pp. 289–325. World Scientific (2011)
6. Auger, A., Hansen, N.: Linear convergence of comparison-based step-size adaptive randomized search via stability of markov chains. SIAM J. Optim. **26**(3), 1589–1624 (2016)
7. Beyer, H.G.: The Theory of Evolution Strategies. Springer, Heidelberg (2013). https://doi.org/10.1007/978-3-662-04378-3
8. Beyer, H.G., Hellwig, M.: The dynamics of cumulative step size adaptation on the ellipsoid model. Evol. Comput. **24**(1), 25–57 (2016)
9. Beyer, H.G., Melkozerov, A.: The dynamics of self-adaptive multirecombinant evolution strategies on the general ellipsoid model. IEEE Trans. Evol. Comput. **18**(5), 764–778 (2014)
10. Chen, Y., Zou, X., He, J.: Drift conditions for estimating the first hitting times of evolutionary algorithm. Int. J. Comput. Math. **88**(1), 37–50 (2011)
11. Ding, L., Kang, L.: Convergence rates for a class of evolutionary algorithms with elitist strategy. Acta Mathematica Scientia **21**(4), 531–540 (2001)
12. Doob, J.L.: Stochastic Processes. Wiley, New York (1953)
13. Droste, S., Jansen, T., Wegener, I.: On the analysis of the (1 + 1) evolutionary algorithm. Theoret. Comput. Sci. **276**(1–2), 51–81 (2002)
14. He, J., Kang, L.: On the convergence rate of genetic algorithms. Theoret. Comput. Sci. **229**(1–2), 23–39 (1999)
15. He, J., Lin, G.: Average convergence rate of evolutionary algorithms. IEEE Trans. Evol. Comput. **20**(2), 316–321 (2016)
16. He, J., Yao, X.: Drift analysis and average time complexity of evolutionary algorithms. Artif. Intell. **127**(1), 57–85 (2001)
17. He, J., Yu, X.: Conditions for the convergence of evolutionary algorithms. J. Syst. Architect. **47**(7), 601–612 (2001)
18. He, J., Zhou, Y., Lin, G.: An initial error analysis for evolutionary algorithms. In: Proceedings of the Genetic and Evolutionary Computation Conference Companion, pp. 317–318. ACM (2017)
19. Huang, H., Xu, W., Zhang, Y., Lin, Z., Hao, Z.: Runtime analysis for continuous (1+1) evolutionary algorithm based on average gain model. SCIENTIA SINICA Informationis **44**(6), 811–824 (2014)
20. Jebalia, M., Auger, A., Hansen, N.: Log-linear convergence and divergence of the scale-invariant (1+1)-es in noisy environments. Algorithmica **59**(3), 425–460 (2011)
21. Meyn, S., Tweedie, R.: Markov Chains and Stochastic Stability. Springer, London (1993). https://doi.org/10.1007/978-1-4471-3267-7
22. Rudolph, G.: Local convergence rates of simple evolutionary algorithms with Cauchy mutations. IEEE Trans. Evol. Comput. **1**(4), 249–258 (1997)
23. Rudolph, G., et al.: Convergence rates of evolutionary algorithms for a class of convex objective functions. Control Cybern. **26**, 375–390 (1997)
24. Varga, R.: Matrix Iterative Analysis. Springer, Heidelberg (2009). https://doi.org/10.1007/978-3-642-05156-2

Optimization Design of Multi-layer Logistics Network Based on Self-Adaptive Gene Expression Programming

Huazhi Zhou, Kangshun Li(✉), Runxuan Xu, Xinyao Qiu, and Zhanbiao Zhu

College of Mathematics and Informatics, South China Agricultural University, Guangzhou 510642, Guangdong, China
723400361@qq.com

Abstract. In order to solve the multi-layer logistics network optimization problem of modern enterprises, a mixed integer programming model with minimum total cost is established by considering the inventory cost and transportation cost and the switching state of the transit logistics nodes. According to the characteristics of multi-layer logistics network optimization problem, the gene expression programming with the characteristics of multi-gene structure is adopted, and the self-adaptive evolution mechanism is introduced to dynamically adjust the genetic operator. A self-adaptive gene expression programming algorithm based on Prüfer coding (SA-GEP) is proposed to solve the model. The algorithm introduces the insertion operator based on the original genetic operator of gene expression programming. The experimental results show that compared with STD-GEP algorithm and EC algorithm, the optimization effect of SA-GEP algorithm is more significant, which greatly improves the performance of the algorithm, and verifies the feasibility of the model and the effectiveness of the algorithm.

Keywords: Gene expression programming · Logistics network · Self-adaptive mutation · Prüfer coding

1 Introduction

How to integrate and optimize the modern logistics network has always been one of the hot topics of enterprise development. In recent years, with the rapid development of the economy, the rise of global supply chain management and the development of e-commerce technology, enterprises are paying more and more attention to the construction and improvement of logistics networks. The integration and optimization of the logistics network structure has become the core to improve the efficiency of the supply chain [1]. Therefore, logistics must be networked to meet the needs of the development of the times [2].

The logistics network structure mainly consists of two components, nodes and paths. Logistics network optimization is the process of optimizing network nodes and all logistics paths in the supply chain [3]. With the improvement of the complexity of logistics network, there are multiple logistics centers to choose from in the process of

K. Li et al. (Eds.): ISICA 2019, CCIS 1205, pp. 46–58, 2020.
https://doi.org/10.1007/978-981-15-5577-0_4

product distribution in modern logistics, and each logistics center has many optional distribution centers, which makes the logistics network optimization problem become a typical NP-hard problem. It is difficult to solve this problem effectively by traditional methods.

For the optimization of logistics network, there are related research at home and abroad. Syarif et al. [4] and Xu Hang et al. [5] used the idea of spanning tree to propose a multi-stage logistics network optimization model based on the evolution algorithm of Prüfer number. Wang et al. [6] studied the problem of two-level logistics distribution area division with the goal of minimizing total cost, and proposed a hybrid algorithm based on extended particle swarm optimization and genetic algorithm (EPSO-GA). Cho SY et al. [7] considered a balanced allocation and vehicle path, proposed a two-stage solution method and established a multi-resource multi-objective mixed integer programming logistics network model. Li Bozhen et al. [8] considered the uncertainty of consumer demand, established a closed-loop logistics network stochastic programming model by stochastic programming method, and proposed a genetic algorithm based on new priority coding to solve the model. Liu Yanqiu et al. [9] aimed to minimize the overall cost, considered dynamically adjusting the node closure state, established an optimization model for describing multi-level distribution network design problems with capacity constraints, and proposed an improved simulated annealing algorithm to solve the model.

Based on the above analysis, considering the inventory cost and transportation cost as well as the switching state of the transit logistics node, this paper builds a mixed integer programming model for the cost optimization problem of the multi-layer logistics network, with the goal of minimizing the total cost generated by the operation logistics network. According to the characteristics of multi-layer logistics network, this paper adopts gene expression programming with the characteristics of multi-gene structure, and introduces self-adaptive evolution mechanism to dynamically adjust genetic operators. A self-adaptive gene expression programming algorithm is proposed to solve the model.

2 Problem Description and Model Construction

2.1 Problem Description

Most of the traditional logistics network models are two-tier logistics networks that are distributed between suppliers and their customers, which is also called simple logistics networks. With the development of China's economy, China's logistics industry is in a period of great development. This paper focuses on the node location problem and product allocation problem of complex multi-layer logistics network, which belongs to the location-allocation problem (LAP) of logistics network planning. The multi-tier logistics network includes supplier node, transit logistics node, and customer node. The multi-layer logistics network optimization problem studied in this paper can be described as: Several customers feed back the product demand to the supplier, and the supplier delivers the product to the corresponding customer without exceeding the maximum supply. During the delivery process, the product passes through one to several layers of transit logistics nodes and finally reaches the customer. Under the premise of meeting the customer's product demand, the cost incurred by the whole process is required to be the least.

2.2 Model Hypothesis

The logistics network optimization model studied in this paper is based on the following assumptions:

(1) Taking a certain type of product in a single cycle as the research object, regardless of environmental factors and social benefits;
(2) The logistics network consists of multiple suppliers, multi-tier transit logistics nodes and multiple customers;
(3) The number of suppliers and the number of customers are known, and the number of transit logistics nodes per layer is known;
(4) The maximum capacity and inventory cost of each transit logistics node (if the node is enabled, the fee is required) is known;
(5) The transportation cost per unit of product between each node of the adjacent logistics network layer is known;
(6) Cross-layer provisioning and horizontal supply between peer nodes are not considered.

2.3 Parameter Definition

K: the number of layers of the multi-layer logistics network (the first layer is the supplier, the Kth layer is the final customer, and the middle is the transit logistics node), $K \geq 3$, k = 1, 2, 3,..., K;
M: number of supplier nodes, m = 1, 2, 3,..., M;
N: number of client nodes, n = 1, 2, 3,..., N;
W_k: the number of k-th layer logistics nodes (W_1 represents the number of supplier nodes when W = 1, i.e., W_1 = M; when k = K, W_k represents the number of client nodes, i.e., W_k = N);
$A_{k,i}$: the maximum processing capacity of the i-th node of the kth layer, i.e. the maximum capacity;
S_i: the maximum supply of products by the i-th supplier;
V_i: The demand for the i-th customer;
$C_{k,i}^{stock}$: inventory cost of the i-th logistics node of the kth layer;
$C_{k,i,j}^{trans}$: unit product transportation cost from the i-th node of the kth layer to the jth node of the k + 1th layer;
Decision variables:
$X_{k,i,j}$: number of product shipments from the i-th node to the j-th node of the kth layer;
$R_{k,i}$: the number of products passing through the i-th transit logistics node of the kth layer, k = 2, 3,... , K − 1;

$$Y_{k,i} = \begin{cases} 1, \text{Enable the i - th logistics node in the k - th layer} \\ 0, \quad \text{Other} \end{cases}$$

$$U_{k,i,j} = \begin{cases} 1, \text{The i - th node of the k - th layer sends the product to the jth node of the k + 1th layer} \\ 0, \quad \text{Other} \end{cases}$$

2.4 Mathematical Model Construction

The cost to be considered in the design of the multi-layer logistics network of this paper mainly includes inventory cost and transportation cost. Starting from the description and assumptions of the model, a mixed integer programming model with the primary goal of minimizing total cost is constructed to meet the needs of customers and reduce the operating costs of the enterprise as much as possible.

Objective function:.

$$\text{Min } C^{total} - C^{stock} + C^{trans} \tag{1}$$

Where C^{total} represents the total cost, C^{stock} represents the inventory cost of the transit logistics node, and C^{trans} represents the transportation cost.

$$C^{stock} = \sum_{k=2}^{K-1} \sum_{i=1}^{W_k} C_{k,i}^{stock} \cdot Y_{k,i} \tag{2}$$

$$C^{trans} = \sum_{k=1}^{K-1} \sum_{i=1}^{W_k} \sum_{j=1}^{W_{k+1}} C_{k,i,j}^{trans} \cdot X_{k,i,j} \tag{3}$$

s.t.

$$\sum_{i=1}^{M} S_i \geq \sum_{i=1}^{N} V_i \tag{4}$$

$$\sum_{i=1}^{W_k} \sum_{j=1}^{W_{k+1}} X_{k,i,j} = \sum_{i=1}^{W_{k+1}} \sum_{j=1}^{W_{k+2}} X_{(k+1),i,j}, \quad \forall \mathrm{k} = 1, 2, \ldots, \mathrm{K} - 2 \tag{5}$$

$$R_{k,i} = \sum_{j=1}^{W_{k+1}} X_{k,i,j} \cdot U_{k,i,j}, \quad \forall \mathrm{k} = 1, 2, \ldots, \mathrm{K} - 1; \ \forall \mathrm{i} = 1, 2, \ldots, W_k \tag{6}$$

$$R_{k,i} \leq A_{k,i} \cdot Y_{k,i}, \quad \forall \mathrm{k} = 1, 2, \ldots, \mathrm{K}; \ \forall \mathrm{i} = 1, 2, \ldots, W_k \tag{7}$$

$$\sum_{i=1}^{W_k} \sum_{j=1}^{W_{k+1}} X_{k,i,j} = \sum_{i=1}^{N} V_i, \ \forall \mathrm{k} = 1, 2, \ldots, \mathrm{K} - 1 \tag{8}$$

$$\sum_{i=1}^{W_k} X_{k,i,j} \cdot U_{k,i,j} = V_i, \quad k = K - 1 \tag{9}$$

$$Y_{k,i}, U_{k,i,j} \in \{0, 1\} \tag{10}$$

$$S_i, V_i, A_{k,i}, X_{k,i,j}, R_{k,i}, C_{k,i}^{stock}, C_{k,i}^{stock} \geq 0, \quad \forall k, i, j \tag{11}$$

Among them, (4) stipulates that the total supply of the supplier must meet the total demand of all customers; (5) represents the flow conservation constraint of the nodes of different logistics layers; (6) represents the different logistics layers. The flow conservation constraint on the node ensures that the supply quantity of the product does not exceed the supply quantity; (7) represents the processing capacity limit of the logistics node, that is, the capacity limitation constraint; (8) and (9) stipulate that the supply quantity must satisfy the customer. The product demand ensures that the quantity of products delivered to the final customer node through the node is equal to the actual demand of the customer. (10) specifies the type of decision variable; (11) represents the non-negative constraint of the variable.

3 Self-Adaptive Gene Expression Programming Algorithm Based on Prüfer Coding

Gene Expression Programming (GEP) [12] is a new self-adaptive evolutionary algorithm based on natural selection and genetic mechanisms. It combines the advantages of genetic algorithm (GA) and genetic programming (GP) while overcoming the disadvantages of the two algorithms. Gene expression programming is characterized by the ability to separate genotypes from phenotypes and to solve complex problems with simple coding. The multi-layer logistics network problem studied in this paper is an NP-hard problem, which is difficult to solve with an accurate algorithm. It is more complicated than the traditional two-tier logistics network. This paper proposes a Self-Adaptive Gene Expression Programming Algorithm based on Prüfer Coding (SA-GEP). According to the multi-gene structure of GEP, in each chromosome, each gene can represent a simple logistics network in a multi-layer logistics network. Multiple simple logistics networks form a complete multi-layer logistics network, so a chromosome composed of multiple genes can represent a complete multi-layer logistics network, which is a major advantage of GEP in solving the logistics network optimization model.

3.1 Gene Coding and Decoding

When using the intelligent optimization algorithm to solve the multi-layer logistics network model, the coding method of the feasible solution is a key issue to be considered. Prüfer coding [10, 11] is an effective method for network coding. According to Cayley's theorem, in a complete graph with n vertices, there are n^{n-2} different label trees. Let the tree T be a label tree with n nodes whose node numbers are {1, 2,… , n}, which can be uniquely represented by an array of lengths n − 2 formed by natural numbers between 1 and n. This arrangement is usually called the Prüfer number. Please refer to literature [4, 5] for the decoding process of Prüfer for logistics network problems.

3.2 Fitness Function

This paper adopts the fitness function based on the objective function formula (1):

$$\text{Fitness} = \delta \cdot \frac{1}{C^{total} + 1} \tag{12}$$

Where C^{total} represents the total cost, δ is the fitness evaluation factor, and δ is a positive number.

3.3 Genetic Operation Design for Logistics Network

The basic genetic operators of GEP include selection, mutation, reversal, interpolation, root insertion, gene transformation, one-point recombination, two-point recombination and gene recombination [13]. The genetic operations designed for multi-layer logistics network optimization problems mainly include selection and copy operation, mutation operation, insertion operation, inversion operation and recombination operation.

(1) Selection and copy operations
The traditional GEP selection operator adopts the Roulette Wheel selection method. According to the fitness value, the individuals with better fitness values in the population are more likely to be selected and copied directly to the next generation to generate new populations. This paper adopts the combination strategy of roulette selection method and elite retention strategy. Firstly, N individuals are selected by roulette wheel selection method to perform a series of genetic operations to generate offspring, and then the parent population and the offspring population are combined to build a temporary population with a size of 2N, and then we carry out the elite retention strategy. By comparing the fitness value of the individual, select the N outstanding individuals with larger fitness values in the temporary population to form the next generation of new populations. The superior individuals in the population replace the poor individuals in the offspring population.
(2) Mutation operation
In view of the multi-layer logistics network optimization problem studied in this paper, the mutation operation first randomly selects a gene on the chromosome according to a certain mutation probability, and then generates a new individual by changing a certain node in the gene.
(3) Insertion operation
For the multi-layer logistics network optimization problem studied in this paper, each chromosome represents a complete multi-layer logistics network. Each gene in the chromosome corresponds to a simple logistics network. In a multi-layer

logistics network, one gene represents the adjacent two-layer logistics network. According to the requirements of Prüfer coding, the length of each gene in the chromosome is related to the number of nodes in the adjacent two-layer logistics network, and the position of each gene in the chromosome cannot be changed, otherwise the wrong result will be obtained, so gene insertion is not suitable for genetic operation of multi-layer logistics network. Since IS and RIS operations remove codes that exceed the length of the gene's head, this may affect the integrity and validity of the gene, and at the same time cause an imbalance in the Prüfer coding sequence. Therefore, this paper introduces a insertion operator [14], which randomly selects a substring from a gene, moves the substring to an arbitrary position of the gene according to a certain insertion probability. The substring of the insertion operation in the gene and its length is random.

(4) Inversion operation
In view of the multi-layer logistics network optimization problem studied in this paper, the inversion operation randomly selects a substring from the gene, and then inverts the characters in the substring in order, hat is to say, the center character of the substring is taken as the symmetry axis, and the symmetrical character positions are interchanged in order. The substring that is inverted in the gene and its length is random.

(5) Recombination operation
Recombination operation include one-point recombination, two-point recombination, and genetic recombination.

3.4 Self-Adaptive Operator Design

The mutation probability of the standard gene expression programming algorithm is usually a fixed constant when it performs mutation operation on individuals. This method of ignoring the individual's superiority and inferiority of the population limits the convergence speed of the algorithm to a certain extent. Therefore, this paper adopts the self-adaptive evolution mechanism to dynamically adjust the mutation probability according to the individual fitness value of the population, reduce the damage to the optimal individual, and improve the convergence speed of gene expression programming algorithm to a certain extent.

In the process of individual evolution, the self-adaptive adjustment formula of the mutation operator is as follows:

$$P_m = \begin{cases} P_1 - \frac{(P_1 - P_2)\cdot(f_{max} - f)}{f_{max} - f_{avg}} & ,f \geq f_{avg} \\ P_1 & ,f < f_{avg} \end{cases} \tag{13}$$

In (13), P_1 is the initial mutation probability, P_2 is the adjustment parameter, P_1 and P_2 are the constants in [0, 1], f_{max} is the maximum fitness value of the population, f_{avg} is the average fitness value of the population, and f is the fitness value of the individual which will be mutated. In the experiment of this paper, $P_1 = 0.3$ and $P_2 = 0.05$ were set.

3.5 Self-Adaptive Operator Design

The SA-GEP algorithm flow proposed in this paper is shown in the Fig. 1.

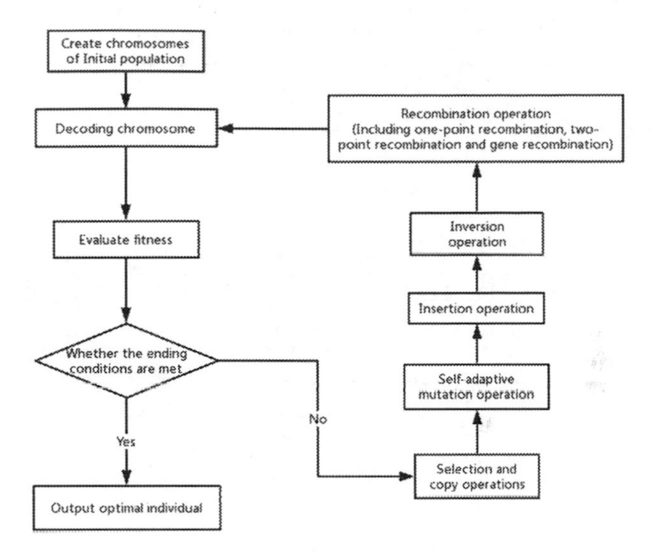

Fig. 1. Algorithm flow chart of SA-GEP.

4 Experimental Simulation and Analysis

4.1 Experimental Data and Algorithm Parameters

This section of the experiment considers the classic four-tier logistics network structure, namely, supplier (S), Logistics Center (LC), Distribution Center (DC), and Customer (C). Suppose a logistics network consists of 6 suppliers, 10 logistics centers, 15 distribution centers and 20 customers. The logistics center and distribution center belong to the transit logistics node mentioned in the model. Tables 1, 2, 3, 4, 5, 6 and 7 are experimental data of a four-layer logistics network, and Table 8 is an algorithm parameter description.

Table 1. Maximum supply of supplier (S)

	1	2	3	4	5	6
Supply	1251	1250	1531	1797	1267	1455

Table 2. Maximum capacity of logistics center (LC) and distribution center (DC)

	1	2	3	4	5	6	7	8	9	10	11	12	13	14	15
LC	928	955	955	757	815	998	985	846	785	918	–	–	–	–	–
DC	613	498	635	561	714	822	506	482	590	682	629	646	862	703	454

Table 3. Customer (C) product demand

	1	2	3	4	5	6	7	8	9	10
Demand	251	350	224	518	583	758	230	217	987	391

	11	12	13	14	15	16	17	18	19	20
Demand	313	198	535	261	514	222	206	482	490	482

Table 4. Inventory costs of logistics center (LC) and distribution center (DC)

	1	2	3	4	5	6	7	8	9	10
LC	1517	1828	1621	1298	1867	1751	1466	1371	1655	1889
DC	1040	1573	1042	1221	1508	1021	1205	1252	1382	1307

	11	12	13	14	15
LC	--	--	--	--	--
DC	1411	1427	1457	1265	1254

Table 5. Unit product transportation costs from supplier (S) to logistics center (LC)

S	LC									
	1	2	3	4	5	6	7	8	9	10
1	8	2	6	5	3	4	1	7	7	6
2	7	8	5	2	4	7	6	5	8	5
3	8	2	7	1	8	6	2	7	8	4
4	2	1	4	7	6	6	5	8	6	6
5	3	5	5	4	2	3	6	3	6	2
6	3	4	5	4	4	3	7	3	5	3

Table 6. Transportation costs per unit of product from the logistics center (LC) to the distribution center (DC)

LC	DC														
	1	2	3	4	5	6	7	8	9	10	11	12	13	14	15
1	5	4	7	5	8	4	4	2	6	2	1	5	6	7	8
2	2	1	3	6	7	9	4	2	4	3	7	3	5	5	6
3	5	5	4	3	5	8	4	7	9	8	5	4	4	5	6
4	1	4	5	7	2	7	5	5	4	2	4	7	6	4	1
5	7	4	3	6	5	4	4	3	2	5	4	6	4	7	6
6	5	7	3	5	4	3	8	7	5	2	5	1	6	3	4
7	7	2	6	3	8	4	3	6	1	2	3	6	7	3	7
8	5	4	4	1	5	9	6	5	7	3	4	7	9	5	1
9	4	5	7	5	4	8	2	4	5	2	1	5	4	7	4
10	8	6	5	6	2	7	9	8	8	5	7	9	5	7	4

Table 7. Unit Product transportation cost from distribution center (DC) to customer (C)

DC	C																			
	1	2	3	4	5	6	7	8	9	10	11	12	13	14	15	16	17	18	19	20
1	1	6	4	7	5	6	7	5	5	1	1	8	4	4	7	9	4	6	6	2
2	2	9	6	4	6	5	7	5	6	5	2	7	5	4	6	7	4	4	5	2
3	7	7	2	2	3	1	1	7	5	8	1	7	6	4	7	7	5	4	3	1
4	4	5	3	3	2	1	2	5	7	5	8	7	5	4	5	6	5	2	3	4
5	2	4	6	6	3	5	4	7	9	9	1	6	3	9	1	3	7	1	2	4
6	7	4	9	3	6	8	1	6	7	1	1	6	7	8	5	1	1	1	8	4
7	1	7	5	8	1	7	6	4	7	7	8	2	4	9	3	6	9	4	4	7
8	9	4	7	2	5	1	8	2	6	9	6	2	1	6	5	4	8	3	8	1
9	1	5	8	2	5	7	5	2	4	5	3	3	6	2	6	6	7	6	9	7
10	8	2	4	9	3	6	9	4	4	7	6	8	2	8	4	3	3	6	2	6
11	2	4	6	6	3	5	4	7	9	9	4	5	9	9	7	9	5	8	3	2
12	1	6	7	5	1	6	7	8	5	1	7	8	1	8	1	8	6	3	2	2
13	4	6	5	7	5	6	5	2	7	5	5	7	5	2	4	5	4	6	1	8
14	3	8	4	3	6	1	2	3	6	7	8	1	7	6	4	7	7	8	2	4
15	5	5	4	5	9	9	3	2	6	4	3	3	1	2	3	3	4	9	7	3

4.2 Experimental Results and Analysis

In order to verify the effectiveness of the algorithm, this paper uses the SA-GEP algorithm, the standard GEP algorithm (STD-GEP) and the traditional evolutionary computation method (EC) of the literature [5] to carry out 100 experiments on the model, and each experiment output the optimal distribution scheme and optimal cost corresponding to the optimal individual in the population. Table 9 shows the performance indicators of the SA-GEP algorithm, the standard GEP algorithm, and the EC algorithm. Figure 2 shows the evolution of the population evolution process of the SA-GEP algorithm, the standard GEP algorithm, and the EC algorithm under the same evolutionary algebra.

It can be seen from Table 9 that the SA-GEP algorithm is better than the STD-GEP and EC algorithms in comparing the indicators related to the objective function of the model. It can be seen from Fig. 2 that in the initial stage of the algorithm (before the 50th generation), the convergence speed of the three algorithms is almost the same, but with the increase of the population evolution algebra, the EC algorithm is easy to fall into the local optimal solution. The self-adaptive operator of SA-GEP algorithm greatly improves the performance of the algorithm, and the optimal value obtained by solving the model is always better than the other two algorithms.

Table 8. Algorithm parameter setting and description

Parameter name	Parameter setting and description
Population size	100
Mutation probability	The initial probability is P_1 = 0.3, self-adaptive adjustment
Insertion probability	0.4
Inversion probability	0.4
One-point recombination probability	0.4
Two-point recombination probability	0.4
Genetic recombination probability	0.4
Fitness evaluation factor δ	10000
Maximum evolution algebra	1000
Number of genes	the number of logistics network layers is reduced by 1, i.e. K − 1

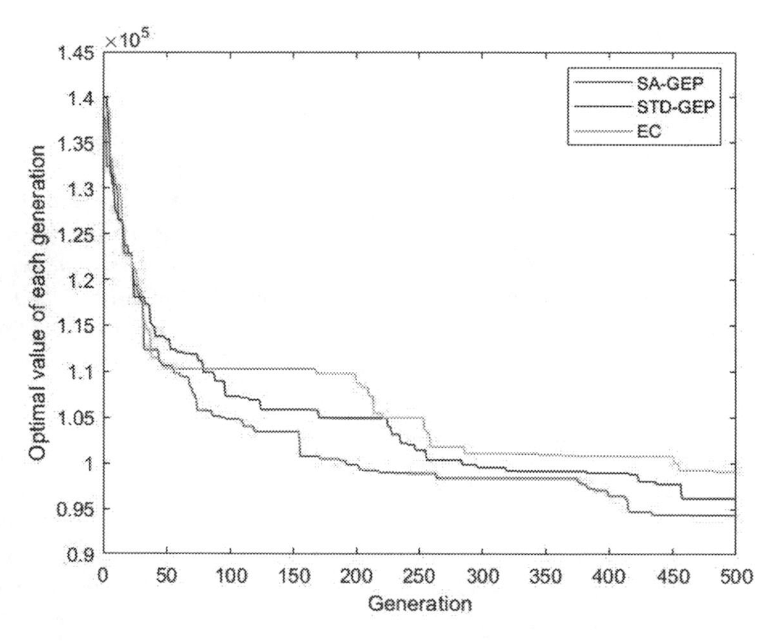

Fig. 2. Comparison of population evolution processes of SA-GEP, STD-GEP and EC under the same evolutionary algebra

Table 9. Comparison of performance indicators of SA-GEP algorithm, standard GEP algorithm and EC algorithm

Algorithm	Optimal cost	Worst cost	Average cost	Standard deviation	Average evolutionary algebra
SA-GEP	92898	108576	99966.01	2869.54	494.97
STD-GEP	97919	113759	104826.63	3582.15	455.70
EC	99819	115246	107659.62	3274.70	305.10

5 Conclusion

Logistics network optimization is the process of optimizing network nodes and all logistics paths in the supply chain. In this paper, a hybrid integer programming model with minimum total cost is constructed for the cost optimization problem of multi-layer logistics network. According to the characteristics of the model, the gene expression programming with the characteristics of multi-gene structure is adopted, and the self-adaptive evolution mechanism is introduced to dynamically adjust the genetic operator. A self-adaptive gene expression programming algorithm based on Prüfer coding is proposed to solve the model. Through experimental comparison and analysis, the optimization effect of SA-GEP algorithm is more significant than the other two algorithms, which verifies the feasibility of the model and the effectiveness of the algorithm.

Acknowledgement. This work is supported by Innovation and Entrepreneurship Training Program for college students of South China Agricultural University with No. 201810564015, the Major Science and Technology Project in Dongguan with No. 2018215121005, National Key R&D Program of China with the Grant No. 2018YFC0831100, the National Natural Science Foundation of China with the Grant No. 61773296, the National Natural Science Foundation Youth Fund Project of China under Grant No. 61703170, Foreign Science and Technology Cooperation Program of Guangzhou with the grant No. 201907010021, the Open Foundation of Key Lab of Data Analysis and Processing of Guangdong Province in Sun Yat-sen University with No. 201901.

References

1. Wang, Z., Zhang, Q., Yang, B., He, W.: 4/R/I/T distribution logistics network 0-1 programming model and application. Comput. Ind. Eng. **55**(2), 365–378 (2008)
2. Tian, Q., Yan, L., Li, Z.: Logistics network design based on transport planning and combined GA. J. Tsinghua Univ. (Sci. Technol. **44**(11), 1441–1444 (2004)
3. Liu, D., Zhu, C., Zhou, W.: Enterprise logistics network optimization based on genetic algorithm. Adv. Mater. Res. **744**, 595–600 (2013)
4. Syarif, A., Yun, Y., Gen, M.: Study on multi-stage logistic chain network: a spanning tree-based genetic algorithm approach. Comput. Ind. Eng. **43**, 299–314 (2002)
5. Xu Hang, H., Xu, W., Ye, Q.: Study on evolutionary computation based on Prüfer number—multi-stage logistic network optimization. Comput. Eng. Appl. **31**, 5–7 (2005)

6. Wang, Y., Ma, X., Maozeng, X., Liu, Y., Wang, Y.: Two-echelon logistics distribution region partitioning problem based on a hybrid particle swarm optimization–genetic algorithm. Expert Syst. Appl. **42**, 5019–5031 (2015)
7. Cho, S.Y., Lee, Y.H., Cho, D.W., Gen, M.: Logistics network optimization considering balanced allocation and vehicle routing. Mariti. Econ. Logistics **18**(1), 41–60 (2016)
8. Li, B., Zhao, G., Ge, Y.: Stochastic programming model of closed loop logistics network based on genetic algorithm. Comput. Integr. Manuf. Syst. **23**(9), 2003–2011 (2017)
9. Liu, Y., Jiao, N., Li, J.: Optimization design of multi-level logistics network based on determined network. Yanggong Univ. Sci. **37**(1), 64–68 (2015)
10. Wang, X., Wu, Y.: Optimal algorithm for coding and decoding Prufer codes. Small Comput. Syst. **4**(4), 687–690 (2008)
11. Ma, C., Ma, C.: Robust model and algorithms for uncertain traveling salesman problem. J. Comput. Appl. **34**(7), 2090–2092 (2014)
12. Peng, Y., Yuan, C., Mai, X., Yan, X.: Survey on theoretical research of gene expression programming. J. Comput. Appl. **28**(2), 413–419 (2011)
13. Yuan, C., Peng, Y., Yan, X., et al.: Principle and Application of Gene Expression Programming Algorithm. Science Press, Beijing (2010)
14. Maojun, L., Richeng, L., Tiaosheng, T.: Analysis on the genetic operators of parthenogenetic algorithm. Syst. Eng. Electron. **23**(8), 84–87 (2001)

Potential Well Analysis of Multi Scale Quantum Harmonic Oscillator Algorithms

Jin Jin[1,2](✉) and Peng Wang[3]

[1] Chengdu Institute of Computer Application, Chinese Academy of Sciences, Chengdu 610041, China
jinjin@nsu.edu.cn
[2] University of Chinese Academy of Sciences, Beijing 100049, China
[3] Southwest Minzu University, Chengdu 610041, China

Abstract. The multiscale quantum harmonic-oscillator algorithm is an intelligent optimization algorithm based on quantum harmonic wave functions. Although it is effective for many optimization problems, an analysis for its performance is still lacking. This paper discusses the harmonic-oscillator potential well, delta-function potential well, and infinite-square potential well in terms of their application in evolutionary algorithms. Of the three, the harmonic-oscillator potential well is considered to give the most precise approximation for complex objective functions. When combined with the harmonic-oscillator potential well, the multiscale quantum harmonic-oscillator algorithm exhibits good adaptability in terms of the convergence of the wave function. To verify its global optimization performance, experiments are conducted using a double-well function to analyze the convergence of the multiscale quantum harmonic-oscillator algorithm and a suite of benchmark functions to compare the performance of different potential wells. The experimental results indicate that the multiscale quantum harmonic-oscillator algorithm with the harmonic-oscillator potential well is a better practical choice than the other two potential well models, and show that the multiscale quantum harmonic-oscillator algorithm is a potential quantum heuristic algorithm for optimization.

Keywords: Multi-scale quantum harmonic oscillator algorithm · Heuristic algorithm · Global optimization · Potential well

1 Introduction

Optimization problems are common in industrial design, automation, control, and other applied disciplines. Many machine learning problems can be formulated in terms of some optimization goal [2]. Direct optimization algorithms are an appealing approach for solving complicated problems. Some heuristics follow a trial-and-error framework [1], with swarm intelligence and evolutionary algorithms (EAs) being notable examples.

K. Li et al. (Eds.): ISICA 2019, CCIS 1205, pp. 59–71, 2020.
https://doi.org/10.1007/978-981-15-5577-0_5

The use of quantum mechanics to construct an optimization algorithm is an effective approach. The simulated annealing (SA) algorithm developed by Kirkpatrick *et al.* was an early exploration using physical mechanisms [9]. SA is a robust and efficient minimization algorithm in which a virtual temperature is slowly decreased to minimize the objective function over a large search space. Through the tunneling effect in quantum mechanics, the SA algorithm can jump over barriers and find the optimal solution. Quantum annealing (QA) was proposed in the 1990s [8]. This can be seen as an offspring of SA in which the particle is assigned an appropriate time-dependent kinetic energy and the optimal solution is determined through the change in the energy state $|\Psi_t\rangle$.

The process of QA in diffusion Monte Carlo sampling, whereby the solution jumps out of the local quantitatively optimal region through quantum tunneling, has been analyzed in the literature [5]. In addition, the optimal performance of classical annealing and quantum annealing has been compared for harmonic-oscillator potential energy, double well potential energy, and parabolic potential energy functions [11].

These quantum heuristic algorithms use the probability interpretation of Born [3], and their performance is improved by model approximation; however, other potential well constraints have not been analyzed in detail.

The multiscale quantum harmonic-oscillator optimization algorithm (MQHOA) is a quantum heuristic and population-based optimization algorithm based on the wave function. The core idea of MQHOA is derived from the continuous convergence process of the quantum wave function from a high-energy state to the ground state.

In MQHOA, we assume that the objective function $f(x)$ is the potential well $V(x)$ in Schrödinger's equation. Because the second-order Taylor approximation of the complex objective function is the harmonic-oscillator potential, the wave function of the quantum harmonic oscillator will reflect the probability density distribution of the optimal solutions [10,11].

Because of the adoption of a wave function corresponding to different potential wells has a profound influence on the performance of the algorithm. This paper studies the performance of the MQHOA with different potential wells and reveals the working mechanism of the wave function under different constraint conditions. To further analyze the influence of the potential well on the algorithm, comparative experiments are conducted and the experimental results are statistically analyzed. The experimental results demonstrate that MQHOA with a harmonic-oscillator potential well model is a better choice, both in theory and in practice, than two other potential well models, and show that MQHOA is a potential quantum heuristic algorithm.

The remainder of this paper is as follows. Section 3 analyzes the different potential well models and summarizes the theoretical framework of the algorithm under these models. Section 4 describes the experiments and presents the results. Finally, the conclusions to this study and ideas for future work are outlined in Sect. 5.

2 Multiscale Quantum Harmonic-Oscillator Algorithm

2.1 Core Concepts of MQHOA

As the probability interpretation is used, the wave function describes the distribution of the global optimal solution of the objective function, which plays an important role in the iterative process. The physical model of MQHOA includes the following core ideas and key points:

1. Equivalent hypothesis: The objective function $f(x)$ is mapped to the potential well $V(x)$ in Schrödinger's equation, transforming the optimization problem into that of solving a constrained ground state wave function.
2. Taylor expansion: Usually, the objective function $f(x)$ is a complicated equation that cannot be easily solved using mathematical methods. Moreover, the objective function, which is equivalent to the potential energy of the wave function, cannot typically be determined by directly solving Schrödinger's equation.
3. Wave function: The wave function indicates the distribution of the current global optimal solution of the objective function. The iterative process of the optimization algorithm aims to maximize the probability of the global optimal solution in the wave function.

2.2 Physical Model of MQHOA

The Taylor expansion of the objective function near the global optimal solution can be written as:

$$f(x) = f(x_0) + f'(x_0)(x - x_0) + \frac{1}{2}f''(x_0)(x - x_0)^2 + \ldots \quad (1)$$

where $f(x_0)$ is a constant, x_0 is the global optimum, i.e., $f'(x_0) = 0$, and $f''(x_0) > 0$ is negligible as long as $(x - x_0)$ is sufficiently small. We then have $f(x) \approx \frac{1}{2}f''(x_0)(x - x_0)^2$, where $f''(x_0)$ is the spring constant k in harmonic motion:

$$f(x) = \frac{1}{2}kx^2 \quad (2)$$

where $f(x)$ is defined as the potential energy of a quantum harmonic oscillator and k is the coefficient of resilience. Substituting Eq. (2) into the time-independent Schrödinger equation:

$$E\psi(x) = \left(-\frac{\hbar^2}{2m}\frac{\partial^2}{\partial x^2} + V(x)\right)\psi(x) \quad (3)$$

The wave function is then as follows:

$$\psi_n(x) = exp(-\frac{m\omega x^2}{\hbar}) \cdot H_n(\sqrt{\frac{m\omega}{\hbar}}x) \quad (4)$$

where n is the energy level. The probability density function is $|\psi_0|^2$. As $n \to 0$, Eq. (4) becomes:

$$|\psi_0|^2 = (\frac{m\omega}{\pi\hbar})^{\frac{1}{2}} \cdot exp(-\frac{m\omega x^2}{\hbar}) \tag{5}$$

which is a form of the Gaussian function:

$$\psi(x) = \frac{1}{\sqrt{2\pi}\sigma} exp(-\frac{(x-\mu)^2}{2\sigma^2}) \tag{6}$$

From the above derivation, it can be inferred that, from a high energy level to the ground state, the wave function of the quantum harmonic oscillator converges from the accumulation of n Gaussian functions in Eq. (4) to the single superposed Gaussian function in Eq. (5). The different energy levels of the wave function indicate the probability distributions of the optimal solution.

2.3 Framework of MQHOA

According to the physical model of MQHOA, we now present the framework for the algorithm. The framework of MQHOA is concise and includes three phases: the energy-level stabilization phase, energy-level transition phase, and scale-reduction phase. Figure 1 illustrates these three processes.

In MQHOA, the wave function at scale σ_s are defined as:

$$\psi(x) = \sum_{i=1}^{n} \psi(x) = \sum_{i=1}^{n} \frac{1}{\sqrt{2\pi}\sigma_s} exp(-\frac{(x-x_i)^2}{2\sigma_s^2}) \tag{7}$$

In this case, x_i is the center of the sampling point and σ_s represents the current scale. In this algorithm, we have exploited the quantum mechanics principle of superposition. The wave function is considered to be the superposition of k Gaussian samplings of the probability distribution, and represents the distribution of the current local minimum of the objective function. As the iterations progress, the probability distribution of the wave function becomes concentrated near the global optimal solution.

3 Analysis of Different Potential Wells

In MQHOA, the wave function corresponds to the probability distribution of the global optimal solution. Thus, the adoption of different potential well wave functions will have a profound impact on the algorithm. Therefore, the next critical step in MQHOA is the choice of a suitable attractive potential field that can guarantee bound states for particles moving in the quantum environment. The wave function is the core problem of quantum mechanics.

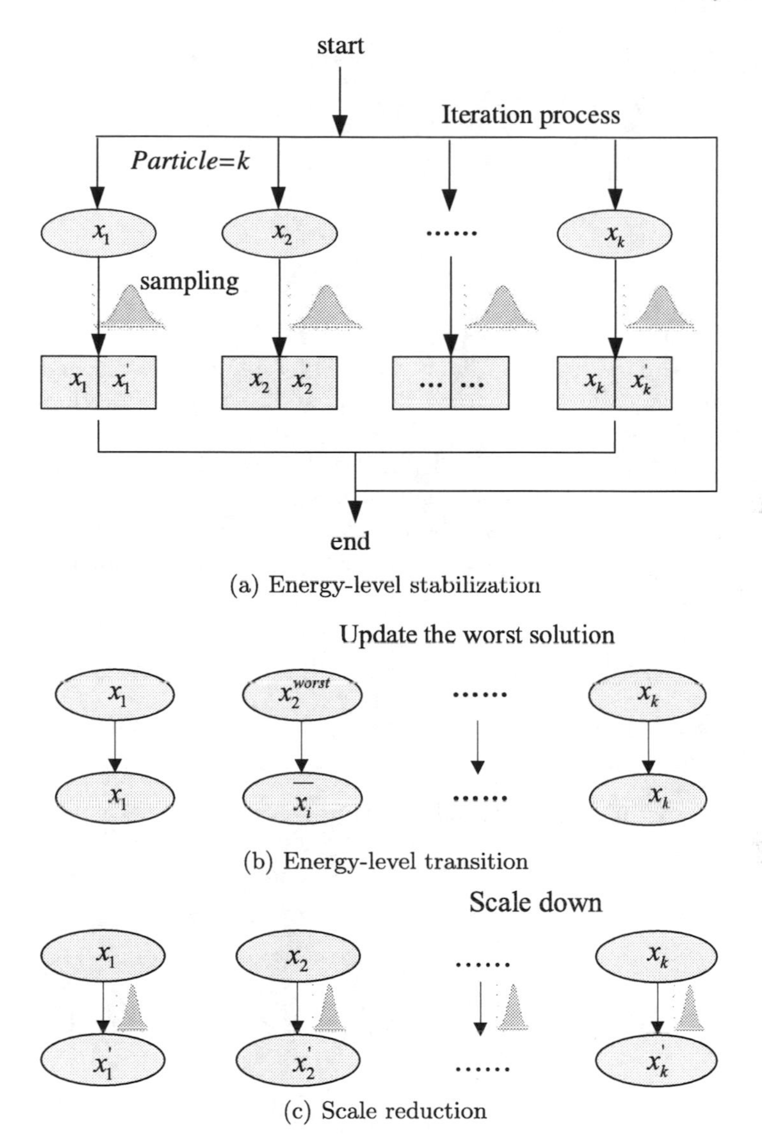

(a) Energy-level stabilization

(b) Energy-level transition

(c) Scale reduction

Fig. 1. Three phases of MQHOA: energy-level stabilization phase, energy-level transition phase, and scale-reduction phase

3.1 Choice of the Potential Well

In addition to the quantum harmonic-oscillator potential well, other formulations can be used as an expansion approximation of the objective function. Equation (1) can be rewritten as follows:

$$f(x) = f(x_0) + M \tag{8}$$

where M represents another potential well model, such as the delta potential well:

$$V(x) = -\alpha\delta(x) \tag{9}$$

where $\delta(x)$ is the Dirac delta function, or the infinite square well:

$$f(x) = \begin{cases} 0 & \text{if } 0 \leqslant x \leqslant W \\ \infty & \text{otherwise} \end{cases} \tag{10}$$

By means of potential well equivalence, the wave function corresponding to different potential well models can be obtained by solving Schödinger's equation. The wave function can fully describe the probability distribution characteristics of the harmonic oscillator.

In this paper, we consider the delta potential well, infinite square potential well, and harmonic-oscillator potential well as the choices for the bound state. Of course, there are other constraints besides these potential wells, but the analytic solutions for these wells are not easy to determine. As we will see, it is very important that the algorithm can find the wave function easily.

3.2 Proposed Algorithm with Different Potential Wells

According to Table 1, MQHOA can be summarized as follows under the different potential wells.

1) Choose an appropriate potential well constraint model (harmonic-oscillator, delta, or square potential well model) to solve Schrödinger's equation and obtain the corresponding wave function. The probability density function of the particles can be obtained using:

$$Q(x) = |\psi(x)|^2 \tag{11}$$

2) Use Monte Carlo simulations to obtain the specific sampling calculation formula.
3) Identify the effects of sampling on the energy-level stabilization process and energy-level adjustment process of the algorithm. The pseudocode of MQHOA is described in Algorithm 1.

In quantum mechanics, the harmonic-oscillator, delta, and square potential wells are the three most common formulations. According to Taylor series theory, the harmonic-oscillator potential well corresponds to the second-order approximation of the Taylor expansion and the other two potential wells correspond to the zero-order approximation.

Table 1. Summary of the probability density functions and update equations of the delta, harmonic-oscillator, and square potential wells, where u is the random distribution, $\alpha = \frac{1}{\sigma}$, σ is the standard scale of the harmonic-oscillator potential well, L is the *characteristic length* of the delta potential well, and W is the *characteristic length* of the square potential well. x_i is the sampling center point of the particle, which is the center point of the wave function in the algorithm.

Potential well model	Probability density function	Sampling points
1. Harmonic oscillator	$Q(x) = \frac{\alpha}{\sqrt{\pi}} \exp(-\alpha^2 x^2)$	$x_i' = x_i \pm \frac{1}{\alpha}\sqrt{\ln \frac{1}{u}}$
2. Delta	$Q(x) = \frac{1}{L} \exp(-2\|x\|L)$	$x_i' = x_i \pm \frac{L}{2} \ln \frac{1}{u}$
3. Square	$Q(x) = \frac{a}{W} cos^2(\frac{\xi}{W} x)$	$x_i' = x_i \pm \frac{W}{\pi} arccos(\sqrt{u}) - \frac{W}{4}$

Algorithm 1: Pseudocode of MQHOA under different potential wells

Initialize k, σ_{min}, σ_s, d_{min}, d_{max}, maxFE
Randomly generate x_i, $(i = 1, ..., k)$ in $[d_{min}, d_{max}]^D$
Evaluate the standard deviation σ_k for all x_i
while $(\sigma_s > \sigma_{min})$ *or* $w <= maxFE$ **do**
  while $(\sigma_k > \sigma_s)$ **do**
    set $Flag_{stable} = 0$
    while $(Flag_{stable} == 0)$ **do**
      $Flag_{stable} = 1$
      Generate x_i' from sampling point of different potential well model:
      $x_i' = x_i \pm \frac{1}{\alpha}\sqrt{\ln \frac{1}{u}}$
      or $x_i' = x_i \pm \frac{L}{2} \ln \frac{1}{u}$
      or $x_i' = x_i \pm \frac{W}{\pi} arccos(\sqrt{u}) - \frac{W}{4}$
      $\forall$ x_i and x_i',
      if $f(x_i') < f(x_i)$;
      $x_i = x_i'$
    end
    update the worst solution: $x^{worst} = x^{mean}$;
  end
  $\sigma_s = \sigma_s / \lambda$
  or $L = L/\lambda$
  or $W = W/\lambda$
end
output x^{best}, $f(x^{best})$

4 Experimental Results and Discussion

To verify the performance of MQHOA under different potential well models, two types of experiments were conducted. The first used the double well function to verify the convergence of MQHOA, and the second used well-defined multidimensional CEC benchmark functions to compare the performance of MQHOA with that of heuristic algorithms. All experiments were executed on a 2.5-GHz Intel(R) Core(TM) i5-7200U CPU with 8-GB RAM using MATLAB R2017a.

4.1 Convergence Under Double Well Function

The double well function is an important model [7] that reflects the tunneling effect of quantum heuristic algorithms [13], and thus their convergence characteristics. The double well function has two extreme points: a local optimum and the global optimum [12]. The wave function is temporarily stable in the ground state, according to the MQHOA wave function representing the convergence process of the algorithm. To demonstrate the convergence of the algorithm, a double well function is proposed.

The double-well function can be written as:

$$f(x) = \nu \frac{(x^2 - a^2)^2}{a^4} + \delta \cdot x \tag{12}$$

where ν, a, and δ ($\delta \geq 0$) are real constants. When $\delta = 0$, the function has two minima located at $x = \pm a$, which are separated by a barrier of height ν. We set $\nu = 6$, $a = 2$, and $\delta = 0.3$. Therefore, the one-dimensional double well function has a global minimum near -2 and a local minimum near 2. Figure 2 shows the double well function and the iteration process of MQHOA using different potential well models.

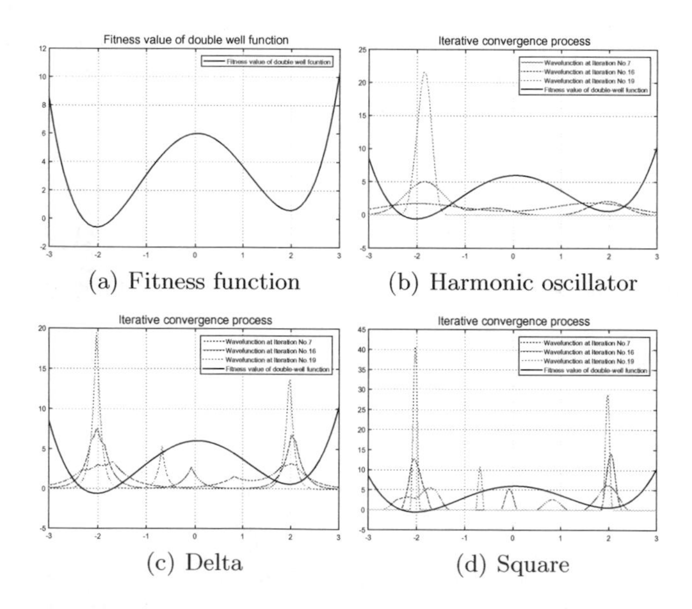

(a) Fitness function (b) Harmonic oscillator

(c) Delta (d) Square

Fig. 2. Particle number set to $k = 8$, search space is $[-3, 3]$, $\nu = 6$, $a = 2$, and $\delta = 0.3$. This figure shows the wave function at iterations 7, 16, and 19 for different potential wells.

The wave function varies from the high-energy state to the ground state at σ_s. The high energy of the wave function corresponds to the entire search domain. As the energy decreases, the wave function gradually reaches the ground state. In the ground states of the different potential wells, the wave function indicates the distribution of the optimal solution. Through the iterative algorithm, the wave function gradually converges to the minimum point from the initial relatively uniform distribution.

In Fig. 2(b), as the probability at the transient position decreases, the wave function gradually moves to the vicinity of -2. In terms of the physical process, in the high-energy state, the quantum fluctuation is very strong, the probability distribution of the solution is almost uniform, and the optimization algorithm has good exploration ability. The evolution then starts, and the quantum fluctuation decreases. As the objective function becomes dominant, the optimization algorithm starts to exploit the process. Finally, the quantum fluctuation ends, and the solution is close to the global minimum. Figure 2(c) shows the process for the delta potential well (recall that the quantum delta potential well has a Laplacian distribution). As the algorithm runs, the particle moves toward the optimal solution and finally reaches the ground state. However, the convergence is slower than that using the harmonic-oscillator potential well. Figure 2(d) shows the wave function movement of the square potential well. Initially, there are particles in both positions. As the iteration progresses, the wave function is superimposed. As can be seen from the vertical coordinate, the value of the function increases significantly to the ground state. However, it can be seen from the double well experiment that the wave function of the delta potential well is still locally optimal. With the same number of iterations, the harmonic oscillator wave functions are distributed around the optimal solution, whereas the delta and square wave functions are also distributed around the locally optimal solution, indicating that they are disturbed by the local minimum.

The critical step in deriving MQHOA is the selection of a suitable potential well. MQHOA converges to the minimum for all three potential wells. However, the wave function of the harmonic-oscillator potential well MQHOA is smoother and can converge completely to a near-global optimum. Thus, the harmonic-oscillator potential well MQHOA outperforms the delta potential well and square potential well algorithms.

4.2 Comparison of Potential Well Models for Global Optimization

Under the MQHOA framework, the Gaussian distribution samples were replaced with Laplacian and cosine distributions. The variance of the normal distribution, the characteristic length of the Laplacian distribution, and the period of the cosine distribution were taken as the scale parameters of the MQHOA. The three potential well algorithms were tested against nine classic benchmark functions.

Benchmark Problems. The benchmark functions presented in Table 2 are test functions for optimization problems [6], and are widely used to evaluate the performance of optimization algorithms. These functions are either unimodal or multimodal.

Table 2. Benchmark test functions

Function	No	Search space
Ackley	f_1	[−32.77, 32.77]
Griwank	f_2	[−100, 100]
Levy	f_3	[−10, 10]
Rotated Hyper Ellipsoid	f_4	[−65.54, 65.54]
Sphere	f_5	[−5.12, 5.12]
Zakharov	f_6	[−5, 10]
Sum of Different Power	f_7	[−100, 100]
Quartic function	f_8	[−1.28, 1.28]
Sum of Squares	f_9	[−10, 10]

Parameter Settings. The benchmark parameters adopted by the different potential well model algorithms were set as follows: particle number $k = 40$, maximum number of iterations $maxFE = (1e5 * DIM)/10$, and termination criterion $\sigma_{min} = 1e - 6$.

Performance Metric. We tested the best value, mean value, and mean iteration time of each algorithm over 51 independent runs [4]. Because high-dimensional problems are highly complex, we introduce the success rate (SR) as a measure of performance. SR represents the rate at which the optimal solution was determined.

Experimental Results and Analysis. Table 3 presents the test results for MQHOA under the three different potential wells for various dimensions of the nine test functions. This experiment tested four different dimensions, namely 2, 20, 30, and 40. The performance was measured in terms of the mean value, best value, and SR. The bold font in the table indicates the best value of the test results.

It can be seen that the three potential wells have similar SR values in the low-dimensional case, whereas the harmonic-oscillator potential well model is superior to the other potential wells when the functions have 20, 30, or 40 dimensions. As the number of dimensions increases, the accuracy of the delta well approaches that of the square potential well.

The experimental results show that the harmonic-oscillator potential well performs well for most test functions in the low-dimensional case. In the two-dimensional case, the optimal solution performance of the delta potential well is slightly better than that of the other two potential well models for f_2, f_6, f_7, f_8. For functions f_1, f_3, f_4, f_9, the three potential well algorithms find the optimal solution quickly and accurately; however, in terms of the precision, number of iterations, and average time, the harmonic-oscillator potential well performs slightly better than the delta potential well, and both performed significantly better than the square potential well.

For dimensions of 30 and 40, the delta potential well and square potential well converged to a local optimum for functions f_1, f_3, f_6, f_7, f_9; however, the harmonic-oscillator potential well model found the optimal solution with higher accuracy. As the dimension increases, the advantages of the harmonic-oscillator potential well become more obvious. In the case of test function f_2, for optimization problems with more than 20 dimensions, the statistical frequency with which the harmonic-oscillator potential well identified the optimal solution within a finite number of iterations is higher than that of the other two potential wells.

Table 3. Detailed computational results obtained by three different potential well models for different dimensions of the nine test functions

	Item	Harmonic-oscillator			Delta			Square		
D	Value	Best	Mean	SR	Best	Mean	SR	Best	Mean	SR
2	f_1	**4.03E−08**	5.74E−07	100%	1.06E−07	**5.33E−07**	100%	1.50E−07	1.21E−06	100%
	f_2	2.93E−05	1.72E−03	45%	**3.83E−12**	**8.23E−04**	**75%**	4.50E−04	3.27E−03	18%
	f_3	**2.46E−17**	**3.11E−15**	100%	2.82E−17	2.42E−15	100%	8.13E−15	1.90E−10	100%
	f_4	**2.64E−15**	**1.38E−13**	100%	4.17E−15	1.72E−13	100%	3.09E−14	8.79E−13	100%
	f_5	**2.40E−16**	**2.04E−14**	100%	3.23E−16	1.91E−14	100%	6.57E−16	1.17E−13	100%
	f_6	6.36E−16	**5.03E−14**	100%	**2.59E−E−16**	1.54E−13	100%	2.91E−14	2.08E−12	100%
	f_7	2.11E−16	**3.59E−14**	100%	**5.82E−18**	5.42E−14	100%	1.60E−15	2.32E−13	100%
	f_8	3.44E−15	7.90E−14	100%	**5.34E−16**	**5.85E−14**	100%	1.70E−14	3.81E−13	100%
	f_9	**1.18E−17**	**2.02E−14**	100%	3.18E−17	3.66E−14	100%	3.47E−12	1.72E−08	100%
20	f_1	**3.89E−06**	**5.58E−02**	96%	7.00E−06	5.80E−01	69%	2.72E+00	5.97E+00	0%
	f_2	**2.10E−12**	**3.20E−02**	**75%**	2.28E−12	2.80E−02	24%	8.91E−08	4.07E−02	18%
	f_3	**5.56E−12**	**8.91E−03**	**98%**	7.33E−12	4.08E−02	78%	8.88E−07	2.95E−01	22%
	f_4	**6.42E−09**	**1.08E−08**	100%	1.62E−08	2.66E−08	100%	1.19E−07	2.02E−07	100%
	f_5	**1.24E−11**	**2.13E−11**	100%	2.80E−11	5.06E−11	100%	1.65E−10	3.67E−10	100%
	f_6	**1.78E−10**	**2.72E−10**	**100%**	2.66E+00	7.26E+00	0%	3.07E+00	1.66E+01	0%
	f_7	**3.70E−10**	2.89E−05	100%	4.13E−09	**1.75E−06**	100%	5.27E−03	1.33E+00	0%
	f_8	**3.08E−09**	**4.29E−09**	100%	4.59E−09	1.03E−08	100%	3.44E−08	7.93E−08	100%
	f_9	**8.50E−10**	**2.49E−08**	100%	6.37E−08	1.78E−07	100%	8.51E−06	2.52E−05	100%
30	f_1	**6.90E−06**	**1.85E−01**	86%	9.52E−06	1.75E+00	22%	3.73E+00	6.66E+00	0%
	f_2	**4.50E−12**	**1.01E−02**	**67%**	5.25E−12	5.07E−03	57%	1.04E−07	1.09E−02	39%
	f_3	**1.86E−11**	**6.92E−02**	**80%**	2.07E−11	1.25E−01	67%	1.59E−05	6.24E−01	6%
	f_4	**4.44E−08**	**6.26E−08**	100%	7.07E−08	1.34E−07	100%	5.26E−07	8.79E−07	100%
	f_5	**3.61E−11**	**5.27E−11**	100%	7.04E−11	1.04E−10	100%	4.49E−10	7.60E−10	100%
	f_6	**6.02E−10**	**1.06E−09**	**100%**	6.35E+00	2.99E+01	0%	5.18E+00	4.37E+01	0%
	f_7	**7.41E−07**	8.40E−03	45%	3.11E−06	**1.94E−03**	**53%**	4.24E−02	5.42E+00	0%
	f_8	**1.14E−08**	**2.46E−08**	100%	3.67E−08	5.25E−08	100%	1.74E−07	3.41E−07	100%
	f_9	**5.09E−09**	**3.17E−08**	100%	7.59E−08	1.76E−07	100%	9.82E−06	2.06E−05	100%
40	f_1	**9.32E−06**	**7.00E−01**	**59%**	1.15E−05	2.36E+00	4%	5.15E+00	7.09E+00	0%
	f_2	**7.69E−12**	**3.57E−03**	**69%**	1.05E−11	4.93E−03	57%	1.64E−06	7.20E−03	51%
	f_3	**7.93E−11**	**1.61E−01**	**53%**	4.32E−11	3.23E−01	33%	8.95E−02	1.10E+00	0%
	f_4	**1.75E−07**	**2.21E−07**	100%	2.76E−07	3.88E−07	100%	1.48E−06	2.60E−06	100%
	f_5	**7.19E−11**	**1.05E−10**	100%	1.34E−10	1.90E−10	100%	7.41E−10	1.28E−09	100%
	f_6	**1.41E−09**	**2.40E−09**	**100%**	9.70E+00	6.62E+01	0%	1.01E+01	9.90E+01	0%
	f_7	**5.01E−06**	**4.26E−02**	**24%**	9.84E−05	3.13E−02	6%	4.50E−01	1.50E+01	0%
	f_8	**6.50E−08**	**8.72E−08**	100%	8.61E−08	1.51E−07	100%	6.20E−07	1.06E−06	100%
	f_9	**9.21E−09**	**3.48E−08**	100%	4.81E−08	1.63E−07	100%	1.15E−05	2.00E−05	100%

In some higher-dimensional cases, the harmonic-oscillator well model algorithm can find the optimal solution within a reasonable number of iterations where both the delta and square potential well models fail.

The above results show that, in terms of optimizing the performance, the MQHOA algorithm performs better when using the harmonic-oscillator potential well than when using the delta or square potential wells. Further, in the low-dimensional optimization problem, the harmonic-oscillator and delta potential wells offer almost the same performance, and both are better than the square potential well, but the delta potential well requires more iterations. The advantage of the harmonic-oscillator is more obvious in the high-dimensional optimization problems. Therefore, in practical applications, if the number of dimensions is low (below 20), the harmonic-oscillator and delta potential wells can be selected, whereas for high-dimensional problems, it is better to choose the harmonic-oscillator potential well.

5 Conclusion

This paper has analyzed the framework of MQHOA in detail. The fundamental idea of MQHOA is to utilize the objective function of an optimization problem to represent the quantum potential well from the perspective of Taylor approximation. Schrödinger's equation can be solved by choosing different attractive potential well models.

The potential well models of MQHOA were analyzed theoretically. The harmonic-oscillator potential well was found to be more suitable for the objective function from the perspective of the Taylor approximation.

Finally, two experiments were conducted. From the first experiment, it was found that the harmonic-oscillator potential well offers the best performance in terms of convergence. Moreover, the test results from multiple complex standard test functions showed that MQHOA based on the harmonic-oscillator potential well model had obvious advantages in terms of its performance.

On the one hand, the experiment results show that the application of the potential well model greatly impacts the performance. Previously, the impacts were never discussed. On the other hand, the results presented in this paper suggest that MQHOA using a quantum harmonic oscillator as the basic quantum model satisfies the theoretical requirements and achieves good optimization performance. The choice of a suitably attractive potential well model is critical in MQHOA. Analysis of distinct potential well models is helpful for further understanding the MQHOA framework.

Note that, although this paper uses the concepts of quantum wave functions and potential wells, MQHOA is not really a quantum algorithm, but a quantum-heuristic algorithm. In the current basic framework, different sampling and learning approaches adopt different wave functions according to the constraints of the potential well, but the statistical characteristics are not analyzed between iterations, and the algorithm is essentially different from true quantum approaches. In future work, a probabilistic statistical interpretation will be examined and further testing for engineering applications will be conducted.

References

1. Beheshti, Z., Shamsuddin, S.M.H.: A review of population-based meta-heuristic algorithms. Int. J. Adv. Soft Comput. Appl **5**(1), 1–35 (2013)
2. Bennett, K.P., Parrado-Hernández, E.: The interplay of optimization and machine learning research. J. Mach. Learn. Res. **7**, 1265–1281 (2006)
3. Born, M.: Statistical interpretation of quantum mechanics. Science **122**(3172), 675 (1955)
4. Derrac, J., García, S., Molina, D., Herrera, F.: A practical tutorial on the use of nonparametric statistical tests as a methodology for comparing evolutionary and swarm intelligence algorithms. Swarm Evol. Comput. **1**(1), 3–18 (2011). https://doi.org/10.1016/j.swevo.2011.02.002
5. Finnila, A., Gomez, M., Sebenik, C., Stenson, C., Doll, J.: Quantum annealing: a new method for minimizing multidimensional functions. Chem. Phys. Lett. **219**(5–6), 343–348 (1994). https://doi.org/10.1016/0009-2614(94)00117-0
6. Jamil, M.: A literature survey of benchmark functions for global optimization problems. Math. Model. Num. Optim. **4**, 150–190 (2013)
7. Jelic, V., Marsiglio, F.: The double-well potential in quantum mechanics: a simple, numerically exact formulation. Eur. J. Phys. **33**(6), 1651 (2012)
8. Kadowaki, T., Nishimori, H.: Quantum annealing in the transverse ising model. Phys. Rev. E **58**(5), 5355 (1998)
9. Kirkpatrick, S., Gelatt, C.D., Vecchi, M.P.: Optimization by simulated annealing. Science **220**(4598), 671–680 (1983). https://doi.org/10.1126/science.220.4598.671
10. Meshoul, S., Batouche, M.: A novel quantum behaved particle swarm optimization algorithm with chaotic search for image alignment. In: IEEE Congress on Evolutionary Computation. IEEE, July 2010. https://doi.org/10.1109/cec.2010.5585954
11. Stella, L., Santoro, G.E., Tosatti, E.: Optimization by quantum annealing: lessons from simple cases. Phys. Rev. B **72**(1) (2005). https://doi.org/10.1103/physrevb.72.014303
12. Stella, L., Santoro, G.E., Tosatti, E.: Monte carlo studies of quantum and classical annealing on a double well. Phys. Rev. B **73**(14), 144302 (2006). https://doi.org/10.1103/PhysRevB.73.144302
13. Wang, L., Zhang, Q., Xu, F., Cui, X.D., Zheng, Y.: Quantum tunneling process for double well potential. Int. J. Quantum Chem. **115**(4), 208–215 (2014). https://doi.org/10.1002/qua.24818

Design and Implementation of Key Extension and Interface Module Based on Quantum Circuit

Chengcheng Wang, Jiahao Sun, Zhijin Guan, Jiaqing Chen, and Yuehua Li(✉)

School of Information Science and Technology, Nantong University, Nantong, Jiangsu, China
lyh@ntu.edu.cn

Abstract. Encryption technology that based on quantum circuit is an important technology in the field of information security and the encryption system designed by this technology can improve the encryption effect, and its anti-attack ability is $(2^n - 1)!$ times of the traditional method. In order to increase the complexity of the key that belongs to the encryption system, we propose a method of constructing the algorithm of key extension based on quantum circuit and finish the design of the key extension module, in the method, we transform the nonlinear transformation into linear transformation which can be constructed by quantum circuit easily, and complete the transformation of operation in quantum logic which can simplify the generation process while maintaining the same performance. In order to increase the practicability of encryption system based on quantum circuit, we analyze the characteristics of the encryption system and complete the design of the interface module. What's more, the key extension module and the interface module are tested under the environment of QUARTUS and real object, and the validity and correctness of the two modules are verified.

Keywords: Quantum circuit · Key extension · Interface · Encryption

1 Introduction

With the booming of new information technologies such as cloud computing and IOT, all kinds of data are expanding rapidly [1]. However, most kinds of emerging technologies are still in the initial stage of development with the immature information protection technology, so there will inevitably be security problems such as Information theft, tampering and disclosure when a large number of sensitive information is processed, transmitted and stored on the network. Therefore, reliable and efficient technology of encryption has become a research hotspot in all walks of life [2]. Encryption technology that based on quantum circuit is an important technology to protect the security of network. It uses quantum circuit to construct encryption algorithm and realizes it with hardware to form encryption system of quantum circuit, the encryption system can not only improve the encryption efficiency, but also increase the encryption complexity by $(2^n - 1)!$ times [3].

K. Li et al. (Eds.): ISICA 2019, CCIS 1205, pp. 72–86, 2020.
https://doi.org/10.1007/978-981-15-5577-0_6

Document [3] proposes a construction method of quantum circuit for multiplication in finite field, and constructs encryption algorithm based on the method, however, it only completes the hardware design of the encrypted part and the decrypted part based on quantum circuit. In reference [4], a quantum circuit implementation method is proposed for S-box transformation in AES algorithm, and the idea of constructing encryption algorithm based on quantum circuit is proposed. Literature [5] proposes Quantum fully homomorphic encryption scheme based on universal quantum circuit and does research on the process of quantum information. In reference [6], authors describe the advantages and feasibility of applying reversible Logic in cryptography and coding theory on the basis of confirmatory study. Document [7] proposes a method of synthesis and optimization for linear nearest neighbor quantum circuits by parallel processing which can reduce the quantum cost for the design and implementation of quantum circuits.

The existing literatures provide a theoretical idea for the research of encryption system based on quantum circuit, and complete the design of encryption and decryption part. However, the encryption and decryption process of the encryption system needs the participation of the key. The complexity and security of the key largely determine the encryption complexity and anti-attack ability of the encryption system, so the design of the key extension and generation part is particularly important. Furthermore, the encryption system must have an interface to interact with users or devices in the practical application, so the design of the interface part is also essential.

For the purpose of improving the complexity of key generation in the encryption system and increasing the security of the key, this paper designed the key extension module based on quantum circuit. And this module can expand the combination types of generated key to $2^n!$, greatly increase the difficulty of cracking. At the same time, in order to improve the practicability of the encryption system, an interface module is designed, and the user or device can implement encryption and decryption operations of information through different interfaces in the module.

2 Relevant Technology

2.1 Quantum Gate

Quantum gates are basic units of quantum circuits], it can be expressed by matrix or vector multiplication which means states of quantum bit [8]. Common quantum gates are CNOT gates and SWAP gates, Fig. 1 shows a CNOT gate, when the control bit B is 1, the target bit A is reversed, and when the control bit B is 0, the target bit A remains unchanged. Figure 2 shows a SWAP gate, which indicates that two connected quantum bits A and B will be exchanged, so it is also called a switching gate.

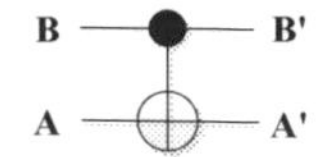

Fig. 1. CNOT gate

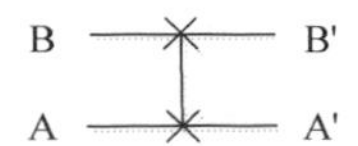

Fig. 2. SWAP gate

2.2 Quantum Circuit

Quantum circuit is a logic circuit with function and is built by a series of quantum gates [8]. Quantum circuit has $2^n!$ kinds of substitutions for the input of n-bits, its input number is equal to the output number and it has no heat loss, no fan-in, no fan-out, no feedback [5]. Figure 3 is a quantum circuit which consists of seven SWAP gates cascaded with three CNOT gates, it represents the multiply 2 operation on a finite field.

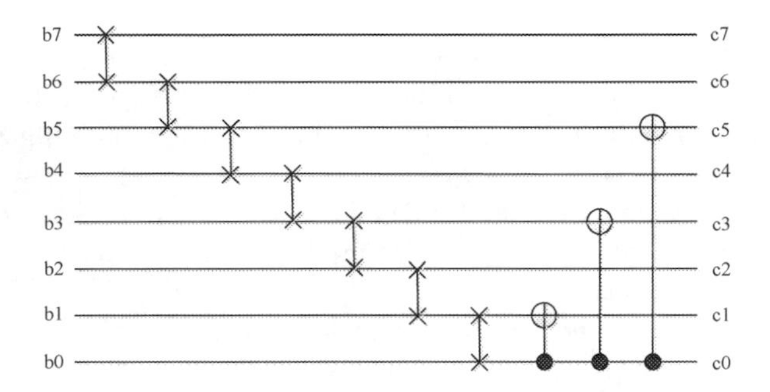

Fig. 3. Quantum circuit

2.3 Encryption Technology and Encryption System of Quantum Circuit

Encryption technology of quantum circuit is to apply the characteristics of non-feedback and high complexity of quantum circuitry to the field of encryption technology, it can design encryption algorithms based on quantum circuit [3]. The encryption system of quantum circuit is designed by encryption technology of quantum circuit, it consists of four parts: encryption module, decryption module, key extension module and interface module, Among them, encryption module, decryption module and key extension module are all constructed based on quantum circuit, meanwhile the interface module is designed by hardware language which can adapt to the characteristics of quantum circuits and work with encryption and decryption module.

Figure 4 is the framework of the encryption system. The key extension module and interface module designed in this paper are used to work with the encryption module and the decryption module to improve the performance of them. The key extension module is connected to the encryption and decryption module in order to provide the required keys for them. Similarly, the interface module is connected to the encryption and decryption module, it can supply convenience for users to exchange information with them and realize the operation of information encryption or decryption. In addition, A switch is used in the system to decide whether the information enters the encryption module or the decryption module, and also decide whether the information is output by the encryption module or the decryption module.

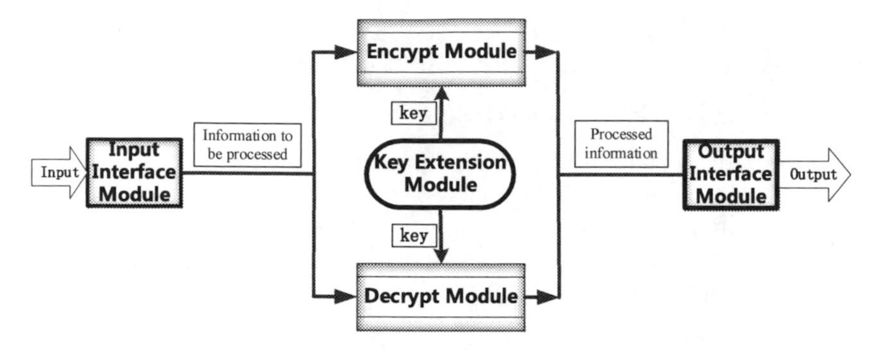

Fig. 4. Structural framework of encryption system

3 Design of Key Extension Module Based on Quantum Circuit

3.1 Key and Key Extension

In cryptography, the key is secret data used to complete cryptographic applications [9]. In the encryption system designed in this paper, the key is used for the encryption and decryption module to complete the encryption and decryption operations.

Key extension refers to the extension of the initial key with a small number of bits into the key with large number of bits [9]. In the encryption system designed in this paper, the initial key has 128 bits, which needs to be extended to 1408 bits, the extended keys are called round key. The necessities of key extension are as follows:

1. The key extension can eliminate the symmetry and similarity of the generation methods for different keys, and increase the complexity of the key, thus improving the encryption complexity of the encryption system.
2. The key extension can expand the difference of the key, therefore, each bit of the initial key can directly or indirectly affect many bits in the process which will make key cracking more difficult.
3. The key extension makes the key sufficiently non-linear to prevent that the difference of the keys completely determined by the initial key, thus the key extended from the initial key can also affect the other extended keys, and it will expand the difference of many steps in encryption process to enhance the encryption effect.

3.2 Introduction of Key Extension Algorithms

Key extension is the process of expanding 128-bit initial key to 1408-bit round key. For convenience, we regard 128-bit initial key as a matrix with four rows and four columns, each element of the matrix has 8-bit data. The key extension process is as follows:

1. The four elements in a column of matrix is counted as W[j], it has 32 bits, therefore, the initial keys are counted as W[0], W[1], W[2], W[3];
2. The extended keys are also expressed in the form of W[j], where j is an integer and the range of it is [4, 43].

3. If j%4 = 0, w[j] = w[j − 4] ⊕ g(w[j − 1]), otherwise, w[j] = w[j − 4] ⊕ w[j − 1]. Where "g()" represents g-function and "⊕" denotes XOR.

The contents of g-function are as follows:

a. Move the corresponding W[j] cycle to the left by 8 bits;
b. S-box substitution for each byte in the corresponding W[j];
c. XOR the corresponding W[j] with constants (Rcon[j/4], 0, 0, 0), Rcon is a one-dimensional array, Rcon[1] = 0 × 01, Rcon[i] = Rcon[i − 1]*(02) (i > 1, "*(02)" represent multiplication over finite fields).

3.3 Theory of Designing Key Extension Algorithms Based on Quantum Circuits

Feasibility and Necessity

1. The step (b) of g-function in key extension algorithm involves S-box replacement, which maps an 8-bit data to another 8-bit data by searching the S-box table. The traditional way to realize S-box replacement is to store the S-box table in read-only memory, and then compare the input 8-bit data with the memory to find the corresponding 8-bit data. Using quantum circuit to realize S-box replacement does not need the tedious operation of looking up tables, it evolves the logic of S-box table into the conversion relationship between finite field and compound field and the result can be obtained only by conversion between different fields. In the whole process, only addition on the finite field is involved which can be implemented by using simple CNOT gates. Therefore, realizing S-box replacement based on quantum circuits can not only accelerate the replacement rate and reduce the delay time, but also reduce power consumption and hardware size [10].
2. The step (c) of g-function in key extension algorithm involves XOR, division and multiplication over finite fields. Traditional methods for hardware to implement division operation is that they first expand the digits of the dividend, and then continuously compare the high digits of the dividend with the divisor to determine the value of quotient until the end of the calculation. The whole process involves many operations such as comparison, subtraction and shift. If we implement division operations with quantum circuits, it generally involves only shift operation which can be implemented by SWAP gates in quantum circuits. In addition, XOR is equivalent to addition over finite fields. Document [3] designs a method for quantum circuits to implement operations over finite fields, and draws a conclusion that the operation rate is higher. Therefore, the implementation of this step based on quantum circuits can simplify the process, make the operation faster and reduce the delay [11].
3. The key extension algorithm contains many operations. The operations of dividing, XOR and g- function in formulas have been introduced in the preceding article, On the step of controlling the XOR of two data according to the value of "j%4", the traditional methods will design the data selector, when the quantum circuit realizes it by using the TOFFOLI gates which have double control bits. So the quantum circuit can realize critical path with high working frequency and reduce the occupancy of resources in hardware.

Advantages of Designing Key Extension Algorithm Based on Quantum Circuit. Designing Key Extension Algorithm based on quantum circuit does not change the algorithm itself and the complexity, what change are the implementation method of the algorithm. Therefore, the characteristics of quantum circuits will be applied to the algorithm, add the following three advantages for the key extension:

1. Quantum circuit's input number is equal to output number, it has no fan-in, no fan-out, no heat loss [6]. Therefore, using quantum circuit to implement key extension algorithm can greatly reduce the energy consumption in the process of key extension, reduce the occupancy rate of hardware resources, and improve the rate at which the key expands.
2. There are $2^n!$ kinds of reversible networks for input and output vectors of quantum circuits, and there are $2^n!$ kinds of permutations for input of n-bit data [6]. Therefore, the implementation of key extension algorithm with quantum circuit can increase the complexity of the key extension process by $(2^n - 1)!$ times and it is difficult to get all the secret keys from the partially extended keys, as a result, it greatly increases the security of key extension.
3. In the encryption system, the key extension module is connected with the encryption and decryption module to work with them, the encryption and decryption module are constructed by quantum circuits, Therefore, implementing the key extension algorithm based on quantum circuits can make the key extension module combine with encryption and decryption module more closely, and improve the ability of the whole cooperative work.

3.4 Implementation of Key Extension Algorithms Based on Quantum Circuit

The implementation of the key extension module is mainly divided into two steps, Firstly, we complete the design of g-function, and then complete the implementation of key extension according to the formula.

Realization of g-Function

1. *Left shift of circulation.* Left shift of circulation is to move the W[j] to the left for 8 bits. Since the operation of left shift is only a change of bit's position, the quantum circuit can be constructed by using the quantum SWAP gate, when converting it to hardware circuit we can change the output sequence of circuit layout.
2. *S-box replacement.* S-box replacement is to replace the four bytes in W[j] with another four bytes. In the implementation, the data W[j] in the finite domain $GF(2^8)$ is transformed into its composite domain $GF\left((2^4)^2\right)$, after inversion in the composite domain, the data is transformed back into the finite domain [10], finally the reversible affine transformation is carried out and S-box replacement is finished.
3. *XOR of Rcon.* XOR of Rcon is an operation that realize XOR of W[j] and 32-bit constant (Rcon[j/4], 0, 0, 0). Since the (Rcon[j/4], 0, 0, 0) is all zero except Rcon, the first step is to construct the circuit to generate the values of Rcon. In addition, using different values in Rcon array for XOR is depending on the value of "j/4", so it is necessary to construct the circuit to generate the value of "j/4".

(1) Circuit Construction of Generating Rcon Values. Rcon is a one-dimensional array, Rcon[1] = 0 × 01, Rcon[i] = Rcon[i − 1]*(02) (i > 1, "*(02)" is the multiplication over finite fields). Since 128-bit seed keys are expressed as W[0] to W [3], the extended secret keys are expressed as W[4] to W[43], therefore the range of "j" in W[j] is [4, 43], and the range of "j/4" is [1, 10], Rcon array only needs to calculate the values of Rcon[1] to Rcon[10]. The quantum circuits of generating Rcon value are shown in Fig. 5. It is necessary to prepare quantum bits with initial state of | 00000001> to initialize the value of Rcon[1], then we use SWAP gate and CNOT gate to realize multiplication of 2 over finite fields, the circuit of multiplication is encapsulated as U device, and after Cascading 9 U devices with 8 CNOT gates [3], the construction of quantum circuit is finished. When Converting it into hardware circuit, we represent quantum auxiliary bit | 1 > with high levels, and quantum auxiliary bits | 0 > with low levels, then cascade them with nine U devices to build hardware circuit. The hardware circuit in Quartus environment is shown in Fig. 6.

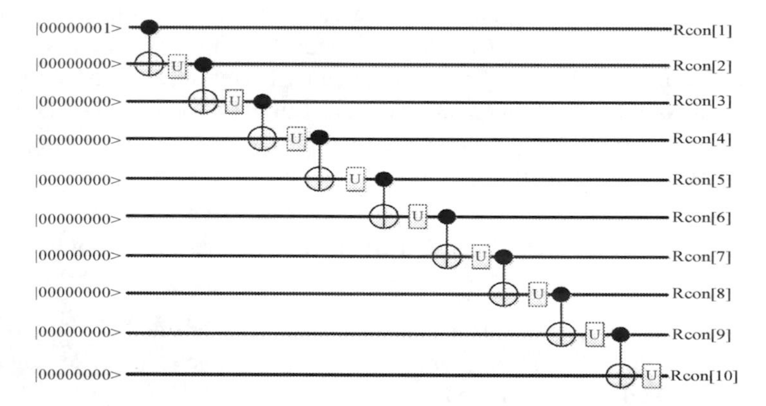

Fig. 5. Quantum circuit graph of value generation of the Rcon array

(2) Circuit Construction of "j/4" Operation. "j/4" means that j is divided by 4 to take an integer. The operation of dividing 8-bits data by 4 can be obtained by shifting the data to the right by two bits, Suppose that the value of "j" is $b_7b_6b_5b_4b_3b_2b_1b_0$, then the two digits (b_1b_0) removed from the low bits are the remainder, and the value ($00b_7b_6b_5b_4b_3b_2$) is the integral value divided by four. When constructing the operation of "j/4" based on quantum circuits, we use auxiliary bits | 0 > to supplement the two high bits [3] the right-shift operation can be implemented by SWAP gates. When converting quantum circuits into hardware circuits, we represent quantum bit |1 > with high level, and quantum bit | 0 > with low level, the shift operation requires changing position during layout. Quantum circuit of dividing four operations designed in this paper is shown in Fig. 7.

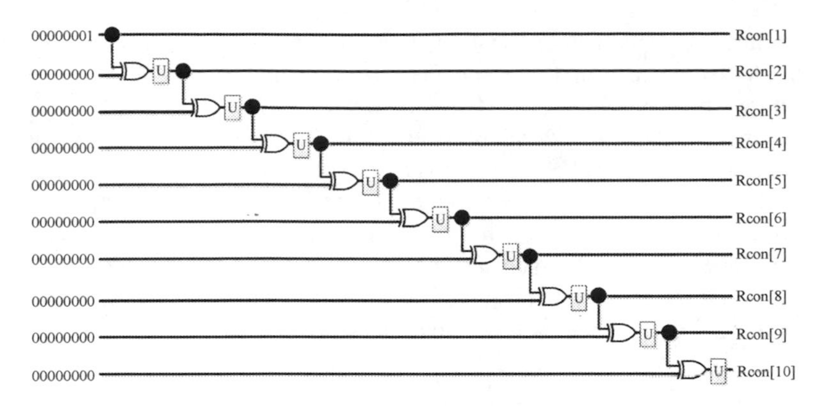

Fig. 6. Hardware circuit graph of value generation of the Rcon array

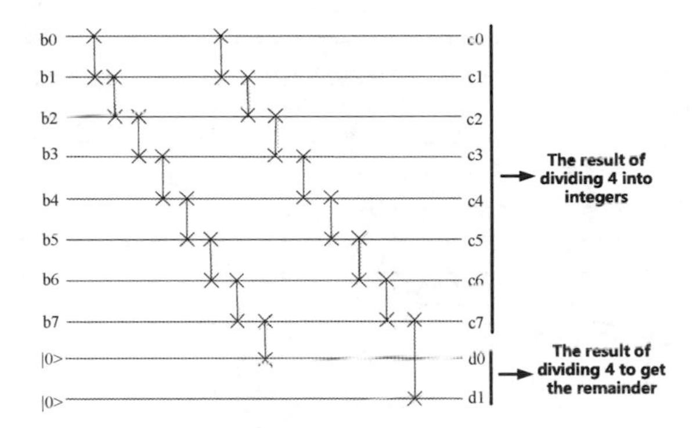

Fig. 7. Quantum circuit graph of the operation of dividing 4

4. *Encapsulation of g-Function circuit.* The construction of key extension based on quantum circuits is complex. In order to introduce the construction of subsequent quantum circuits, we simplify the circuits of g-Function logically, encapsulate the g-Function circuits in three steps 1, 2 and 3 as G devices, as shown in Fig. 8.

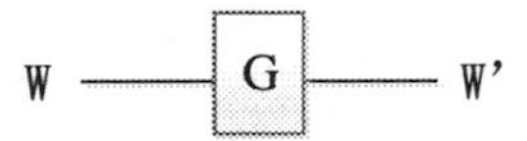

Fig. 8. G device

Overall Design of Key Extension Algorithms

The core content of key extension algorithm is as follows: If $j\%4 = 0$, $w[j] = w[j-4] \oplus g(w[j-1])$, otherwise, $w[j] = w[j-4] \oplus w[j-1]$; Since the whole quantum circuit of key extension is complex and every four lines is a construction cycle, we only shows the partial quantum circuit of the key extension in Fig. 9. In the figure, the left side is the input, the right side is the output, "| 0000… >" represents | 0 > auxiliary bits of 32 bits.

When implementing the algorithm, XOR can be implemented by CNOT gates, the g-function has been realized before which is represented by a G device. After cascading 80 CNOT gates and 10 G devices according to the rules, we finish the overall design of key extension based on quantum circuit. When converting it to hardware circuit, we can use logical gate XOR to implement CNOT gate. The partial hardware circuit in the Quartus environment is shown in Fig. 10, the left side is the input, the right side is the output, and the "0000…" means 0 bit which length is 32-bit.

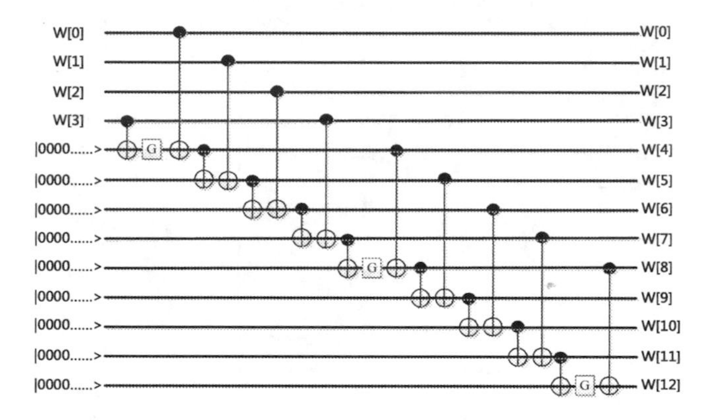

Fig. 9. Quantum circuit diagram of key extension module

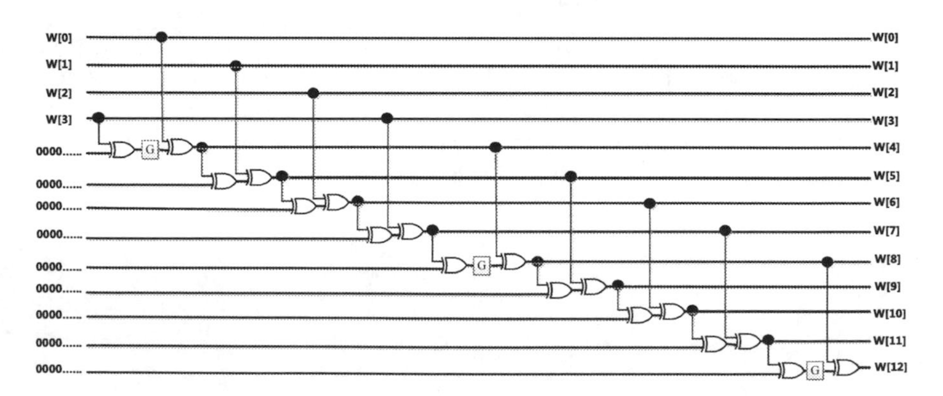

Fig. 10. Hardware circuit diagram of key extension module

4 The Design of Interface Module

In order to make it easier for users to use the encryption module and the decryption module, we research on the characteristics of the encryption and decryption module and design the interface module to match them. The interface module has multiple types of interface, such as UART, SPI, and the different interfaces can meet different needs of users. The following takes the SPI interface as an example to describe how the interface module is designed.

4.1 Introduction of the SPI Interface

SPI is a high-speed, full-duplex, synchronous communication bus [13] with four IO pins: SCLK, MOSI, MISO, CS. When the SPI data is transmitted, the host selects the slave to communicate according to "CS", and then sends data according to the clock "SCLK" via the "MOSI" line, the slave reads the data through the line. After that the slave sends data from the MISO line, and the host reads data from the same line [13], this process loops until the data transfer is complete. In addition, the SPI interface needs to set the phase and polarity, the polarity (CPOL) sets the level of the idle clock and the phase (CPHA) sets the clock edge for transmission of data. The phase and polarity of the master and slave must be the same.

4.2 The Design of SPI Interface

The encryption and decryption module perform operations on the data sent by the user, so the encryption and decryption module is a slave device, and the interface designed in this paper is an interface of the slave device. Since the encryption and decryption module perform operations on every 128 bits of data, a cache is needed to temporarily store data received and to be transmitted from the device. We set the phase and polarity to zero, in addition, when the operations of encryption and decryption are finished, the processed information needs to be output, so the interface module is divided into an input interface module and an output interface module. This article uses the Verilog language to implement the design of the SPI interface.

Input Interface Module. The users or devices can input data into the encryption or decryption module through the input interface module. Figure 11 is a logical representation of the SPI input interface in the Quartus environment, in the figure, "SPI_in_8" implements the function of receiving one byte of data and it stores the 8-bit data received each time into the buffer "Buffer_128". When the buffer is full of 128-bit data, the 128-bit data is output to the corresponding encryption module or decryption module. The main implementation steps are as follows:

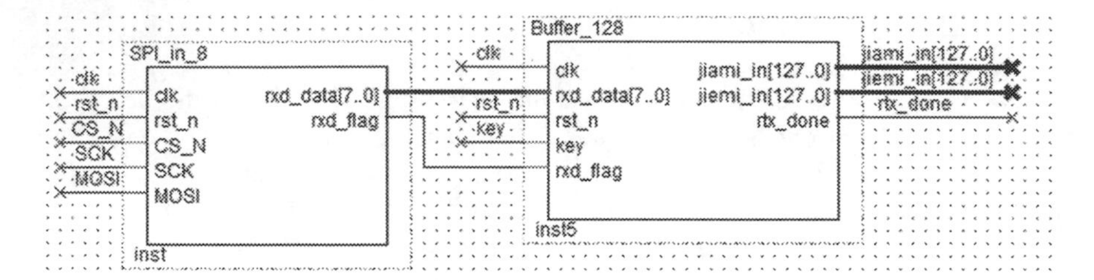

Fig. 11. Logic diagram of input interface module

1. *Capture the Trigger Edge.* Each step of the SPI must be performed after the trigger edge of the clock is generated and this article sets the rising edge as the trigger edge. "sck_r1" is used to record the high or low of the previous clock, and "sck_r0" is used to record the high or low of the next clock. If "sck_r1" is low and "sck_r0" is high, the rising edge is captured.
2. *Receive and Store 8-bit Data.* The first step is to determine whether the user wants to send data for encryption and decryption, this step can be achieved by determining whether the selection signal is low level. If the selection signal is low, the value obtained by the pin "MOSI" will be stored in the register under the clock. When the rising edge is captured and the select signal is 0, the data will be received. Meanwhile, the number of the received data is recorded and when the number is full of 8 bits, the value of corresponding flag is set to 1 and it indicates the completion of data transmission in one byte.
3. *Setting 128-bit buffer.* The data of the register in step 2 will be input into the buffer, and when the buffer is stored in 128-bit data, the data will be output to the encryption module or the decryption module. In implementation, we can define memory data which is 16 bits long and 8 bits wide to store 8-bit data received at a time. In addition, a switch is needed to determine whether the data in buffer enter the encryption module or decryption module.
4. *Data Filling.* When the number of bits needed to be encrypted is not a multiple of 128, it is necessary to fill in the data to meet the requirements of data size in buffer. We record the number of bytes entering the buffer, if the number less than 16, all the bytes in the vacancy are set at 0-bit. When the data is restored, the original data can be obtained only by clearing all the parts that is set to 0-bit.

Output Interface Module. The data processed by encryption or decryption module is output by output interface module. The output interface module and the input interface module are inverse processes and their implementation methods are similar, therefore, this article does not make a superfluous introduction for it.

5 Verification and Testing

In order to verify the validity of the key extension module and the interface module, we verify the key generation of the key extension module and carry out the transmission test of the interface module. Then the analysis results are obtained.

5.1 Verification of the Key Extension Module

The key extension module expands the original key of 128-bit into the key of 1408-bit according to the rule, each 32-bit key is a group and there are 44 groups in total. The process of verification is as follows:

1. The 128-bit original key (hexadecimal) entered is " 3E 19 3D 71 25 56 6A 49 4F 4B 3F 7C 2A 8C 62 23".
2. The 1408-bit extended key (hexadecimal) generated is shown in Table 1.

Table 1. 1408-bit Extended key

Group name	Group data	Group name	Group data
W[0]	3E 19 3D 71	W[22]	3A 9E 49 5B
W[1]	25 56 6A 49	W[23]	79 93 58 98
W[2]	4F 4B 3F 7C	W[24]	02 A2 09 5F
W[3]	2A 8C 62 23	W[25]	CA 6F 2F 42
W[4]	C8 B6 CC E9	W[26]	F0 F1 66 19
W[5]	ED E0 A6 A0	W[27]	89 62 3E 81
W[6]	A2 AB 99 DC	W[28]	ED FB 77 C1
W[7]	88 27 FB FF	W[29]	27 94 58 83
W[8]	03 16 D0 72	W[30]	D7 65 3E 9A
W[9]	EE F6 76 D2	W[31]	5E 07 00 1B
W[10]	4C 5D EF 0E	W[32]	BC FB BB A0
W[11]	C4 7A 14 F1	W[33]	9B 6F E3 23
W[12]	D7 8F F3 A8	W[34]	4C 0A DD B9
W[13]	39 79 85 7A	W[35]	12 0D DD A2
W[14]	75 24 6A 74	W[36]	46 03 95 0A
W[15]	B1 5E 7E 85	W[37]	DD 6C 76 29
W[16]	BE 0E 80 48	W[38]	91 66 AB 90
W[17]	87 77 05 32	W[39]	83 6B 76 32
W[18]	F2 53 6F 46	W[40]	AF B9 07 8A
W[19]	43 0D 11 C3	W[41]	72 D5 71 A3
W[20]	4f BA 23 2F	W[42]	E3 B3 DA 33
W[21]	C8 CD 26 1D	W[43]	60 D8 AC 01

The above keys are shown in hexadecimal form, every two characters represents hexadecimal number of 8-bit. The extended keys in Table 1 represent a set of 32-bit keys from top to bottom, the first list shows W[0] to W[21], the second list shows W[22] to W[43]. The number of output key is 1408 bits, and the content of them conforms to the logic of the key extension algorithm, therefore Verification passed.

5.2 Transfer Test of Interface Module

Taking SPI interface as an example, this paper builds a communication environment including master device and slave device to verify the correctness of the interface.

Test Preparation.

1. *SPI Upper Computer.* The upper computer is the host equipment, which is used to communicate with the slave interface designed in this paper. The right of Fig. 12 shows a USB-to-SPI device, Users can set the length of transmission data and see the data sent and received through the corresponding software.
2. *SPI Lower computer.* The SPI lower computer is a slave device, which can communicate with the upper computer. Firstly, we connect the two output pins of the input module directly with the corresponding input pins of the output module, so that the information received by the input module will be directly transmitted to the output module for output, which is convenient for testing. Secondly, the connected lines are burned into the core board of the FPGA which is shown in the left of Fig. 12, and after setting the pin parameters, the FPGA has the function of SPI interface designed in this paper [14], which is used as the lower computer.
3. *Connection of Upper and Lower Computer.* Connecting the prepared upper computer with the lower computer in the way shown in Fig. 12 when the upper computer connects the PC.

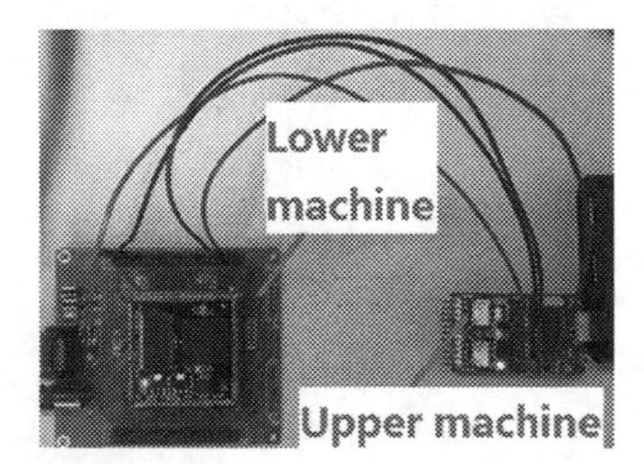

Fig. 12. Physical connection diagram of upper and lower machine

Testing Process

1. *Input the data to be transmitted.* As shown in Fig. 13, we input hexadecimal data of 8-bit in each address, totaling 16 addresses, so the total length is 128 bits.

Fig. 13. Input data

2. *Receive the data returned by the lower computer.* The data read by the host computer from MISO is shown in Fig. 14.

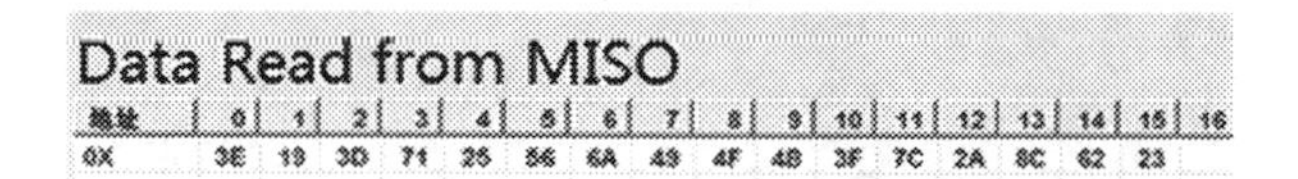

Fig. 14. Received data

As can be seen from the above two graphs, the data transmitted from the host computer to the FPGA is exactly the same as the data received from the FPGA, and the verification is successful.

6 Summary

This paper designs and implements the key extension module based on quantum circuit and interface module based on logic language, it also verify the correctness and validity of the two modules. This design can be used in the encryption system of quantum circuit to improve the encryption complexity, cracking difficulty and practicability. In the further work, we will design the two modules into the encryption system even better, and construct a complete encryption system to protect the hot technologies in urgent need of security, such as narrowband Internet of Things.

Acknowledgements. This work was partially supported by National Natural Science Foundation of China (NO. 61402244), Natural Science Foundation of Jiangsu Province (NO. BK20151274) and Next Generation Project of Technology Innovation for CERNET Internet (NO. NGII20180310).

References

1. Xu, L., Jiang, C., Wang, J.: Information security in big data: privacy and data mining. IEEE Access **2**, 1149–1176 (2014)
2. Natalia, M., Alexander, T.: Internet of Things: information security challenges and solutions. Cluster Comput. **22**, 103–106 (2018)
3. Chen, J., Guan, Z., Chen, X.: Design and hardware implementation of quantum reversible logic encryption algorithms. J. PLA Univ. Sci. Technol. (Nat. Sci. Ed.) **0**(0), 1–3 (2018)

4. Liu, Y., Zhu, F.: Quantum circuit implementation of S-Box transform in AES. Inf. Secur. Commun. Secrecy **5**, 92–94 (2008)
5. Liang, M.: Quantum fully homomorphic encryption scheme based on universal quantum circuit. Quantum Inf. Process. **14**(8), 2749–2754 (2015)
6. Datta, K., Sengupta, I.: Applications of reversible Logic in cryptography and coding theory. In: International Conference on Vlsi Design and International Conference on Embedded Systems, pp. lxvi–lxvii. IEEE Computer Society. Washington DC (2013)
7. Zhang, Z., Guan, Z., Zhan, H.: A method of synthesis and optimization for linear nearest neighbor quantum circuits by parallel processing. Quantum Inf. Comput. **18**(13), 1095–1114 (2018)
8. Li, Z., Chen, S., Song, X.: Quantum Circuit synthesis using a new quantum logic gate library of NCV quantum gates. Int. J. Theor. Phys. **56**(4), 1023–1038 (2017)
9. Du, Z., Xu, Q., Zhang, J.: The design of a key extension algorithm based on dynamic dislocation counts. In: 11th International Conference on Computational Intelligence and Security (CIS), pp. 345–349. IEEE, Shenzhen (2015)
10. Guan, Z., Chen, J., Chen, X.: Implementation method of quantum byte replacement hardware module for AES hardware encryption system: CN108322305A[P], 24 July 2018
11. Nayak, A., Sen, P.: Invertible Quantum Operations and Perfect Encryption of Quantum States. Quantum Inf. Comput. **7**(1–2), 103–110 (2012)
12. Bossuet, L., Datta, N., Mancillas-Lopez, C.: ELmD: a pipelineable authenticated encryption and its hardware implementation. IEEE Trans. Comput. **65**(11), 3318–3329 (2016)
13. Wei, P., Zhang, C., Huang, X.: Design of a high-performance and dynamic reconfigurable SPI IP core with master and slave mode. Appl. Electron. Tech. **44**(3), 15–18 (2018)
14. Xie, Z.: A design of PCI target interface based on FPGA. Adv. Mat. Res. **186**, 342–347 (2011)

Research on Atmospheric Data Assimilation Algorithm Based on Parallel Time-Varying Dual Compression Factor Particle Swarm Optimization Algorithm with GPU Acceleration

Ke Chen[1], Yadong Liu[1], Liyuan Liu[1], Yidong Yu[1], Yiqi Dong[1], and Yala Tong[1,2](✉)

[1] School of Science, Hubei University of Technology, Wuhan 430068, China
1339734057@qq.com
[2] Hubei Engineering Technology Research Center of Energy Photoelectric Device and System, Wuhan 430068, China

Abstract. Intelligent optimization algorithms such as particle swarm optimization (PSO) have been introduced into four-dimensional variational assimilation of atmospheric data to solve complex optimization problems. The time-varying double compression model can solve the problem of accuracy well. But when confronted with the problem of high accuracy, the long training time will become the weakness. Parallelization acceleration is one of the effective ways to solve the conundrum. And applying Graphic Processing Unit (GPU) to accelerate PSO algorithm in parallel has the advantage of low hardware cost. In this paper, a parallel time-varying double compression factor PSO algorithm based on GPU acceleration is proposed. The parallel operation of particle swarm optimization algorithm is carried out by GPU, in which the time can be improved with the same precision kept. Compared with the dynamic inertia weight algorithm and time-varying double compression factor algorithm, the experimental results display that the accuracy is better than the former and the consuming time is shorter than the latter, which proves that the method can process the prediction in a faster and more accurate way.

Keywords: Data assimilation · Graphic Processing Unit · Time-varying double compression factor PSO algorithm · Weather forecast

1 Introduction

Atmospheric data assimilation results from objective analysis of the formation of initial numerical weather forecasts: In terms of functions, the most likely values on a predetermined grid point are obtained by combining the irregular distribution of observation data with the characteristics of weather dynamics and observation data in a computer program running with a certain algorithm (without manual intervention). In other words, the purpose is to provide the initial value conditions of the numerical prediction model by analyzing the atmospheric state as accurate as possible.

K. Li et al. (Eds.): ISICA 2019, CCIS 1205, pp. 87–96, 2020.
https://doi.org/10.1007/978-981-15-5577-0_7

Four-dimensional variational assimilation is one of the most practical methods. Variational assimilation problem is a nonlinear optimization problem. Due to the large amount of computation, complex mode and the fact that the assimilation results depend on the selection of initial speculative value, the convergence and accuracy of the algorithm are in highly demanding. In 1949, Panofsky [1] first proposed the data assimilation method of polynomial interpolation. Subsequently, the data assimilation method went through the stages of gradual revision [2], optimal interpolation [3, 4], three-dimensional variational [6, 7], kalman filter and set kalman filter [8]. One of the most representative Wang Shunfeng [9], Bai Chen [10] design such as genetic algorithm GA assimilation scheme, as well as Zheng Qin [11] through the design appropriate weight drop strategy and configured with the appropriate parameter learning factor, put forward a kind of effective solution containing discontinuous switching process of variational data assimilation problems of dynamic inertia weight PSO algorithm [11]. Particle Swarm Optimization (PSO), as intelligent Optimization algorithms, has been introduced into four dimensional variational assimilation to resolve sophisticated optimization problems.

PSO algorithm is an intelligent optimization algorithm proposed by Kennedy and Eberhart in 1995 [14]. It is derived from the research on the foraging behavior of birds which can be detected a bunch of the characteristics of intelligence. It has been successfully applied to many fields as a result of the simple concept, easy implementation and strong convergence and global searching ability.

Facing optimization problems with large complexity, PSO algorithm usually requires a tedious convergence time and even cannot obtain satisfactory results. Therefore, it is of great significance to find a method with high-efficiency [13]. It is an algorithm based on a model of group, in which the behavior of each individual contains essential parallelism. Consequently, the ideas of parallelizing the algorithm to accelerate the calculation have been put forward successively. In 2003, Schutte et al. made the first attempt to parallelize the PSO algorithm. At present, there are four ways of parallel PSO algorithm: The first is the parallel computing method based on computer cluster, MPI parallel computing environment and master-slave parallel mode are commonly employed. The second is the parallel computing method based on multi-core CPU, which generally adopts the multi-thread parallel mode of OpenMPI Shared storage. The third is the parallel computing method based on FGPA and other professional parallel devices, and the last one is the parallel computing based on GPU technology. On the one hand, parallel computing based on computer cluster or multi-core CPU generates a lot of communication overhead and complex management loss, On the other hand, it requires exorbitant hardwares; As for FGPA, the cost is even more expensive. Compared with the other three ways, GPU mode has a dominant advantage of low hardware cost, further as the GPU technology developed rapidly, GPU currently has strong parallel computing capability. At the same time, the single-precision floating point computing capability is more than 10 times that of the contemporary CPU [12]. In recent years, GPU-based general computing has become more popular and to be applied in various scientific computing fields, showing great potentials. Therefore, GPU can be used in assimilation system to solve problems faster and more effectively.

2 Particle Swarm Optimization

PSO algorithm is a random search algorithm based on group intelligence, which is inspired by the bird foraging model, and has been widely used in various optimization problems. When using PSO algorithm, each particle can be regarded as a search individual in the d-dimensional search space, The current position of the particle is a candidate solution of the corresponding optimization problem, and the flight process of the particle is the search process of the individual. The flight speed of particles can be dynamically adjusted according to the optimal position of particle history and the optimal position of population history.

Let's suppose a D dimensional search space, there are N particles in a population, the i(i < N) particle can be described by two indices in t^th^ generation: The position can be represented as the D vector [14] of $X_i^t = \left(x_{i1}^t + x_{i2}^t, \ldots, x_{ij}^t, \ldots, x_{iD}^t\right)$. The flight speed can be expressed as $V_i^t = \left(v_{i1}^t + v_{i2}^t, \ldots, v_{ij}^t, \ldots, v_{iD}^t\right)$. If the optimal position of individual history is $P_i = (p_{i1} + p_{i2}, \ldots, p_{ij}, \ldots, p_{iD})$, here the ith particle is searched to the tth generation, historical optimal position of the whole particle swarm at generation t is $P_g = (p_{g1} + p_{g2}, \ldots, p_{gj}, \ldots, p_{gD})$, for the $t + 1^{th}$ generation, the iterative updating formula of velocity and position of the ith particle in the dimension j is as follows:

$$v_{ij}^{t+1} = \omega\, v_{ij}^t + c_1 r_1 (p_{ij}^t - x_{ij}^t) + c_2 r_2 (p_{gj}^t - x_{gj}^t) \tag{1}$$

$$x_{ij}^{t+1} = x_{ij}^t + v_{ij}^t \tag{2}$$

Where, ω is the inertia weight, which measures the impact of the speed at the previous moment on the next movement, c_1 and c_2 are the learning factor, r_1 and r_2 are the internal random $[0, 1]$.

3 A Particle Swarm Optimization Algorithm with Time-Varying Compression Factor Based on GPU Acceleration

3.1 Introduction to Particle Swarm Optimization with Time-Varying Dual Compression Factors

Particle Swarm Optimizer with Time Varying Compression Factor (PSOTVCF) replaces the original inertia weights by compression factors, to balance the contradictory relationship between global and local search. The compression factors are calculated from the acceleration factors:

$$C_1 = (C_{1N} - C_{1M}) \frac{ITER}{MAXITER} + C_{1M} \tag{3}$$

$$C_2 = (C_{2N} - C_{2M}) \frac{ITER}{MAXITER} + C_{2M} \tag{4}$$

Here, ITER is current evolution, MAXITER is the maximum number of iterations, C_{1N}, C_{1M}, C_{2N}, C_{2M}, are the initial maximum and minimum values of the first and second accelerators respectively.

The velocity update model of PSOTVCF is transformed into:

$$v(t+1) = \chi(v(t) + \phi y(t)) \tag{5}$$

$$y(t+1) = \chi x(t) + (1 - \chi\phi)y(t) \tag{6}$$

To meet the system matrix:

$$M = \begin{bmatrix} \chi & \chi\varphi \\ -\chi & 1 - \chi\varphi \end{bmatrix} \tag{7}$$

$$\varphi = C_1 + C_2 \tag{8}$$

To satisfy the conditions $\varphi > 4$, the compression factor is defined by the convergence criterion:

$$\chi = \frac{2}{\left|2 - \varphi - \sqrt{\varphi^2 - 4\varphi}\right|} \tag{9}$$

Of which, χ is Positive real Number, simultaneously $\chi \in (0, 1)$.

Due to the limited ability of a single compression factor to balance global and local search, as bad as the low accuracy in solving high-dimensional multi-peak functions such as four-dimensional variational assimilation, the dual compression factor should be adopted.

$$\begin{aligned} Vel_{i,j}(k+1) = \chi_1(&Vel_{i,j}(k) + C_1 Rand(P_{i,j} - X_{i,j}(k) \\ &+ C_2 Rand(G + X_{i,j}(k)))) \end{aligned} \tag{10}$$

$$Vel_{i,j}(k+2) = \chi_2 \otimes (Vel_{i,j}(k+1)) \tag{11}$$

Set the number of cycles as M, and then simplified particle velocity can be obtained:

$$Vel_{i,j}(M) = (\chi_2 \otimes \chi_2)^{M-1}(Vel_{i,j}(k+1) + \phi Y) \tag{12}$$

Above these PSOTVCF algorithm synthetically uses time-varying compression factors and dual compression factors to maximize the global and local search ability.

3.2 Design Principle of Assimilation Algorithm of PSOTVCF Based on GPU Acceleration

As is introduced, PSOTVCF algorithm replaces the original inertia weight with compression factors to balance the contradiction between global and local search. On the basis of the same number of iterations, higher precision can be obtained. Nevertheless, when faced with the situation of high complexity, it usually takes a long calculating time, the convergence and the calculation of data assimilation is particularly a problem of high complexity. PSOTVCF modeling needs a large amount of particle number and particle dimension, along with the growing number of samples, excessive computational load make computing time increase sharply, sometimes even arriving an unbearable level. Hence parallelized PSOTVCF algorithm acceleration is a simple efficiency improvement.

The existing method of parallel computation acceleration of PSOTVCF mainly utilizes the parallelism of particle storage, most of which is completed through computer clustering. The current mainstream GPU floating point processing power has reached more than 10 times of the CPU in the same period. Unlike traditional parallel computing methods such as computer cluster, GPU accelerated parallel algorithm has the advantages of low hardware cost and strong programmability. PSO algorithm naturally has the parallelism of group individual behavior. For complex problems, hundreds or even more particles are needed. By making use of the parallelism of particle behavior in the group, the parallel computing ability can be given full play.

3.3 PSOTVCF Algorithm Based on GPU Acceleration

The task of a GPU is to synthesize and display images of millions of pixels on the screen. That is to say, there are millions of tasks that need to be processed in parallel. Therefore, the GPU is designed to handle multiple tasks in parallel, instead of completing a single task like the CPU.

The parallelism of PSO based GPU algorithm is originated from the parallelism of individual behaviors in the group, which is reflected in the following three aspects:

(1) Particle velocity and position updates are parallelizable;
(2) The computation of particle fitness is parallelizable;
(3) The updating of the optimal position of individual particles is in parallel;
(4) The flow chart of using GPU to transform PSO and improve its parallelism:

According to above, PSOTVCF based GPU acceleration can be drawn as follow (Fig. 1):

In the figure, when a large number of particle position velocity updates are carried out, GPU is faster than CPU in calculation speed, and the repeated behavior of particle update is processed in parallel to achieve acceleration effect, thus improving the efficiency of particle update. The assimilation algorithm of PSOTVCF is designed as followed:

Step 1: initialize the particle swarm, and generate the particle directly on the GPU, name its position as $q_i^{(k)}$ and flight speed as $q_i^{(k)}$, $k = 0$, $i \in I = \{1, \ldots\ldots, m\}$; Set the maximum number of iterations.

Step 2: perform operations on gpu-generated particles to calculate the data assimilation cost function of each particle $J(q_i^{(k)})$.

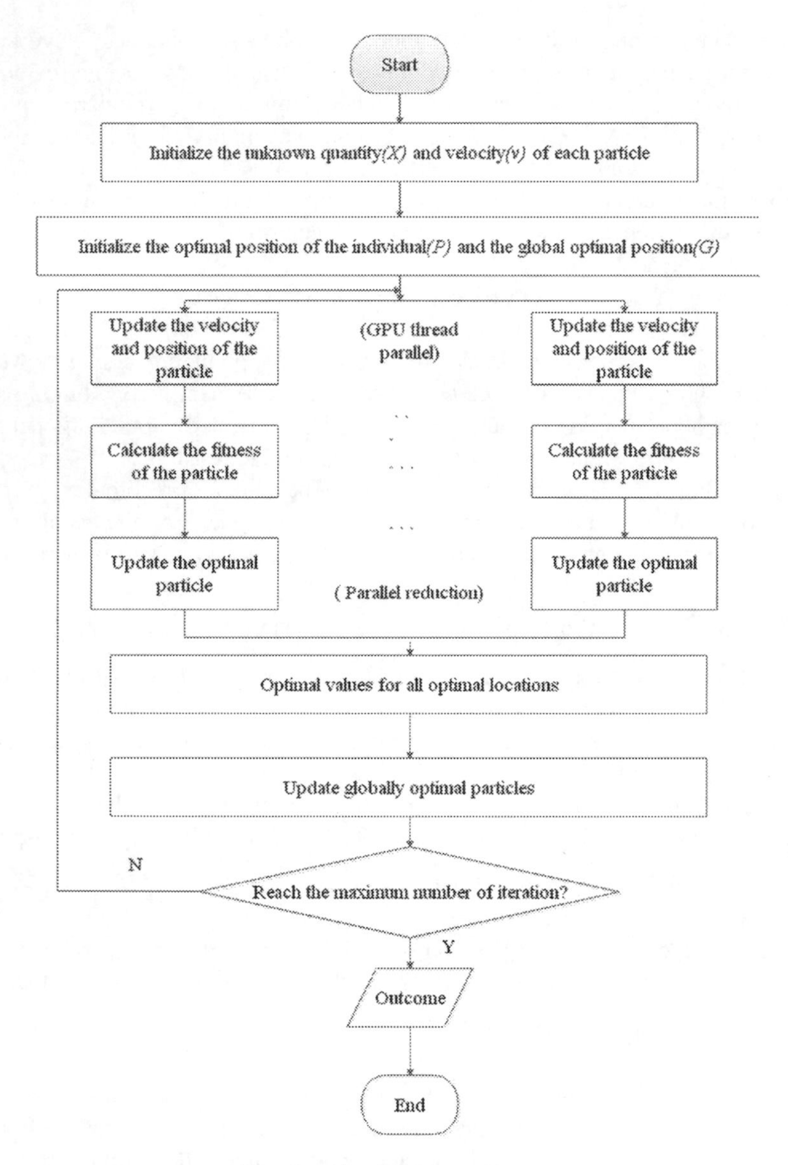

Fig. 1. Algorithm flow diagram

Step 3: use time-varying dual compression factor to find the best historical position $g^{(k)}$ of each particle and the best historical position $p_i^{(k)}$ of the whole particle group. If $J(q_i^{(k)}) < J(p_i^{(k-1)})$, $p_i^{(k)} = q_i^{(k)}$, else $p_i^{(k)} = p_i^{(k-1)}$.
Step 4: pass the data on the GPU back to the CPU and use the gather function to check if it works. If the condition $k > N_{iter}$ is satisfied, stop the algorithm and output $g^{(k)}$; Otherwise, go to step 5.

Step 5: use equation $Vel_{i,j}(M) = (\chi_2 \otimes \chi_1)^{M-1} \times (Vel_{i,j}(k+1) + \varphi \times Y)$ (12) to update the particle swarm $k = k+1$, and go to step 2.

4 Numerical Test Results and Analysis

To verify the effect of GPU acceleration, we compare the assimilation results of this algorithm with PSODIWAF(PSO with dynamic inertia weight) and PSOTVCF without GPU, Experimental data and experimental analysis methods in literature [11] are borrowed, The parameter of PSOTVCF: the first compression factor is constant, $C_1 = 2.6$, $C_2 = 1.2$; The second compression factor is time-varying, $C_{1N} = 2.88$, $C_{1M} = 2.68$, $C_{2N} = 1.45$, $C_{2M} = 1.25$. Convergence condition is $N_{iter} = 50/100/150/200$. Test environment: hardware Intel Core i7, software MATLAB R2017b.

4.1 Convergence Accuracy

In order to contrast easily, the accuracy is represented by the logarithm lgJ of the cost function. Figure 2, Fig. 3, Fig. 4 and Fig. 5 show the comparison results of 200 assimilation tests when the number of iterations is 50, 100, 150 and 200 respectively. In the figures, red line represents PSODIWAF (PSO with dynamic inertia weight). The blue line represents gpu-based PSOTVCF, and the horizontal line represents the average value. The y-coordinate is the logarithm of convergence precision, the smaller the lgJ is, the closer the J is to 0, that is the initial value is closer to the observed value after assimilation.

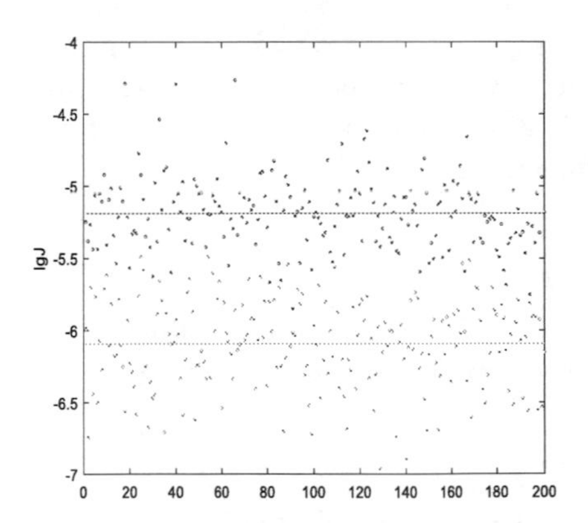

Fig. 2. Comparison of convergence accuracy after Iter = 50 (Color figure online)

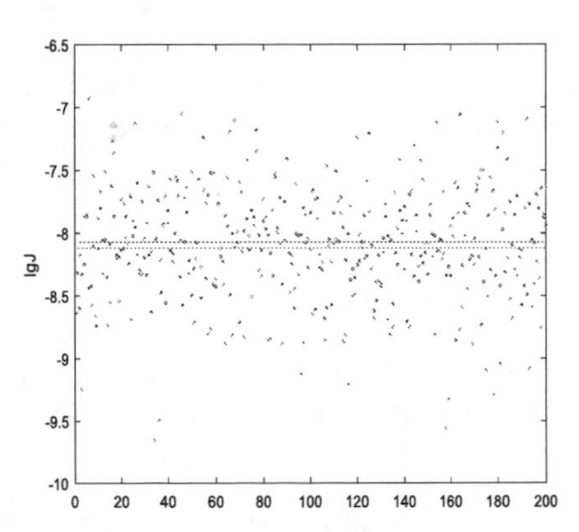

Fig. 3. Comparison of convergence accuracy after Iter = 100 (Color figure online)

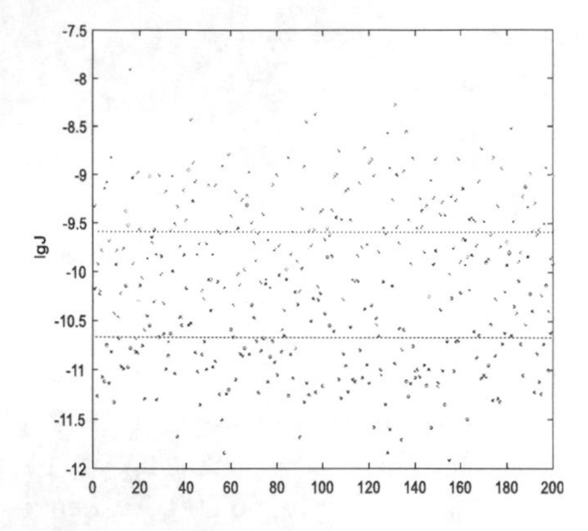

Fig. 4. Comparison of convergence accuracy after Iter = 150 (Color figure online)

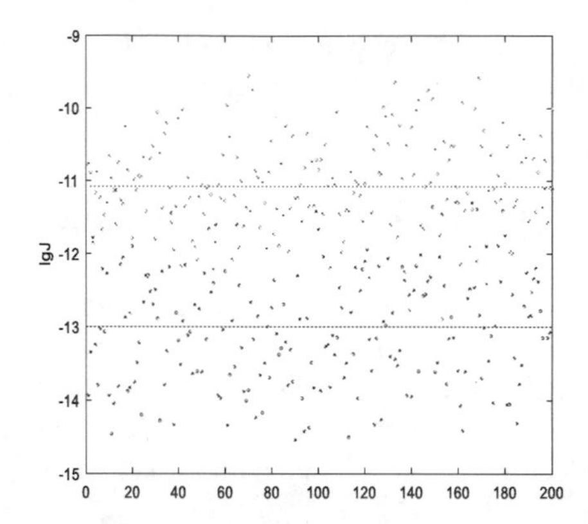

Fig. 5. Comparison of convergence accuracy after Iter = 200 (Color figure online)

As can be seen from the figure, the quality of results of PSODIWAF is better than that of PSOTVCF algorithm based on GPU acceleration at the beginning of assimilation (at the time of 50 iterations); By the mid assimilation stage (100 iterations), the assimilation capacity of the two is almost equal; During the later period of assimilation (150 iterations), PSOTVCF algorithm based on GPU acceleration came from behind, which is obvious that the assimilation ability of PSOTVCF is enhanced, and the accuracy of assimilation results lgJ was far beyond and could reach −10.75. After 200 iterations, PSODIWAF can only converge to −11, while PSOTVCF algorithm based on GPU can converge to −13 on average, which improves the accuracy by two orders of magnitude. Because of the action of time-varying accelerators and double compression factors, particles conducted self-learning in combination with the assimilating environment, and the communication among groups gained further strengthening, Therefore, the quality of assimilation results were greatly improved and almost all of them were corresponding to the observed values.

4.2 Assimilation Time

Table 1 shows the assimilation time consumed by different iterations. In order to avoid contingency, a total of 6 sets of assimilation tests were carried out for each assimilation window, and the data in the table are their average values.

Table 1. Comparison of iteration time of different algorithms

T_{200}/s ITER	50	100	150	200
Based on GPU acceleration time – PSO with varying dual compression factor	121. 273093	243. 593469	346. 460145	500. 339676
PSO with Time-varying dual compression factor	134. 390372	270. 694395	395. 541064	552. 676933
PSO with Dynamic inertia weight	129. 418112	256. 3248432	372. 042627	521. 157825

It can be calculated that after 150–200 interactions, although the accuracy of psotvcf is higher than that of psodiwaf by two orders of magnitude, the average time of single assimilation is slightly longer than that of psodiwaf by 0.1 s. The reason is that learning factors need to updated after each iteration of assimilation time, which makes the time of assimilation slightly increase. But the psotvcf accelerated by GPU shortens the assimilation time on the original basis. The reason is that through GPU parallel way, particles can find the optimal location together, rather than one by one. It saves a lot of time for particles to update their location at the same time, the accuracy of gpu-based PSOTVCF not only exceeds PSODIWAF by two orders of magnitude, but also accelerate the time by about 10% through parallel operation, and the improved PSOTCF also consumes less time of 4% than PAODIWAF.

5 Conclusions and Prospects

In this paper, gpu-end parallel PSOTVCF algorithm is adopted to have a more rapid data assimilation problem solving. By processing each particle in a computational parallel way on the gpu-end, the convergence speed of the whole particle swarm is accelerated, thus reducing the computation time. For PSOTVCF algorithm, to substantially raise the number of particles in GPU segment, the error can be greatly reduced when the running time is extremely limited, The experimental results demonstrate that compared with the serial PSOTVCF at the CPU end, the parallel PSOTVCF of GPU segment wins to reduce the computation time under the premise of consistent optimization stability, and the outcomes obtained are obviously superior to those in existing literatures. However, the acceleration time is not significantly shortened, thus there is still room for improvement. Additionally, the perfection of particle size will also ameliorate the parallel acceleration.

Funding Information. This research was financially supported by the Application of improved particle swarm optimization in data assimilation (20191050013), Teaching research project of Hubei provincial department of education (2016294, 2017320), humanities and social science research project of Hubei province (17D033).

References

1. Panofsky, H.: Objective weather map analysis. J. Met. **6**, 386–392 (1949)
2. Gilchrist, B., Cressman, G.P.: An experiment in objective analysis. Tellus A **6**, 309–318 (1954)
3. Bergthorsson, P., Dose, B.: Numerical weather map analysis. Tellus A **7**, 329–340 (1955)
4. Rutherford, I.D.: Data assimilation by statistical interpolation of forecast error fields. J. Atmos. Sci. **29**, 809–815 (1972)
5. Sasaki, Y.: Numerical variational analysis with weak constraint and application to surface analysis of severe storm gust. Mon. Wea. Rev. **98**, 899–910 (1970)
6. Fisher, M., Andersson, E.: Developments in 4D-Var and Kalman filtering. ECMWF Tech. Memo. (347) (2001). (Available from European Centre for Medium-Range Weather Forecasts, Shinfield Park, Reading, Berkshire RG2 9AX, UK)

7. Cao, X., Huang, S., Du, H.: A new method of orthogonal wavelet simulation for horizontal error function in variational assimilation. J. Phys. **57**, 1984–1989 (2008)
8. Guan, Y.H., Zhou, G.Q., Lu, W.S., et al.: Theory development and application of data assimilation methods. Meteorol. Disaster Reduction Res. **30**, 938–950 (2007)
9. Bai, C., Wu, C., Wu, L.: A four-dimensional assimilation method based on the combination of genetic algorithm and conjugate gradient method. J. Nanjing Inst. Meteorol. **29**(6), 850–854 (2006)
10. Zheng, Q., Ye, F., Sha, J., Wang, Y.: The application of dynamic weight particle swarm algorithm in four-dimensional variational data assimilation with switch. Weather Sci. Technol. **41**(2), 286–293 (2013)
11. Chen, F.: Research and application of particle swarm optimization neural network based on GPU. Jiangsu University of Science and Technology (2015)
12. Zhang, C.X.: Particle swarm optimization based on time varying constrict factor. Comput. Eng. Appl. **51**(23), 59–64 (2015)
13. Xia, X.W., Liu, J.N., Gao, K.F., et al.: Particle swarm optimization algorithm with reverse-learning and local-learning behavior. J. Software **38**(7), 1398–1409 (2015)
14. Kennedy, J., Eberhart, R.: Particle swarm optimization. In: IEEE International Conference on Neural Networks, pp. 1942–1948 (1995)

A Parallel Gene Expression Clustering Algorithm Based on Producer-Consumer Model

Lei Yang[1(✉)], Xin Hu[1], Kangshun Li[1], Wensheng Zhang[2], Yaolang Kong[1], Rui Xu[1], and Dongya Wang[3]

[1] College of Mathematics and Informatics, South China Agricultural University, Guangzhou 510642, China
yanglei_s@scau.edu.cn

[2] Institute of Automation, Chinese Academy of Sciences, Beijing 100190, China

[3] College of Engineering, Mathematics and Physical Sciences, University of Exeter, Exeter EX4 4QF, UK

Abstract. Clustering is one of the important tasks of machine learning. Gene Expression Programming (GEP) is used to solve clustering problems because of its strong global searching ability. In order to solve the limitation of lower rate of convergence and easy falling into optimal local solution in the traditional GEP clustering process, this paper proposes a parallel GEP clustering algorithm based on the producer-consumer model (PGEPC/PCM), which parallelizes the time-consuming operations such as fitness calculation, recombination, and mutation in GEP clustering analysis to speed up, improves the calculation method of fitness function to enable it to cluster automatically. This algorithm can fast calculate accurate clustering center points in parallel. Extensive experiments on four widely used benchmark Iris, Wine, Soybean and Seeds from the UCI machine learning data sets are conducted to investigate the influence of algorithmic component and results are compared with traditional GEP clustering algorithm. These comparisons demonstrate the competitive efficiency of the proposed algorithm.

Keywords: Gene Expression Programming · Clustering algorithm · Producer-consumer model · Clustering center

1 Introduction

Data mining and knowledge discovery have become an important research topic with the rapid expansion of data. Clustering is an important task of data mining. Because the accurate clustering requires corresponding algorithms, it is particularly important to design an efficient and precise algorithm. The design of the clustering algorithm depends on the type of data and the purpose of cluster analysis. The traditional clustering methods have made related researches in clustering techniques from different perspectives, but many existing algorithms

K. Li et al. (Eds.): ISICA 2019, CCIS 1205, pp. 97–111, 2020.
https://doi.org/10.1007/978-981-15-5577-0_8

have difficulty in automatically clustering data without prior knowledge of data. Evolutionary algorithms are applied to cluster analysis because of their high degree of parallelism, randomization, adaptive search, etc. Murthy and Chowdhury [13] proposed a clustering algorithm based on genetic algorithm (GA) that uses binary coding. Each data unit occupies one bit of an individual chromosome, due to the chromosome length limitation, it can only solve the clustering problem of small data sets. Bandyopadhyay and Maulik [2] proposed an improved GA clustering algorithm that uses cluster center points instead of individual coding schemes in chromosomes, which enormously reduced the chromosome length and can handle larger data sets. Yu Chen, Changjie Tang et al. [3] proposed an auto-clustering algorithm based on gene expression programming. Daihong Jiang and Sanyou Zhang et al. [11] combine the advantages of two kinds of algorithms. From the analysis of gene expression programming and K-Means algorithm, which realized the algebraic operation of gene clustering under the condition of unknown cluster division information, ensuring all grammar of chromosome evolution through this algorithm is correct. It not only avoids the large use of computational resources for editing illegal chromosomes, but also allows the use of chromosomes for modification, the experimental results are better than the traditional K-means clustering algorithm. Yifei Zheng, Lixin Jia, Hui Cao [15] proposed a multi-objective gene expression clustering algorithm, which can automatically determine the number of data sets and appropriate partitions. Hongguo Cai, Changan Yuan [8] proposed a serial clustering algorithm based on gene expression programming, which makes use of the parallelism advantage of Gene Expression Programming (GEP) to combine with the existing serial clustering DBSCAN algorithm. It makes the following program parallelized and improves the efficiency of the algorithm. Clustering algorithms based on gene expression programming have achieved some results. However, these algorithms also have some shortcomings, such as the low speed of operating, the requirement of prior knowledge when clustering, and falling into local optimum when it is disturbed by noises. Therefore, how to solve the problems existing in the traditional GEP algorithm in cluster analysis and obtain more accurate results will be the focus of this paper.

Aiming at the slow running speed and instability of the traditional gene expression programming, and also it is easy to fall into local optimal. This paper proposes a parallel GEP clustering algorithm based on producer-consumer model (PGEPC/PCM). This algorithm adopts the producer-consumer model to parallelize. It performs multi-thread parallelization on the calculation of fitness, selection, recombination, mutation, etc. in GEP clustering analysis to increase the speed and improves the calculation method of fitness function to enable it to cluster automatically. In this paper, Extensive experiments carried out to compare the average accuracy, the highest accuracy, the running time and other related indicators of the parallel gene expression clustering algorithm based on producer-consumer model and traditional gene expression programming clustering algorithm in Iris, Wine, Soybean, and Seeds from UCI public data sets. The results show that the parallel GEP clustering algorithm based on producer-consumer model (PGEPC/PCM) has a significant improvement in the efficiency of the algorithm.

The remainder of this study is organized as follows. Section 2 presents a discussion of the traditional Clustering analysis based on GEP. Section 3 presents our new parallel gene expression clustering algorithm based on producer-consumer model. Section 4 presents the experimental results and discussion. Finally, Sect. 5 concludes this study and presents future research directions.

2 Clustering Algorithm Based on Gene Expression Programming

2.1 Gene Expression Programming

The gene expression programming proposed by Ferreira [6] is an extension of Genetic Programming (GP), which combines the simple and fast characteristics of Genetic Algorithm's fixed-length linear coding and the flexible and diverse advantages of Genetic programming tree structure. It can solve complex problems with simple coding. The concept of functions in GEP is quite extensive, it includes the intermediate structure of any other non-terminal in the system, the set of functions can include the arithmetic symbols of the problem domain related to the application. Gene expression programming is widely used in symbol regression [16,17], classification [10,14], clustering [12], forecast [5,7,9], combinatorial optimization [1], modeling [4] and resource management [18] because of the fast random searching ability, potential parallelism, scalability, and robustness.

Gene expression programming uses a particular description method different from the genetic algorithm, which primarily uses a generalized hierarchical computer program to describe the problem. The formal description of this generalized hierarchical computer program requires two types of symbols, the terminators and the functions, which are meta-languages for constructing a program in gene expression programming. The chromosome of GEP consists of one or more genes through junction function and terminals, each gene consists of head, and tail, the head of a gene consists of a set of terminals and a set of functions, the tail of the gene only consists of the set of terminals. For each problem, the length of the head is selected in advance, and the length t of the tail is a function of the head lengths h and n, where n is the number of arguments to the function with the most variables, t is obtained by Eq. 1 below.

$$t = h(n - 1) + 1 \quad (1)$$

A gene is transformed into an expression by parsing, Moreover, the functions and terminals of the gene are filled from top to bottom in a hierarchical traversal. The basic GEP genetic operators include mutations, insertion sequences, and recombination. Mutation can introduce new nodes into genes, which is the most efficient operator among all the operators with modification ability. To ensure the gene can be expressed generally after mutation. The category of nodes would be different, according to the place where mutation happened. When a mutation occurs in the head of the gene, the node after mutation can be a function or terminal. If a mutation happened in the tail, the node only could be the terminal. Mutations can take place anywhere in the gene, without regard to the type of the original type. Some of the nodes at the position of mutation, which can be modified randomly based on the above rules. The insertion sequence elements of GEP are fragments of the gene locus that can be activated and jump to another location of the chromosome. There are three kinds of transposable elements in GEP: (1) Short fragments whose the beginning position is function or terminal are transposed to the head of genes except for the root (insertion sequence elements or IS elements); (2) Short fragments with a function at the first position are transposed to the root of genes (root IS elements or RIS elements); and (3) entire genes are transposed into the beginning of chromosomes. There are three recombinations in GEP: single-point recombination, two-point recombination, and genetic recombination. In all cases, two randomly selected parent chromosomes pair and interacted with each other. Gene recombination produces different arrangements of existing genes. The evolutionary ability of GEP is not only based on gene rearrangement, but also based on the continuous generation of new genetic material, which is caused by mutation and insertion sequence.

2.2 Cluster Analysis Based on Gene Expression Programming

Cluster analysis is the process of grouping a collection of physical or abstract objects into multiple classes of similar objects. The goal of cluster analysis is to measure the similarities between different data sources and to classify data sources into different clusters. Clustering can be used as an independent tool to obtain the distribution of data, which also can observe the characteristics of each cluster of data, and focus on further analysis of specific clusters. Cluster analysis can also be used as a preprocessing step for other algorithms. The clustering algorithm based on gene expression programming does not need any prior knowledge of the data sets, which can automatically divide clusters and complete cluster analysis.

Clustering Algorithm Based on GEP

(1) Encoding. Due to the particularity of the clustering problem, the encoding mode uses the single-gen encoding, which consists of a head and a tail. The range of the header encoding is || or &&, and the tail encoding is an individual instance. According to the uniform distribution, the data sequence numbers are randomly extracted from the data set to form a tail sequence. The length of the head is n, and the length of the tail is $n + 1$.

(2) Cluster Fitness Function. The fitness function is the driving force of GEP population evolution, which can make the algorithm evolve in the required direction. In most cases, an individual's fitness evaluation consumes most of the GEP running time. Choosing different fitness functions will affect the quality of the

evolutionary results directly, if the fitness function is selected improperly, it will make a consequence that the iteration does not converge or converge to an unrelated solution. The GEP clustering analysis draws on the principle of "survival of the fittest" in nature to give the fitness function of the corresponding gene expression programming. In other words, the closer the genetic expression is to the actual observation, the higher the fitness will be. The fitness function in the GEP clustering algorithm is defined as follows in Eqs. 2 and 3:

$$f = \frac{1}{1+E} \tag{2}$$

$$E = \sum_{i=1}^{k} \sum_{p \in C_i} |p - m_i|^2 \qquad m_i = \frac{1}{n_i} \sum_{x_j \in C_i} x_j \tag{3}$$

In this formula, the k is the number of clusters in the data set, and the p is the point in space. n_i is the number of data points in the cluster C_i. m_i is the average of cluster C_i. It is calculated by taking the arithmetic mean of the dimensions of all the elements in the cluster. After decoding the chromosomal information, the coordinates of the center points of each possible cluster are obtained, and then the data points of the data set are sequentially assigned to the nearest cluster according to the Euclidean distance from each cluster center. Recalculate the center coordinates of each cluster after clustering, and calculate the sum of squared errors E of all data in the data set. The smaller the E value is, the smaller the cluster will be as compact and independent as possible.

(3) Procedure of GEP Clustering Algorithm. The procedure of the clustering algorithm based on GEP is shown below. Firstly, the data is normalized and preprocessed when the initial training set data is imported. Then, the GEP Algorithm is used to cluster the training set data and get the cluster center point. Then determine whether the fitness of the best individual is ranked in the top ten of the historical operation record. If so, put the resulting individuals in this calculation into the best individuals to compile statistics. Otherwise, discard it. Finally, this algorithm integrates the best individuals in the statistical fitness set to derive rules.

3 Parallel Gene Expression Programming Clustering Algorithm Based on Population Migration Strategy (PGEPC/PCM)

In order to study the shortcomings of the basic gene expression programming clustering algorithm and the necessity of improving the GEP clustering algorithm, in this section, we applied the basic gene expression clustering algorithm to four common data sets, study the performance of the basic gene expression clustering algorithm on the four public data sets and analyze its inadequacies. After that, the parallel gene expression clustering algorithm based on producer-consumer model was proposed.

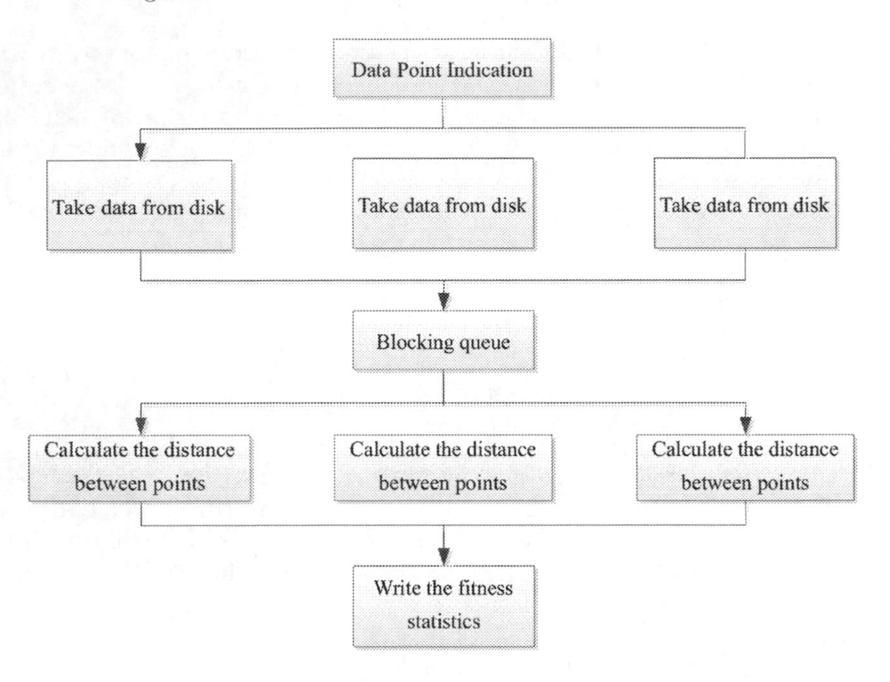

Fig. 1. Fitness calculation parallelization process

3.1 Inadequacies of Basic Gene Expression Programming Clustering Algorithm

Speed of the Algorithm. The main steps of GEP clustering operation are calculating fitness values, recombination, mutation, Furthermore, loop the steps above. The individual fitness calculation process is as follows, for each discrete point, the nearest center point is selected as the cluster to which it is assigned, each point does not need to rely on the calculation results of other points to calculate the center point of the cluster to which it will be assigned. The algorithm costs the most time while calculating the fitness. And it is the main module that limits the efficiency of the program. It needs to adjust the method of individual fitness calculation to improve the efficiency of the algorithm.

Automatic Clustering. After the GEP completes the clustering of each data point, the algorithm determines whether there are clusters that can merge in the cluster set and merges them. Head coding does not have to be involved in recombination and mutation operations, the traditional individual fitness function does not perform automatic clustering well.

3.2 Parallel GEP Clustering Algorithm Based on Producer-Consumer Model (PGEPC/PCM)

The Clustering Algorithm with Parallel Gene Expression Programming based on producer-consumer model proposed in this paper will optimize the basic GEP clustering algorithm from two aspects: The first is parallelizing the time-consuming operations in the algorithm to promote the speed of calculation; The second is improving the fitness function to enable automatic clustering.

Algorithm Local Parallelization

Parallel Method Mining. The calculation of fitness value is the main factor limiting the efficiency of GEP clustering algorithm, Parallel analysis of the phase is needed to improve the efficiency of the algorithm. Design to create a separate thread for each point to calculate the point to which it will be assigned. However, in the case of an enormous amount of data, assuming that there are thousands of pieces of data, it is necessary to open another thread to calculate. Therefore, the system resources will be consumed tremendously.

To solve this problem, we must analyze the critical resources, with the sum of discrete points and distances firstly. Assuming that there are n workers (threads) randomly selecting a point from thousands of points to classify, and then gather statistics of the sum of the distances within the classes. We can use the producer-consumer model to realize the algorithm parallelism. The producer-consumer model is significant in the operating system, which describes a mechanism of waiting and notifying. The producer-consumer is a classic problem in the thread model: producers and consumers share the same storage space during the same period. The producer is responsible for taking out data and stuffing it into the blocking queue. The consumer is responsible for taking out the statistical fitness results from the queue. This parallel computing model is adopted to reduce system overhead and improve system operation efficiency. The algorithm flow is shown in Fig. 1.

This algorithm's description can be divided into the following steps: Randomly take a data point and get the data from the disk with the label; Put the extracted point data into the blocking queue; Calculate the point of the thread taken from the blocking queue and calculate its distance from the center cluster point; Write the calculated data to statistical fitness.

BWP Fitness

Definition 1. *Definition of baw(j,i): Let* $K = \{X, R\}$ *be the cluster space, where* $X = \{x_1, x_2, \ldots, x_n\}$ *. Assuming that n sample objects are clustered into class c, the cluster distance baw(j, i) of the i-th sample defining the j-th class is the sum of the minimum inter-class distance and the intra-class distance of the sample. That is:*

$$baw(j,i) = b(j,i) + w(j,i) = \min_{1 \le k \le c, k \ne j} \left(\frac{1}{n_k} \sum_{p=1}^{n_k} \left\| x_p^k - x_i^j \right\|^2 \right) + \frac{1}{n_j - 1} \sum_{q=1, q \ne i}^{n_j} \left\| x_q^j - x_i^j \right\|^2 \tag{4}$$

Definition 2. *Definition of bsw(j,i): Let* $K = \{X, R\}$ *be the cluster space, where* $X = \{x_1, x_2, \dots, x_n\}$. *Assuming that n sample objects are clustered into class c, the cluster separation distance bsw(j, i) of the i-th sample defining the j-th class is the difference between the minimum inter-class distance and the intra-class distance of the sample. That is:*

$$baw(j,i) = b(j,i) - w(j,i) = \min_{1 \le k \le c, k \ne j} \left(\frac{1}{n_k} \sum_{p=1}^{n_k} \left\| x_p^k - x_i^j \right\|^2 \right) - \frac{1}{n_j - 1} \sum_{q=1, q \ne i}^{n_j} \left\| x_q^j - x_i^j \right\|^2 \tag{5}$$

According to $baw(j,i)$ and $bsw(j,i)$, Let $K = X, R$ be the cluster space, where $X = \{x_1, x_2, \dots, x_n\}$. Assuming that n sample objects are clustered into class c, defining between-withing proportion (BWP) of the i-th sample of class j. The index $BWP(j,i)$ is the ratio of the clustering dispersion distance and the clustering distance of the sample. That is:

$$BWP(j,i) = \frac{bsw(j,i)}{baw(j,i)} = \frac{b(j,i) - w(j,i)}{b(j,i) + w(j,i)} = \frac{\min_{1 \le k \le c, k \ne j} \left(\frac{1}{n_k} \sum_{p=1}^{n_k} \left\| x_p^{(k)} - x_i^{(j)} \right\|^2 \right) - \frac{1}{n_j - 1} \sum_{q=1, q \ne i}^{n_j} \left\| x_q^{(j)} - x_i^{(j)} \right\|^2}{\min_{1 \le k \le c, k \ne j} \left(\frac{1}{n_k} \sum_{p=1}^{n_k} \left\| x_p^{(k)} - x_i^{(j)} \right\|^2 \right) + \frac{1}{n_j - 1} \sum_{q=1, q \ne i}^{n_j} \left\| x_q^{(j)} - x_i^{(j)} \right\|^2} \tag{6}$$

When the BWP fitness calculation is running, the distance to each center point needs to be calculated for each point, and the shortest distance to the points in the other clusters is calculated after the classification. The calculation amount is several times larger than the previous calculation of the distance within the group. When the operating environment is officially running the algorithm, it takes a high time cost, and eventually the data volume is too large to run. However, the main time-consuming part of the program operation is the calculation of the distance between groups. When transplanting into GEP clustering, a compromise method is adopted, and it is not necessary to accurately calculate the distance between groups, and the main idea is extracted to perform an approximate calculation.

Because the GEP chromosomes are limited in length, they only contain a limited number of points as cluster centers. It is not necessary to cluster all the points and then calculate the distance between the clusters and the BWP value for measuring the fitness of a chromosome. Therefore, the $b(i, j)$ of the new algorithm design proposed in this paper keeps the original calculation method unchanged, and $w(i, j)$ is simplified to calculate the shortest distance from the center of the original cluster to the center of other clusters. If the clustering effect is good enough and the distance within the group is tight, the distance between the clusters to the other clusters is approximately equal to the distance between the discrete points closest to the other clusters to the other clusters.

Table 1. Info of data sets

Data sets	Number of attributes	Number of categories	Number of instances
Iris	4	3	150
Wine	13	3	178
Soybean	35	4	47
Seeds	7	3	210

Table 2. Parameter table for clustering experiments of Iris

Parameter name	Value	Parameter name	Value
Running times	10	Function set	||, &&
Number of generations	100	Terminal set	x1–x150
Size of groups	100	Mutation probability	0.4
Head length	2	Single point recombination probability	0.7
Tail length	3	Double point recombination probability	0.7

4 Experiment and Result Analysis

The parallel GEP clustering algorithm based on producer-consumer model (PGE PC/PCM) proposed in this paper is implemented by Java language. This paper compares the parallel GEP clustering algorithm based on producer-consumer model (PGEPC/PCM) and the basic gene expression programming clustering algorithm. And it also uses these two algorithms to compare the average accuracy, the highest accuracy, the running time, the average correct clustering document, the highest correct clustering document and other evaluation indicators of the four data sets.

4.1 Data Sets

The experiment in this paper uses 4 data sets. The data sets are derived from the UCI database for machine learning (http://archive.ics.uci.edu/ml/) proposed by the University of California Irvine. The four data sets are Iris, Wine, Soybean, and Seeds. The main info of four data sets is listed in Table 1.

4.2 Parallel GEP Clustering Algorithm Based on Producer-Consumer Model (PGEPC/PCM)

In order to analyze the influence of the parallelization method based on the producer-consumer model on the GEP clustering algorithm, this paper carried out the parallel GEP clustering algorithm based on producer-consumer model (PGEPC/PCM) at first, and cluster for the above 4 data sets. The algorithm running environment is windows operating system and Java 1.7, using Redis as a tool. The experimental parameter settings are shown in Tables 2, 3, 4 and 5.

Table 3. Parameter table for clustering experiments of Wine

Parameter name	Value	Parameter name	Value
Running times	10	Function set	||, &&
Number of generations	100	Terminal set	x1–x178
Size of groups	100	Mutation probability	0.4
Head length	2	Single point recombination probability	0.7
Tail length	3	Double point recombination probability	0.7

Table 4. Parameter table for clustering experiments of Soybean

Parameter name	Value	Parameter name	Value
Running times	10	Function set	||, &&
Number of generations	100	Terminal set	x1–x47
Size of groups	100	Mutation probability	0.4
Head length	3	Single point recombination probability	0.7
Tail length	4	Double point recombination probability	0.7

Table 5. Parameter table for clustering experiments of Seeds

Parameter name	Value	Parameter name	Value
Running times	10	Function set	||, &&
Number of generations	100	Terminal set	x1–x210
Size of groups	100	Mutation probability	0.4
Head length	2	Single point recombination probability	0.7
Tail length	3	Double point recombination probability	0.7

Table 6. Clustering experiment results of Iris

Data set	Running times	Highest accuracy		Run time/s	
		GEP	PGEPC/PCM	GEP	PGEPC/PCM
Iris	1	89.33	89.33	445	234
	2	88	89.33	435	245
	3	88	90.66	479	250
	4	90.66	90.66	470	247
	5	88	89.33	430	240
	6	90.66	90.66	447	255

The four data sets were clustered based on the above parameter settings, and the experimental results are shown in Tables 6, 7, 8 and 9.

The comparison of the running time and the highest accuracy of the parallel GEP clustering algorithm based on producer-consumer model (PGEPC/PCM) clustering and the basic GEP clustering is shown in Figs. 2, 3, 4 and 5.

It can be compared from Fig. 2, 3, 4 and 5, the parallel GEP clustering algorithm based on producer-consumer model (PGEPC/PCM) has no visible accuracy improvement in each data set, but it has improved a lot in the running

Table 7. Clustering experiment results of Wine

Data set	Running times	Highest accuracy		Run time/s	
		GEP	PGEPC/PCM	GEP	PGEPC/PCM
Wine	1	87.07	87.64	683	308
	2	87.64	87.07	678	309
	3	83.70	87.07	664	310
	4	87.07	87.07	679	305
	5	87.07	87.64	684	312
	6	87.64	87.07	664	315

Table 8. Clustering experiment results of Soybean

Data set	Running times	Highest accuracy		Run time/s	
		GEP	PGEPC/PCM	GEP	PGEPC/PCM
Soybean	1	85.10	85.10	85.10	85.10
	2	85.10	87.23	87.23	87.23
	3	85.10	87.23	87.23	87.23
	4	87.23	87.23	87.23	87.23
	5	87.23	87.10	87.10	87.10
	6	87.10	87.10	87.10	87.10

Table 9. Clustering experiment results of Seeds

Data set	Running times	Highest accuracy		Run time/s	
		GEP	PGEPC/PCM	GEP	PGEPC/PCM
Seeds	1	78.57	78.57	803	515
	2	79.04	78.57	789	504
	3	78.57	78.57	794	512
	4	78.57	78.57	782	518
	5	79.04	79.04	795	519
	6	78.57	78.57	785	514

speed. It can be concluded that this parallel GEP clustering algorithm based on producer-consumer model (PGEPC/PCM) improves the running speed of the algorithm without affecting the change of the overall fitness, thus improving the performance of the algorithm, and the improvement is visible.

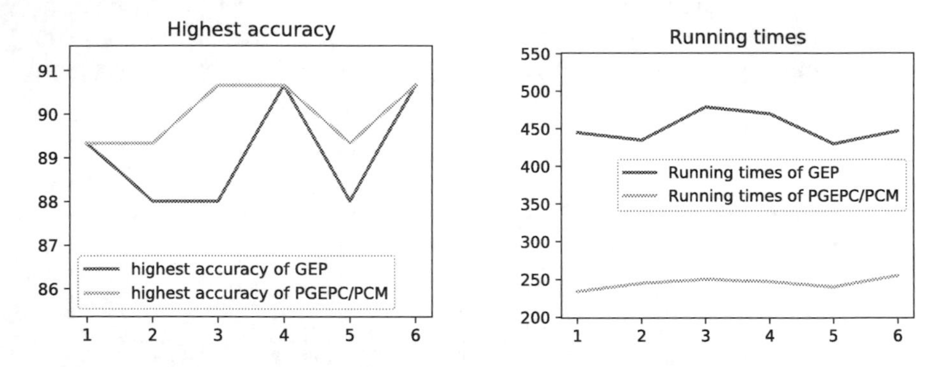

Fig. 2. Clustering experiment result comparison figure of Iris

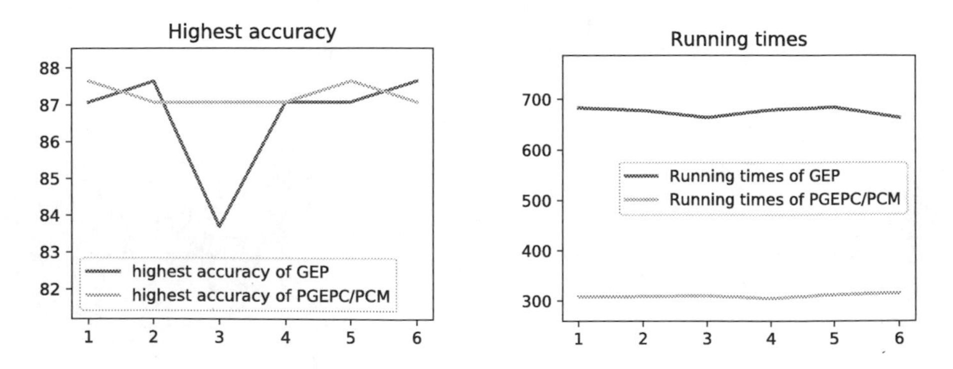

Fig. 3. Clustering experiment result comparison figure of Wine

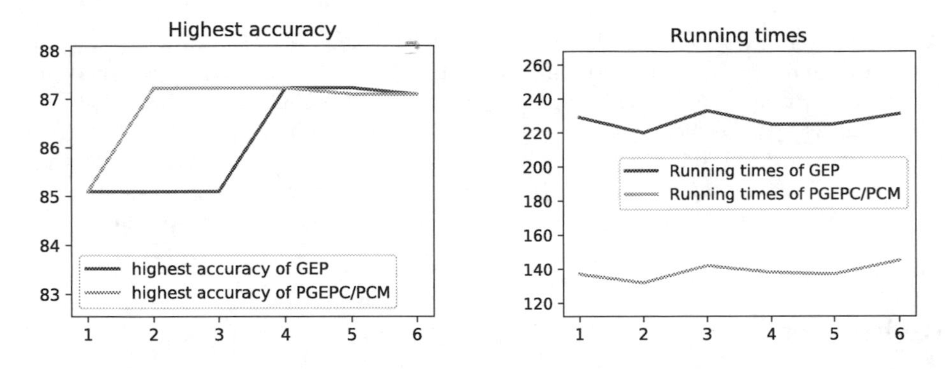

Fig. 4. Clustering experiment result comparison figure of Soybean

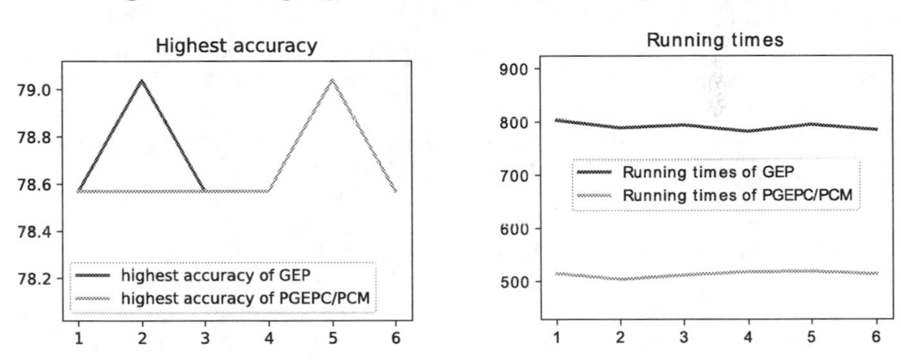

Fig. 5. Clustering experiment result comparison figure of Seeds

5 Conclusion and Future Work

In order to get better clustering results, in this paper, we have conducted analysis and research to optimize the evolutionary operators and algorithm flows based on GEP for cluster analysis. Aiming at the shortcomings of the basic GEP clustering algorithm, a parallel GEP clustering algorithm based on producer-consumer model (PGEPC/PCM) proposed, which parallelizes the time-consuming operations such as fitness calculation, recombination and mutation in GEP clustering analysis to speed up, improves the calculation method of fitness function to enable it to cluster automatically. Extensive experiments on 4 widely used benchmark Iris, Wine, Soybean and Seeds from the UCI machine learning data sets are conducted to investigate the influence of algorithmic components and results are compared with the traditional GEP clustering algorithm. These comparisons demonstrate the competitive efficiency of the proposed algorithm.

Then there are several areas worth studying. For example, The method of fitness calculation has not been perfected and the problem of local optimization has not been solved, which will be the follow-up work of this paper, how to set the head coding of chromosome reasonably and the selection of weights needs further study.

Acknowledgments. This work was partially supported by the National Natural Science Foundation of China (Grant Nos. 61573157 and 61703170), Science and Technology Project of Guangdong Province of China (Grant Nos. 2018A0124 and 2017A020224004), Science and Technology Project of Tianhe District of Guangzhou City (Grant No. 201702YG061), Science and technology innovation project for College Students (Grant No. 201910564129). The authors also gratefully acknowledge the reviewers for their helpful comments and suggestions that helped to improve the presentation.

References

1. Ferreira, C.: Gene Expression Programming. Springer, Heidelberg (2006). https://doi.org/10.1007/3-540-32849-1
2. Bandyopadhyay, S., Maulik, U.: An evolutionary technique based on k-means algorithm for optimal clustering in RN. Inf. Sci. **146**(1–4), 221–237 (2002). https://doi.org/10.1016/s0020-0255(02)00208-6
3. Chen, Y., Tang, C., Ye, S.Y., Li, C., Liu, Q.H.: An auto-clustering algorithm based on gene expression programming **39**, 107–112 (2007)
4. Colbourn, E., Roskilly, S., Rowe, R., York, P.: Modelling formulations using gene expression programming - a comparative analysis with artificial neural networks. Eur. J. Pharm. Sci. **44**(3), 366–374 (2011). https://doi.org/10.1016/j.ejps.2011.08.021
5. Deng, S., Zhou, A.H., Yue, D., Hu, B., Zhu, L.P.: Distributed intrusion detection based on hybrid gene expression programming and cloud computing in a cyber physical power system. IET Control Theory Appl. **11**(11), 1822–1829 (2017). https://doi.org/10.1049/iet-cta.2016.1401
6. Ferreira, C.: Gene expression programming: a new adaptive algorithm for solving problems. ArXiv cs.AI/0102027 (2001)
7. Gholami, A., Bonakdari, H., Zeynoddin, M., Ebtehaj, I., Gharabaghi, B., Khodashenas, S.R.: Reliable method of determining stable threshold channel shape using experimental and gene expression programming techniques. Neural Comput. Appl. **31**(10), 5799–5817 (2018). https://doi.org/10.1007/s00521-018-3411-7
8. Hongguo, C., Chanan, Y.: Research on GEP-based cluster algorithm for serial program to be parallelized. J. S.-Cent. Univ. Nationalities (Nat. Sci.Ed.) **4**, 112–115 (2017)
9. Huang, Z., Li, M., Chousidis, C., Mousavi, A., Jiang, C.: Schema theory-based data engineering in gene expression programming for big data analytics. IEEE Trans. Evol. Comput. **22**(5), 792–804 (2018). https://doi.org/10.1109/tevc.2017.2771445
10. Jedrzejowicz, J., Jedrzejowicz, P., Wierzbowska, I.: Implementing gene expression programming in the parallel environment for big datasets' classification. Vietnam J. Comput. Sci. **06**(02), 163–175 (2019). https://doi.org/10.1142/s2196888819500118
11. Jiang, D.H., Zhang, S.Y.: K-means auto-clustering algorithm based on gene expression programming. Comput. Simul. **27**, 216–220 (2010)
12. Jiang, Z., Li, T., Min, W., Qi, Z., Rao, Y.: Fuzzy c-means clustering based on weights and gene expression programming. Pattern Recogn. Lett. **90**, 1–7 (2017). https://doi.org/10.1016/j.patrec.2017.02.015
13. Murthy, C., Chowdhury, N.: In search of optimal clusters using genetic algorithms. Pattern Recogn. Lett. **17**(8), 825–832 (1996). https://doi.org/10.1016/0167-8655(96)00043-8

14. Yang, L., Li, K., Zhang, W., Zheng, L., Ke, Z., Qi, Y.: Optimization of classification algorithm based on gene expression programming. J. Ambient Intell. Humanized Comput. (2017). https://doi.org/10.1007/s12652-017-0563-8
15. Zheng, Y., Jia, L., Cao, H.: Multi-objective gene expression programming for clustering. Inf. Technol. Control **41**(3) (2012). https://doi.org/10.5755/j01.itc.41.3.1330
16. Zhong, J., Feng, L., Ong, Y.S.: Gene expression programming: a survey [review article]. IEEE Comput. Intell. Mag. **12**(3), 54–72 (2017). https://doi.org/10.1109/mci.2017.2708618
17. Zhong, J., Ong, Y.S., Cai, W.: Self-learning gene expression programming. IEEE Trans. Evol. Comput. **20**(1), 65–80 (2016). https://doi.org/10.1109/tevc.2015.2424410
18. Aytac, G., Ali, A.: New approach for stage-discharge relationship: gene-expression programming. J. Hydrol. Eng. **14**(8), 812–820 (2009). https://doi.org/10.1061/(ASCE)HE.1943-5584.0000044

Evolutionary Multi-objective and Dynamic Optimization

Decomposition-Based Dynamic Multi-objective Evolutionary Algorithm for Global Optimization

Qing Zhang[1], Ruwang Jiao[2(✉)], Sanyou Zeng[2], and Zhigao Zeng[3]

[1] School of Computer Science, Huanggang Normal University, Huanggang 438000, China
zq3652000@yahoo.com.cn

[2] School of Mechanical Engineering and Electronic Information, China University of Geosciences, Wuhan 430074, China
ruwangjiao@gmail.com, sanyouzeng@gmail.com

[3] College of Computer, Hunan University of Technology, Zhuzhou 412007, Hunan, China
zzgzzg99@163.com

Abstract. This paper proposes a decomposition-based dynamic multi-objective evolutionary algorithm for addressing global optimization problems. In the proposed method, the niche count function is regarded as an additional objective to maintain the population diversity. The niche count function is controlled by niche radius. The niche radius is dynamically reduced from a large value to zero, which can provide a tradeoff between exploration and exploitation. In each generation, the proposed algorithm decomposes the transformed bi-objective optimization problem into a number of simple bi-objective subproblems, and then solving these bi-objective subproblems in a collaborative way. Experiments have been conducted on benchmark test instances and compared with four state-of-art algorithms. The experimental results have demonstrated its effectiveness in solving complex global optimization problems.

Keywords: Global optimization · Multi-objective optimization · Decomposition · Diversity and convergence · Exploration and exploitation

1 Introduction

In many academic researches and engineering applications, the problems to locate the global optimum are usually encountered. Such problems are known as global optimization problems. Many approaches have been proposed to deal with global optimization problems. Nevertheless, when the objective is nonlinear, non-convex or non-differentiable, conventional mathematical methods may become inefficient, or even fail to work. Evolutionary algorithm (EA) is a kind of population based iterative heuristic optimization paradigm. Over the last two

K. Li et al. (Eds.): ISICA 2019, CCIS 1205, pp. 115–126, 2020.
https://doi.org/10.1007/978-981-15-5577-0_9

decades, EAs have been widely studied and applied for many scientific and real-world optimization problems with promising results [1]. However, the basic EAs are easy to fall into some local optima when tackling complicated global optimization problems with a large number of local optima. Apparently, the key to solving global optimization problems using EAs is how to handle the relationship between exploration and exploitation.

The exploration refers to the capability of an EA to investigate undiscovered regions in the search space to find more potential solutions, while the exploitation means the ability to apply the knowledge of the discovered promising solutions to further improve their quality. Usually, high exploitation rate leads to faster convergence. In practice, exploration and exploitation are conflicting with each other. Thus, in order to achieve good performances on global optimization problems, the exploration and exploitation abilities should be well balanced [2]. It is extensively recognized that an EA should put more emphasis on exploration at the early evolution stage and high exploitation at the later evolution stage.

Dynamic multi-objective technique [3] is a very recent work which can provide a well balance between exploration and exploitation. It converts a single-objective optimization problem into an equivalent dynamic multi-objective optimization problem (DMOP) which considers the niche count function as an additional objective. The niche count function is controlled by the niche radius. At the early evolution stage, a large niche radius enables the search pay more attention to the exploration to investigate more promising regions. At the later evolution stage, the niche radius gradually decreased to zero as the environmental changes, which allows an EA focus on exploitation to guarantee the convergence.

Simultaneously, a large number of multi-objective evolutionary algorithms (MOEAs) have been proposed in the past two decades, which can generally fall into three types: Pareto domination-based methods [4], indicator-based methods [5], and decomposition-based methods [6]. Decomposition is a well-known method in traditional multi-objective optimization. Decomposition-based MOEAs decompose a multi-objective optimization problem into a number of scalar optimization subproblems and optimize them in a collaborative manner using an EA [6–13]. Among them, MOEA/D-M2M [7] is an improved version of MOEA/D [6]. Different from most decomposition-based MOEAs which decompose a MOP into certain single-objective subproblems, MOEA/D-M2M decomposes a MOP into a number of simple multi-objective subproblems, and then solving these multi-objective subproblems in a cooperative way.

In this paper, we make an attempt to use decomposition-based dynamic multi-objective evolutionary algorithm to solve complex global optimization problems. Firstly, a global optimization problem is transformed into a dynamic bi-objective optimization problem: the original objective and the niche count function objective. Subsequently, a decomposition-based multi-objective evolutionary algorithm: MOEA/D-M2M, is adopted to address the converted dynamic bi-objective optimization problem. The dynamic version of MOEA/D-M2M is termed DMOEA/D-M2M. DMOEA/D-M2M decomposes the transformed bi-objective optimization problem into a set of bi-objective optimization subproblems which are easier to solve. Throughout the entire evolution stage, the niche radius of the niche count function is dynamically reduced from an initial value to zero, which pushes the population from diverse directions to the global optimum.

The rest of this paper is organized as follows. In Sect. 2, we briefly presents some preliminary. In Sect. 3, a global optimization problem is transformed into a dynamic bi-objective optimization problem. The details of the proposed DMOEA/D-M2M for solving global optimization problems are presented in Sect. 4. Experimental settings and comparisons of DMOEA/D-M2M with four algorithms are presented in Sect. 5. Finally, conclusions are drawn in Sect. 6.

2 Preliminary

2.1 Global Optimization Problem

Without loss of generality, a minimization global optimization problem can be represented as follows:

$$\begin{aligned} \min \quad & y = f(\boldsymbol{x}) \\ \text{where } & \boldsymbol{x} = (x_1, ..., x_n) \in \mathbf{X} \\ & \mathbf{X} = \{\boldsymbol{x} | \boldsymbol{l} \leq \boldsymbol{x} \leq \boldsymbol{u}\} \\ & \boldsymbol{l} = (l_1, ..., l_n), \boldsymbol{u} = (u_1, ..., u_n), \end{aligned} \tag{1}$$

where $\boldsymbol{x}$ is the solution vector and $\mathbf{X}$ denotes the solution space, $\boldsymbol{l}$ and $\boldsymbol{u}$ are the lower and upper bounds of the solution space. The aim of the global optimization problem is to find the global optimum while satisfying the bound constraints $(\boldsymbol{l} \leq \boldsymbol{x} \leq \boldsymbol{u})$.

2.2 Multi-objective Optimization Problem

A minimization MOP can be formulated as follows:

$$\begin{aligned} \min \quad & y = \boldsymbol{f}(\boldsymbol{x}) = (f_1(\boldsymbol{x}), ..., f_m(\boldsymbol{x})) \\ \text{where } & \boldsymbol{x} = (x_1, ..., x_n) \in \mathbf{X} \\ & \mathbf{X} = \{\boldsymbol{x} | \boldsymbol{l} \leq \boldsymbol{x} \leq \boldsymbol{u}\} \\ & \boldsymbol{l} = (l_1, ..., l_n), \boldsymbol{u} = (u_1, ..., u_n), \end{aligned} \tag{2}$$

where $\boldsymbol{x}$ is the solution vector, $\mathbf{X}$ represents the solution space, $\boldsymbol{l}$ and $\boldsymbol{u}$ denote the lower bound and upper bound of the solution space. The objective vector $\boldsymbol{f}(\boldsymbol{x})$ consists of m objective functions.

For any two solution vectors $\boldsymbol{a}$ and $\boldsymbol{b}$, $\boldsymbol{a}$ is said to dominate $\boldsymbol{b}$ ($\boldsymbol{a} \prec \boldsymbol{b}$), if $f_i(\boldsymbol{a}) \leq f_i(\boldsymbol{b})$ for all $i \in \{1, ..., m\}$, and $f_i(\boldsymbol{a}) < f_i(\boldsymbol{b})$ for at least one $i \in \{1, ..., m\}$. A solution $\boldsymbol{x}^*$ is called Pareto optimal if there is not another solution $\boldsymbol{x}$ satisfying $\boldsymbol{x} \prec \boldsymbol{x}^*$. The *Pareto set* (*PS*) is the set of all the Pareto optimal solutions, the *Pareto front* (*PF*) is the set of the images of the solutions in the *PS*. The aim of the MOP is to find a set of non-dominated solutions which are evenly distributed on the *PF*. Note that when $m = 1$, problem (2) is equal to problem (1).

3 Conversion of a Global Optimization Problem to a Dynamic Multi-objective Optimization Problem

The idea of using dynamic multi-objective technique for dealing with unconstrained and constrained single-objective optimization problems borrows from [3,14–17]. An additional objective, niche count function, is introduced to tradeoff the exploration and exploitation and, hence, avoid an EA from being trapped in local optima.

Definition 1. ***(Niche-count)*** *Suppose the combination of parent population and offspring population* $\mathbf{R} = \{\boldsymbol{x}_1, \boldsymbol{x}_2, \cdots, \boldsymbol{x}_{2NP}\}$, *the niche-count function for* $\boldsymbol{x} \in \mathbf{R}$*:*

$$nc(\boldsymbol{x}|\mathbf{R}, \sigma) = \sum_{i=1, \boldsymbol{x}_i \neq \boldsymbol{x}}^{2NP} sh(\boldsymbol{x}, \boldsymbol{x}_i), \tag{3}$$

where the sharing function between $\boldsymbol{x}_1$ *and* $\boldsymbol{x}_2$*:*

$$sh(\boldsymbol{x}_1, \boldsymbol{x}_2) = \begin{cases} 1 - (\frac{d(\boldsymbol{x}_1, \boldsymbol{x}_2)}{\sigma}), & d(\boldsymbol{x}_1, \boldsymbol{x}_2) \leq \sigma \\ 0, & otherwise \end{cases} \tag{4}$$

$d(\boldsymbol{x}_1, \boldsymbol{x}_2)$ is the Euclidean distance between $\boldsymbol{x}_1$ and $\boldsymbol{x}_2$, σ is the niche radius.

A ***dynamic multi-objective optimization problem (DMOP)*** is a sequence of MOPs $\{MOP^{(s)}\}$, $s = 0, 1, \cdots, S$, defined as follows

Definition 2. ***(DMOP)***

$$\begin{array}{ll} MOP^{(0)}: & min\ \boldsymbol{y} = (f(\boldsymbol{x}), nc(\boldsymbol{x}|\mathbf{R}, \sigma^{(0)})) \\ MOP^{(1)}: & min\ \boldsymbol{y} = (f(\boldsymbol{x}), nc(\boldsymbol{x}|\mathbf{R}, \sigma^{(1)})) \\ \ldots\ldots & \ldots\ldots \\ MOP^{(S)}: & min\ \boldsymbol{y} = (f(\boldsymbol{x}), nc(\boldsymbol{x}|\mathbf{R}, \sigma^{(S)})) \\ & where\ \boldsymbol{x} \in \mathbf{X} = \{\boldsymbol{x}|\boldsymbol{l} \leq \boldsymbol{x} \leq \boldsymbol{u}\}, \end{array} \tag{5}$$

where $\sigma^{(0)} > \sigma^{(1)} > \cdots > \sigma^{(S)} = 0$. S is the maximum environmental state which is equal to the maximum generation, $\sigma^{(s)}$ represents the niche radius value at state s. An environment change can be seen as a shrinkage of the niche radius σ from state s to $s+1$.

The initial niche radius $\sigma^{(0)}$ is obtained by $\sigma^{(0)} = \frac{1}{2}\sqrt[n]{\frac{2n\prod_{i=1}^{n}(u_i - l_i)}{2NP\pi}}$, which considers the average space occupied by each individual in the initial population. The final niche radius $\sigma^{(S)} = 0$, which guarantees the convergence of the population. The reduction of the niche radius σ for each environment change is based on:

$$\sigma^{(s)} = Ce^{-(\frac{s}{D})^2} - \varepsilon \tag{6}$$

Algorithm 1. DMOEA/D-M2M

Input: K: the number of subproblems;
K unit direction vectors: $v^1, ..., v^K$;
NS: the size of subpopulation.
Output: The nondominated solutions.

1: Uniformly and randomly generate NP $(K \times NS)$ solutions.
2: Initializing the niche radius $\sigma = \sigma^{(0)}$, MOP$^{(0)}$, environmental state $s = 0$.
3: Computing the objective values of the initial population and then use them to set $P_1, ..., P_K$.
4: **while** the halting criterion is not satisfied **do**
5: Reducing $\sigma = \sigma^{(s+1)}$ according to Eq. (6);
6: Updating MOP$^{(s+1)}$ according to Eq. (5);
7: Set $R = \emptyset$;
8: **for** k = 1 **to** K **do**
9: Using DE to generate new subpopulation P'_k by P_k;
10: $R = R \bigcup P'_k$;
11: **end**
12: $Q = R \bigcup (\bigcup_{k=1}^{K} P_k)$;
13: Using Q to allocate subpopulations $P_1, ..., P_K$;
14: $s = s + 1$;
15: **end**
16: **return** nondominated solutions.

where ε is a given positive close-to-zero number (ε = 1e−8). C and D can be calculated in terms of the initial niche radius $\sigma^{(0)}$ and the final niche radius $\sigma^{(S)} = 0$.

As the niche radius σ shrinks from $\sigma^{(0)}$ to $\sigma^{(S)} = 0$, the MOPs change gradually from the initial MOP$^{(0)}$ to the final MOP$^{(S)}$ which has the form:

$$MOP^{(S)} : \min \boldsymbol{y} = (f(\boldsymbol{x}), 0). \tag{7}$$

It is obvious that the global optima solution of $MOP^{(S)}$ in Eq. (7) is equal to the original optima solution of the global optimization problem in Eq. (1). So we can say the global optimization problem is converted into the DMOP is an equivalent conversion.

4 The Proposed DMOEA/D-M2M Algorithm

Algorithm 1 shows the pseudo code of DMOEA/M2M algorithm. At first, after generate an initial population, we need to set the initial niche radius, then decomposes the initial transformed bi-objective optimization problem into K bi-objective optimization subproblems. At each generation, the niche radius is dynamically reduced according to Eq. (6). DMOEA/D-M2M maintains K subpopulations: $P_1, ..., P_K$, where P_k is for subproblem k. Each subpopulation is consist of NS solutions. In this paper, DE/rand/1/bin operator [18] is used to generate NS offspring solutions for each subpopulation.

Algorithm 2. Allocation of solutions to each subpopulation

Input: A set of solutions in Q.
Output: Subpopulations $P_1, ..., P_K$.
1: **for** $k = 1$ **to** K **do**
2: Initialize P_k as the solutions in Q whose objective values are in its subspace;
3: **if** $|P_k| \leq NS$ **then**
4: randomly choose $NS - |P_k|$ solutions from Q and add them to P_k;
5: **else**
6: Select NS solutions from P_k using nondominated sorting;
7: **end**
8: **end**
9: **return** sub-populations $P_1, ..., P_K$.

In line 13 of Algorithm 1, the union of the offspring population and the parent population is used to allocate solutions to each subpopulation. Algorithm 2 guarantees that each subpopulation P_k has NS individuals at each generation and, hence, enhances the population diversity during the search [7].

In terms of the search of the DMOEA/D-M2M, optimizing the original objective $f(\boldsymbol{x})$ is beneficial to the population convergence, while minimizing the additional objective, the niche-count function $nc(\boldsymbol{x}|\mathbf{R}, \sigma)$, is helpful to the population diversity. Through the entire evolution process, DMOEA/D-M2M can provide a sound balance between the exploration and the exploitation by dynamically reducing the niche radius from an initial value to zero, from the diverse population to the global optimum.

5 Experimental Studies

5.1 Investigation of the Population Diversity

In this paper, the standard deviation of the distribution of population is used to measure the population diversity [19], which is calculated as follows:

$$D(\boldsymbol{X}) = \frac{1}{NP} \sum_{i=1}^{NP} \sqrt{\sum_{j=1}^{n} (x_i^j - \bar{x}^j)^2}, \tag{8}$$

$$\bar{x}^j = \frac{1}{NP} \sum_{i=1}^{NP} x_i^j, \tag{9}$$

where $\bar{\boldsymbol{x}}$ is the mean position of the population. Apparently, a big value of $D(\boldsymbol{X})$ denotes the population distribution is scattered, while a small value of $D(\boldsymbol{X})$ indicates the population distribution is dense. A good optimization algorithm should keep the bigger value of $D(\boldsymbol{X})$ at the early evolution stage in order to avoid the algorithm falling into some local optima, while maintain the smaller value of $D(\boldsymbol{X})$ at the later stage to seek fast convergence.

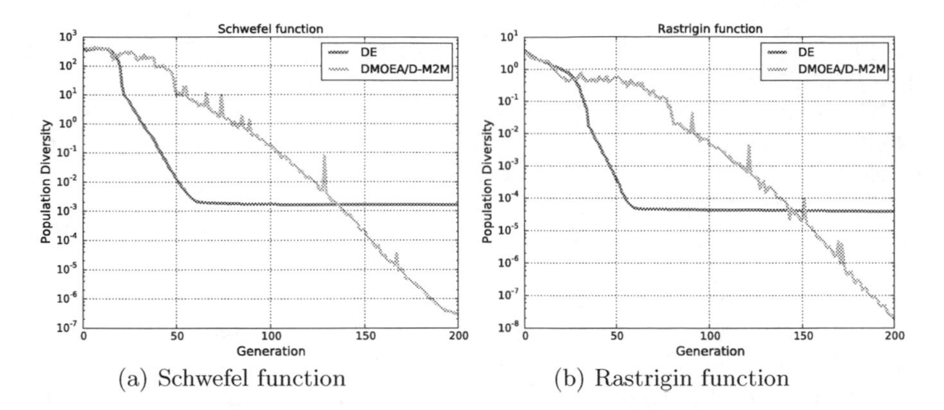

(a) Schwefel function (b) Rastrigin function

Fig. 1. Population diversity comparison among DE and DMOEA/D-M2M on Schwefel function and Rastrigin function, respectively.

We adopt Schwefel function and Rastrigin function as the test bench, which is highly multi-modal:

Schwefel function:

$$\min\ f(\boldsymbol{x}) = 418.9829n - \sum_{j=1}^{n} x_j sin(\sqrt{|x_j|})$$

$$\text{where}\quad -500 \leq x_j \leq 500, j = 1, ..., n.$$

Rastrigin function:

$$\min\ f(\boldsymbol{x}) = 10n + \sum_{i=1}^{n}(x_i^2 - 10cos(2\pi x_i))$$

$$\text{where}\quad -5.12 \leq x_j \leq 5.12, j = 1, ..., n.$$

First, to verify that the proposed DMOEA/D-M2M can preserve good diversity and convergence and could tradeoff these two abilities properly, we conduct comparison experiments among DMOEA/D-M2M and DE on solving Schwefel and Rastrigin problems. The proposed DMOEA/D-M2M also adopts DE to generate offspring population, the main differences between DE and DMOEA/D-M2M is DMOEA/D-M2M regards niche-count as the second objective, and employs the decomposition-based DMOEA to environment selection. When solving these two functions, the population size and the maximum number of function evaluations were set to 100 and 20000, respectively.

Figure 1 depicts the diversity of DE and DMOEA/D-M2M on these two functions. It is clear that Fig. 1(a) and Fig. 1(b) have a similar variation tendency. For DE, the population diversity keeps unchanged since 60th generation, this is because that all individuals of DE have converged to a small region both on Schwefel and Rastrigin problems. By contrast, our proposed DMOEA/D-M2M can provide more diversity than DE in the early evolution stage, while it has the capability to provide better exploitation in the later evolution stage.

Overall, the above experimental results validate that the proposed DMOEA/D-M2M can preserve good exploration and exploitation abilities and can compromise these two well to search the space during the evolution.

5.2 Benchmark Test Functions

The performance of DMOEA/D-M2M is tested by the IEEE CEC2014 test problems [20]. IEEE CEC2014 test suite contains three unimodal functions (F01-F03), thirteen simple multimodal functions (F04-F16), and the remaining test problems are hybrid or composition functions (F17-F30). This test suite exhibit various complex characteristics, such as strong nonlinearity, rotated landscape. Therefore, it can provide a systematic assessment on the performance of the proposed DMOEA/D-M2M.

In this paper, four state-of-the-art global optimization algorithms are adopted to compare the proposed DMOEA/D-M2M: MOMPSO [21], LX-BBO [22], MERDE [23], and M-PSO-MA [24].

To detect the differences of different algorithms for statistical significance, the Wilcoxon rank sum test is adopted with a 0.05 significance level is performed between DMOEA/D-M2M and each compared peer algorithm, where symbols '+', '−', and '≈' represent that the proposed DMOEA/D-M2M is significantly better than, worse than, and equivalent to the corresponding algorithm, respectively. In addition, the Friedman test is chosen to sort all algorithms on all test instances.

5.3 Parameter Settings

In the experiments, all test functions are with variable dimension $D = 10$. The maximal number of function evolutions for all algorithms are set to 100,000 on all test problems. For DMOEA/D-M2M, the population size is $NP = 100$. The number of subproblems $K = 10$, the size of each subpopulation $NS = 10$, which are the same as MOEA/D-M2M in [7]. In DE operator, $CR = 0.9$, F is randomly chosen in [0.0, 1.0]. In addition, 31 independent runs were performed on each test instance.

The results of MOMPSO, LX-BBO, MERDE, and M-PSO-MA are directly taken from their original source [21–24].

5.4 Comparisons with State-of-the-Art Algorithms

Table 1 summarizes the average objective function error value and standard deviation derived from the five compared algorithms over 31 independent runs. The objective function error value represents the absolute value of difference between obtained objective value at the termination of each algorithm and the known optimal value. In these tables, the best results of each test problem were highlighted.

Table 1. The average and standard deviation error value obtained by five algorithms for 10-dimentional IEEE CEC2014 benchmark problems. Better results are highlighted in bold.

Prob	MOMPSO	LX-BBO	MERDE	M-PSO-MA	DMOEA/D-M2M
F01	1.41e+04±1.77e+04	1.61e+03±1.14e+03	**1.58e+00±7.61e+00**	4.01e+01±3.17e+01	3.47e+02±7.14e+02
F02	8.43e+03±3.80e+03	5.80e+03±2.27e+03	**6.31e-05±1.12e-04**	3.17e-02±4.63e-02	6.06e+02±1.27e+03
F03	1.21e+04±1.18e-04	4.47e+03±5.52e+03	1.35e-03±1.22e-03	**0.00e+00±0.00e+00**	2.25e-07±6.31e-07
F04	6.18e+00±1.07e+01	1.72e+00±4.19e-03	**0.00e+00±0.00e+00**	6.18e+00±1.07e+01	1.74e+01±1.98e+01
F05	2.00e+01±3.81e-02	**1.01e+00±2.81e-01**	1.90e+01±2.71e+00	2.00e+01±5.56e-03	1.57e+01±7.80e+00
F06	3.53e+00±1.77e+00	3.45e+00±1.52e+00	8.93e-01±2.81e-01	1.22e+00±1.32e+00	**6.90e-02±1.21e-01**
F07	1.17e-01±6.19e-02	2.56e-01±1.40e-01	**1.83e-02±1.51e-02**	5.37e-02±2.81e-02	6.09e-02±2.62e-02
F08	1.05e+01±5.36e+00	**0.00e+00±0.00e+00**	**0.00e+00±0.00e+00**	3.12e-01±1.08e+00	**0.00e+00±0.00e+00**
F09	1.28e+01±8.22e+00	1.10e+01±4.27e+00	5.58e+00±1.74e+00	**5.08e+00±2.02e+00**	5.71e+00±2.33e+00
F10	3.14e+02±2.03e+02	9.01e+02±5.45e−02	**3.67e-02±3.98e-02**	9.68e+01±9.89e+01	2.57e-01±7.24e-02
F11	4.70e+02±2.54e+02	1.12e+03±5.93e+02	7.55e+01±7.63e+01	**2.52e+00±1.76e+02**	5.40e+01±8.40e+01
F12	1.90e-01±1.23e-01	1.00e-01±4.20e-17	1.17e-01±6.93e-02	**5.29e-02±3.61e-02**	1.07e-01±9.01e-02
F13	**7.31e-02±4.04e-02**	3.12e-01±1.50e-01	1.17e-01±4.43e-02	1.02e-01±4.80e-02	7.99e-02±1.97e-02
F14	2.20e-02±8.27e-03	2.39e-01±2.22e-01	9.37e-02±2.73e-02	**2.09e-02±7.87e-03**	7.13e-02±2.41e-02
F15	6.85e-01±1.93e-01	1.51e+00±7.88e-01	6.72e-01±2.18e-01	**6.46e-01±1.58e-01**	7.61e-01±2.28e-01
F16	2.69e+00±4.30e-01	2.37e+00±4.16e-01	1.53e+00±4.63e-01	1.47e+00±5.68e-01	**7.72e-01±2.83e-01**
F17	1.08e+03±4.01e+02	5.66e+03±6.81e+03	**7.93e+00±9.59e+00**	2.38e+02±1.37e+02	4.54e+01±1.08e+02
F18	7.82e+02±1.20e+03	7.02e+03±7.18e+03	2.72e+00±1.29e+00	4.63e+02±5.90e+02	**1.68e+00±9.29e-01**
F19	2.75e+00±1.43e+00	3.69e+00±7.27e+00	5.10e-01±1.76e-01	9.29e-01±6.98e-01	**1.42e-01±9.44e-02**
F20	3.95e+01±3.89e+01	1.61e+04±2.06e+04	1.70e+00±7.50e-01	2.50e+00±1.71e+00	**2.34e-01±4.46e-01**
F21	3.64e+02±2.63e+02	6.26e+03±7.23e+03	8.54e+00±2.66e+01	6.11e+01±7.63e+01	**3.13e-01±2.48e-01**
F22	5.35e+01±6.99e+01	7.59e+01±7.45e+01	3.24e+00±3.96e+00	1.50e+01±9.03e+00	**1.13e+00±3.24e+00**
F23	3.29e+02±1.38e-12	2.44e+02±5.65e+01	3.29e+02±2.68e-11	3.29e+02±1.38e-12	**2.18e+02±4.10e+01**
F24	1.20e+02±6.74e+00	1.01e+03±8.54e+02	1.15e+02±2.45e+00	**1.14e+02±4.30e+00**	1.17e+02±3.61e+00
F25	1.98e+02±1.60e+01	1.78e+02±1.50e+01	**1.36e+02±8.34e+00**	1.54e+02±4.17e+01	1.63e+02±1.79e+00
F26	1.00e+02±3.34e-02	**3.71e-03±6.58e-03**	1.00e+02±4.18e-02	1.00e+02±3.30e-02	1.05e+02±6.17e-05
F27	3.16e+02±1.71e+02	**1.05e+01±7.68e+00**	2.87e+01±7.60e−01	2.90e+02±1.57e+02	1.88e+02±4.33e+01
F28	4.73e+02±1.07e+02	5.29e+02±1.14e+02	3.36e+02±7.37e+00	4.52e+02±7.24e+01	**3.19e+02±3.70e-01**
F29	3.63e+02±9.23e+01	3.53e+05±7.54e+05	3.17e+02±5.48e+01	2.93e+02±4.57e+01	**2.05e+02±1.67e+00**
F30	7.02e+02±2.63e+02	6.31e+04±6.97e+04	5.34e+02±6.06e+01	6.09e+02±1.78e+02	**2.31e+02±4.93e+00**

For Table 1, we can see that from the angle of the number of obtained best results, DMOEA/D-M2M has the best performance on 12 test instances, while its peer competitors MOMPSO, LX-BBO, MERDE, and M-PSO-MA obtains 1, 4, 8, and 7 best results, respectively. For unimodal test functions, DMOEA/D-M2M has poor performance than its competitors, this can be attributed to the fact that DMOEA/D-M2M put more emphasis on investigating more regions in the search space to discover the potential solutions, so it has slower convergence speed than its competitors on unimodal test functions. However, on complex hybrid or composition functions, DMOEA/D-M2M outperforms the compared algorithms. These test instances involve multi-modality and strong nonlinearity, so an algorithm is easy to fall into some local optima. However, a larger niche radius at the early evolution stage can maintain the population diversity and makes DMOEA/D-M2M more powerful in global exploration, so it has better performance than its competitors on these problems.

In terms of the multiple-problem Wilcoxon's signed rank test in Table 2, DMOEA/D-M2M provides higher R+ values than R− values in all cases. According to the Friedman's test in Fig. 2, DMOEA/D-M2M ranks the first among the five compared algorithms.

Table 2. Statistical test results of DMOEA/D-M2M and four peer algorithms by the Multiple-Problem Wilcoxon's test for IEEE CEC2014.

DMOEA/D-M2M vs	R+	R−	$\alpha = 0.05$
MOMPSO	433.0	32.0	+
LX-BBO	381.0	54.0	+
MERDE	233.0	202.0	≈
M-PSO-MA	302.0	163.0	≈

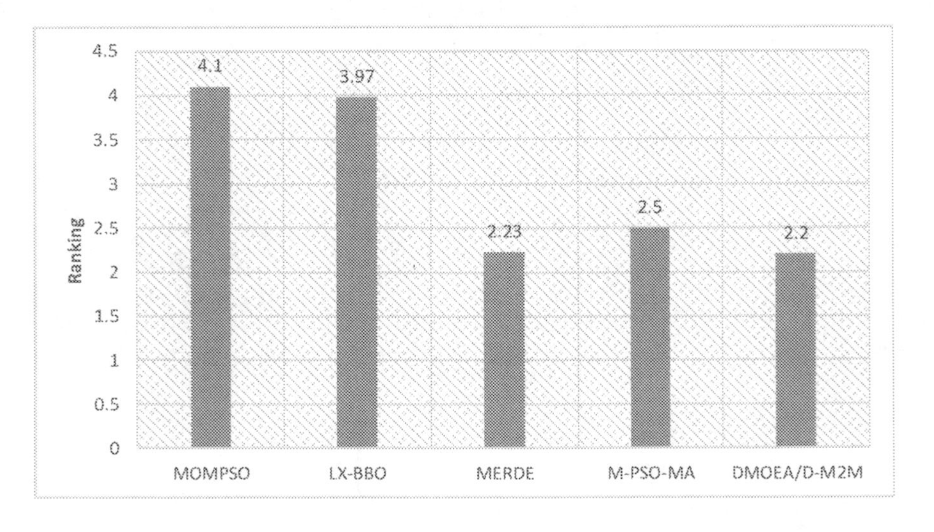

Fig. 2. Ranking of 5 algorithms for CEC2014 by Friedman's test.

The above comparison demonstrates that DMOEA/D-M2M exhibits better performance than the four compared algorithms on the 30 test problems with 10D from IEEE CEC2014.

6 Conclusion

To solve complicated global optimization problems, a decomposition-based dynamic bi-objective evolutionary algorithm embedded in the framework of MOEA/D-M2M was proposed. The proposed algorithm named DMOEA/D-M2M, which converts a global optimization problem to an equivalent dynamic bi-objective optimization problem: one is the original objective and the another is the niche count function. During the evolution stage, DMOEA/D-M2M decomposes the converted bi-objective optimization problem into a set of simple bi-objective optimization subproblems, and these simple bi-objective subproblems are solved in a collaborative way simultaneously. In addition, the niche count function can provide a balance between exploration and exploitation by dynamically reducing the niche radius from an initial value to zero. Experimental results suggest that the proposed DMOEA/D-M2M is effective for solving global optimization problems.

In the future, the proposed decomposition-based dynamic multi-objective evolutionary algorithm will be adopted to solve real-world antenna design optimization problems.

Acknowledgments. This work was supported in part by the Major Project for New Generation of AI under Grant 2018AAA0100400, in part by the Fundamental Research Funds for National Universities, China University of Geosciences (Wuhan), in part by the high-performance computing platform of China University of Geosciences.

References

1. Del Ser, J., et al.: Bio-inspired computation: where we stand and what's next. Swarm Evol. Comput. **48**, 220–250 (2019)
2. Singh, A., Deep, K.: Exploration–exploitation balance in Artificial Bee Colony algorithm: a critical analysis. Soft Comput. **23**(19), 9525–9536 (2018). https://doi.org/10.1007/s00500-018-3515-0
3. Jiao, R., Zeng, S., Alkasassbeh, J.S., Li, C.: Dynamic multi-objective evolutionary algorithms for single-objective optimization. Appl. Soft Comput. **61**, 793–805 (2017)
4. Deb, K., Pratap, A., Agarwal, S., Meyarivan, T.: A fast and elitist multiobjective genetic algorithm: NSGA-II. IEEE Trans. Evol. Comput. **6**(2), 182–197 (2002)
5. Bader, J., Zitzler, E.: HypE: an algorithm for fast hypervolume-based many-objective optimization. Evol. Comput. **19**(1), 45–76 (2011)
6. Zhang, Q., Li, H.: MOEA/D: a multiobjective evolutionary algorithm based on decomposition. IEEE Trans. Evol. Comput. **11**(6), 712–731 (2007)
7. Liu, H.L., Gu, F., Zhang, Q.: Decomposition of a multiobjective optimization problem into a number of simple multiobjective subproblems. IEEE Trans. Evol. Comput. **18**(3), 450–455 (2014)

8. Zhou, A., Zhang, Q.: Are all the subproblems equally important? resource allocation in decomposition-based multiobjective evolutionary algorithms. IEEE Trans. Evol. Comput. **20**(1), 52–64 (2015)
9. Yuan, Y., Xu, H., Wang, B., Zhang, B., Yao, X.: Balancing convergence and diversity in decomposition-based many-objective optimizers. IEEE Trans. Evol. Comput. **20**(2), 180–198 (2015)
10. Trivedi, A., Srinivasan, D., Sanyal, K., Ghosh, A.: A survey of multiobjective evolutionary algorithms based on decomposition. IEEE Trans. Evol. Comput. **21**(3), 440–462 (2017)
11. Ma, X., Zhang, Q., Tian, G., Yang, J., Zhu, Z.: On Tchebycheff decomposition approaches for multiobjective evolutionary optimization. IEEE Trans. Evol. Comput. **22**(2), 226–244 (2017)
12. Li, K., Zhang, Q.: Decomposition multi-objective optimisation: current developments and future opportunities. In: Proceedings of the Genetic and Evolutionary Computation Conference Companion, pp. 1002–1031. ACM (2019)
13. Li, H., Deb, K., Zhang, Q., Suganthan, P.N., Chen, L.: Comparison between MOEA/D and NSGA-III on a set of novel many and multi-objective benchmark problems with challenging difficulties. Swarm Evol. Comput. **46**, 104–117 (2019)
14. Zeng, S., Jiao, R., Li, C., Li, X., Alkasassbeh, J.S.: A general framework of dynamic constrained multiobjective evolutionary algorithms for constrained optimization. IEEE Trans. Cybern. **47**(9), 2678–2688 (2017)
15. Jiao, R., Sun, Y., Sun, J., Jiang, Y., Zeng, S.: Antenna design using dynamic multi-objective evolutionary algorithm. IET Microw. Antenna. Propag. **12**(13), 2065–2072 (2018)
16. Zeng, S., Jiao, R., Li, C., Wang, R.: Constrained optimisation by solving equivalent dynamic loosely-constrained multiobjective optimisation problem. Int. J. Bio-Insp. Comput. **13**(2), 86–101 (2019)
17. Jiao, R., Zeng, S., Li, C.: A feasible-ratio control technique for constrained optimization. Inf. Sci. **502**, 201–217 (2019)
18. Storn, R., Price, K.: Differential evolution-a simple and efficient heuristic for global optimization over continuous spaces. J. Glob. Optim. **11**(4), 341–359 (1997)
19. Yang, Q., Chen, W.N., Da Deng, J., Li, Y., Gu, T., Zhang, J.: A level-based learning swarm optimizer for large-scale optimization. IEEE Trans. Evol. Comput. **22**(4), 578–594 (2018)
20. Liang, J.J., Qu, B.Y., Suganthan, P.N.: Problem definitions and evaluation criteria for the CEC 2014 special session and competition on single objective real-parameter numerical optimization. Technical report, pp. 1–32 (2013)
21. Singh, G., Deep, K.: Effectiveness of new multiple-PSO based membrane optimization algorithms on CEC 2014 benchmarks and iris classification. Nat. Comput. **16**, 473–496 (2017)
22. Garg, V., Deep, K.: Performance of Laplacian biogeography-based optimization algorithm on CEC 2014 continuous optimization benchmarks and camera calibration problem. Swarm Evol. Comput. **27**, 132–144 (2016)
23. Qu, B.Y., Liang, J.J., Xiao, J.M., Shang, Z.G.: Memetic differential evolution based on fitness Euclidean-distance ratio. In: 2014 IEEE Congress on Evolutionary Computation, pp. 2266–2273. IEEE (2014)
24. Singh, G., Deep, K., Nagar, A.K.: Cell-like P-systems based on rules of particle swarm optimization. Appl. Math Comput. **246**, 546–560 (2014)

A Novel Multi-objective Evolutionary Algorithm Based on Space Partitioning

Xiaofang Wu[1,2], Changhe Li[1,2(✉)], Sanyou Zeng[3], and Shengxiang Yang[4]

[1] School of Automation, China University of Geosciences, Wuhan 430074, China
{wuxiaofang,lichanghe}@cug.edu.cn

[2] Hubei Key Laboratory of Advanced Control and Intelligent Automation for Complex Systems, Wuhan 430074, China

[3] School of Mechanical Engineering and Electronic Information, China University of Geosciences, Wuhan 430074, China
sanyouzeng@gmail.com

[4] School of Computer Science and Informatics, De Montfort University, Leicester LE1 9BH, United Kingdom
syang@dmu.ac.uk

Abstract. To design an effective multi-objective optimization evolutionary algorithms (MOEA), we need to address the following issues: 1) the sensitivity to the shape of true Pareto front (PF) on decomposition-based MOEAs; 2) the loss of diversity due to paying so much attention to the convergence on domination-based MOEAs; 3) the curse of dimensionality for many-objective optimization problems on grid-based MOEAs. This paper proposes an MOEA based on space partitioning (MOEA-SP) to address the above issues. In MOEA-SP, subspaces, partitioned by a k-dimensional tree (kd-tree), are sorted according to a bi-indicator criterion defined in this paper. Subspace-oriented and Max-Min selection methods are introduced to increase selection pressure and maintain diversity, respectively. Experimental studies show that MOEA-SP outperforms several compared algorithms on a set of benchmarks.

Keywords: Multi-objective optimization · kd-tree space partitioning · Max-Min method

1 Introduction

Multi-objective optimization problems (MOPs) widely exist in engineering practice. There is no single optimal solution, but a set of trade-off optimal solutions, because of the conflict between objectives. Multi-objective evolutionary algorithms (MOEAs), with the ability to obtain a set of approximately optimal solutions in a single run, have become a useful tool to solve MOPs.

Over the past decades, with the development of evolutionary multi-objective optimization (EMO) research, domination-based and decomposition-based algorithms have attracted many researchers.

K. Li et al. (Eds.): ISICA 2019, CCIS 1205, pp. 127–142, 2020.
https://doi.org/10.1007/978-981-15-5577-0_10

One of the most classical domination-based MOEAs is the nondominated sorting genetic algorithm (NSGA-II) [1]. NSGA-II selects offspring according to elitist nondominated sorting and density-estimation metrics. However, it takes too many resources on the convergence during evolution, which results in loss of diversity.

One of the most classical decomposition-based MOEAs is the MOEA based on decomposition (MOEA/D) [2], which decomposes a MOP into multiple single-objective subproblems. Although it simplifies the problem, it still suffers several issues. Decomposition approaches are very sensitive to the shapes of Pareto front (PF). For example, Weighted Sum (WS) can not find Pareto optimal solutions in the nonconvex part of PF with complex shape. On extremely convex PF, a set of solutions obtained by Tchebycheff (TCH) are not uniformly distributed.

To address the sensitivity issue discussed above, some researchers have proposed the grid-based MOEAs, such as the grid-based evolutionary algorithm (GrEA) [3] and the constrained decomposition approach with grids for MOEA (CDG-MOEA) [4]. The two grid-based algorithms evenly partition each dimension (i.e., objective), which is beneficial to maintain diversity. However, with the increase in the number of objectives, the number of grids will increase exponentially, resulting in the curse of dimensionality.

In this paper, an MOEA based on space partitioning (MOEA-SP) is proposed to address the above issues. In MOEA-SP, subspaces, partitioned by a k-dimensional tree (kd-tree), are sorted according to a bi-indicator criterion defined in this paper. The two indicators refer to dominance degree and the niche count that measure convergence and diversity, respectively. The introduction of historical archive pushes the population toward the true PF and distributed evenly.

The rest of this paper is organized as follows. Section 2 describes the related work of MOEAs based on decomposition and grid decomposition. Section 3 gives the details about MOEA-SP. Section 4 presents the experimental studies. Finally, Sect. 5 concludes this paper.

2 Related Work

This section discusses related work regarding the improvement of decomposition-based and grid-based MOEAs.

The weight vector generation method in MOEA/D makes the population size inflexible, and the distribution of the generated weight vector is not uniform. In order to address these issues, researchers have proposed some improved methods. For example, Fang *et al.* [5] proposed the combination of transformation and uniform design (UD) method. Deb *et al.* [6] proposed the two-layer weight vector generation method, including the boundary and inside layer, where the weight vectors of the inside layer are shrunk by a coordinate transformation, finally, the weight vectors of the boundary and inside layer are merged into a set of weight vectors.

The three decomposition methods of MOEA/D are sensitive to the shapes of PF, and it is difficult to solve the PF with the shape of nonconvex or extremely convex. For Penalty-based Boundary Intersection (PBI), it is also difficult to set the penalty parameters. The improved methods include inverted PBI (IPBI) [7], new penalty scheme [8], and angle penalized distance (APD) decomposition method [9]. Although these methods improve the performance of the algorithms for some MOPs, they are still sensitive to the shape of PF and difficult to set the penalty parameters. Besides, to overcome the shortcomings of traditional decomposition methods, Liu *et al.* [10] proposed a new alternative decomposition method, i.e., using a set of reference vectors to divide the objective space into multiple subspaces and assign a subpopulation to evolve in each subspace. This method does not need traditional decomposition methods.

The neighborhood could have a significant impact on generating offspring and environment selection, so the improper definition of the neighborhood relationship may mislead algorithms. To define the proper neighborhood, Zhao *et al.* [11] proposed to dynamically adjust the neighborhood structure according to the distribution of the current population. To address the issue of the loss of diversity due to the large update area in the selection of offspring, Li *et al.* [12] introduced the concept of the maximum number of a new solution to replace the old solutions. Zhang *et al.* [13] proposed a greedy-based replacement strategy, which calculates the improvement of the new solution for each subproblem, and then replaces the solutions of the two subproblems with the new solution with the best improvement.

Although there are many improved methods, they can not fundamentally overcome the above limitations. In order to overcome the limitations of the sensitivity for the shapes of PF and loss of diversity in decomposition-based method, Cai *et al.* [4] proposed a constrained decomposition approach with grids for MOEA (CDG-MOEA), which uniformly divides the objective space into $M \times K^{M-1}$ (where K is the grid decomposition parameter, and M is the number of objectives) subproblems, and choose offspring according to a decomposition-based ranking and lexicographic sorting method. CDG-MOEA has great advantages to maintain diversity. However, with the increasing number of objectives, the number of subproblems will increase exponentially, resulting in the curse of dimensionality.

3 MOEA Based on Space Partitioning

This section introduces the design process of the proposed algorithm. The major idea of MOEA-SP is to partition the objective space into a set of subspaces, then select offspring according to the rank of subspaces after sorting.

3.1 The Framework of MOEA-SP

Algorithm 1 gives the framework of MOEA-SP in detail. MOEA-SP starts with initialization, then repeat generate offspring, partition objective space, and select offspring (i.e., environmental selection) until the termination conditions are satisfied. The environmental selection mainly includes nondominated sorting of the subspaces, subspace-orient selection, and Max-Min selection.

Algorithm 1. MOEA-SP

Input:

N: the population size of P;

K: number of subspace based on kd-tree partitioning method.

Output: A solution set P.

Step 1: Initialization

1.1 Initialize a population $P_0 = \{x^1, \ldots, x^N\}$ randomly;

1.2 $A = P_0$; // A is historic archive

1.3 Set $t = 0$.

Step 2: Reproduction

2.1 Update z^*, z^{nad} by P_t;

2.2 Build a kd-tree based on objective space formed by $[z^*, z^{nad}]$;

2.3 Obtain neighborhood based on kd-tree: $N(x), x \in P_t$;

2.4 Generate an empty set $Q_t = \emptyset$;

for all $x \in P_t$ **do**

2.5 Fill mating pools of x

$$M(x) = \begin{cases} N(x) & rand < \delta \ \ and \ \ |N(x)| > 3, \\ P_t & otherwise. \end{cases}$$

2.6 Select three solutions x^1, x^2 and x^3 randomly from $M(x)$, then generate offspring y by DE operator [12], and put y to Q_t.

end for

2.7 $A = A \cup Q_t$.

Step 3: Environmental selection

3.1 $P_{t+1} = \emptyset$.

3.2 Find nondominated solutions A^r, and $A^d = A \backslash A^r$;

3.3

if $|A^r| < N$ **then**

$P_{t+1} = P_t \cup A^r$;

Update z^*, z^{nad} by A;

Partition the objective space formed by $[z^*, z^{nad}]$ with a kd-tree;

$R =$ SSBB (kd-tree); // Algorithm 2

$P_{t+1} =$ DSOS (R, A^d, P_{t+1}); // Algorithm 4

Update A, delete some solutions from each subspace;

else

Update z^*, z^{nad} by A^r;

Partition the objective space formed by $[z^*, z^{nad}]$ with a kd-tree;

$R =$ SSBB (kd-tree);

$P_{t+1} =$ NSOMMS (R, A^r, P_{t+1}); // Algorithm 5

Update A, this is $A = A^r$;

end if

3.5 $t = t + 1$.

Step 4: Termination

If the stopping criterion is satisfied, terminate the algorithm. Otherwise, go to Step 2.

3.2 Subspace-Oriented Domination and Sorting

In MOEA-SP, the multiple subspaces are obtained after the objective space is partitioned by kd-tree method. The set of subspaces is denoted as $R = \{s_1, s_2, \ldots, s_n\}$, where each subspace may contain a number of individuals. As shown in Fig. 1a, the objective space is partitioned into twenty subspaces, eleven of them, which contain individuals, are marked with $s_1, \ldots, s_{11}$.

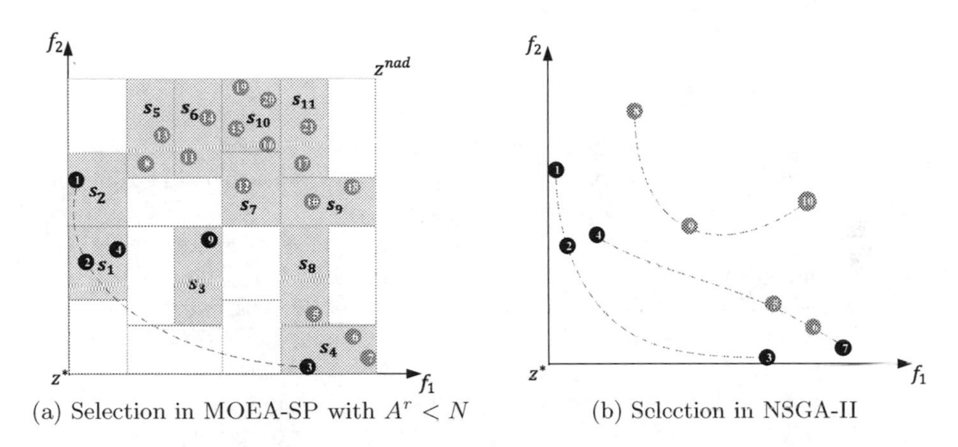

(a) Selection in MOEA-SP with $A^r < N$ (b) Selection in NSGA-II

Fig. 1. (a) Selection in MOEA-SP with $A^r < N$. (b) Selected in NSGAII. Solid circles represent individuals, where deeper-color circles represent individuals selected. Individuals connected by dotted lines are the same level of individuals after nondominated sorting. Note that in (a), there are individuals distributed in the gray subspaces (i.e, $s_1, \ldots, s_{11}$), and the total number of individuals in the historical archive is twenty-one.

This paper defines the concept of subspace-oriented: neighborhood $N(x)$, niche count (nc), dominance ratio (dr), dominance matrix (D) and dominance degree (dd).

The neighborhood $N(x)$ of individual x is defined as the set of individuals in the neighborhood of the subspace to which it belongs. It is worth noting that a subspace itself belongs to its own neighborhood. As shown in Fig. 1a, for x_1 of belonging to s_2, the neighborhood of subspace s_2 are subspaces s_1, s_2, s_5, which contains five individuals: x_1, x_2, x_4, x_8, x_{13}, i.e., $N(x_1) = \{x_1, x_2, x_4, x_8, x_{13}\}$.

The niche count (nc) of the subspace is defined as the sum of all individuals in its adjacent subspaces in historical archive A. The niche count of all subspaces is recorded as $NC = \{nc(s_1), ..., nc(s_n)\}$. As shown in lines 9 to 12 of Algorithm 2. For example, the neighborhood of subspace s_2 includes subspaces s_1, s_2, and s_5, and they consist of five individuals with a fold of x_1, x_2, x_4, x_8, and x_{13}, so $nc(s_2) = 5$. In this example, the niche count of all subspaces can be obtained, i.e., $NC = \{3,\ 5,\ 2,\ 4,\ 5,\ 9,\ 11,\ 7,\ 6,\ 9,\ 9\}$.

The dominance ratio (dr) is defined as

$$dr(s_1, s_2) = \frac{NO(s_1 \prec s_2) - NO(s_2 \prec s_1)}{M}. \tag{1}$$

where $NO(s_1 \prec s_2)$ denotes the number of objectives subspace s_1 dominates s_2, and M is the number of objectives. $dr(s_1, s_2)$ denotes the ratio of the difference between the number of objectives s_1 dominates s_2 and the number of objectives s_2 dominates s_1 to M, obviously, $dr(s_1, s_2) = -dr(s_2, s_1)$. For example, subspaces s_1 and s_3 in the bi-objective MOP shown in Fig. 1a, s_1 dominates s_3 on the second objective, and there is no comparability on the first one, so the difference in objectives number of s_1 dominance s_3 are 1, so $dr(s_1, s_3) = 1/M = 1/2$. If subspaces s_1 and s_2 do not dominate each other, then $dr(s_1, s_2) = 0$, such as subspaces s_1 and s_4 are nondominated; if subspace s_1 completely dominates s_2, then $dr(s_1, s_2) = 1$, for example, subspace s_1 completely dominates s_7.

The dominance matrix of the subspaces is defined as $D_{n \times n}$, where $D[i][j] = dr(s_i, s_j)$. In this example, as shown in Algorithm 2, the dominance matrix D is calculated as

$$D = \begin{pmatrix}
0 & \frac{1}{2} & \frac{1}{2} & 0 & 1 & 1 & 1 & \frac{1}{2} & 1 & 1 & 1 \\
-\frac{1}{2} & 0 & 0 & 0 & \frac{1}{2} & \frac{1}{2} & \frac{1}{2} & 0 & \frac{1}{2} & 1 & \frac{1}{2} \\
-\frac{1}{2} & 0 & 0 & 0 & 0 & \frac{1}{2} & 1 & \frac{1}{2} & 1 & 1 & 1 \\
0 & 0 & 0 & 0 & 0 & 0 & 0 & \frac{1}{2} & \frac{1}{2} & 0 & \frac{1}{2} \\
-1 & -\frac{1}{2} & 0 & 0 & 0 & \frac{1}{2} & \frac{1}{2} & 0 & 0 & \frac{1}{2} & \frac{1}{2} \\
-1 & -\frac{1}{2} & -\frac{1}{2} & 0 & -\frac{1}{2} & 0 & \frac{1}{2} & 0 & 0 & \frac{1}{2} & \frac{1}{2} \\
-1 & -\frac{1}{2} & -1 & 0 & -\frac{1}{2} & -\frac{1}{2} & 0 & 0 & \frac{1}{2} & \frac{1}{2} & \frac{1}{2} \\
-\frac{1}{2} & 0 & -\frac{1}{2} & -\frac{1}{2} & 0 & 0 & 0 & 0 & \frac{1}{2} & 0 & \frac{1}{2} \\
-1 & -\frac{1}{2} & -1 & -\frac{1}{2} & 0 & 0 & -\frac{1}{2} & -\frac{1}{2} & 0 & 0 & \frac{1}{2} \\
-1 & -1 & -1 & 0 & -\frac{1}{2} & -\frac{1}{2} & -\frac{1}{2} & 0 & 0 & 0 & \frac{1}{2} \\
-1 & -\frac{1}{2} & -1 & -\frac{1}{2} & -\frac{1}{2} & -\frac{1}{2} & -\frac{1}{2} & -\frac{1}{2} & -\frac{1}{2} & -\frac{1}{2} & 0
\end{pmatrix}.$$

Algorithm 2. Subspaces Sorting Based on Bi-indicator (SSBB)

Input: A kd-tree (K subspaces).
Output: New sorting of subspaces R.
1: $R = \{s \mid \exists$ solution $x \in s$th subspace$\}$, $n = |R|$;
Dominance matrix: $D_{n\times n}$; Dominance degree: $DD = \{dd(s_1), \ldots, dd(s_n)\}$;
Neighbor of sth subspace: $NR(s)$; Niche count: $NC = \{nc(s_1), \ldots, nc(s_n)\}$;
/* Calculate dominance degree DD */
2: Calculate $D_{n\times n}$ according to dominance ratio in equation (1)
3: **for** $s = 1$ to n **do**
4: $GP(l) = \{l|D[s][l] > 0\}$,$GN(l) = \{l|D[s][l] < 0\}$;
5: $val_s = \sum_{l\in GP(l)}^{n} D[s][l]$;
6: $val\_ed_s = \sum_{l\in GN(l)}^{n} D[s][l]$;
7: $dd(s) = val_s/val\_ed_s$;
8: **end for**
/* Calculate niche count NC */
9: **for** $s = 1$ to n **do**
10: $NR(s) = \{s_1, s_2, ...\}$;
11: $nc(s) = \sum_{i\in NR(s)} |subspace(s)_{ind}|$;
12: **end for**
13: $Rank(s) = Nondominated\text{-}sorting(DD, NC)$; // maximum dd and minimum nc
14: $R = sort(R, Rank(s))$. // ascending sort R according to rank

The dominance degree (dd) of subspace denotes the degree of convergence, calculated according to the dominance matrix D. The dominance degree of all subspaces is recorded as $DD = \{dd(s_1), dd(s_2), ..., dd(s_n)\}$. As shown in lines 3 to 8 of Algorithm 2, according to the domination matrix, the dominating value val_s and the dominated value val_ed_s of each subspace can be obtained, i.e., $val_s = \sum_{l\in GP(l)}^{n} D[s][l]$ ($GP(l) = \{l \mid D[s][l] \geq 0\}$), $val\_ed_s = \sum_{l\in GN(l)}^{n} |D[s][l]|$ ($GN(l) = \{l \mid D[s][l] \leq 0\}$), and then the dominance degree of the subspace is calculated by $dd(s) = val_s/val\_ed_s$. Note that when val or val_ed is 0, dd is assigned to $+1e5$ or $-1e5$, respectively. In this example, $val_2 = D[2][5] + D[2][6] + D[2][7] + D[2][8] + D[2][9] + D[2][10] = 7/2$, $val\_ed_2 = |D[2][1]| = 1/2$, then $dd(s_2) = val_2/val\_ed_2 = 7$. Finally, DD can be obtained, i.e., $DD = \{+1e5,\ 7,\ 10,\ +1e5,\ 4/3,\ 3/5,\ 3/7,\ 2/3,\ 1/8,\ 1/9,\ -1e5\}$.

In summary, the dominance degree dd and niche count nc can measure the convergence and diversity, respectively and push the population to these two directions. This paper uses the two indicators as two objectives, i.e.,, maximizing dd, minimizing nc, and then performs nondominated sorting by these two indicators. As shown in lines 13 to 14 of Algorithm 2, a new sorting subspaces set R is obtained according to the rank value of nondominated sorting. In this example, according to the DD and NC values calculated above, the eleven subspaces can be divided into eight ranks, $\{s_1, s_3\}, \{s_4\}, \{s_2\}, \{s_5\}, \{s_8, s_9\}, \{s_6\}, \{s_7, s_{10}\}, \{s_{11}\}$ by nondominated sorting, i.e., $R = \{s_1, s_3, s_4, s_2, s_5, s_8, s_9, s_6, s_7, s_{10}, s_{11}\}$.

3.3 Environmental Selection

This paper uses the historical archive to find the nondominated individuals for the selection of offspring, where the set of nondominated individuals is recorded as A^r, and the set of other individuals is A^d.

Algorithm 3. Subspace-oriented Selection (SOS)

Input:
R: Sorted subspaces;
A': A set of candidate solutions;
P: A set of solutions;
num_max: The maximum selected from each subspace;
Output:
A set of solutions P;
The index of subspace s.
1: Let $s = 1$;
// N denotes the population size
2: **while** $|P| + min\{|R(s)_{inds}|, num\_max\} \leq N$ **do**
3: **if** $|R(s)_{ind}| \leq num\_max$ **then**
4: $P = P \cup R(s)_{inds}$;
5: **else**
6: $Nondominated\ selection\ (R(s)_{inds})$;
7: $P = P \cup R(s)_{inds}[1 : (num\_max)]$; // select num_max solutions from $R(s)_{inds}$
8: **end if**
9: $s = s + 1$;
10: **end while**

The environmental selection is divided into two cases according to the size of $|A^r|$, as shown in step 3.3 of Algorithm 1. IF $|A^r| < N$, then besides A^r, $N - |A^r|$ individuals need to be selected from A^d for the next-generation population; Otherwise, N individuals of the next-generation population need to be selected from A^r.

In either case of the above, the next-generation population is selected according to the sorted subspaces. Algorithm 3 shows subspace-oriented selection in detail, where num_max is the maximum number of individuals selected from each subspace. Firstly, we select the individuals from the first subspace in R, and proceed in sequence but not more than N (population size). To maintain diversity, this algorithm sets the maximum number of individuals (num_max) selected from each subspace. If the number of individuals in a subspace is not greater than num_max, then all the individuals in this subspace are selected; Otherwise, num_max individuals are selected according to the NSGAII [1].

Selection from Dominated Subspaces. In the first case, the number of nondominated individuals is smaller than N ($|A^r| < N$). $N - |A^r|$ individuals need to be selected from A^d for next-population, as mentioned earlier. The subspaces set R sorted was obtained by Algorithm 2. The next step is to select individuals from A^d distributed in dominated subspaces, the detailed procedure is shown in Algorithm 4. Note that if individuals in each subspace are not evenly distributed, the number of selected individuals will be less than N. At this time, we need to select the remaining individuals from the unselected A^d according to the NSGA-II [1], as shown in lines 6 to 7 of Algorithm 4.

As shown as Fig. 1a, the result of environment selection for MOEA-SP is $P = \{x_1, x_2, x_3, x_4, x_9\}$, compared with the result $P = \{x_1, x_2, x_3, x_4, x_7\}$ of NSGA-II in Fig. 1b. From the results of the two selection, the diversity of MOEA-SP is better than that of NSGA-II during evolution.

Algorithm 4. Dominated Subspaces-oriented Selection (DSOS)

Input:
  R: Sorted subspaces;
  A^d: A set of candidate solutions;
  P: A set of solutions.
Output: A set of solutions P.
1: Let $s = 1$, $num\_max = 5$;
2: $(P,\ s) = \text{SOS}\ (R,\ A^d,\ P,\ num\_max)$; // Algorithm 3
3: **if** $(s \leq |R|)\&\&(|P| < N)$ **then**
4:   *Nondominated selection* $(R(s)_{inds})$;
5:   $P = P \cup R(s)_{inds}[1 : (N - |P|)]$;
6: **else if** $s > |R|\&\&|P| < N$ **then**
7:   Select $(N - |P|)$ solutions from the unselected solution from A^d;
8: **end if**

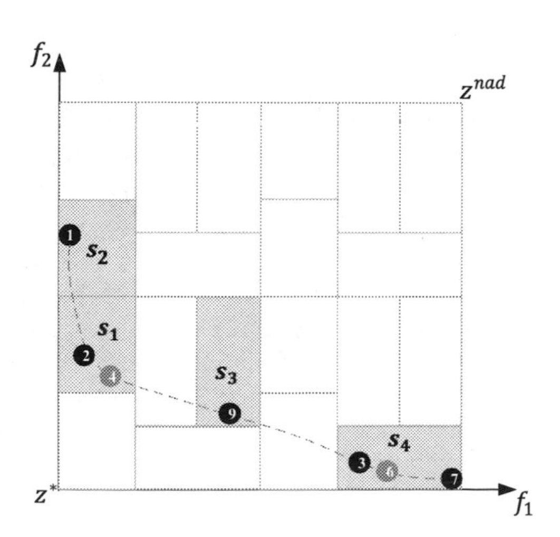

Fig. 2. Nondominated Subspaces based Max-Min Selection

Selection from Nondominated Subspaces. In the second case, the number of nondominated individuals is larger than N ($A^r \geq N$). N individuals of next-generation population need to be selected from A^r. As shown in Fig. 2, $|A^r| = 7 > N = 5$. In this case, the population normally enters into the late stage of the evolution. This means diversity should be paid more attention than convergence.

Algorithm 5. Nondominated Subspaces-oriented Max-Min Selection (NSOMMS)

Input:

R: Sorted subspaces;
A^r: A set of candidate solutions;
P: A set of solutions;

Output: A set of solutions P.

1: Let $num\_max = \left\lfloor \frac{N}{|R|} \right\rfloor$;
2: $P = \text{SOS}\ (R,\ A^d,\ P,\ num\_max)$; // Algorithm 3
3: $A^r = A^r \setminus P$
/* Select lacked individuals by *Max-Min* method */
4: **while** $|P| < N$ **do**
5: Let $Dist_{1\times|A^r|}$, $i = 1$;
6: **for all** $x \in A^r$ **do**
7: $Dist[i] = \min_{y\in P}\ distance\_Minkowski(x, y)$;
8: $i = i + 1$;
9: **end for**
10: $q = arg\ \max_{x\in A^r} Dist[x_{index}]$;
11: $P = P \cup \{q\}$;
12: $A^r = A^r \setminus \{q\}$;
13: **end while**

Different from Algorithm 4, in Algorithm 5, num_max is the average number of individuals in each subspace, which is set to $num\_max = \left\lfloor \frac{N}{|R|} \right\rfloor$. In this way, the value of num_max will be adaptively adjusted according to the distribution of the current population. The setting of num_max ensures that individuals will be selected from each nondominated subspace, which is conducive to maintaining the diversity of the population.

Max-Min selection is described with lines 4 to 13 of Algorithm 5. The major idea is to select an individual each time that its minimum distance to the set of selected individuals is the largest. In other words, an individual, who is as far away as possible from the selected individual, is expected to select. Firstly, the minimum Minkowski distance between each individual in A^r and individual in P is calculated and stored in *Dist*. Then, the individual in A^r corresponding to the maximum value of *Dist* is found and selected into P, which proceeds in turn until the number of individuals in P reaches N. The method is conducive to maintaining diversity greatly. Here the Minkowski distance is calculated as

$$distance\_Minkowski\ (x, y) = \left(\sum_{i=1}^{n} |x_i - y_i|^p \right)^{\frac{1}{p}} \quad (0 < p < 1). \tag{2}$$

where n is the dimension of x or y, p is a parameter. The reason why the Minkowski distance $(0 < p < 1)$ is used is that it can enlarge the difference and facilitate the comparison of distances.

For example in Fig. 2, $num\_max = \left\lfloor \frac{5}{4} \right\rfloor = 1$, and individuals x_1, x_2, x_9, x_3 are selected from s_1, s_2, s_3, s_4, respectively, i.e., $P = \{x_1, x_2, x_9, x_3\}$. It is necessary to select an individual from $\{x_4, x_6, x_7\}$ according to the following Max-Min method because of $5 - |P| = 1$. Obviously, individual x_7 in $\{x_4, x_6, x_7\}$ is the farthest away from all the selected individuals. So individual x_7 is selected, i.e., $P = \{x_1, x_2, x_9, x_3, x_7\}$.

3.4 Historical Archive Update

The reason for introducing the historical archive is that it can guide the direction of population evolution and facilitate the convergence and diversity of the population. The historical archive participates in the calculation of the aforementioned two indicators and environmental selection. In the early stage of evolution, due to the pressure and information of historical individuals, it promotes the convergence of population and accelerates the evolution speed and efficiency. In the late stage, it can increase diversity to select offspring from the set of all nondominated historical individuals according to the Max-Min method. However, if the number of historical individuals is too large, computing resources will be overtaken, so we need to delete some individuals from the historical archive.

The update of the historical archive in this paper is divided into two cases according to the size of nondominated individuals. In the first case, the number of nondominated individuals is smaller than the size of the population, which needs historical individuals to guide algorithm search. Therefore, the historical archive reserves some representative individuals in each sampled subspace. In the second case, when the number of nondominated individuals is larger than the size of the population, the historical archive only preserves nondominated individuals.

4 Experimental Studies

To verify the validity of the proposed algorithm MOEA-SP, this paper makes some comparative experiments on a set of benchmarks.

4.1 Benchmark Functions and Performance Metric

GLT1-GLT6 [14] benchmark problems are used for testing the performance of algorithms. The shape of these functions has a variety of forms, including convex, nonconvex, extremely convex, disconnected, nonuniformly distributed. The dimensions of the decision variables of GLT1-GLT6 are set to 10.

In this paper, the Inverted Generational Distance (IGD) is used as a performance metric to measure the quality of a solution set P, which represents the average distance from a set of reference points P^* on true PF to the solved population P. The IGD metric is defined as

$$IGD(P^*, P) = \frac{\sum_{v \in P^*} dist(v, P)}{|P^*|}. \tag{3}$$

where $dist(v, P)$ is the minimum Euclidean distance from the solution v in P^* to solution in P. The IGD metric can measure the convergence and diversity of a set of solutions P, and the smaller the value of the IGD is, the better the algorithm performs.

4.2 Peer Algorithms and Parameter Settings

The compared algorithms with MOEA-SP are CDG-MOEA [4], NSGA-II [1] and MOEA/D [2]. These MOEAs belong to grid-based, dominance-based and decomposition-based methods, respectively. The population size is set to 200. Here the population size, in MOEA/D, is set to 210 (the closest integer to 200) for three objective problems because of the same as the number of weight vectors.

The number of evaluations of each test function is: 300,000; $\delta = 0.9$; In the DE operator: $CF = 1$, $F = 0.5$, $\eta = 20$, $p_m = 1/D$, where D is the dimensions of the decision variables. Each of the above four MOEAs runs 30 times independently on each test problem.

4.3 Experimental Results

Table 1 gives the mean and standard deviation of the IGD-metric values of the four algorithms (i.e., MOEA-SP, CDG-MOEA, NSGA-II, and MOEA/D) on GLT1-GLT6 test problems in 30 independent runs, where the IGD-metric value with the best mean for each test problem is highlighted in boldface. MOEA-SP performs the best on GLT5 and GLT6 test problems, compared with the other three MOEAs. For GLT1-GLT4 test problems, the other three MOEAs have their own strong points. NSGA-II performs the best on GLT2 and GLT3. CDG-MOEA and MOEA/D perform the best on GLT1 and GLT4, respectively.

The reason for the best performance of CDG-MOEA on GLT1 is that the grid partitioning is very favorable for GLT1 with the linear shape of PF, compared with the kd-tree partitioning in the MOEA-SP. Although NSGA-II performs the best on GLT3 with the extremely concave PF, MOEA-SP does not perform poorly. As shown in Fig. 3, we give the distribution of the nondominated solutions

Table 1. The IGD-metric values comparisons of the MOEA-SP with the other three MOEAs on GLT test problems in terms of the mean and standard deviation values, where the IGD-metric value with the best mean for each test problem is highlighted in boldface.

		MOEA-SP	CDG-MOEA	NSGA-II	MOEA/D
GLT1	mean	4.30E-03	**1.55E-03**	1.76E-03	1.78E-03
	std	7.25E-04	**6.45E-04**	1.71E-04	5.58E-06
GLT2	mean	3.64E-02	1.03E-01	**1.66E-02**	1.45E-01
	std	1.54E-03	6.63E-02	**4.41E-04**	1.83E-02
GLT3	mean	5.96E-03	1.03E-01	**5.48E-03**	9.18E-03
	std	5.16E-04	8.45E-02	**6.22E-03**	1.42E-04
GLT4	mean	1.91E-02	6.59E-03	9.79E-03	**4.78E-03**
	std	4.06E-02	3.56E-03	3.37E-02	**8.17E-06**
GLT5	mean	**2.91E-02**	2.52E-01	4.18E-02	7.86E-02
	std	**5.05E-04**	1.18E-01	1.87E-03	3.98E-04
GLT6	mean	**2.93E-02**	1.90E-01	2.97E-02	4.56E-02
	std	**5.32E-03**	4.88E-02	1.65E-03	5.21E-04

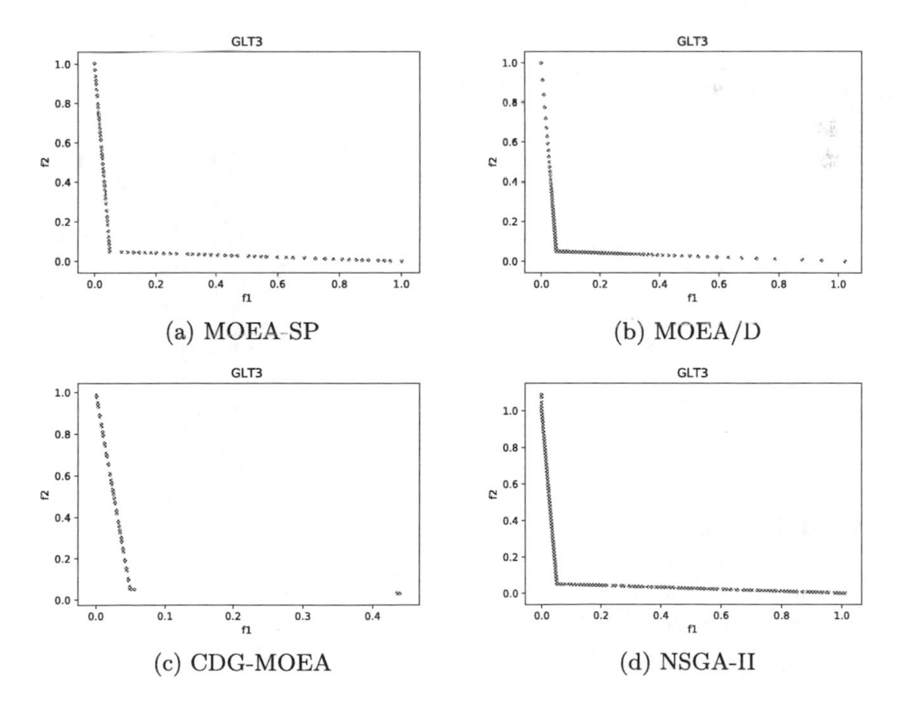

Fig. 3. The final nondominated solution set obtained by the four algorithms on GLT3

sets obtained by the four algorithms on GLT3. In terms of the uniformity, the distribution of the nondominated solutions set obtained by MOEA-SP is better than MOEA/D. The performance of MOEA-SP on GLT5 and GLT6 is superior to the other three MOEAs. The reason is that, in MOEA-SP, the subspace sorting and the subspace-oriented selection increase selection pressure and are conducive to solving GLT5 and GLT6 with three objectives. As shown in Fig. 4, we give the distribution of the nondominated solutions sets obtained by the four algorithms on GLT6. The distribution of nondominated solutions set obtained by MOEA-SP on GLT6 is superior to the other three algorithms in terms of both the convergence and diversity.

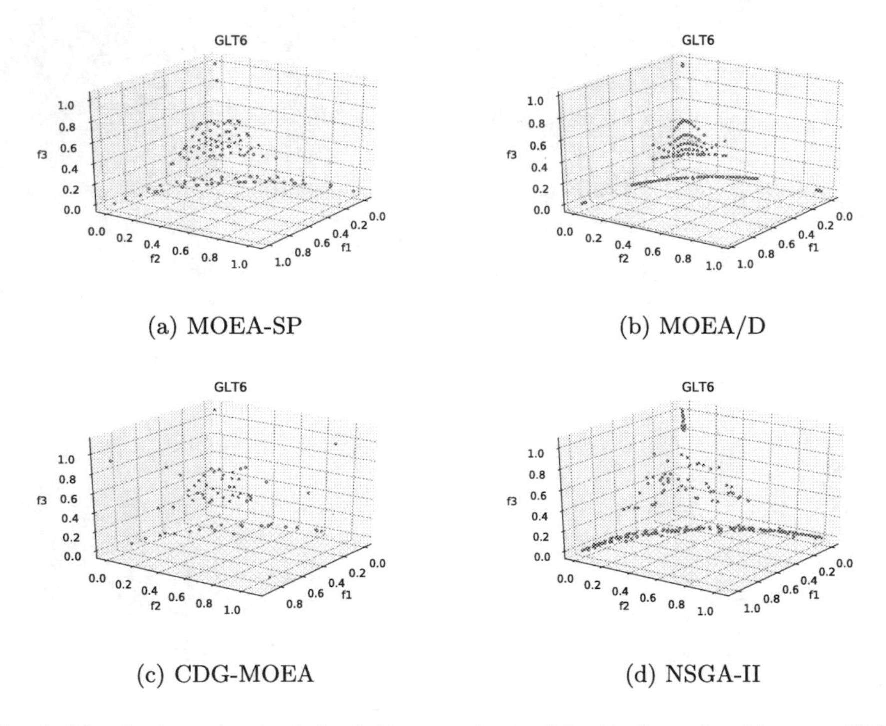

(a) MOEA-SP (b) MOEA/D

(c) CDG-MOEA (d) NSGA-II

Fig. 4. The final nondominated solution set obtained by the four algorithms on GLT6

5 Conclusion

This paper proposes a novel MOEA based on space partitioning (MOEA-SP). The proposed MOEA-SP transforms a MOP into a bi-objectives optimization problem to sort the subspaces, then selects offspring by the subspace-oriented selection, which simplifies the complexity of the problem. The *k*d-tree space partitioning method overcomes the limitation of shape sensitivity to PF and the curse of dimensionality. The subspaces sorting and the Max-Min selection are used for pushing the population convergence and maintaining the diversity, respectively, which has great potential to solve MaOPs. MOEA-SP is compared with three MOEAs on GLT test suite. The experimental results show that MOEA-SP outperforms the compared algorithms in some benchmarks.

Although MOEA-SP overcomes some of the above issues, it still faces great challenges. If the number of partitioning subspaces and the value of *num_max* are not set properly, the performance of the algorithm will be affected to some extent. In addition, the definition of the bi-indicator criterion needs to fine-tune to the more accurate description of the convergence and the diversity. Adaptively adjusting the *num_max* and furtherly modifying the definition of the bi-indicator criterion what needs to be done in the future.

Acknowledgments. This work was supported in part by the National Natural Science Foundation of China under Grant No. 61673355, in part by the Fundamental Research Funds for the Central Universities, China University of Geosciences (Wuhan) under Grant (CUG170603, CUGGC02).

References

1. Deb, K., Pratap, A., Agarwal, S., Meyarivan, T.: A fast and elitist multiobjective genetic algorithm: NSGA-II. IEEE Trans. Evol. Comput. **6**(2), 182–197 (2002)
2. Zhang, Q., Hui, L.: MOEA/D: a multiobjective evolutionary algorithm based on decomposition. IEEE Trans. Evol. Comput. **11**(6), 712–731 (2007)
3. Yang, S., Li, M., Liu, X., Zheng, J.: A grid based evolutionary algorithm for many objective optimization. IEEE Trans. Evol. Comput. **17**(5), 721–736 (2013)
4. Cai, X., Mei, Z., Fan, Z., Zhang, Q.: A constrained decomposition approach with grids for evolutionary multiobjective optimization. IEEE Trans. Evol. Comput. **22**(4), 564–577 (2017)
5. Fang, K., Yang, Z.: On uniform design of experiments with restricted mixtures and generation of uniform distribution on some domains. Stat. Probab. Lett. **46**(2), 113–120 (2000)
6. Deb, K., Jain, H.: An evolutionary many-objective optimization algorithm using reference-point-based nondominated sorting approach, part I: solving problems with box constraints. IEEE Trans. Evol. Comput. **18**(4), 577–601 (2014)
7. Sato, H.: Analysis of inverted PBI and comparison with other scalarizing functions in decomposition based MOEAs. J. Heuristics **21**(6), 819–849 (2015). https://doi.org/10.1007/s10732-015-9301-6
8. Yang, S., Jiang, S., Jiang, Y.: Improving the multiobjective evolutionary algorithm based on decomposition with new penalty schemes. Soft Comput. **21**(16), 4677–4691 (2016). https://doi.org/10.1007/s00500-016-2076-3
9. Cheng, R., Jin, Y., Olhofer, M., Sendhoff, B.: A reference vector guided evolutionary algorithm for many-objective optimization. IEEE Trans. Evol. Comput. **20**(5), 773–791 (2016)
10. Liu, H., Gu, F., Zhang, Q.: Decomposition of a multiobjective optimization problem into a number of simple multiobjective subproblems. IEEE Trans. Evol. Comput. **18**(3), 450–455 (2014)
11. Zhao, S., Suganthan, P., Zhang, Q.: Decomposition-based multiobjective evolutionary algorithm with an ensemble of neighborhood sizes. IEEE Trans. Evol. Comput. **16**(3), 442–446 (2012)

12. Li, H., Zhang, Q.: Multiobjective optimization problems with complicated pareto sets, MOEA/D and NSGA-II. IEEE Trans. Evol. Comput. **13**(2), 284–302 (2009)
13. Zhang, H., Zhang, X., Gao, X., Song, S.: Self-organizing multiobjective optimization based on decomposition with neighborhood ensemble. Neurocomputing **173**, 1868–1884 (2016)
14. Gu, F., Liu, H., Tan, K.: A multiobjective evolutionary algorithm using dynamic weight design method. Int. J. Innov. Comput. Inf. Control IJICIC **8**(5), 3677–3688 (2012)

Neural Architecture Search Using Multi-objective Evolutionary Algorithm Based on Decomposition

Weiqin Ying[1(✉)], Kaijie Zheng[1], Yu Wu[2(✉)], Junhui Li[1], and Xin Xu[3,4]

[1] School of Software Engineering, South China University of Technology, Guangzhou 510006, China
yingweiqin@scut.edu.cn

[2] School of Computer Science and Cyber Engineering, Guangzhou University, Guangzhou 510006, China
wuyu@gzhu.edu.cn

[3] School of Physics and Information Engineering, Minnan Normal University, Zhangzhou 363000, China

[4] College of Information and Engineering, Jingdezhen Ceramic Institute, Jingdezhen 333000, Jiangxi, China

Abstract. NSGA-Net is a popular method for neural architecture search (NAS). It conducts the improved non-dominated sorting genetic algorithm (NSGA-II) during its search procedure. In this paper, a NAS method using the multi-objective evolutionary algorithm based on decomposition (MOEA/D-Net) is proposed to heighten the running efficiency of NSGA-Net. MOEA/D-Net aims to minimize the number of floating-point operations (FLOPs) and error rate of neural architectures through the multi-objective evolutionary algorithm based on decomposition (MOEA/D) during the search process. It selects parents within the neighborhoods of a subproblem and conducts multi-point crossover and mutation to generate offspring individuals at every generation. Experiment results on the CIFAR-10 image classification dataset indicate that MOEA/D-Net obtains architecture networks with less FLOPs and MOEA/D-Net outperforms NSGA-Net in terms of running efficiency.

Keywords: Neural architecture search · Image classification · Multi-objective optimization · Evolutionary algorithms · Decomposition

1 Introduction

With the development of deep learning techniques, the importance of architecture is gradually noticed. In general, the deeper architecture causes the better result. There are many deep convolutional neural networks (CNN) that can achieve good performances for the image task, such as VGG [1], GoogLeNet [2], ResNet [3] and DenseNet [4]. Networks such as MobileNet [5], XNOR-Net [6] and BinaryNets [7] need less hardware resources in practice. However, these

K. Li et al. (Eds.): ISICA 2019, CCIS 1205, pp. 143–154, 2020.
https://doi.org/10.1007/978-981-15-5577-0_11

architectures are all designed by hand, which needs professional knowledge, experience and repeated debugging. People have to spend a lot of time to get the best architecture when the task is challenging and difficult. In a word, designing deep learning architecture should take both the performance and complexity into account. The latter means its actual application value in the real world. Therefore, people begin to conceive whether it is possible to automate the process of designing the neural network architecture through algorithms.

Neural architecture search (NAS) approaches aim to design neural network architectures automatically. Most of NAS procedures consist of four steps: (1) defining the search space and encoding method, (2) using a search strategy to find candidate network structures, (3) evaluating different network structures, (4) executing the next iteration base on the feedback.

There are three main search strategies in NAS field: (1) reinforcement learning (RL) methods, (2) gradient descent methods, (3) evolutionary algorithms. Q-Learning method has been widely used in reinforcement learning. MetaQ-NN [8] is a meta-modeling algorithm based on reinforcement learning, which adopts ε-greedy exploration strategy, experience replay and sets the validation accuracy as reward for Q-learning. However, the network ignores the skip connection. It just consists of convolution, pooling and fully-connected layers. Basing on MetaQNN, Zhong et al. presented BlockQNN [9]. It gets the optimal network blocks firstly and then stacks them. So this method reduces the search space of the network structure greatly. NASNet [10] transfers the search structure from network architecture to two kinds of cells, Normal Cell and Reduction Cell. In addition, it incorporates skip connection in the cell structure. It adopts a RNN controller to get architectures and updates the controller by policy gradient.

The network structure in DiffRNN [11] is differentiable for using gradient descent methods. It can solve simple problems, for example the sequence prediction problems. DiffRNN indicates that gradient descent methods can adjust the size of a recurrent network efficiently according to the change of problem. DARTS [12] also uses a gradient descent method to search network architectures. It conducts the softmax function to select candidate operations and then connects them for getting the best cell structures.

Apart from those, evolutionary algorithms can be a kind of search strategy for its iterative and selective features. In general, the accuracy in a validation set represents the fitness [13]. After the worst pair of parents being obsoleted, two offspring individuals generated from the best pair of parents are added into the population. The Genetic CNN [14] presents a new encoding method, a fixed-length binary string, to represent a network architecture. It also defines selection, crossover and mutation operators. But it just adopts genetic algorithms to search new structures. The regularized evolution for image classifier architecture search [15] proposes an age property which prefers younger architectures in the selection process. In addition, it makes a comparison between the evolutionary algorithms and the reinforcement learning algorithms for NAS. The result concludes that the former is faster under the same hardware condition.

However, the methods as described above are all based on the single objective problems. They only take account of the accuracy in the validation set. In order to consider both the performance and speed, NSGA-Net [16,17] introduces multi-objective evolutionary algorithms for NAS. It selects architectures near the Pareto front for balancing the performance and resource issues.

In this paper, a NAS method using the multi-objective evolutionary algorithm based on decomposition, called MOEA/D-Net, is presented to further reduce the running time of NSGA-Net. MOEA/D-Net incorporates the multi-objective evolutionary algorithm based on decomposition (MOEA/D) in NAS. MOEA/D has solved a variety of multi-objective optimization problems successfully. MOEA/D-Net makes use of the coevolution between subproblems. In each iteration, NSGA-Net needs to select parents from the whole population while MOEA/D-Net does in the neighborhoods of a subproblem. Therefore, the latter searches faster. Meanwhile, MOEA/D-Net has great diversity and fitness allocation, because it optimizes the N scalar subproblems at the same time instead of directly solving the whole multi-objective optimization problems.

2 MOEA/D-Net Method

For the limitation of hardware resources, we must usually balance conflicting objectives, such as accuracy and computation complexity, for NAS tasks. Evolutionary algorithms are effective for multi-objective problems and MOEA/D-Net is a NAS method using the multi-objective evolutionary algorithm based on decomposition. MOEA/D-Net automatically generates different network architectures and selects architectures that approximate the Pareto front to do the next iteration.

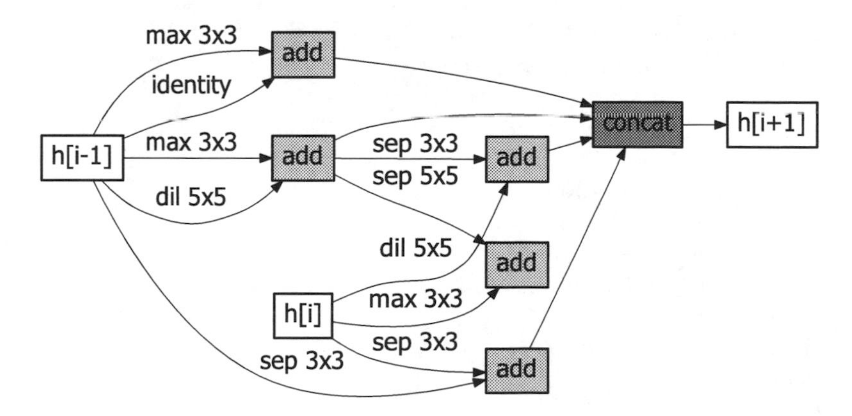

Fig. 1. Normal Cell = [2 0 1 0 7 0 1 0 7 1 3 3 1 1 4 3 3 1 3 0]

2.1 Search Space

As suggested by NASNet [10] and DiffRNN [11], all convolutional networks in the search space are composed of cells with the same structure and different weights. In this paper, MOEA/D-Net also searches the best cell structures and stacks them to get a complete network.

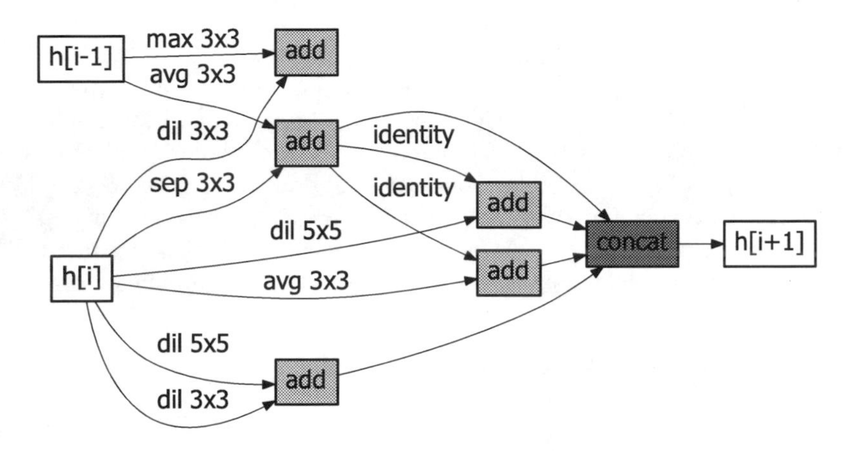

Fig. 2. Reduction Cell = [0 0 3 1 6 1 1 0 2 2 7 1 2 2 0 1 7 1 6 1]

There are two kinds of convolution cell structures, Reduction Cell and Normal Cell. Putting feature map into cells, the output size of Reduction Cell is half of the input and the other cell keeps it as the same as the input size. MOEA/D-Net performs the convolution operations in the order of ReLU-Conv-BN. As suggested by NASNet [10] and NSGA-Net [17], MOEA/D-Net includes the candidate basic operations as below:

3×3 max pooling
3×3 average pooling
3×3 separable convolution
3×3 dilated separable convolution
5×5 separable convolution
5×5 dilated separable convolution
7×7 separable convolution
7×1 then 1×7 convolution
skip connection

2.2 Operation Encoding

In MOEA/D-Net, each cell adopts an encoding method suggested by Juefei-Xu et al. [18] to represent the connection relationship among basic operations. Assuming that a cell is composed of B blocks and each block corresponds to a combination of two operations. Firstly, MOEA/D-Net creates a list for nine basic operations as follows: [$'avg\_pool\_3 \times 3$, $'max\_pool\_3 \times 3'$, $'skip\_connect'$, $'sep\_conv\_3{\times}3'$, $'sep\_conv\_5{\times}5'$, $'sep\_conv\_7{\times}7'$, $'dil\_conv\_3{\times}3'$, $'dil\_conv\_5{\times}5'$, $'conv\_7 \times 1\_1 \times 7'$]. And MOEA/D-Net uses 2 bits (a_i, b_j) to represent every basic operation. The notation a_i indicates the index of the operation in the list and b_j denotes that the operation is performed on the output of the b_j block. Specifically, a cell has two input nodes and one output node. And the former is just the output results of two previous cells. The values b_j for these previous cells are 0 and 1, respectively.

Algorithm 1: Framework of MOEA/D-Net

Input: P: the size of the population ; n_{off}: the number of the offpsring ; n_{gen}: the number of generation; T: the training epoch ;
Output: N: Net Architectures

1 Initialize the population and generate the genotypes of P individuals randomly;
2 **for** i=1,$\cdots$,n_{gen} **do**
3 | Decode genotypes into different Neural network architectures;
4 | **for** *epoch*=1,$\cdots$,T **do**
5 | | Train architectures on the training dataset by GPU;
6 | **end**
7 | Acquire the error rate and FLOPs on the validation dataset by GPU;
8 | **for** h=1,$\cdots$,n_{off} **do**
9 | | Generate an integer j from 1 to P randomly;
10 | | Select two individuals, x_k and x_l, from the neighborhood of subproblem j;
11 | | Apply multi-point crossover and polynomial mutation to generate offspring individuals;
12 | | Decode genotypes of offspring individuals into different neural network architectures;
13 | | **for** *epoch*=1,$\cdots$,T **do**
14 | | | Train these architectures on the training dataset by GPU;
15 | | **end**
16 | | Acquire the error rate and FLOPs on the validation dataset by GPU;
17 | | Update the ideal point Z;
18 | | Calculate the decomposed scalar values for each neighbor by the PBI aggregation method;
19 | | **if** $FV_{off} < FV_j$ **then**
20 | | | Replace the current individual j by the offspring
21 | | **end**
22 | **end**
23 **end**
24 Decode genotypes into different neural network architectures N;
25 **return** neural network architectures N;

For the multi-point crossover operator, MOEA/D-Net selects sub-strings in the same position from the parent encoding strings randomly, and swaps them to get offspring individuals. Figure 3 presents an example of the multi-point crossover. In order to increase the diversity of population and avoid the local optimal trouble, MOEA/D-Net also adopts the polynomial mutation [18] for offspring individuals. The mutation probability $P_m = 1/n$ and the distribution index $\eta_m = 3$.

X = [[3 0 6 1 1 2 6 2 7 1 3 2 0 1 5 1 7 2 1 4 1 1 5 1 1 2 5 0 5 3 2 0 4 0 3 0 5 4 3 3]
[7 0 6 1 7 0 7 0 1 0 1 0 4 1 5 2 6 1 1 4 7 0 6 1 1 0 0 0 5 3 1 2 0 2 6 2 7 3 1 2]]

a new offspring = [[3 0 6 1 1 2 6 2 7 1 1 0 4 1 5 2 6 1 1 4 7 0 6 1 1 0 0 0 5 3 1 2 4 0 3 0 5 4 3 3]
[7 0 6 1 7 0 7 0 1 0 3 2 0 1 5 1 7 2 1 4 1 1 5 1 1 2 5 0 5 3 2 0 0 2 6 2 7 3 1 2]]

Fig. 3. An example of crossover. X represents parents. Yellowish shade and greyish shade denote the sub-strings to be operated. (Color figure online)

Here is an example for this encoding. Presuming the number of blocks $N = 5$, so we need forty bits to represent a group of Reduction Cell and Normal Cell. In this paper, we set the first twenty bits to represent the Normal Cell and the others to represent the Reduction Cell. A block adds 2 basic operations and an operation is represented by 2 bits. For example, the bits from one to four mean the operations for the output of block $b_j = 0$. Adding these operations composes block $b_j = 2$. In the same way, the next four bits means the composed operations of block $b_j = 3$. The concrete encodings of Normal Cell and Reduction Cell are presented in Fig. 1 and Fig. 2, respectively. Thus the complete encoding is a combination of Normal Cell and Reduction Cell, i.e., [2 0 1 0 7 0 1 0 7 1 3 3 1 1 4 3 3 1 3 0 0 0 3 1 6 1 1 0 2 2 7 1 2 2 0 1 7 1 6 1]. The operation represented by each edge, i.e., the line with an arrow, is annotated above the line.

2.3 Search Process

MOEA/D-Net is an iterative evolutionary process. It initializes a set of weight vectors and determines the neighbor set for each sub-problem according to the euclidean distance of vectors. Afterwards, MOEA/D-Net generates a population of initial architectures randomly and initializes the ideal point. In every iteration, it selects parents in the neighborhood and conducts the multi-point crossover and polynomial mutation to generate offspring individuals. For the evaluation of individuals, MOEA/D-Net needs to train different architectures on GPU to obtain the error rate and FLOPs. At last, MOEA/D-Net selects an architecture from the Pareto front artificially to train from scratch and evaluate it on the validation dataset. The framework of MOEA/D-Net is given in Algorithm 1.

Because the performance ranking when networks converges is almost as the same as the ranking in the early training stage, we only set the training epoch $T = 20$ and then evaluate it on the validation set. Furthermore, MOEA/D-Net also sets the number of offspring to be a half of the population size. These measures all help MOEA/D-Net to obtain architectures more faster during the search process.

3 Experimental Results

3.1 Training Details

We choose the CIFAR-10 dataset to compare MOEA/D-Net with NSGA-Net on image classification. Consisting of 50,000 training color images and 10,000 testing color images, this dataset is divided into 10 categories and the size of all images is 32 * 32 * 3. In our experiment, we split the original training set in an 8:2 ratio to create a training set and a validation set for NAS. The CIFAR-10 original testing set is used at the end of the search to acquire the accuracy of the final model. The hyper-parameters of MOEA/D-Net are similar to NSGA-Net.

Table 1. The comparison between MOEA/D-Net and NSGA-Net in FLOPs and error rate

Architectures		NSGA-Net	MOEA/D-Net
Search Cost (GPU-days)		3.5	3
Best	Error rate	11.78%	13.94%
	FLOPs (M)	28.15	28.46
Mean	Error rate	13.50%	14.43%
	FLOPs (M)	64.85	31.16
Worst	Error rate	16.06%	15.21%
	FLOPs (M)	142.54	34.53

We set the number of blocks in one cell to 5 and the number of basic operations is set to 9. The probability of crossover and mutation operations are set to 0.9 and 0.02 respectively. We conduct the penalty boundary intersection (PBI) aggregation function to measure the fitness of the solution associated with a reference vector. In order to speed up the search process, we set the number of offspring to be a half of the population size. In our experiment, the population size is 20 and MOEA/D-Net generates 10 offspring individuals in each generation. And The number of generation is also 20. Therefore, MOEA/D-Net can search 220 network architectures.

In the training phase, we adopt the stochastic gradient descent (SGD) algorithm and a cosine annealing learning rate schedule [19]. The initial learning rate is 0.025 and the number of epochs is 20. And the batch-size is set to 128.

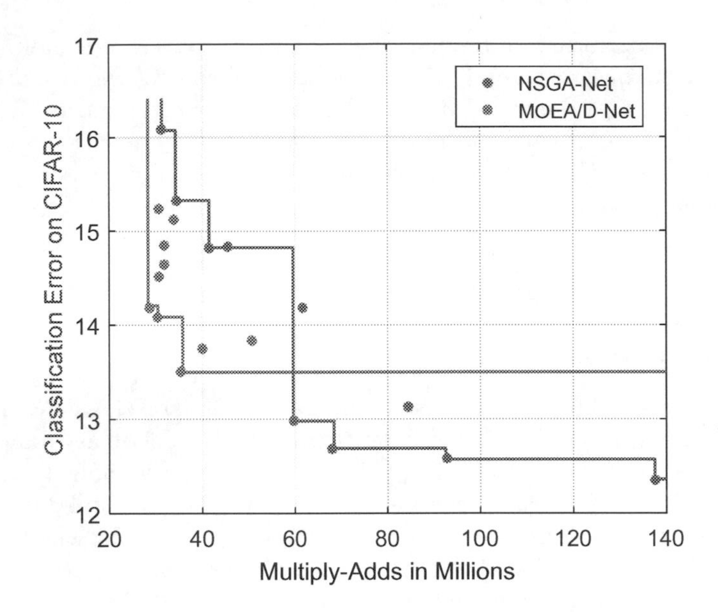

Fig. 4. The Pareto front obtained, respectively, by MOEA/D-Net and NSGA-Net

Architecture search and training are conducted on a server with 8 Nvidia 1080Ti GPUs. It takes around 20 min to train each network architecture on one GPU. According to NSGA-Net [17], we also consider the classification error rate and FLOPs as two objectives.

After the search procedure, we choose the optimization model on the Pareto front to train from scratch. The epoch is set to 120 and the batch-size is 96. We also adopt two kinds of data pre-processing technologies, regularization and cutout [20]. In addition, an auxiliary head classifier is added to avoid the gradient loss when the networks is too deep. Hence, the loss function value is the sum of the loss from the auxiliary head classifier and the loss of the original architecture during the training. The other hyper-parameters stay the same with those adopted by NSGA-Net [17].

3.2 Result Analysis

The results obtained by MOEA/D-Net and NSGA-Net on CIFAR-10 are presented in Table 1. It shows that the mean value of MOEA/D-Net is better than NSGA-Net in terms of FLOPs. Note that the best validation error rate in the table is not corresponding to the FLOPs. The lowest error rate and FLOPs constitute the ideal point. When achieving the minimum error rates, the corresponding FLOPs obtained by NSGA-Net and MOEA/D-Net are 140.80M and 30.56M FLOPs, respectively. Although the minimum error rate obtained by NSGA-Net is lower than that by MOEA/D-Net, the corresponding FLOPs by NSGA-Net is 4.6 times as much as that by MOEA/D-Net. Therefore, MOEA/D-Net actually performs better.

Figure 4 presents the Pareto fronts obtained, respectively, by MOEA/D-Net and NSGA-Net. The architectures acquired by MOEA/D-Net have the lower FLOPs and the Pareto front by MOEA/D-Net seems better. It can be concluded that though NSGA-Net gains the architectures with the lower classification error, the FLOPs of the architectures by NSGA-Net is obviously greater than that by MOEA/D-Net.

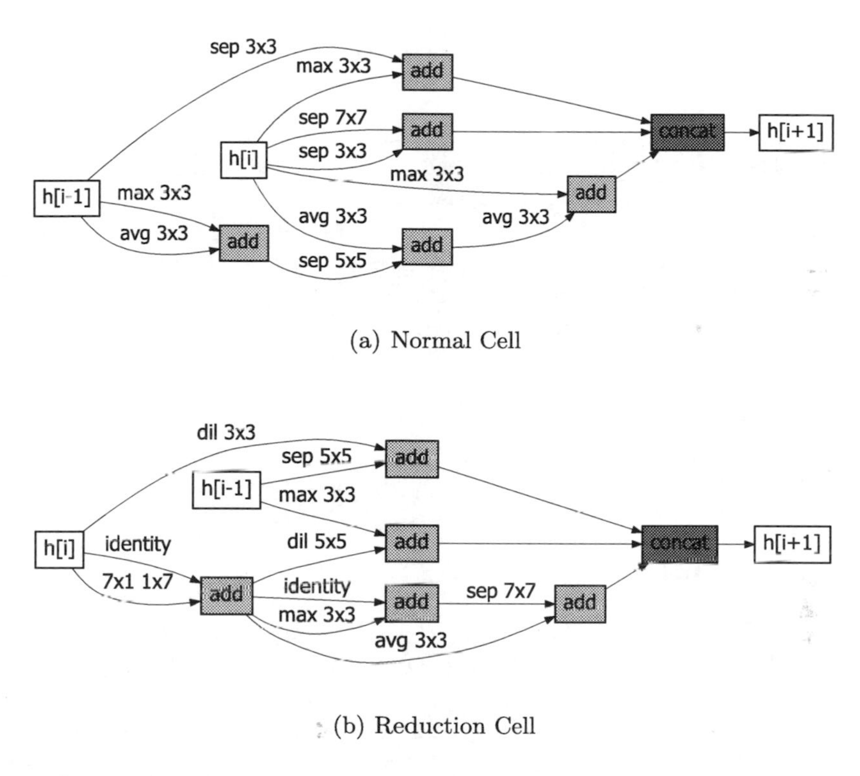

(a) Normal Cell

(b) Reduction Cell

Fig. 5. The selected architectures to train from scratch for NSGA-Net.

Considering two competing objectives, we choose the obtained architecture for NSGA-Net whose classification accuracy is 87.02% and FLOPs is 59.623M FLOPs to train from scratch. Similarly, we choose the obtained architecture for MOEA/D-Net whose classification accuracy is 86.06% and FLOPs is 30.56M FLOPs. The concrete architectures are presented in Fig. 5 and Fig. 6, respectively. On the final result, the accuracy achieved by NSGA-Net is 95.08%, and the number of parameters is 2.114 MB. The accuracy by MOEA/D-Net is 94.64%, and the number of parameters is 1.113 MB. Compared with NSGA-Net, the number of parameters by MOEA/D-Net is halved. Moreover, the accuracy difference is 0.96% initially, but it decreases to 0.4% after the complete training.

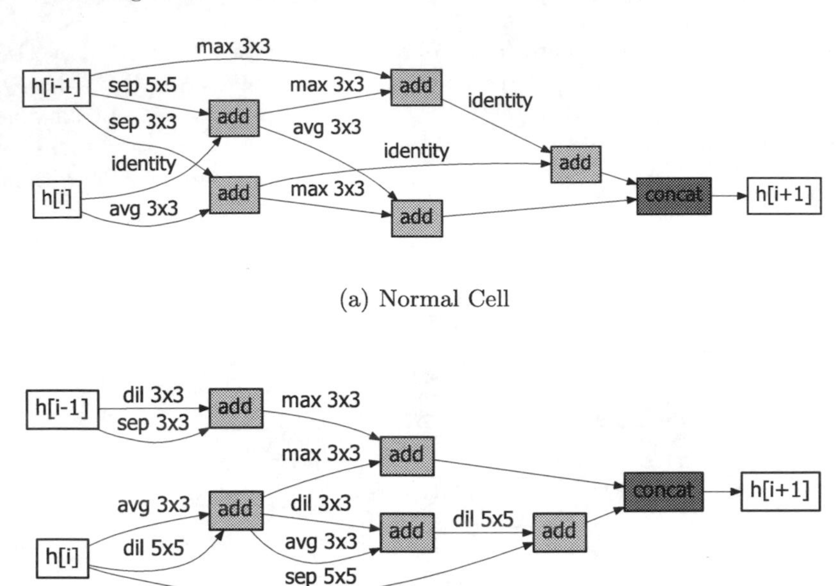

(a) Normal Cell

(b) Reduction Cell

Fig. 6. The selected architectures to train from scratch for MOEA/D-Net.

4 Conclusion

In this paper, a NAS method using the multi-objective evolutionary algorithm based on decomposition, MOEA/D-Net, is presented. By incorporating MOEA/D in the neural architecture search procedure, MOEA/D-Net can trade off among two competing objectives and obtains a variety of network architectures which approximate the Pareto front. MOEA/D-Net also has the lower complexity than NSGA-Net. Experiment results on the CIFAR-10 image classification dataset show that MOEA/D-Net achieves architecture networks with less FLOPs than NSGA-Net and MOEA/D-Net outperforms NSGA-Net in terms of running efficiency.

In the future, we plan to design the parallel version of MOEA/D-Net to further reduce the running time of NAS by training different networks architectures on multiple GPUs. In addition, we will also conduct the hyper-parameter optimization for MOEA/D-Net.

Acknowledgments. This work was supported in part by the Natural Science Foundation of Guangdong Province, China, under Grant 2015A030313204, in part by the Pearl River S&T Nova Program of Guangzhou under Grant 2014J2200052, in part by the National Natural Science Foundation of China under Grant 61203310, Grant 61503087 and Grant 61702239, in part by the Fundamental Research Funds for the Central Universities, SCUT, under Grant 2017MS043, in part by the Science Foundation of Jiangxi Provincial Department of Education under Grant GJJ170765 and Grant GJJ170798, and the Project of Jingdezhen Science and Technology Bureau under Grant 20161GYZD011-011.

References

1. Simonyan, K., Zisserman, A.: Very deep convolutional networks for large-scale image recognition. arXiv preprint arXiv:1409.1556 (2014)
2. Szegedy, C., Liu, W., Jia, Y., et al.: Going deeper with convolutions. In: IEEE Conference on Computer Vision and Pattern Recognition, pp. 1–9. IEEE (2015)
3. He, K., Zhang, X., Ren, S., et al.: Deep residual learning for image recognition. In: IEEE Conference on Computer Vision and Pattern Recognition, pp. 770–778. IEEE (2016)
4. Huang, G., Liu, Z., Van Der Maaten L., et al.: Densely connected convolutional networks. In: IEEE Conference on Computer Vision and Pattern Recognition, pp. 4700–4708. IEEE (2017)
5. Howard, A. G., Zhu, M., Chen, B., et al.: MobileNets: efficient convolutional neural networks for mobile vision applications. arXiv preprint arXiv:1704.04861 (2017)
6. Rastegari, M., Ordonez, V., Redmon, J., Farhadi, A.: XNOR-Net: ImageNet classification using binary convolutional neural networks. In: Leibe, B., Matas, J., Sebe, N., Welling, M. (eds.) ECCV 2016. LNCS, vol. 9908, pp. 525–542. Springer, Cham (2016). https://doi.org/10.1007/978-3-319-46493-0_32
7. Courbariaux, M., Hubara, I., Soudry, D., et al.: Binarized neural networks: training deep neural networks with weights and activations constrained to +1 or -1. arXiv preprint arXiv:1602.02830 (2016)
8. Baker, B., Gupta, O., Naik, N., et al.: Designing neural network architectures using reinforcement learning. arXiv preprint arXiv:1611.02167 (2016)
9. Zhong, Z., Yan, J., Wu, W., et al.: Practical block-wise neural network architecture generation. In: IEEE Conference on Computer Vision and Pattern Recognition, pp. 2423–2432. IEEE (2018)
10. Zoph, B., Vasudevan, V., Shlens, J., et al.: Learning transferable architectures for scalable image recognition. In: IEEE Conference on Computer Vision and Pattern Recognition, pp. 8697–8710. IEEE (2018)
11. Miconi, T.: Neural networks with differentiable structure. arXiv preprint arXiv:1606.06216 (2016)
12. Liu, H., Simonyan, K., Yang, Y.: DARTS: differentiable architecture search. arXiv preprint arXiv:1806.09055 (2018)
13. Real, E., Moore, S., Selle, A., et al.: Large-scale evolution of image classifiers. In: International Conference on Machine Learning, pp. 2902–2911. PMLR (2017)
14. Xie, L., Yuille, A.: Genetic CNN. In: IEEE International Conference on Computer Vision, pp. 1379–1388. IEEE (2017)
15. Real, E., Aggarwal, A., Huang, Y., et al.: Regularized evolution for image classifier architecture search. In: AAAI Conference on Artificial Intelligence, pp. 4780–4789. AAAI (2019)

16. Chu, X., Zhang, B., Ma, H., et al.: Fast, accurate and lightweight super-resolution with neural architecture search. arXiv preprint arXiv:1901.07261 (2019)
17. Lu, Z., Whalen, I., Boddeti, V., et al.: NSGA-Net: a multi-objective genetic algorithm for neural architecture search. arXiv preprint arXiv:1810.03522 (2018)
18. Felix, J.-X., Naresh Boddeti, V., Savvides, M.: Local binary convolutional neural networks. In: IEEE Conference on Computer Vision and Pattern Recognition, pp. 19–28. IEEE (2017)
19. Loshchilov, I., Hutter, F.: SGDR: stochastic gradient descent with warm restarts. arXiv preprint arXiv:1608.03983 (2016)
20. DeVries, T., Taylor, G.W.: Improved regularization of convolutional neural networks with cutout. arXiv preprint arXiv:1708.04552 (2017)

A Collaborative Evolutionary Algorithm Based on Decomposition and Dominance for Many-Objective Knapsack Problems

Hainan Huang[1], Weiqin Ying[1(✉)], Yu Wu[2], Kaijie Zheng[1], and Shaowu Peng[1]

[1] School of Software Engineering, South China University of Technology, Guangzhou 510006, China
yingweiqin@scut.edu.cn

[2] School of Computer Science and Cyber Engineering, Guangzhou University, Guangzhou 510006, China
wuyu@gzhu.edu.cn

Abstract. Multi-objective evolutionary algorithms (MOEAs) are popular for solving many-objective knapsack problems. Among various MOEAs, the multi-objective evolutionary algorithm based on decomposition (MOEA/D) behaves well. However, MOEA/D often retains multiple copies of one individual in the population, which might hamper the diversity of the population. To overcome the disadvantage, a collaborative evolutionary algorithm based on decomposition and dominance, called MOEA/D-DDC, is presented in this paper. It mainly adopts a decomposition-dominance collaboration mechanism. The mechanism consists of a decomposition-based population and a dominance-based archive. The decomposition-based population collects elite individuals for the dominance-based archive. Meanwhile the dominance-based archive assists to repair the decomposition-based population and heighten the diversity. The experiment results show that MOEA/D-DDC obtains the better set of solutions than MOEA/D for many-objective knapsack problems with 4 to 8 objectives.

Keywords: Knapsack problems · Many-objective optimization · Evolutionary algorithms · Decomposition · Dominance

1 Introduction

In recent years, many-objective knapsack problems have attracted a lot of attention in the field of evolutionary optimization [1]. In general, the method developed by Hisao et al. [2] is adopted for the generation of test cases of many-objective knapsack problems. This method constructs many-objective knapsack

K. Li et al. (Eds.): ISICA 2019, CCIS 1205, pp. 155–166, 2020.
https://doi.org/10.1007/978-981-15-5577-0_12

problems on the basis of a bi-objective knapsack problem. The mathematical formula of this bi-objective knapsack problem can be described as follows [3]:

$$maximize\ f(\mathbf{x}) = (f_1(\mathbf{x}), f_2(\mathbf{x})) \tag{1}$$

$$subject\ to \sum_{j=1}^{500} p_{ij}x_j \leq c_i,\ i = 1, 2 \tag{2}$$

$$x_j = 0\ or\ 1,\ j = 1, 2, \ldots, 500 \tag{3}$$

$$where \quad f_i(\mathbf{x}) = \sum_{j=1}^{500} q_{ij}x_j,\ i = 1, 2. \tag{4}$$

In the above formula, $\mathbf{x}$ is a binary vector of length 500. Thus, this bi-objective knapsack problem with 500 items is also called 2-500 knapsack problem for short. In addition, q_{ij} and p_{ij} are, respectively, the profit and weight of item j for knapsack i, and c_i is the capacity of this knapsack. The value of c_i is numerically equal to half of the sum of all weights related to knapsack i. Moreover, the values of q_{ij} and p_{ij} are integers randomly generated within interval $[10, 100]$.

On the basis of the 2-500 knapsack problem, $m - 2$ more objective functions can be constructed as follows [2]:

$$f_i(\mathbf{x}) = \sum_{j=1}^{500} q_{ij}x_j,\ i = 3, 4, \ldots, m, \tag{5}$$

where the variables have the same meanings as described for the 2-500 knapsack problem. Thus an m-objective knapsack problem with 500 items, written as $m - 500$ test case, is generated according to the method developed by Hisao et al. [2] as follows:

$$maximize\ \ f(\mathbf{x}) = (f_1(\mathbf{x}), f_2(\mathbf{x}), \ldots, f_m(\mathbf{x})) \tag{6}$$

$$subject\ to\ \ (2)\ and\ (3). \tag{7}$$

Essentially, many-objective knapsack problems are a kind of multi-objective optimization problems (MOPs). They have not only the basic properties of MOPs but also several other important characteristics. One the one hand, the dimension of decision vectors of many-objective knapsack problems is very high. One the other hand, the number of objectives is relatively large. Furthermore, these problems are also a kind of constrained optimization problems [4] with the natures of the combination blast and discrete variables. These characteristics present some challenges for solving many-objective knapsack problems. Both dominance-based [6,7] and hypervolume-based [5,8] evolutionary algorithms encounter many difficulties when dealing with these many-objective knapsack problems. In contrast, the multi-objective evolutionary algorithm based on decomposition (MOEA/D) is more effective, and its effectiveness has been widely verified for many multi-objective evolutionary algorithms [2,9,10].

However, the original version of MOEA/D still has a few weaknesses when dealing with many-objective knapsack problems. For instance, the original version of MOEA/D will be at risk of diversity degradation when a locally better solution leads to the generation of many copies of itself. In fact, there usually exist a large number of locally optimal solutions for many-objective knapsack problems because of their discrete, non-differentiable and highly non-linear characteristics. This often causes MOEA/D the diversity degradation. In this paper, a collaborative evolutionary algorithm based on decomposition and dominance, referred to as MOEA/D-DDC, is proposed for many-objective knapsack problems. It adopts a decomposition-dominance collaboration mechanism (DDC) to heighten the diversity through the cooperation between the decomposition-based population and dominance-based archive.

The remainder of this paper is organized as follows. Section 2 describes the decomposition-dominance collaboration mechanism. The procedure of MOEA/D-DDC is described in Sect. 3. The experimental results are given in Sect. 4. Finally, Sect. 5 concludes this paper.

2 Decomposition-Dominance Collaboration Mechanism

The DDC mechanism designed in this paper aims to overcome the troubles in the original version of MOEA/D. As shown in Fig. 1, this mechanism consists of two parts. The left subgraph of Fig. 1 represents a decomposition-based population, and the right subgraph indicates a dominance-based archive.

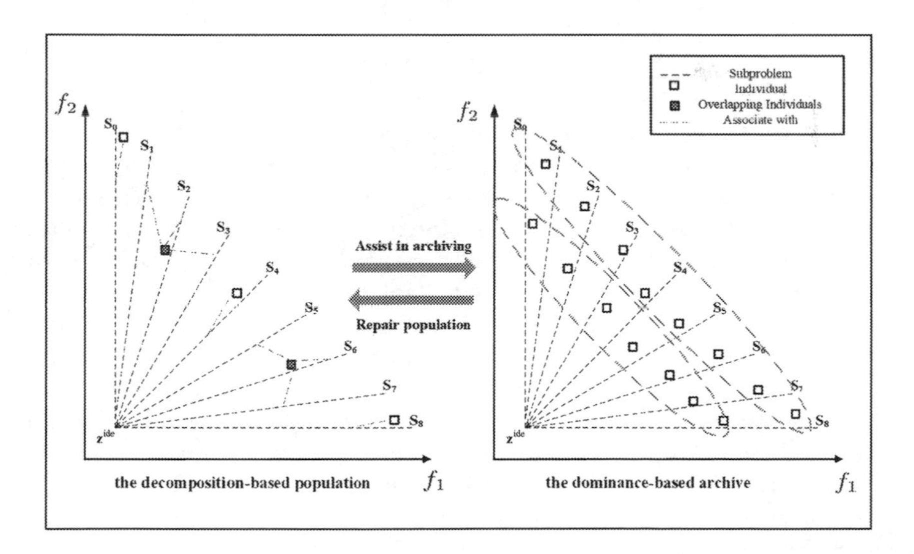

Fig. 1. Decomposition-dominance collaboration mechanism.

The arrow from left to right indicates the generation of the dominance-based archive with the help of the decomposition-based population. It is worth mentioning that the diversity of the decomposition-based population is maintained by a set of uniformly distributed weight vectors. And the individuals in the dominance-based archive are collected from the population. Therefore, the diversity of the archive can be guaranteed to a certain extent. At the same time, the individuals collected in the archive have good convergence and speed up the convergence of the archive. Next, the archive will retain more elite individuals by the use of the non-dominated sorting method and corner solution selection.

The arrow from right to left represents the repair of the decomposition-based population with the help of the dominance-based archive. In the right subgraph of Fig. 1, individuals included in an ellipse are at the same non-dominance level. The non-dominated individuals in the archive are utilized to repair the population. After the repair operation, the same copies of any individual in the population will be removed. Moreover, the trouble of the mismatches between solutions and subproblems has been solved.

2.1 Generation and Update of the Archive

This subsection focuses on the generation and update process of the dominance-based archive. The archive is specially designed for solving many-objective knapsack problems. The processes of generation and update of the archive are under the help of the decomposition-based population and corner solutions. And the corner solutions of a solution set P' are defined in the following formulas [11]:

$$P_1 = \{\mathbf{x} | \mathbf{x} = \arg\min_{\mathbf{x} \in P'} \; dist^{\perp}(\mathbf{f}(\mathbf{x}), \mathbf{e}^i), i = 1, 2, \cdots, m\} \quad (8)$$

$$P_2 = \{\mathbf{x} | \mathbf{x} = \arg\min_{\mathbf{x} \in P'} f_i(\mathbf{x}), i = 1, 2, \cdots, m\} \quad (9)$$

In the above formulas, m is the number of objectives, $dist^{\perp}(\mathbf{v}_1, \mathbf{v}_2)$ indicates the perpendicular distance from the point $\mathbf{v}_1$ to a direction vector $\mathbf{v}_2$ and $\mathbf{e}^i$ represents the unit direction vector along the i-th objective axis. The solutions in P_1 are closest to the coordinate axis in the objective space while each solution in P_2 has the lowest value along each of the coordinate axes. The set of corner solutions, written as P_c, is composed of the solutions in P_1 and P_2 [11–13].

The generation of the archive occurs during the neighbor update procedure. When an individual associated with one neighboring subproblem is inferior than the offspring $\mathbf{y}$ according to the scalar aggregate function such as the weighted sum or Tchebycheff [9], it is replaced by the offspring $\mathbf{y}$. That is, it is discarded by the population P. But if it is not dominated by the offspring $\mathbf{y}$, the individual will be collected into the archive A to avoid the loss of valid information.

The update process of the archive aims to eliminate solutions with poor qualities from the archive A, as shown in Algorithm 1. In line 1, the non-dominated sorting of archive A is performed and the first non-dominated level R_1 is obtained. The ideal point z^* in line 2 consists of the best value for each objective of the archive A. And the nadir point z^{nad} in line 4 is composed of the worst value for each objective of the set of corner solutions, P_c.

Function ABSD-SELECTION in line 6 represents the process of selecting elite individuals according to the sum of the absolute value of difference. It is noting that this function is similar to the procedure of the angle-based selection

Algorithm 1: Update of the archive

Input: A: the current archive; N: the size of archive.
Output: A': the updated archive.
1 $R_1, R_2, ..., R_l$ = NondominatedSorting(A);
2 Update z^*;
3 $P_c \leftarrow$ Choose the corner solutions from R_1;
4 Update z^{nad};
5 **if** $|R_1| > N$ **then**
6 | A' = ABSD-SELECTION(R_1, P_c, z^*, z^{nad});
7 **else**
8 | $A' = R_1$;
9 **end**
10 **return** A';

[12]. The main difference is that this function is based on the sum of the absolute difference. The difference along each coordinate axis between two individuals in the objective space is considered to calculate the sum of absolute difference as follows:

$$SumAbs(\mathbf{x}^p, \mathbf{x}^q) = \sum_{i=1}^{m} |f_i(\mathbf{x}^p) - f_i(\mathbf{x}^q)| \tag{10}$$

where $\mathbf{x}^p$ and $\mathbf{x}^q$ represent two different individuals and m is the number of objectives.

2.2 Repair of the Population

In the DDC mechanism, the population needs to be repaired by the archive. The process of repairing the population consists of the following two strategies: a redundancy removal (RR) strategy and a re-association (RA) strategy.

As discussed previously, there are often multiple same copies of one individuals, in the population of MOEA/D. The existence of these copies will lead to the degradation of the diversity of the population. In fact, individuals who are identical in all objectives are called redundant individuals. The RR strategy aims to remove these redundant individuals and maintain the diversity of the population. Since the elite individuals in the archive may be more suitable for certain subproblems, it is necessary to utilize them to re-associate the subproblems for the population. The RA strategy needs to mix the population with the archive. It is worth reminding that the RA strategy is different from the association strategy of MOEA/D-AGR [10]. Specifically, this strategy finds the most suitable individual for each subproblem from the mixed population, while MOEA/D-AGR finds the most suitable subproblem for each offspring individual.

3 MOEA/D-DDC

3.1 Basic Framework

The proposed algorithm, MOEA/D-DDC, adopts the above DDC mechanism. Algorithm 2 presents the basic framework of MOEA/D-DDC. First, the initialization is preformed from line 1 to 4. Next, the current number t of generation of the population is initialized to 0 in line 5. In line 7, the archive, A, is initialized by the non-dominated solutions of P. The main loop of the algorithm starts on line 8 and ends on line 18. In line 19, P represents the final population by combining the population and archive. Finally, it returns and output the final population P. The implementation detail of each component in MOEA/D-DDC will be explained in the following subsections.

Algorithm 2: MOEA/D-DDC Framework

Input: T: the neighborhood size; m: the number of objectives; H_1, H_2: the parameters for generating weight vectors.
Output: P: The population.
1 Generate the uniformly distributed weight vectors [14,15], $W = \{w^1, \cdots, w^N\}$;
2 Initialize T neighbor indexes $B^i = \{j_1, j_2, \cdots, j_T\}$ for each weight vector $w^i \in W$, where $w^{j_1}, w^{j_2}, \cdots, w^{j_T}$ are the T closest reference directions to w^i;
3 Create N initial solutions $P = \{\mathbf{y}^1, \mathbf{y}^2, \cdots, \mathbf{y}^N\}$ by uniformly randomly sampling from the decision space Ω and then evaluate them;
4 Initialize the ideal point $z^{ide} = (z_1^{ide}, z_2^{ide}, \cdots, z_m^{ide})$ where $z_j^{ide} = \max_{\mathbf{y} \in P} f_j(\mathbf{y})$;
5 $t = 0$;
6 $R_1, R_2, ..., R_l$ = NondominatedSorting(P);
7 $A = R_1$ //the archive;
8 **while** $t < t_{max}$ **do**
9   **for** $i = 1$ *to* N **do**
10   A pair of individuals, *pp*, are randomly selected as parents from the neighbors of the current subproblem S_i;
11         $\mathbf{y} \leftarrow$ **reproduction**(*pp*);
12   Apply $\mathbf{y}$ to update the current ideal point $\mathbf{z}^{ide}$;
13   P,A $\leftarrow$ **updateNeighbors**($\mathbf{y}$,$\mathbf{z}^{ide}$,P,A,B^i,w^i);
14   **if** t % 10 == 0 **then**
15   P,A $\leftarrow$ **repairPopulation**(P,A);
16   **end**
17   **end**
18 **end**
19 $P \leftarrow$ **combiningPopulations**(P,A);
20 **return** P;

3.2 Reproduction and Greedy Repair

In this stage, an offspring $\mathbf{y}$ is generated by the uniform crossover operator and bit-flip mutation operator and then it is evaluated. For each gene c_i, the procedure of generation is to randomly take a probability p from the interval [0, 1]. However, the offspring obtained by the crossover operator and mutation operator may be an infeasible solution. Thus the method for repairing the infeasible solution in MOEA/D-DDC refers to the greedy repair method [3]. All the items are first sorted in terms of the ratio r_j between profit and weight, as shown in formula 11. Then the corresponding item with the lowest ratio is deleted until the constraint condition is satisfied.

$$r_j = max\{q_{ij}/p_{ij}|i = 1, 2\}, j = 1, 2, ..., 500. \tag{11}$$

3.3 Combining Population and Archive

In this procedure, P and A are first mixed to obtain a temporary population M. Then, the individuals in M are sorted to obtain the first non-dominated level. Next the redundant individuals are removed to obtain another temporary population M'. Population P' further collects the intersection from population M' and population P. In fact, the individuals in population P' are the non-dominated solutions in population P without redundant individuals. Similarly, the individuals in population A' are the non-dominated solutions in population A without redundant individuals. If the size of P' is less than the size of P, the remaining solutions are selected from A', and then MOEA/D-DDC returns population P' with these selected solutions.

4 Numerical Experiments

4.1 Performance Metric

To verify the performance of the proposed algorithm, MOEA/D-DDC, three many-objective knapsack problems, 4-500 test case, 6-500 test case, and 8-500 test case, are chosen in our experiment. The hardware environment is composed of 2 CPUs (Intel Xeon CPU E5-2620 v4 2.10 GHz) and 128 GB memory. All algorithms are implemented in the Java language and performed on a 64-bit Linux OS.

A widely used indicator, hypervolume, was adopted to assess the performances of the algorithms in our experiments. And the indicator can be calculated by Eq. 12:

$$HV(S) = VOL(\bigcup_{\mathbf{x}\in S} [f_1(\mathbf{x}), z_1^r] \times \cdots \times [f_m(\mathbf{x}), z_m^r]) \tag{12}$$

where $VOL(\cdot)$ represents the Lebesgue measure, m represents the number of objectives, and z^r is the reference point defined in the objective space.

Table 1. The relative hypervolume results (best, mean, worst) obtained by the algorithms for all test cases.

m	MOEA/D-DDC:W	MOEA/D-DDC:T	MOEA/D:W	MOEA/D:T	NSGA-II
4	$1.01494_{(2)}$	$\mathbf{1.02072}_{(1)}$	$1.00695_{(3)}$	$1.00354_{(4)}$	$0.87624_{(5)}$
	$1.00774_{(2)}$	$\mathbf{1.01304}_{(1)}$	$1.00000_{(3)}$	$0.99529_{(4)}$	$0.86029_{(5)}$
	$0.99641_{(2)}$	$\mathbf{1.00680}_{(1)}$	$0.99360_{(3)}$	$0.98805_{(4)}$	$0.84798_{(5)}$
6	$\mathbf{1.01560}_{(1)}$	$0.98398_{(3)}$	$1.00730_{(2)}$	$0.95360_{(4)}$	$0.79742_{(5)}$
	$\mathbf{1.00817}_{(1)}$	$0.97565_{(3)}$	$1.00000_{(2)}$	$0.94146_{(4)}$	$0.78241_{(5)}$
	$\mathbf{0.99705}_{(1)}$	$0.96279_{(3)}$	$0.99107_{(2)}$	$0.93013_{(4)}$	$0.76635_{(5)}$
8	$\mathbf{1.03141}_{(1)}$	$0.95627_{(3)}$	$1.01162_{(2)}$	$0.94719_{(4)}$	$0.75450_{(5)}$
	$\mathbf{1.01447}_{(1)}$	$0.93692_{(3)}$	$1.00000_{(2)}$	$0.92884_{(4)}$	$0.72338_{(5)}$
	$\mathbf{1.00053}_{(1)}$	$0.92211_{(3)}$	$0.98376_{(2)}$	$0.91461_{(4)}$	$0.69979_{(5)}$

4.2 Parameter Settings

In our experiment, two algorithms, NSGA-II and MOEA/D, are chosen as comparison algorithms of MOEA/D-DDC for solving many-objective knapsack problems. It is worth noting that MOEA/D-DDC is an improved version of MOEA/D for many-objective knapsack problems. Thus the comparison between MOEA/D and MOEA/D-DDC is primary in our experiments. And it is also meaningful to compare MOEA/D-DDC with NSGA-II since the DDC mechanism is designed by taking advantage of Pareto dominance. The general experimental parameter settings are summarized below:

1. The number of evaluations: 400 000
2. The crossover probability: 0.8
3. The mutation probability: 0.004
4. The 30 independent runs of each algorithm on each test case were conducted.
5. The neighborhood size for MOEA/D and MOEA/D-DDC: 10
6. The scalar aggregate functions for MOEA/D and MOEA/D-DDC: weighted sum and Tchebycheff
7. The reference point: $1.1z^*$
8. The numbers of divisions for the boundary and inside layers of weight vectors are set to (H_1, H_2) = (7, 0), (4, 0), and (3, 2), respectively, for the 4-500, 6-500, and 8-500 test cases.
9. The population size is set to 120, 126, and 156, respectively, for the 4-500, 6-500, and 8-500 test cases.

4.3 Experimental Results

The experimental results of relative hypervolume value obtained by all algorithms for all test cases are listed in Table 1 with the best value, the mean value, and the worst value. In Table 1, the origin is used as a reference point for the calculation of hypervolume. Then, the relative hypervolume of each test case is calculated based on the mean value of MOEA/D:W among the results of 30 runs. MOEA/D:W and MOEA/D:T represent two versions of MOEA/D, respectively, with the weighted sum function and the Tchebycheff function. The same is true for MOEA/D-DDC:W and MOEA/D-DDC:T. In addition, all values obtained by the algorithms are ordered and the rank of each value is represented as a subscript.

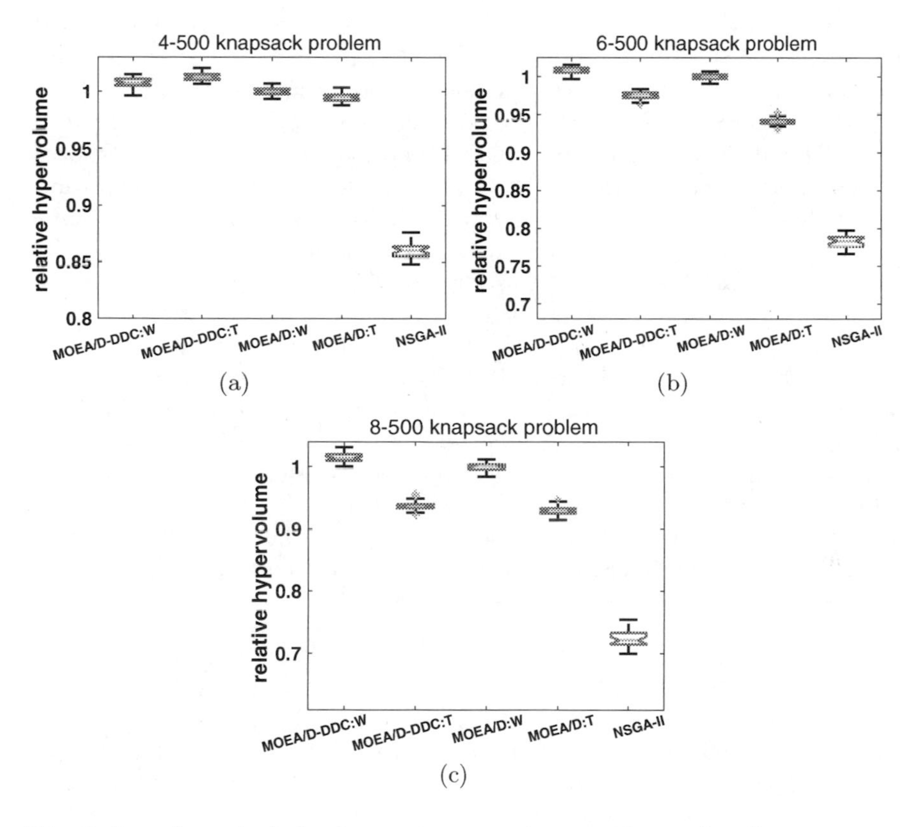

Fig. 2. Box plots of relative hypervolume results over 30 runs for all test cases.

Table 1 shows that, in all test cases, the performance of MOEA/D-DDC is better than that of MOEA/D, no matter in terms of the weighted sum function and the Tchebycheff function. Even for some test cases, the performances of both MOEA/D-DDC:W and MOEA/DDC:T can surpass the highest performance of MOEA/D using the weighted sum function or Tchebycheff function. For example, in the 4-500 test cases, MOEA/D-DDC:T and MOEA/D-DDC:W respectively achieve the first- and second-best performances. Moreover, it can be observed that, on the higher-dimensional many-objective knapsack problems such as the 6-500 and 8-500 test cases, the weighted sum function is usually better than the Tchebycheff function for both MOEA/D and MOEA/D-DDC. This may be that the Pareto fronts are widely distributed and convex in the higher-dimensional objectives spaces, and the weighted sum function is suitable for the cases with such characteristics.

In addition, it can be shown that as the number of objectives increases, the performance gap between MOEA/D-DDC and the other algorithms is gradually opened. In particular, the dominance-based evolutionary algorithm, NSGA-II, is getting worse and worse. This fully demonstrates the superiority of decomposition-based evolutionary algorithms for solving many-objective knapsack problems, while dominance-based evolutionary algorithms face poor selection pressure in high-dimensional objective spaces and lead to poor performances.

On the whole, MOEA/D-DDC can effectively improve the performance of MOEA/D and the degree of improvement depends on the test case. For example, the improvement for the 8-500 test case is significant. It may be owing to the fact that the number of redundant individuals increases in the higher dimensional objective space. This greatly affects the performance of MOEA/D and the effectiveness of DDC is more obviously reflected.

Further, the performance of each algorithm is more intuitively presented through box plots. Figure 2 shows the box plots of relative hypervolume results obtained by each algorithm for all test cases. It can be concluded from Fig. 2 that most algorithms are relatively stable for all test cases, and MOEA/D-DDC achieves the best results. For the 4-500 test case, MOEA/D-DDC with the Tchebycheff function behaves the best, while for the other test cases, MOEA/D-DDC with the weighted sum function performs the best. This also reveals that the performances of algorithms are closely related to the scalar aggregate functions and the DDC mechanism can usually improve the performance of MOEA/D effectively for many-objective knapsack problems.

5 Conclusion

In this paper, MOEA/D-DDC is proposed to solve many-objective knapsack problems efficiently. The performance of the MOEA/D-DDC benefits from the design of the DDC mechanism. As discussed previously, the original version of MOEA/D has some disadvantages. On the one hand, it may often lose some excellent individuals that are not suitable for the current subproblems. On the other hand, the neighbor update procedure in MOEA/D may lead to multiple redundant individuals in the population. These disadvantages might hamper the diversity and convergence of the population.

However, MOEA/D-DDC can retain elite individuals to avoid the degradation of population diversity and convergence. First, the population can help the archive to collect elite individuals. Second, the archive assists the population to remove redundant individuals. The re-association strategy in the DDC mechanism utilizes the elite individuals in the archive to help the population to find the most suitable individual for each subproblem. The DDC mechanism takes advantage of the collaborative relationship between the decomposition-based population and dominance-based archive. As a result, it can play a key role in MOEA/D-DDC for many-objective knapsack problems.

In the future researches, we will expand the MOEA/D-DDC algorithm to other complex many-objective optimization problems to further verify the effect of the decomposition-dominance collaboration mechanism.

Acknowledgments. This work was supported in part by the Natural Science Foundation of Guangdong Province, China, under Grant 2015A030313204, in part by the Pearl River S&T Nova Program of Guangzhou under Grant 2014J2200052, in part by the National Natural Science Foundation of China under Grant 61203310, Grant 61503087 and Grant 61702239, in part by the Fundamental Research Funds for the Central Universities, SCUT, under Grant 2017MS043, and the Project of Foshan Science and Technology Innovation Foundation under Grant 2016AG100291.

References

1. Ishibuchi, H., Kaige, S., Narukawa, K.: Comparison between Lamarckian and Baldwinian repair on multiobjective 0/1 knapsack problems. In: Coello Coello, C.A., Hernández Aguirre, A., Zitzler, E. (eds.) EMO 2005. LNCS, vol. 3410, pp. 370–385. Springer, Heidelberg (2005). https://doi.org/10.1007/978-3-540-31880-4_26
2. Ishibuchi, H., Akedo, N., Nojima, Y.: Behavior of multiobjective evolutionary algorithms on many-objective knapsack problems. IEEE Trans. Evol. Comput. **19**(2), 264–283 (2015)
3. Zitzler, E., Thiele, L.: Multiobjective evolutionary algorithms: a comparative case study and the strength Pareto approach. IEEE Trans. Evol. Comput. **3**(4), 257–271 (1999)
4. Wang, Y., Cai, Z.: A constrained optimization evolutionary algorithm based on multiobjective optimization techniques. In: IEEE Congress on Evolutionary Computation, pp. 1081–1087. IEEE (2005)
5. Bader, J., Zitzler, E.: HypE: an algorithm for fast hypervolume-based many-objective optimization. Evol. Comput. **19**(1), 45–76 (2011)
6. Köppen, M., Yoshida, K.: Substitute distance assignments in NSGA-II for handling many-objective optimization problems. In: Obayashi, S., Deb, K., Poloni, C., Hiroyasu, T., Murata, T. (eds.) EMO 2007. LNCS, vol. 4403, pp. 727–741. Springer, Heidelberg (2007). https://doi.org/10.1007/978-3-540-70928-2_55
7. Deb, K., Pratap, A., Agarwal, S., Meyarivan, T.: A fast and elitist multiobjective genetic algorithm: NSGA-II. IEEE Trans. Evol. Comput. **6**(2), 182–197 (2002)
8. Beumea, N., Emmerich, M.: SMS-EMOA: multiobjective selection based on dominated hypervolume. Eur. J. Oper. Res. **181**(3), 1653–1669 (2007)
9. Zhang, Q., Hui, L.: MOEA/D: a multiobjective evolutionary algorithm based on decomposition. IEEE Trans. Evol. Comput. **11**(6), 712–731 (2007)
10. Wang, Z., Zhang, Q., Zhou, A., Gong, M., Jiao, L.: Adaptive replacement strategies for MOEA/D. IEEE Trans. Cybern. **46**(2), 474–486 (2017)
11. Singh, H.K., Isaacs, A., Ray, T.: A Pareto corner search evolutionary algorithm and dimensionality reduction in many-objective optimization problems. IEEE Trans. Evol. Comput. **15**(4), 539–556 (2011)
12. Cai, X., Yang, Z., Fan, Z., Zhang, Q.: Decomposition-based-sorting and angle-based-selection for evolutionary multiobjective and many-objective optimization. IEEE Trans. Cybern. **47**(9), 2824C–2837 (2017)

13. Freire, H., de Moura Oliveira, P.B., Solteiro Pires, E.J., Bessa, M.: Many-objective optimization with corner-based search. Memetic Comput. **7**(2), 105–118 (2015). https://doi.org/10.1007/s12293-015-0151-4
14. Das, I., Dennis, J.E.: Normal-boundary intersection: a new method for generating the Pareto surface in nonlinear multicriteria optimization problems. SIAM J. Optim. **8**(3), 631–657 (1998)
15. Li, K., Deb, K., Zhang, Q., Kwong, S.: An evolutionary many-objective optimization algorithm based on dominance and decomposition. IEEE Trans. Evol. Comput. **19**(5), 694–716 (2014)

A Many-Objective Algorithm with Threshold Elite Selection Strategy

Shaojin Geng, Di Wu, Penghong Wang, and Xingjuan Cai(✉)

Complex System and Computational Intelligence Laboratory,
Taiyuan University of Science and Technology, Taiyuan, China
xingjuancai@163.com

Abstract. The study of many-objective evolutionary algorithm (MaOEA) has become particularly important, especially with the increasing complex engineering optimization problems. Considering that the convergence and diversity of the population are two important indicators to measure the performance of the algorithm, a many-objective evolutionary algorithm with threshold elite selection strategy (MaOEA-TES) are proposed in this paper. The algorithm adopts the balanceable fitness estimation strategy and the reference-point based non-dominated sorting strategy to balance the convergence and diversity of the solution. An adaptive penalty distance boundary intersection strategy is designed to dynamically adjust the impact of convergence and diversity on the algorithm. In addition, a dynamic threshold selection strategy is proposed to ensure that the algorithm emphasizes diversity at an early stage, emphasizes convergence at a later stage, and ensures that the result is closer to the real non-dominant front. The DTLZ test suite is used to evaluate the performance of MaOEA-TES. The experimental results show that the MaOEA-TES has the best performance comparing with three other state-of-the-art algorithms on many-objective optimization.

Keywords: Many-objective optimization · Balanceable fitness estimation · Reference points · Elite selection strategy

1 Introduction

There are many optimization problems in the real society, and these problems are often composed of multiple objectives that conflict and affect each other, which are called multi-objective optimization Problems (MOPs) [1]. Since the early 1960s, MOPs have attracted more and more researchers from different backgrounds [2,3], and it has very important scientific and practical significance to solve multi-objective optimization problems. However, with the continuous development of society, many engineering problems have become more and more complicated, and the mathematical model is no longer a simple multi-objective optimization model. The many-objective optimization problems (MaOPs) [4] (when the objectives of optimization reach four or more) have been appeared in

K. Li et al. (Eds.): ISICA 2019, CCIS 1205, pp. 167–179, 2020.
https://doi.org/10.1007/978-981-15-5577-0_13

real life, such as wing design problem [5], water distribution system [6] and car engine calibration problem [7].

Whether in scientific research [8] or engineering applications [9], optimization issues generally involve multiple conflicting objective functions. Therefore, there are a set of optimal solution sets, which composed of numerous Pareto optimal solutions in MaOPs. In recent years, many-objectives evolutionary algorithms (MaOEAs) [10] have been researched by many scholars for MaOPs. And these evolutionary algorithms can be divided into the following three categories.

(1) Many-objective evolutionary algorithm based on pareto-dominance. For example, NSGA-III [11], NSGAIII-NE [12], etc. The disadvantage of this approach is that there are many parameters in the algorithm, which requires adjust heuristically.
(2) Many-objective evolutionary algorithm based on decomposition [13]. The main idea is that the algorithm decomposes the complex MaOPs into a series of sub-problems, and then solves them one by one. This method effectively overcomes the diversity maintenance difficulties, but it is still in its infancy.
(3) Many-objective evolutionary algorithm based on performance indicators [14]. For example, a density estimation strategy that uses a simple coordinate transformation to put the solution with poor convergence into crowded area.

Reference-point based non-dominated sorting strategy uses widely distributed reference points (one reference direction can be associated with multiple solutions) to maintain diversity. However, as the number of objective increases, the pareto-dominance makes the selection pressure of the strategy insufficient. Therefore, in order to effectively improve the performance of the algorithm, this paper combines the balanceable fitness estimation strategy and the adaptive penalty distance boundary intersection strategy to increase the selection pressure. And on this basis, a dynamic threshold selection strategy is proposed. The specific principle is as follows:

1) Balanceable fitness estimation strategy is used to balance the solution of convergence and diversity.
2) Reference-point based non-dominated sorting strategy is used to achieve the goal of distributing the non-dominated solution in the objective space as uniform as possible.
3) Adaptive penalty distance boundary intersection strategy is designed to dynamically adjust the impact of convergence and diversity on the algorithm.
4) Dynamic threshold selection strategy is proposed to ensure that the algorithm emphasizes diversity in the early stage and convergence in the late stage to make sure that the results are closer to the real non-dominant front.

The structural of this paper is organized as follows: Sect. 2 introduces related works about balanceable fitness estimation and reference-point based non-dominated sorting strategy. Section 3 describes the proposed algorithm in detail.

The simulation of several algorithms has been experimented in Sect. 4. Section 5 gives the conclusion of this paper.

2 Related Works

This section describes the balanceable fitness estimation strategy and the reference-point based non-dominated sorting strategy, which are important components of MaOEA-TES.

2.1 Balanceable Fitness Estimation Strategy (BFE)

Balanceable fitness estimation (BFE) [15] is a strategy which combined the diversity and convergence distance to balance the convergence and diversity for each solution in objective space.

Suppose that the population $P = \{p_1, p_2, ..., p_N\}$ includes N individuals. Each individual has the position X_i. For each individual p_i, the value of BFE $BFE(p_i, P)$ is consisted of two components: the diversity distance and the convergence distance, the equation is as follows:

$$BFE(p_i, P) = \alpha \times Cd(p_i, P) + \beta \times Cv(p_i, P) \tag{1}$$

where $Cd(p_i, P)$ and $Cv(p_i, P)$ represent the diversity and convergence distances of p_i, respectively. Both α and β are the weight factors, when calculating the value of BFE, each objective of population p_i will be normalized firstly by using the maximum and minimum values of the corresponding objective. This normalization approach helps to eliminate the impact of different amplitudes on multiple objectives. The normalized objective $f_k'(p_i)$ of p_i is obtained with the following equation:

$$f_k'(p_i) = \frac{f_k(p_i) - f_k min}{f_k max - f_k min} \tag{2}$$

where $f_k max$ and $f_k min$ are the maximum and minimum values of the $k-th$ objective obtained from the non-dominated solutions available in the external archive, respectively. And the objective $f_k'(p_i)$ is normalized to $[0, 1]$. Then, the normalized diversity distance $Cd(p_i, P)$ is showed as follows:

$$Cd(p_i, P) = \frac{SDE(p_i) - SDE_{\min}}{SDE_{\max} - SDE_{\min}} \tag{3}$$

where $SDE_{\max}$ and $SDE_{\min}$ are the maximum and minimum SDE distances in the population, respectively. $SDE(p_i)$ is the original SDE distance defined in [4], which uses the shifted euclidean distance to the nearest neighbor, the equation is showed as follow:

$$SDE(p_i) = \min_{p_j \in P, j \neq i} \sqrt{\sum_{k=1}^{m} sde(f_k'(p_i), f_k'(p_j))^2} \tag{4}$$

$$sde(f'_k(p_i), f'_k(p_j)) = \begin{cases} f'_k(p_j) - f'_k(p_i) & \text{if } f'_k(p_j) > f'_k(p_i) \\ 0 & otherwise \end{cases} \tag{5}$$

And the convergence distance $Cv(p_i, P)$ is used to reflect the convergence ability of $f'_k(p_i)$ ($k = 1, 2, ..., m$) with respect to the ideal point z^*. The equation is calculated as follows:

$$Cv(p_i, P) = 1 - \frac{dis(p_i)}{\sqrt{m}} \tag{6}$$

where $dis(p_i)$ denotes the euclidean distance from $f'_k(p_i)$($k = 1, 2, ..., m$) to the ideal point z^*. It can be computed as follows:

$$dis(p_i) = sqrt(\sum_{k=1}^{m} f'_k(p_i)^2) \tag{7}$$

The larger the value of $Cd(p_i, P)$, the further away the neighborhood is from p_i. The larger the value of $Cv(p_i, P)$, the closer the distance between the $f'_k(p_i)$($k = 1, 2, ..., m$) and the ideal point z^*. To minimize all the objectives, individuals with larger convergence distances should be prioritized to increase selection pressure, and selected individuals should move toward ideal points when external archive are updated. At the same time, in order to balance the diversity distance and the convergence distance, the two weight factors α and β can be used to adjust the individual weights adaptively on the basis of their original diversity distance and convergence distance.

2.2 Reference-Point Based Non-dominated Sorting Strategy (RNS)

In reference-point based non-dominated sorting strategy, the diversity maintenance is achieved by initializing a set of reference points [12]. Supposed that the initial population is P_t and the size is N ($N \approx H$), the offspring population is Q_t, and the combined population $S_t = P_t \cup Q_t$. Then, the population S_t selects individuals of different non-domination levels $(F_1, F_2, ..., F_l, ...)$ at the same time, the termination condition is that the size of S_t is equal to or larger than N. Suppose that the current level is the $l-th$ level. Select F_{l-1} individuals from S_t and put them into the next generation population P_{t+1}. The rest individuals $N - number(S_t(F_{l-1}))$ are chosen from F_l by the reference points strategy. The reference points strategy needs to normalize the objective function values. Also, the ideal point of the population z^* is defined as $(0, ..., 0)$ and reference points just lie on this normalized hyper-plane. Then, the perpendicular distance between each individual in S_t with reference line is calculated and the individual of minimum distance belongs to the niche corresponding to the reference point. The ones associated with the reference points whose niche counts are small have better chances to be selected. The procedure is presented in Algorithm 1.

3 The Proposed Algorithm

In this section, the detailed process of the proposed MaOEA-TES algorithm is described in Sect. 3.1. And then, designed strategies are given in Sect. 3.2, 3.3.

Algorithm 1. Selection Operator

1: Input:H structured reference points, Initial population P_t
2: Output: P_{t+1}
3: **While (stop criterion is met)**
4: Q_t= Crossover (P_t) + Mutation (P_t)
5: $S_t = P_t \cup Q_t$
6: $(F1, F2, \cdots)$=Non-dominated sort (S_t)
7: $S_t = S_t \cup F_i$
8: **Until** $|S_t| \geq N$
9: Last front to be included:$F_l = F_i$
10: **If** $|S_t| = N$, then
11: $P_{t+1} = S_t$, break
12: **else**
13: $K = N - number(S_t(F_{l-1}))$
14: Normalize objective function values
15: Choose K members from F_l
16: According to the reference point to construct P_{t+1}
17: **End If**
18: **End**

3.1 Many-Objective Evolutionary Algorithm Based on Threshold Elite Selection Strategy (MaOEA-TES)

In this section, a many-objective evolutionary algorithm with threshold elite selection strategy (MaOEA-TES) is proposed. A reference-point based non-dominated sorting strategy is used to select good individuals. The non-dominated sorting strategy can be layered according to the level of individual non-dominated solutions, and the pareto optimal solution can be quickly searched. Furthermore, based on the reference point and adaptive penalty distance boundary intersection strategy, individuals with uniform distribution in the objective space can be selected to enhance the diversity of the population, and the convergence information can be combined to ensure convergence and distribution of the population. Furthermore, the dynamic threshold selection strategy is adopted to ensure that the BFE method can replace the general environment selection mechanism in the later stage of the algorithm, so that the algorithm can balance the convergence and distribution well in the whole group evolution process. The pseudo code is shown in Algorithm 2.

3.2 Adaptive Penalty Distance Boundary Intersection Strategy (APDBI)

As shown in Fig. 1, it can be seen that the perpendicular distance between the individual and the reference line is d_2 in the non-dominated sorting strategy based on the reference point. However, d_2 represents the diversity of individuals and does not balance well with the relationship between convergence and diversity.

Algorithm 2. The framework of MaOEA-TES

1: Input:H structured reference points, Initial population P_t and calculate the function minimum $Zmin$
2: Output: P_{t+1}
3: $t = 0$
4: **While (stop criterion is met)**
5: $t = t + 1$
6: $rhm = rand$
7: **If** $rhm > t/100$
8: Q_t= Crossover (P_t) + Mutation (P_t)
9: $S_t = P_t \cup Q_t$
10: $(F1, F2, \cdots)$=Non-dominated sort (S_t)
11: $S_t = S_t \cup F_i$
12: **Until** $|S_t| \geq N$
13: Last front to be included:$F_l = F_i$
14: **If** $|S_t| = N$, then
15: $P_{t+1} = S_t$, break
16: **else**
17: $K = N - number(S_t(F_{l-1}))$
18: Normalize objective function values
19: Choose K members from F_l
20: Calculate the adaptive penalty distance of boundary intersection between each individuals and references point to construct P_{t+1}
21: **End If**
22: **Else**
23: $O_t = BFE(P_{t+1}, Population)$
24: $R_t = Crossover(O_t) + Mutation(O_t)$
25: $Population = BFE(O_t, R_t)$
26: According to the reference point to construct P_{t+1}
27: **End If**
28: **End**

Therefore, $d(x) = d_{j,1}(x) + \theta \times d_{j,2}(x)$ is used to replace d_2, and then the value of θ can dynamically balance adaptive penalty distance of boundary intersect with the number of iterations increases, which makes the solution closer to PF.

And the equations are showed as follows:

$$d(x) = d_{j,1}(x) + \theta \times d_{j,2}(x) \tag{8}$$

$$d_{j,1}(x) = \frac{\left\| (w - F(x))^T \sigma \right\|}{\|\sigma\|} \tag{9}$$

$$d_{j,2}(x) = \|f(x) - (w - d_1 \sigma)\| \tag{10}$$

$$\theta = \sigma \cdot e^{M-1 \cdot \frac{gen+1}{maxgen} \cdot \|w^j\|} \tag{11}$$

where w represents the ideal point, M is the number of objective, gen represents the current iterations number, and the maximum iterations number is $maxgen$,

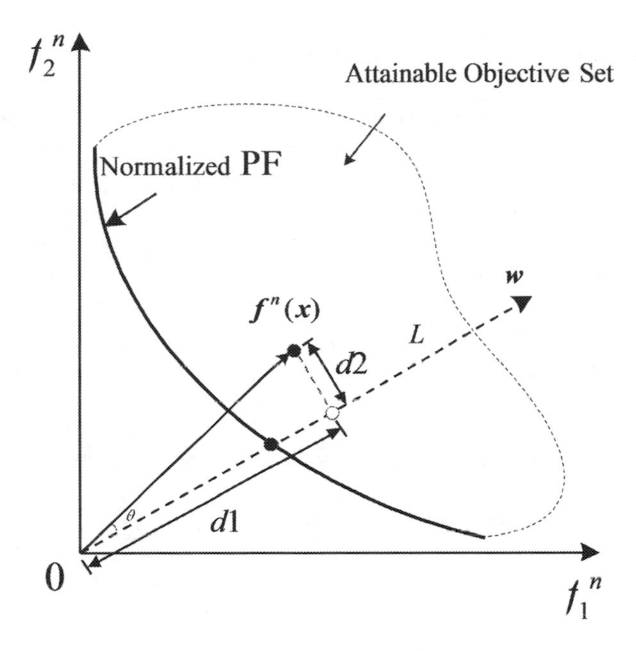

Fig. 1. The description of APDBI

σ is the reference point and θ is the adaptively defined penalty parameter. Also, $d_{j,1}(x)$ represents the distance between the point that the individual maps to the reference line and the ideal point, $d_{j,2}(x)$ is the distance between the individual and reference line.

3.3 Dynamic Threshold Selection Strategy (DTS)

Because of the different proportions of convergence and diversity in the different evolution stages of the algorithm, dynamic threshold selection strategy is adopted to balance them in the evolution process dynamically. And a threshold is set to select different strategy in the algorithm. When the condition is met, the algorithm uses the reference point strategy to select the offspring, otherwise balanceable fitness estimation strategy is used to select offspring. The equation for this threshold is as follows:

$$Thre = \frac{t+1}{MaxIt} \tag{12}$$

where t represents the current generation, $MaxIt$ represents the maximum generation, and the threshold $Thre$ increases with the number of iterations. A random number $rand$ within $[0, 1]$ is generated. If $rand$ is larger than the threshold, the reference-point based non-dominated sorting strategy is selected, otherwise, the balanceable fitness estimation strategy is selected.

4 Experimental Results and Analysis

This section will evaluate the performance of MaOEA-TES and discuss results. Firstly, we will describe the test problems DTLZ [16]. Meanwhile, the performance indicator is used in the experiment. Then we will introduce three most advanced algorithms for comparison and corresponding parameter design. Finally, the experimental results are discussed. The simulation results of MaOEA-TES on three to fifteen objective optimization problems are provided. Seven problems are used as test problems in DTLZ. DTLZ problems are non-convex, multi-modal, non-connected and non-uniform Pareto front. These benchmark issues are challenging to evaluate the performance of MaOEAs.

Table 1. Population size corresponding to different objective numbers

Population size	Number of objectives
91	3
210	5
156	8
275	10
135	15

In the evaluation criterion, the inverse intergenerational distance (IGD) [17] is chose as the performance indicator to evaluate the quality of the solution set, as a separate measure, which can provide a combination of information about the convergence and diversity of the obtained solutions. By calculating the average Euclidean distance and standard deviation between the Pareto optimal solution set and the obtained optimal solution set (in parentheses), the formulas are as follows:

$$IGD(A, Z) = \frac{1}{|Z|} \sum_{i=1}^{|z|} \min_{j=1}^{|A|} d\left(z_i, a_j\right) \tag{13}$$

where $d\left(z_i, a_j\right) = \|z_i - a_i\|_2$. The IGD value is smaller, the solution set obtained by the algorithm is closer to the PF. If the IGD value is large, it proves that no solution related to the reference point has been found.

For each algorithm, the number of iterations is 10000 generations, running 30 times independently, and the best, median and worst IGD performance values are reported. For all algorithms, performance indicators are computed using the final solution set. Table 1 shows the different population sizes for different objective sizes.

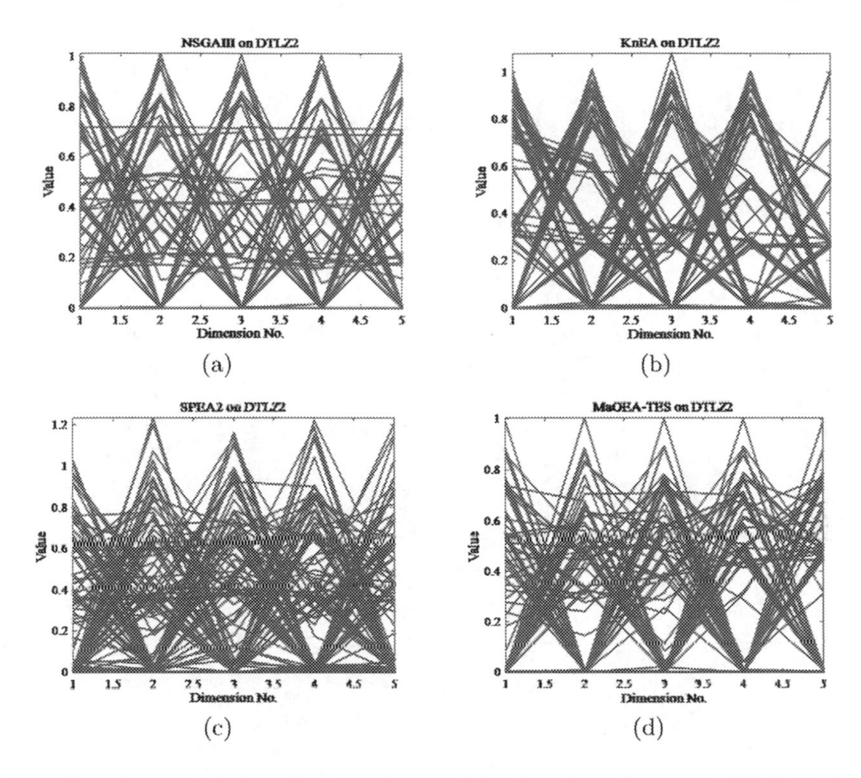

Fig. 2. Coordinates of the solutions obtained by five algorithms on DTLZ2 with five objectives. (a) NSGAIII (b) KnEA (c) SPEA2 (d) MaOEA-TES

It can be seen from Table 2 that the average value of the algorithm MaOEA-TES is better than other algorithms. The feasibility and accuracy of the algorithm are verified. It can be seen from DTLZ1 that MaOEA-TES is significantly better than other algorithms. For DTLZ2, the reason of the proposed is slightly worse than KnEA is that KnEA adopts the inflection point strategy which is suitable for solving the concave problem. And for DTLZ3, MaOEA-TES receives the best performance on 8, 10, 15 objectives, which is caused by the balanceable fitness estimation strategy. At the same time, by comparing the IGD value and standard deviation (in parentheses) on DTLZ4-7, it can be concluded that the MaOEA-TES algorithm has a significant role in promoting population convergence and diversity. Observing from the pareto front of the algorithm, and the results are shown in Fig. 2. It can be seen that the solution set of MaOEA-TES is at the center of the pareto, which can maintain a better diversity.

Table 2. IGD comparisons the four algorithms on DTLZ test suite

Problem	N	M	D	NSGA-III	KnEA	SPEA2	MaOEATESS
DTLZ1	91	3	7	2.3058×10^{-1} $(2.23 \times 10^{-1})-$	1.5470×10^{-1} $(1.83 \times 10^{-1})=$	1.5804×10^{-1} $(1.80 \times 10^{-1})=$	1.2757×10^{-1} (1.43×10^{-1})
	210	5	9	1.6254×100 $(4.81 \times 10^{-1})-$	7.0935×10^{-1} $(3.22 \times 10^{-1})=$	1.5155×10^{0} $(7.72 \times 10^{-1})-$	8.6358×10^{-1} (3.43×10^{-1})
	156	8	12	2.0535×10^{0} $(8.89 \times 10^{-1})-$	3.9979×10^{0} $(2.04 \times 10^{0})-$	9.5366×10^{1} $(2.97 \times 10^{1})-$	6.0595×10^{-1} (3.14×10^{-1})
	275	10	14	4.2252×10^{0} $(1.77 \times 10^{0})-$	2.6308×10^{0} $(1.30 \times 10^{0})-$	9.7740×10^{1} $(2.68 \times 10^{1})-$	1.8022×10^{0} (1.11×10^{0})
	135	15	19	1.5061×10^{0} $(8.52 \times 10^{-1})-$	7.2083×10^{0} $(6.51 \times 10^{0})-$	1.5876×10^{2} $(4.03 \times 10^{1})-$	7.4380×10^{-1} (5.14×10^{-1})
DTLZ2	91	3	12	5.4957×10^{-2} $(1.67 \times 10^{-4})+$	7.5937×10^{-2} $(4.64 \times 10^{-3})=$	5.7424×10^{-2} $(6.21 \times 10^{-4})+$	7.7083×10^{-2} (2.51×10^{-3})
	210	5	14	1.8553×10^{-1} $(4.04 \times 10^{-3})-$	1.7727×10^{-1} $(3.72 \times 10^{-3})+$	2.3063×10^{-1} $(1.05 \times 10^{-2})-$	1.8270×10^{-1} (3.55×10^{-3})
	156	8	17	3.9090×10^{-1} $(7.00 \times 10^{-2})-$	3.7864×10^{-1} $(1.01 \times 10^{-2})-$	1.4962×10^{0} $(1.75 \times 10^{-1})-$	3.6421×10^{-1} (2.85×10^{-2})
	275	10	19	5.9934×10^{-1} $(9.67 \times 10^{-2})-$	4.5085×10^{-1} $(1.48 \times 10^{-2})=$	1.4107×10^{0} $(1.49 \times 10^{-1})-$	4.6149×10^{-1} (6.32×10^{-2})
	135	15	24	7.5601×10^{-1} $(5.93 \times 10^{-2})-$	6.1126×10^{-1} $(1.39 \times 10^{-2})+$	2.4955×10^{0} $(2.57 \times 10^{-2})-$	6.3724×10^{-1} (4.61×10^{-2})
DTLZ3	91	3	12	1.0504×10^{1} $(3.13 \times 10^{0})-$	6.4702×10^{0} $(3.22 \times 10^{0})=$	9.0596×10^{0} $(5.26 \times 10^{0})=$	6.5991×10^{0} (3.50×10^{0})
	210	5	14	5.8276×10^{1} $(1.03 \times 10^{1})-$	3.2447×10^{1} $(8.71 \times 10^{0})+$	5.0996×10^{1} $(1.39 \times 10^{1})-$	4.0504×10^{1} (1.12×10^{1})
	156	8	17	8.4808×10^{1} $(2.54 \times 10^{1})-$	1.2374×10^{2} $(3.40 \times 10^{1})-$	7.9994×10^{2} $(1.21 \times 10^{2})-$	3.0523×10^{1} (9.66×10^{0})
	275	10	19	1.5685×10^{2} $(3.89 \times 10^{1})-$	1.1612×10^{2} $(3.52 \times 10^{1})-$	8.6838×10^{2} $(1.43 \times 10^{2})-$	8.3597×10^{1} (2.42×10^{1})
	135	15	24	1.5216×10^{2} $(5.28 \times 10^{1})-$	2.8794×10^{2} $(1.66 \times 10^{2})-$	1.4128×10^{3} $(2.04 \times 10^{2})-$	2.7982×10^{1} (1.09×10^{1})
DTLZ4	91	3	12	2.3080×10^{-1} $(2.63 \times 10^{-1})-$	2.0242×10^{-1} $(2.34 \times 10^{-1})=$	2.9428×10^{-1} $(2.92 \times 10^{-1})-$	2.1563×10^{-1} (2.17×10^{-1})
	210	5	14	1.9176×10^{-1} $(5.55 \times 10^{-3})-$	1.9539×10^{-1} $(3.99 \times 10^{-2})=$	2.4515×10^{-1} $(1.16 \times 10^{-2})-$	1.9132×10^{-1} (4.19×10^{-2})
	156	8	17	4.2521×10^{-1} $(7.39 \times 10^{-2})-$	4.4162×10^{-1} $(2.85 \times 10^{-2})-$	1.3066×10^{0} $(1.22 \times 10^{-1})-$	3.9986×10^{-1} (6.80×10^{-2})
	275	10	19	5.6600×10^{-1} $(5.14 \times 10^{-2})-$	5.3957×10^{-1} $(7.12 \times 10^{-3})-$	1.2556×10^{0} $(8.95 \times 10^{-2})-$	4.6221×10^{-1} (3.50×10^{-2})
	135	15	24	8.0334×10^{-1} $(8.61 \times 10^{-2})-$	6.4400×10^{-1} $(7.42 \times 10^{-3})+$	2.4389×10^{0} $(2.17 \times 10^{-1})-$	6.6414×10^{-1} (2.74×10^{-2})

(*continued*)

Table 2. (*continued*)

Problem	N	M	D	NSGA-III	KnEA	SPEA2	MaOEATESS
DTLZ5	91	3	12	1.1778×10^{-2} $(1.11 \times 10^{-3})+$	2.5055×10^{-2} $(1.50 \times 10^{-2})-$	5.4912×10^{-3} $(2.54 \times 10^{-4})+$	1.4119×10^{-2} (3.09×10^{-3})
	210	5	14	2.1423×10^{-1} $(4.67 \times 10^{-2})-$	2.8699×10^{-1} $(1.26 \times 10^{-1})-$	2.2368×10^{-1} $(4.63 \times 10^{-2})-$	1.0968×10^{-1} (3.12×10^{-2})
	156	8	17	2.6442×10^{-1} $(1.09 \times 10^{-1})=$	3.9955×10^{-1} $(1.11 \times 10^{-1})-$	1.0693×10^{0} $(2.82 \times 10^{-1})-$	2.2965×10^{-1} (5.01×10^{-2})
	275	10	19	2.1507×10^{-1} $(5.10 \times 10^{-2})=$	3.8203×10^{-1} $(1.02 \times 10^{-1})-$	1.0059×10^{0} $(2.41 \times 10^{-1})-$	2.2076×10^{-1} (6.34×10^{-2})
	135	15	24	3.2934×10^{-1} $(6.20 \times 10^{-2})+$	5.7370×10^{-1} (1.39×10^{-1})	1.5772×10^{0} $(5.17 \times 10^{-1})-$	4.1069×10^{-1} (1.03×10^{-1})
DTLZ6	91	3	12	1.8179×10^{-2} $(3.07 \times 10^{-3})-$	1.8622×10^{-2} $(1.05 \times 10^{-2})=$	4.6313×10^{-3} $(5.99 \times 10^{-4})+$	1.5668×10^{-2} (2.80×10^{-3})
	210	5	14	3.6214×10^{0} $(7.13 \times 10^{-1})-$	1.3820×10^{0} $(4.53 \times 10^{-1})-$	4.3482×10^{0} $(8.94 \times 10^{-1})-$	9.3977×10^{-1} (5.54×10^{-1})
	156	8	17	5.5482×10^{0} $(8.34 \times 10^{-1})-$	3.1498×10^{0} $(5.49 \times 10^{-1})-$	9.3170×10^{0} (2.42×10^{-1})	1.0758×10^{0} (7.27×10^{-1})
	275	10	19	6.8329×10^{0} $(5.60 \times 10^{-1})-$	3.1085×10^{0} $(6.03 \times 10^{-1})+$	8.9506×10^{0} $(3.95 \times 10^{-1})-$	4.3246×10^{0} (9.61×10^{-1})
	135	15	24	6.9227×10^{0} $(6.49 \times 10^{-1})-$	3.9146×10^{0} $(6.34 \times 10^{-1})-$	9.6993×10^{0} $(1.15 \times 10^{-1})-$	1.7440×10^{0} (5.82×10^{-1})
DTLZ7	91	3	22	9.4297×10^{-2} $(8.27 \times 10^{-3})+$	1.1232×10^{-1} $(1.00 \times 10^{-1})=$	8.6602×10^{-2} $(5.28 \times 10^{-2})=$	1.4807×10^{-1} (1.35×10^{-1})
	210	5	24	7.5161×10^{-1} $(1.33 \times 10^{-1})-$	4.1326×10^{-1} $(1.16 \times 10^{-1})+$	5.0430×10^{-1} $(5.84 \times 10^{-2})=$	5.1477×10^{-1} (1.07×10^{-1})
	156	8	27	5.1510×10^{0} $(1.06 \times 10^{0})-$	2.4092×10^{0} $(1.19 \times 10^{0})+$	3.5575×10^{0} $(1.12 \times 10^{0})=$	3.2171×10^{0} (1.51×10^{0})
	275	10	29	1.3637×10^{1} $(1.47 \times 10^{0})-$	8.7958×10^{0} $(2.16 \times 10^{0})+$	7.9420×10^{0} $(2.05 \times 10^{0})+$	1.0206×10^{1} (1.36×10^{0})
	135	15	34	1.7003×10^{1} $(1.97 \times 10^{0})-$	1.8376×10^{1} $(3.22 \times 10^{0})-$	1.2369×10^{1} $(2.99 \times 10^{0})=$	1.3221×10^{1} (2.13×10^{0})
+/−/=				4/29/2	8/18/9	4/25/6	—

5 Conclusion

Due to the non-dominated sorting strategy based on the reference-point lacks the pareto selection pressure and the diversity maintenance mechanism is insufficient in the late stage of algorithm. In order to solve this problem, a many-objective evolutionary algorithm with threshold elite selection strategy MaOEA-TES is proposed in this paper. The proposed algorithm combines balanceable fitness estimation with reference-point based non-dominated sorting strategy. Meanwhile, dynamic threshold selection strategy is designed to better balance the diversity and convergence of population. The experiment results show that the proposed MaOEA-TES is superior to other advanced algorithms.

Acknowledgments. This work is supported by the National Natural Science Foundation of China under Grant No.61806138, Natural Science Foundation of Shanxi Province under Grant No. 201801D121127, Taiyuan University of Science and Technology Scientific Research Initial Funding under Grant No. 20182002. Postgraduate education Innovation project of Shanxi Province under Grant No. 2019SY493.

References

1. Coello, C.C., Cortés, N.C.: An approach to solve multiobjective optimization problems based on an artificial immune system. In: First International Conference on Artificial Immune Systems (ICARIS 2002), pp. 212–221 (2002)
2. Cui, Z., Li, F., Zhang, W.: Bat algorithm with principal component analysis. Int. J. Mach. Learn. Cybernet (2018). https://doi.org/10.1007/s13042-018-0888-4
3. Cai, X., (ed.): Bat algorithm with triangle-flipping strategy for numerical optimization. Int. J. Mach. Learn. Cybern. **9**(2), 199–215 (2018)
4. Li, M., Yang, S., Liu, X.: Shift-based density estimation for Pareto-based algorithms in many-objective optimization. IEEE Trans. Evol. Comput. **18**(3), 348–365 (2014)
5. Wickramasinghe, U.K., Carrese, R., Li, X.: Designing airfoils using a reference point based evolutionary many-objective particle swarm optimization algorithm. In: IEEE Congress on Evolutionary Computation, pp. 1–8 (2010). https://doi.org/10.1109/CEC.2010.5586221
6. Fu, G., Kapelan, Z., Kasprzyk, J.R., Reed, P.: Optimal design of water distribution systems using many-objective visual analytics. J. Water Resour. Plann. Manag. **139**, 624–633 (2013)
7. Lygoe, R., Cary, M., Fleming, P.: A real-world application of a many-objective optimisation complexity reduction process. In: Purshouse, R., Fleming, P., Fonseca, C., Greco, S., Shaw, J. (eds.) EMO 2013. LNCS, vol. 7811, pp. 641–655. Springer, Heidelberg (2013). https://doi.org/10.1007/978-3-642-37140-048
8. Fleming, P.J., Purshouse, R.C., Lygoe, R.J.: Many-objective optimization: an engineering design perspective. In: Coello Coello, C.A., Hernández Aguirre, A., Zitzler, E. (eds.) EMO 2005. LNCS, vol. 3410, pp. 14–32. Springer, Heidelberg (2005). https://doi.org/10.1007/978-3-540-31880-4_2
9. Mkaouer, M.W., (ed.): High dimensional search-based software engineering: finding tradeoffs among 15 objectives for automating software refactoring using NSGA-III. In: Proceedings of the 2014 Annual Conference on Genetic and Evolutionary Computation, Vancouver, BC, Canada, 12–16 July, pp. 1263–1270 (2014)
10. Cui, Z.H., et al.: A pigeon inspired optimization algorithm for many-objective optimization problems. Sci. China Inf. Sci. **62**(7), 070212 (2019)
11. Deb, K., Jain, H.: An evolutionary many-objective optimization algorithm using reference-point-based nondominated sorting approach, Part I: solving problems with box constraints. IEEE Trans. Evol. Comput. **18**(4), 577–601 (2014)
12. Bi, X., Wang, C.: A niche-elimination operation based NSGA-III algorithm for many-objective optimization. Appl. Intell. **48**(1), 118–141 (2017). https://doi.org/10.1007/s10489-017-0958-4
13. Li, K., Deb, K., Zhang, Q., Kwong, S.: An evolutionary many-objective optimization algorithm based on dominance and decomposition. IEEE Trans. Evol. Comput. **19**(5), 694–716 (2015)

14. Zitzler, E., Künzli, S.: Indicator-based selection in multi-objective search. In: Yao, X., et al. (eds.) PPSN 2004. LNCS, vol. 3242, pp. 832–842. Springer, Heidelberg (2004). https://doi.org/10.1007/978-3-540-30217-9_84
15. Lin, Q.Z., et al.: Particle swarm optimization with a balanceable fitness estimation for many-objective optimization problems. IEEE Trans. Evol. Comput. **22**(1), 32–46 (2018)
16. Deb, K., (ed.): Scalable test problems for evolutionary multi-objective optimization. In: Evolutionary Multi-objective Optimization, pp. 105–145 (2005)
17. Zhang, Q., (ed.): Multi-objective optimization test instances for the CEC 2009 special session and competition, p. 264. Technical report, University of Essex, Colchester, UK and Nanyang technological University, Singapore, special session on performance assessment of multi-objective optimization algorithms (2008)

Multi-objective Optimization Algorithm Based on Uniform Design and Differential Evolution

Jinrong He[1,2,3], Dongjian He[1,3(✉)], Aiqing Shi[1,3], and Guoliang He[4]

[1] College of Mechanical and Electronic Engineering, Northwest A&F University, Yangling 712100, China
hdjl68@nwsuaf.edu.cn
[2] College of Mathematics and Computer Science, Yan'an University, Yan'an 716000, China
[3] Key Laboratory of Agricultural Internet of Things, Ministry of Agriculture and Rural Affairs, Yangling 712100, China
[4] School of Computer, Wuhan University, Wuhan 430070, China

Abstract. The multi-objective optimization problem is an important research direction in the field of optimization. Because the traditional mathematical programming method often cannot achieve the optimal global solution, the researchers introduced the heuristic method into the multi-objective optimization problem. The heuristic method is a method of searching based on empirical rules, which can get the optimal solution or solution set of problems in the limited search space. In this paper, we proposed a multi-objective evolutionary algorithm based on uniform design and differential evolution, which use the uniform design table to construct the weight vector and utilize the crossover in differential evolution and mutation process to replace the simulated binary intersection and the simulated polynomial variation. Compared with the classical algorithm, the experimental results show that the improved algorithm is superior to the original algorithm.

Keywords: Multi-objective optimization · Heuristic method · Evolutionary algorithm · Uniform design · Differential evolution

1 Introduction

In general, optimization problem with more than one optimization goals need to be processed at the same time is called multi-objective optimization problem (MOP, [1]). A typical approach to directly solve a multi-objective optimization problem is to convert the multi-objective optimization problem into a single-objective optimization problem in a certain way (aggregation by weight or transformation multiple objectives into constraint conditions), to obtain an optimal solution. However, this kind of traditional method can only be applied to a small set of problems and has poor generalization. Secondly, most algorithms can only get a locally optimal solution. If the neighborhood is enlarged, the complexity of the algorithm will be increased.

K. Li et al. (Eds.): ISICA 2019, CCIS 1205, pp. 180–193, 2020.
https://doi.org/10.1007/978-981-15-5577-0_14

In 1967, Rosenberg proposed that genetic algorithms could be used to solve multi-objective optimization problems [2]. Eighteen years later, Schafferf proposed a vector-based fitness calculation method, called Vector Evaluated Genetic Algorithm (VEGA, [3]), which is an extension of SGA (Simple Genetic Algorithms). A multi-objective genetic algorithm combined with genetic algorithms is proposed for the first time, creating a precedent for multi-objective heuristic optimization algorithms. Since 1967, multi-objective optimization problems have begun to attract the attention of researchers in various fields. By the end of the 20th century, as a new method to solve multi-objective optimization problems, the evolutionary algorithm of multi-objective heuristic optimization algorithms has attracted much attention, and some have successfully applied to the project.

Multi-objective optimization algorithms based on evolutionary algorithms are mostly characterized by non-dominated selection and diversity preservation strategies based on shared functions. In 1993, Fonseca and Fleming proposed the Multi-Objective Genetic Algorithm (MOGA, [4]). Srinivas and Deb proposed the Non-Dominated Sorting Genetic Algorithm (NSGA, [5]). Horn and Nafpliotis proposed the Niched Pareto Genetic Algorithms (NPGA, [6]), which are commonly called the first-generation evolutionary multi-objective optimization algorithms. In 2002, Zitzler and Thiele proposed the Strength Pareto Evolutionary Algorithm (SPEA, [7]). In 2000, Knowles and Corne proposed the Pareto Archived Evolution Strategy (PAES, [8]), and in 2002, Deb et al. proposed the NSGA-II, which is still widely used up to now, obtained by improving NSGA [9]. In 2003, Moore and Chapman proposed the first method to solve the multi-objective optimization problem using PSO, in which they introduced the Pareto preference to change records and use individual optimal locations [10]. Ray et al. introduced the Pareto sorting strategy into PSO to solve multi-objective optimization problem, and adopted crowding degree to maintain the diversity of particles [11]. Coello and Lechuga introduced the Pareto archiving evolution strategy into PSO to obtain MOPSO [12], and based on artificial immunity, they proposed the multi-objective immune system algorithm (MISA, [13]), which first applied an artificial immune system to solve the multi-objective optimization problems. Luh, Chueh, and Liu proposed the multi-objective immune algorithm (MOIA, [14]), in which the antibody adopts binary string encoding. Based on distributed estimation, Khan et al. proposed the multi-objective Bayesian optimization algorithm (mBOA, [15]), which combined the non-dominant selection strategy in NSGA-II with Bayesian Optimization Algorithm (BOA) to solve the spoofing multi-objective optimization problem. Laumanns et al. also proposed a multi-objective optimization algorithm based on Bayesian optimization, which combined SPEA2 and BOA to solve the multi-objective knapsack problem [16].

Domestic researchers have also proposed a variety of multi-objective heuristic optimization algorithms. In response to the multi-objective optimization problem with constraints, Prof. Cai Zixing and Dr. Wang Yong from Central South University proposed a new evolutionary algorithm based on the existing multi-objective optimization technology solving the optimization problem with constraints and getting good results [17]. In the research of multi-objective algorithm based on particle swarm, Li Xiaodong introduced the non-dominant sorting mechanism of NSGA-II into PSO and adopted niche strategy to ensure the diversity of solutions, achieving excellent results [18]. In the research on multi-objective Optimization Algorithm based on artificial immunity, Prof. Jiao licheng, Prof. Gong Maoguo, and Dr. Shang Ronghua et al. proposed Immune Dominance Clone Multi-objective Optimization Algorithm (IDCMA, [19]). In 2008, Prof. Gong Maoguo and Prof. Jiao Licheng et al. presented the nonsorted neighbor immune algorithm (NNIA, [20]), which used an individual selection method based on non-dominant neighbors. In the study of multi-objective optimization algorithm based on density estimation, Zhang Qingfu and Zhou Aimin et al. proposed Regularity Model based Multi-objective Estimation of Distribution Algorithm (RM-MEDA) which is a classic algorithm combined with distributed estimation algorithm and multi-objective optimization algorithm [21]. In the study of multi-objective optimization algorithm based on co-evolution, Liu Jing proposed a multi-objective evolutionary algorithm based on co-evolution, in which a crossover operator and three co-evolution operators were designed to maintain the diversity of individuals in the population and accelerate the convergence rate [22]. Tan Kaichen et al. proposed a new distributed cooperative co-evolutionary algorithm (DCCEA, [23]) based on the idea of distributed cooperative coevolution. In 2010, Prof. Tan Ying et al. from Peking University proposed the fireworks algorithm, which is a new idea of optimization problem study [24]. In 2013, Dr. Zheng Yujun et al. proposed the application of multi-target fireworks algorithm to the variables in oil crop production [25]. In 2016, Prof. Xie Chengwang et al. proposed to introduce an elite reverse learning mechanism into multi-objective fireworks algorithm to make the algorithm better, which provided a new research direction for multi-objective optimization [26].

This paper proposes MOEA/D based on uniform design and differential evolution. According to the existing multi-objective heuristic optimization algorithm combined with existing mathematical knowledge and differential evolution algorithm, the improvement of MOEA/D algorithm was proposed, and compared with the original algorithm in experimental analysis.

2 Multi-objective Evolutionary Algorithm

2.1 Problem Formulation of Multi-objective Optimization

In general, the multi-objective optimization problem consists of n decision variables, M objective functions, and K constraints, which can be formulated as follows.

$$\begin{cases} \min \ \vec{y} = f(\vec{x}) = [f_1(\vec{x}), f_2(\vec{x}), \cdots, f_M(\vec{x})] \\ s.t. \quad g_i(\vec{x}) \leq 0 \qquad i = 1, 2, \ldots, p \\ \qquad\ \ h_i(\vec{x}) = 0 \qquad i = 1, 2, \ldots, q \end{cases} \tag{1}$$

where $\vec{x} = (x_1, x_2, \ldots, x_n) \in D$ is the decision vector; $\vec{y} = (f_1, f_2, \ldots, f_M) \in Y$ represents the target vector; D is the decision space formed by the decision vector; Y represents the target space formed by the target vector; $g_i(\vec{x}) \leq 0 (i = 1, 2, \ldots, p)$ defines p inequality constraints; $h_i(\vec{x}) = 0 (i = 1, 2, \ldots, q)$ defines q equality constraints.

Definition 1 (Pareto dominance). If $x_1, x_2 \in X_f$ is two feasible solutions to the multi-objective optimization problem (1), then x_1 Pareto dominates x_2 if and only if

$$\forall i = 1, 2, \ldots, M,\ f_i(x_1) \leq f_i(x_2) \wedge \exists j \in \{1, 2, \ldots, M\},\ f_i(x_1) < f_i(x_2) \quad (2)$$

For short, x_1 dominates x_2, which can be denoted as $x_1 \succ x_2$, also called compared with x_2, x_1 is Pareto dominant.

Definition 2 (Pareto optimal). Solution $x^* \in X_f$ is Pareto optimal (or non-dominated), if and only if

$$!\exists\, x \in X_f : x \succ x^* \quad (3)$$

Definition 3 (Pareto frontier). The target vector corresponding to all Pareto optimal solutions in the Pareto optimal set P_s constitutes the surface, which can be represented as:

$$PF = \{F(x^*) = (f_1(x^*), f_2(x^*), \cdots, f_M(x^*))^T \,|\, x^* \in P_s\} \quad (4)$$

2.2 Multi-objective Evolutionary Algorithm Based on Decomposition (MOEA/D)

MOEA/D decomposes the multi-objective optimization problem into N scalar sub-problems, which can be solved simultaneously by evolving the population of solutions. In each generation, the population is a collection of optimal solutions for each sub-problem selected from all generations. The degree of association of two adjacent sub-problem keys is determined by the distance between their aggregation coefficient vectors. For two adjacent sub-problems, the optimal solution should be very similar. For each sub-question, it can be optimized with information about its adjacent sub-problems. The details of MOEA/D are summarized in Algorithm 1.

Algorithm 1. Decomposition-based multi-objective evolutionary algorithm

Step 1. Uniformly generate the initial population $\{x^1, x^2, ..., x^N\}$ with size N in the feasible space.

Step 2. Calculate the Euclidean distance between any two weight vectors, and find the T weight vectors nearest to each weight vector. For any $i = 1,2,...,N$, let $B_i = \{i_1, i_2, ..., i_t\}$ and $\lambda^{i_1}, \lambda^{i_2}, ..., \lambda^{i_T}$ is the nearest T weight vectors to λ^i.

Step 3. Initialize $z = \{z_1, z_2, ..., z_m\}^T$, where $z_i = \min\{f_i(x^1), f_i(x^2), ..., f_i(x^N)\}$; Set EP to null.

Step4. Genetic recombination and improvement. Select two random serial Numbers k, l from B_i, use genetic operator to generate a new solution y from x^k, x^l, and use the repair and improvement of y based on the test problem to inspire the generation of y'.

Step 5. Update z. For any $j = 1,2,...,m$, if $z_i < f_i(y')$, then $z_i = f_i(y')$.

Step 6. Update neighborhood solution. For any $j \in B_i$, if $g^{te}(y' | \lambda^j, z) \le g^{te}(x^j | \lambda^j, z)$, then $x^j = y', FV^j = F(y')$.

Step 7. Update *EP*. Remove all vectors that are dominated by $F(y')$ from *EP*. If the vector in *EP* does not dominate $F(y')$, add $F(y')$ to *EP*.

Step 8. Determine the termination condition. If it terminates, output *EP* as a result, otherwise go to **Step 3**.

3 MOEA/D Based on Uniform Design and Differential Evolution

3.1 Generate a Weight Vector Using Uniform Design Methods

In MOEA/D, the simplex lattice point design is used to set the weight vector, but the generated weight vector points are not distributed uniformly, and there are too many points on the boundary, so that the weight of some targets in the sub-question is zero. We use a uniform design method to generate uniformly distributed weight vectors. The basic idea behind uniform design is to make the experimental points in the factor space have better uniform dispersion, and to minimize some uniformity measure. Let $U_n(q^s)$ denotes the uniform design table, where *U* denotes a uniform design table, *S* denotes the number of factors, *q* is the number of experimental levels, and *n* denotes the number of experiments.

In this paper, we use the good lattice point method to construct a uniform design table $U_N(N^{m-1})$ and use the reliable deviation (CD_2) as the uniformity measure to generate *N* weight vectors with uniform distribution of *m* dimensions. The use of good lattice point method to construct a uniform design table involves the following two definitions. The results of uniform design in this paper are discussed on the basis of the following definitions.

Definition 4. If each column element of a $n \times s$ matrix $U = (u_{ij})$ is a permutation of a set $\{1, 2, \ldots, q\}$, it is called a U – matrix and is denoted as $U(n, n^s)$. Transform u_{ij} to $x_{ij} = \frac{2u_{ij}-1}{2n}$, the matrix consisting of x_{ij} is denoted $X_u = (x_{ij})_{n \times s}$, the n rows of X_u can be regarded as n points above $C^s = [0, 1]^s$, and X_u is the set of points on C^s.

The n points determined by the U-matrix or the corresponding X_u are not necessarily evenly distributed, but finding the most uniform one among them can be used as a uniform design. In this paper, the centralization L_2 – deviation is used to measure the uniformity of the point set X_u. The centralization L_2 – deviation of the point set X_u is "0", and the calculation method is

$$CD_2(X_u) = \left[\left(\frac{13}{12}\right)^s - \frac{2^{1-s}}{n} \sum_{k=1}^{n} \sum_{i=1}^{s} \left(2 + \left|x_{ki} - \frac{1}{2}\right| - \left|x_{ki} - \frac{1}{2}\right|^2\right) + \frac{1}{n^2} \right. \\ \left. + \frac{1}{n^2} \sum_{k,l-1}^{n} \sum_{i=1}^{s} \left(1 + \frac{1}{2}\left|x_{ki} - \frac{1}{2}\right| + \frac{1}{2}\left|x_{li} - \frac{1}{2}\right| - |x_{ki} - x_{li}|\right)^{\frac{1}{2}} \right] \tag{5}$$

Definition 5. The U-matrix U^* of size $n \times s$ is called a uniform design, if and only if its corresponding X_{U^*} has the smallest CD_2 – value in all X_U of the same type, and U^* is denoted as $U(n^s)$.

3.2 Differential Evolution Method

The evolutionary process of MOEA/D mainly adopts simulated polynomial variation (PM) and simulated binary crossover (SBX). In this paper, mutation and crossover strategies in differential evolutionary algorithm (DE) are adopted to evolve population. The basic idea of differential variability is to randomly select two individuals from the population to obtain the difference vector, and to weight the difference vector and then sum the third individual according to certain rules to produce the variant individual. The basic idea of differential crossing is to mix the mutated individuals with a predetermined objective individual to generate test individuals. The specific calculation method is as follows:

The new solution obtained by DE mutation is $u' = (u'_1, u'_2, \ldots, u'_n)$, where the component u'_k is calculated by:

$$u'_k = x_{r_1}(g) + F \cdot (x_{r_2}(g) - x_{r_3}(g)), k \neq r_1 \neq r_2 \neq r_3 \tag{6}$$

where $k = 1, 2, \ldots, n$, F is the control parameter of the mutation process.

The new solution generated by DE crossover is $\overline{y} = (\overline{y_1}, \overline{y_2}, \ldots, \overline{y_n})$, where the component $\overline{y_k}$ is calculated by:

$$\overline{y_k} = \begin{cases} v_k & if \quad rand_1 < CR,\ k = rand_2, \\ x_k^i & otherwise. \end{cases} \tag{7}$$

in which

$$v_k = x_k^i + F \times (x_k^{r_1} - x_k^{r_2}) \tag{8}$$

where $k = 1, 2, \ldots, n$, and $rand_1 \in [0, 1]$, $rand_2 \in \{1, \ldots, n\}$ are two random numbers, and CR is the control parameter of the crossing process.

3.3 Algorithm Description

UDEMOEA/D, like the general idea of MOEA/D, decomposes the multi-objective optimization problem into sub-problems of n scalars, which solves all sub-problems simultaneously by evolving a solution population. For each generation of populations, the population is a collection of optimal solutions for each sub-question selected from all generations. The degree of association of two adjacent sub-problem keys is determined by the distance between their aggregation coefficient vectors. For two adjacent sub-problems, the optimal solution should be very similar. For each sub-question, just optimize it with information about its neighboring sub-problems. The specific algorithm steps are shown in Algorithm 2.

Algorithm 2. Decomposition multi-objective evolutionary algorithm based on uniform design and differential evolution

Step 1. Uniformly generate the initial population $\{x^1, x^2, \ldots, x^N\}$ with size *N* in feasible space.

Step 2. Calculate the Euclidean distance between any two weight vectors, and find the *T* weight vectors nearest to each weight vector. For each $i = 1, 2, \ldots, N$, let $B_i = \{i_1, i_2, \ldots, i_t\}$ and $\lambda^{i_1}, \lambda^{i_2}, \ldots, \lambda^{i_T}$ is the nearest *T* weight vectors to λ^i.

Step 3. Initialize the uniformly distributed weight vector $z = \{z_1, z_2, \ldots, z_m\}^T$, and let $z_i = \min\{f_i(x^1), f_i(x^2), \ldots, f_i(x^N)\}$; set *EP* to null.

Step 4. Genetic recombination and improvement. Three serial numbers j, k, l are randomly selected from B_i, and selects a suitable genetic operator according to the objective function to generate a new solution y from x^j, x^k, x^l, and generates a y' based on the repair and improvement of the test problem.

Step 5. Update z. For any $j = 1, 2, \ldots, m$, if $z_i < f_i(y')$, then $z_i = f_i(y')$.

Step 6. Update the neighborhood solution. For any $j \in B_i$, if $g^{te}(y' | \lambda^j, z) \le g^{te}(x^j | \lambda^j, z)$, then $x^j = y', FV^j = F(y')$.

Step 7. Update *EP*. Remove all vectors that are dominated by $F(y')$ from EP. If the vector in *EP* does not dominate $F(y')$, then add $F(y')$ to *EP*.

Step 8. Determine the termination condition. If it terminates, output *EP* as a result, otherwise go to **Step 3.**

In Algorithm 2, Step 1 uses a uniform design method of good lattice points to construct the weight vector. The specific steps are shown in Algorithm 3.

Algorithm 3. Good lattice point uniform design method

Step 1. Given the number n of experiments, look for an integer h smaller than n and let the greatest common divisor of n and h be 1. A positive integer that meets this condition constitutes a vector $h = (h_1, h_2, \ldots, h_m)$, where m is determined by the Euler function $\varphi(n)$.

Step 2. The jth column of the uniform design table is generated by calculating $u_{ij} = ih[\text{mod} \quad n]$, and then U_{ij} is generated according to the recursion formula.

Step 3. Select the first of all generated uniform design tables as the weight vector.

In Algorithm 3, the $[\text{mod} \quad n]$ mentioned in Step 2 represents the congruence operation. If ih exceeds n, it is subtracted by an appropriate multiple of n to make the difference fall in $[1, n]$. At the same time, the recursion formula used to generate U_{ij} is:

$$\begin{cases} u_{1j} = h_j \\ u_{i+1j} = \begin{cases} u_{ij} + h_j & if \quad u_{ij} + h_j \leq n \\ u_{ij} + h_j - n & if \quad u_{ij} + h_j > n \end{cases} \end{cases} \quad i = 1, 2, \ldots, n-1 \tag{9}$$

4 Experiment Results

4.1 Experimental Settings

In the experiment, the initial population size is set to 100, the number of iterations is 1000, the simulated binary cross-distribution index used in the process is 15, the simulated polynomial variation distribution index is 20, the cross-index of differential evolution is 0.9, the variation index of differential evolution is 0.5. All simulation experiments were performed on a PC with Intel Core i3-2350M 2.30 GHz and 2G memory.

To evaluate the performance improvement of UDEMEA/D compared to MOEA/D, the coverage metric I_C is used to evaluate the optimal Pareto solution set. Coverage metric [7] is a relatively common measure of approximation that is used to quantitatively evaluate the dominance of two sets. Assume A and B are the two approximate Pareto optimal solution sets in the target space, the coverage index $I_C(A, B)$ is calculated as follows:

$$I_C(A, B) \stackrel{def}{=} \frac{|b \in B; \exists a \in A : a \rhd b|}{|B|} \tag{10}$$

where $\rhd$ means that Pareto is not inferior, and $I_C(A, B) \in [0, 1]$, $I_C(A, B) = 1$ means that for all points in the set B, at least one point that is not inferior to it can be found in the set A. Instead, $I_C(B, A) = 0$ means that for all points in the set A, at least one point that is not inferior to A can be found in the set B. This indicator considers both $I_C(A, B)$ and $I_C(B, A)$, since $I_C(A, B)$ is not necessarily equal to $1 - I_C(B, A)$.

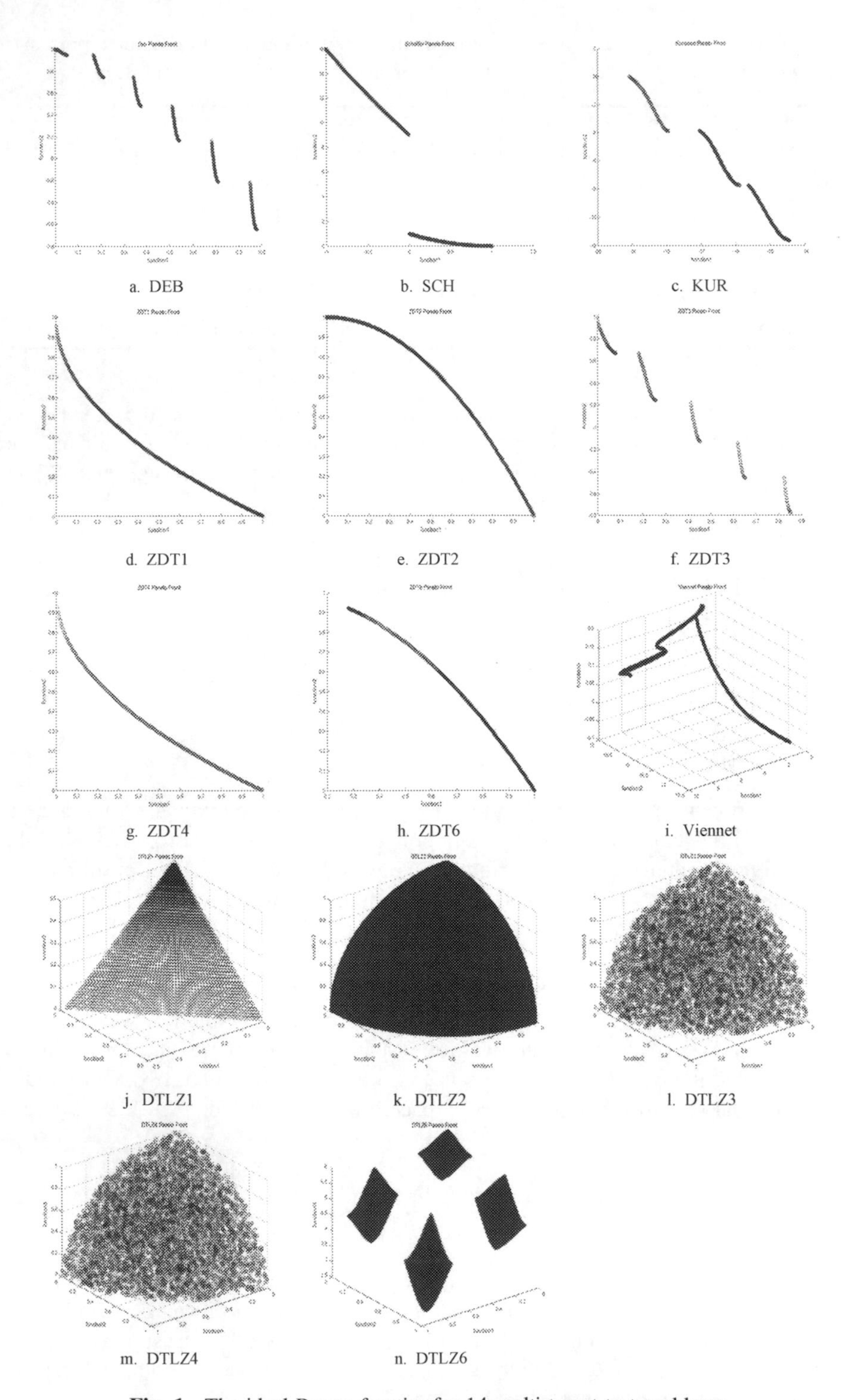

Fig. 1. The ideal Pareto frontier for 14 multi-target test problems

4.2 Benchmark Problems

The test functions used in the experiment were selected from the widely used test functions in the field of multi-objective optimization, including DEB, SCH, KUR, five ZDT problems, Viennet issues and five DTLZ issues [9]. Among them, the test functions DEB, KUR and SCH are two-objective optimization problems, and in five ZDT problems, ZDT1, ZDT2, ZDT3 have 30 decision variables, ZDT4 and ZDT6 have 10 decision variables, and Viennet problem is three-objective optimization problems, with 2 decisions Variables. The number of decision variables in five DTLZ problems and targets can be extended to any number. This article will set the value of k and $|x_k|$ according to Deb's suggestion. For DTLZ1, $k = 3$, $|x_k| = 5$ For DTLZ2, DTLZ3 and DTLZ4, $k = 3$, $|x_k| = 10$; for DTLZ6, $k = 3$, $|x_k| = 20$. Figure 1 is an ideal Pareto front for 14 test questions.

4.3 Experimental Results

UDEMOEA/D and MOEA/D were evaluated according to coverage index, and Table 1 and Fig. 2 were obtained respectively.

Table 1. UDEMOEA/D and MOEA/D coverage index evaluation box diagram

Coverage indicator value	UDEMOEA/D			MOEA/D		
	Min	Average	Max	Min	Average	Max
DEB	0	**0.1493**	0.5000	0	0.0028	0.0500
SCH	0	0.0498	0.2500	0	**0.0912**	0.2000
KUR	0.0450	**0.4360**	0.5000	0	0	0
ZDT1	0.4950	**0.4997**	0.5000	0	0	0
ZDT2	0.5000	**0.5000**	0.5000	0	0	0
ZDT3	0.5000	**0.5000**	0.5000	0	0	0
ZDT4	0.5000	**0.5000**	0.5000	0	0	0
ZDT6	0	**0.3665**	0.5000	0	0	0
Viennet	0	**0.0016**	0.0067	0	0.0012	0.0033
DTLZ1	0.3067	**0.3321**	0.3333	0	0	0
DTLZ2	0.0267	**0.0850**	0.2600	0	0	0
DTLZ3	0.0033	**0.2924**	0.3333	0	0.0046	0.1167
DTLZ4	0	**0.1439**	0.2767	0	0	0
DTLZ6	0	**0.2984**	0.3333	0	0	0

Table 1 shows the minimum, average, and maximum coverage of the solutions obtained by running the algorithm independently for 30 times. Figure 2 is the coverage index box graph of the two algorithms, in which MOEA/D represents the coverage of the solution obtained by MOEA/D for UDEMOEA/D, and UDEMOEA/D represents the coverage of the solution obtained by UDEMOEA/D to the solution obtained by MOEA/D. The display in Fig. 2 shows that in DEB, KUR, 5 ZDT problems, Viennet and 5 DTLZ problems, the box diagram of UDEMOEA/D is higher than the

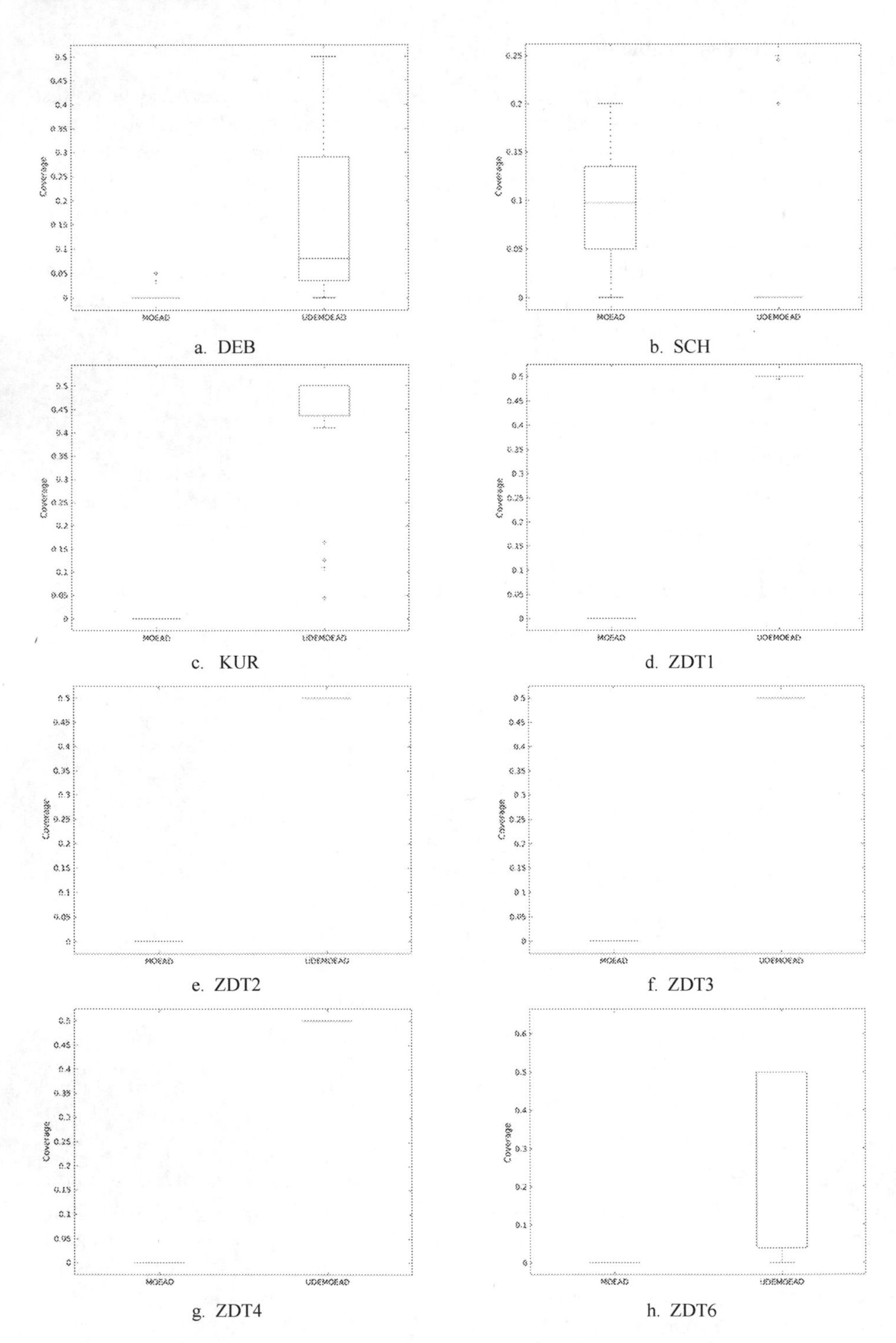

Fig. 2. MOEA/D and UDEMOEA/D cover the coverage index of the 14 test questions

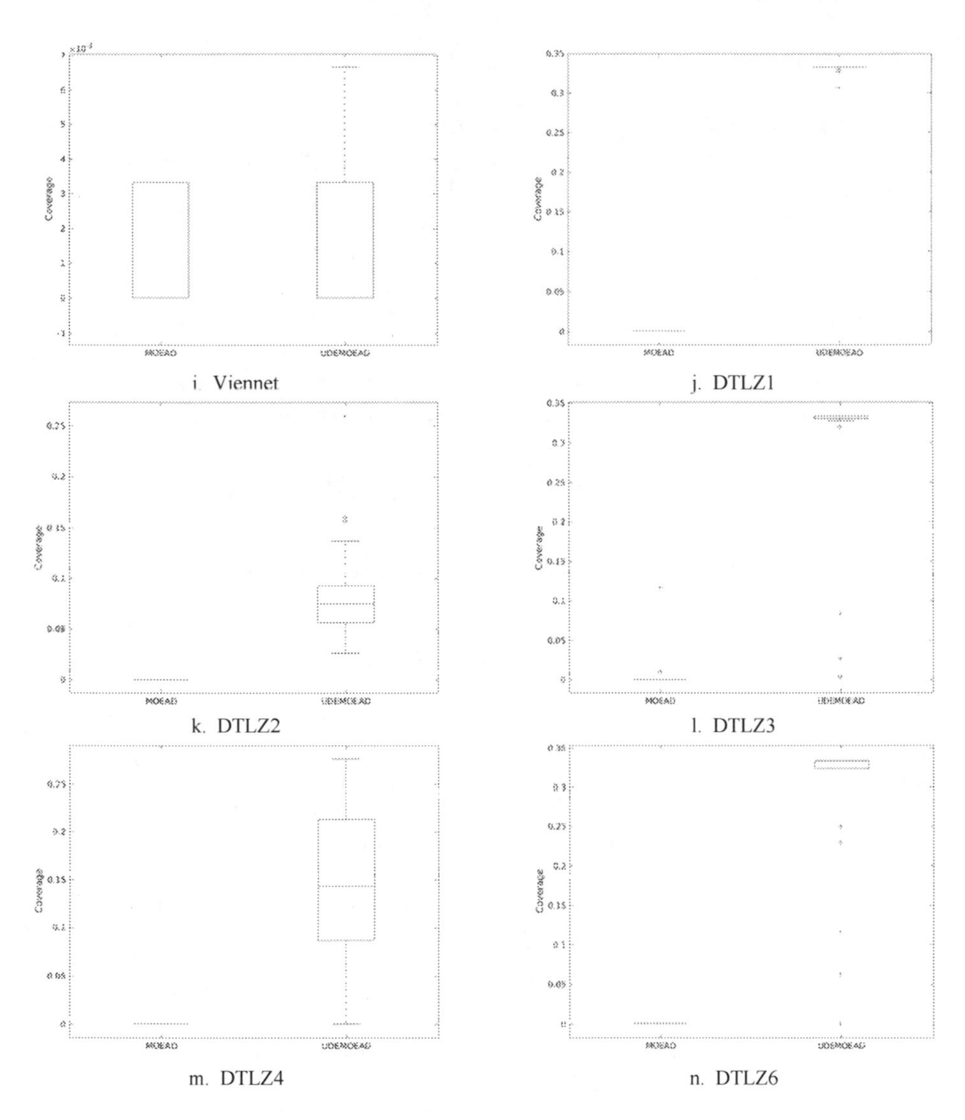

Fig. 2. (*continued*)

corresponding box diagram of MOEA/D, only MOEA in KUR/The box diagram of D is higher than the corresponding box diagram of UDEMOEA/D. At the same time, as can be seen from Table 1, among the four problems of ZDT1, ZDT2, ZDT3, and ZDT4, the coverage of the optimal Pareto solution set obtained by UDEMOEA/D for the optimal Pareto solution set of MOEA/D is 0.5. It can be concluded that under the same parameter setting, the optimal Pareto solution set obtained by UDEMOEA/D in solving 13 other test problems except SCH problem is better than the optimal Pareto solution set obtained by MOEA/D, and the solution set obtained by UDEMOEA/D is closer to the ideal Pareto frontier. UDEMOEA/D is superior to MOEA/D only considering approximation performance.

5 Conclusions

Based on that the distribution of weight vector points generated by the decomposition-based multi-objective evolutionary algorithm is not uniform, which leads to the fact that the final solution cannot be close to the ideal Pareto front, this paper proposed to use the uniform design table to construct the weight vector and utilize the crossover in differential evolution and mutation process to replace the simulated binary intersection and the simulated polynomial variation, and an improved algorithm is obtained—a decomposition-based multi-objective evolutionary algorithm based on uniform design and differential evolution. After the improved algorithm is implemented, it is compared with the original algorithm, and the obtained results are calculated to obtain the coverage index between the two algorithms to solve the test problems. By comparing the minimum, average and maximum of the two algorithms coverage index value, it can be seen that the improved algorithm is superior to the original algorithm in approximation performance.

Acknowledgement. This work was partially supported by National Natural Science Foundation of China (61902339, 61876136), the China Postdoctoral Science Foundation (2018M633585), Natural Science Basic Research Plan in Shaanxi Province of China (No. 2018JQ6060), and Google Supported Industry-University Cooperation and Education Project, Doctoral Starting up Foundation of Yan'an University (YDBK2019-06).

References

1. Anirban, M., Ujjwal, M., Sanghamitra, B., Carlos, A.C.C.: A survey of multi-objective evolutionary algorithms for data mining. IEEE Trans. Evol. Comput. **18**(1), 20–35 (2014)
2. Rosenberg, R.S.: Simulation of genetic populations with biochemical properties. Ph.D. thesis, University of Michigan, Michigan (1967)
3. Schaffer, J.D.: Multiple objective optimization with vector evaluated genetic algorithms. In: Proceedings of the International Conference on Genetic Algorithms and Their Applications, pp. 93–100. L. Erlbaum Associates, Hillsdale (1985)
4. Fonseca, C.M., Fleming, P.J.: Genetic algorithm for multi-objective optimization: formulation, discussion and generation. In: Forrest, S., (ed.) Proceedings of the 5th International Conference on Genetic Algorithms, pp. 416–423. Morgan Kauffman Publishers, San Mateo (1993)
5. Srinivas, N., Deb, K.: Multi-objective optimization using non-dominated sorting in genetic algorithms. Evol. Comput. **2**(3), 221–248 (1994)
6. Horn, J., Nafpliotis, N., Goldberg, D.E.: A niched Pareto genetic algorithm for multi-objective optimization. In: Fogarty, T.C., (ed.) Proceedings of the 1st IEEE Congress on Evolutionary Computation, pp. 82–87. IEEE, Piscataway (1994)
7. Zitzler, E., Laumanns, M., Thiele, L.: SPEA2: improving the strength Pareto evolutionary algorithm. In: Giannakoglou, K., Tsahalis, D.T., Périaux, J., Papailiou, K.D., Fogarty, T., (eds.) Evolutionary Methods for Design, Optimization and Control with Applications to Industrial Problems, pp. 95–100. Springer, Berlin (2002)
8. Knowles, J.D., Corne, D.W.: Approximating the non-dominated front using the Pareto archived evolution strategy. Evol. Comput. **8**(2), 149–172 (2000)

9. Deb, K., Pratap, A., Agarwal, S., Meyarivan, T.: A fast and elitist multi-objective genetic algorithm: NSGA-II. IEEE Trans. Evol. Comput. **6**(2), 182–197 (2002)
10. Moore, J., Chapman, R.: Application of particle swarm to multi-objective optimization. In: International Conference on Computer Science and Software Engineering (2003)
11. Ray, T., Liew, K.M.: A swarm metaphor for multi-objective design optimization. Eng. Optim. **34**(2), 141–153 (2002)
12. Coello, C.C.A., Pulido, G.T., Lechuga, M.S.: Handing multiple objectives with particle swarm optimization. IEEE Trans. Evol. Comput. **8**(3), 256–279 (2004)
13. Coello, C.C.A., Cortes, N.C.: Solving multi-objective optimization problems using an artificial immune system. Genet. Program. Evolv. Mach. **6**(2), 163–190 (2005). https://doi.org/10.1007/s10710-005-6164-x
14. Luh, G.C., Chueh, C.H., Liu, W.: MOIA: multi-objective immune algorithm. Eng. Optim. **35** (2), 143–164 (2003)
15. Khan, N., Goldberg, D.E., Pelikan, M.: Multi-objective Bayesian optimization algorithm. In: Proceedings of the Genetic and Evolutionary Computation Conference, p. 684. Morgan Kaufmann, New York (2002)
16. Laumanns, M., Ocenasek, J.: Bayesian optimization algorithms for multi-objective optimization. In: Guervós, J.J.M., Adamidis, P., Beyer, H.-G., Schwefel, H.-P., Fernández-Villacañas, J.-L. (eds.) PPSN 2002. LNCS, vol. 2439, pp. 298–307. Springer, Heidelberg (2002). https://doi.org/10.1007/3-540-45712-7_29
17. Cai, Z., Wang, Y.: A multi-objective optimization based evolutionary algorithm for constrained optimization. IEEE Trans. Evol. Comput. **10**(6), 658–675 (2006)
18. Li, X.: A non-dominated sorting particle swarm optimizer for multiobjective optimization. In: Cantú-Paz, E., et al. (eds.) GECCO 2003. LNCS, vol. 2723, pp. 37–48. Springer, Heidelberg (2003). https://doi.org/10.1007/3-540-45105-6_4
19. Jiao, L., Gong, M., Shang, R., Du, H., Lu, B.: Clonal Selection with Immune Dominance and Anergy Based Multiobjective Optimization. In: Coello Coello, Carlos A., Hernández Aguirre, A., Zitzler, E. (eds.) EMO 2005. LNCS, vol. 3410, pp. 474–489. Springer, Heidelberg (2005). https://doi.org/10.1007/978-3-540-31880-4_33
20. Gong, M.G., Jiao, L.C., Du, H.F., et al.: Multi-objective immune algorithm with non-dominated neighbor-based selection. Evol. Comput. **16**(2), 225–255 (2008)
21. Zhang, Q.F., Zhou, A.M., Jin, Y.: RM-MEDA: a regularity model based multi-objective estimation of distribution algorithm. IEEE Trans. Evol. Comput. **12**(1), 41–63 (2007)
22. Liu, J.: Research on Organizational Coevolutionary Algorithm and its Applications. Ph.D. thesis. Xidian University Xi'an (2004)
23. Tan, K.C., Yang, Y.J., Goh, C.K.: A distributed cooperative evolutionary algorithm for multi-objective optimization. IEEE Trans. Evol. Comput. **10**(5), 527–549 (2006)
24. Tan, Y., Zhu, Y.: Fireworks algorithm for optimization. In: Tan, Y., Shi, Y., Tan, K.C. (eds.) ICSI 2010. LNCS, vol. 6145, pp. 355–364. Springer, Heidelberg (2010). https://doi.org/10.1007/978-3-642-13495-1_44
25. Zheng, Y.J., Song, Q., Chen, S.Y.: Multi-objective fireworks optimization for variable-rate fertilization in oil crop production. Appl. Soft Comput. **13**(11), 4253–4263 (2013)
26. Xie, C., Xu, L., Xia, X., Wei, B., et al.: Multi-objective fireworks optimization algorithm using elite opposition-based learning. Acta Electronica Sinica **44**(5), 1180–1188 (2016)

Research on Optimization of Multi-target Logistics Distribution Based on Hybrid Integer Linear Programming Model

Jinfeng Wang[1(✉)], Jiayan Tang[1], Linqi He[1], Zirong Ji[1], Zhenyu He[1], Cheng Yang[1], Rongliang Huang[1], and Wenzhong Wang[2(✉)]

[1] College of Mathematics and Informatics, South China Agricultural University, Guangzhou 510642, China
wangjinfeng@scau.edu.cn

[2] College of Economics and Management, South China Agricultural University, Guangzhou 510642, China
wangwenzhong@scau.edu.cn

Abstract. At present, it is difficult to complete scheduling efficiently and cost-effectively in the process of logistics and distribution in the complex background, such as multi-target transportation, multi-model and multi-storage. Based on the detailed analysis of this problem, a multi-objective mixed integer linear distribution cost optimization model is established under multi-resource constraints. The model adopts the method of expanding the scale of calculation to enrich the vehicle distribution scheme. A linear representation of nonlinear cost is proposed with the mutual restraint way of four constraint modules which includes actual capacity, dynamic balance, daily running time and cost constraint. It can resolve the practical problems including customer loading queue, time consumption of path selection, and daily idleness of vehicles. Through the simulation experiments, the results show that the model can not only obtain the optimal distribution scheme with the fewest cost, but also have strong stability, wide practical application and scalability.

Keywords: Multi-resource constraints · Hybrid integer linear distribution model · Cost optimization · Dynamic balance

1 Introduction

Production outsourcing has become a common phenomenon since 1990, and technological updates have been involved in more manufacturing industry markets. Nowadays, the optimization of distribution scheme under resource constraint is the problem of optimizing the total cost of distribution under the constraint stake of cost, volume, route and so on. The costs can be divided into production and distribution cost, total quality cost and opportunity cost. At this time, in order to better calculate the cost,

This work is supported by the Technology Planning Project of Guangdong Province (No.: 2017A04-0406023) and the Technology Planning Project of Guangzhou City (No.: 201804010353).

K. Li et al. (Eds.): ISICA 2019, CCIS 1205, pp. 194–208, 2020.
https://doi.org/10.1007/978-981-15-5577-0_15

Anna N. et al. proposed that Euler method adjust the discrete time to deal with the industry flow, the time equation of price to realize the problem of cost minimization, control equilibrium conditions [1]. This model provided the best purchase or manufacturing and contractor selection decisions for each original company. In the actual user distribution problem of large enterprises, the distance between users and manufacturers is an inevitable cost factor. Therefore, Marcel T. et al. analyze the geographical location of the lead's dispersed location to measure the expected logistics cost problem sits geographically by averaging the location distance [2]. Meanwhile, a theoretical cost optimization model is also put forward for the construction of this kind of distribution logistics system to obtain the best distribution solution [3], which is similar to the hypothesis of this article. But it used the empirical research of shale gas data to solve the multi-target scheme of transportation cost optimization and reducing goods loss rate. Valentina C. et al. proposed a complete model solution for this distribution system [4]. They set up an integer linear planning model to optimize the time of the double constraint sq. for train scheduling, which provides the solution idea for the integer linearity. On the model selection of distribution problems, we refer to the nonlinear planning algorithm, which only enhances the general characteristics of problem solution characteristics [5]. It is more complex to implement than the linear model. So, we choose the linear model as the framework for this article.

In this article, the best distribution scheme aims at minimizing the total cost function. In this way, we need to consider the type of distribution goods required, the destination point, the maximum load of the vehicle, the matching degree of the vehicle and the user, the scheduling problem, route selection and other factors. An effective model is established in order to meet the different requirements of customers and suppliers, greatly reduce the distribution loss and maximize the benefits of both parties.

Firstly, this article proposes a multi-target 0–1 hybrid integer linear programming model [6] combining with the realistic nonlinear cost function to solve the complex distribution problem of multi-resource constraints. Meanwhile, a block-constrained model is established to obtain the best distribution scheduling problem with minimized cost. This model takes cost minimization, distribution scheme and time maximization as the targets, not only for meeting the cost loss phenomenon caused by the supplier's manual dispatch problem, thus saving the cost, but solving the problem of customer loading queue and time consumption of path selection. Combining the real data of the supplier *W* with the actual operation, the simulation result shows that our model reduces the total cost significantly compared with the original cost data of manual scheduling.

Finally, with half an hour for the schedule restrictions, it greatly shortens the loading and unloading time and improves the efficiency of logistics operation in the factory.

The full text structure is scheduled as follows: The second section describes the background of the problem, analyzes the problem and gives out necessary assumptions. In the next section, the model construction is presented in detail. The section four starts with data pre-processing, determines the target function and constraints to construct a hybrid cost optimization model. Then, experiments are implemented for verifying the effectiveness of the new model. Finally, the conclusions and prospects are summarized.

2 Model Construction Background

2.1 Description of the Problem

The problem to be solved in this article comes from the problem in practical application. Large-scale suppliers are operated by decentralized managing and centralized supplying, which involves the delivery of vehicles by different service departments to the users under their jurisdiction. A supplier hires the vehicle to deliver the goods, and needs to design a distribution system. It is necessary to consider comprehensively various factors such as the type of goods to be delivered, the destination location, maximum load of the vehicles, the matching degree between vehicles and users, the scheduling problem, route selection [7], load cost and transportation unit price [8]. It aims to minimize the total cost of the supplier's distribution and improve the efficiency of the transport vehicle and promote the revenue of distributors and the users accordingly.

The parsing diagrams of the problems solved in this article is as follows (Fig. 1):

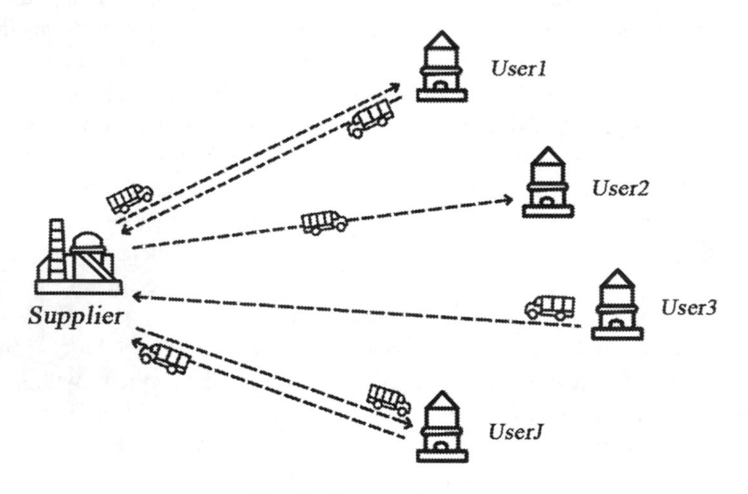

Fig. 1. Parsing diagrams

2.2 Basic Assumptions

The goal of establishing the distribution cost optimization model is to reduce the total cost of distribution, improve the quality of system service and increase customer satisfaction under guaranteed mutual constraints. Before building a model, the relevant conditions need to be summarized and appropriate assumptions need to be given to facilitate systematic processing.

The corresponding assumptions are listed below.

1) A single shipment can only be shipped to one user, allowing multiple shipments that can truck different orders for the same user;

2) The total cost considered by the supplier is related to distance (H km bounded) and actual capacity loading rate;
3) The loading warehouse has a fixed number of loading pipes, and the shift of the loading needs to be adjusted through the queue;
4) The user's destination has practical restrictions such as unloading pipe length, truck width, truck height, length et al.;
5) Vehicles for different shifts of the same user appear as little as possible in the case of queues at the destination for discharge;
6) The departure time of the vehicle's frequency is related to the distance, path selection, and actual loading and unloading time of the single shipment;
7) The commute time of all workers who involved in the system is 8:00–21:00;
8) The suppliers need to consider the loss of time for vehicles to wait for loading in the warehouse due to factors such as loading ports.

3 Model Construction

In this article, the main study focus on the cost optimization problem of reasonable distribution of multiple types of truck (trucking volumc) from the supplier's fixed points to multiple fixed-point (destination) with multi-position vehicles under multi-resource constraints in multi-distribution scenarios. Some factors in model construction are restricted mutually, such as the vehicle conditions, load cost, traffic cost, truck loss time and others.

3.1 Basic Thoughts

In order to control the total cost of loading and transportation, the supplier wants to maximize the use of the worker's effective work length to complete the delivery of goods demand. The cost calculation involved in the completion of a vehicle delivery is a fixed numerical ratio to the actual loading capacity and the transportation distance. It is a nonlinear function. If the nonlinear programming algorithm is directly adopted, the stability is not as good as the linear model [9]. Therefore, the modeling process introduces 0–1 variables to establish a hybrid integer linear model to improve the stability of the model. There are many constraints involved in this article. The following sections will give detailed descriptions of the model one by one module in order to express the model comprehensively.

In order to be able to clearly explain the model construction process, before introducing the submodules separately, the relevant variables used in the model are defined in a unified manner. The corresponding explanations are listed for understanding easily, as detailed in Table 1.

The overall implementation process of the model is shown in Fig. 2.

The model proposed in this article needs to solve the unavoidable truck condition restrictions in reality, which will lead to time-consuming problems such as temporary replacement of trucks, reloading materials, road congestion, and thus increasing the total cost of suppliers. Therefore, the model should give priority to the truck width,

Table 1. Symbol description

Symbol	Symbol description
C	Represents the total cost of shipment for all users within the supplier
i	i for the No. of the ith truck each supplier has a total of I trucks
j	j for the No. of the jth user for a vendor, with a total of J households
b_i	Indicates that the truck i has b_i loading scheme
q_{jf}	Indicates the demand of the user j for the material f
q_{jzf}	Indicates the actual delivery volume of the material f when the scheme z is used to deliver the material for the user j
f	f for the No. of the fth material, with a total of F kinds of transportation materials
a_{zf}	Indicates the maximum load of the truck that delivers the material f with the scheme z
M_{jz}	Indicates the total quantity of all actual shipments delivered by the user j when the shipment is delivered using the scheme z
c_{jz}	Indicates the actual distribution cost when the user j distributes the material with the scheme z
$\bar{c}_{jz}$	Indicates that the user j distributes the material with the scheme z, and the corresponding delivery cost when the truck is fully loaded, that is, the single maximum cost when the truck is delivered the users j
$\bar{q}_z$	The maximum load of the truck indicating the scheme z
$Kilo_{jz}$	Indicates the round-trip distance for the user j to deliver the material with the scheme z
$Time_{jz}$	Indicates the round-trip transit time for the user j to deliver the shipment with the scheme z
y_{jz}	0–1 variable
g_{jzk}	0–1 variable (k = 1: No load; k = 2: when the loading rate does not exceed u; k = 3: When the loading rate exceeds u)
ω_{jz1}	Actual load rate not exceeding u
ω_{jz2}	Actual load rate over u
d_i	Times of transportations of the trucks i
z	z represents the scheme z, and there are Z schemes corresponding to the total number of trips for all trucks

unloading tube length, length, truck height and other factors to avoid the phenomenon of mismatch with the user's destination.

$$Toweru_i \leq CarTu_i(j = 1, \cdots, J; i = 1, \cdots, I) \quad \text{(Unloading Pipe Length Matching)} \tag{1}$$

$$CarWidth_i \leq TowerWidth_j(j = 1, \cdots, J; i = 1, \cdots, I) \quad \text{(Truck Width Matching)} \tag{2}$$

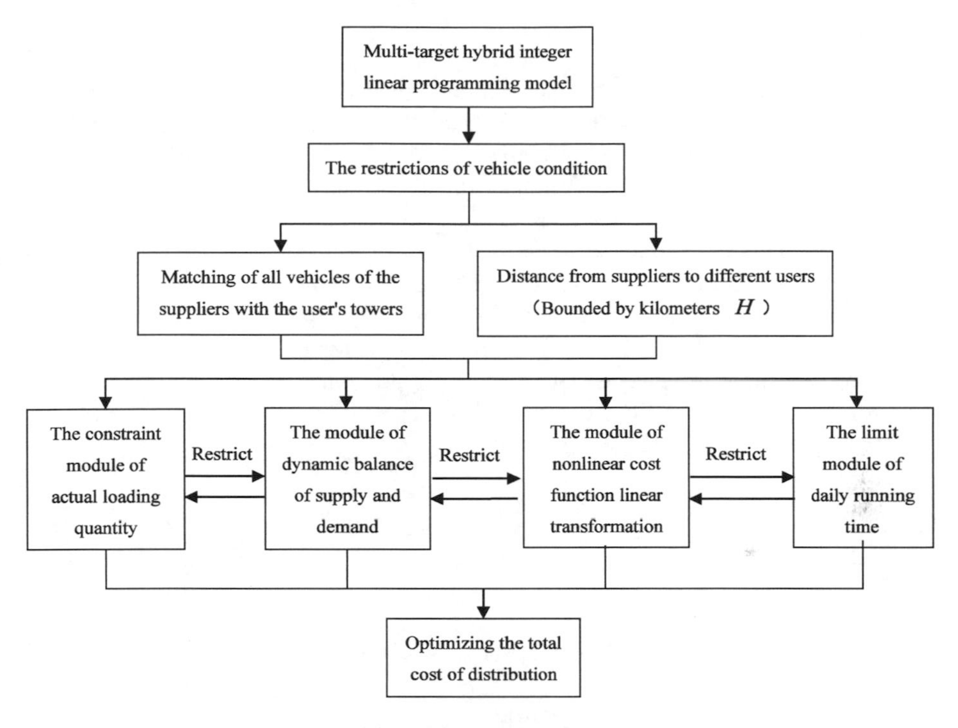

Fig. 2. Model flowchart

$$CarLength_i \leq TowerLength_j (j = 1, \cdots, J; i = 1, \cdots, I) \quad \text{(Truck Length Matching)} \tag{3}$$

$$CarHeight_i \leq TowerHeight_j (j = 1, \cdots, J; i = 1, \cdots, I) \quad \text{(Truck Height Matching)} \tag{4}$$

Symbolic Note

Toweru: Minimum unloading pipe length required by the user
TowerWidth: Maximum truck width allowed by destination
TowerLength: Maximum truck length allowed by destination
TowerHeight: Maximum truck height allowed by destination
CarTu: Unloading pipe length of the truck i
CarWidth: The width of the truck i
CarLength: The length of the truck i
CarHeight: The height of the truck i

In addition, if the supplier's distance to each user is greater than the specified threshold of H km $\left(Kilo_{jz} \geq H\right)$, then $y_{jz} = 1$; otherwise, $y_{jz} = 0$ means that the actual round-trip distance does not exceed H km $\left(Kilo_{jz} \leq H\right)$, this setting is convenient for the algorithm to realize the cost calculation of the post-text transportation.

3.2 Model Descriptions

It is still difficult to directly achieve the overall optimal solution in traditional nonlinear programming problems. The stability is weaker than linear solution taking the large amount of data into account. The realization of nonlinear model overall optimal solution needs more techniques to promote the algorithm [10].

Therefore, the model proposed in this article is constructed on the premise of expanding the number of schemes. The total number of schemes for a single truck is the product of number of schemes multiplied by the maximum number of runs per day. This treatment replaces the multi-dimensional nonlinear expression of the same single truck in the same scheme, which ensures that all constraints in the model are linear expression. It cannot change the nature of linear programming and improve ease of operation [11].

The target function is defined as follows:

$$\min C = \sum_{j=1}^{J}\sum_{z=1}^{Z} c_{jz} = \sum_{j=1}^{J}\sum_{z=1}^{Z} u \times \bar{c}_{jz} \times g_{jz2} + \sum_{j=1}^{J}\sum_{z=1}^{Z} \left[\bar{c}_{jz} \times \omega_{jz2} + \left(Kilo_{jz} - H\right) \times y_{jz} \times M_{jz} \times a\right] \tag{5}$$

The model is parsed from different modules below.

3.2.1 Actual Load Constraint Module

In the optional scheme of any truck, the warehouse has certain limits on the actual load of any kind of cargo. First of all, it can be allocated by the scheme to meet the basic condition of not being greater than the maximum load of any kind of truck, as shown in Eq. (6).

$$0 \le q_{jzf} \le a_{zf}\ (j = 1, \cdots J; z = 1, \cdots, Z; f = 1, \cdots, F) \tag{6}$$

On the contrary, the reality does not allow for the actual load exceeding the maximum load. Failure to do so can result in overflowing of shipments or overcapacity waste to the destination.

3.2.2 Dynamic Balance Module Between Supply and Demand

Based on the choice of distribution scheme, the dynamic balance model of supply and demand can effectively avoid the problem of excess load waste, and also ensure that the demand of customers and suppliers are equal to supply and demand.

In the actual transport process, the balance between the supply of goods and the demand is to ensure that the actual load of all programmers allocated to any type of shipment is equal to the actual demand of all users for any shipment (balance of supply and demand). This condition is not only a limitation of the previous constraint, but also a necessary solution condition for cost function calculation. The cumulative sum of the supply and demand balance of all transportation materials equals to the actual load of a truck with any scheme. Its loading rate should be used for the calculation of load cost:

$$\sum_{z=1}^{Z} q_{jzf} = q_{jf} (j = 1, \cdots, J; f = 1, \cdots, F), M_{jz} = \sum_{f=1}^{F} q_{jzf} \text{ (Actual load)} \quad (7)$$

3.2.3 Linear Conversion of Nonlinear Cost Function

In actual transportation, it is usually stipulated that when the transport distance is less than *H km*. It is paid by the actual carrying capacity by unit price. If distance exceeds *H km*, it is considered as a long-distance transport. When the total cost function involved in this model can be divided into the load cost of the actual load and the transportation cost of the delivery (Fig. 3):

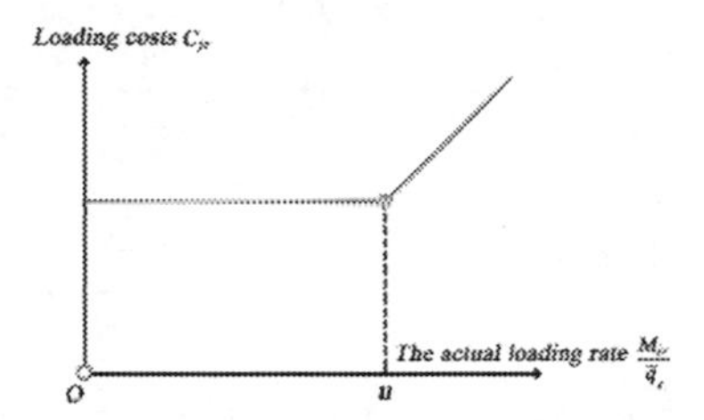

Fig. 3. Actual load cost function

$$c_{jz} = c_{jz1} + c_{jz2}$$

(1) Load Cost Function

$$\bar{c}_{jz} = c \times \bar{q}_z$$
$$\text{(Vehicle full load cost of } C\, yuan/ton \times \textit{full load weight (tons)}) \quad (8)$$

The actual load cost function is a piecewise function, and there is a discontinuity point (actual load rate = 0). It is difficult for such a function to deal with by nonlinear programming. The model construction introduces 0–1 decision variables, i.e. logical variables, which can be discussed in the same problem in a multi-class situation. So the construction of the 0–1 integer programming model can be used to solve the three types of loading cases defined in this article: no load, no overload *u* load rate and more than *u* load rate. Then, nonlinear function can be converted to linear expression:

$$c_{jz1} = 0 \times \bar{c}_{jz} \times g_{jz1} + u \times \bar{c}_{jz} \times g_{jz2} + \bar{c}_{jz} \times \omega_{jz2} \quad (9)$$

Here, $\frac{M_{jz}}{\bar{q}_z} = \omega_{jz1} + \omega_{jz2}$ represents the actual load rate, and $g_{jz1} + g_{jz2} + g_{jz3} = 1$ is used to constrain the range of ω_{jz1} values and ω_{jz2} values for $0 \le \omega_{jz1} \le u \times g_{jz3}$, $u \times g_{jz3} \le \omega_{jz2} \le g_{jz3}$ in which ω_{jz1} value range is $[0, u]$; ω_{jz2} value range is $\{[u, 1], 0\}$. g_{jz1} represents an empty load, g_{jz2} represents that the loading rate is not above *u* and g_{jz3} represents a load rate of more than *u*.

(2) Transportation Cost Function

Assuming that the unit price for transport ingress over *H km* is specified as *a yuan/kilometer/ton of actual loading*, the transportation cost calculation formula is:

$$c_{jz2} = \left(Kilo_{jz} - H\right) \times M_{jz} \times a \tag{10}$$

As described in the overview, the total cost calculation can be summarized into the four cost scenarios in list (Table 2):

Table 2. Costing classification table

Cost of a single shipment	Delivery distance not exceeding *H km*	Delivery distance of more than *H km*
The actual load rate is not exceeded u	$u \times \bar{c}_{jz}$	$u \times \bar{c}_{jz} + \left(Kilo_{jz} - H\right) \times M_{jz} \times a$
Actual load rate exceeds u	$\bar{c}_{jz} \times \omega_{jz2}$	$\bar{c}_{jz} \times \omega_{jz2} + \left(Kilo_{jz} - H\right) \times M_{jz} \times a$

Variable y_{jz} is introduced to classify the distance between starting point to any user's distance. At the same time, if shipping time difference does not exceed that the user can accept and the actual traffic cost difference do not exceed the fixed difference, then the path with shorter time is chosen; on the contrary, path with less cost is preferred. In the total cost function, where y_{jz} is a determining value, the final expression of the total cost can be converted into a linear equation.

Therefore, the target function of cost is defined as follows:

$$c_{jz} = 0 \times \bar{c}_{jz} \times g_{jz1} + u \times \bar{c}_{jz} \times g_{jz2} + \bar{c}_{jz} \times \omega_{jz2} + \left(Kilo_{jz} - H\right) \times y_{jz} \times M_{jz} \times a \tag{11}$$

3.2.4 Daily Run Time Limit Module

The data preparation in the database is to arrange all the vehicles in order, and the total number of schemes is also arranged in order. The truck i has b_i kinds of charging schemes. It is known that the loading speed of the supplier is v minutes/ton. Assuming that the time ratio of loading and unloading materials is 1:2, there are:

$$\sum_{j=1}^{J} \sum_{z=1+\sum_{n=1}^{i} b_{(n-1)}}^{z=\sum_{n=1}^{i} b_n} \left(g_{jz2} + g_{jz3}\right) \times Time_{jz} + \frac{M_{jz}}{10^3} \times v \times 3 \leq (13 - t) \times 60 \tag{12}$$

in which $g_{jz2} + g_{jz3} = 1$ means loading (using a truck), $g_{jz2} + g_{jz3} = 0$ means no load. $Time_{jz}$ represents the traffic time if the truck departs to the delivery destination.

In this case, the normal working time of the staff is 8:00–21:00. The total working time is 13 h, and t hours is reserved as the total time for a truck to wait for loading (not counting the time for loading and unloading the truck), so the time is limited to $13 - t$ hours. In addition, the time unit of this model is minutes, and then $13 - t$ hours can be converted into $(13 - t) \times 60$ min.

3.2.5 Scheduling Module

This model can be used for a single vendor and multi-user distribution problem, while the result of multi-objective mutual constraint module is to minimize the total cost. So there can be multiple cost scenarios. Here, the theoretical model cannot be deviated from the actual discussion. If the following conditions occur in the results, the model error can be confirmed [12].

1) With the same actual carrying capacity of the truck, but a long distance of highway or low-speed road is chosen as the trip path;
2) There are even-numbered shifts for the same truck to be delivered to the same user, and the shift with a loading rate exceeding u is even-numbered minus one;
3) The same truck is loaded with different users' cargo at any frequency;

If the above contradicts the basic assumptions, it can prove that the results are wrong. Comparing the actual data of the supplier into the actual model operation and comparing the resulting cost data with the original manual scheduling cost data, it is obvious that the total cost has been reduced.

The scheduling problem in this article uses a random planning model to schedule every half hour, for the setting of this article, consider the limited number of loading ports, still need to ensure that all shifts queue loading time is the smallest target to meet the following basic conditions: (See Fig. 4 for details).

1) Trucks with different shifts on the same order may not be blocked at the destination;
2) Different departures of the same truck cannot be scheduled at the same time;

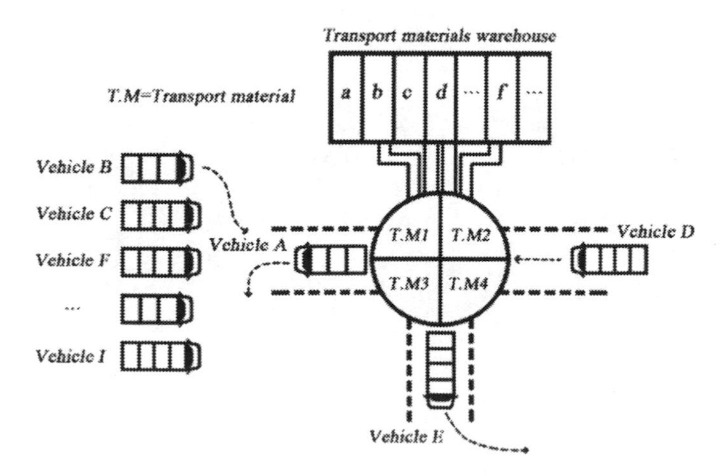

Fig. 4. Queuing diagram

4 Simulations and Result Analysis

In order to verify the validity of the optimization model, the real data of supplier *W* is applied to this model. *Python3.0* is adopted as the programming language in the *Window10* operating system platforms to complete the implementation of the optimization model. Comparative analysis with actual cost is presented similar to [13].

In the simulation data of this article, given loading rate $u = 80\%$, threshold $H = 100$, unit price of actual load $c = 71\, yuan/ton$, $a = 0.22\, yuan/ton \times kilometer$ part of the order data sample is listed in Table 3 and available vehicle information is presented in Table 4. Order information includes supplier name, user name, order number, transport material type and quantity. Among the selected orders, there are four types of assignable transport materials: *a, b, c* and *d*. The quantity of materials required for each user's order is not equal. According to the above information, a basic database is established.

Table 3. Order information

Current order information				
Supplier name	User name	Order number	Transport material type	Transport material quantity
W	User 1	201910141	*a*	16000
W	User 2	201910142	*b*	8000
W	User 3	201910143	*a*	12000
W	User 4	201910144	*c*	10000
W	User 5	201910145	*d*	15000
W	User 6	201910146	*b*	10000
W	User 7	201910147	*c*	9000
W	User 8	201910148	*c*	7000

Table 4. Vehicle information

Vehicle information of supplier *W*			
Supplier name	Registration number	Loading quantity of vehicle bins	Number of runs
W	Vehicle *A*	12000, 8000	2
W	Vehicle *B*	13000, 7000	2
W	Vehicle *C*	2500, 3000, 4000, 4500	2
W	Vehicle *D*	5750, 5750, 5750, 5750	2
W	Vehicle *E*	10000, 10000	2

The distribution scheme outputted by the model is as follows:

① Registration V.A → Transport material a → Load on 12t → High speed → User 3
② Registration V.A → Transport material c → Load on 9t → High speed → User 7
③ Registration V.B → Transport material a → Load on 16t → Low speed → User 1
④ Registration V.B → Transport material d → Load on 15t → High speed → User 5
⑤ Registration V.C → Transport material b → Load on 8t → High speed → User 2
⑥ Registration V.C → Transport material b → Load on 10t → High speed → User 6
⑦ Registration V.D → Transport material a → Load on 7t → High speed → User 8
⑧ Registration V.E → Transport material a → Load on 12t → High speed → User 4

The vehicle distribution scheme is related to the order demand and route selection. Because the traffic cost of different routes at high and low speed is within the difference acceptable to the supplier, the vehicle driving scheme will give priority to the route with less driving time. At the same time, in the vehicle arrangement, there is a reasonable phenomenon that different types of materials are distributed to the same user. Another reasonable phenomenon is that the same vehicle transports multiple users in different plans on the same day. At the same time, the actual loading rate in the transportation cost of a single vehicle is close to 80%. It is to realize the goal result of minimum load cost. Maximum time efficiency and the total cost are as small as possible.

The total cost of this scheme is ¥12376.

In this article, the running cost of the model is compared with the manual scheduling cost of supplier W. The simulation results of the model not only reduce the total cost and improve the time utilization rate, but also generate a visual chart as Table 5 through a simple queuing model [14]. It is used for the vehicle time travel arrangement provided by the supplier to the outsourcer. The vehicle travel table includes registration number, loading time, route selection, distribution destination and ideal cost under actual conditions.

According to the above simulation results, there are many factors to be considered in the actual vehicle distribution process. The model constructed in this article can solve multi-objective optimization problems under the basic constraints, such as goods damage time, transportation cost, loading cost, time efficiency and so on. At the same time, the simulation results show that multiple orders can be delivered at the same time, which can be reasonably arranged when the number of available vehicles is sufficient. All of them can be arranged reasonably. On the contrary, the system will provide suppliers with prompt information such as adding vehicles, and visual window settings will enable suppliers and outsourcers to understand information intuitively and ensure the feasibility of operation.

Table 5. Visual chart of vehicle arrangement

Result generation				
Loading time	Registration number	User name	Loading scheme	Recommend scheme and notes
8:00	*B*	User 1	{*a* 16000}	**Low speed** (High speed distance 198.0 km, ¥2816; Low speed distance 102.0 km, ¥1143)
8:00	*C*	User 2	{*b* 8000}	**High speed** (High speed distance 350.0 km, ¥1235; Low speed distance 386.0 km, ¥1298)
8:00	*A*	User 3	{*a* 12000}	**High speed** (High speed distance 292.0 km, ¥1642; Low speed distance 264.0 km, ¥1568)
8:30	*E*	User 4	{*c* 10000}	**High speed** (High speed distance 328.0 km, ¥1637; Low speed distance 314.0 km, ¥1606)
8:30	*D*	User 8	{*c* 7000}	**High speed** (High speed distance 448.0 km, ¥1842; Low speed distance 362.0 km, ¥1709)
12:30	*B*	User 5	{*d* 15000}	**High speed** (High speed distance 328.0 km, ¥1888; Low speed distance 314.0 km, ¥1842)
14:00	*C*	User 6	{*b* 10000}	**High speed** (High speed distance 224.0 km, ¥1608; Low speed distance 196.0 km, ¥1006)
14:00	*A*	User 7	{*c* 9000}	**High speed** (High speed distance 224.0 km, ¥1381; Low speed distance 196.0 km, ¥1326)

5 Conclusions and Perspectives

5.1 Conclusions

(1) The multi-target 0–1 hybrid integer linear programming model proposed in this article can solve the optimal distribution scheduling problem with cost minimization under the mutual constraints of cost, traffic volume, route and time.
(2) The constraint model of actual loading quantity is based on the random distribution of the system load and the fixed maximum vehicle load. By setting the restricted conditions, the problem of waste caused by overload in this kind of model is solved.
(3) The linear transformation of nonlinear cost function adopts the 0–1 variable to effectively express the three loading situations of vehicles, and sets the loading rate according to the requirements of the logistics party, so as to calculate the cost function. By transforming the nonlinear cost function into the linear function, the multiple linear programming of the model is constructed to realize the cost minimization with the restricted constraints.
(4) The limit model of daily running time limits the working hours of vehicles within the working range. The number of vehicles and schemes are arranged in order to solve the problem that goods cannot be delivered to customers on time during the working period.

5.2 Perspectives

In this article, the multi-target 0–1 hybrid integer linear programming model is applied to the optimization of distribution scheme under resource constraints. The feasibility of the model in solving this kind of problem is verified by the comparison of actual data. The model established in this article can be used for the distribution problem of one supplier and multi-user. The available field of the model should be further expanded to realize the multi-supplier to multi-user distribution scheduling.

In addition, referred to [15], the extensive application of the model is not limited only to the material transportation and distribution system. It can be used for small-scale in plant scheduling, merchant distribution, large-scale flight control to the airport, large-scale processing plant distribution problems, and so on.

References

1. Anna, N., Li, D.: A supply chain network game theory model with product differentiation, outsourcing of production and distribution, and quality and price competition. Ann. Oper. Res. **226**(1), 479–503 (2015). https://doi.org/10.1007/s10479-014-1692-5
2. Marcel, T., Andreas, K.: Demand dispersion and logistics costs in one-to-many distribution systems. Eur. J. Oper. Res. **223**(2), 499–507 (2012)
3. Li, H.J., An, H.Z., Fang, W., et al.: A theoretical cost optimization model of reused flowback distribution network of regional shale gas development. Energy Policy **100**, 359–364 (2017)

4. Valentina, C., Fabio, F., Martin, P.K.: Approaches to a real-world Train Timetabling Problem in a railway node. Omega **58**, 97–110 (2016)
5. Ademir, A.R., Mael, S., Sandra, A.S.: On the augmented subproblems within sequential methods for nonlinear programming. Comput. Appl. Math. **36**(3), 1255–1272 (2017). https://doi.org/10.1007/s40314-015-0291-7
6. Shipra, A., Wang, Z.Z., Ye, Y.Y.: A dynamic near-optimal algorithm for online linear programming. Oper. Res. **62**(4), 876–890 (2014)
7. Tao, Y., Chew, E.P., Lee, L.H., et al.: A column generation approach for the route planning problem in fourth party logistics. J. Oper. Res. Soc. **68**(2), 165–181 (2017)
8. Mario, C.S.J., Thiago, E.P.: Logistics in the road modal: a costs approach in function of the distance of transport and type of vehicle. Semina Ciências Exatas e Tecnológicas **30**(1), 63 (2009)
9. Jose, L., Andrade, P., David, C., Pedro, L.G-R.: On modelling non-linear quantity discounts in a supplier selection problem by mixed linear integer optimization. Ann. Oper. Res. **258**(2), 301–346 (2017). https://doi.org/10.1007/s10479-015-1941-2
10. Zhou, Z.Y., Yu, B.: The flattened aggregate constraint homotopy method for nonlinear programming problems with many nonlinear constraints. In: Abstract and Applied Analysis 2014 (2014)
11. Janne, E., Tomi, S., Juuso, T., et al.: Multiple-method analysis of logistics costs. Int. J. Prod. Econ. **137**(1), 29–35 (2012)
12. Gu, Y., Dong, S.J.: Logistics cost management from the supply chain perspective. J. Serv. Sci. Manage. **09**(03), 229–232 (2016)
13. Wiljar, H., Inger, B.H., Knut, V.: Logistics costs in Norway: comparing industry survey results against calculations based on a freight transport model. Int. J. Logist. Res. Appl. **17** (6), 485–502 (2014)
14. Sun, L., Atul, R., Mark, H.K., et al.: Transportation cost allocation on a fixed route. Comput. Ind. Eng. **83**, 61–73 (2015)
15. Karol, W., Jacek, W., Bogusław, S.: The concept of computer software designed to identify and analyse logistics costs in agricultural enterprises. J. Agribus. Rural Dev. **2**(12), 267–278 (2009)

Research of Strategies of Maintaining Population Diversity for MOEA/D Algorithm

Wenxiang Wang(✉), Xingzhen Tao, Lei Deng, and Jun Zeng

College of Information Engineering, Jiangxi College of Applied Technology, Ganzhou 341000, Jiangxi, China
406873165@qq.com

Abstract. In recent years, MOEA/D algorithm has been recognized by the industry for its inherent advantages in dealing with super multi objective optimization problems, and its application is also very extensive. However, MOEA/D algorithm also has the problem of lack of population diversity during the later stage of evolution, resulting in slow convergence speed. In this paper, it makes a research on the strategy of maintaining population diversity based on MOEA/D algorithm and proposes three population diversity maintenance strategies, namely SBX-DE operator competition, mutation probability adaptive modulation, and double-faced mirrors theory boundary processing. The experiments' result shows that all of these three strategies can effectively improve the diversity of the MOEA/D algorithm in the late evolutionary population, and contribute to the convergence speed of the MOEA/D algorithm.

Keywords: MOEA/D · Population diversity · Convergence speed

1 Introduction

Multi Objective Optimization Problems (MOPs) play an important role in scientific research, production practice, and engineering applications. The research on MOP issues has important theoretical and academic value. Compared with single-objective optimization, the goal of MOP is no longer a single optimal solution, but a set of mutually constrained solution sets, that is, the improvement of one optimization target may be accompanied by the deterioration of other optimization goals. To this end, the academic community generally uses the Pareto solution set to represent a set of compromise solution sets of various objectives in multi-objective optimization, and the projection of Pareto optimal solution sets in the target domain is called Pareto Front (PF) [1].

The early multi objective evolutionary algorithms are evolutionary algorithms based on the Pareto dominance relationship, such as SPEA [2], SPEA-II [3], NSGA-II [4], PESA [5], PESA-II [6] and so on. Although these evolutionary algorithms have achieved remarkable results in the field of multi-objective optimization, they face challenges when facing the problem of ultra-multi-objective optimization (the number of objectives is greater than or equal to 3). In 2007, Zhang [7] and others proposed a decomposition-based multi-objective evolutionary algorithm (MOEA/D), which decomposes multi-objective optimization problems into a set of single-objective

K. Li et al. (Eds.): ISICA 2019, CCIS 1205, pp. 209–221, 2020.
https://doi.org/10.1007/978-981-15-5577-0_16

optimization problems, and then uses evolutionary algorithm to solve these decomposition problems simultaneously. At the same time, MOEA/D algorithm also introduces neighbor relation to share evolution information. In addition, the MOEA/D algorithm is directly applicable to the super multi-objective optimization problem, and overcomes the limitations of the evolutionary algorithm based on the Pareto dominance relation. Therefore, the MOEA/D algorithm framework has become a research hotspot in recent years.

Although MOEA/D algorithm has made great progress in the field of multi-objective optimization, there are also some problems such as the lack of diversity and slow convergence rate in the late evolution. For this reason, scholars have proposed various improvement schemes of MOEA/D algorithm, mainly in the following aspects: improvement of weight vector, improvement of decomposition method, improvement of evolutionary operator, improvement of matching selection, improvement of replacement selection. However, these studies have not done much to solve the problem of population diversity loss in MOEA/D algorithm, and also lack systematic research programs for diversity loss.

In order to solve the problem of missing population diversity of MOEA/D algorithm, this paper systematically studied the population diversity maintenance strategy of MOEA/D algorithm, and carried out research on the improvement strategies of three aspects: one is to adopt a competitive employment evolution strategy combined with simulated binary crossover (SBX) [8] and Differential Evolution (DE) [9]. The second is to adopt the mutation probability adaptive adjustment strategy. The mutation probability will be adjusted for adaptive growth according to the diversity change of the population in the evolution process. At the same time, if the evolution falls into the stagnation state, the algorithm will temporarily increase the mutation probability for breaking through the stagnation state. Third, the boundary processing problem of double specular reflection principle and the problem of overcoming the boundary aggregation in the process of population evolution. Simulation experiments show that the above three population diversity maintenance strategies can enhance the diversity of MOEA/D algorithm populations in the late evolution stage and improve the evolution rate of MOEA/D algorithm.

2 Related Work

2.1 Multi Objective Optimization Problems and the Related Concepts

Taking the minimization of multi objective problems as an example, the MOP problem can be described as:

The decision variable $X = (x_1, x_2, \ldots, x_n)$ should satisfy the constraint:

$$\begin{cases} g_i(X) \geq 0 & i = 1, 2, \ldots, k \\ h_i(X) = 0 & i = 1, 2, \ldots, k \end{cases} \tag{1}$$

With m objective functions, the optimization objective is expressed as

$$F(X) = (f_1(X), f_2(X), \ldots, f_n(X))^T \quad (2)$$

Then the MOP problem is to find a set of solution sets $X^* = (x_1^*, x_2^*, \ldots, x_n^*)$ in the decision space to make $F(X^*)$ minimum under the premise of satisfying the constraints. On this basis, the following definitions are given:

(1) Feasible solution set X_f: a set of decision variables x satisfying constraint condition (2). That is $X_f = \{x \in X | g(x) \geq 0, h(x) = 0\}$

(2) Pareto dominance:
Suppose p and q are any two different individuals in the evolutionary group, if the following relationship is satisfied: ① All sub-targets of p are no worse than q, namely $f_k(p) \leq f_k(q)(k - 1, 2, \ldots, n)$; (2) p at least has one sub-goal that is better than q, namely $\exists l \in \{1, 2, \ldots, n\}, f_k(p) < f_k(q)$, then p dominates q, p dominates and q is dominated.

(3) Pareto's optimal solutions:
Suppose $x^* \in X_f$, if there is no other solution $x'^* \in X_f$ to make $f_i(x'^*) \leq f_i(x^*)(i = 1, 2, \ldots, m)$, and at least one is a strict inequality, then x^* is said to be the Pareto's optimal solution of this MOP problem.

(4) Pareto's optimal solution set:
The set of all Pareto optimal solutions of MOP problem is Pareto Set (PS).

(5) Pareto Front:
The projection of the Pareto optimal solution set on the target space is the Pareto Front (PF), namely $PF = \{F(x) | x \in PS\}$.

2.2 Introduction to MOEA/D Algorithm

MOEA/D algorithm uses aggregation function to decompose MOPs into N subproblems, and optimizes these subproblems at the same time. The algorithm divides neighborhood for each subproblem by calculating Euclidean distance between weight vectors. Each subproblem realizes coevolution through neighborhood. MOEA/D algorithm flow is as follows:

Step 1: Initialization

a) Initialize the weight vector $\lambda = \{\lambda_1, \lambda_2, \ldots, \lambda_N\}$ and divide it into neighborhoods. Assuming the individual $x_i \in X$, then $B(i) = \{i_1, i_2, \ldots, i_T\}$ is the neighborhood of the No. i weight vector.
b) Initialize the population $pop = \{x_1, x_2, \ldots, x_N\}$ and calculate its target value $FV_i = F(x_i)$.
c) Calculate the ideal point $z = \{z_1, z_2, \ldots, z_m\}^T$, where m is the target number and z_i is the optimal value of f_i in the current generation.

Step 2: Update

a) Evolution: Firstly, two individuals x_j and x_k are randomly selected from the neighborhood $B(i)$, and a new individual x_d is generated by using a simulated binary crossover (SBX) algorithm. Then, x_d is subjected to polynomial mutation to obtain a mutant individual y, and finally, the mutation individual y is improved to y' according to the constraint of the problem.

b) Update the ideal point: for $j = 1, 2, \ldots, m$, if $f_j(y') < z_j$, then $z_j = f_j(y')$.

c) Update the neighborhood: For $j \in B(i)$, if the aggregate function value of y' is not greater than x_j, then $x_j = y'$, and the corresponding target value $FV_j = F(y')$ is updated at the same time.

Step 3: Stop iteration

If the stop condition is satisfied, stop iteration and output the best solution set; Otherwise, continue to Step 2.

3 Study on Population Diversity Maintenance Strategy of MOEA/D Algorithm

According to the above MOEA/D algorithm flow, the simulation experiment of the test function was carried out on the MOEA/D algorithm. The experiments show that it's easy to have the problems of premature convergence and lack of diversity.

Taking the test function DTLZ3 as an example, MOEA/D algorithm is used to solve the evolution of the test function DTLZ3. The test function DTLZ3 is propagated for 200 generations. The average of the Pareto optimal solution sets of the three optimization targets of each generation is calculated to reflect the evolution degree of the MOEA/D algorithm. The variation curve of the average value of Pareto optimal solution sets is shown in Fig. 1. It can be seen from Fig. 1 that after the MOEA/D population has multiplied to 40 generations, the average reduction trend of Pareto optimal solution sets of the three optimization objectives of the algorithm is significantly slower, indicating that the evolution speed of the MOEA/D algorithm has slowed down.

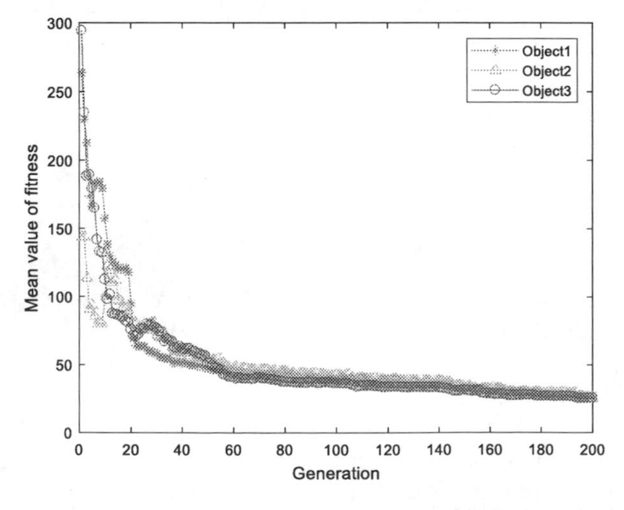

Fig. 1. Pareto optimal solution set mean curve of MOEA/D algorithm

The reason why MOEA/D algorithm's evolution speed slows down is the lack of diversity after the population has multiplied to a certain extent, which leads to the algorithm falling into a local optimal solution. Variance can well reflect the diversity of the population. The larger the variance is, the better the diversity of the population, and vice versa. For this reason, the mean value of the variance of each dimension in each generation is further studied and counted, as shown in Fig. 2. It can be seen that the significant decrease of variance mean in all dimensions of the population is basically synchronous with the slow evolution of the algorithm, which also proves that the slow evolution of MOEA/D algorithm is indeed related to the lack of population diversity.

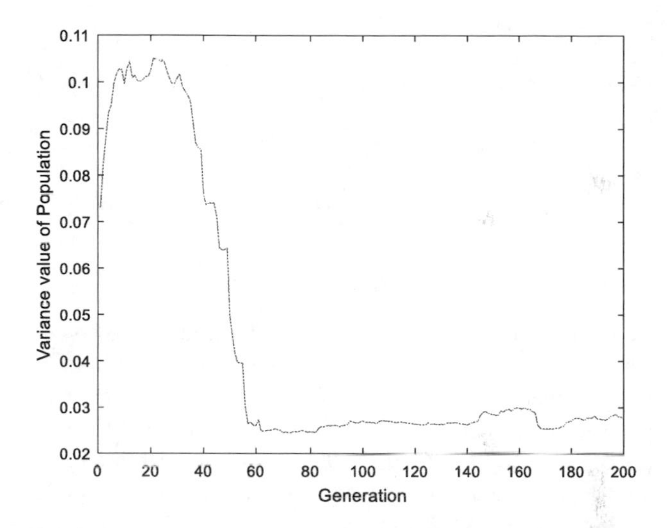

Fig. 2. Population variance mean curve of MOEA/D algorithm

The study also found that the Pareto optimal solution set was unevenly distributed in the simulation process of MOEA/D algorithm, which showed that the solution set was aggregated near the boundary while thc other regions were sparsely distributed, as shown in Fig. 7. This will also affect the population diversity of MOEA/D algorithm.

The reason of uneven distribution of Pareto optimal solution set of MOEA/D algorithm is that the algorithm adopts the following methods when dealing with boundary problems:

$$y = \begin{cases} X_{\min}, & y < X_{\min} \\ X_{\max}, & y > X_{\max} \\ y, & others \end{cases} \tag{3}$$

Where y is an new individual obtained by evolution, $X_{\min}$ is its lower boundary limit, and $X_{\max}$ is its upper boundary limit. This method will make all the trans boundary individuals to gather and project onto the boundary.

In order to solve the problem of lack of population diversity in MOEA/D algorithm, this paper tries three strategies to maintain population diversity.

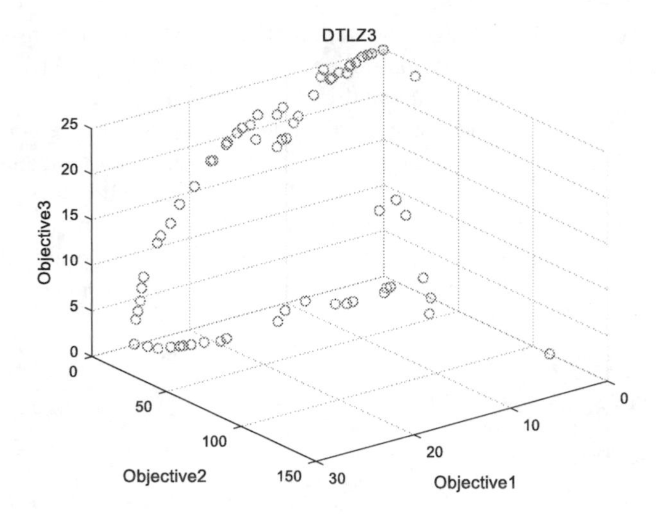

Fig. 3. Pareto optimal solution set distribution of MOEA/D algorithm

3.1 The SBX-DE Operator Competition Strategy

SBX-DE operator competition strategy is that the algorithm starts to use DE algorithm to perform crossover operation, and counts the number of neighbor updates in each round of evolution in real time. When the number of neighbor updates is significantly reduced in five consecutive rounds of evolution, it indicates that the evolution speed is slowing down. SBX algorithm is used to replace DE algorithm to perform crossover operation at this time, and then this competition polling mechanism is used to alternately use the two evolutionary algorithms.

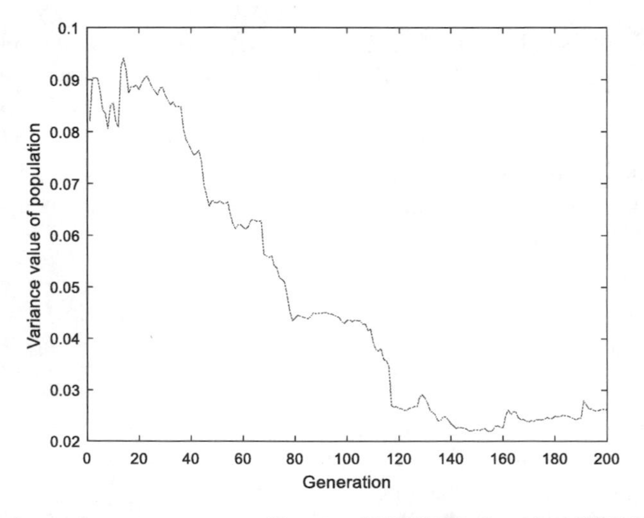

Fig. 4. Population variance mean curve of improved MOEA/D algorithm (SBX-DE competition induction strategy)

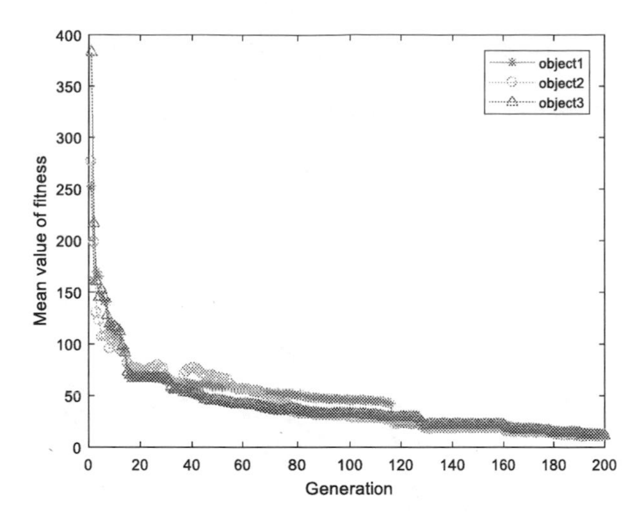

Fig. 5. Pareto optimal solution set mean curve of improved MOEA/D algorithm (SBX-DE competition induction strategy)

The following simulation and verification experiments are carried out. The DTLZ3 test function is taken as the test object, the number of decision variables is set to 12, the dimension of the problem is set to 3, and the population reproduction algebra is set to 200 generations. The variance mean curve of each dimension of MOEA/D algorithm population was obtained after adopting the SBX-DE operator competition strategy. As shown in Fig. 4, it can be seen that the population variance decrease smoothly, while it's has a sudden decrease in Fig. 2, indicating that the SBX-DE operator competition strategy can enrich the diversity of the MOEA/D algorithm population.

After enriching the diversity of the population, in order to further study the influence of the strategy on the convergence speed of MOEA/D algorithm, the experiment continued to count the Pareto optimal solution set mean of the improved algorithm, and obtained the Pareto optimal solution set mean curve of the improved MOEA/D algorithm as shown in Fig. 5. Compared with Fig. 1, it can be seen that the Pareto optimal solution set mean continues to evolve to a better value after 40 generations of reproduction. It shows that SBX-DE operator competition strategy can improve the convergence speed of MOEA/D algorithm.

3.2 Adaptive Adjustment Strategy of Mutation Probability

Another solution to maintain population diversity is to increase the probability of mutation. The adaptive adjustment strategy of mutation probability is to make the individual mutation probability adapt to the evolution process, and the adjustment rule is to make the mutation probability change in a negative exponential growth law with the evolution process. The variation curve of mutation probability is shown in Fig. 6. The population diversity performs very well at the beginning of evolution process, and the mutation probability is low. But with the evolution process the population diversity decrease, which need to increase the mutation probability to improve it. However, the mutation probability should tend to be stable after reach a certain value to avoid the problem of reducing convergence speed caused by the high mutation probability. The mutation probability computation formula is as follows:

$$Muta\_rate = 0.2 * (-e^{-\frac{gen}{\chi}} + 1) \tag{4}$$

Where *gen* is the population evolution algebra and χ is the scaling factor, here set to 400.

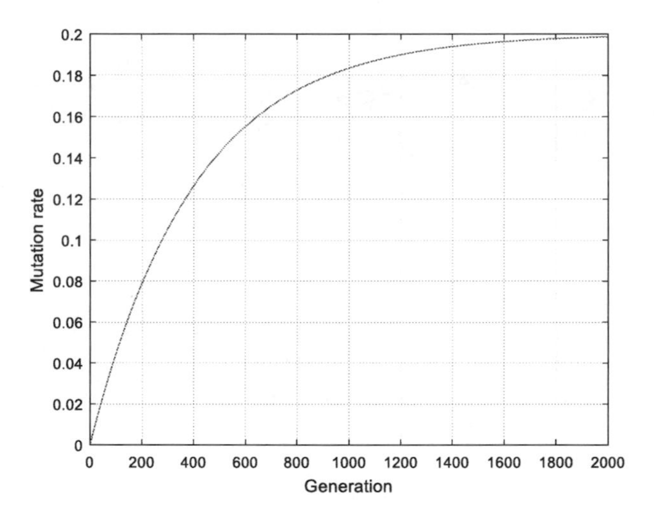

Fig. 6. Variation probability adaptive adjustment curve

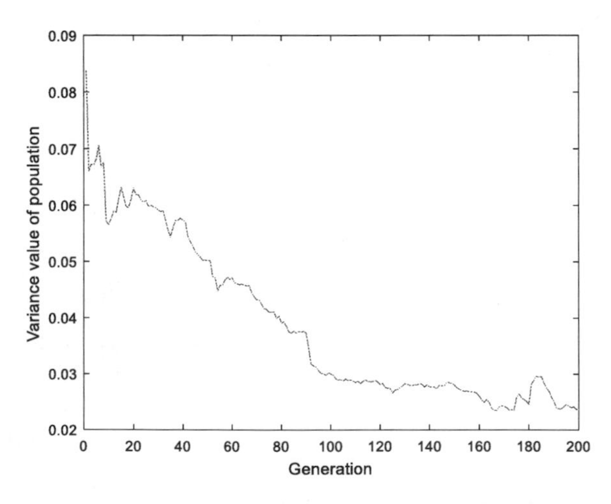

Fig. 7. Population variance mean curve of improved MOEA/D algorithm (adaptive adjustment strategy of mutation probability)

In addition, by recording the number of neighbor updates, when it is detected that the algorithm is stuck in a stagnant state (the number of neighbor updates counted for 5 generations of continuous evolution is less than a set threshold), the mutation probability will be temporarily increased to seek a breakthrough in the stagnant state. Figure 7 is the mean variance curve of every population dimension of improved MOEA/D algorithm by adopting adaptive adjustment strategy of mutation probability. It can be seen that the strategy can also mitigate the attenuation trend of the MOEA/D algorithm population diversity.

In terms of convergence, Fig. 8 shows the mean curve of Pareto optimal solution set of improved MOEA/D algorithm by adopting adaptive adjustment strategy of mutation probability. Compared with Fig. 1, it can be seen that the evolution rate is faster and does not fall into the evolutionary stagnation state.

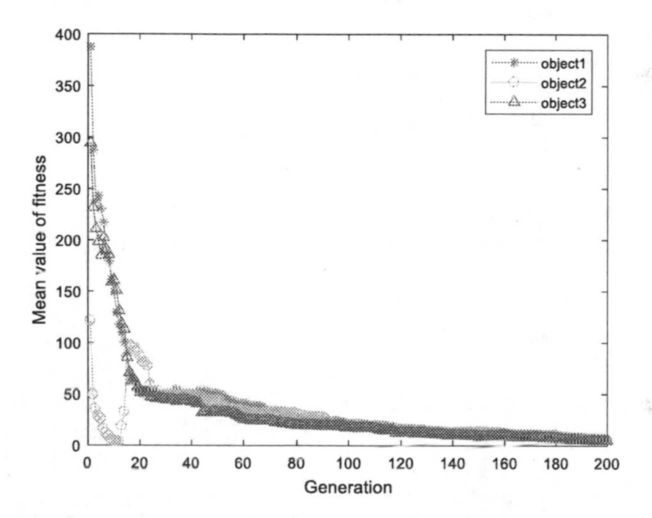

Fig. 8. Pareto optimal solution set mean value curve of improved MOEA/D algorithm (adaptive mutation probability)

3.3 Double-Faced Mirrors Theory Boundary Processing Strategy

The principle of double-faced mirrors theory: as shown in Fig. 9, the boundary is $[X_{\min}, X_{\max}]$, y is set as the crossing point, where the upper and lower boundaries are two mirrors. y is the propagating beam and the size of y indicates the light intensity. The intermediate medium has optical loss. Then, after a large amount of double-faced mirrors reflection, the light will eventually be exhausted at a certain point y' in the boundary due to the intermediate medium loss. If there are several beams of continuous intensity values, the final dissipation point of these beams after double-faced mirrors reflection will be evenly distributed in the boundary region, as shown in Fig. 10.

If the cross-boundary value is regarded as a beam of light, the cross-boundary problem can be dealt with by the principle of double-faced mirrors reflection. The boundary's value will eventually fall to a certain point in the boundary after double-mirrors reflection. Continuous and uniform coverage in the boundary area can be

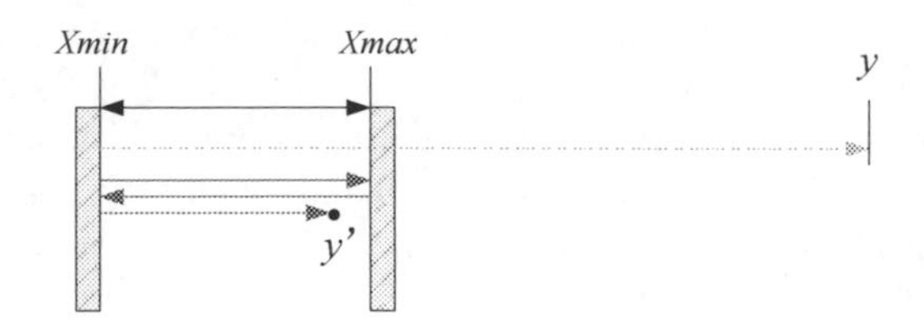

Fig. 9. The principle of double specular reflection

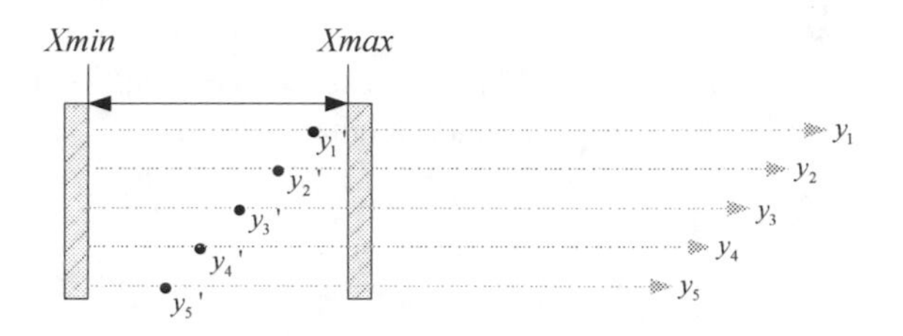

Fig. 10. Processing effect of the double-faced mirrors reflection principle

realized for multiple random cross-boundary values. The formula for dealing with boundary problems by the principle of double-faced mirrors reflection is as follows:

$$y' = \begin{cases} X_{\max} - \mathrm{mod}(y - X_{\max}, X_{\max} - X_{\min}), & y > X_{\max} \\ X_{\min} + \mathrm{mod}\,(X_{\min} - y, X_{\max} - X_{\min}), & y < X_{\min} \end{cases} \tag{5}$$

Apply the principle of the double-faced mirrors reflection to the boundary processing of MOEA/D algorithm and observe the distribution of Pareto optimal solution set in the evolution process, as shown in Fig. 11. Compared with Fig. 3, there is no Pareto optimal solution sets gathering at the boundary. Meanwhile, the mean variance of each dimension of the improved MOEA/D algorithm using the double-faced mirrors reflection boundary processing test was calculated, as shown in Fig. 12. It can be seen that the decrease trend of variance was slower than that in Fig. 2, indicating that the double-faced mirrors reflection boundary processing strategy can also effectively improve the diversity of the algorithm population.

In terms of convergence, Pareto optimal solution set mean curve of the improved MOEA/D algorithm using the double-faced mirrors theory boundary processing strategy is shown in Fig. 13, which is better than the overall evolution speed of Fig. 1, but there are also some problems of slow convergence of some optimized goals.

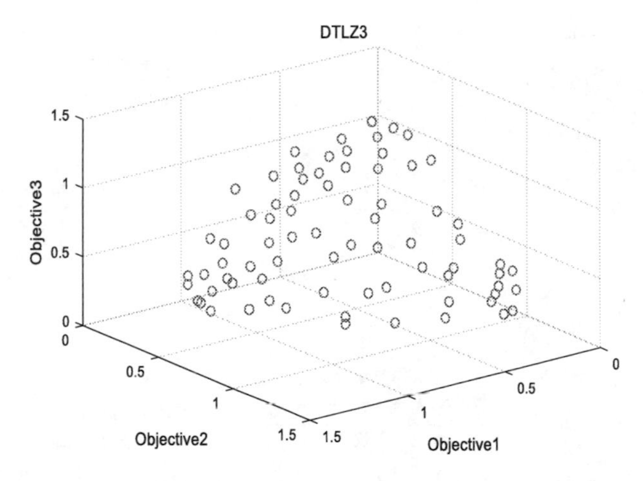

Fig. 11. Pareto optimal solution set distribution diagram of improved MOEA/D algorithm (double-faced mirrors reflection boundary processing strategy)

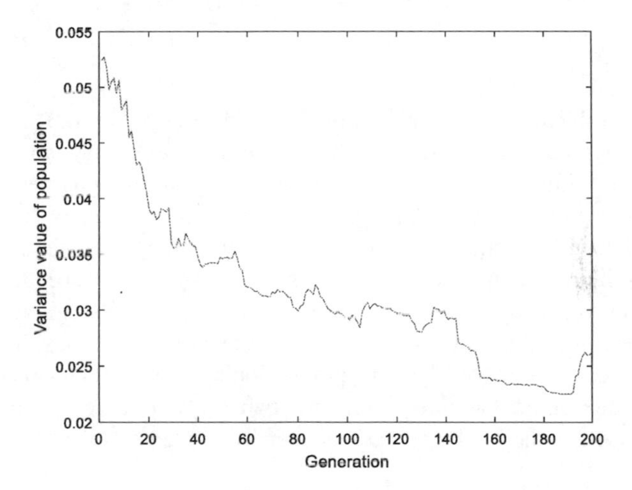

Fig. 12. Population variance mean change curve of improved MOEA/D algorithm (double-faced mirrors reflection boundary processing strategy)

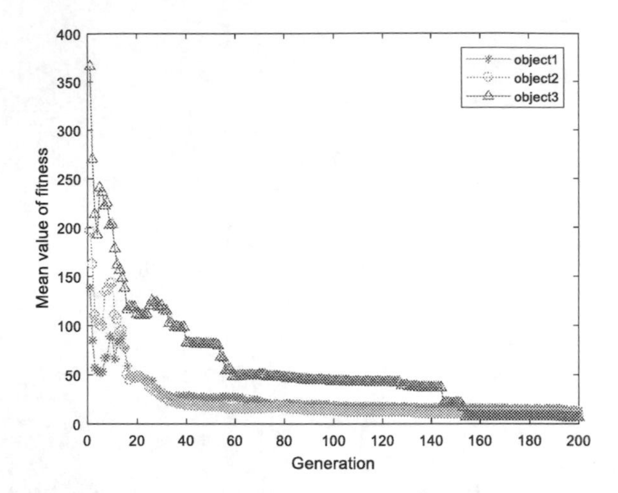

Fig. 13. Pareto optimal solution set mean curve of improved MOEA/D algorithm (double-faced mirrors theory boundary processing strategy)

4 Conclusion

In view of the problem that the loss of population diversity of MOEA/D algorithm in the late evolution process which leads to the slowdown or even stagnation of evolution, this paper studies three population diversity maintenance strategies of MOEA/D algorithms. Among them, the SBX-DE operator competitive strategy and the mutation probability adaptive adjustment strategy can improve the population diversity of MOEA/D algorithm and meanwhile improving its convergence speed. Although the double-faced mirrors theory boundary processing strategy can also improve population diversity, there is also the problem of slow convergence of some evolutionary goals. Due to the different positions of the three population diversity maintenance strategies in the MOEA/D algorithm, the three strategies can be used together in the practical application process, which will achieve better improvement results.

Acknowledgement. This work was supported by the Key Research and Development Project of Ganzhou, the name is "Research and Application of Key Technologies of License Plate Recognition and Parking Space Guidance in Intelligent Parking Lot".

References

1. Sofokleous, A.A., Angelides, M.C.: Dynamic selection of a video content adaptation strategy from a pareto front. Comput. J. **52**(4), 413–428 (2009)
2. Reznick, M.D.: Effects of larval density on postmetamorphic spadefoot toads (Spea hammondii). Ecology **82**(2), 510–522 (2001)

3. Wahid, A., Gao, X., Andreae, P.: 2015 IEEE International Conference on Data Science and Advanced Analytics (DSAA) - Multi-objective Clustering Ensemble for High-Dimensional Data Based on Strength Pareto Evolutionary Algorithm (SPEA-II), Campus des Cordeliers, Paris, France, 19–21 Oct 2015. IEEE International Conference on Data Science and Advanced Analytics, pp. 1–9. IEEE (2015)
4. Deb, K., Pratap, A., Agarwal, S., et al.: A fast and elitist multiobjective genetic algorithm: NSGA-II. IEEE Trans. Evol. Comput. **6**(2), 182–197 (2002)
5. Meniru, G.: Studies of percutaneous epididymal sperm aspiration (PESA) and intracytoplasmic sperm injection. Hum. Reprod. Update **4**(1), 57–71 (1998)
6. Gadhvi, B., Savsani, V., Patel, V.: Multi-objective optimization of vehicle passive suspension system using NSGA-II, SPEA2 and PESA-II. Procedia Technol. **23**, 361–368 (2016)
7. Zhang, Q., Li, H.: MOEA/D: a multi-objective evolutionary algorithm based on decomposition. IEEE Trans. Evol. Comput. **11**(6), 712–731 (2007)
8. Agrawal, R.B., Deb, K., Agrawal, R.B.: Simulated binary crossover for continuous search space. Complex Syst. **9**(3), 115–148 (2000)
9. Das, S., Suganthan, P.N.: Differential evolution: a survey of the state-of-the-art. IEEE Trans. Evol. Comput. **15**(1), 4–31 (2011)

Intelligent Multimedia Systems

Farm Characteristics, Social Dynamics and Dairy Farmers' Conversions to Organic Farming

Qing Xu[1,2](✉), Sylvie Huet[3](✉), and Wei Li[1](✉)

[1] Jiangxi University of Science and Technology, Ganzhou 341000, China
{xuqing, liwei}@jxust.edu.cn
[2] UCA, AgroParisTech, Inra, Irstea, VetAgro Sup, Territoires, Clermont-Ferrand, France
[3] UCA Lapsco, Irstea Lisc – Irstea, Aubière, France
sylvie.huet@irstea.fr

Abstract. This work aims to study the interaction effect of farm characteristics and social dynamics on the conversion to organic farming with an agent-based model. In the model, an agent's decision on the conversion to organic farming is based on the comparison between the satisfaction with its current situation and the potential satisfaction with an alternative farming strategy. A farmer agent's satisfaction is modeled with the Theory of Reasoned Action. It is computed by comparing the agent's outcomes over time and comparing its outcomes against those of other agents to whom it lends great credibility ('important others'). We initialize the model with the Agricultural Census' data about the farm characteristics in 27 French "cantons". The analysis of the simulations shows that farm characteristics are very important to predict the conversion. However, their distribution in the population can also strongly influence the simulation results due to social dynamics. Indeed the interaction between farms having similar or different characteristics can decrease or increase the adoption rate compared to what can be inferred only from farm characteristics.

Keywords: Organic farming · Decision making · Major change · Theory of reasoned action · Agent-based model · Social influence · Credibility

1 Introduction

How to disentangle the role of farm characteristics from the one of social dynamics in the process of organic farming adoption? The position varies with the disciplinary approach chosen to consider the question in this strong debate. Agronomists argue for the farm characteristics [1–3], economists argue for prices and funding [4–6], and social psychologists for subjective norm and professional identities [7–11]. Our work tries to address this issue by studying a simple agent-based model involving farm characteristics and social dynamics. It is implemented with dairy farms in different French regions. The dairy production is particularly interesting. Indeed, the numerous dairy crises and an increasing knowledge about how to farm organically have made the conversion interesting from an economical point of view [12, 13]. This allows our model to put apart the economic issue and to focus on farm characteristics and social dynamics.

K. Li et al. (Eds.): ISICA 2019, CCIS 1205, pp. 225–241, 2020.
https://doi.org/10.1007/978-981-15-5577-0_17

The literature identifies a broad range of factors associated with the adoption and non-adoption of organic farming [7, 14–16]. A strand of research takes a behavioral approach [17] suggesting that motives, values and attitudes determine farmers' decision-making processes. Recently, the conversion to organic farming has been qualified as a major change [18] or a transformational adaptation [19, 20], as well as a social movement [2, 14, 18, 21]. Conversion often implies strong changes in a farmer agent's worldview and social network, and generally begins with a strong need for change [18, 22]. Such a change that engages a number of social processes involving the farm, the farmer agent, its peers and its environment has rarely been studied [23].

Agent-based modeling [24] or individual-based modeling [25] appear relevant and well-geared to help identifying the main drivers that can explain the observed dynamics. However, as pointed out in [26, 27], none of current agent models is well fit to represent the decision process about a major change that is at stake in the conversion to organic farming. This is the reason why our model, argued and described with details in relation to the literature [27], has proposed a dynamic version of the Theory of Reasoned Action (TRA); and showed its potential to explain why farmers do not convert to organic farming in a prototypical farmer population.

In [27], the model is studied with prototypical farm populations and shows different reasons for the lack of conversion to organic farming. [28] shows how to implement the model with real French dairy farms' characteristics given by the French Agricultural Census 2000. It outlines the importance of farm characteristics for the adoption. The present work goes further by implementing with more numerous French "cantons" (27 "cantons") varying in their farm type distributions. It aims to better understand the role of social dynamics compared to farm characteristics in the conversion to organic farming. In general, final adoption rates depend strongly on farm characteristics. However, the conversion explanation is more complex since farmers' social interactions can decrease or increase the adoption rate compared to what can be inferred from farms' characteristics. After presenting the model's principles, we outline the model's behaviors and some explanations before going on to synthesis, and discuss our conclusions.

2 Materials and Methods

2.1 The Model

Basic Elements

Farmer

The model studies the evolution of a population with N farmer agents. Each agent is characterized by its farm; its farming strategy; its performance defined on several dimensions i to evaluate a practice; the importance W_i given to each dimension of practice; the credibility $(C(f,v))$ it lends to each other agent; its memory of applied strategies and performances during the last M periods; its satisfactions with current farming strategy (I_C) and with an alternative one (I_A); and its duration for staying with a strategy (DC) and for being dissatisfied for a strategy (DD).

DC* and *DD capture the duration between two events related to the decision process. *DD* counts an agent's dissatisfaction duration with its current strategy. In the model, an agent has to be dissatisfied long enough with its current strategy to change it. *DC* counts the duration since last strategy change. An agent cannot consider changing strategy again even if the agent is dissatisfied with it during the confirmation period. This is consistent with the theory of innovation diffusion [29] in which an agent has a confirmation period just after adopting a new strategy. Both durations are necessary to account for an agent's stability and consistency. The delayed action of both can only occur when the corresponding duration is above the parameter *TD*.

Except for W_i, all these attributes of a farmer agent are dynamic during the simulation, and their changing rules are described in detail below.

Credibility. Each agent *f* gives a credibility *C(f, v)* to another agent *v* by comparing their outcomes. Credibility is between 0 (not credible at all) and 1 (very credible).

Satisfaction. Each agent has a satisfaction with its current farming strategy (I_C) that corresponds to an evaluation of that strategy. It may also evaluate an alternative strategy in certain cases and have a satisfaction for it (I_A). Satisfaction with a farming strategy lies between 0 (not satisfied at all) and 1 (very satisfied).

If an agent is satisfied with its current farming strategy, it does not consider an alternative. Otherwise, its satisfactions with its current farming strategy (I_C) and with an alternative one (I_A) are computed and compared. If I_A is higher enough than I_C, the agent will change its farming strategy. I_C is thus computed at every iteration, whereas I_A is only computed when a stable agent is dissatisfied with its current farming strategy.

In accordance with TRA, the satisfaction I_S with a farming strategy *S* depends on two elements: attitude A_S and subjective norm SN_S toward *S*. In the original theory, the interaction between these two elements varies with different agents facing different situations. In order to keep the model simple, satisfaction is assumed as the average value of these two elements.

$$I_S = \frac{A_S + SN_S}{2} \quad (1)$$

Both attitude and subjective norm lie between −1 (very negative attitude/subjective norm concerning the farming strategy to evaluate) and 1 (very positive attitude/subjective norm concerning the farming strategy to evaluate). They are computed with farms' outcomes, farmers' strategies, and credibility. See the section "Farmers' dynamics" for the computation details.

Considering the value range of attitude and subjective norm towards a farming strategy, the satisfaction should also lie between −1 and 1. However, to facilitate other calculations, the satisfaction is normalized between 0 and 1.

Performance. The performance is evaluated over two dimensions: the level of the output production (i.e. the productivity impact, in our case milk production), and the level of environmental amenities production (i.e. the environmental impact), respectively called productivity performance (P_0) and environmental performance (P_1) in the paper. We assume P_0 and P_1 lie between 0 (very bad on this performance dimension) and 1 (very good on this performance dimension).

Importance Given to Each Dimension. The importance given to productivity dimension is termed W_0, and the one given to environmental dimension is termed W_1. W_0 and W_1 lie between 0 (not important at all) and 1 (most important). They sum to 1.

$$W_0 + W_1 = 1 \quad (2)$$

Importance defines an agent's personal values. An agent uses its own lens to judge the information it receives and the other agents it meets. In this model, both W_0 and W_1 are kept constant if an agent does not change its farming strategy.

Farming Strategy. It is defined by the importance that a farmer gives to each dimension of practice. Two farming strategies are considered: organic and conventional. The organic strategy lends more importance to environmental dimension and less to productivity dimension, whereas the conventional strategy lends more importance to productivity dimension and less to environmental dimension. It is assumed that when a farmer agent changes its strategy, it changes accordingly the importance given to different dimensions.

Farm

Afarm has three attributes: its farming total production (productivity outcome) T_0, its environmental amenities outcome T_1 and its reference R. R is the maximum possible productivity performance considering a farm's all characteristics and evolution. Interviews and experts' arguments show that conventional farms' references are grounded on the negotiations with dairy enterprises (often expressed by "quota" in Europe in the past). Organic farms have more constraints in terms of reference due to stricter regulations. So a conventional farm f's reference is considered as its farmer f's initial productivity performance $P_0^f(t = 0)$ and that for an organic farm is a function l of $P_0^f(t = 0)$.

$$R^f = \begin{cases} P_0^f(\mathrm{t} = 0) \text{ if } f \text{ is a conventional farm/farmer} \\ l\left(P_0^f(\mathrm{t} = 0)\right) \quad \text{otherwise} \end{cases} \quad (3)$$

The implementation of a farm may need more attributes for different use cases. The detailed computation of T_0 and T_1 are defined in the model's implementation (see Sect. 2.2).

Media

When an agent is dissatisfied with its current farming strategy and looks for an alternative, it first searches in the population for other agents having similar characteristics but applying an alternative farming strategy. If it cannot find one, it has access to the media for an alternative farming strategy's model. These models are given by some functions depending on the farm's current outcomes.

Dynamics

Overview of a Farmer's Dynamics over Years

One time-step (iteration) t $\rightarrow$ t + 1 represents one year, i.e. once a year, farmers decide their farming strategies, their performances and so on. During an iteration, farmers' update order is picked up at random by a uniform law.

As shown in Fig. 1 and the pseudocode of Algorithm 1, during each iteration, an agent evaluates its satisfaction with its current farming strategy. If the agent is in a stable period and is satisfied with its current strategy, it does not consider a change. Otherwise, the agent looks for an alternative and evaluates it. If the agent has been dissatisfied for long enough and the alternative is good enough, it will change. Otherwise, the agent stays with its current farming strategy. It will then update its credibility given to other agents and its performance. See the detail in the following.

```
For each iteration {
  Generate the order of the population
  For each agent f in the population {
      Compute I_C
      If DC>TD and I_C <TA, compute I_A
         If DD>TD and I_A > I_C +TO, change strategy and update W_0, W_1
      For each agent v that is different from agent f in the population, compute C (f, v)
      Compute P_0, P_1  } }
```

Algorithm 1— Population updating loop. I_C is the satisfaction with a current strategy; I_A is the satisfaction with an alternative one. *DC* is an agent's confirmation duration; *DD* is an agent's dissatisfaction duration. *TD* is the minimum time of dissatisfaction before considering the alternative. *TA* is threshold of I_C to consider an alternative. *TO* is the threshold of I_A to change strategy. W_0 is the importance given to productivity performance, W_1 is the importance given to environmental performance. $C\ (f,\ v)$ is the credibility that agent *f* gives to agent *v*. P_0 is the productivity performance, P_1 is the environmental performance.

Credibility Update

Every relationship between two agents is characterized by the credibility one gives to another, and depends on an agent's personal view of its difference to another in outcome (i.e. total production). For agent *f*, its difference to agent *v* is the sum of differences on each outcome dimension weighted by the importance given to that dimension.

$$D_v^f = \sum_{i=1}^{2} \left(W_i^f \left(T_i^v - T_i^f \right) \right) \tag{4}$$

The credibility that agent *f* gives to agent *v* is calculated with *f*'s difference to *v*:

$$C_v^f = \frac{1}{1 + e^{-\alpha D_v^f}} \tag{5}$$

with α the parameter of the logistic function.

In Fig. 2, agent *f*'s difference and credibility to *v* are respectively plotted on the x-axis and y-axis. When the difference is negative, it means that *v* has a worse outcome

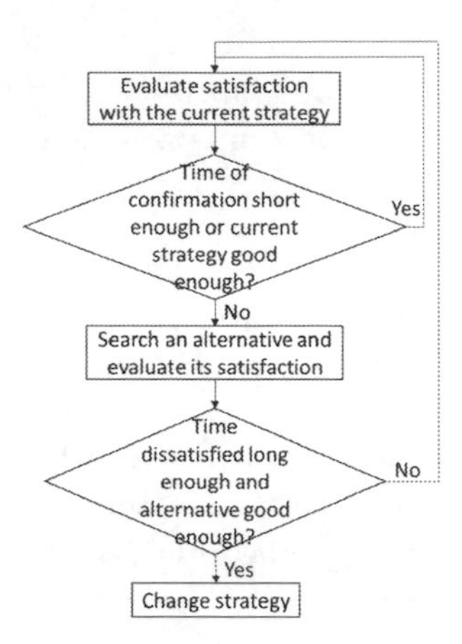

Fig. 1. Overview of the farmer's update

than *f*, thus *f* gives little credibility to *v*. When the difference is positive, *v* has a better outcome than *f*, thus *f* gives big credibility to *v*.

The lines with different colors represent α variations to characterize the bias degree that an agent has for others with better outcomes. When α is small, the bias is small. An agent tends to give same credibility to others, whether or not with better outcomes. If α is big, the bias is strong. Only others with better outcomes are credible.

In the model, every two agents are connected. Credibility depends on an agent's perceived difference in outcome to another and it is then used to update the agent's outcome which can change the perceived difference. Thus, the relationship associated to credibility is dynamic.

Farming Strategy Change

An agent changes its farming strategy according to its satisfaction evaluations with its current farming strategy (I_C) and with the alternative one (I_A). If an agent is in a stable state (its confirmation duration $DC >$ threshold TD) and it is still dissatisfied with its current farming strategy ($I_C <$ threshold TA), it will consider an alternative one. If the agent is dissatisfied long enough (its dissatisfaction duration since being stable $DD > TD$) and its satisfaction evaluation of the alternative is better enough than that with its current one ($I_A > I_C$ (1 + threshold TO)), it will change farming strategy. As stated in Eq. (1), satisfaction I with a strategy is the average sum of the related attitude A and subjective norm SN.

In Eq. (1), attitude (A_S) represents an agent's personal view of the difference between its experience and the (potential) outcome of evaluated strategy S. The agent f's experience is its average outcome on the farm $\left(\overline{T_{C,0}}, \overline{T_{C,1}}\right)$ with its current farming strategy (S^f) in memory (M). It is computed like this:

$$\overline{T_{C,\iota}} = \frac{\sum_{t \text{ and } S^t = S^f}^{M} T_i^t}{Nb(S^t = S^f)} \tag{6}$$

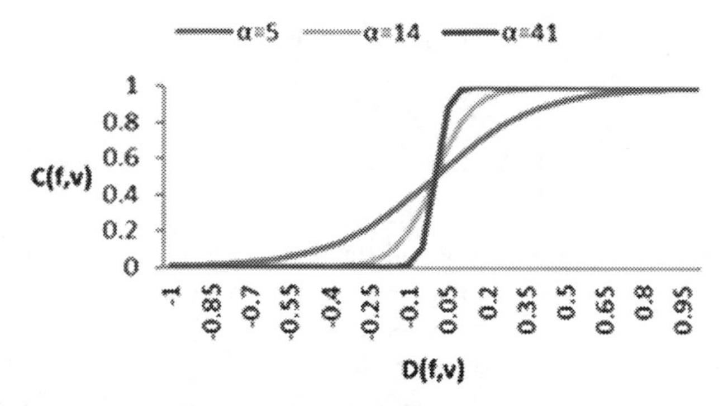

Fig. 2. The credibility (y-axis) agent f gives to v depends on f's difference (x-axis) to v for three values of parameter α (different-colored lines) (Color figure online)

The evaluated outcome depends on the strategy to be evaluated. For agent f's current farming strategy evaluation, the evaluated outcome is f's current outcome $\left(T_i^f\right)$.

Agent f's attitude toward the current farming strategy (A_C) is like this:

$$A_C^f = \sum_{i=1}^{2}\left(W_i^f\left(T_i^f - \overline{T_{C,\iota}^f}\right)\right) \tag{7}$$

If an agent's outcome on its farm changes and this change is considered better than its experience, A_C will be positive and strengthen the agent's decision to keep its current strategy. Otherwise, A_C will be negative and may influence I_C. Then the agent may be dissatisfied and evaluate I_A.

For agent f's evaluation of an alternative farming strategy, the evaluated outcome is the average outcome $\left(\overline{T_{A,0}^f}, \overline{T_{A,1}^f}\right)$ of other agents having similar characteristics as f but applying the alternative. The similarity is defined by a function Y over the farm's characteristics and is compared with a distance threshold (*simi*). Y is designed in the model's implementation (see Sect. 2.2). The evaluated outcome is computed as follows:

$$\overline{T_{A,\iota}^f} = \frac{\sum_{Y_v^f < simi\,and\,S^v \neq S^f}^{N} T_i^v}{Nb(Y_v^f < simi\,and\,S^v \neq S^f)} \tag{8}$$

If there is no corresponding peer (no other agents similar to f and applying the alternative farming strategy), agent f will search the media for a stereotypical farm to evaluate the alternative.

$$\overline{T_{A,\iota}^f} = T_{model,i}^f \tag{9}$$

Therefore, agent *f*'s attitude toward an alternative is:

$$A_A^f = \begin{cases} 0\ if\ (I_C^f > TA) \\ \sum_{i=1}^{2} (W_i^f \left(\overline{T_{A,\iota}^f} - \overline{T_{C,\iota}^f} \right)) & otherwise \end{cases} \tag{10}$$

Another component of satisfaction is the subjective norm, which represents how an agent considers others' opinions on the evaluated farming strategy through outcomes, i.e. the strategy's implementation results. It is thus an agent's perceived difference between the outcome to be evaluated and the average of other agents' outcomes.

For agent *f*'s evaluation of current farming strategy, the subjective norm is:

$$SN_C^f = \sum_{i=1}^{2} \left(W_i^f \left(T_i^f - \frac{\sum_{v \neq f}^{N} (C_v^f T_i^v)}{\sum_{v \neq f}^{N} C_v^f} \right) \right) \tag{11}$$

An agent will be socially satisfied if it perceives that other agents, especially those to whom it lends great credibility ("important others"), consider it as a 'good farmer'. The agent may be so satisfied to have a good social image that it will never consider a major change. Otherwise, if the agent feels socially bad, it may try to become more similar to others in the group or to change of group. This can be done with a change of strategy.

For the evaluation of an alternative farming strategy, the subjective norm is:

$$SN_A^f = \begin{cases} 0\ if\ (I_C^f > TA) \\ \sum_{i=1}^{2} \left(W_i^f \left(\overline{T_{A,\iota}^f} - \frac{\sum_{v \neq f}^{N} (C_v^f T_i^v)}{\sum_{v \neq f}^{N} C_v^f} \right) \right) & otherwise \end{cases} \tag{12}$$

If in other agents' opinions, especially those to whom agent *f* lends great credibility ("important others"), the alternative is not better, then it is judged not good enough to improve the situation. Agent *f* will tend to keep its current strategy. Otherwise, the agent's subjective norm strengthens its intention to change its strategy.

If an agent changes its strategy, it also changes the importance given to each dimension of practice. The parameter *W* is the conventional farmers' initial importance given to the productivity dimension.

$$\text{For conventional agents: } W_0 = W;\ W_1 = 1 - W \tag{13}$$

$$\text{For organic agents: } W_0 = 1 - W;\ W_1 = W \tag{14}$$

Performance Update

As farmers co-construct their practices [26], at each time *t*, a farmer agent updates its performance by copying the practices of its credible peers with a similar farm.

$$\Delta P_i^f = \frac{\sum_{v \neq f\ and\ Y_v^f < simi}^{N} C_v^f \left(P_i^v - P_i^f \right)}{\sum_{v \neq f\ and\ Y_v^f < simi}^{N} C_v^f} \tag{15}$$

Both performance dimensions are between 0 and 1. A farmer's productivity performance is also limited by the reference on its farm.

$$P_0^{f,t+1} = \begin{cases} 0 \; if \left(P_0^{f,t} + \Delta P_0^f < 0 \right) \\ R^f \; if(\left(P_0^{f,t} + \Delta P_0^f \right) > R^f) \\ P_0^{f,t} + \Delta P_0^f \; otherwise \end{cases} \tag{16}$$

A special case: if agent *f* looks for an alternative and cannot find a similar peer applying an alternative strategy, it will look for an alternative in the media. If agent *f* adopts the alternative found in the media after evaluation, then it will also copy the performance.

2.2 The Design of Use Cases Based on Agricultural Data

The model is implemented with data from Agricultural General French Census (RGA 2000). In each of the three main dairy production French regions, we have selected one or two "départements" and their related "cantons" having at least 60 dairy farmers, on average 113 farms with a standard-deviation of 34.6. 27 French "cantons" are simulated. For sake of comparison, each of our virtual cantons has 100 farmers. These "cantons" have strong variations on practice intensity and homogeneity of farm types. According to expertise and literature [30, 31], a farmer's farm is defined by three variables: the utilized agricultural area (*UAA*), the number of dairy cows (*NC*), and the quota (*Q*) which is a synthetic indicator of the farm's maximum milk production. A farm's intensity can be measured by the average *UAA*/the average *NC*. The implementations of a farm, a farmer and the media are presented in the following.

Farmer

A farmer agent is designed by its practice with productivity performance P_0 and environmental performance P_1 designed from data. P_0 is directly deduced from the farm's initial characteristics and corresponds to the normalized average milk volume produced by one cow in one year. For farmer/farm *f*, at the initial time $t = 0$, $P_0^f = Q^f / NC^f$. At every time *t*, $P_1^f = T_1^f / UAA^f$, the Eq. (15) is only used to update P_0^f.

The *Y* function telling how two farmers are judged similar is based on a similarity of their farms' characteristics regarding *UAA* and *NC*. For agent *f*, agent *v* is a similar peer if $\frac{|UAA^f - UAA^v|}{UAA^f} < simi$ and $\frac{|NC^f - NC^v|}{NC^f} < simi$. The value of threshold *simi* is fixed at 0.1 in the model.

Farm

As shown above, each farm is initialized by the crossed distribution of discretized utilized agricultural areas (*UAA*) and quotas (*Q*) of its "canton" from the RGA 2000. They remain constant all along the simulation. The number of dairy cows (*NC*) is computed from a law extracted from data (with a regression $r^2 = 0.9563$):

$$NC = 0.2463\,UAA + 0.0001106\,Q \tag{17}$$

From databases regarding farmers' production and various sources[1], a law is built to compute the potential maximum milk production of an organic farm starting with the conventional strategy, knowing its initial productivity performance P_0 after the normalization and P'_0 before the normalization. A farm f's normalized reference R and reference R' before the normalization are computed as follows:

$$\text{For a conventional farm: } R = P_0(t = 0) \tag{18}$$

$$\text{For an organic farm: } R' = 0.6046\ P'_0(t = 0) + 1913\,NC \tag{19}$$

The environmental amenities outcome T_1 is computed at every time by an aggregated function of literature [30, 31]. It considers mineral impacts and energy consumption related to the total milk production and the farm's agricultural surface:

$$\text{For a conventional farm: } T_1 = (53\ UAA + 2.918\,T_0)/2 \tag{20}$$

$$\text{For an organic farm: } T_1 = (-10\ UAA + 2.588\,T_0)/2 \tag{21}$$

Using French dairy farms' database in RGA 2000, R' is to be normalized between 0 (very low production) and 1 (very high production). 53 UAA, 2.918 T_0, −10 UAA and 2.588 T_0 are normalized values between 0 and 1. The normalization is:

$$x = (x' - min)/(max - min) \tag{22}$$

With: *min* = minimum real value in the database; *max* = maximum real value in the database; x is the normalized value of real value x'.

The Media

We use laws extracted from data to design farmers' alternative models. When a conventional farmer *f* wants to evaluate the organic strategy at time $t + 1$, it computes $T'_0(t+1)$ as follows and its related $T'_1(t+1)$ with the Eq. (20). Noting that T'_0 is the farmer's real productivity outcome on the farm (before the normalization).

$$T'_0(t+1) = 0.6046\,T'_0(t) + 1913\,NC \tag{23}$$

When an organic farmer *f* wants to evaluate the conventional strategy at time $t + 1$, it computes $T'_0(t+1)$ as follows and its related $T'_1(t+1)$ with the Eq. (21).

$$T'_0(t+1) = \left(T'_0(t) - 1913\,NC\right)/0.6046 \tag{24}$$

[1] http://www.cantal.chambagri.fr/fileadmin/documents/Internet/Autres%20articles/pdf/2014/Bio/ABBL2008-2012.pdf.
http://www.tech-n-bio.com/.
http://www.agrobio-bretagne.org/.

2.3 Experimental Design

[27] has studied a large set of 625 parameter values identifying all the qualitative behaviors of the model explaining the absence of conversion. This work aims to study conversion in different French "cantons". A larger interval of parameter variation with 81 parameter sets gives the same conclusion as the 625 parameter sets in [27]. The main parameters to vary are: α (slope of logistic function) with values: 5, 23 and 41, *TA* (threshold to consider an alternative) with values: 0.41, 0.45 and 0.49, *TO* (threshold to consider an alternative) with values: 0.01, 0.05 and 0.09, *W* (importance given to the dimension representing farming strategy) with values: 0.6, 0.8 and 1. *TD* (threshold for two counters of duration) is kept constant with the value 5. Agents' memory is also kept constant at 10 years, as well as the distance for similarity, *simi*, valued 0.1. The evolution of each "canton" with a population of 100 conventional farmers (no organic farming at the beginning) is simulated for 30 years and replicate 100 times.

3 Model Behaviors

[28] shows that farm characteristics are very important for farmers' decisions towards the conversion to organic farming. In the following study, we distinguish:

- Virtual conversions which would only be explained by initial farm characteristics (at time 0), or after the first 5 years (value of the "waiting" period *TD*) during which a conversion is not possible but farms' performances can be changed due to the model dynamics. Virtual conversions indicate the **potential adoptions** in a "canton".
- Real conversions which can be explained by farm characteristics, but also by the social dynamics of interacting farms. These real conversions are measured at time 6, the first date at which a farm can adopt an alternative strategy, and at time 29, the end of the simulation.

The measures presented in what follows are averages over the 81 parameter sets and their related 100 replicas. The details corresponding to the variations due to a particular parameter set are not investigated here since they have been presented further in [27]. These averages are used as global indicators of the coupled dynamics instead of particular indicators of a particular dynamics. This approach is relevant to compare what is explained by farm characteristics to what is explained by social dynamics.

3.1 Adoption Rate Evolution in 27 "Cantons"

In Fig. 3, different colors represent "cantons" in different "départements". The first three figures show the adoption rates at time 0 (initial farm characteristics), 5 (initial practice evolution effect), 6 (decision effect) and 29 (complete effect) considering average values over 100 replications of each parameter set. The fourth one shows the adoption rates at time 29 considering average values over all the 81 parameter sets and their related 100 replicas.

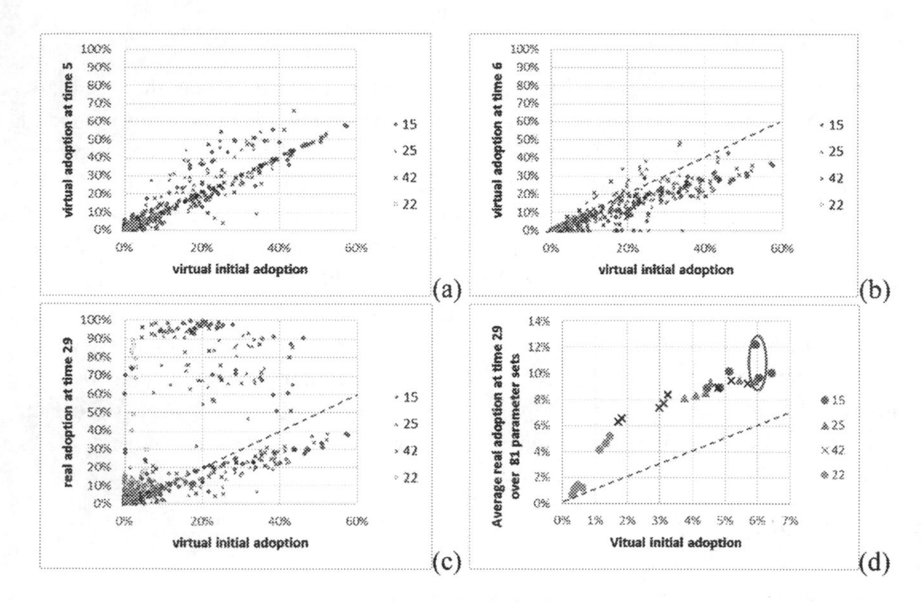

Fig. 3. (a–c) The average virtual/real adoption rates of each parameter set over its 100 replicas at time 5, 6 and 29 comparing to the virtual adoption rates at time 0, (d) the average real adoption rates over all the 81 parameter sets and their related 100 replicas. Purple lines represent the virtual adoption rates at time 0. Points with different colors represent different "cantons" from the four different "départements" to which they belong (15 in blue, 25 in green, 42 in purple and 22 in orange). (Color figure online)

As shown in Fig. 3, all the conversion rates at these times keep the same tendency as the virtual conversion at time 0 depending only on the farms' characteristics. "Département" 22 converts the least, 25 and 42 have nearly the same conversion rates, 15 converts the most. This indicates that the principal determinant of the conversion to organic farming is probably the initial farm characteristics.

In some cases, the social dynamics can favor adoption until a 100% adoption rate; in some cases, the social dynamics can impede the adoption and give the adoption rate inferior to the virtual one at time 0. Indeed, at time 5, the virtual adoption rates are slightly higher than the initial rates in most "cantons". The adoption rates measured at time 6 are lower than at time 5. However, the adoption rates at time 29 are finally better than at other times. This indicates that the adoption rate depends also on social dynamics.

During the first 5 times, an agent cannot change its farming strategy. The variation of adoption rate is only due to the performance evolution. Less productive farms cannot increase their performances much due to the maximum production R while more productive ones can decrease theirs. Then more productive farms will have less total productions and negative personal evaluations compared to their experiences, and eventually worse satisfactions with the current situation. Thus, the virtual adoption rate at time 5 is larger than at time 0, especially for more productive farms.

At time 6, less productive farms convert immediately since they can gain in total production (Eq. 18 to Eq. 19). This can make the alternative less interesting for other agents searching an alternative. For first adopters, referring to the media, the alternative evaluation depends directly on their own farm characteristics. Then they can be referred to by other "similar" farms (on farm characteristics) which are looking for an alternative. Since these first adopters are rather less productive farms, they give a worse evaluation of the alternative than that given by the media. Therefore, the adoption rate at time 6 is lower than at time 5.

Later, with the adoption diffusion, there are more and more adopters. At time 29, the adoption rates are larger than initial virtual rates due to the social dynamics.

For the selected "cantons", the conversion to organic farming depends mainly on farm characteristics. However, social dynamics also have a strong impact on the decision. Indeed, as shown by the red circles in Fig. 3(d), some "cantons" have similar initial virtual adoption rates among which some have also similar real final ones while others have different real final ones. They correspond to "cantons" 1512, 1515 and 2517 which have a similar virtual adoption rate at time 0. At time 29, 1515 and 2517 they still have a similar final adoption rate while the adoption rate in 1512 is much higher than in other "cantons". In the following, we aim at understanding these variations.

3.2 Understand the "Cantons" Impact on the Adoption Rate

To better understand the detailed situations, the three typical "cantons" (1512, 1515 and 2517) outlined in Fig. 3 are studied through two points: (1) farm characteristics of each "canton"; (2) temporal adoption rates of different farm types in each "canton".

Farm Characteristics of Each "Canton" from RGA

As shown in [27, 28], different farms in terms of quota have quite different behaviors. According to each farm's quota (Q), three farm types are classified for the following behavior analysis. Q inferior to 133000 L are considered as small farms, between 133000 and 257000 L as medium farms, superior to 257000 L as large farms. As shown in Fig. 4, "cantons" 1512 and 1515 are similar in farm characteristics and type distributions, while 2517 is quite different from others.

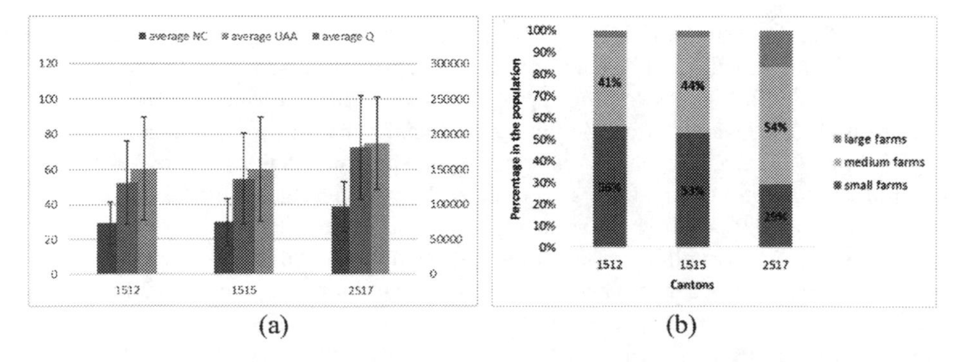

(a) (b)

Fig. 4. (a) On the left: farm characteristics of each "canton" in terms of average number of cows (NC), average utilized agricultural area (UAA) in are and average quota (Q) in liters, NC and UAA use the left dimension and quota Q uses the right dimension. Error bars indicate standard deviation. (b) On the right: distribution of farm types in terms of quota Q in each "canton".

Temporal Adoption Rates for Each Farm Type in Each "Canton"
Figure 5 shows the average virtual/real adoption rates for each farm type in each "canton" at time 0, 5, 6 and 29 (averaged over the 81 parameter sets and their replicas). In general, small farms convert the most except for the "canton" 2517. Large farms convert the least in all cases.

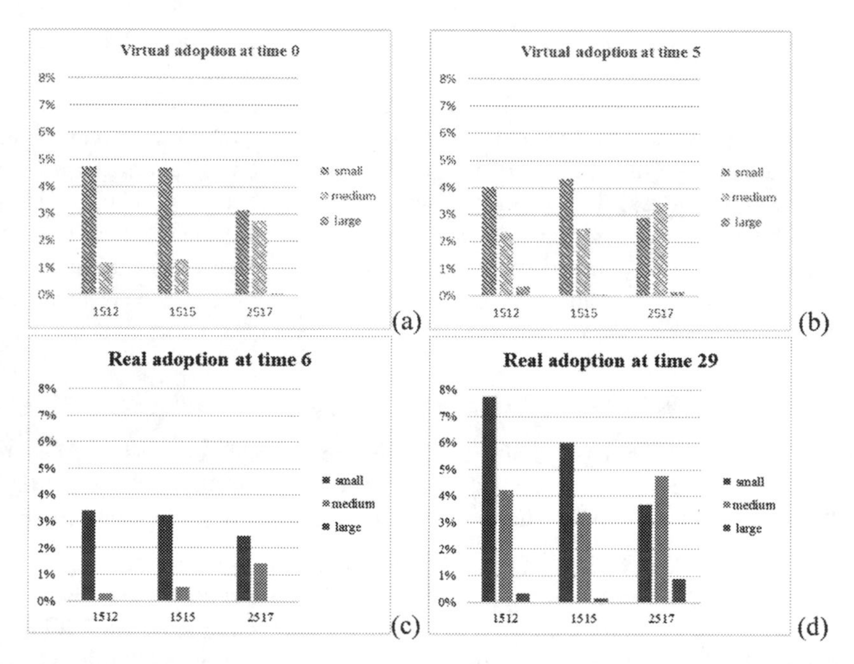

Fig. 5. Average temporal adoption rates over the 81 parameter sets for each farm type in each "canton".

4 Conclusion Discussion

[26] and [27] argue that the organic farming adoption is strongly influenced by the imitation of the practices of "important others" and is very sensitive to the distance between farms. [28] shows the impact of population characteristics on the adoption rate. In this paper, we focus on the impact of social dynamics for different farm type distributions (initialization of 27 "cantons" from the data of the French 2000 Agricultural Census) on the adoption rate predicted by an agent-based model.

At the individual farm level, as in [28], we notice the importance of farm types. Indeed, small farms with low total milk production and high environmental total production convert more than others. Such farms have extensive practices and are more likely to convert according to the experts. Accordingly, [32, 33] also argue that large farms intensify more than small farms, developing characteristics preventing them from converting.

At the population level, we globally notice the same impact of characteristics on the level of adoption, but the adoption can be impeded or favored by the social dynamics of interacting farms.

Indeed, some "cantons" have more small and medium farms convert more than others. "Cantons" that are more intensive (those located in the Brittany region) have almost no conversion compared to extensive ones (those located in mountainous areas).

However, "cantons" having rather similar characteristics and just a tiny difference of their type distributions can have very different final adoption rates. This is due to the effect of the interaction among farmers, which are very sensitive to farm type distributions and can change the alternative referred to in the population. Another particular remark, populations with different characteristics can also have a similar final global adoption rate but different rates regarding each farm type. This is also often due to different population distributions. Overall, we conclude that despite their low social value, small farms, if they are numerous enough, can have a slow repeated impact on larger farms. They can change their practices in such way they are interested in adopting organic farming. However, they can also change their practice without adopting. These two possible evolutions are in accordance with [18, 34] and depend on the farm type distributions of the population. This is also due to constraint in terms of level of total production, as the agronomic characteristics of the soil and/or regions, the size of the farm and the quota, which limit the total possible production.

We cannot compare our adoption rates for the chosen French regions to real adoption rates: firstly because the data is not so easy to obtain; secondly very important dynamics are not presented in the model and make the comparison hardly relevant. Adding to the model external factors such as the economy crashing, but also demographic evolution implying an increasing average size of farms and consequently of maximum possible production, are the next steps for research about this model. Indeed, as already said, most of the results obtained in this study are due to the fact that the maximum total production is limited.

Acknowledgements. We acknowledge funding from the Scientific Research Foundation of Jiangxi University of Science and Technology (Grant No. JXXJBS19012).

References

1. Neumeister, D., Fourdin, S., Dockès, A.-C.: Etude des freins et motivations des éleveurs laitiers au passage en agriculture biologique dans les zones de piémont et de montagne. Rencontres Autour Rech. Sur Rumin. (2011)
2. Pavie, J., Dockès, A.-C., Echevarria, L.: Etude des freins à la conversion à l'agriculture biologique des exploitations laitières bovines. Institut de l'Elevage, Paris Cedex 12 (2002)
3. De Buck, A.J., Van Rijn, I., Roling, N.G., Wossink, G.A.A.: Farmers' reasons for changing or not changing to more sustainable practices: an exploratory study of arable farming in The Netherlands. J. Agric. Educ. Ext. **7**(3), 153–166 (2001)
4. Kaufmann, P., Stagl, S., Franks, D.W.: Simulating the diffusion of organic farming practices in two New EU Member States. Ecol. Econ. **68**(10), 2580–2593 (2009)
5. Kerselaers, E., De Cock, L., Lauwers, L., Van Huylenbroeck, G.: Modelling farm-level economic potential for conversion to organic farming. Agric. Syst. **94**(3), 671–682 (2007)

6. Latruffe, L., Nauges, C., Desjeux, Y.: Le rôle des facteurs économiques dans la décision de conversion à l'agriculture biologique. Innov. Agron. **32**, 259–269 (2013)
7. Lamine, C., Bellon, S.: Conversion to organic farming: a multidimensional research object at the crossroads of agricultural and social sciences. A review. Agron. Sustain. Dev. **29**(1), 97–112 (2009)
8. Padel, S., et al.: Conversion to organic farming: a typical example of the diffusion of an innovation? Sociol. Rural. **41**(1), 40–61 (2001)
9. Burton, R.J.F., Wilson, G.A.: Injecting social psychology theory into conceptualisations of agricultural agency: towards a post-productivist farmer self-identity? J. Rural Stud. **22**(1), 95–115 (2006)
10. Mzoughi, N.: Farmers adoption of integrated crop protection and organic farming: do moral and social concerns matter? Ecol. Econ. **70**(8), 1536–1545 (2011)
11. Stock, P.V.: 'Good farmers' as reflexive producers: an examination of family organic farmers in the US Midwest. Sociol. Rural **47**(2), 83–102 (2007)
12. Dedieu, M.-S., Lorge, A., Louveau, O., Marcus, V.: Les exploitations en agriculture biologique: quelles performances économiques? (2017)
13. Sainte-Beuve, J., Bougherara, D., Latruffe, L.: Performance économique des exploitations biologiques et conventionnelles: Levier économique à la conversion, 10 p. Transversalités L'agriculture Biol. Strasbg. (2011)
14. Darnhofer, I., Schneeberger, W., Freyer, B.: Converting or not converting to organic farming in Austria: farmer types and their rationale. Agric. Hum. Values **22**(1), 39–52 (2005)
15. Fairweather, J.R.: Understanding how farmers choose between organic and conventional production: results from New Zealand and policy implications. Agric. Hum. Values **16**(1), 51–63 (1999)
16. Rigby, D., Young, T., Burton, M.: The development of and prospects for organic farming in the UK. Food Policy **26**, 599–613 (2001)
17. Burton, R.J.F.: Reconceptualising the 'behavioural approach' in agricultural studies: a socio-psychological perspective. J. Rural Stud. **20**(3), 359–371 (2004)
18. Sutherland, L.-A., Burton, R.J.F., Ingram, J., Blackstock, K., Slee, B., Gotts, N.: Triggering change: towards a conceptualisation of major change processes in farm decision-making. J. Environ. Manag. **104**, 142–151 (2012)
19. Rickards, L., Howden, S.M.: Transformational adaptation: agriculture and climate change. Crop Pasture Sci. **63**(3), 240 (2012)
20. Dowd, A.-M., Marshall, N., Fleming, A., Jakku, E., Gaillard, E., Howden, M.: The role of networks in transforming Australian agriculture. Nat. Clim. Change **4**(7), 558–563 (2014)
21. Fairweather, J.R., Hunt, L.M., Rosin, C.J., Campbell, H.R.: Are conventional farmers conventional? Analysis of the environmental orientations of conventional New Zealand farmers. Rural Sociol. **74**(3), 430–454 (2009)
22. Barbier, C., Cerf, M., Lusson, J.-M.: Cours de vie d'agriculteurs allant vers l'économie en intrants: les plaisirs associés aux changements de pratiques. Activités **12**(12–2), (2015)
23. Burton, M., Rigby, D., Young, T.: Modelling the adoption of organic horticultural technology in the UK using duration analysis. Aust. J. Agric. Resour. Econ. **47**(1), 29–54 (2003)
24. Goldstone, R.L., Janssen, M.A.: Computational models of collective behavior. Trends Cogn. Sci. **9**(9), 424–430 (2005)
25. Grimm, V.: Ten years of individual-based modelling in ecology: what have we learned and what could we learn in the future? Ecol. Model. **115**(2), 129–148 (1999)
26. Huet, S., Rigolot, C., Xu, Q., De Cacqueray-Valmenier, Y., Boisdon, I.: Toward modelling of transformational change processes in farm decision-making. Agric. Sci. **09**(03), 340–350 (2018)

27. Xu, Q., Huet, S., Poix, C., Boisdon, I., Deffuant, G.: Why do farmers not convert to organic farming? Modelling conversion to organic farming as a major change. Nat. Resour. Model. **31**, e12171 (2018)
28. Xu, Q., Huet, S., Perret, E., Boisdon, I., Deffuant, G.: Population characteristics and the decision to convert to organic farming (2018)
29. Rogers, E.M.: Diffusion of Innovations, 3rd edn. Free Press/Collier Macmillan, New York/London (1983)
30. Pavie, J., Chambaut, H., Moussel, E., Leroyer, J., Simonin, V.: Evaluations et comparaisons des performances environnementales, économiques et sociales des systèmes bovins biologiques et conventionnels dans le cadre du projet CedABio. Renc Rech Rumin. **19**, 37–40 (2012)
31. Chambaut, H., et al.: Profils environnementaux des exploitations d'élevage bovins lait et viande en agriculture biologique et conventionnelle: enseignements du projet CedABio. In: Rencontres Autour Rech. Sur Rumin., pp. 53–56 (2011)
32. Bos, J.F.F.P., Smit, A.(B.)L., Schröder, J.J.: Is agricultural intensification in The Netherlands running up to its limits? NJAS Wagening. J. Life Sci. **66**(Suppl. C), 65–73 (2013)
33. Groeneveld, A., Peerlings, J., Bakker, M., Heijman, W.: The effect of milk quota abolishment on farm intensity: shifts and stability. NJAS Wagening. J. Life Sci. **77**(Suppl. C), 25–37 (2016)
34. Sutherland, L.-A.: 'Effectively organic': environmental gains on conventional farms through the market? Land Use Policy **28**(4), 815–824 (2011)

AnimeGAN: A Novel Lightweight GAN for Photo Animation

Jie Chen[1], Gang Liu[2(✉)], and Xin Chen[2]

[1] School of Civil Engineering, Wuhan University, Wuhan 430072, China
cjjjack@163.com
[2] School of Computer Science, Hubei University of Technology, Wuhan 430072, China
lg0061408@126.com, ghj9527@163.com

Abstract. In this paper, a novel approach for transforming photos of real-world scenes into anime style images is proposed, which is a meaningful and challenging task in computer vision and artistic style transfer. The approach we proposed combines neural style transfer and generative adversarial networks (GANs) to achieve this task. For this task, some existing methods have not achieved satisfactory animation results. The existing methods usually have some problems, among which significant problems mainly include: 1) the generated images have no obvious animated style textures; 2) the generated images lose the content of the original images; 3) the parameters of the network require the large memory capacity. In this paper, we propose a novel lightweight generative adversarial network, called AnimeGAN, to achieve fast animation style transfer. In addition, we further propose three novel loss functions to make the generated images have better animation visual effects. These loss function are grayscale style loss, grayscale adversarial loss and color reconstruction loss. The proposed AnimeGAN can be easily end-to-end trained with unpaired training data. The parameters of AnimeGAN require the lower memory capacity. Experimental results show that our method can rapidly transform real-world photos into high-quality anime images and outperforms state-of-the-art methods.

Keywords: Generative adversarial networks · Neural style transfer · Computer vision

1 Introduction

Anime is a common artistic form in our daily life. This artistic form is widely used in several fields including advertising, film and children's education. Currently, the production of animation mainly relies on manual implementation. However, manually creating anime is very laborious and involves substantial artistic skills. For animation artists, creating high-quality anime works requires careful consideration of lines, textures, colors and shadows, which means that it is difficult and time consuming to create the works. Therefore, the automatic techniques that can automatically transform real-world photos to high-quality

K. Li et al. (Eds.): ISICA 2019, CCIS 1205, pp. 242–256, 2020.
https://doi.org/10.1007/978-981-15-5577-0_18

animation style images are very valuable and necessary. It not only allows the artists to focus on more creative work and save time, but also makes it easier for ordinary people to implement their own animation.

Currently, image-to-image translation based on deep learning [25] has achieved great results. Recently, learning-based style transfer methods [5–8] have become the common image-to-image translation methods. These methods can learn the style of the reference image (style image) and apply the learned style to the input image (content image) to generate a new image which fuses the content of the content image and the style of the style image. These methods primarily uses the correlations between deep features and an optimization-based method to encode the visual style of an image.

Generative adversarial networks (GANs) [9,28] have been applied for style transfer and achieved great results. Many researchers have proposed many GAN-based style transfer methods [3,30]. Although these methods have achieved some success in anime style transfer, they still have many obvious problems. These important problems mainly include: 1) the generated images have no obvious animated style textures; 2) the generated images lose the content of the original photos; 3) a large number of the parameters of the network require more memory capacity.

In order to solve the above problems, we propose a novel lightweight GAN, called AnimeGAN, which rapidly transforms real-world photos into high-quality anime images. The proposed AnimeGAN is a lightweight generative adversarial model with fewer network parameters and introduces Gram matrix [16] to get more vivid style images. Our method takes a set of photos and a set of anime images for training. To produce high quality results while making the training data easy to obtain, we use the unpaired data for training, which means that the photos and anime images in the training set are not related in content.

To further improve the animation visual effects of the generated images, we propose three new simple yet efficient loss functions. The proposed loss functions are the grayscale style loss, the color reconstruction loss, and the grayscale adversarial loss. In the generative network, the grayscale style loss and the color reconstruction loss make the generated image have more obvious anime style and preserve the color of the photos. The grayscale adversarial loss in the discriminator network makes the generated image have vivid color. In the discriminator network, we also used the edge-promoting adversarial loss proposed by the literature [3] for preserving clear edges.

In addition, in order to make the generated images have the content of the original photos, we introduce the pre-trained VGG19 [26] as the perceptual network to obtain the L1 loss of the deep perceptual features of the generated images and original photos. Before AnimeGAN starts training, we perform an initialization training on the generator to make the training of AnimeGAN easier and more stable. A large number of experimental results show that our AnimeGAN can quickly generate higher-quality anime style images and outperform state-of-the-art methods.

The remainder of this paper is organized as follows. The related work are described in Sect. 2. The architecture of AnimeGAN is presented in Sect. 3. Experimental datasets and results are reported in Sect. 4. Finally, some conclusions are given in Sect. 5.

2 Related Work

Neural Style Transfer. A variety of Neural style transfer (NST) algorithms [5–8,13] have been developed to synthesize a novel image by combining the content of one image with the style of another image (typically a painting) based on matching the Gram matrix statistics of deep features extracted from a pre-trained convolutional network. Gatys et al. [5–8] have achieved many impressive results in a series of style transfer tasks, but their network models often have a large number of parameters, which makes the training process quite time-consuming. Furthermore, their methods focus on the style transfer of the paintings. For other style transfer tasks, such as animation style transfer, photography style transfer, etc, they still need to be improved. Huang et al. [11] proposed a simple but effective method, called AdaIN, which aligns the mean and variance of the content image features with those of the style image features. The method does not use complex Gram matrix. Li et al. [17] proposed a novel approach called Universal Style Transfer (UST), which used Whitening and Coloring transform (WCT) to directly match the features covariance in the content image to those in the given style image. In order to achieve photorealistic image stylization, Li et al. [18] presented a novel fast photography style transfer method, which consists of a stylization step and a photorealistic smoothing step. The stylization step transfers the style of the reference photo to the content photo, the smoothing step ensures spatially consistent stylizations. Each of the steps has a closed-form solution and can be computed efficiently. But their method requires additional semantic label maps as supervision to help style transfer between corresponding semantic regions. In a word, these existing neural style transfer methods are usually only suitable for specific style transfer task. When they are used for animation style transfer, they tend to obtain unsatisfactory results.

Image-to-Image Translation with GANs. Image-to-image (I2I) translation, which is a research hotspot in the field of computer vision, refers to the task of mapping an image from a source domain to a target domain, e.g. semantic maps to real images, grayscale images to color images, low-resolution images to high-resolution images, and so on. Generative adversarial networks (GANs) have become a research focus of artificial intelligence, which is inspired by two-player zero-sum game. A GAN often comprises a generator and a discriminator that learn simultaneously and these two networks are optimized using a min-max game. Recently, the image-to-image translation methods based on GANs have achieved many results. Isola et al. [12] proposed the pix2pix to use the conditional GAN (cGAN) and U-Net neural network to achieve general image-to-image transfer. They demonstrated that this approach is effective at synthesizing photos from label maps, reconstructing objects from edge maps, and colorizing images, among other tasks. Wang et al. [29] proposed the pix2pixHD approach on the basis of the pix2pix approach for synthesizing high-resolution photo-realistic images from semantic label maps using conditional generative adversarial networks. CycleGAN [30] is an approach for learning to translate an image from a source domain X to a target domain Y in the absence of paired examples. Almahairi et al. [2] proposed Augmented CycleGAN, which learns many-to-many mappings between domains in an unsupervised way. Their model can learn mappings which produce a diverse set of outputs for each input and can learn mappings across substantially different domains.

It can be seen that the style transfer methods based on GANs or convolutional neural networks (CNNs) [15] have been extensively studied and many excellent results have been achieved. In recent years, animation style transfer has become a new research direction. Chen et al. in the literature [3] presented style transfer method based on Generative Adversarial Networks that seems to work really well in terms of photo cartoonization problem. Maciej et al. in the literature [23] proposed a solution to transform a video into a comics. They build two stages to transform input video into a comics. Their main contribution is to propose a keyframes extraction algorithm that selects a subset of frames from the video to provide the most comprehensive video context. Their network structure is the same as the literature [3] and some other training strategies are used in their method.

As a comparison, we propose a more lightweight AnimeGAN to learn the mapping between photo and anime manifolds using unpaired training data. We also proposed three dedicated loss functions to synthesize high quality anime images.

3 Our Method

3.1 AnimeGAN Architecture

In this paper, we present a simpler and more efficient generative adversarial network called AnimeGAN. AnimeGAN consists of two convolution neural networks: One is the generator G which is used to transform the photos of real-world scenes into the anime images; the another is the discriminator D which discriminates whether the images are from the real target domain or the output produced by the generator. The architecture of AnimeGAN can be seen in Fig. 1.

In Fig. 1, the generator of AnimeGAN can be considered as a symmetrical encoder-decoder network, which is mainly composed of the standard convolutions, the depthwise separable convolutions [4], the inverted residual blocks (IRBs) [24], the upsampling and downsampling modules. In the generator, the last convolutional layer with 1×1 convolution kernels does not use the normalization layer and is followed by the tanh nonlinear activation function.

The structure of Conv-Block, DSConv and the inverted residual blocks are shown in Fig. 2. Conv-Block is composed of the standard convolution with 3×3 convolution kernels, the instance normalization layer [27] and the LRelu activation function [19]. DSConv is composed of the depthwise separable convolution

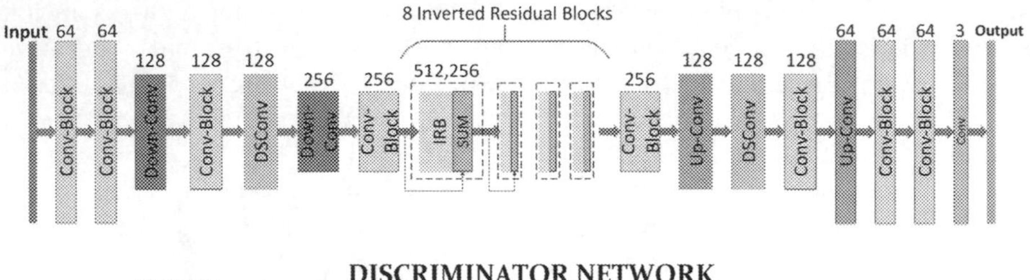

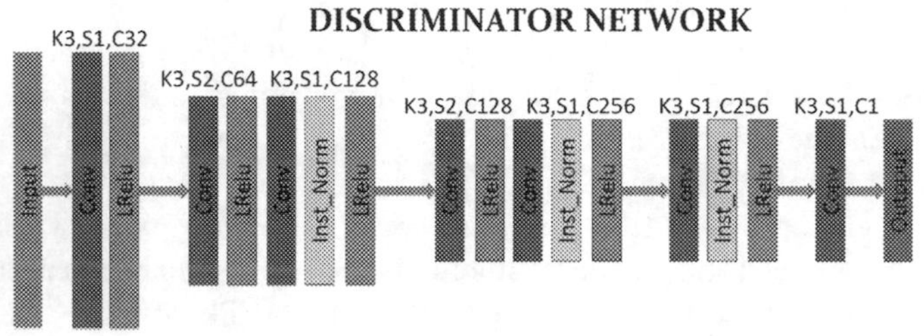

Fig. 1. The architecture of the generator and the discriminator in the proposed AnimeGAN. In the generator, the numbers on all boxes represent the number of channels and SUM means the element-wise sum. In the discriminator, "K" is the kernel size, "C" is the number of the feature maps, "S" is the stride in each convolutional layer and Inst_Norm indicates the instance normalization layer.

with 3×3 convolution kernels, the instance normalization layer and the LRelu activation function. The inverted residual block contains Conv-Block, the depthwise convolution, the pointwise convolution and the instance normalization layer.

In order to avoid the loss of the feature information caused by max-pooling, the proposed Down-Conv is used as the downsampling module to reduce the resolution of the feature maps. The architecture of Down-Conv is shown in Fig. 2. It contains the DSConv module with stride 2 and the DSConv module with stride 1. In Down-Conv, the feature maps are resized to half the size of the input feature maps. The output of the Down-Conv module is the sum of the output of the DSConv module with stride 2 and the DSConv module with stride 1.

The proposed Up-Conv is used as the upsampling module to increase the resolution of the feature maps. The architecture of Up-Conv is also shown in Fig. 2. In Up-Conv, the feature maps are resized to 2 times the size of the input feature maps. The Up-Conv module is used instead of the fractionally strided convolutional layer with stride $\frac{1}{2}$ used in the literature [3]. The reason is the upsampling method used in the literature [3] can cause the checkerboard artifacts [22] in the synthesized images and affect the quality of the images.

In order to effectively reduce the number of the parameters of the generator, we use 8 consecutive and identical IRBs in the middle of the network. Compared with the standard residual blocks [10], IRBs can significantly reduce the number of the parameters and computational workload of the network. IRB used in the generator consists of the pointwise convolution with 512 kernels, the depthwise

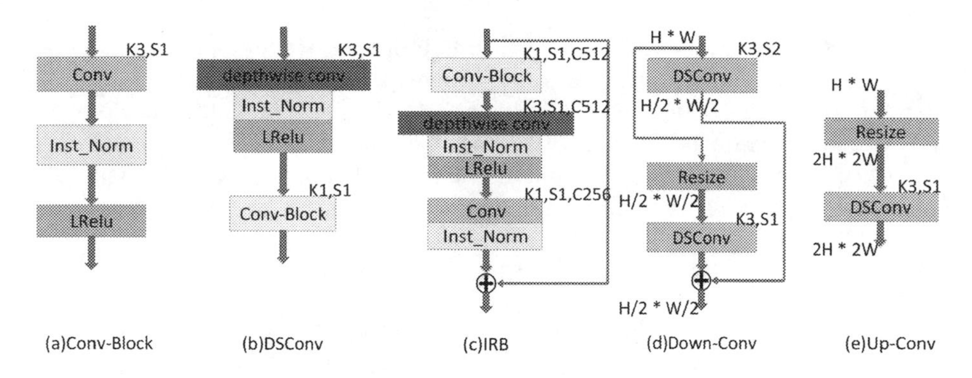

Fig. 2. The architectures of Conv-Block, DSConv, IRB, Down-Conv and Up-Conv in AnimeGAN. The "K" is the kernel size, "C" is the number of feature maps, "S" is the stride in each convolutional layer. The "H" is the height of the feature map, "W" is the width of the feature map. Resize means the interpolation method to set the size of the feature maps. The ⊕ indicates the element-wise addition.

convolution with 512 kernels and the pointwise convolution with 256 kernels. It is worth noting that the last convolution layer does not use activation function.

Complementary to the generator network, we use the same discriminator network as in the literature [3]. The architecture of the discriminator is shown in Fig. 1. All convolutional layers in the discriminator are the standard convolutions. For the weight of each convolutional layer, the spectral normalization [21] is used to make the network training more stable.

3.2 Loss Function

We formulate the approach of transforming real-world photos into animation images as an image-to-image mapping model which maps the photo domain P to the animation domain A. AnimeGAN is trained using unpaired training data $S_{data}(p) = \{p_i \mid i = 1, \cdots, N\} \subset P$ and $S_{data}(a) = \{a_i \mid i = 1, \cdots, M\} \subset A$, where N and M are the numbers of the photos and the animation images in the training set, respectively.

In our task, the texture of the animation images instead of the color is transferred into the photo images. We use the grayscale Gram matrix to make the generated images have the texture of the anime images instead of the color of the anime images. Therefore, it is necessary to transform the animation images into the grayscale images for eliminating color interference. The color animation image a_i in $S_{data}(a)$ is transformed to the grayscale image x_i. The grayscale animation training data $S_{data}(x) = \{x_i \mid i = 1, \cdots, M\} \subset X$ corresponding to the color animation domain $S_{data}(a)$ is obtained. In addition, as mentioned in the reference [3], the training data $S_{data}(e) = \{e_i \mid i = 1, \cdots, M\} \subset E$ is constructed by removing the edges of the animation images in the training data $S_{data}(a)$. In our method, we further process the images in $S_{data}(e)$ into the grayscale images to obtain $S_{data}(y)$. The reason is to avoid the influence of the color of the images in $S_{data}(e)$ on the color of the generated images.

In order to enable AnimeGAN to generate the higher quality images and make the training of the entire network more stable, the least squares loss function in LSGAN [20] is employed as the adversarial loss $L_{adv}(G, D)$. The loss function $L(G, D)$ used in AnimeGAN can be simply expressed as follows:

$$L(G,D) = \omega_{adv} L_{adv}(G,D) + \omega_{con} L_{con}(G,D) + \omega_{gra} L_{gra}(G,D) + \omega_{col} L_{col}(G,D) \tag{1}$$

where $L_{adv}(G, D)$ is the adversarial loss that affects the animation transformation process in the generator G, $L_{con}(G, D)$ is the content loss which helps to make the generated image retain the content of the input photo, $L_{gra}(G, D)$ represents the grayscale style loss which makes the generated images have the clear anime style on the textures and lines. Since the use of the grayscale style loss can easily cause the generated image to be displayed as the grayscale image, $L_{col}(G, D)$ is used as the color reconstruction loss to make the generated images have the color of the original photos. ω_{adv}, ω_{con}, ω_{gra} and ω_{col} are the weights to balance four given loss functions. In all our experiments, we set $\omega_{adv} = 300$, $\omega_{con} = 1.5$, $\omega_{gra} = 3$ and $\omega_{col} = 10$ which achieves a good balance of style and content preservation.

The loss functions used in the generator include the least squares loss function, the content loss function, the grayscale loss function and the color reconstruction loss function. For the content loss $L_{con}(G, D)$ and the grayscale style loss $L_{gra}(G, D)$, the pre-trained VGG19 is used as the perceptual network to extract the high-level semantic features of the images. $L_{con}(G, D)$ and $L_{gra}(G, D)$ can be expressed as:

$$L_{con}(G,D) = E_{p_i \sim S_{data}(p)}[\| VGG_l(p_i) - VGG_l(G(p_i)) \|_1] \tag{2}$$

$$L_{gra}(G,D) = E_{p_i \sim S_{data}(p)}, E_{x_i \sim S_{data}(x)}[\| Gram(VGG_l(G(p_i))) - Gram(VGG_l(x_i)) \|_1] \tag{3}$$

where $VGG_l(\bullet)$ refers to the feature maps of the lth layer in VGG and "$\bullet$" indicates the input. In our method, the lth layer is "conv4-4" in VGG. Gram represents the Gram matrix of the features. $G(p_i)$ means the generated image and p_i means the input photo. The $l1$ sparse regularization is used to calculate $L_{con}(G, D)$ and $L_{gra}(G, D)$.

In order to make the image color reconstruction better, we convert the image color in RGB format to the YUV format to build the color reconstruction loss $L_{col}(G, D)$. In $L_{col}(G, D)$, L1 loss is used for the Y channel and Huber Loss is used for the U and V channels. Formally, $L_{col}(G, D)$ can be defined as:

$$L_{col}(G,D) = E_{p_i \sim S_{data}(p)}[\| Y(G(p_i)) - Y(p_i) \|_1 + \| U(G(p_i)) - U(p_i) \|_H + \| V(G(p_i)) - V(p_i) \|_H] \tag{4}$$

where $Y(p_i)$, $U(p_i)$, $V(p_i)$ represent the three channels of the image p_i in the YUV format, respectively, and H represents Huber Loss. Finally, the loss function of the generator can be expressed as follows:

$$L(G) = \omega_{adv} E_{p_i \sim S_{data}(p)}[(G(p_i) - 1)^2] + \omega_{con} L_{con}(G, D) + \omega_{gra} L_{gra}(G, D) + \omega_{col} L_{col}(G, D) \tag{5}$$

For the loss function used by the discriminator, in addition to introducing edge-promoting adversarial loss [3] to make the images generated by AnimeGAN have clearly reproduced edges, a novel grayscale adversarial loss is also used to prevent the generated image from being displayed as a grayscale image. Finally, the loss function of the discriminator is expressed as follows:

$$L(D) = \omega_{adv} [E_{a_i \sim S_{data}(a)}[(D(a_i) - 1)^2] + E_{p_i \sim S_{data}(p)}[(D(G(p_i)))^2] + E_{x_i \sim S_{data}(x)}[(D(x_i))^2] + 0.1 E_{y_i \sim S_{data}(y)}[(D(y_i))^2]] \tag{6}$$

where $E_{y_i \sim S_{data}(y)}[(D(y_i))^2]$ represents the edge-promoting adversarial loss and $E_{x_i \sim S_{data}(x)}[(D(x_i))^2]$ represents the grayscale adversarial loss. 0.1 is the scaling factor. The purpose of setting the scaling factor of the edge-promoting adversarial loss to 0.1 is to avoid the edges of the generated image being too sharp.

3.3 Training

The proposed AnimeGAN can be easily end-to-end trained with unpaired training data. Since the GAN model is highly nonlinear, with random initialization, the optimization can be easily trapped at suboptimal local minimum. The literature [3] suggests that the pre-training of the generator helps to accelerate GAN convergence. Hence, the generator network G is pre-trained with only the content loss $L_{con}(G, D)$. The initialization training is performed for one epoch and the learning rate is set to 0.0001.

For the training phase of AnimeGAN, the learning rates of the generator and discriminator are 0.00008 and 0.00016, respectively. The training epochs for AnimeGAN is 100 and the batch size is set to 4. Adam optimizer [14] is used to minimize the total loss. AnimeGAN is trained on a Nvidia 1080ti GPU using the Tensorflow [1].

4 Experiments

4.1 Data

The proposed AnimeGAN can be easily end-to-end trained to generate the high-quality anime style images with the unpaired data. The real-world photos as the content images and the anime images as the style images are used as the training data, and the test data only includes the real-world photos. The resolution of all training images is set to 256×256. For the content images, 6,656 real-world photos are employed for training and these photos have been used as the training data for CycleGAN [30]. For the style images, since different animation artists have their own unique animation creation styles, in order to obtain a series of the animation images with the same style, we use the key frames of the animated films drawn and directed by the same artist as the style images. In our experiments, 1792 animation images from the movie "The Wind Rises" directed by Miyazaki Hayao are used for training the Miyazaki Hayao style model, 1650 animation images from the movie "Your Name" directed by Makoto Shinkai are used for training the Makoto Shinkai style model and 1553 animation images from the movie "Paprika" directed by Kon Satoshi are used for training the Kon Satoshi style model.

4.2 Results

We first compare AnimeGAN with the two state-of-the-art anime style transfer methods, namely, CartoonGAN [3] and ComixGAN [23]. It is worth noting that CartoonGAN and ComixGAN have the same network structure and loss function. The difference between them is that ComixGAN uses different training strategies and applies animation style transfer to video. Furthermore, the architecture of the discriminator used by AnimeGAN is the same as that used by CartoonGAN. The discriminator used by AnimeGAN is the lightweight convolutional neural network with a model size of $4.30M$. For different generators, the comparisons between them in term of model size, the number of the parameters and computational cost are shown in Table 1. In Table 1, the size of each input photo is 256×256 for inference.

Table 1. Comparisons of the performance of the different generators

Network	Params	Model size	FLOPs	Inference time
CartoonGAN	12253152	46.74M	108.98B	51 ms/image
AnimeGAN	**3956096**	**15.09M**	**38.66B**	**43 ms/image**

As can be seen from Table 1, compared with CartoonGAN, AnimeGAN significantly reduces the number of parameters and computational cost, and has the faster inference speed.

The qualitative results of CartoonGAN, ComixGAN and AnimeGAN are shown in Fig. 3. It can be clearly seen that the three methods can effectively capture the anime style. However, the image generated by CartoonGAN produces obvious colorful artifacts in the local areas and loses the color of the original content images. The images generated by ComixGAN are easily excessively stylized in the local areas, which causes the generated images to lose the content of the original photos. The images generated by AnimeGAN can effectively preserve the content of the photos and the color of the corresponding areas in the photos. Compared with CartoonGAN and ComixGAN, AnimeGAN can produce the higher quality animated visual effects.

It is worth noting that the adversarial loss of AnimeGAN is much smaller than the content loss and the grayscale style loss. In order to balance the influence

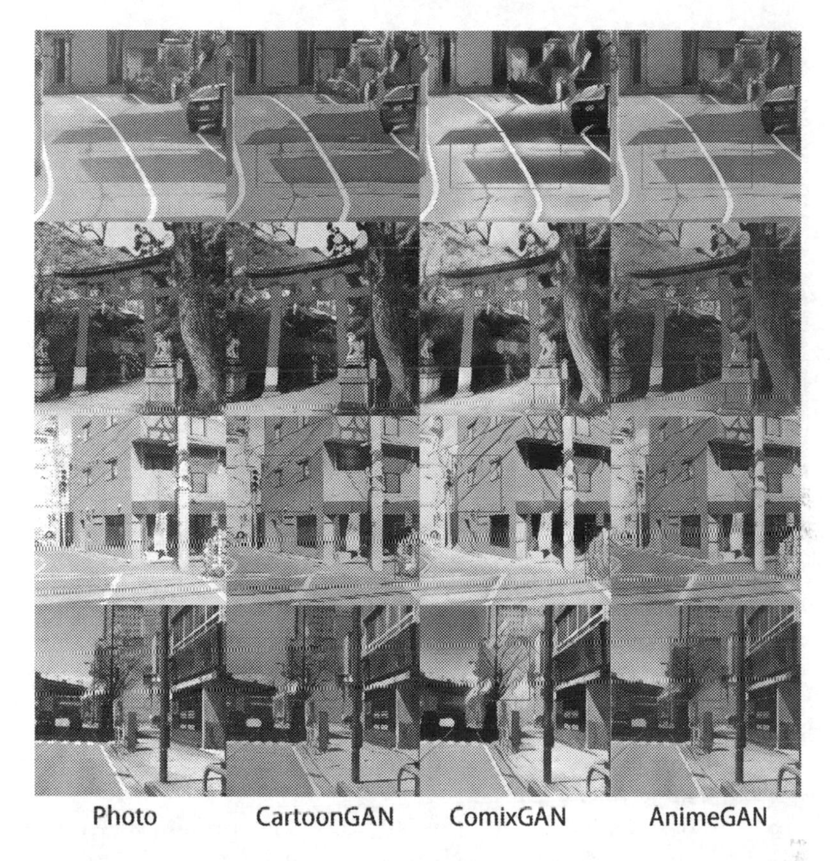

Fig. 3. Qualitative comparison of CartoonGAN, ComixGAN and AnimeGAN.

of different losses on the performance of the network, we need to reasonably weight different losses. If the weight of the content loss is too large, the generated images will be very close to the real photo. If the weight of the grayscale style loss is too large, the generated images will lose the content of the original photos. In the experiments, the weights of the content loss and grayscale style loss are set to 1.5 and 3.0, respectively. These two losses are larger than other losses and must be given the small weights.

To further investigate the effect of the weights on the quality of the generated images, we compared the effects of the weight of the adversarial loss and the weight of the color reconstruction loss on the generated images. The comparison results are shown in Fig. 4. Compared to AnimeGAN with $\omega_{col} = 10$ and $\omega_{adv} = 300$, the images generated by AnimeGAN with $\omega_{col} = 10$ and $\omega_{adv} = 250$ are closer to the input photo, and the generated images have no obvious animation style. Although the images generated by AnimeGAN with $\omega_{col} = 10$ and $\omega_{adv} = 350$ have obvious animation style, the hue of the generated images will be affected by the style of the training data. As shown in the third column of Fig. 4, the images generated by AnimeGAN with $\omega_{col} = 10$ and $\omega_{adv} = 350$ have obvious dark green.

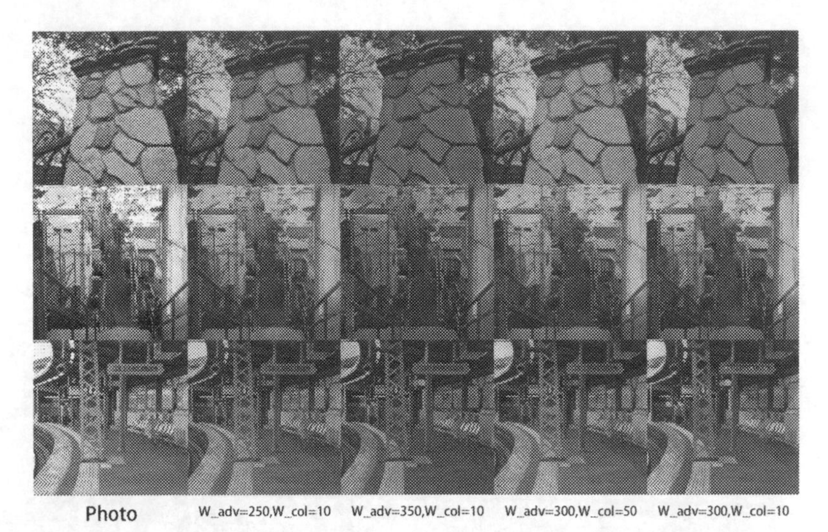

Fig. 4. Qualitative comparison of different weights. (Color figure online)

As can be seen from the last two columns in Fig. 4, compared to AnimeGAN with $\omega_{col} = 10$, the images generated by AnimeGAN with $\omega_{col} = 50$ have more realistic content but the animation style of the images is not obvious. Therefore, when $\omega_{col} = 10$ and $\omega_{adv} = 300$, the images generated by AnimeGAN have the satisfactory animated visual effects.

The color reconstruction loss is essentially the comparison between the pixels of the generated images and the pixels of the input photos. If its weight is set too large, the generated images are closer to the input photos.

To further study the effect of the grayscale images on the results, we conduct the ablation experiment to analyze the effectiveness of the grayscale adversarial loss and the blurred grayscale edge images $S_{data}(y)$ used for the edge-promoting adversarial loss. In Fig. 5, "A" indicates the inference results generated by AnimeGAN without using the grayscale adversarial loss, "B" indicates the inference results generated by AnimeGAN using the blurred color edge images in $S_{data}(e)$ for the edge-promoting adversarial loss, and "C" indicates the inference results generated by AnimeGAN. The results in A and B are similar. The difference between C and B is that the edge-blurred color images in $S_{data}(e)$ are processed into the grayscale images in $S_{data}(y)$ to avoid the color effects of the edge-blurred images on the generated results. Compared with C, the colors in A are not rich enough and saturated. Hence, the grayscale adversarial loss can promote the generated images to have a more saturated color.

Compared with B, the use of the grayscale data from $S_{data}(y)$ for the edge-promoting adversarial loss can not only enable the generated images to have clear edges but also make the generated images have more saturated color.

In order to fully verify the ability of AnimeGAN for photo animation, a large number of experiments were conducted on three animation style datasets.

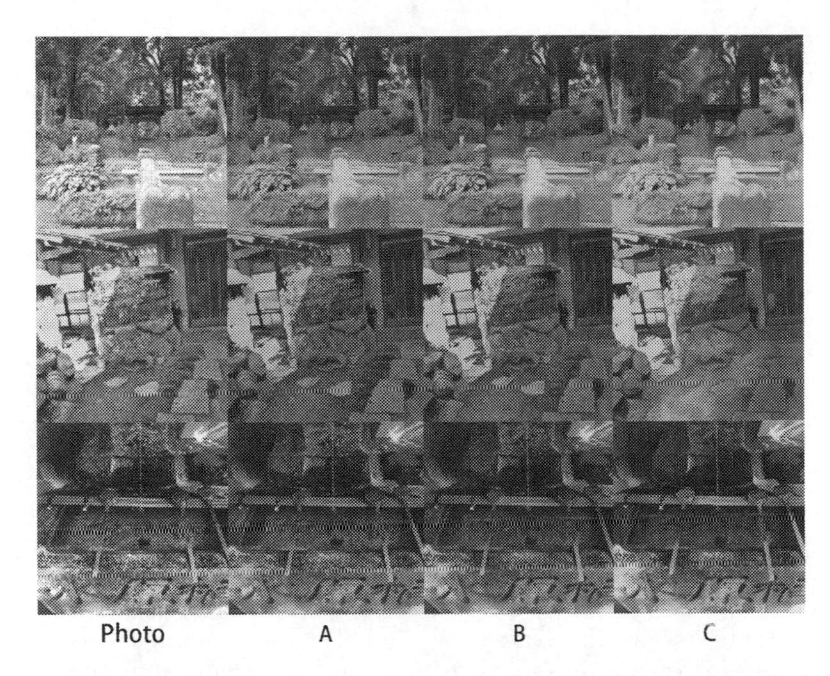

Fig. 5. Qualitative comparison of the grayscale adversarial loss and the different edge-promoting adversarial losses.

Fig. 6. The results produced by AnimeGAN for 3 different styles.

The experimental results are shown in Fig. 6. It can be seen that AnimeGAN successfully learns the animation styles of three different artists and generates high-quality animation visual images.

5 Conclusions

In this paper, we proposed AnimeGAN which is the lightweight generative adversarial network to fast transform real-world photos to the high-quality anime style images. The main contributions of this paper are as follows: 1) the novel grayscale style loss for transforming the anime style textures and lines; 2) the novel color reconstruction loss to preserve the color of the content images; 3) the novel grayscale adversarial loss for preventing the generated images from being displayed as the grayscale images; 4) a lightweight generator using depthwise separable convolutions and inverted residual blocks to achieve faster transfer. The experiments show that AnimeGAN can transform photos of real-world scenes to anime style images with high-quality animated visual effects and significantly outperforms the state-of-the-art stylization methods. In the experiments we also analyzes the effects of the grayscale adversarial loss and the weights on the quality of the generated images. We find that the appropriate weights and the grayscale adversarial loss can effectively improve the quality of generated images. In the future work, in order to achieve a faster animation style transfer, we will further study the more lightweight GAN to achieve real-time applications on small mobile devices. Furthermore, it is also considered to integrate the inference phase of AnimeGAN into the video processing pipeline to achieve the animation style transfer of videos.

Acknowledgment. The work described in this paper was support by National Natural Science Foundation of China Foundation No. 61300127. Any conclusions or recommendations stated here are those of the authors and do not necessarily reflect official positions of NSFC.

References

1. Abadi, M., et al.: TensorFlow: a system for large-scale machine learning. CoRR abs/1605.08695 (2016). http://arxiv.org/abs/1605.08695
2. Almahairi, A., Rajeswar, S., Sordoni, A., Bachman, P., Courville, A.: Augmented CycleGAN: learning many-to-many mappings from unpaired data. In: Proceedings 35th International Conference on Machine Learning, ICML 2018, Stockholm, Sweden, pp. 300–309 (2018)
3. Chen, Y., Lai, Y.K., Liu, Y.J.: CartoonGAN: generative adversarial networks for photo cartoonization. In: Proceedings 31st Meeting of the IEEE/CVF Conference on Computer Vision and Pattern Recognition, CVPR 2018, Salt Lake City, UT, United States, pp. 9465–9474 (2018)
4. Chollet, F.: Xception: deep learning with depthwise separable convolutions. In: Proc. 30th IEEE Conference on Computer Vision and Pattern Recognition, CVPR 2017, Honolulu, HI, United States, pp. 1800–1807 (2017)

5. Gatys, L.A., Ecker, A.S., Bethge, M.: A neural algorithm of artistic style. CoRR abs/1508.06576 (2015). http://arxiv.org/abs/1508.06576
6. Gatys, L.A., Ecker, A.S., Bethge, M.: Texture synthesis using convolutional neural networks. In: Proceedings 29th Annual Conference on Neural Information Processing Systems, NIPS 2015, Montreal, QC, Canada, pp. 262–270 (2015)
7. Gatys, L.A., Ecker, A.S., Bethge, M.: Image style transfer using convolutional neural networks. In: Proceedings 29th IEEE Conference on Computer Vision and Pattern Recognition, CVPR 2016, Las Vegas, NV, United States, pp. 2414–2423 (2016)
8. Gatys, L.A., Ecker, A.S., Bethge, M., Hertzmann, A., Shechtman, E.: Controlling perceptual factors in neural style transfer. In: Proceedings 30th IEEE Conference on Computer Vision and Pattern Recognition, CVPR 2017, Honolulu, HI, United States, pp. 3730–3738 (2017)
9. Goodfellow, I.J., et al.: Generative adversarial nets. In: Proceedings 28th Annual Conference on Neural Information Processing Systems 2014, NIPS 2014, Montreal, QC, Canada, pp. 2672–2680 (2014)
10. He, K., Zhang, X., Ren, S., Sun, J.: Deep residual learning for image recognition. In: Proceedings 29th IEEE Conference on Computer Vision and Pattern Recognition, CVPR 2016, Las Vegas, NV, United States, pp. 770–778 (2016)
11. Huang, X., Belongie, S.: Arbitrary style transfer in real-time with adaptive instance normalization. In: Proceedings 16th IEEE International Conference on Computer Vision, ICCV 2017, Venice, Italy, pp. 1510–1519 (2017)
12. Isola, P., Zhu, J.Y., Zhou, T., Efros, A.A.: Image-to-image translation with conditional adversarial networks. In: Proceedings 30th IEEE Conference on Computer Vision and Pattern Recognition, CVPR 2017, Honolulu, HI, United States, pp. 5967–5976 (2017)
13. Johnson, J., Alahi, A., Fei-Fei, L.: Perceptual losses for real-time style transfer and super-resolution. In: Leibe, B., Matas, J., Sebe, N., Welling, M. (eds.) ECCV 2016. LNCS, vol. 9906, pp. 694–711. Springer, Cham (2016). https://doi.org/10.1007/978-3-319-46475-6_43
14. Kingma, D.P., Ba, J.: Adam: a method for stochastic optimization. In: Proceedings the 3rd International Conference for Learning Representations, ICLR 2015, San Diego, CA, United States, pp. 1–15, May 2015
15. Krizhevsky, A., Sutskever, I., Hinton, G.E.: ImageNet classification with deep convolutional neural networks. In: Proceedings 26th Annual Conference on Neural Information Processing Systems 2012, NIPS 2012, Lake Tahoe, NV, United States, pp. 1097–1105 (2012)
16. Li, Y., Fang, C., Yang, J., Wang, Z., Lu, X., Yang, M.H.: Diversified texture synthesis with feed-forward networks. In: Proceedings 30th IEEE Conference on Computer Vision and Pattern Recognition, CVPR 2017, Honolulu, HI, United States, pp. 266–274 (2017)
17. Li, Y., Fang, C., Yang, J., Wang, Z., Lu, X., Yang, M.H.: Universal style transfer via feature transforms. In: Proceedings 31st Annual Conference on Neural Information Processing Systems, NIPS 2017, Long Beach, CA, United States, pp. 386–396 (2017)
18. Li, Y., Liu, M.-Y., Li, X., Yang, M.-H., Kautz, J.: A closed-form solution to photorealistic image stylization. In: Ferrari, V., Hebert, M., Sminchisescu, C., Weiss, Y. (eds.) ECCV 2018. LNCS, vol. 11207, pp. 468–483. Springer, Cham (2018). https://doi.org/10.1007/978-3-030-01219-9_28

19. Maas, A.L., Hannun, A.Y., Ng, A.Y.: Rectifier nonlinearities improve neural network acoustic models. In: ICML Workshop on Deep Learning for Audio, Speech and Language Processing (2013)
20. Mao, X., Li, Q., Xie, H., Lau, R.Y., Wang, Z., Smolley, S.P.: Least squares generative adversarial networks. In: Proceedings 2017 IEEE International Conference on Computer Vision, ICCV 2017, Venice, Italy, pp. 2813–2821 (2017)
21. Miyato, T., Kataoka, T., Koyama, M., Yoshida, Y.: Spectral normalization for generative adversarial networks. CoRR abs/1802.05957 (2018). http://arxiv.org/abs/1802.05957
22. Odena, A., Dumoulin, V., Olah, C.: Deconvolution and checkerboard artifacts. Distill (2016). https://doi.org/10.23915/distill.00003. http://distill.pub/2016/deconv-checkerboard
23. Pesko, M., Svystun, A., Andruszkiewicz, P., Rokita, P., Trzcinski, T.: Comixify: transform video into a comics. CoRR abs/1812.03473 (2018). http://arxiv.org/abs/1812.03473
24. Sandler, M., Howard, A., Zhu, M., Zhmoginov, A., Chen, L.C.: MobileNetV2: inverted residuals and linear bottlenecks. In: Proceedings 31st Meeting of the IEEE/CVF Conference on Computer Vision and Pattern Recognition, CVPR 2018, Honolulu, HI, United States, pp. 4510–4520 (2018)
25. Schmidhuber, J.: Deep learning in neural networks: an overview. Neural Netw. **61**, 85–117 (2015)
26. Simonyan, K., Zisserman, A.: Very deep convolutional networks for large-scale image recognition. CoRR abs/1409.1556 (2014). http://arxiv.org/abs/1409.1556
27. Ulyanov, D., Vedaldi, A., Lempitsky, V.S.: Instance normalization: The missing ingredient for fast stylization. CoRR abs/1607.08022 (2016). http://arxiv.org/abs/1607.08022
28. Wang, K., Gou, C., Duan, Y., Lin, Y., Zheng, X., Wang, F.Y.: Generative adversarial networks: introduction and outlook. IEEE/CAA J. Autom. Sinica **4**(4), 588–598 (2017)
29. Wang, T.C., Liu, M.Y., Zhu, J.Y., Tao, A., Kautz, J., Catanzaro, B.: High-resolution image synthesis and semantic manipulation with conditional gans. In: Proceedings 31st Meeting of the IEEE/CVF Conference on Computer Vision and Pattern Recognition, CVPR 2018, Salt Lake City, UT, United States, pp. 8798–8807 (2018)
30. Zhu, J.Y., Park, T., Isola, P., Efros, A.A.: Unpaired image-to-image translation using cycle-consistent adversarial networks. In: Proceedings 16th IEEE International Conference on Computer Vision, ICCV 2017, Venice, Italy, pp. 2242–2251 (2017)

BERT-BiLSTM-CRF for Chinese Sensitive Vocabulary Recognition

Yujuan Yang[1], Xianjun Shen[1(✉)], and Yujie Wang[2]

[1] School of Computer, Central China Normal University, Wuhan, Hubei, China
xjshen@mail.ccnu.edu.cn

[2] Collaborative & Innovation Center, Central China Normal University, Wuhan, Hubei, China

Abstract. There are tens of thousands of unstructured textual data generated from social media platforms every day, as well as vast sensitive words that may spread harmful information and even insult with each other. Therefore, effectively identifying sensitive words on the network, can not only filter out these inappropriate remarks, but also build a healthy and clean network environment. Recently, there are many researchers devote themselves into the study of emotional vocabulary, just fewer studies draw attention to the identification of sensitive vocabulary on the network. Furthermore, traditional methods of vocabulary recognition need a lot of manpower to make rules and extract features. In this paper, we firstly construct the dataset of uncivilized language (ULN dataset), which is acquired through web crawler on the website. Secondly, we propose a method to identify Chinese sensitive words on the network, which combine the Bidirectional Encoder Representations from Transformers with Bidirectional Long Short Term Memory Network and Conditional Random Field (BERT-BiLSTM-CRF). We use three models to identify sensitive words in ULN dataset. The experimental results show that, the model proposed in this paper has excellent performance, compared with the classical Bidirectional Long Short Term Memory Network (BiLSTM-CRF) and Convolutional Neural Network (CNN).

Keywords: Sensitive vocabulary · BERT-BiLSTM-CRF model · Chinese sensitive lexicon

1 Introduction

There are a lot of vulgar and insulting information on social media, which is extremely harmful to human life, social harmony and national stability. Chinese is very different from other languages. First, Chinese language is composed of Chinese characters and Pinyin. Second, Chinese characters have radicals, and Pinyin has four tones. Third, Chinese contains a large number of polyphonic characters, homophonic words. Therefore, there are many variations of words, such as misspelled, mixture of Chinese and English, homophone substitution and so on. It brings a lot of trouble to the research in social media. The short text is the most important unstructured data in social media. How to quickly and accurately identify sensitive information in short text data is an enormous challenge for Natural Language Processing (NLP).

K. Li et al. (Eds.): ISICA 2019, CCIS 1205, pp. 257–268, 2020.
https://doi.org/10.1007/978-981-15-5577-0_19

At present, there are two main methods for studying sensitive information on the network at home and abroad. One is to transform the text on the network into classification problem. Aggarwal [1] proposed the BERT and Two-Vote Classification method to classify offensive language on the network. Altin [2] used the BiLSTM model to identify offensive languages in Twitter. The other one is to identify harmful information by keyword list matching. Li [3] proposed Chinese keyword matching algorithm for spam filtering. The classifier of feature-based method cannot recognize the variations of words in short texts. The traditional method based on keyword matching can only accurately match the text in the keyword table. It is difficult to quickly and accurately identify sensitive information on the network from massive data. In order to extract valuable information from massive data, researchers have constructed a variety of emotional dictionaries, such as HowNet Lexicons [4], Affective Lexicon Ontology [5], and these lexicons are helpful to analyze emotions of the text.

Identifying the sensitive words in the short text is a key step in building the dictionary. There are three main methods of word recognition research. One is to identify new words by constructing rules. Li [6] improved Apriori Algorithm to process the corpus and generate association rules, and then used association rules to extract new professional vocabulary. Sasano [7] leveraged derivation rules and onomatopoeia patterns, and correctly recognized certain types of unknown words. This rule-based method needs to consume human resources to build a rule base, which has strong pertinence, and it is difficult to apply to all fields. Another method needs select new words by calculating various statistics. Li [8] proposed a new word discovery method based on the degree of word internal cohesion and boundary freedom. Su [9] analyzed the performance of classical statistical methods in extracting new words from microblog texts and proposed an improved algorithm of adjacency entropy to extract new words. Although this method is not limited by the field, and it is fast and easy to implement, it needs large-scale corpus training. It also cannot recognize words that are low-frequency. The other one is a combination of statistics and rules. Xiang [10] proposed a method to identify Chinese named entities, which used Hidden Markov model for Part-Of-Speech tagging, and then modified and transformed the results of the first step by using matching rules with priority. Although the combination of statistics and rules can improve the recognition effect of new words to a certain extent, it still needs to extract features manually and cannot be applied to various fields.

The deep neural network provides a new method for word recognition. Maryam [11] proposed the LSTM-CRF model for biomedical named entity recognition, and achieved excellent performance. Dong [12] proposed a character-based LSTM-CRF model with radical-level features for Chinese named entities recognition.

Inspired by the above, we try to build scientific and relatively complete sensitive words. In this paper, we propose the BERT-BiLSTM-CRF method to identify the sensitive vocabulary on the network. It mainly identifies abusive words, heresy words and sensitive words in the ULN dataset. There are three main contributions in this paper. Firstly, we use the crawler technology to obtain the short text on the network, and then annotate the text in BIO format to build the ULN dataset. Secondly, we use BERT-base Chinese model to generate word vectors, which are more suitable for the context language environment. Finally, we compare and analyze the lexical recognition effects of the BERT-BiLSTM-CRF model, BiLSTM-CRF model and CNN model.

2 Method

This paper proposes the BERT-BiLSTM-CRF model for harmful words recognition, and the structure of the model is shown in Fig. 1. The model we have introduced is an end-to-end language model, which can catch the syntactic and semantic characteristics automatically and understand contextual associations better. There are three parts in the model, the input embedding layer, the bidirectional LSTM layer and the CRF layer. The functions and principles of each layer are as follows.

2.1 Input Layer

It is well known that the computer can understand text only by processing it into numbers, vectors or matrices. Word embedding is to map words into vector space and use vectors to represent words. There are many researchers devote themselves into the study of appropriate word representations. In the traditional One-Hot vector representation, the position of the corresponding words is expressed as 1, and the other words are expressed as 0. Although this method is easy to understand, its usability is bigger problems. When the dictionary is large, the corresponding dimension of word vector is very large, and the semantic relationship of the text cannot be obtained. However, the basic idea of Word2Vec is to represent each word as a k-dimension vector, mapping similar words into different parts of the vector space [13]. It can learn the relationship between words, but it cannot solve the problem of polysemy. No matter what the context is, each word has only one word vector, and the word vector cannot change with the change of the contextual environment.

Peters [14] proposed the ELMO model, which is a splicing of a forward LSTM language model and a backward LSTM language model. The word vector is related to a specific task and can understand different ambiguous words. It has achieved excellent performance in six natural language processing problems. But it is essentially a unidirectional language model. Compared with Word2Vec, this model represents the context of words perfectly.

Vaswani [15] proposed a transformer model that can obtain a longer range of contextual information. Radford [16] proposed the Open AI GPT model, which is trained through two processes: unsupervised pre-training and supervised fine tuning. The pre-training language model can be applied to various natural language processing tasks without changing the basic structure of the model. The experimental results show that the transformer encoder used in this model is better than the LSTM network in ELMO. However, the Open AI GPT model is still a unidirectional language model, which uses a left to right limited transformer encoder. If we can learn a real bidirectional and universal method of word representation, it will contribute to promote the development of natural language research.

The BERT [17] model proposed by Google effectively solves the above problems. The training process of BERT is similar to that of Open AI GPT, and it is divided into two tasks: pre-training and fine tuning. BERT uses two methods of mask language model and next sentence prediction [18] in the pre-training task. It not only can obtain word and sentence level representation, but also utilize a large amount of unlabeled data. The model of BERT uses a transformer encoder. Based on the model, the upper and lower layers are fully connected. It is a real bidirectional language model, which effectively solves the problem of long dependence in NLP. In this paper, the pre-trained BERT-base model [17] is used instead of Word2Vec to generate the word vector of the text. This model can generate different word vector representations for the same word according to the context of text. The output of this layer will be used as the input to the next layer.

2.2 BiLSTM Layer

The traditional Full Connected Neural Network has no connection between the nodes of the same layer, so it cannot deal with the problems of time or space dependence. Recurrent Neural Network captures the dynamic information of the sequence through loops in the node network, which can retain the state information [19] in any long context window. In the actual training process, however, it is easy to appear the problem of gradient disappearance and gradient explosion [20]. The Long Short Term Memory Network is a variant of RNN. The mainly difference from RNN is that it adds a "processor" to the algorithm to judge whether the information is useful or not. The architecture of processor is called gate. There are three gates in an LSTM unit, which are called the forget gate, the input gate, and the output gate. Gate is a way to choose information to pass. In other words, the forget gate determines which information is discarded from the gate state, the input gate determines which information is to be updated, and the output gate controls the output information. These three gates work in coordination to jointly control the status of the LSTM unit. The unit of LSTM updating formula at a certain time t is:

$$i_t = \sigma(w_{xi}x_t + w_{hi}h_{t-1} + w_{ci}c_{t-1} + b_i) \tag{1}$$

$$f_t = \sigma(w_{xf}x_t + w_{hf}h_{t-1} + w_{cf}c_{t-1} + b_f) \tag{2}$$

$$c_t = (1 - i_t) \odot c_{t-1} + i_t + \odot \tan h(w_{xc}x_t + w_{hc}h_{t-1} + b_c) \tag{3}$$

$$o_t = \sigma(w_{xo}x_t + w_{ho}h_{t-1} + w_{ho}h_{t-1}) \tag{4}$$

$$h_t = o_t \odot \tan h(c_t) \tag{5}$$

i_t represents the input gate, f_t represents the forget gate, c_t represents the LSTM unit state at the current t time, o_t represents the output gate, h_t represents the hidden layer output at the current time, x_t indicates that the current input. w_{xi}, w_{hi}, w_{ci}, w_{xo}, w_{ho}, w_{co} represent the weight matrix of the corresponding gate. b_i, b_c, b_f represent the corresponding offsets.

Since LSTM preserves information from front to back, it is possible to remember long dependencies. The meaning of Chinese words is rich in language environment, so the computer must understand the contextual information to judge the exact meaning of words. Dyer [21] proposed a bidirectional LSTM model composed of forward LSTM and backward LSTM, which can represent contextual information excellently. For example, if the output to the forward network is h_{t1} at a certain time t and the backward network output is h_{t2}, the output information h_t of the bidirectional LSTM model at this time can be expressed as $h_t = [h_{t1}; h_{t2}]$.

2.3 CRF Layer

In fact, we can use the output of bidirectional LSTM to predict the probability value of each word, but this is not reasonable. Because the possible sequence of output is B-ABU, I-HER, but the sequence of correct is B-ABU, I-ABU. Our task is the identification of multiple types of vocabulary, and there are strong dependencies between the output tags of each category [12]. Therefore, after bidirectional LSTM layer, we add CRF layer to adjust the possible output tags. Different from the traditional CRF, it is not affected by the characteristic function and the corresponding weight value. It is only learning the state transition matrix between tags.

Suppose $X = (x_1, x_2, \cdots, x_n)$ is the sequence of input sentence, where x_i represents the input vector of the ith word. $y = (y_1, y_2, \cdots y_n)$ represents the sequence of predictions of X. P represents the output of $n \times k$ matrix of the bidirectional LSTM model, where k represents the number of the tags. p_{ij} represents the score of the jth label of the ith word in the sentence.

The score for this sentence is:

$$s(X, y) = \sum\nolimits_{i=0}^{n} A_{y_i, y_{i+1}} + \sum\nolimits_{i=1}^{n} p_{i, y_i} \tag{6}$$

In the above formula, A represents the label state transition matrix, which is obtained during the training model. $A_{y_i, y_{i+1}}$ represents the transition probability from the label y_i to the label y_{i+1}.

After the softmax function, the probability of the sequence y is:

$$p(y|X) = \frac{e^{s(X,y)}}{\sum_{\widetilde{y} \in Y_X} e^{s(X, \widetilde{y})}} \tag{7}$$

Then take the logarithm to maximize the probability of the sequence $p(y|X)$:

$$\log(p(y|X)) = s(X, y) - \log \sum\nolimits_{\widetilde{y} \in Y_X} e^{s(X, \widetilde{y})} \tag{8}$$

Finally, the maximum score for the possible of output sequence is:

$$y^* = arg\max_{\widetilde{y} \in Y_X} s(X, \widetilde{y}) \tag{9}$$

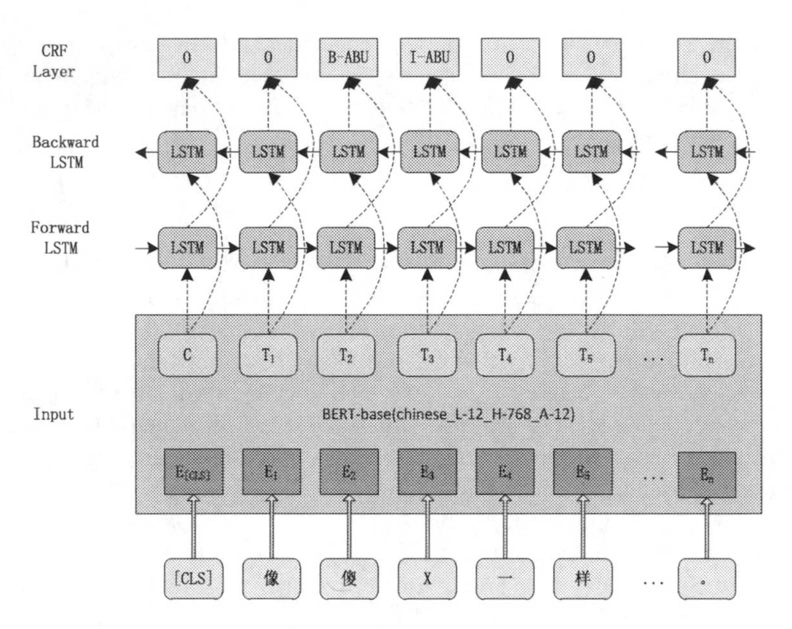

Fig. 1. BERT-BiLSTM-CRF model diagram

3 Experiment

3.1 Dataset

We construct the dataset of uncivilized language (referred to as ULN dataset). First and foremost, the data comes from YouTube comments on hot issues, which is acquired through web crawler on the website. Through careful reading and analyzing, more than 9,600 texts were selected. In addition, we remove invalid characters in the text, such as HTML tags, emoticons. Last but not least, the text is used as the initial corpus for manual annotation. We divide the vocabulary into three categories: abusive words (ABU), political sensitive words (SEN) and heresy words (HER). These words are marked in BIO format, where "B" stands for the beginning of the vocabulary and "I" stands for the middle or end of the vocabulary, "O" indicates other information without sensitive information. For example, "脑残" is labeled as "B-ABU I-ABU". Due to the special combination of sensitive words on the network, there is no strict standard of language structure. In the process of annotation, different people have different understanding of the text, so the boundary of the same word may be different. Therefore, to ensure accuracy, we asked three people to annotate the same text. We will further study the texts with different opinions when labeling.

In the dataset, there are 37912 tags, which are divided into train set, test set, and validation set in a reasonable proportion.

Table 1. Number of labels in the ULN dataset

	TRAIN	TEST	DEV	TOTAL
ABU	22599	4886	4073	31558
SEN	1460	457	345	2262
HER	2318	986	788	4092
TOTAL	26377	6329	5206	37912

3.2 Evaluation

We use the precision, recall, and F_1 value to evaluate performance of the model. Precision and recall are important evaluate indicators of the model. However, the two measures are contradictory, and F_1 combine the results of the precision rate and the recall rate. The closer the F_1 value is to 1, the better the performance of the model. The calculation formula is as follows:

$$precision = \frac{identify\ the\ correct\ number\ of\ entities}{number\ of\ entities\ identified} \times 100\% \tag{10}$$

$$recall = \frac{identify\ the\ correct\ number\ of\ entities}{number\ of\ entities\ in\ the\ sample} \times 100\% \tag{11}$$

$$F_1 = \frac{2 * precision * recall}{precision + recall} \times 100\% \tag{12}$$

3.3 Experimental Settings

We use TensorFlow to implement the model structure and run it in a Linux environment. During the training, we update the parameters of the model using a Stochastic Gradient Descent Algorithm. We use the dropout [22] technique and early stopping [23] technique to prevent overfitting on the train set. After experimental analysis, the parameters of the model are shown in Table 2.

Table 2. Experimental parameter setting

Parameter	BERT-LSTM-CRF
train_batch_size	32
learning_rate	5e−5
num_train_epoch	100
dropout	0.5
lstm_size	128
num_layers	1

3.4 Model Comparison One

The BiLSTM-CRF model showed excellent performance in different fields. Chalapathy [24] proposed bidirectional LSTM-CRF model for clinical concept extraction. Zeng [25] proposed LSTM-CRF for drug named entity recognition. Chen [26] proposed BiLSTM-CRF model to classify sentences and then analyze sentiment. In the comparative experiment, the BiLSTM-CRF model is used to identify sensitive words on the network. The input vector of the model is 100-dimensional, trained by word2vec. The variable of early stoppping is set to 3. The learning rate is set to 0.001. The hidden layer size is 100. The dropout size is set to 0.5. The decay rate is set to 0.9.

3.5 Model Comparison Two

Prechelt [27] proposed Convolutional Neural Network in 1998. The CNN model has achieved excellent results in the field of image recognition and text classification. Zha [28] proposed CNN for unconstrained video classification. Kim et al. [29] proposed using CNN for sentence-level classification tasks. In this paper, CNN is used to identify sensitive words on the network. The word vector of the model is the same as the BiLSTM-CRF.

4 Experimental Results

4.1 Overall Evaluation

As is shown in Table 1, the distribution of labels for each category is extremely unbalanced. The number of abuse labels is very large, almost 10 times for the heresy labels. To evaluate the performance of the model, we use the mirco-F1 method to calculate the F_1 value. The experimental results are shown in Table 3.

Table 3. Experimental results of each model on the validation set and test set

Set	Model	Precision (%)	Recall (%)	F_1 (%)
DEV	CNN	83.94	78.26	81.00
	BiLSTM-CRF	82.34	79.67	80.98
	BERT-BiLSTM-CRF	87.27	92.03	87.66
TEST	CNN	81.02	71.67	76.06
	BiLSTM-CRF	79.22	75.14	77.13
	BERT-BiLSTM-CRF	79.33	79.99	**79.66**

Table 3 shows the result of overall evaluation of the three models on the validation set and the test set. In the validation set, it can be found that the best effect is BERT-BiLSTM-CRF model, and there is no significant difference between CNN and BiLSTM-CRF. However, in the test set, we can find that the BERT-BiLSTM-CRF model achieved the highest F_1 value.

4.2 Entity Evaluation

There are three kinds of entity tags in the dataset. So, we did an entity level assessment. The experimental results are shown in Table 4.

Table 4. Entity level evaluation of models on test set

Tag	Model	Precision (%)	Recall (%)	F_1 (%)
ABU	CNN	78.75	71.17	74.76
	BiLSTM-CRF	76.93	74.69	75.80
	BERT-BiLSTM-CRF	77.38	78.85	**78.11**
SEN	CNN	80.70	33.33	47.18
	BiLSTM-CRF	74.67	40.29	52.34
	BERT-BiLSTM-CRF	77.98	65.38	**71.13**
HER	CNN	94.69	90.72	92.66
	BiLSTM-CRF	94.55	92.31	93.41
	BERT-BiLSTM-CRF	94.68	95.36	**95.02**

From Table 4, we can find that in the three models, the F_1 value of the HER class is above 92. For the abusive words, the F_1 value of the BERT-BiLSTM-CRF model is 3% higher than the CNN and 2% higher than the BiLSTM-CRF. It is worth mentioning that in the category of politically sensitive words, the results of the three models are quite different. For the SEN category words, the best results are obtained by the BERT-BiLSTM-CRF model. Its F_1 value is 18% higher than the BiLSTM-CRF model and 24% higher than the CNN model.

4.3 Comparative Analysis of Vocabulary Labels

We counted the number of three types of vocabulary for test set and predicative text of the model. Then, we removed the duplicate words from the text, and the results are shown in Table 5 and Table 6. There are 2554 words in the test set, and the predicative text predict a total of 2582 words (we will not discuss whether the possible output tags of the words are correct). It the test set, there are 23 HER words after removing duplicate words. There are 27 words after removing duplicate words in the predicative text. It indicates that the vocabulary of HER class in the ULN data set has few categories. Meanwhile, The SEN class vocabulary removes only a few repetitions in both test and predictive text. This situation indicates that there are varieties of SEN class vocabularies in the ULN dataset. For these words, the results of prediction of CNN and BiLSTM-CRF model are not ideal, while the F_1 value of BERT-BiLSTM-CRF model reach 71.13. It further indicates that the model proposed in this paper has better performance.

Table 5. Statistics of the labels in the test text

	ABU	SEN	HER
Word	2072	141	341
Remove duplicate	613	83	23

Table 6. Statistics of the labels in the predicative text

	ABU	SEN	HER
Word	2173	127	282
Remove duplicate	623	74	27

4.4 Error Analysis

The mainly reasons for the error were analyzed by observing the predicative text. Firstly, one reason is that the boundaries of the words are not clearly defined. For example, the tag is "太贱" and the result of the model is "太贱了". Secondly, the vocabulary is too long to fully identify. At last, the model cannot fully understand the human language environment. For example, an old couple may call the other party "死老头", it is not contain abusive information. But the computer is unable to understand it.

5 Conclusion

In this paper, we try to build a sensitive lexicon by identifying the harmful words in short texts on the network to provide data support for the network monitoring of the National Language Commission. The web crawlers are utilized to acquire more text owing to the deficiency of dataset. In the meanwhile, we have preprocessed and manually annotated the raw data, the precision and dependability can be ensured. Comparing performance of the three models, the F_1 value of the BERT-BiLSTM-CRF model that we have proposed in this paper is 2.53% better than BiLSTM-CRF. Furthermore, our model has improved 3.6% compared with CNN. In terms of the recognition results of the three types of vocabulary, the performance of label of heresy is superior to the others. We might devote to optimizing the effect of the other two categories in further study. In the future work, a standardized language annotation rules need to be constructed. And we will update and integrate more sensitive vocabulary to assure quality and quantity of the data. Moreover, other advanced computing model such as semi-supervised method will be utilized for word recognition to reduce manual annotation.

Acknowledgements. This research is supported by the National Language Commission Key Research Project (ZDI135-61), the National Natural Science Foundation of China (No.61532008 and 61872157), and the National Science Foundation of China (61572223).

References

1. Aggarwal, P., Horsmann, T., Wojatzki, M., Zesch, T.: LTL-UDE at SemEval-2019 Task 6: BERT and two-vote classification for categorizing offensiveness. In: 2019 Proceedings of the 13th International Workshop on Semantic Evaluation, pp. 678–682 (2019)
2. Altin, L.S.M., Serrano, À.B., Saggion, H.: LaSTUS/TALN at SemEval-2019 Task 6: identification and categorization of offensive language in social media with attention-based Bi-LSTM model. In: 2019 Proceedings of the 13th International Workshop on Semantic Evaluation, pp. 672–677 (2019)
3. LI-Lin, F., Xiao-Dong, W.: An algorithm of Chinese keys matching used in filtering junk mail. J. HenNan Univ. Sci. Technol. (Nat. Sci.) **27**(5), 35–37 (2006)
4. Dong, Z., Dong, Q.: HowNet-a hybrid language and knowledge resource. In: International Conference on Natural Language Processing and Knowledge Engineering, 2003, pp. 820–824. IEEE (2003)
5. Linhong, X., Hongfei, L., Yu, P., Hui, R., Jianmei, C.: Constructing the Affective Lexicon Ontology. J. China Soc. Sci. Tech. Inf. **27**(2), 180–185 (2008)
6. Liming: new words discovery research for specific areas. Nanjing University of Aeronautics and Astronautics (2012)
7. Sasano, R., Kurohashi, S., Okumura, M.: A simple approach to unknown word processing in Japanese morphological analysis. In: 2013 Proceedings of the Sixth International Joint Conference on Natural Language Processing, pp. 162–170 (2013)
8. Wenkun, L., Yangsen, Z., Ruoyu, C.: New word detection based on inner combination degree and boundary freedom degree of word. Appl. Res. Comput. **32**(8), 2302–2304 (2015)
9. Su, Q., Liu, B.: Chinese new word extraction from MicroBlog data. In: 2013 International Conference on Machine Learning and Cybernetics, 2013. IEEE, pp. 1874–1879 (2013)
10. Xiaowen, X., Xiaodong, S., Hualin, Z.: Chinese named entity recognition system using statistics-based and rules-based method. Comput. Appl. **25**(10), 2404–2406 (2005)
11. Habibi, M., Weber, L., Neves, M., Wiegandt, D.L., Leser, U.: Deep learning with word embeddings improves biomedical named entity recognition. Bioinformatics **33**(14), i37–i48 (2017)
12. Dong, C., Zhang, J., Zong, C., Hattori, M., Di, H.: Character-based LSTM-CRF with radical-level features for Chinese named entity recognition. In: Lin, C.-Y., Xue, N., Zhao, D., Huang, X., Feng, Y. (eds.) ICCPOL/NLPCC -2016. LNCS (LNAI), vol. 10102, pp. 239–250. Springer, Cham (2016). https://doi.org/10.1007/978-3-319-50496-4_20
13. Goldberg, Y., Levy, O.: word2vec Explained: deriving Mikolov et al.'s negative-sampling word-embedding method. arXiv preprint arXiv:1402.3722 (2014)
14. Peters, M.E., et al.: Deep contextualized word representations (2018)
15. Vaswani, A., et al.: Attention is all you need. In: 2017 Advances in Neural Information Processing Systems, pp. 5998–6008 (2017)
16. Radford, A., Narasimhan, K., Salimans, T., Sutskever, I.: Improving language understanding by generative pre-training. https://s3-us-west-2.amazonaws.com/openai-assets/researchcovers/languageunsupervised/languageunderstandingpaper.pdf (2018)
17. Devlin, J., Chang, M.W., Lee, K., Toutanova, K.: BERT: pre-training of deep bidirectional transformers for language understanding (2018)
18. Bengio, Y., Ducharme, R., Vincent, P., Jauvin, C.: A neural probabilistic language model. J. Mach. Learn. Res. **3**(Feb), 1137–1155 (2003)
19. Lipton, Z.C., Berkowitz, J., Elkan, C.: A critical review of recurrent neural networks for sequence learning. arXiv preprint arXiv:1506.00019 (2015)

20. Bengio, Y., Simard, P., Frasconi, P.: Learning long-term dependencies with gradient descent is difficult. IEEE Trans. Neural Netw. **5**(2), 157–166 (1994)
21. Dyer, C., Ballesteros, M., Ling, W., Matthews, A., Smith, N. A.: Transition-based dependency parsing with stack long short-term memory. arXiv preprint arXiv:1505.08075 (2015)
22. Hinton, G.E., Srivastava, N., Krizhevsky, A., Sutskever, I., Salakhutdinov, R.R.: Improving neural networks by preventing co-adaptation of feature detectors. arXiv preprint arXiv:1207. 0580 (2012)
23. Prechelt, L.: Automatic early stopping using cross validation: quantifying the criteria. Neural Netw. **11**(4), 761–767 (1998)
24. Chalapathy, R., Borzeshi, E.Z., Piccardi, M.: Bidirectional LSTM-CRF for clinical concept extraction (2016)
25. Zeng, D., Sun, C., Lin, L., Liu, B.: LSTM-CRF for drug-named entity recognition. Entropy-Switz. Entropy-Switz **19**(6), 283 (2017)
26. Chen, T., Xu, R., He, Y., Wang, X.: Improving sentiment analysis via sentence type classification using BiLSTM-CRF and CNN. Expert Syst. Appl. **72**, 221–230 (2017)
27. LeCun, Y., Bottou, L., Bengio, Y., Haffner, P.: Gradient-based learning applied to document recognition. P IEEE **86**(11), 2278–2324 (1998)
28. Zha, S., Luisier, F., Andrews, W., Srivastava, N., Salakhutdinov, R.: Exploiting image-trained CNN architectures for unconstrained video classification. arXiv preprint arXiv:1503. 04144 (2015)
29. Kim, Y.: Convolutional neural networks for sentence classification. arXiv preprint arXiv: 1408.5882 (2014)

The Classification of Chinese Sensitive Information Based on BERT-CNN

Yujie Wang[2], Xianjun Shen[1](✉), and Yujuan Yang[1]

[1] School of Computer, Central China Normal University, Wuhan, Hubei, China
xjshen@mail.ccnu.edu.cn
[2] Collaborative & Innovation Center, Central China Normal University, Wuhan, Hubei, China

Abstract. The traditional classification method of Chinese sensitive information mainly relies on the frequency of the co-occurrence of sensitive words and keywords. However, it is difficult to detect the meaning and context relationship of some complex statements. In this paper, a new model is proposed to classify Chinese network sensitive information. The model, which is based on CNN (Convolutional Neural Network) and latest pre-trained BERT (Bidirectional Encoder Representation from Transformers), is called the BERT-CNN deep learning model. Firstly, nearly 20,000 comments of corresponding videos are collected from YouTube, after data cleaning and labeling, the texts are divided into four categories to build the SLD (sensitive language detection) dataset. Finally, the word vectors from the pre-trained BERT are used as the input of the CNN model, constructing the Chinese sensitive information classifying BERT-CNN model. The result reveals that, compared with the traditional neural network model, the BERT-CNN model proposed in this paper improves the generalization ability of word embedding, and can effectively achieve the recognition and classification of the network sensitive information in short text dataset. The experimental results perform better than the classic CNN and RNN (Recurrent Neural Network) models.

Keywords: Sensitive information · Short text classification · BERT · CNN

1 Introduction

In recent years, the international situation has been complex and volatile, such as trading war between China and the United States, demonstrations and disturbances around the world. There is a large amount of data in various social platforms, which can be used for public opinion analysis. In the current international environment, the classification of Chinese sensitive information is of great significance.

Text classification is a key link in natural language processing (NLP). It paved the way for the following tasks, such as search engine, network information filtering and public opinion analysis [1]. With the development of network society, more and more sensitive information is flooded in various social platforms and video websites. The main representatives of social platforms are twitter and microblog, while the main representatives of video websites are YouTube and YOUKU.

K. Li et al. (Eds.): ISICA 2019, CCIS 1205, pp. 269–280, 2020.
https://doi.org/10.1007/978-981-15-5577-0_20

Because YouTube does not need any threshold to upload video and has no high technical requirements, it is convenient for people of all ages to upload video and pictures. Therefore, YouTube is the fastest growing and most used video website in the world. The total number of video uploads reached 114 billion minutes, with a billion users and billions of comments. Compared with social media, YouTube tends to spread non-textual content, which will contain more information. The comment under video has more information for us to analyze.

At present, research on sensitive information mainly focuses on the research and analysis of hate and aggression [2], cyberbullying [3]. Since 2019, the conflicts on the internet have become more and more intense. A variety of statements containing sensitive information are increasing in the network platforms. The research on the network sensitive information has become more and more detailed, such as the research on network sensitive information based on OLID data set, it contains three subtasks, one of which, task A, is to classify offensive and non-offensive texts [4]. Since the definition of the task is a little vague and repetitive in related subtasks, it is also necessary to distinguish hate speech from general offensive speech [5]. Deep learning algorithm is based on large-scale manually labeled data, however, it lacks uniform standards, and everyone has different definitions of sensitive information, leading to the problem of semantic ambiguity. Some studies only consider the sub-task of hatred, not abusive bullying or other directions [6].

The method based on constructing a sensitive word dictionary can deal with unstructured data. Therefore, it is widely used for detecting sensitive information. However, with the rapid development of the Internet, the expansion speed of sensitive words is faster than the filling speed of dictionaries. Therefore, the method based on word frequency statistics can hardly be perfectly applied. At the same time, if we want to determine whether a sentence is sensitive or not, it may have nothing to do with the presence of sensitive words and keywords. We need to understand the meaning of the sentence from the context.

With the rapid development of the Internet, the expansion speed of sensitive words is faster than the filling speed of dictionaries. Therefore, the method based on word frequency statistics can hardly be perfectly. At the same time, if we want to determine whether a sentence is sensitive or not, it may have nothing to do with the presence of sensitive words and keywords. We need to understand the meaning and context relationship of the sentence.

In this paper, the model we proposed fuses the convolutional neural network and BERT, which can learn semantic and grammatical information of the context. BERT-CNN can understand meaning and context relationship of some complex statements and effectively improve the accuracy of identifying and classifying sensitive information on the Internet. The comments under the video were crawled through Google YouTube API. After data cleaning and labeling, we build SLD dataset. The experimental results show that, compared with the traditional neural network model, the BERT-CNN deep learning model can effectively classify dataset, which consist of sensitive information on the Internet, and is significantly superior to the classical CNN and RNN models.

2 Related Work

In this part, we will give a description of the earlier work of the short texts classification and public opinions analysis based on social network platform in the field. Normally, a network sensitive information detection includes specifying the problem, labeling the harmful information as aggression, cyberbullying or hate speech, and then classifying them using different algorithm. As for the traditional classification of short texts, the main task contains three themes: characteristics engineering, characteristics selection and the machine learning algorithm selection based on the former two themes. In 2012, junmingxu [7] researched about the bully action existing in the network social media, they confirmed several key problems and arranged the problems into NLP tasks: context classification, character labeling and emotion analysis using topic model to identify related topics [8]. Homophobia and socialism may be easily labeled as hate speech [9], yet some kinds of gender aggression can be difficultly classified, because it is hard to define a large dictionary to accurately classify them. By the method of traditional machine learning, they classified nearly 24,000 tweet data into three categories. The Germ Eval released 8,500 labeled German twitters which were used for coarse-grained binary classification - offensive and unoffensive. Meanwhile, they carried about another task to subdividing the subset of the offensive set. The task included multiple hierarchical tasks where the binary classification was firstly finished and then the fine-grained classification was launched in the certain set of the offensive set from the former classification [10]. The idea made a pleasant effort in the realistic application. All the papers mentioned above used only some traditional machine learning algorithms, but just as was mentioned in the last parts of the papers, the problem is still far from settled although the algorithms, which have problems such as sparse matrix, are effective in some ways.

In recent years, the occurrence of DNN (deep neural network) settled the sparse matrix problem. Additionally, DNN model performs better than the traditional words-frequency statistical method, whose limitation leads to a difficulty learning the relationship of the context, in learning the grammar and meanings of the words, understanding the context profoundly and classifying contexts. Institute of Automation, Chinese Academy of Sciences [10] proposed a Recurrent Convolution Neural Network applying in context classification in 2015. The new neural network model, using cycle structure to study as more context information as possible and CNN to construct context representatives, can have less noises than the CNN. Some work compared the word2vec with the traditional n-gram model, and construct a twitter hate speech context classifying system based on deep learning. Four CNN models were trained based on word2vec at the same dataset to construct words embedding of meaning information, generate words embedding including ones combined with n-g characters randomly. Maxpooling was used to narrow the characteristics set in the net, and Tweet was classified by softmax function. After ten cross validations, the model based on word2vec embedding proved to be the best. The latest works were mainly about improving classification performance with perceptual network, Mikolov [11] employed a new continuous Skip-gram for effective embedding performance. Kalchbrenner and Blunsom [12] proposed a new dialect behavior classification cycle network. Collobert [13] introduced a CNN which is used for meaning character markings.

3 Method

In this paper, the model that we proposed is BERT-CNN. The model structure is shown in Fig. 1. The BERT-CNN model is applied to the dataset of Chinese sensitive information. In this model, the word and sentence embedding are learned by utilizing the pre-trained BERT model. After that, the embedding vectors are extracted from BERT as the input of the CNN. In other words, we treat the embedding extracted from BERT as the conventional embedding layer.

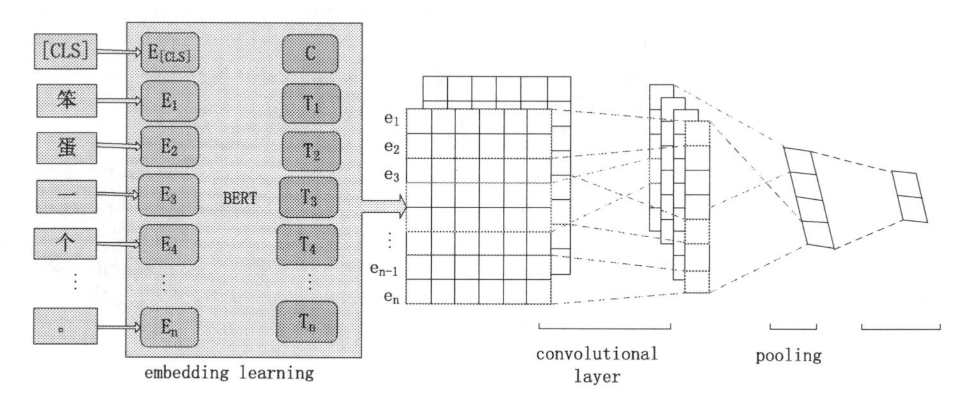

Fig. 1. The architecture of BERT-CNN

3.1 BERT

Compared with the ELMO [14] model proposed by Google in 2018, BERT [15] has changed language model from BiLSTM to Transformer [16], and realized the concept of bidirectional encoding in a real sense. model structure of BERT is shown in Fig. 2.

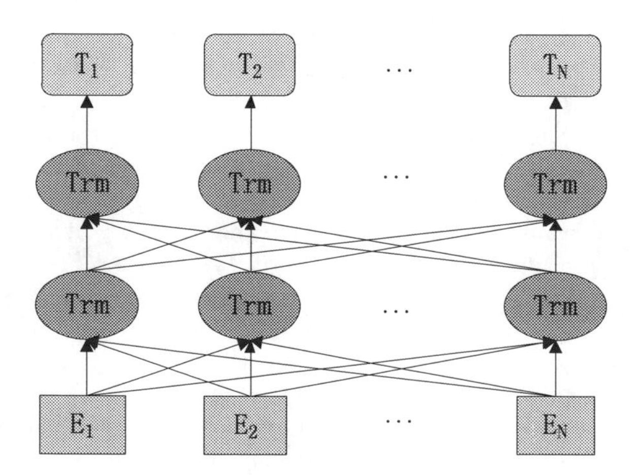

Fig. 2. The architecture of BERT

In the past, the language model was the probability distribution of a long statement, specifically, it is the probability value of the text of length of m. In order to facilitate calculation, the N-Gram model was proposed. The formula of N-Gram model is as follows:

$$p(S) = p(w_1 w_2 \cdots w_n) = p(w_1)p(w_2|w_1) \cdots p(w_n|w_{n-1}w_{n-2} \cdot \cdot w_2 w_1) \quad (1)$$

N-Gram is used to calculate the probability that the current word appears for the first n words. At that time, the N-Gram model would have the problem of sparse of matrix as n increased. In order to solve the problem of sparse of matrix, the solution given by the neural network is to optimize the model parameters based on the BP algorithm, through the random initialization of the network model parameters. The latest pre-training idea is that the value of parameter is no longer random initialization, the task was trained to obtain a fixed parameter value, which was used to initialize the model, and then the model was trained.

BERT model further enhanced the generalization ability of word embedding, and more fully learned the feature of character level, word level and sentence level. In the process of language training, they abandoned to predict the model from left to right, rather than using random masking. The method of masking randomly masks off 15% token in the corpus, and then hidden layer which contain masked word treat as a input of the softmax to predict.

Some models can only learn the features of token level but also need the features of sentence level. BERT borrowed the idea of negative sampling from word2vec [17] and used it for sentence level sampling to make dichotomy of sentence level to judge whether a sentence is noise. BERT can directly obtain the unique embedding of an entire sentence. It adds a special mark (CLS) in every input, and then let the Transformer to deep encode. Because the Transformer can ignore the space and distance of the global information, and the highest hidden layer as the presentation of sentences or words, it can directly connect with softmax, through the BP (back propagation), BERT can learn the whole feature of the upper input.

Compared with GPT [18], BERT realizes the concept of bidirectional encoding in a real sense. Because of self-attention, the transformer used by BERT realizes full connection of the model. The transformer can be regarded as a multilevel Encoder-Decoder model. Each layer of Encoder has two operations, which are Self-Attention and Feed Forward. In the other hand, each layer of Decoder has three operations, which are Self-Attention, Encoder-Decoder Attention and Feed Forward operation. Self-Attention and Encoder-Decoder Attention both use Multiheaded Attention mechanism. The dot product of Attention is scale. The input includes Q and K with dimension d_k, and V with dimension d_v. Take the dot product of Q with all K and divide by $\sqrt{d_k}$. Then use a softmax function to get the weight of V.

$$Attention(Q, K, V) = softmax\left(\frac{QK^t}{\sqrt{d_k}}\right)V \quad (2)$$

3.2 TextCNN

TextCNN [19] is a convolutional neural network proposed in 2014 for text classification. Convolutional neural network is a common neural network model. The general structure of convolutional neural network includes convolutional layer, pooling layer and full connection layer. Compared with traditional neural network, it contains convolution layer and pooling layer. Neurons between the convolutional layers are no longer connected in full connection, and neurons in the next layer only receive part of the input results from neuron nodes in the previous layer. The incoming input of the next layer of neurons is determined by the output of the previous layer of neurons, while the output of the previous layer of neurons is determined by the convolution kernel. Compared with full connection, the convolution layer realizes parameter sharing greatly reduces the number of parameters. Pooling layer is the pooling of the results of the convolution layer, also known as sub-sampling. Once again, the pooling layer carries out feature screening for the convolution layer, and removes useless features, and further optimizes the parameters of the convolutional neural network.

In the convolutional process of TextCNN, we embed the result of last layer of BERT as embedding layer. If the dimension of the word vector is d and the number of words is s, the matrix $A \in R^{s \times d}$ can be obtained. Then we use the convolutional neural network to extract the features. Since the correlation degree of the words in the short text is very high, the 1-dimensional convolution kernel can be used. Assuming that there is convolutional kernel $W \in R^{h \times d}$, we have $h * d$ parameters that need to be updated, and the result obtained by the last layer of BERT goes through embedded layer to obtain matrix A. The formula of convolution layer can be expressed as:

$$o_i = w \cdot A[i : i + h - 1], i = 1, 2, \ldots, s - h + 1 \tag{3}$$

Add offset b, the characteristic is obtained under the activation of activation function f, and the specific formula is as follows:

$$c_i = f(o_i + \mathrm{b}) \tag{4}$$

In the process of pooling, because there are multiple convolution kernels of different sizes, the number of features obtained is generally different. Therefore, pooling layer is added to select the maximum feature value of each convolution kernel, and then cascade to obtain the final feature result. Then, the feature result inputs into softmax layer to complete classification of short text.

4 Experiment

4.1 Dataset

With comments greater than 1000 and view counts greater than 10000 as thresholds, we chose 20 hot Chinese videos from YouTube. In addition, the comments below video were crawled through Google YouTube V3 API, and the total number of comments was calculated as 20000. We stored the 20000 comments, then we cleaned the data, and converted the traditional Chinese character in the comments into simplified Chinese character, and filtered out the useless data with messy codes and HTML tags. Finally, we got 18,707 comments. All the 18707 comments were manually labeled by two people. Before manually labeling the data, we formulated the labeling standard according to the actual situation. The labels used 4 highly different words (un-sensitive, abuse, political sensitive, heresy). Finally, there are 8603 pieces of non-sensitive data, 8228 pieces of abuse data, 1362 pieces of political sensitive data and 514 pieces of heresy data. In our paper, we collated all these comments into a new SLD (Sensitive Language Detection) dataset. Finally, SLD dataset was divided into train set, test set and validation set according to 6:2:2. The dataset partition is shown in Table 1.

Table 1. Dataset distribution

Data	Train set	Test set	Val set	Total
Un-sensitive	5357	1700	1546	8603
Abuse	5075	1660	1493	8228
Sensitive	841	303	218	1362
Heresy	327	96	91	514
Total	11600	3759	3348	18707

4.2 Experiment Setting

In process of training CNN and RNN and BERT-CNN, we use the cross entropy as loss function. CNN uses Adam as the optimizer, and its parameters are set as the learning rate: 0.001. At the same time, we added dropout technology and early-stop technology in the training process. The principle of dropout technology is to randomly abandon a certain proportion of nodes in the training process to prevent the occurrence of over fitting. In the end, dropout is set as 0.5. Early-stop technology evaluates the performance of the model on the validation set after each iteration. When the evaluation result of the validation set is no longer improved in N consecutive rounds, the iterative process is truncated and the training of the model is stopped. When N is set too small, the iteration may not converge. When N is set too large, the time consumed will increase. The hyper parameters setting in this paper is shown in Table 2.

Table 2. Hyper parameter setting

Parameter	Value
Embedding dim	64
Learning rate	1e−3
Train epoch	100
Dropout	0.9
Batch size	64
Epoch	10

4.3 Evaluation Metrics

The formula of evaluation standard is as follows. P value is the precision rate, indicating the proportion of texts which are correctly predicted in the predicted text, and the R value is the Recall rate, indicating the proportion of correctly predicted text in the all positive text. The value F_1 is the harmonic mean of P and R. TP is the number of successfully predicted positive cases, FP is the number of falsely predicted positive cases, and FN is the number of falsely predicted negative cases.

$$P = \frac{TP}{TP + TF} \tag{5}$$

$$R = \frac{TP}{TP + FN} \tag{6}$$

$$F_1 = \frac{2 \cdot P \cdot R}{P + R} \tag{7}$$

5 Results and Discussion

5.1 Performance Evaluation

In order to verify the performance of classification of the BERT-CNN we proposed. We conducted comparative experiments on three models of CNN, RNN and BERT-CNN under the same hyper parameters and same dataset.

The experiment results are shown in Table 3. Table 3 shows the results of P value, Recall value and F value of the three models on the same validation set.

Table 3. The performance of comparison

Approach	Tag	Precision	Recall	F_1
CNN	UN-SEN	95.0	93.0	94.0
	ABU	95.0	97.0	96.0
	SEN	43.0	33.0	37.0
	HER	79.0	92.0	85.0
RNN	UN-SEN	91.0	94.0	93.0
	ABU	96.0	95.0	95.0
	SEN	32.0	10.0	15.0
	HER	80.0	86.0	83.0
BERT- CNN	UN-SEN	94.0	94.0	94.0
	ABU	96.0	97.0	97.0
	SEN	**79.0**	**72.0**	**75.0**
	HER	81.0	90.0	85.0

The results of the classification of different labels are shown in Table 3. In particular, un-sensitive labels and abuse labels have a significantly better results than the other two labels, and F_1 value of these two types of labels is higher than 90%. In the non-sensitive label, the F_1 value of BERT-CNN is 1% higher than the RNN. On the Abuse tag, F_1 value of the BERT-CNN is 1% higher than CNN and 2% higher than RNN. Compared with the traditional neural network model, BERT-CNN has a considerable improvement in sensitive labeling. In the label of heresy, all three models performed similarly. BERT-CNN had the same F_1 value as CNN on the heresy label, which is 2% higher than RNN. The results of abuse label shown in Fig. 3.

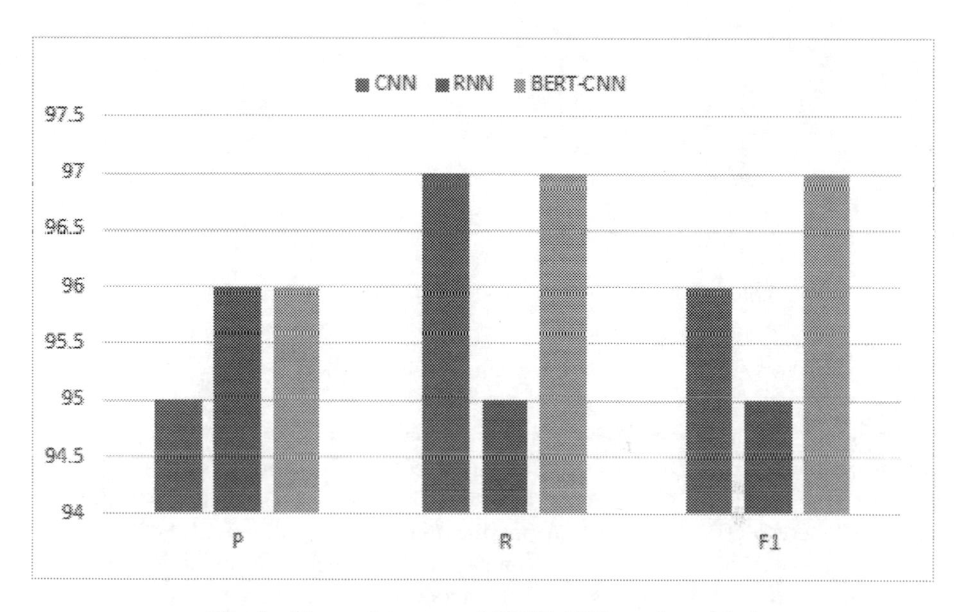

Fig. 1. The performance of BERT-CNN on abuse label

In the classification of sensitive information, it can be clearly seen from Fig. 4 that BERT-CNN is far better than the other two comparison of models in P value, Recall value and F_1 value.

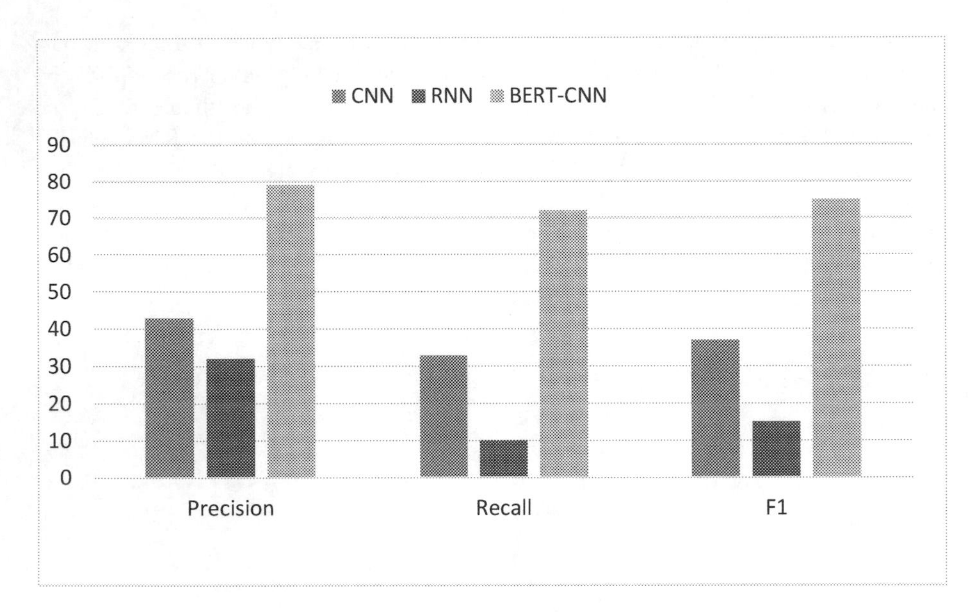

Fig. 4. The performance of BERT-CNN on sensitive label

6 Conclusion

Detecting the specific meaning and context relationship of some complex sentences in the dataset of unstructured short texts based on emotion dictionary shows high difficulty. In This paper, we build a new dataset SLD (Sensitive Language Detection) to achieve Chinese sensitive information classification. We propose a BERT-CNN model based on pre-trained model. BERT improved the generalization ability of word embedding, and added features of characters, words and sentences levels. The embedding learned from BERT was used as the input layer of TextCNN for Chinese sensitive information classification. Dropout technology was applied during the training to prevent over-fitting.

The result reveals that, compared with the traditional neural network model, the BERT-CNN model proposed in this paper improves the generalization ability of word embedding, and can effectively achieve the recognition and classification of the network sensitive information in short text dataset. and is significantly superior to the classical CNN and RNN models.

Although BERT-CNN has a significant improvement compared with CNN and RNN in the tetrad classification, it is still can be improved in the sensitive labeling. Since the SLD data set has a certain imbalance of negative and positive examples, the prediction effect of short text is influenced. In addition, comments under YouTube videos are not independent and have a strong interactive relationship, that is to say, comments do not exist independently and will have a certain logical relationship with other comments. How to classify the text of sensitive information in a more fine-grained way, taking other relevant comments of the same video into consideration as context, will be the future direction of this article.

Acknowledgement. This research is supported by the National Language Commission Key Research Project (ZDI135-61), the National Natural Science Foundation of China (No. 61532008 and 61872157), and the National Science Foundation of China (61572223).

References

1. Aggarwal, C.C., Zhai, C.X.: A Survey of Text Classification Algorithms. In: Aggarwal, C., Zhai, C. (eds.) Mining Text Data, pp. 163–222. Springer, Boston (2012). https://doi.org/10.1007/978-1-4614-3223-4_6
2. Zampieri, M., Malmasi, S., Nakov, P., Rosenthal, S., Farra, N., Kumar, R.: SemEval-2019 task 6: identifying and categorizing offensive language in social media (OffensEval) (2019)
3. Dinakar, K., Reichart, R., Lieberman, H.: Modeling the detection of textual cyberbullying 2011 (2011)
4. Zampieri, M., Malmasi, S., Nakov, P., Rosenthal, S., Farra, N., Kumar, R.: Predicting the type and target of offensive posts in social media (2019)
5. Davidson, T., Warmsley, D., Macy, M., Weber, I.: Automated hate speech detection and the problem of offensive language (2017)
6. Gambäck, B., Sikdar, U.K.: Using convolutional neural networks to classify hate-speech. In: Proceedings of the First Workshop on Abusive Language Online 2017, pp. 85–90 (2017)
7. Xu, J.M., Jun, K.S., Zhu, X., Bellmore, A.: Learning from bullying traces in social media. In: Conference of the North American Chapter of the Association for Computational Linguistics: Human Language Technologies 2012 (2012)
8. Blei, D.M., Ng, A.Y., Jordan, M.I.: Latent dirichlet allocation. J. Mach. Learn. Res. **3**, 993–1022 (2003)
9. O'Connor, B., Balasubramanyan, R., Routledge, B.R., Smith, N.A.: From tweets to polls: linking text sentiment to public opinion time series 2010 (2010)
10. Wiegand, M., Siegel, M., Ruppenhofer, J.: Overview of the germeval 2018 shared task on the identification of offensive language (2018)
11. Lai, S., Xu, L., Liu, K., Zhao, J.: Recurrent convolutional neural networks for text classification. In: Twenty-Ninth AAAI Conference on Artificial Intelligence 2015 (2015)
12. Mikolov, T., Sutskever, I., Chen, K., Corrado, G.S., Dean, J.: Distributed representations of words and phrases and their compositionality. In: Advances in Neural Information Processing Systems 2013, pp. 3111–3119 (2013)
13. Kalchbrenner, N., Blunsom, P.: Recurrent continuous translation models. In: Proceedings of the 2013 Conference on Empirical Methods in Natural Language Processing 2013, pp. 1700–1709 (2013)
14. Collobert, R., Weston, J., Bottou, L., Karlen, M., Kavukcuoglu, K., Kuksa, P.: Natural language processing (almost) from scratch. J. Mach. Learn. Res. **12**, 2493–2537 (2011)
15. Devlin, J., Chang, M., Lee, K., Toutanova, K.: Bert: pre-training of deep bidirectional transformers for language understanding. arXiv preprint arXiv:1810.04805 (2018)
16. Peters, M.E., Neumann, M., Iyyer, M., Gardner, M., Clark, C., Lee, K., Zettlemoyer, L.: Deep contextualized word representations. arXiv preprint arXiv:1802.05365 (2018)
17. Vaswani, A., et al.: Attention is all you need. In: Advances in Neural Information Processing Systems 2017, pp. 5998–6008 (2017)
18. Mikolov, T., Chen, K., Corrado, G., Dean, J.: Efficient estimation of word representations in vector space. arXiv preprint arXiv:1301.3781 (2013)

19. Radford, A., Narasimhan, K., Salimans, T., Sutskever, I.: Improving language understanding by generative pre-training. https://s3-us-west-2.amazonaws.com/openai-assets/researchcovers/languageunsupervised/languageunderstandingpaper.pdf (2018)
20. Kim, Y.: Convolutional neural networks for sentence classification. arXiv preprint arXiv: 1408.5882 (2014)

RASOP: An API Recommendation Method Based on Word Embedding Technology

Bin Zhang[1], Lihua Sheng[3], Lei Jin[1], and Wanzhi Wen[1,2](✉)

[1] School of Information Science and Technology, Nantong University, Nantong, Jiangsu, China
wenwanzhi@126.com

[2] Key Laboratory of Safety-Critical Software, Nanjing University of Aeronautics and Astronautics, Ministry of Industry and Information Technology, Nanjing, China

[3] Modern Educational Technology Centre, Nantong University, Nantong, Jiangsu, China

Abstract. Users' demand for the function of the software is increasingly affluent, and the scale of software is getting larger and larger. The structure of software presents the characteristics of complexity. In the process of software development, developers are likely to face a lot of difficulties, so they need to query the appropriate APIs. However, finding the right APIs can be time-consuming and laborious. It's especially difficult for developers who don't have much programming experience. In this paper, to solve the problems developers may face in the actual development process and improve the development efficiency, we propose RASOP (Recommendation APIs by Stack Overflow posts and Java Packages), an API recommendation approach leveraging word embedding technique and the information crawling from Stack Overflow posts and Java core packages, to recommend appropriate APIs for developers. Furthermore, RASOP also provides developers with label words, similar questions and relevant code. To evaluate the effectiveness of RASOP, we decided to analyze our system by simulating an instance. By testing a problem encountered during development, the API and tags and other recommendations from the RASOP output can indeed solve our problem. RASOP shows great results not only in the effect of API recommendation but also in the content of practicability.

Keywords: API recommendation · Code search · Stack Overflow · Word embedding

1 Introduction

Research shows that software developers spend an average of a fifth of their development time in web search [1], mainly looking for relevant code snippets for their tasks. Developers often query the API they need through a search engine. However, using search engines usually requires accurate functional descriptions [2]. This method is very effective in querying the functionality of a particular API, but it may not be useful in finding the API that implements a specific function, and the search results

K. Li et al. (Eds.): ISICA 2019, CCIS 1205, pp. 281–295, 2020.
https://doi.org/10.1007/978-981-15-5577-0_21

even need to be manually selected by the developer. Another approach is to use API documentation. For example, in a project implemented in the Java language, developers can use the Java SE 8 API specification. API documentation contains the Java language a large number of commonly used API specification, include its built-in method and code instance. This method also can be a perfect solution to learn a specific API function, but in realizing the purpose of searching API is hard to achieve. There is another way to get help from experienced developers, but this is very subjective and inefficient. Therefore, we can greatly assist developers in code search by collecting developer questions and then using a technique to find a set of related API classes or functions for developers [1, 3].

A large number of approaches focusing on API recommendation have developed. These techniques have come up with their unique strategies to varying degrees of improved recommendation effectiveness. McMillan et al. [2] came up with a method called Portfolio, which implemented a recommended API for a given code search query and demonstrated it in a large codebase. Chan et al. [4] improved the Portfolio approach by adding sophisticated graphics mining and text similarity techniques to develop recommendations. Rahman et al. [5] proposed another method, RACK, which constructs a keyword-API mapping database, extracts keywords questions from Stack Overflow, and collects mapping APIs from corresponding high-score answers. Qiao Huang et al. [6] proposed a method called BIKER, which used word embedding technology in API recommendation for the first time.

Although all of these technologies perform well in particular problems, they have some limitations of their factors and hardly solve our research problems. Firstly, these technologies use keyword matching to select the recommended API, so different key matching technologies have a high impact on the recommendation results of an API [7]. Besides, developers should have rich development experience and understand the meaning of certain specific keywords so that they can use these technologies [8]. It is not particularly friendly for a novice developer. For example, if you have a question: "How do I read/convert an InputStream into a String in Java?" you may have to understand the role of InputStream. Secondly, the names of API keywords and the keywords used in search queries are very likely to be different. Different developers may use different words to express some concepts. "Read" and "convert" have the same meaning above the question. Different developers may use different words. In fact, the concepts of this term are very similar. It's what we usually call word mismatch problem [3]. And some identifiers in the code can also be a problem for some keyword queries [9].

Based on the above problems, we should improve the accuracy and efficiency of searching order to solve the issues that developers encounter in querying the appropriate API. In this paper, we propose an API recommendation technique RASOP (Recommendation APIs by Stack Overflow posts and Java Packages), an API recommendation approach leveraging word embedding technique and the information from Stack Overflow posts and Java core packages, to recommend appropriate APIs for developers. The purpose of this technology is to integrate multiple source information to make API recommendation for developers. It uses a method of word embedding for data problem set and establishes a relational database of questions and answers [10]. On one hand, the question-and-answer information on Stack Overflow and the existing

API help documents are comprehensively used to solve the problem of a single source of API resources [11, 12]. On the other hand, we also use the LDA model and the LSA model to help us extract the topic words and solve the semantic bias problem. By calculating the similarity of the questions and word vectors, the appropriate API is finally recommended to the developers. At the same time, to meet the needs of development, we can provide multiple solutions to the same problem, and finally choose an optimal solution to apply to the development process.

To evaluate the effectiveness of RASOP, we decided to analyze our system by simulating an instance. Analyzing the correlation between recommendation API and target query is an effective method.

The main contributions of this paper are as follows:

1. We propose RASOP (Recommendation APIs by Stack Overflow posts and Java Packages), an API recommendation approach leveraging word embedding technique and the information from Stack Overflow posts and Java core packages, to recommend appropriate APIs for developers.
2. By simulating an instance, we analyze the accuracy of API recommendation and the practicability of content.

The rest of this paper is organized as follows. The preliminaries are introduced in Sect. 2. Section 3 presents our approach. Section 4 shows us an instance. Section 5 discusses the conclusion and our future work.

2 Preliminaries

In this paper, RASOP mainly uses word embedding technology to recommend appropriate APIs to developers by extracting data from Stack Overflow and Java SE 8 API Specification. This section introduces three aspects of background knowledge: Stack Overflow, Word Embedding Technology, LSA and LDA topic models.

2.1 Stack Overflow

Stack Overflow is a program-related IT technology Q&A website. Users can submit questions, browse questions, and index-related content free of charge on the website. When searching for APIs, many developers choose to browse through a large number of Stack Overflow posts and pick out useful APIs based on discussions, so Stack Overflow is often used as a link between programming tasks and APIs. Because Stack Overflow is a platform that discusses problems in different task environments and different languages, it can provide developers with help recommendations from others. It is such a brainstorming website that it is welcomed by developers. In Rahman's experiment, the title of the Stack Overflow problem contains a large number of keywords for actual code search, which matches to a large extent the keywords we collect for code search [5]. Some studies show many developers choose to use Stack Overflow when looking for the right API. Because they can find similar questions, and the answers to these questions usually include the relevant APIs to be used [6, 24].

2.2 Word Embedding Technology

In Natural Language Processing (NLP), we collectively refer to language model and representation learning techniques as word embedding techniques. Conceptually, it uses a multi-dimensional vector to represent words, and the vector representation of words with similar semantics is very close in spatial position. In this paper, we mainly use Word2vec model [10]. Word2vec is a software tool for training word vectors. Word2vec can transform a word into vector form quickly and effectively according to the given corpus through the optimized training model, which provides a new tool for the applied research in the field of natural language processing. Word2vec relies on skip-grams or continuous word bag to build an embedding module. Word2vec involves two algorithms: CBOW algorithm and Skip-Gram algorithm. CBOW algorithm is used to predict the headword in a given context. Skip-Gram algorithm predicts the context through the headword. Both of them need only one layer of hidden layer. Because RASOP integrate all kinds of information and combine with some data sets for API recommendation. So we choose the CBOW algorithm to implement the system.

2.3 LSA&LDA

LSA (Latent Semantic Analysis) [13] is a new indexing and retrieval method. This method uses vectors to represent terms and documents, and uses the relationship between vectors to determine the relationship between words and documents. The LSA maps words and documents to the latent semantic space, thereby removing some of the noise in the original vector space and improving the accuracy of information retrieval. This mapping must be strictly linear and based on matrix SVD (singular value decomposition).

LDA (Latent Dirichlet Allocation) [14] is a topic generation model based on the Bayesian model. It consists of a three-tier structure of words, topics, and documents. The topics of each document in the document set can be given in the form of a probability distribution, so that by analyzing some documents to extract their themes distributions, topic clustering or text categorization can be performed according to the theme distribution. LDA is mainly used to identify hidden topic information in the corpus.

In order to solve the problems that developers encounter in querying appropriate APIs, we should improve the accuracy and efficiency of searching [5–8]. We design a system RASOP that can facilitate developers to query APIs more efficiently.

3 Approach

The Fig. 1 shows the detailed framework of RASOP. RASOP design an interactive system with users. The system is divided into two modules: the offline training module and the online recommendation module.

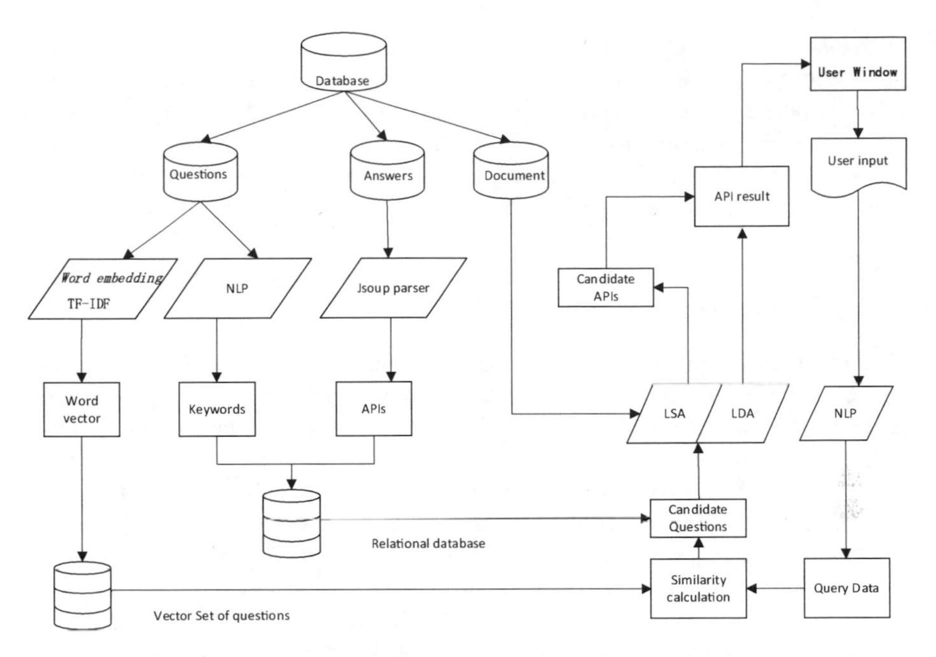

Fig. 1. Process of RASOP

In the offline training module, we have to preprocess the collected data. The collected data is divided into two parts. We need to download the Java SE 8 API specification, and another part is to download the Q&A dataset for the Java API from Stack Overflow. The second step is to pre-process the collected data. We divide the data set obtained into problem data set and answer data set from Stack Overflow, and collect all the words in the problem data set to build a corpus. We need to segment all the content in the corpus, remove the treatment of stop words and transform words into roots. Then, the word embedding modeling is performed on the data in the corpus, that is, all the words are converted into corresponding word vectors to generate a word model. Create an IDF (Inverse Document Frequency) dictionary that sets a higher weight for words that may contain important information. The operation of word segmentation, deletion of stop words and conversion roots is performed on all questions in the problem data set to generate a keyword list. We then extract the data in the answer data set, such as the API, and the code, and save the problem and the data extracted from the answer set in a relational database.

In the online recommendation module, we collect developer input questions through the UI interface. By using the method of processing the corpus in offline processing, we need to segment all the content that develop input in the corpus, remove the treatment of stop words and transform words into roots. We use the word model and the IDF dictionary to calculate the similarity between the user's problem and the problem data set. Through calculations, we can obtain candidate problem sets sorted by similarity. We use the LDA model to extract themes from these candidate problem sets, and get the keywords of the related questions. We treat these keywords as labels for such issues and provide them to users. Then, we use the LSA model to parse the candidate problem set, and filter out the similarity of the problem by filtering the cosine similarity of these problems. Through these questions, we get the corresponding API in the database. Then combining with API documentation, we can get candidate APIs. RASOP also provides developers with tag words, similar questions and related code for the questions to be queried. Finally, we return this content to the developer via the interface.

3.1 Data Collection

We divide the data collection into two parts, and one is the Java SE8 API documentation, the other is the Stack Overflow Question and Answer Data Set on the Java API. The text of the Java SE8 API file has more resources on the Internet and is easy to obtain. It can be downloaded directly from the Internet. Since there are many versions on the Internet, the version chosen is downloaded from Oracle. The other is the dataset on the Java API on Stack Overflow, which needs to be retrieved from his website. Stack Overflow uploads the data on its website to its official website on time. They provide data dump to researchers doing data analysis. The data dump version used in this experiment was released in December 2017. Because the purpose of RASOP is to provide API recommendations, the part of the Java API that we obtained from data dump is extracted as our data set.

3.2 Data Preprocessing

After obtaining the two data sets of the previous step, we need to do a data preprocessing step. Firstly, we divide the data set of Java API and add the content of question attribute and answer attribute to the two data sets respectively, which are question data set and answer attribute. Collect all the words in the question data set to form a corpus.

3.2.1 Processing Corpus

Our experiment uses NLTK (Natural Language Toolkit) class library for text processing. NLTK [15] is the most commonly used Python Library in the field of natural language processing. It provides easy-to-use interfaces through which more than 50 corpuses and lexical resources (such as WordNet) can access, as well as a set of text processing libraries for classification, tagging, stemming, parsing and semantic reasoning. Algorithm shows the algorithm of segmentation, deletion of stop words, and processing of data for extracting stems.

```
Algorithm 1- Analyzing vocabulary corpus
1:  procedure 1  RASOP(Q)    //Q: DataSet_Questions
2:    R ← { }
3:    //collecting every word from DataSet_Questions
4:    for i in DataSet_Questions:
5:      for j in i:
6:        List_word  ← WordPunctTokenizer( j )
10:     end for
11:   end for
12:   //removing stop words
13: if stopwords.words('English') not in List_word:
14:   List_OutStopWords ← n
15: end if
16: //stemming words
17: for word in List_OutStopWords:
18:   List_StemmerWords ← snowballStemmer().stem ( word )
19: end for
20: //return preprocessed data
```

3.2.2 Building Language Model

After collecting the word build corpus in all question data sets, we need to perform word embedding modeling. We use the word2vec model for word embedding through the packages in the Gensim library [17]. Each word in the lexicon is converted into a word vector by using the word 2vec method.

There are two aspects to consider when building a language model. The first part is the algorithm. After selecting the word2vec training data set, the model involves two algorithms: CBOW algorithm and Skip-Gram algorithm. Because RASOP integrate all kinds of information and combine with some data sets for API recommendation. So we choose the CBOW algorithm to implement the system. After using COBW algorithm, we need to consider the dimension of vectors. The dimension of the vector can be achieved by tuning the parameters. Based on our extensive data set, we set the size the parameter of word2vec method to 100. The reason is to make all the words of the data in the corpus uniquely represented, thus improving the accuracy of the recommendation [10, 17].

3.2.3 Establishing TF-IDF Dictionary

Establish a TF-IDF Dictionary of a word. The TF-IDF value of a word indicates the reverse rate of the word in corpus. If the TF-IDF value of a word is lower, the lower the frequency of the word in corpus, the more likely it will contain important information. TF-IDF is used to weigh the similarity of words. The TF-IDF dictionary is also computed using the data set TextCorpus obtained in the previous step. The calculation of TF-IDF can be divided into word frequency calculation and reverse document frequency calculation.

$$\mathrm{TF-IDF} = \mathrm{tf}_{i,j} \cdot \mathrm{idf}_i = \frac{n_{i,j}}{\sum_k n_{k,j}} \cdot \log \frac{|\mathrm{D}|}{\left|\{j : t_i \in d_j\}\right| + 1} \tag{1}$$

In formula (1), $n_{i,j}$ in the formula represents the number of occurrences of the target word t_i in the file d_i while denominator represents the sum of occurrences of all words in the whole file d_i. D is the total number of files in the data set TextCorpus, $\left|\{j : t_i \in d_j\}\right|$ is the number of the file containing the target word t_i. In order to ensure that the target word must be in the data set, preventing divisions from being zero, we usually use $\left|\{j : t_i \in d_j\}\right| + 1$.

3.2.4 Building Relational Database

Stack Overflow provides a user rating system for the answers to each question, so we can filter question data sets by its score. We should recommend high-scoring answers to users. Then, we extract the target API from the answer data sets. The Jsoup parser is used to extract API and extract all code segments containing <code> tags from the answer data sets. However, some code segments may be tagged by other tags or without tag information, so RASOP also uses a set of regular expressions to assist in extracting Java class APIs with hump writing.

A relational database is constructed by combining the question data sets and the content of answer data sets. The primary keys in this mapping database include the serial number of the problem, the question, the keyword and the API of the corresponding problem extracted from answer set.

3.3 Top-N Questions and Top-APIs

After collecting the query that developers enter through the UI interface, we need to calculate the similarity between this query and the data in our problem set. We calculate the similarity by using the previously established word model and the TF-IDF dictionary. Finally, we get a list of candidate questions based on the similarity ranking. This list contains our top-n questions.

After getting top-n questions, we can first extract the API information corresponding to these problems. Then we use these questions to compare with the description of the function of the method in the java API documentation, or use the similarity calculation method to calculate the function description with higher similarity. We combine the above two methods to get the API and form a top-API according to the corresponding similarity ranking.

3.4 Recommendation

3.4.1 Using LSA to Exclude Low Similar Items

We use the LSA model to exclude corpora that do not meet the target characteristics. LSA is a machine learning-based indexing and retrieval method. By mapping words and documents to potential semantic space, some "noise" in the original vector space is removed, and the accuracy of information retrieval is improved. This mapping is done by performing singular value decomposition (SVD) on the matrix. We also achieve dimensionality reduction of the matrix through singular value decomposition. We usually arrange the elements of Σ in descending order, and the singular value has a characteristic, which is especially fast, and usually the sum of the singular values of the first 10% or even 1% accounts for more than 99% of the sum of all the singular values:

$$X_{m*n} = U_{m*m}\Sigma_{m*n}V_{n*n}^{T} \approx U_{m*k}\Sigma_{k*k}V_{k*n}^{T} \quad (2)$$

In formula (2), U is an $m * m$ matrix; Σ is a semi-definite $m * n$ diagonal matrix; and V^T is a conjugate transpose of V. The latter approximation is the result of our reduction of X_{m*n} to k dimension. That is to say, we can use the k singular values appearing above and the corresponding left and right singular vectors to approximate the description matrix.

In RASOP, we generate the word matrix for the candidate problem and reduce the dimension by singular value decomposition. Then we measure the similarity of these problems by calculating the vector angle of these problems, and screen out the problems of those angles. We calculate cosine similarity of words, where $cos<i,j>$ is calculated by the following formula: $\frac{\sum_{k=1}^{n} i{\cdot}j}{\sqrt{\sum_{k=1}^{n} i^2}\cdot\sqrt{\sum_{k=1}^{n} j^2}}$. This way we have ruled out some problems with large semantic differences.

3.4.2 Generating Query Labels

We use the LDA model to generate themes for documents and process them into tags to help users find APIs more easily. The process of the model is basically as follows: For each document, a topic is extracted from the topic distribution, a word is extracted from the word distribution corresponding to the extracted topic, and the process is repeated until each word in the document is traversed.

LDA is an unsupervised machine learning technique that uses a bag of words approach that treats each document as a word frequency vector, transforming the text information into digital information that is easy to model. However, the word bag method does not consider the order between words and words, which simplifies the complexity of the problem and also provides an opportunity for the improvement of the model. Therefore, we also need to assist in generating tags by counting the word frequency of the candidate problem set. Those that are generated by both the LDA model and the target words in the previous section of the word frequency statistics are selected as our labels.

3.4.3 Supplying Recommended Content

RASOP use the list of candidate APIs with similarity as the sort to get the recommended APIs. However, as a recommendation system, it is not enough for RASOP to recommend APIs. So we decided to add other recommendations. When multiple APIs with a high degree of similarity are found, other content can be supported at the same time: (a) a functional description in the API help documentation (b) a similar problem with the user input (c) a code segment related to the query API. By supplementing these additional recommendations, RASOP can provide users with a list of related issues to help developers compare. Users can also learn more about the functionality of these APIs, and they can learn to use these methods through code snippets.

4 Empirical Study

4.1 Simulating an Instance

We simulate an instance to test the recommended effect of RASOP. The API recommendation results will obtained through RASOP by inputting problems encountered by the user in actual development. We found a post on the Java tag on Stack Overflow and selected a post that was recently published. The question we get is: How to initialize an array in Java with a constant value efficiently?

Before using RASOP, we need to process the data first. This part of the data comes from the Java SE 8 API Specification and the Q&A data set about Java API on Stack Overflow. We divide the data set obtained from Stack Overflow into a problem data set and an answer data set, and collect all the words in the problem data set to build a word library. Then, the word embedding modeling is performed on the data in the word library, that is, all the words are converted into corresponding word vectors to generate a word model. Create TF-IDF (Inverse Document Frequency) dictionary that sets a higher weight for words that may contain important information. The Fig. 2 shows the matrix representation and idf of Array and Value. The matrix dimension in our model is 100. In the paper we show the first 20 dimensions.

word	matrix	idf
Array	[1.12746218e-01 3.10163074e-02 -1.10002900e-03 3.464567 69e-04 3.95832744e-02 3.16977051e-03 1.68261725e-01 -1.41162743e-01 7.63311130e-02 -2.08550958e-01 -6.02869057e-02 1.66583753e-01 3.85899648e-02 1.07580325e-01 -2.22708942e-02 -7.05205782e-03 8.93873694e-02 5.83802817e-02 1.59950751e-01 -5.15916022e-02......]	2.83905779
Value	[9.13028949e-03 9.99668760e-02 -5.47133560e-02 6.78238728e-02 -1.17038134e-01 1.28493716e-01 1.35517367e-02 -7.79208122e-02 1.15511037e-01 -2.25283003e-02 -6.32656275e-02 4.02586759e-03 1.34669212e-02 -3.68029686e-02 -1.37853577e-01 -1.67833999e-01 3.19488890e-02 9.43804662e-02 5.23366106e-02 -3.55417347e-02......]	13.95190859

Fig. 2. Matrix representation and idf of Array and Value

A relational database is constructed by combining the word library and the API extracted from the corresponding problems in question data set. The primary keys in this mapping database include the serial number of the problem in question data set, the problem, the word library and the API of the corresponding problem extracted from answer data set.

We enter the question we want to query at the user input of RASOP: How to initialize an array in Java with a constant value efficiently? We use the method of online module to preprocess, segment, delete stop words and extract stem words. The results are shown in Table 1.

Table 1. Data preprocessing

Step	Result
Query	'How to initialize an array in Java with a constant value efficiently?'
Segment	'how', 'to', 'initialize', 'an', 'array', 'in', 'java', 'with', 'a', 'constant', 'value', 'efficiently'
Delete stopwords	'how', 'initialize', 'array', 'java', 'constant', 'value', 'efficiently'
Extract stem	'how', 'initialis', 'array', 'java', 'constant', 'valu', 'effici'

Then, the pre-processed data is combined with the word model in offline processing and TF-IDF dictionary to calculate the similarity. A list of APIs and similarities is obtained after calculating all the problems in the database, as shown in Table 2. The 10 questions in Table 2 are the most relevant to our instance that is derived from Stack Overflow. In fact, we calculated the 50 most relevant questions. For the convenience of display, we selected the top ten.

Table 2. Top 10 similarity questions

Rank	Question
1	Java Static Class Variable Initialisation Efficiency
2	Most efficient way to append a constant string to a variable string in Java?
3	How do I declare a variable with an array element?
4	How to declare an ArrayList of objects inside a class in java
5	How to assign a value to byte array?
6	Java - Efficient way to access an array
7	Initialising array names in java
8	How to initialize byte array in Java?
9	More efficient for an Java Array
10	Java How to use class object for List initialisation?

We use the LSA model to exclude corpora that do not meet the target characteristics. In this step, we take the 10 questions extracted in Table 2 as an example. Through the LSA model, these ten questions are converted into vector form. In the vector space, we filtered out the options that are too large for the current query problem. We also conducted a semantic filtering through this step. The Fig. 3 shows 10 problem vector space locations in the LDA model. The vector angle between D9 and D1 is obviously larger than the angle between other problems, so we think that D9 is the one with the biggest semantic difference, we need to eliminate it. Among all 50 questions, we calculated the 10 largest angles of the vector and excluded them.

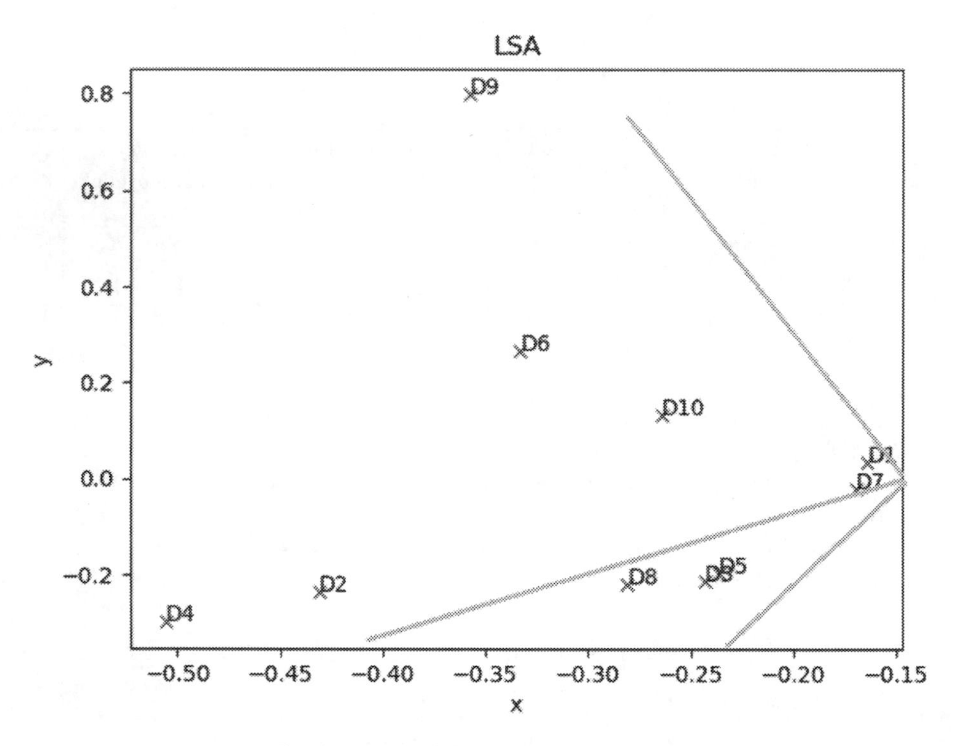

Fig. 3. Result of the API recommendation.

Next, we use the LDA model to generate document themes and process them into tags to help users find the API more easily. After passing the excluded problems, we use the LDA model to extract keywords for the remaining similar problems. We treat these keywords as labels for such issues and provide them to users.

Our query is a problem of initializing arrays, so what we need is an API for arrays. We analyze the first API: java.util. Arrays.fill which is shown in Fig. 4. We found in the JAVA documentation that the Arrays class in java.util contains various methods for manipulating arrays such as sorting and searching. Second, the fill() method in the class assigns the specified data type to an array of the specified data type. Obviously, this is an initialization assignment problem with data types, which is relevant to our question. Our analysis was also confirmed by the functional description in the recommended results. Through the list of similar questions, we found that the problem is focused on what data type to initialize an array, and these problems can also help us solve the problem. In the final code snippet, an example of the use of the fill() method is given. Through those codes, we know the parameters and return values of this method. In summary, through the above, the recommended content can help us solve this problem.

Recommend 1	Tags: array, initialize, string, create, value
API: java.util.Arrays.fill	
Function: Assigns the specified long value to each element of the specified array of longs.	
Similar questions: 1.How to initialize all the elements of an array to any specific value in java 2.How to initialize each element of an array with a specific value at declaration? 3.initializing a boolean array in java	
Relevant code : Arrays.fill(array, -1);	

Fig. 4. Result of the API recommendation

4.2 Result Analysis

RASOP is a synthesis of several traditional API query methods [4–7], which can bring more efficient and accurate API recommendation to developers. Word embedding technology overcomes the problem of semantic deviation and solves the problem that developers query the corresponding API when inputting problems. By constructing language model, we get the most relevant question list and API recommendation, which ensures that every question can get the most relevant result content recommendation, and reduce the time for developers to compare queries. In addition, RASOP also provides developers with tag words, similar questions and related code for the questions to be queried, which significantly improves the programming efficiency of developers in the development process.

5 Conclusion and Future Work

In this paper, we propose RASOP. Furthermore, RASOP also provides developers with similar problems and code snippets, respectively. API recommendation in our instance demonstrates that RASOP shows better effect not only in the accuracy of API recommendation but also in the content of practicability.

In our work, RASOP combines the question and answer information on Stack Overflow with the information on Java API help document [23, 24]. In future work, we can consider merging other databases. For example, we can combine GitHub history submission information, code records, and other content to make comprehensive recommendations [20]. Make our data more diversified, more reference value. In terms of accuracy of recommendation, the accuracy of RASOP needs to be further improved. Combining with many keyword matching techniques, an attempt is made to enhance the accuracy of recommendation further. From the perspective of word embedding technology, many mainstream word embedding technologies have been proposed, such as FastText technology and Elmo technology (the most advanced context-based word embedding technology at present) [21, 22]. These technologies can also be considered in our recommendation system. In terms of the construction time of RASOP recommendation system, the primary time of our system spend on the construction of the language model. How to save this time is also a factor we need to consider. Overall, in future work, so we will discuss more factors to recommend and continuously improve the API recommendation system.

Acknowledgements. This work is supported partially by Natural Science Foundation of China under Grants no. 61602267, partially by the Open Project of Key Laboratoriesm of Ministry of Industry and Information Technology for Software Development and Verification Technology of High Safety Systems at Nanjing University of Aeronautics and Astronautics under Grants no. NJ2018014, partially by Natural Science Foundation of Nantong University under Grants no. 15Z14.

References

1. Brandt, J., Guo, P.J., Lewenstein, J., Dontcheva, M., Klemmer, S.R.: Two studies of opportunistic programming: interleaving web foraging, learning, and writing code. In: Proceedings SIGCHI, pp. 1589–1598 (2009)
2. McMillan, C., Grechanik, M., Poshyvanyk, D., Xie, Q., Fu, C.: Portfolio: finding relevant functions and their usage. In: Proceedings ICSE, pp. 111–120 (2011)
3. Kevic, K., Fritz, T.: Automatic search term identification for change tasks. In: Proceedings ICSE, pp. 468–471 (2014)
4. Chan, W.-K., Cheng, H., Lo, D.: Searching connected API subgraph via text phrases. In: Proceedings of the ACM SIGSOFT 20th International Symposium on the Foundations of Software Engineering, pp. 10:1–10:11. ACM (2012)
5. Rahman, M.M., Roy, C.K., Lo, D.: RACK: automatic API recommendation using crowdsourced knowledge. In: 2016 IEEE 23rd International Conference on Software Analysis, Evolution, and Reengineering (SANER), vol. 1, pp. 349–359. IEEE (2016)
6. Huang, Q., Xia, X., Xing, Z., Lo, D., Wang, X.: API method recommendation without worrying about the task-API knowledge gap. In: Proceedings of the 2018 33rd ACM/IEEE International Conference on Automated Software Engineering, ASE 2018, pp. 292–303 (2018)
7. Chan, W., Cheng, H., Lo, D.: Searching connected API subgraph via text phrases. In: Proceedings FSE, pp. 10:1–10:11 (2012)
8. Bajracharya, S.K., Lopes, C.V.: Analyzing and mining a code search engine usage log. Empirical Softw. **17**(4–5), 424–466 (2012)

9. Haiduc, S., Marcus, A.: On the effect of the query in IR-based concept location. In: Proceedings ICPC, pp. 234–237 (2012)
10. Mikolov, T., Sutskever, I., Chen, K., Corrado, G.S., Dean, J.: Distributed representations of words and phrases and their compositionality. In: Advances in Neural Information Processing Systems, pp. 3111–3119 (2013)
11. Ye, X., Shen, H., Ma, X., Bunescu, R., Liu, C.: From word embeddings to document similarities for improved information retrieval in software engineering. In: Proceedings of the 38th International Conference on Software Engineering, pp. 404–415. ACM (2016)
12. Treude, C., Robillard, M.P.: Augmenting API documentation with insights from stack overflow. In: 2016 IEEE/ACM 38th International Conference Software Engineering (ICSE), pp. 392–403. IEEE (2016)
13. https://github.com/laserwave/topic_models/tree/master/LSA-demo
14. https://github.com/laserwave/lda_gibbs_sampling
15. Bird, S., Loper, E.: NLTK: the natural language toolkit. In: Proceedings of the ACL 2004 on Interactive Poster and Demonstration Sessions, p. 31. Association for Computational Linguistics (2004)
16. Stopword List. https://code.google.com/p/stop-words
17. Řehůřek, R., Sojka, P.: Software framework for topic modelling with large corpora. In: Proceedings of the LREC 2010 Workshop on New Challenges for NLP Frameworks, pp. 45–50 (2010)
18. Manning, C.D., Raghavan, P., Schütze, H.: Introduction to Information Retrieval. Cambridge University Press, Cambridge (2008)
19. Thung, F., Wang, S., Lo, D., Lawall, J.: Automatic recommendation of API methods from feature requests. In: Proceedings ASE, pp. 290–300 (2013)
20. Yan, M., Zhang, X., Yang, D., Xu, L., Kymer, J.D.: A component recommender for bug reports using discriminative probability latent semantic analysis. Inf. Softw. Technol. **73**, 37–51 (2016)
21. Joulin, A., Grave, E., Bojanowski, P., et al.: FastText.zip: compressing text classification models. Under Review as a Conference Paper at ICLR, pp. 1–13 (2016)
22. Athiwaratkun, B., Wilson, A.G., Anandkumar, A.: Probabilistic FastText for multi-sense word embeddings (2018)
23. Regehr, J., Reid, A., Webb, K.: Eliminating stack overflow by abstract interpretation. ACM Trans. Embed. Comput. Syst. **4**(4), 751–778 (2005)
24. Vasilescu, B., Filkov, V., Serebrenik, A.: StackOverflow and GitHub: associations between software development and crowdsourced knowledge. In: 2013 International Conference on Social Computing, vol. 35, pp. 188–195. IEEE Computer Society (2013)

Application of Improved Collaborative Filtering Algorithm in Personalized Tourist Attractions Recommendation

Yujie Liang[1], Xin Li[1], Jiali Lin[2](✉), and Dazhi Jiang[1,3]

[1] Department of Computer Science, Shantou University, Shantou 515063, China
{19yjliang, lixin, dzjiang}@stu.edu.cn
[2] Business School, Shantou University, Shantou 515063, China
jllin@stu.edu.cn
[3] Key Laboratory of Intelligent Manufacturing Technology (Shantou University), Ministry of Education, Shantou 515063, China

Abstract. With the advent of the era of information globalization, information in tourism industry is growing at an explosive rate in the Internet. More and more people hope to get high-quality tourism resources information more quickly and effectively. This trend is promoting the development of personalized tourism recommendation. In this context, the appropriate recommendation algorithm can make better use of tourism information and make the recommendation model have the ability to discover the potential value of users. However, tourist attractions are difficult to characterize because they contain many information such as geographical location, transportation route, cost information and user evaluation. Item-based Collaborative Filtering Algorithm (ItemCF) can skillfully avoid this situation, therefore we used ItemCF to build a personalized recommendation model for tourist attractions and apply it to real tourism dataset. The improvement of ItemCF is proposed in this paper, which can effectively improve the coverage of recommendation model. The biggest disadvantage of collaborative filtering algorithm is the cold start problem. This paper proposes a strategy to alleviate this problem. The experimental results show that the improved collaborative filtering algorithm has a good performance in the application of personalized tourist attractions recommendation.

Keywords: Personalized recommendation · Tourism · Collaborative filtering algorithm

1 Introduction

With the rapid development of the Internet industry, tourism commercial websites have emerged, which collect a large number of tourist attractions information. In this context, the problem that users face is no longer how to obtain the information of tourist attractions, but how to quickly and accurately obtain useful information of tourist attractions. Search technology can solve this problem to some extent, but it is still far from enough. Personalized recommendation of tourist attractions can solve this

K. Li et al. (Eds.): ISICA 2019, CCIS 1205, pp. 296–306, 2020.
https://doi.org/10.1007/978-981-15-5577-0_22

problem to a great extent. It can make a list of tourist attractions recommendation for each user by using the user behavior data retained by the website.

Personalized tourism recommendation method based on user geographical location has been proposed [1–4]. It has rich topological and temporal information, which helps to mine user preferences and stimulate the validity of tourist attractions recommendation. Researchers extracted the behavioral patterns of tourists on social media and explored the attitudes of tourists to scenic spots [5, 6]. The purpose was to analyze the useful tendencies of specific tourist spots comprehensively and recommend scenic spots with appropriate emotions for the relevant scenic spots. In [7], from the perspective of user and context, authors uses context theory and machine learning method to realize the user model of tourism activity recommendation. User-based collaborative filtering algorithm is applied to scenic spot recommendation [8]. Authors calculate the similarity between each user and generate scenic spot recommendation based on the visiting history of neighbor users. In [9], authors presented an improved algorithm of association rules based on Apriori, which takes into account the nature, behavior and situation of tourists, establishes a tourist behavior preference model and constructs a personalized recommendation model for scenic spots.

The above-mentioned documents mainly focus on user geographical location, user social networks, user contexts and the correlations between users. More or less, the correlations between tourist attractions are neglected. Tourist attractions are difficult to characterize because they contain many information such as geographical location, transportation route, cost information and user evaluation. Therefore, it is obviously a difficult task to calculate the similarity between scenic spots from their content attributes. The Item-based Collaborative filtering algorithm can skillfully avoid this problem by calculating the similarity between scenic spots by analyzing user historical tourism records. The algorithm considers that scenic spot A and scenic spot B have a great similarity due to most of the users who have visited scenic spot A have also visited scenic spot B. Just as the famous Amazon e-commerce website has used this algorithm, its recommendation explanation to users is interpreted as "users who have bought a commodity have also bought B commodity".

This paper proposes the idea of combining the improved ItemCF algorithm with the personalized recommendation of tourist attractions, so as to get a higher coverage rate of tourist attractions. For the cold start problem in collaborative filtering algorithm, we first analyze the characteristics of new user, classify new user, and then the most visited scenic spots of old users that belong to new user class are recommended to new user. This strategy alleviates the problem to a certain extent. In order to test the real effectiveness of the personalized recommendation model, we crawled a large number of real data from Baidu Tourism, a large domestic tourism website, and applied the personalized recommendation model to it.

The rest of this paper is organized as follows. Section 2 summarizes the related work on recommendation algorithm and personalized tourism recommendation. Section 3 introduce the steps of ItemCF algorithm in detail and the improvement strategy of recommendation formula is proposed. In Sect. 4, we will show the real data set we have collected and introduce in detail the experimental steps and results of establishing a personalized recommendation model for tourist attractions. And finally, Sect. 5 give the conclusion of this paper.

2 Related Works

2.1 Recommendation Algorithm

Content-based recommendation algorithm is a simple but important recommendation idea [10]. The algorithm first constructs the project attribute vector, then set up the user interest preference vector by analyzing the user historical behavior records. By calculating the similarity between the user interest preference vector and the attribute vector of each item without evaluation, it generates the item prediction score or top-N recommendation for the target user. The problem of association rules was put forward by Agrawal et al. [11] in 1993. A classification random walk algorithm based on association rule mining is proposed, which uses association rules to mine the association between user attributes and items [12]. A personalized knowledge recommendation method based on constructivist learning theory is proposed by introducing the nearest neighbor first candidate knowledge selection strategy [13]. A hybrid recommendation algorithm combining user clustering and scoring preferences is proposed, which utilizes data mining knowledge and collaborative filtering algorithm [14]. In order to solve the problem of data sparsity and scalability, a collaborative filtering hybrid recommendation algorithm based on user feature clustering and Slope One filling is proposed [15]. In the application of movie recommendation, Hadoop distributed platform and Mahout tool are combined to design and implement the combined recommendation algorithm, which is applied to the movie recommendation [16]. Collaborative filtering recommendation is proposed by Goldberg et al. [17] in 1992. User-based collaborative filtering algorithm integrates user historical behavior data, calculates the similarity between users, establishes neighbor users according to similarity and finally selects items from neighbor user set for recommendation.

2.2 Personalized Tourism Recommendation

Personalized tourism recommendation is an important part of recommendation application. Therefore, compared with personalized recommendation systems for movies and music, personalized tourism recommendation provides more diverse content. Scholars from various countries have done a lot of research on personalized tourism recommendation system. Natural Language Processing (NLP) is used to assess the emotions of user reviews and extract implicit features. Explicit and implicit feedback ratings are used for hotel recommendation [18]. In order to provide tourists with real-time and accurate scenic spot recommendation, the weather conditions are used as the related attributes of scenic spot recommendation by using context awareness [19]. The discrete time is introduced to represent the traffic condition in the basic static network [20]. Based on dynamic programming, the main problems are transformed into recursive equations to get the recommendation of tourism routes. An expert system framework is designed, which uses data mining technology to classify users accurately and predict them separately so as to provide users with complete travel recommendation [21]. Data mining technology is used to mine chat text information between tourism service providers and users, and then the system extracts recommendation content from the database using the information mined to form a complete travel recommendation [22]. However, in personalized tourism recommendation, tourist attractions are rich in content information and have a large number of content characteristics. Traditional recommendation algorithm has poor adaptability, while Item-based Collaborative filtering algorithm has strong adaptability but faces the cold start problem. When the system has a new user or item, for other users and items, the new user has no behavior data, and the new item has no behavior with users. They are blank and can not be analyzed. This makes it impossible for the system to generate recommendations for new users and new items can not to be recommended to system users.

3 Methodology

3.1 Item-Based Collaborative Algorithm

Item-based collaborative filtering algorithm can recommend items that similar to the set of items that users have behaved before. For example, this algorithm will recommend "Database Design and Implementation" due to you have purchased "Database Principles". However, this algorithm does not calculate the similarity between items according to the attributes of the content of the item, but mainly calculates the similarity between items by recording and analyzing the user behavior data. If the algorithm considers that there is a high similarity between Item X and Item Y, that is because most users who are interested in Item X are also interested in Item Y. The concept map of the Item-based collaborative filtering recommendation algorithm is shown in Fig. 1.

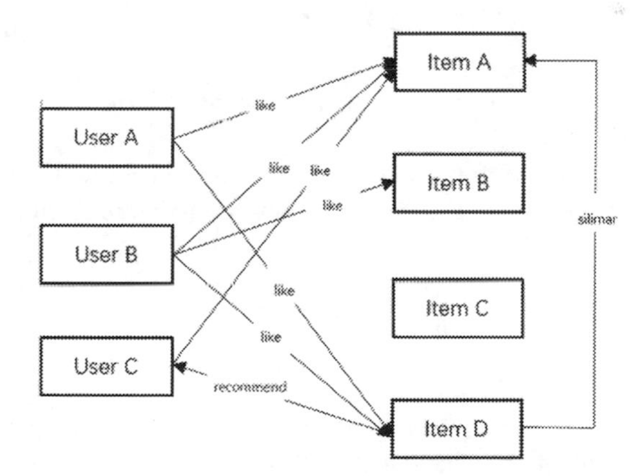

Fig. 1. The concept map of item-based collaborative algorithm

Through the above description, ItemCF can be divided into the following two steps.

(1) Computing the similarity between items.
(2) Combining and analyzing the historical behavior data generated by users and the similarity matrix of the constructed items to build the recommendation list for users.

ItemCF also collects user behavior data, constructs it into user-item inversion table, and uses user-item inversion table to calculate the similarity between items. In the user-item inversion table, two items in each user item list are added to the co-occurrence matrix C by 1. Figure 2 is an example of calculating item co-occurrence matrix. The table on the left side of the figure records a list of items that each user has behaved. The first step is to get the number of co-occurrences between two items according to the item list, and a matrix will be obtained. In the second step, the matrix C is obtained by adding all the matrices calculated in the first step. C[i][j] is the number of users who have acted on both item i and item j. Similarity matrix W can be obtained by co-occurrence matrix C.

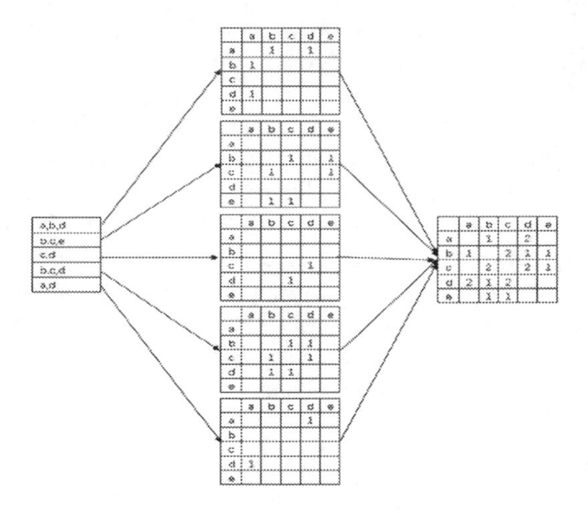

Fig. 2. Computing item co-occurrence matrix

From Amazon classic recommendation, "users who have purchased product *i* have also purchased *j*", the following formula can be defined to calculate the similarity between items:

$$Wij = \frac{|N(i) \cap N(j)|}{|N(i)|} \tag{1}$$

Equation (1) seems reasonable, but there is a problem. For example, if item j is very popular and most people have behaved, $|N(i) \cap N(j)|$ tends to approach $|N(i)|$, so W_{ij} is closer to 1. So using this formula, it is very likely that any item will have a high degree of similarity with popular items. This is not a good property for recommendation systems that can mine long tails. In order to prevent popular goods from becoming more popular, the following formulas can be used:

$$Wij = \frac{|N(i) \cap N(j)|}{\sqrt{|N(i)| * |N(j)|}} \tag{2}$$

The Eq. (2) penalizes the weight of item j, thus reducing the similarity between most commodities and popular commodities.

$$Wij = \frac{wij}{\max\limits_{j} wij} \tag{3}$$

After getting the item similarity matrix, we normalize the similarity by Eq. (3). The advantage of normalization is not only to increase the accuracy of recommendation, but also to improve the coverage and diversity of recommendation.

After calculating the similarity of the item, the recommended item is selected and recommended to the target user. Equation (4) shows the interest of the target user u in the item j.

$$P(u,j) = \sum\nolimits_{i \in S(j,K) \cap N(u)} wjirui \tag{4}$$

In Eq. (4), $N(u)$ denotes all item sets favored by the target user u, S (j, K) is the K item sets most similar to j, W_{ij} denotes the similarity between item j and item i, and R_{ui} denotes the user u preference for item i. If the target user u has ever acted on item i, then we suppose R_{ui} is 1. Equation (4) can be translated as: if some items are more similar to those in the historical behavior of the target user, then when generating the user recommendation list, its ranking will be higher. According to experimental analysis [24], when K is defined as 20, ItemCF has the best effectiveness.

3.2 Improvement of Recommendation Formula for Scenic Spots

Now it can generate a reasonable recommendation list for users, but the experimental results show that the coverage is not very ideal. Coverage is also the most concerned indicator for content providers [24], which shows the ability of a recommendation system to discover long tail items. By referring to and analyzing the improvement strategy of ItemCF in [25], this paper proposes an improved method for the formula of recommendation degree of scenic spots in order to improve the coverage rate of the recommended scenic spots. The improvement is to add the thermal parameters to the recommendation formula and punish them. The purpose is to make the cold scenic spots more likely to be recommended to users. The following formulas can be used:

$$P(u,j) = \sum\nolimits_{i \in S(j,K) \cap N(u)} wjirui * \frac{1}{log(1 + Fi)} \tag{5}$$

F_i represents the heat of a scenic spot. In this paper, if the number of user reviews for a scenic spot is n, the heat of the scenic spot is n. Popular attractions are easily recommended many times, which is tedious for users. Therefore, we add heat parameters to the recommendation formula to punish the hot spots, which not only reduces the recommendation of hot spots, but also improves the novelty of the recommendation model, so as to improve the coverage of the recommendation model. The experimental result show that after improving the recommendation formula, the coverage increases from 22.29% to 25.55%.

4 Dataset and Experiment

4.1 Dataset

In order to test the applicability of our personalized scenic spot recommendation model, we crawled a lot of data from large domestic tourism websites (Baidu Tourism), including 5582 scenic spots, 81282 users and 673811 scenic spots comment data. These real data can enable collaborative filtering algorithm to analyze a large number of real user behavior data, so that the recommendation model can more accurately predict the user recommendation list. Figure 3 and Fig. 4 show the occupational share of users and the proportion of provinces where scenic spots are located in the data set, respectively.

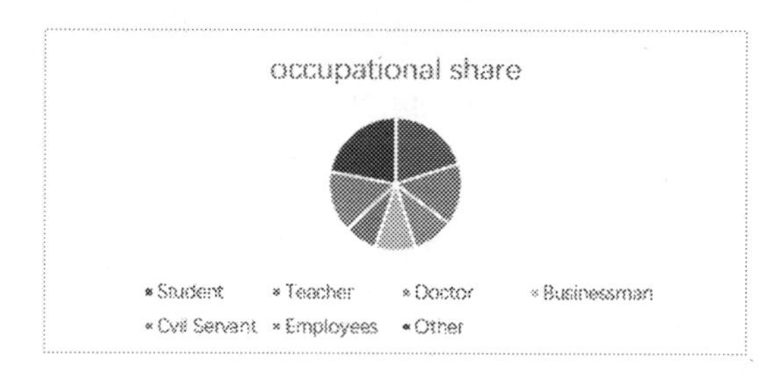

Fig. 3. The occupational share of users

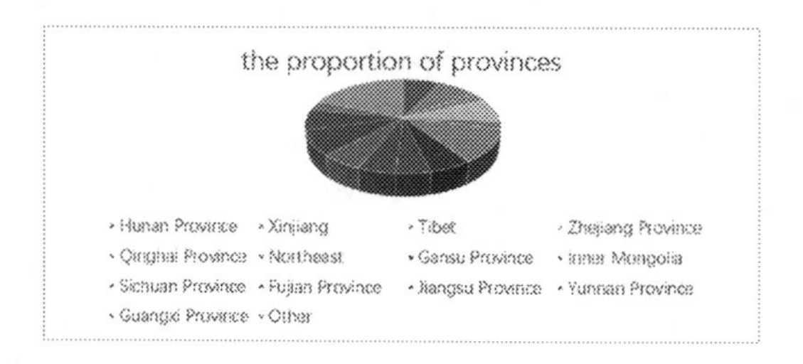

Fig. 4. The proportion of provinces where scenic spots are located

4.2 Experiment

Firstly, according to Baidu tourism dataset, a user-attraction inversion table is established for all users (that is, a list of the attractions that he has visited is established for each user). Then, for each user, add 1 to the co-occurrence matrix C by adding two or two scenic spots in the list of scenic spots, and finally get the complete co-occurrence matrix C. Figure 5 shows some data in the co-occurrence matrix C.

Scenic Spot 1	Scenic Spot 2	Number of simultaneous visits
The Imperial Palace	Qin Terra-Cotta Warriors and Horses Figurines	1258
Yulong Snowmountain	Dali Old City	1235
The Great Hall of the People	Shanghai Bund	1195
Lugu Lake	Erhai	1154
Beijing Zoo	Old Summer Palace	1078
Sun Yat sen University	WuHan University	1036
Wangfujing	Ancient culture street	994
Gulangyu Islet	Xuanwu Lake	962
Ocean Park	Shenzhen window of the world	921
Mount Yuelu.	Hanshan Temple	898

Fig. 5. Frequency of scenic spots co-occurrence

According to the co-occurrence matrix C, similarity Eq. (2) and normalization Eq. (3), the similarity matrix W among all scenic spots is obtained by using the ItemCF. Figure 6 shows some data in the similarity matrix W.

Scenic Spot 1	Scenic Spot 2	Similarity
Temple Street, Hong Kong	Lan Kwai Fong, Hong Kong	0.528
Temple Street, Hong Kong	Central Hongkong	0.464
Temple Street, Hong Kong	Repulse Bay	0.461
Temple Street, Hong Kong	Hong Kong Science Museum	0.456
Temple Street, Hong Kong	Hong Kong Golden Bauhinia Square	0.448
Wuhan Donghu Suspicious Beach Bath	Jinligou, Wuhan	0.654
Wuhan Donghu Suspicious Beach Bath	Pomegranate Red Village, Wuhan	0.645
Wuhan Donghu Suspicious Beach Bath	Yunwu Mountain, Wuhan	0.626
Wuhan Donghu Suspicious Beach Bath	Wuhan Ink Lake	0.618
Wuhan Donghu Suspicious Beach Bath	Wuhan Mulan Grassland	0.593
Coconut Dream Gallery	Guanyin on the Sea in Nanshan, Sanya	0.34
Coconut Dream Gallery	Sanya Luhuitou	0.334
Coconut Dream Gallery	Sanya Big and Small Cave Sky	0.325
Coconut Dream Gallery	Butterfly Valley in Sanya	0.316
Coconut Dream Gallery	Nanshan Temple, Sanya	0.313
Zhongshan Mausoleum, Nanjing	Nanjing Confucius Temple	0.404
Zhongshan Mausoleum, Nanjing	Nanjing Presidential Palace	0.299
Zhongshan Mausoleum, Nanjing	Nanjing the Qinhuai River	0.295
Zhongshan Mausoleum, Nanjing	Xiaoling Tomb of Ming Dynasty in Nanjing	0.291
Zhongshan Mausoleum, Nanjing	Nanjing Xuanwu Lake	0.286

Fig. 6. Similarity of scenic spots

We extract the scenic spots that a real user A has visited from Baidu tourism dataset: Tsim Sha Tsui, Shang Xia Jiu Pedestrian Street, Macau Tourist Tower, Baiyun Mountain, Gulangyu, Jinli, Wu Caichi, Wuhou Temple, Huacheng Square, Longji Terrace, Diwang Building and Changhai. The recommended list generated for him through the ItemCF is shown in Fig. 7.

	Scenic Spot	Recommendation Degree
1	Tap Seac Gallery	0.78
2	Happy Valley	0.68
3	Skyline 100 viewing platform	0.67
4	Macao Science Museum	0.64
5	Nanhai Temple	0.58
6	Chimelong Water Park	0.54
7	Gully mirror cliff	0.48
8	Hong Kong Observatory	0.43
9	Sun Yat-sen University	0.39
10	Baicheng Beach	0.37

Fig. 7. Recommendation list for user A

5 Conclusion

In this paper, we analyses the significance of personalized tourism recommendation in the era of explosive growth of tourism information and the applicability of ItemCF in the context of tourism information. In order to test the accuracy of personalized tourism recommendation model, a large number of real data from large domestic tourism websites are crawled and used in personalized tourism recommendation model. The experimental steps of establishing personalized scenic spot recommendation model by using ItemCF are introduced in detail. The recommendation formula in ItemCF is improved. The experimental results show that the improved recommendation model improves the recommendation coverage. Aiming at the defect of collaborative filtering algorithm—cold start problem, a solution strategy is proposed, which alleviates the problem of cold start to a certain extent.

In this paper, it is a bold attempt to introduce the improved collaborative filtering algorithm into personalized tourism recommendation, but this attempt ignores many aspects of tourism. In the next research, we will consider the travel notes and strategies published by users on tourism websites as user behavior data, extract key information such as attitudes and feelings, and analyze the user preference for scenic spots. Geographical location of tourism users can determine the scope of their activities, so it is also an important information, which can play a key role in personalized tourism recommendation. Therefore, in the future, we will consider adding these two information to the current personalized scenic spot recommendation model to improve the accuracy of recommendation.

Acknowledgements. The authors would like to thank anonymous reviewers for their very detailed and helpful review. This work was supported by National Natural Science Foundation of China (61902232, 61902231), Natural Science Foundation of Guangdong Province (2019A-1515010943), Key Project of Basic and Applied Basic Research of Colleges and Universities in Guangdong Province (Natural Science) (2018KZDXM035), The Basic and Applied Basic Research of Colleges and Universities in Guangdong Province (Special Projects in Artificial Intelligence) (2019KZDZX1030), The Characteristic Innovation Project in Higher Education Institutions of Guangdong Province (2016WTSCX035).

References

1. Ajantha, D., Vijay, J., Sridhar, R.: A user-location vector based approach for personalised tourism and travel recommendation. In: International Conference on Big Data Analytics and Computational Intelligence (ICBDAC), Chirala, India (2017)
2. Zhu, Z., Cao, J., Weng, C.: Location-time-sociality aware personalized tourist attraction recommendation in LBSN. In: IEEE 22nd International Conference on Computer Supported Cooperative Work in Design ((CSCWD)), Nanjing, China (2018)
3. Kesorn, K., Juraphanthong, W., Salaiwarakul, A.: Personalized attraction recommendation system for tourists through check-in data. IEEE Access **5**, 26703–26721 (2017)
4. Jiang, K., Wang, P., Yu, N.: ContextRank: personalized tourism recommendation by exploting context information of geotagged web photos. In: Sixth International Conference on Image and Graphics, Hefei, Anhui, China (2011)
5. Shiranthika, C., Premakumara, N., Fernando, S., Sumathipala, S.: Personalized travel spot recommendation based on unsupervised learning approach. In: International Conference on Advances in ICT for Emerging Regions (ICTer), Colombo, Sri Lanka (2018)
6. Shao, X., Tang, G., Bao, B.K.: Personalized travel recommendation based on sentiment-aware multimodal topic model. IEEE Access **7**, 113034–113052 (2019)
7. Chang, W., Ma, L.: Personalized e-tourism attraction recommendation based on context. In: International Conference on Service Systems and Service Management, Hong Kong, China (2013)
8. Jia, Z., Yang, Y., Gao, W., Chen, X.: User-based collaborative filtering for tourist attraction recommendations. In: International Conference on Computational Intelligence & Communication Technology, Ghaziabad, India (2015)
9. Xi, Y., Yuan, Q.: Intelligent recommendation scheme of scenic spots based on association rule mining algorithm. In: International Conference on Robots & Intelligent System (ICRIS), Huai'an, China (2017)
10. Lops, P., de Gemmis, M., Semeraro, G.: Content-based recommender systems: state of the art and trends. In: Ricci, F., Rokach, L., Shapira, B., Kantor, Paul B. (eds.) Recommender Systems Handbook, pp. 73–105. Springer, Boston, MA (2011). https://doi.org/10.1007/978-0-387-85820-3_3
11. Agrawal, R., Imieliński, T., Swami, A.: Mining association rules between sets of items in large databases. In: Proceedings of the 1993 ACM SIGMOD International Conference on Management of Data, pp. 207–216. ACM, New York, (1993)
12. Shi, H.: Random-walk classification algorithm with association rules mining. Chin. Acad. J. Electron. Publishing House **27**(9), 7–11 (2017)
13. Xie, Z., Jin, C., Liu, Y.: Personalized knowledge recommendation model based on constructivist learning theory. J. Comput. Res. Dev. **55**(1), 125–138 (2018)

14. Gao, M., Duan, Y.: Recommendation algorithm based on user clustering and rating preference, 21 July (2017)
15. Gong, M., Deng, Z., Huang, W.: Collaborative recommendation algorithm based on user clustering and Slope One filling. Comput. Eng. Appl. **54**(22), 139–143 (2018)
16. Han, J.: Research of the Personalized Recommendation Algorithm Based on Distributed Platform, pp. 43–60. Chang'an University, Xian (2017)
17. Glodberg, D., Nichols, D., Okib, M., et al.: Using collaborative filtering to weave an information tapestry. Commun. ACM **35**(12), 61–70 (1992)
18. Ebadia, K.A.: A hybrid multi-criteria hotel recommender system using explicit and implicit feedback. In: proceedings of the International Conference on Applied Science in Information Systems and Technology (2016)
19. Braunhofer, M., Elahi, M., Ricci, F., Schievenin, T.: Context-aware points of interest suggestion with dynamic weather data management. In: Xiang, Z., Tussyadiah, I. (eds.) Information and Communication Technologies in Tourism 2014, pp. 87–100. Springer, Cham (2013). https://doi.org/10.1007/978-3-319-03973-2_7
20. Hasuike, T., Katagiri, H., Tsuda, H.A.: Framework of route recommendation system for sightseeing from subjective and objective evaluation of tourism data. In: Proceedings of the Iiai International Congress on Advanced Applied Informatics (2016)
21. Uar, T.: Developing an intelligent trip recommender system by data mining methods. **6**(01), 119–127 (2016)
22. Devasanthiya, C., Vigneshwari, S., Vivek, J.: An enhanced tourism recommendation system with relevancy feedback mechanism and ontological specifications. In: Suresh, L.P., Panigrahi, B.K. (eds.) Proceedings of the International Conference on Soft Computing Systems. AISC, vol. 398, pp. 281–289. Springer, New Delhi (2016). https://doi.org/10.1007/978-81-322-2674-1_28
23. Wu, J.: Research and Implementation of Personalized Travel Recommendation System based on Collaborative Filtering Recommendation. Jiaotong University, Beijing (2017)
24. Liang, X.: Recommendation System Practice. People's Posts and Telecommunications Publishing House, Beijing (2012)
25. Mu, L.: Research on a Recommendation Algorithm for Tourism Strategy Based on Hybrid Collaborative Filtering. Jilin University (2017)

Research on Partner Selection in Virtual Enterprises Based on NSGA-II

Haixia Gui(✉), Banglei Zhao, Xiangqian Wang, and Huizong Li

School of Economics and Management,
Anhui University of Science and Technology, Huainan 232001, China
guihaixia18@sohu.com

Abstract. Partner selection is a key issue for the success of a virtual enterprise. The research on the selection of existing virtual enterprise partners is performed serially by a single project, and the virtual enterprise partner selection model of multiple projects concurrently is constructed. At the same time, the multi-objective optimization method is applied to the virtual enterprise partner selection, and the design is based on NSGA-II multi-target. The optimized virtual enterprise partner selection algorithm is compared by experiment and single objective optimization. Experiments show that the multi-objective optimization of virtual enterprise partner selection can obtain multiple Pareto optimal solutions, which can provide decision makers with more decision-making solutions.

Keywords: Virtual enterprise · Partner selection · NSGA-II · Multi-objective optimization · Pareto optimal solution

1 Introduction

With the rapid development of the Internet, information technology and economic globalization, it is difficult for traditional enterprises to adapt to the rapid changes in the market. In order to adapt to market demand and competition, companies must seek for superior resources outside the enterprise to cooperate. In this way, Virtual Enterprise (VE) [1, 2] came into being. The so-called virtual enterprise [3, 4] is that when some enterprises with different resources and advantages face new opportunities and challenges in the market, through information network technology, sharing resources and technology, together, they can jointly establish a market response and enhance competition. A major feature of the formation of virtual enterprises is how to choose partners, and this feature is also the decisive factor of the overall competitiveness of virtual enterprises and their market adaptability [5, 6]. In each selection process, due to the different resources and advantages of the partners, the competitiveness and survivability of the integrated alliance formed by the partners are closely related to the partners. Therefore, partner selection is a key focus of relevant research at home and abroad [7, 8].

Related research at home and abroad is generally a series of individual projects, each project is completed by several partners or divided into several sub-projects, each sub-project is completed by one or several candidate enterprises. Sadigh et al. [9]

K. Li et al. (Eds.): ISICA 2019, CCIS 1205, pp. 307–319, 2020.
https://doi.org/10.1007/978-981-15-5577-0_23

proposed a multi-agent hybrid partner selection algorithm based on ontology for multi-agent virtual enterprises. Through different types of agents to collaborate and compete, the unqualified or inefficient enterprises are removed from the enterprise pool and the winning company is selected by the algorithm. Huang et al. [10] proposed a new method to consider engineering start-up time, completion time, transportation time and cost uncertainty by using grey system theory, and proposed a new chaotic particle swarm optimization algorithm to solve virtual enterprise partner selection. Nikghadam et al. [11] used fuzzy analytic hierarchy process to determine four main indicators: unit price, on-time delivery reliability, enterprise past performance and service quality, and evaluated the optimal partner through weighting method. Zhang et al. [12] proposed a trust-based approach to corporate partner selection. This method combines real-coded genetic algorithm and nonlinear learning algorithm, and uses the hybrid learning algorithm of fuzzy cognitive map to provide decision-makers with a multi-view and interactive potential partner based on the dynamic characteristics of the trusted index. Jia [13] aimed to select a time-cost compromised virtual enterprise partner, using the task-resource allocation map as the scheduling model, and established a project deployment diagram that uniformly describes the virtual enterprise processes and resources, and implemented it using an iterative heuristic algorithm. Relative cost effectiveness solving algorithm. Andrade et al. [14] solved the virtual enterprise partner selection based on the game theory method and obtained a list of virtual enterprise rankings by accepting any set of key factors.

In addition, some related researches have proposed multi-objective models, which are generally achieved by corresponding methods or multi-objectives into single goals. This single-objective approach is simply to pursue a certain goal, which leads to the solution of the solution deviating from other goals. The actual demand has certain subjectivity. Xiao et al. [15] used traditional optimization methods to consider various factors such as operating cost, reaction time, and operational risk. An adaptive quantum group evolutionary algorithm with time-varying acceleration coefficients is proposed to solve the problem of partner selection optimization. Nikghadam et al. [16] proposed a method based on fuzzy inference system for evaluating and selecting potential enterprises. The evaluation is based on four indicators: unit price, delivery date, quality and performance. The output of the model is calculated using fuzzy rules, that is, partners. Dong et al. [17] combined the ideal solution similarity sorting technique with the multidimensional preference analysis linear programming technique, defined the fuzzy consistency and inconsistency indicators with relative closeness, and derived the weight vector of the decision maker according to the relative entropy. A new fuzzy linear programming model is constructed by using trapezoidal fuzzy numbers to estimate the index weights. By constructing a multi-objective allocation model, a scheme set sorting matrix is generated. Han et al. [18] proposed a multi-objective optimization virtual enterprise partner selection model, which is based on the shortest completion time, the lowest cost and the highest credibility. The AHP method is used to determine the weight of the evaluation factor. Su et al. [19] proposed an immune-based agile virtual enterprise partner selection algorithm, introduced multi-objective constraints, and designed an adaptive vaccine extraction strategy and a two-dimensional binary coding method for agile virtual enterprise partner selection algorithm.

Based on the above, this paper proposes the virtual enterprise partner selection problem of multiple projects concurrently, and applies the multi-objective optimization method to the virtual enterprise partner selection. The advantage of multi-objective optimization is to maintain a good trade-off between multiple targets, namely, Pareto optimal solution, to find a solution that tries to achieve consistency on each target. Moreover, multi-objective optimization can give a set of Pareto optimal solutions, which increases the choice space of decision makers. Decision makers can choose a reasonable solution that meets their needs, while single-objective optimization can only give a single solution.

The innovation of this paper is to construct a mathematical model of virtual enterprise partner selection for multiple projects concurrently. At the same time, based on the idea of multi-objective optimization, the virtual enterprise partner selection algorithm based on NSGA-II is designed and optimized by experiment and single target. The virtual enterprise partners choose to conduct comparative analysis. In this paper, the multi-objective optimization of virtual enterprise partner selection proposed by multiple projects in this paper can obtain multiple decision-making schemes superior to single-objective optimization.

2 Mathematical Model of Virtual Enterprise Partner Selection Problem

2.1 Mathematical Model

Let $Y = \{y_i | i \in [1, m]\}$ be the set of m projects that the virtual enterprise needs to complete; each project can be divided into n links, Expressed by

$$H = \begin{bmatrix} h_{11} & \cdots & h_{1j} & \cdots & h_{1n} \\ \vdots & \cdots & \vdots & \cdots & \vdots \\ h_{i1} & \cdots & h_{ij} & \cdots & h_{in} \\ \vdots & \cdots & \vdots & \cdots & \vdots \\ h_{m1} & \cdots & h_{mj} & \cdots & h_{mn} \end{bmatrix} \tag{1}$$

among them, $i \in [1, m]$, $j \in [1, n]$. For example, each project can be divided into five parts: design, procurement, manufacturing, distribution, and logistics. $h_{\bullet j}$ is the set of candidate enterprises that are willing to participate in the link $E_j = \{e_{jk} | k \in [1, r_j]\}$ (where $h_{\bullet j}$ refers to the link of all the rows in the j th column), r_j is the number of candidate enterprises of the link $h_{\bullet j}$, $P_{e_{jk}} = \{t_{e_{jk}}, q_{e_{jk}}, c_{e_{jk}}, \ldots\}$ is the set of performance parameters of each link in the $h_{\bullet j}$ of the enterprise e_{jk}, wherein $t_{e_{jk}}$ is time, $q_{e_{jk}}$ is quality, $c_{e_{jk}}$ is cost, and other parameters can be selected according to actual needs; T, Q and C are the time, quality, and cost of completing all projects respectively.

The completion time of the link h_{ij} cannot be greater than $t_{\max ij}$, the quality cannot be lower than $q_{\min ij}$, and the cost cannot be higher than $c_{\max ij}$. The matrix R_i can be used to represent the requirements of all items $t_{\max ij}$, $q_{\min ij}$ and $c_{\max ij}$. This matrix is called the project condition constraint matrix and can be expressed as:

$$R_i = \begin{bmatrix} \ldots t_{\max ij} \cdots \\ \ldots q_{\min ij} \cdots \\ \ldots c_{\max ij} \cdots \end{bmatrix} \tag{2}$$

The optimization goal of the virtual enterprise partner selection problem is to select the m group enterprise $F_i = \{S_1, S_2, \ldots, S_n\}$, and $S_j \cap E_j \neq \varnothing$, satisfy

$$\left.\begin{array}{l} \min T \\ \max Q \\ \min C \\ \text{s.t.} \\ \qquad \max\limits_{e_{jk} \in S_j} t_{e_{jk}} < t_{\max ij} \\ \qquad \sum\limits_{e_{jk} \in S_j} q_{e_{jk}} > q_{\min ij} \\ \qquad \sum\limits_{e_{jk} \in S_j} c_{e_{jk}} < c_{\max ij} \end{array}\right\} \tag{3}$$

among them,

$$T = \sum_{i=1}^{m} \sum_{j=1}^{n} \max_{e_{jk} \in S_j} \frac{t_{e_{jk}}}{t_{\max}} u_{jk} \tag{4}$$

$$Q = \sum_{i=1}^{m} \sum_{j=1}^{n} \sum_{k=1}^{r_j} \left(1 - \frac{q_{e_{jk}}}{q_{\max}}\right) u_{jk} \tag{5}$$

$$C = \sum_{i=1}^{m} \sum_{j=1}^{n} \sum_{k=1}^{r_j} \frac{c_{e_{jk}}}{c_{\max}} u_{jk} \tag{6}$$

$$u_{jk} = \begin{cases} 1 & \text{select } e_{jk} \text{ to join} \\ 0 & \text{don't select } e_{jk} \text{ to join} \end{cases} \tag{7}$$

$t_{\max}$, $q_{\max}$ and $c_{\max}$ are the maximum values of all the performance parameters of all the candidate companies, such as $q_{\max} = \max\limits_{j} \max\limits_{k} q_{e_{jk}}$.

2.2 Model Analysis and Multi-objective Optimization Problems

It can be seen from the above model that to solve the problem of virtual enterprise partner selection, it is necessary to consider the three objective functions of time, quality and cost of all projects, and to satisfy the constraints in Eq. (3), so it can be solved by multi-objective optimization. Virtual enterprise partner selection. There is a conflict between each optimization objective function in the multi-objective optimization problem. It is generally difficult to find an optimal solution to make all the optimization objective function values optimal. One solution is optimal for one of the optimization objective functions, and it is possible to Other optimization objective functions are not optimal, even worst. In view of the multi-objective optimization problem proposed by the above model, the three objective functions also conflict with each other. That is to say, it is difficult to make the three objective function values reach the optimal at the same time, and only the coordination compromise processing can be performed. The three objective function values may be optimally maximized. Therefore, solving the multi-objective optimization problem is to find a set of solutions that have a good constraint relationship with each other. These solutions are generally not easy to compare with each other. Pareto defines this solution as the Pareto optimal solution (also called non-inferior solution) [20, 21].

The multi-objective optimization problem [20, 21] can be expressed in the following mathematical form. Here, taking the minimization as the solution target, for example,

$$\left.\begin{array}{l} \min F(x) = \left(f_1(x), f_2(x), \cdots, f_p(x)\right)^T \\ s.t. \\ \qquad h_s(x) \geq 0, s = 1, 2, \cdots, q \end{array}\right\} \tag{8}$$

Where $x = \left(x_1, x_2, \cdots x_d\right)^T$ is the d-dimensional decision vector, $f_1(x), f_2(x), \cdots, f_p(x)$ total p objective functions, and $h_1(x), h_2(x), \cdots, h_q(x)$ have a total of q constraints.

If there is a decision vector x that satisfies q constraints at the same time, then x is called a feasible solution. The set of all feasible solutions is called the feasible solution set and is denoted as X. For any two feasible solutions x_1 and x_2, If $\forall l = 1, 2, \cdots, p$, can get $f_l(x_1) \leq f_l(x_2)$, $\exists l^* = 1, 2, \cdots, p$, satisfy $f_{l^*}(x_1) < f_{l^*}(x_2)$, then x_1 is said to dominate x_2 and is denoted as $x_1 \succ x_2$.

If there is no solution X in the feasible solution set x^*, then x^* is called a Pareto optimal solution in X. The set of all Pareto optimal solutions is called the Pareto optimal solution set, denoted as X_p. The Pareto frontier is a set of objective function vectors corresponding to the Pareto optimal solution set.

The multi-objective optimization problem is composed of multiple objective functions, and a solution cannot be obtained to optimize the values of all objective functions. Solving the multi-objective optimization problem is to find the Pareto optimal solution set or Pareto frontier of the approximation problem. Therefore, the criterion for evaluating the merits of a multi-objective optimization algorithm is whether it can find the Pareto optimal solution set or approach the optimal solution set [20, 21].

The traditional multi-objective optimization method is essentially a single-objective optimization solution. In the pre-processing stage, certain rules are used to transform multiple targets into a single target solution method, and true multi-objective optimization cannot be achieved. In recent years, researchers have proposed various multi-objective optimization methods based on the definition of Pareto's solution set, such as non-dominated sorting genetic algorithm (NSGA) [21], second-generation non-dominated sorting inheritance. Algorithm (NSGA-II) [22], multi-objective particle swarm optimization [23] and so on. Among them, NSGA-II is improved on the basis of NSGA, and it is solved by adding non-dominated sorting based on standard genetic algorithm. Because it is relatively simple and widely used, it has been tested function [24], power grid planning [25] and supply chain network design [26] and other experimental and practical applications have shown good performance, and become the standard choice for solving multi-objective optimization problems. The method for solving the virtual enterprise partner selection problem is based on NSGA-II.

3 Solving Virtual Enterprise Partner Selection Based on NSGA-II

3.1 Second Generation Non-dominated Sorting Genetic Algorithm (NSGA-II)

In 2000, Srinivas and Deb [27] proposed the second generation non-dominated sorting genetic algorithm (NSGA-II), which is an improved algorithm based on the non-dominated sorting genetic algorithm NSGA. The main flow of the algorithm: In order to generate the progeny population Q_t, the initial population is first used as the parent population P_t for genetic operators (selection, crossover, mutation), and the parent population and the progeny population are merged into a new population R_t, then the new The population size is twice that of the parent or progeny population, namely, the size is $2N$. Then, the new population R_t is quickly non-dominated, and the crowding degree of each individual in the population is calculated, and the non-dominated set F_i is obtained. R_t contains parent and child individuals, then the individual in non-dominated set F_1 is the best in R_t, so first put F_1 into the new parent population P_{t+1}. If the number of individuals in F_1 exceeds the population N, then Select N from F_1 according to the size of the crowd; if the number of individuals in F_1 does not reach the population N, then select the same method from the next layer F_2 until the number of individuals in P_{t+1} is N. Finally, a new progeny population Q_{t+1} is generated by the genetic operator.

3.2 Virtual Enterprise Partner Selection Algorithm Based on NSGA-II

Chromosome Coding

In NSGA-II, the traditional chromosome coding adopts one-dimensional coding. This coding structure is not suitable for the problem solved in this paper. For this reason, this paper extends the NSGA-II chromosome to two-dimensional binary coding. As shown in Fig. 1, each line of the chromosome code represents a project y_i, each column represents the participation of each link partner, and u_{jk} is used to represent each gene position of the chromosome. If $u_{jk} = 1$ indicates that the enterprise participates in the corresponding link, conversely, If $u_{jk} = 0$ indicates that the company is not involved in the corresponding link.

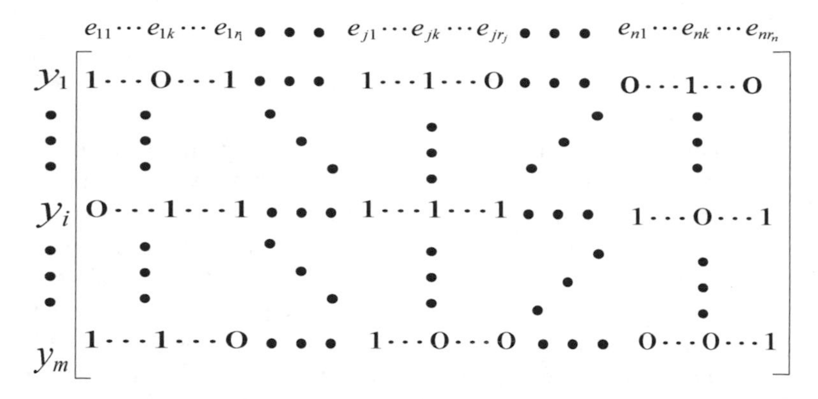

Fig. 1. Two-dimensional binary chromosome encoding

3.3 Virtual Enterprise Partner Selection Algorithm Based on NSGA-II

The steps for solving the virtual enterprise partner selection algorithm based on NSGA-II are as follows:

Step 1. First, initialize the maximum number of iterations T, population size N, crossover probability, mutation probability, randomly generate initial populations of N individuals, and calculate each objective function value of the initial population according to formulas (4), (5), and (6), first assume the iterative algebra $t = 1$.
Step 2. If $t = T$, go to Step 6, otherwise go to Step 3.
Step 3. In the initial population, individuals who meet the constraints are selected according to the constraints in Eq. (3) to perform crossover and mutation operations to generate new populations, and each objective function value is calculated according to Eqs. (4), (5), and (6).
Step 4. Combine the two populations, and obtain the non-dominated level through rapid non-dominated sorting, and generate a new generation of initial population through the elite strategy according to the non-dominated level and the crowding distance.
Step 5. Let $t = t + 1$ go to Step 2.
Step 6. Output the result.

4 Experimental Results and Analysis

Consider three projects concurrently, each of which has been divided into five parts: design D, procurement P, manufacturing M, distribution S, and logistics L. The number of candidates for each link is 21, 18, 17, 16 and 16, respectively. The performance parameters, $t_{e_{jk}}$, $q_{e_{jk}}$ and $c_{e_{jk}}$ of each candidate in the candidate are initialized, and the constraint matrix of each item is R_1, R_2 and R_3, respectively.

$$R_1 = \begin{bmatrix} 8.0 & 5.3 & 6.4 & 4.3 & 4.5 \\ 0.78 & 0.78 & 0.75 & 0.74 & 0.74 \\ 440 & 420 & 450 & 420 & 400 \end{bmatrix}, R_2 = \begin{bmatrix} 8.9 & 5.2 & 9.0 & 4.5 & 4.0 \\ 0.82 & 0.76 & 0.77 & 0.76 & 0.75 \\ 480 & 460 & 440 & 460 & 400 \end{bmatrix}, R_3 = \begin{bmatrix} 8.2 & 7.0 & 6.5 & 7.9 & 8.0 \\ 0.79 & 0.74 & 0.76 & 0.78 & 0.74 \\ 460 & 440 & 400 & 430 & 420 \end{bmatrix}.$$

The NSGA-II was used to perform multi-objective optimization and immune algorithm (IA) [19] single-objective optimization experiments. The immune algorithm draws on the idea of the literature [19], that is, the multi-objective optimization is turned into a single target through the weight method, but here three items are serially processed in sequence. Generally speaking, for an existing optimization algorithm, the performance of the algorithm may change due to different values of the parameters during the execution of the algorithm. At present, the commonly used method for determining parameters is to use an experimental method, that is, a combination of existing work [19, 28–30] and a large number of experimental tests to obtain a relatively good set of parameters. The details are as follows: (1) NSGA-II: population size 100, iteration number 500, crossover probability 0.7, mutation probability 0.05. (2) Immune algorithm: population size 100, iteration number 500, vaccination probability 0.72, crossover probability 0.7, mutation probability 0.05.

Each test sample was randomly generated according to the size and constraints of the problem and run independently 30 times.

Figure 2 shows the CPU runtime (in seconds) for 30 independent experiments of the two algorithms. As can be seen from the above figure, this paper proposes that multi-objective optimization of virtual enterprise partner selection is significantly larger than single-target optimized CPU runtime. This is because multi-objective optimization is solved by multiple objective functions not being dominated, which takes a lot of time in each iteration. This paper is the three objective functions, which are the time T, quality Q and cost C of all projects, which require the least time, the highest quality and the lowest cost. They are mutually exclusive and get a set of solutions that satisfy the

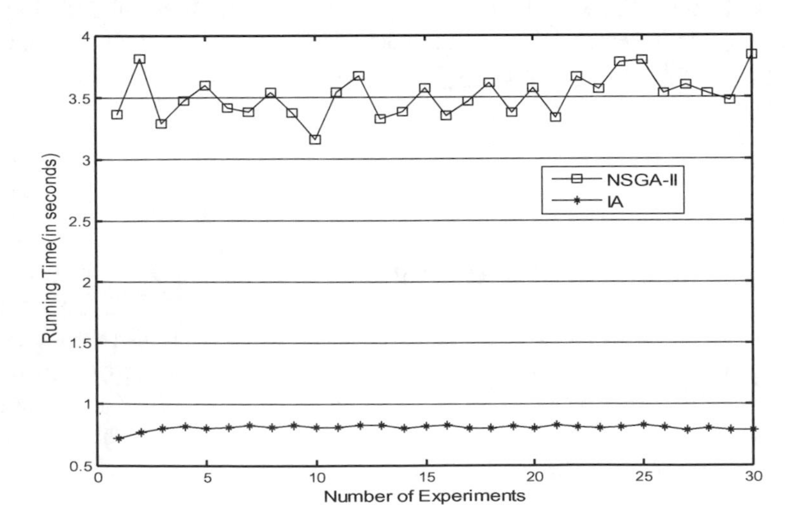

Fig. 2. Running time of two algorithms

constraints. The single-objective optimization is to convert multiple objective functions into one objective function by the weight method, so that it takes little time in each iteration.

Figure 3 shows the Pareto optimal solution set obtained by NSGA-II multi-objective optimization and immune algorithm single-objective optimization. It can be seen from Fig. 3 that NSGA-II multi-objective optimization can search for multiple Pareto optimal solutions. The distribution is more dense and uniform, and the immune target single-objective optimization can only obtain an optimal solution, so that the decision maker has no choice in evaluating the decision, and the NSGA-II multi-objective optimization can give more optimal solutions, the decision maker can choose a reasonable combination plan as needed. The solution idea of the single-objective optimization problem to be compared is similar to that of the literature [19]. The three projects are executed serially in sequence, and the weighting method is used to convert the multi-objective optimization problem into the solution of the single-objective optimization problem, and an objective function value is obtained. The weights corresponding to the objective functions are: 0.38, 0.26, and 0.36, respectively. In this paper, we use NSGA-II to solve the multi-objective optimization problem and obtain a set of solutions with three objective function values.

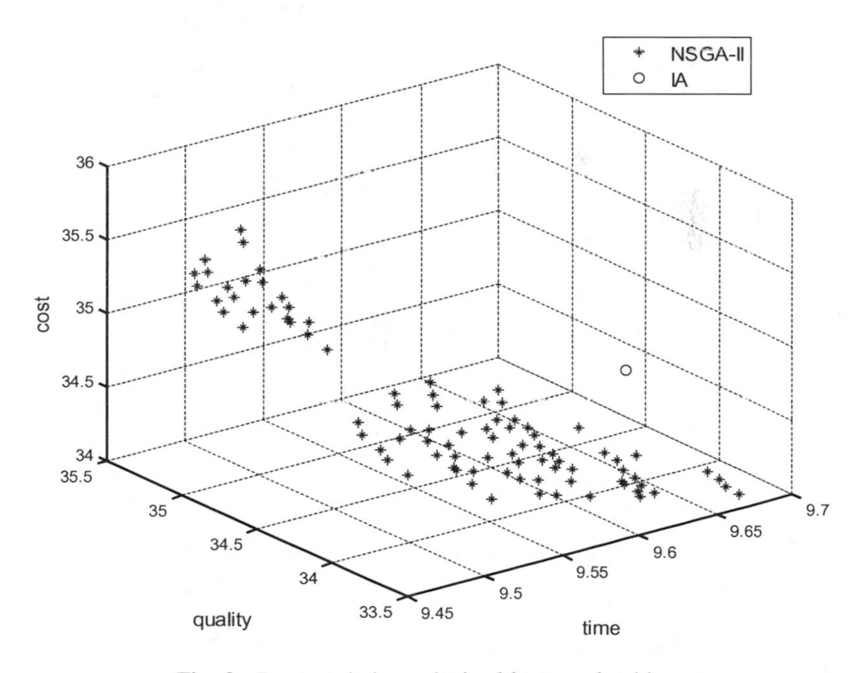

Fig. 3. Pareto solutions obtained by two algorithms

In order to find a solution superior to single-objective optimization from the Pareto optimal solution obtained from multi-objective optimization in this paper, since the single-objective optimization is performed by three items in sequence, the optimal value of the objective function corresponding to the single-objective optimization is

Table 1. Superior Pareto optimal solution and its corresponding partner selection

Algorithm	Better Pareto optimal solution (T, Q, C)	Partner selection	
NSGA-II	(9.600000, 34.456522, 34.541667)	Project 1	$\{D_4, D_{16}\}$, $\{P_5, P_{15}\}$, $\{M_7, M_9, M_{12}, M_{14}, M_{15}\}$, $\{S_6, S_{13}\}$, $\{L_3, L_4, L_{13}\}$
		Project 2	$\{D_6, D_8, D_{17}, D_{20}\}$, $\{P_3, P_4, P_6, P_8, P_{16}\}$, $\{M_{15}, M_{17}\}$, $\{S_3, S_{13}\}$, $\{L_8, L_9, L_{12}\}$
		Project 3	$\{D_6, D_{12}, D_{20}\}$, $\{P_9, P_{16}\}$, $\{M_{10}, M_{17}\}$, $\{S_1, S_{11}\}$, $\{L_3, L_8, L_{12}\}$
	(9.560000, 34.282609, 34.447917)	Project 1	$\{D_3, D_{16}\}$, $\{P_5, P_{15}\}$, $\{M_5, M_9, M_{12}, M_{14}\}$, $\{S_6, S_{13}\}$, $\{L_3, L_4, L_{12}\}$
		Project 2	$\{D_6, D_8, D_{17}, D_{21}\}$, $\{P_3, P_4, P_6, P_8, P_{16}\}$, $\{M_{14}, M_{17}\}$, $\{S_3, S_{13}\}$, $\{L_1, L_3, L_{12}\}$
		Project 3	$\{D_6, D_{12}, D_{21}\}$, $\{P_9, P_{15}\}$, $\{M_{10}, M_{17}\}$, $\{S_1, S_{11}\}$, $\{L_3, L_{11}, L_{13}\}$
	(9.560000, 34.456522, 34.541667)	Project 1	$\{D_3, D_{14}\}$, $\{P_4, P_{15}\}$, $\{M_7, M_9, M_{11}, M_{14}, M_{15}\}$, $\{S_6, S_{13}\}$, $\{L_3, L_4, L_{12}\}$
		Project 2	$\{D_6, D_8, D_{18}, D_{20}\}$, $\{P_3, P_4, P_6, P_8, P_{17}\}$, $\{M_{15}, M_{17}\}$, $\{S_3, S_{13}\}$, $\{L_8, L_9, L_{11}\}$
		Project 3	$\{D_7, D_{12}, D_{20}\}$, $\{P_9, P_{16}\}$, $\{M_{10}, M_{17}\}$, $\{S_1, S_{11}\}$, $\{L_3, L_8, L_{13}\}$
	(9.600000, 34.347826, 34.385417)	Project 1	$\{D_3, D_{14}\}$, $\{P_2, P_8, P_{11}\}$, $\{M_5, M_9, M_{14}, M_{17}\}$, $\{S_5, S_{12}\}$, $\{L_3, L_4, L_{11}\}$
		Project 2	$\{D_5, D_7, D_{17}, D_{20}\}$, $\{P_4, P_6, P_8, P_{16}\}$, $\{M_8, M_{14}\}$, $\{S_3, S_{13}\}$, $\{L_8, L_{11}, L_{12}\}$
		Project 3	$\{D_8, D_{11}, D_{20}\}$, $\{P_9, P_{15}\}$, $\{M_{10}, M_{17}\}$, $\{S_1, S_9\}$, $\{L_3, L_8, L_{12}\}$
	(9.533333, 34.347826, 34.510417)	Project 1	$\{D_3, D_{14}\}$, $\{P_2, P_8, P_{11}\}$, $\{M_5, M_{14}, M_{17}\}$, $\{S_5, S_{13}\}$, $\{L_3, L_{11}, L_{16}\}$
		Project 2	$\{D_1, D_2, D_{17}, D_{21}\}$, $\{P_2, P_4, P_6, P_9, P_{16}\}$, $\{M_5, M_8, M_{14}\}$, $\{S_3, S_{11}\}$, $\{L_8, L_{11}, L_{13}\}$
		Project 3	$\{D_8, D_{16}, D_{17}\}$, $\{P_3, P_{13}\}$, $\{M_2, M_5, M_{17}\}$, $\{S_1, S_{10}\}$, $\{L_3, L_8, L_{14}\}$
	(9.600000, 34.434783, 34.468750)	Project 1	$\{D_3, D_{16}\}$, $\{P_2, P_8, P_{10}\}$, $\{M_5, M_{14}, M_{17}\}$, $\{S_5, S_{13}\}$, $\{L_3, L_4, L_{11}\}$
		Project 2	$\{D_1, D_2, D_{16}, D_{21}\}$, $\{P_3, P_4, P_6, P_7, P_{16}\}$, $\{M_5, M_9, M_{14}\}$, $\{S_3, S_{13}\}$, $\{L_8, L_9, L_{12}\}$
		Project 3	$\{D_6, D_{12}, D_{20}\}$, $\{P_9, P_{16}\}$, $\{M_1, M_5, M_{17}\}$, $\{S_1, S_{11}\}$, $\{L_3, L_8, L_{12}\}$

(*continued*)

Table 1. *(continued)*

Algorithm	Better Pareto optimal solution (T, Q, C)	Partner selection	
	(9.520000, 34.282609, 34.520833)	Project 1	$\{D_4, D_{17}\}$, $\{P_6, P_{15}\}$, $\{M_7, M_9, M_{11}, M_{14}, M_{15}\}$, $\{S_6, S_{13}\}$, $\{L_3, L_4, L_{11}\}$
		Project 2	$\{D_6, D_8, D_{17}, D_{19}\}$, $\{P_2, P_4, P_6, P_8, P_{16}\}$, $\{M_5, M_9, M_{13}\}$, $\{S_3, S_{13}\}$, $\{L_8, L_{11}, L_{13}\}$
		Project 3	$\{D_6, D_{12}, D_{20}\}$, $\{P_9, P_{16}\}$, $\{M_1, M_4, M_{17}\}$, $\{S_1, S_{13}\}$, $\{L_3, L_8, L_{10}\}$
	(9.586667, 34.326087, 34.437500)	Project 1	$\{D_3, D_{14}\}$, $\{P_1, P_8, P_{10}\}$, $\{M_7, M_9, M_{12}, M_{14}\}$, $\{S_7, S_{13}\}$, $\{L_3, L_4, L_{12}\}$
		Project 2	$\{D_6, D_8, D_{20}\}$, $\{P_3, P_4, P_6, P_8, P_{16}\}$, $\{M_5, M_{15}, M_{17}\}$, $\{S_5, S_{13}\}$, $\{L_8, L_{11}, L_{12}\}$
		Project 3	$\{D_8, D_{16}, D_{17}\}$, $\{P_3, P_{13}\}$, $\{M_2, M_5, M_{17}\}$, $\{S_1, S_9\}$, $\{L_3, L_8, L_{14}\}$
	(9.506667, 34.391304, 34.531250)	Project 1	$\{D_3, D_{14}\}$, $\{P_1, P_8, P_{11}\}$, $\{M_7, M_{11}, M_{12}, M_{14}\}$, $\{S_6, S_{13}\}$, $\{L_3, L_4, L_{11}\}$
		Project 2	$\{D_6, D_8, D_{17}, D_{20}\}$, $\{P_3, P_4, P_6, P_8, P_{16}\}$, $\{M_{15}, M_{17}\}$, $\{S_3, S_{13}\}$, $\{L_8, L_{11}, L_{13}\}$
		Project 3	$\{D_8, D_{15}, D_{17}\}$, $\{P_8, P_{16}\}$, $\{M_1, M_4, M_{17}\}$, $\{S_1, S_{13}\}$, $\{L_3, L_8, L_{13}\}$
IA	(9.666667, 34.282609, 4.572917)	Project 1	$\{D_{11}, D_{14}, D_{19}\}$, $\{P_7, P_8\}$, $\{M_7, M_{10}, M_{11}\}$, $\{S_4, S_{13}\}$, $\{L_3, L_{11}, L_{12}\}$
		Project 2	$\{D_1, D_{13}, D_{14}, D_{20}\}$, $\{P_7, P_{10}\}$, $\{M_4, M_{10}, M_{14}\}$, $\{S_1, S_2, S_{12}, S_{14}\}$, $\{L_1, L_5, L_6\}$
		Project 3	$\{D_5, D_8\}$, $\{P_{14}, P_{16}\}$, $\{M_3, M_6, M_{11}\}$, $\{S_4, S_5, S_{13}, S_{16}\}$, $\{L_6, L_{11}, L_{16}\}$

The three objective function values, T, Q and C of the project are sequentially added to obtain a solution containing three objective function values, and this solution is compared with a set of solutions containing three objective function values obtained by multi-objective optimization. According to formula (3), the three objective function values in the optimal solution must satisfy the time value of the solution in which all the items are completed less than or equal to the single target optimization, and the mass is greater than or equal to the quality value of the solution in the single target optimization, and the cost is less than or equal to Cost value in the solution of single-objective optimization. Table 1 is a solution to the single-objective optimization of the optimal solution-dominated immune algorithm obtained by NSGA-II multi-objective optimization and its corresponding partner selection. It can be seen from Table 1 that the virtual enterprise partner selection based on NSGA-II multi-objective optimization can search for nine sets of single-objective optimization solutions, which can provide decision makers with nine decision-making schemes when evaluating decisions. Optimization has only one option.

5 Conclusions

In this paper, a virtual enterprise partner selection algorithm based on overlapping alliance and NSGA-II multi-objective optimization is studied, which realizes the partner selection of multiple projects concurrently processing multiple links. Applying multi-objective optimization ideas to virtual enterprise partner selection is an innovation in this paper. Experiments show that this method can obtain multiple Pareto optimal solutions, which provides decision makers with a variety of decision-making schemes, and decision makers can choose the best one.

Acknowledgement. This work was supported in part by the National Natural Science Foundation of China under (Grant No. 61703005, 51474007, 51874003, 61873004), the Natural Science Foundation of Anhui Province of China (NO. 1808085MG221) and the Master and Doctoral Science Foundation for Anhui University of Science and Technology under Grant 12059.

References

1. Xiao, J., Liu, B., Huang, Y., et al.: An adaptive quantum swarm evolutionary algorithm for partner selection in virtual enterprise. Int. J. Prod. Res. **52**(6), 1607–1621 (2014)
2. Nikghadam, S., Ozbayoglu, A.M., Unver, H.O., et al.: Design of a customer's type based algorithm for partner selection problem of virtual enterprise. Procedia Comput. Sci. **95**(9), 467–474 (2016)
3. Nyongesa, H.O., Musumba, G.W., Chileshe, N.: Partner selection and performance evaluation framework for a construction-related virtual enterprise: a multi-agent systems approach. Architectural Eng. Des. Manage. **13**(5), 344–364 (2017)
4. Xiao, J.H., Niu, Y.Y., Chen, P., et al.: An improved gravitational search algorithm for green partner selection in virtual enterprises. Neurocomputing **217**(12), 103–109 (2016)
5. Arrais, C.A., Varela, M.L., Putnik, G.D., et al.: Ollaborative framework for virtual organisation synthesis based on a dynamic multi-criteria decision model. Int. J. Comput. Integr. Manuf. **31**(9), 857–868 (2018)
6. Polyantchikov, I., Shevtshenko, E., Karaulova, T.: Virtual enterprise formation in the context of a sustainable partner network. Ind. Manage. Data Syst. **117**(7), 1446–1468 (2017)
7. Zhao, Y.P., Zhang, P.P., Li, S.: Partner selection model and method of virtual enterprise in virtual cultivation environment. Stat. Decis. **8**(50), 173–176 (2015)
8. Jin, Y., Zhang, Y.W.: Method for determining comprehensive weight of partner selection index in virtual enterprise. Stat. Decis. **10**(32), 72–74 (2013)
9. Sadigh, B.L., Nikghadam, S., Ozbayoglu, A.M., et al.: An ontology-based multi-agent virtual enterprise system (OMAVE): part 2: partner selection. Int. J. Comput. Integr. Manuf. **30**(10), 1072–1092 (2017)
10. Huang, B., Bai, L.H., Roy, A., et al.: A multi-criterion partner selection problem for virtual manufacturing enterprises under uncertainty. Int. J. Prod. Econ. **196**(2), 68–81 (2018)

11. Nikghadam, S., Sadigh, B.L., Ozbayoglu, A.M., et al.: A survey of partner selection methodologies for virtual enterprises and development of a goal programming & ndash based approach. Int. J. Adv. Manuf. Technol. **85**(5–8), 1713–1734 (2016)
12. Zhang, W., Zhu, Y.C., Zhao, Y.: Fuzzy cognitive map approach for trust-based partner selection in virtual enterprise. J. Comput. Theor. Nanosci. **13**(1), 349–360 (2016)
13. Jia, X.: Selection algorithm of virtual enterprise partner based on task-resource assignment graph. Int. J. Grid Distrib. Comput. **9**(9), 185–192 (2016)
14. Andrade, J., Ares, J., Garcia, R.: A game theory based approach for building holonic virtual enterprises. IEEE Trans. Syst. Man Cybern. Syst. **45**(6), 291–302 (2015)
15. Xiao, J.H., Liu, B.L., Huang, Y.F.: An adaptive quantum swarm evolutionary algorithm for partner selection in virtual enterprise. Int. J. Prod. Res. **52**(6), 1607–1621 (2014)
16. Nikghadam, S., LotfiSadigh, B., Ozbayoglu, A.M., et al.: Evaluation of partner companies based on fuzzy inference system for establishing virtual enterprise consortium. Commun. Comput. Inf. Sci. **577**(1), 104–115 (2015)
17. Dong, J.Y., Wan, S.P.: Virtual enterprise partner selection integrating LINMAP and TOPSIS. J. Oper. Res. Soc. **67**(10), 1288–1308 (2016)
18. Han, J.H., Wang, M.F., Ma, X.S., et al.: Partner selection solving based on a self-adaptive genetic algorithm for virtual enterprise. Comput. Integr. Manuf. Syst. **14**(1), 118–123 (2008)
19. Su, Z.P., Jiang, J.G., Xia, N., et al.: A partner selection algorithm based on immunity for agile virtual enterprises. China Mech. Eng. **19**(8), 925–929 (2008)
20. Wang, P.F., Yan, X., Zhao, F.Y.: Multi-objective optimization of control parameters for a pressurized water reactor pressurizer using a genetic algorithm. Ann. Nucl. Energy **124**(2), 9–20 (2019)
21. Mithilesh, K., Chandan, G.: The elitist non-dominated sorting genetic algorithm with inheritance and its jumping gene adaptations for multi-objective optimization. Inf. Sci. **382–383**(3), 15–37 (2017)
22. Mehmet, B.E., Jonathan, L.G.: Design and implementation of a general software library for using NSGA-II with SWAT for multi-objective model calibration. Environ. Model Softw. **84** (10), 112–120 (2016)
23. Vahid, B., Mohamad, M.K., Azuraliza, A.B.: Multi-objective PSO algorithm for mining numerical association rules without a priori discretization. Expert Syst. Appl. **41**(9), 4259–4273 (2014)
24. Ye, C.J., Huang, M.X.: Multi-objective optimal power flow considering transient stability based on parallel NSGA-II. IEEE Trans. Power Syst. **30**(2), 454–461 (2015)
25. Li, Y., Lu, X., Kar, N.C.: Rule-based control strategy with novel parameters optimization using NSGA-II for powersplit PHEV operation cost minimization. IEEE Trans. Veh. Technol. **63**(7), 3051–3061 (2014)
26. Zahra, A.A., Mohammad, M.P., Seyed, H.N., et al.: A meta-heuristic approach supported by NSGA-II for the design and plan of supply chain networks considering new product development. J. Ind. Eng. Int. **14**(1), 95–109 (2018)
27. Deb, K., Agrawal, S., Pratap, A., et al.: A fast elitist nondominated sorting genetic algorithm for multi-objective optimization: NSGA-II. In: Proceedings of the Parallel Problem Solving from Nature VI Conference, Paris, pp. 849–958 (2000)
28. Zhang, G.F., Wang, Y.Q., Su, Z.P., et al.: Modeling and solving multi-objective allocation-scheduling of emergency relief supplies. Control Decis. **32**(1), 86–92 (2017)
29. Camara, M.V.O., Ribeiro, G.M., Tosta, M.D.R.: A pareto optimal study for the multi-objective oil platform location problem with NSGA-II. J. Petrol. Sci. Eng. **169**(10), 258–268 (2018)
30. Xu, F.Q., Liu, J.C., Lin, S.S., et al.: A multi-objective optimization model of hybrid energy storage system for non-grid-connected wind power: A case study in China. Energy **163**(8), 585–603 (2018)

Research on Big Data System Based on Cultural Tourism in Dongguan

Ding Li[1,2] and Kangshun Li[1,2(✉)]

[1] School of Computer Science, Guangdong University of Science and Technology, Dongguan 523083, Guangdong, People's Republic of China
792046639@qq.com, likangshun@sina.com

[2] School of Mathematics and Informatics, South China Agricultural University, Guangzhou 510642, China

Abstract. This paper proposes a big data platform based on cultural tourism in Dongguan, Guangdong, China, which mainly includes data acquisition system, storage system and intelligent analysis system. Using web search data to analyze the relevance of tourism destinations, construct a search index to measure tourism demand, and analyze the tourists' demand for characteristic cultural tourism through field research, and provide corresponding countermeasures for tourism management departments; The tourism information intelligent collaborative method analyzes the tourism information intelligent collaborative system under big data and analyzes the results based on time series network data. The tourism information intelligent collaborative method based on big data analysis has strong anti-interference ability and strong timeliness. It has high research and judgment strategy for massive data analysis, and can quickly and accurately classify tourism information and extract the tourism products needed by users. The design meets the analysis information and processing requirements of the tourism system for big data.

Keywords: Big data system · Cultural tourism · Tourism demand · Big data analysis · System design

1 Introduction

1.1 Overview of Dongguan's Cultural and Tourism Big Data Construction

Dongguan is located in the Big-Bay District of Guangdong, Hong Kong and Macao, on the east bank of the Pearl River Estuary, a famous historical and cultural city in Guangdong. It is surrounded by 10 county-level administrative regions including Guangzhou, Shenzhen and Huizhou, and has a history and cultural city of more than 1,700 years. At the same time, as an excellent tourist city, it has rich tourism resources. In 2018, Dongguan City received a total of 44.338 million passengers, a year-on-year increase of 7.05%; the total tourism revenue was 52.937 billion Yuan, an increase of 8.28%. There are 25 national tourist attractions in the city, including 15 AAAA-level tourist attractions. However, the contribution rate of tourism to Dongguan's GDP is only 8%, which is lower than the national average of 10% [1]. This shows the realistic

K. Li et al. (Eds.): ISICA 2019, CCIS 1205, pp. 320–330, 2020.
https://doi.org/10.1007/978-981-15-5577-0_24

existence of the low factor contribution rate of the cultural tourism industry in Dongguan as a historical and cultural city. In essence, the low-element contribution rate of Dongguan's cultural tourism industry stems from the lack of precise management and sustainable planning mechanisms in the cultural tourism industry, including the functional orientation of the local cultural tourism industry and the insufficiency of connotation mining and weak extension. In recent years, with the wide application of mobile Internet technology in the development of tourism, the big data application is the core to improve the modernization level of the tourism industry. Big data technology integrates transportation, environmental protection, land and resources, urban and rural construction, business, aviation, postal, telecommunications, meteorology and other related aspects of tourism related data, while with Baidu, Google, Taobao and other major network search engines and travel electronic operators Cooperation, the establishment of social data and tourism and related departments data integration of tourism big data resources, providing new development opportunities and driving force for the characteristic cultural tourism industry.

With the continuous updating of network technology, big data analysis has been more and more recognized by people, and big data information is largely integrated into the Internet [2, 3]. In the past decade or so, the Internet has been inseparable from people's lives and work. People put a lot of time and energy into the Internet, and the information in big data analysis contains a lot of tourism products. The traditional tourism information intelligent method occupies a large amount of network resources, causing congestion of the operating system and reducing the retrieval efficiency of tourism information [4, 5]. Big data analysis is the basic means of modern large data volume information processing system, big data analysis the cloud storage computing system can not only be used for data storage, but also can deeply mine and judge data [6, 7]. With the adjustment of traditional industrial structure and the application of new technologies, the cultural tourism industry has broken through the traditional industrial model and gradually evolved into a multi-faceted, multi-dimensional new industry [8]. With the advent of the era of big data, the real-time data collection of cultural tourism scenic spots, creative, related enterprises and users has led to the explosive growth and accumulation of spatial location information. The traditional index structure has not been applied to large-scale data space. Distributed parallel computing complicates the creation of indexes. Cultural tourism data has formed a diversified and massive data space. These massive data not only need to be stored and shared, but also need to be analyzed, compared, and explored to find out the common law and value and apply to the entire industry of cultural tourism. The core of cultural tourism big data lies in the value of data. How to obtain valuable information from big data is the core content of big data analysis and processing. Traditional data analysis methods include data mining, machine learning, and statistical analysis. In the aspect of big data mining technology, most of them use association analysis to search data and find patterns with higher probability, or analyze the similarity of cultural tourism data through data clustering and classification, and provide decision support for decision makers. The integration of culture and tourism has quietly begun, but it takes a long time to truly integrate ideas, integrate functions, and integrate products. Establishing a relatively uniform evaluation parameter system can better promote the integration of the two. The evaluation system construction provides a good foundation for data analysis and mining, which in turn affects the analysis and decision-making, and plays a guiding role in the development of the business and industry. With the development of the Internet and big data and cloud computing, new formats have emerged as the times require. The previous evaluation index system can be used with fewer new formats and new technologies. The system of separate evaluation of culture and tourism is not conducive to the integration and development of cultural tourism. This paper refers to international standards, combined with China's actual situation, and joins the impact of new technologies on the business of tourism, and initially establishes a big data framework platform that includes the documentary evaluation parameters system.

2 Tourism Demand and Evaluation Parameter

2.1 Tourism Demand Description

Accurate analysis of tourism demand through big data is an important research topic in the field of tourism management. It not only helps the tourism management department to establish the expected judgment on the bearing capacity and tourist flow of the tourism destination, but also formulates corresponding policies based on the analysis results. Improve the tourism management system, thereby improving the economic benefits of local tourism.

At present, the contents of China's tourism evaluation mainly include: the three major tourism markets [9], namely the inbound tourism market, the outbound tourism market, the domestic tourism market. The number of tourist, income, expenses, source of tourists, destination; the number of hotels and travel agencies, income, taxes, employment rate, etc. in these 3 markets respectively. Similar to cultural evaluation, it is mainly measured from the dimension of economy, but the evaluation at the social level is relatively weak.

As a socialized industry, social and public services are equally important. In terms of basic evaluation information, it mainly evaluates from two aspects: tourism demand and tourism supply, including basic concept system, relevant classification standards, corresponding parameter system and evaluation result table. Abiding by the principles of national accounts, a set of global standards and definitions have been set to measure the contribution of tourism to GDP, employment, capital investment, taxation, etc., and the important role of tourism in the national balance of payments.

The evaluation methods of culture and tourism are limited by the current development of science. In the past, they mainly used sample surveys, departmental reports and special surveys. The evaluation survey project shall be jointly formulated by the administrative department and the local administrative department. The local meeting will appropriately increase the regional evaluation content according to the characteristics of the region and actual needs. However, technologies such as the Internet of Things, mobile internet, big data, cloud computing, can shorten the frequency of data updates, adding new methods for evaluation in user portraits and precision marketing.

UNESCO defines culture in this way: culture is the unique spiritual, material, intellectual and emotional characteristics of a society (social group). The scope includes: art and literature, lifestyle, settlement methods, value systems, traditions and beliefs [10]. In the era of "Internet+", China's 2018 cultural industry classification has added Internet entertainment platforms and wearable cultural equipment. This indicates that the evaluation parameter system should be adjusted accordingly with the development of society and industry.

2.2 Big Data Analysis Management System

Cultural tourism data includes spatial data, visitor data and background data. Spatial data includes spatial data of cultural tourism scenic spots, creative cultural tourism data, tourist data including tourist's smart device interaction data, active collection and sharing of data, and background data including tourism platform data. Database construction refers to the acquisition and analysis of these massive multi-type data and design into a standardized database.

Database construction includes research and set up the standards of content in cultural tourism database and meta-database, and the classification and the grading specification of tourism data according to development of tourism industry, and the classification data collection, cataloging and archiving of tourism data. Divide all data from the cultural tourism GIS and resources to realize digital editing and distribution system to establish a cultural tourism big data resource pool. Cultural tourism resources cover all information points and take part in the region platform server to collect data and operate the support system and geographic information in the cultural tourism industry chain from multiple heteroid or isostructural business systems. The system collects data including cultural product information, tourist attractions information, hotel information, store information, service information, etc., covering documents, text, images, XML, HTML, reports, images, audio and video, 3D and so on.

The big data analysis management system adopts the Hadoop architecture which is mainly divided into collection, storage, analysis and present to form a 4-layer architecture and use Hadoop to combine business architecture. The cultural tourism big data analysis management system can combine the data of the cultural tourism industry to form an industry-type big data platform mainly including big data collection system, big data real-time storage system, big data intelligent analysis system. Management and analysis of integration of user activity data, transaction data, and participation data based on the entire platform. Use specialized technology to capture, manage, process and organize into valuable data in a reasonable amount of time.

It is the key to organize the tourism space information and the tourist behavior information reasonably so that the commonly used queries can be executed efficiently. For example a query for a number of hotels closing a famous scenic spot, the system needs to provide an efficient index to optimize the operation of the query, speed up the query response speed while reducing system overhead, and bring a good experience for the user. Therefore, the technical support for cultural tourism big data management and analysis is to provide efficient data organization and access algorithms. Figure 1 is a technical architecture diagram of the big data analysis management system.

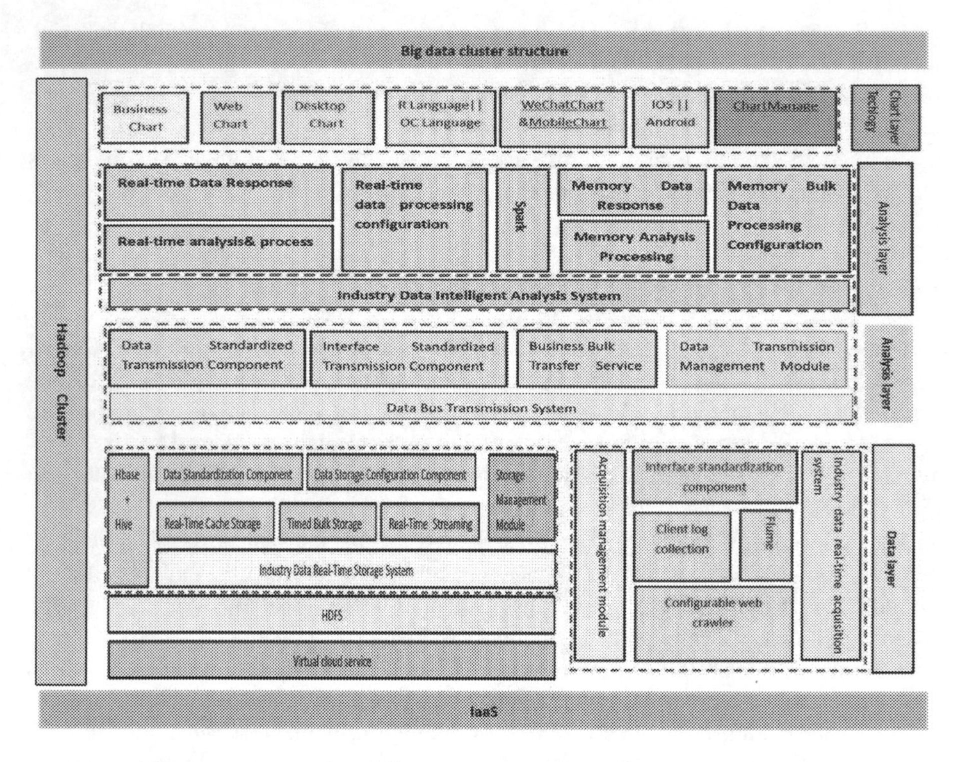

Fig. 1. The technical architecture diagram of big data analysis management system

For citations of references, we prefer the use of square brackets and consecutive num Big data real-time acquisition system: data collection through the customized collection service and interface with the massive log collection system Flume's Source component, the data can be efficiently obtained in batches and then stored in the data cache service by using memory storage and disk store two ways to temporarily store the collected data on the client. In the process of transferring to the cluster center server, the data standardization component DSC is used to customize the standardized interface, and the data is standardized into various storage formats such as json, xml, txt, db, etc., and then the data is collected by the local data cluster component of the client, and finally the data is carried out. The steps of cleaning and decrypting push the data to the cluster center, and the cluster center performs distributed data storage.

Big Data Real-Time Storage System: When the storage is initiated by the data cluster host, the data is first filtered and grouped, and stored in real time storage, batch storage, and streaming storage. At the same time, by cooperating with HBase, the table structure is the configuration file is stored in the cloud server, and then HBase is used to add, modify, delete, and query basic operations. When using data, combined with Hive to query and retrieve SQL statements, it is convenient to program and use.

Big Data Intelligent Analysis System: The system can configure both streaming data computing and memory data computing. Both methods require related response components. The flow calculation responds and starts the analysis by receiving and popping. The analysis process first injects the configuration and the algorithm into the algorithm component, uses the Storm framework to perform the flow analysis, and outputs the calculation result. Memory calculation in the response process to perform memory control, the data is stored in memory in an orderly manner, and then combined with the configuration, algorithm, the data is analyzed through the Spark framework, and finally getting the calculation results.

2.3 Core Technology

Data Indexing Technology: Cultural tourism big data includes spatial data and text data. The text indexing technique adopts the method of inverted indexing, which is an efficient indexing method for mass text retrieval and is a mapping of text data storage locations. The spatial indexing technique divides the spatial region by orthogonal grids and uses corresponding objects for any object in the space. The mapping function maps the object to a specific grid unit, combines the inverted index and the spatial grid division to form an inverted grid index, and implements an algorithm of the index structure on the MapReduce distributed model.

Big data analysis is an important source of tourism information and the first step in obtaining tourism information. How to analyze and study the information in big data analysis is the basis for system design. For the big data tourism information intelligent collaborative method, its main one of the problems is to give accurate travel data and describe the data information that needs to be obtained by iterative calculation.

Assume that $f: R \rightarrow R^m$ is the data information mapping in big data analysis. Under the condition f, the information in big data analysis can be iterated, and the problem algorithm to be solved can be continuously iterated. $f|a = f|a$. Assuming that the nonlinear time sequence in the big data analysis is $\{x(t), t = 1, 2, \ldots, N\}$, the nonlinear coordinate space reconstruction with network data information can obtain the tourism information data coordination parameters:

$$x(n) = [x(n), x(n-t), \ldots, x(n-(m-1)t)]^T \tag{1}$$

Among them, m is the cooperative data embedded information parameter, and t is the network data analysis travel information delay time. From linear time analysis in big data analysis, there is a smooth mapping of tourism information, as in formula (2). Show:

$$X(n+T) = \psi[X(n)] \tag{2}$$

According to the different synthetic methods of tourism in the big data analysis, the collaborative algorithm of tourism information nonlinear time series is different for the global tourism information support prediction.

For the big data analysis, the nonlinear time series calculation and analysis of tourism coordination, we must fit all the big data, analyze the law of the data needed for tourism, and find the mapping law of network data [11]. According to the data law, the characteristics of collaborative tourism are analyzed and calculated by given tourist information, namely:

$$\sum_{i=0}^{N}[X(t+1)-\psi(X(t))]^{2} \tag{3}$$

The big data in big data analysis is coordinated in the form of tourism information, and the minimum value of the calculation time is $\psi = R \rightarrow R^{m}$.

The vector of coordinate values in big data analysis is represented by c_{fg}. The center of the f cluster of the tourism information collaboration center has g-dimensional analysis attributes. G_d represents the quadratic error of the tourism information data cluster, and the optimal relationship between the tourism information collaborative objective function and the big data analysis design change vector is:

$$\min G_d = \sum_{g=1}^{d}\sum_{f=1}^{m_g} \| a_f^{(g)} - c_{fg} \|^2 \tag{4}$$

When all the data information objects in the big data analysis have the attributes of the tourism information data, the collaborative centers of the various tourism features in the network data information cluster are expressed by the formula:

$$Z_g = \frac{\sum_{d,\in fg} d_g}{|R_g|} \quad g \in [1, f] \tag{5}$$

Among them, $|R_g|$ representing the number of tourism feature collaborative data based on the gth data information cluster in big data analysis. When the tourism feature collaborative center continuously calculates the information data in the big data analysis until the tourism cooperation does not change, the travel algorithm ends.

When there are few types of data information in big data analysis, a higher tourism information data collaboration method is adopted, and all network data information is analyzed globally. When there are many types of data information in big data analysis, the analysis and calculation method is computationally intensive. In order to facilitate tourism search and speed up the operation speed of tourism information intelligent collaborative method, regression nonlinear calculation is performed by formula (6):

$$x(t+1) = \sum_{i=0}^{m-1} ax(t-i) + A \tag{6}$$

In the formula, A represents the non-known tourist type of tourism feature in the nonlinear calculation of data information in big data analysis. $a_i(i = 0, 2, \ldots, m-1)$ can be calculated by analyzing the calculation of information sequence in big data analysis by nonlinear time [12, 13], so that the data information of big data analysis is the shortest in part A in the process of tourism feature coordination. The travel feature error obtained by the nonlinear time series of data information is minimized.

$$Q_d = \sum_{t=m}^{N} [x(t) - \sum ax(t-i-1)]^2 \tag{7}$$

To solve the tourism information in the big data analysis, there are

$$\sum_{i=0}^{m-1} a(\sum x(t-i)x(t-j)) = \sum_{t=m}^{N} [x(t) - \sum ax(t-j)]^2, j = 0, \ldots, m-1 \tag{8}$$

The analysis of nonlinear time data information sequence can produce different results of large tourism feature synergy. Tourism research and judgment synergy is the simplest way to coordinate tourism information, and tourism support is close to 100%. Further information on the big data analysis of the nonlinear time series, expressed as a model formula:

$$x(N+T) = a + \sum_{i=1}^{k} a(Nx(N-(i-1)t)) = a + A(N)X(N) \tag{9}$$

Where $A(N) = [a_1, a_2, \ldots a_m]$, Big data analysis information is obtained by simplifying nonlinear coefficients:

$$F(N) = \sum_{i=1}^{k} |x(N+T) - a - A(N)X(N)|^2 \tag{10}$$

Based on the construction of the above big data system, our analysis and application of Dongguan tourist attractions are as follows:

Dongguan has 13 national-level scenic spots and more than 100 provincial-level scenic spots, including Keyuan, Qifengshan, Songshan Lake, Linzexu Yanyanchi, Weiyuan Fort, Yinpingshan Forest Park, Shuiyushan Forest Park. A large number of tourist attractions such as the Opium War Museum and Xiabafang have become the tourist and leisure destinations of countless tourists at home and abroad. Using BD search engine tools to study the tourism needs of three scenic spots, namely, Keyuan, Qifengshan and Songshan Lake, which are currently hot. See Fig. 1 for BD search data based on 3 attraction keywords.

From Fig. 2, it can be concluded that the hot spots of Songshan Lake in January and October of 2019 are relatively high. This is related to the Chinese New Year and National Day holiday. There will be a 7-day holiday in China's Spring Festival and National Day. In March, April and May of 2019, attention to the hot spots of Songshan Lake also showed a small peak, which is related to China's Ching Ming Festival and Labor Day holiday. There are three days of small holidays in China's Ching Ming Festival and Labor Day. It can be seen that the 7-day long vacation and the 3-day long vacation have a significant impact on the heat of the park. It can be concluded from Fig. 1 that the heat of the three scenic spots of Keyuan, Qifengshan and Songshan Lake has been cyclical since the beginning of 2019. Tourists have the highest heat in the

Songshan Lake Scenic Area, and the Garden and Qifeng Mountain are separated by two or three. The blue line represents Songshan Lake, yellow represents the Ke garden, and the green line represents Qifeng Mountain. as shown in Fig. 2.

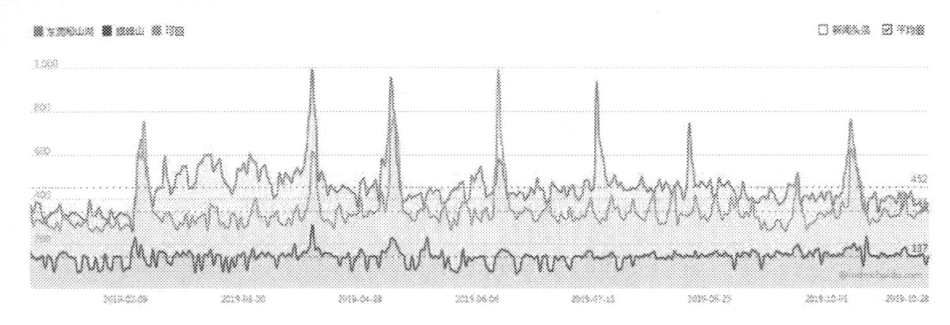

Fig. 2. Search and comparison of Songshan Lake, Keyuan and Qifeng Mountain from the beginning of 2019 to the present

3 Analysis of Local Cultural Tourism

3.1 Analysis Method

This paper uses the joint analysis method to empirically analyze the local cultural tourism demand in Dongguan. The research team conducted a questionnaire survey on the three scenic spots of Dongguan Keyuan, Qifengshan and Songshan Lake. A total of 3,500 samples were distributed, and invalid samples were excluded. A total of 2,689 valid samples were obtained. The questionnaire mainly includes six demand factors such as local characteristics of tourist attractions, per capita daily consumption, characteristics of tourist souvenirs, degree of dietary characteristics, maturity of tourist attractions, and convenience of transportation facilities. Perform attribute level design on 6 demand factors, as shown in Table 1.

Table 1. Tourism demand factors and their attribute levels

NO.	Demand factor	Attribute level
1	Degree of local characteristics of tourist attractions	Unique = 1; high degree of feature = 2; Characteristic level = 3; Poor feature level = 4
2	Per capita daily spending limit	Below 500 yuan = 1; 500 to 1000 = 2; 1000 to 2000 = 3
3	Characteristics of tourist souvenirs	Unique = 1; high degree of feature = 2; characteristic degree = 3; poor degree of character = 4
4	Dietary characteristics	Unique = 1; high degree of feature = 2; characteristic degree = 3; poor degree of character = 4
5	Tourism area maturity	More mature = 1; generally mature = 2; immature = 3
6	Convenience of transportation facilities	Good = 1; generally mature = 2; poor = 3

Using Spearman's rho test and Kendall's tau test method, the goodness-of-fit test was performed on the joint model, and the results obtained are shown in Table 2.

Table 2. Correlation coefficient test

Method	Value	Sig.
Spearman' rho	0.976	0.000
Kendall' tau	0.953	0.000

From Table 2, Spearman's rho test and Kendall's tau test are significant, indicating that the established model fitting accuracy is relatively high. The basic assumptions made in this paper and the analysis of tourism demand in Dongguan using the joint analysis method are Reasonable, that is, the sample of this study can reflect the tourism demand structure of the three scenic spots of Dongguan Keyuan, Qifengshan and Songshan Lake.

The utility analysis in the joint analysis method uses the utility value to reflect the degree of tourists' demand for the tourist destination. The greater the utility value of a demand factor, the higher the expectation of the tourist's demand for the tourist destination. According to the value of the various demand factors of the tourists, the relative weight values of each demand factor to attract tourists can be obtained, as shown in Table 3.

Table 3. Relative weight values of demand factors for tourism demand

Demand factor	Local characteristics of tourist attractions	Per capita daily spending	Tourist souvenirs	Dietary characteristics	Tourism area maturity	Convenience of transportation facilities
Relative weight value	25.21	10.54	12.73	22.58	15.65	13.31

4 Conclusion

Based on big data technology, this paper analyzes the tourism needs of three scenic spots in Dongguan Keyuan, Qifengshan and Songshan Lake. The joint analysis method is used to analyze the demand of tourists for tourist destinations, and the weight of each demand factor is calculated. The research results show that the tourist demand for the three major scenic spots of Dongguan Keyuan, Qifengshan and Songshan Lake is higher in the degree of local characteristics and dietary characteristics of the tourist attractions. This study provides a theoretical support for the relevant tourism management departments to use the big data platform to improve the quality of tourism scenic spots, improve the competitiveness and attractiveness of tourist attractions, promote the urbanization of tourism in a healthy and orderly manner, and improve the regional economic level.

The research on cultural tourism big data will effectively promote the development of cultural tourism industry chain, and then drive the development of food, housing, travel, tourism, shopping and entertainment enterprises in the tourism industry. Future research will focus on the research of the income of chain wholesale, retail enterprises, hotels, hotels, restaurants and attractions, as well as art performance groups, cultural centers, public libraries, museums and other cultural industry chain enterprises. The big data analytics management system is not only for the cultural tourism industry studied in this paper, but also for other industries with similar data structures.

Acknowledgements. The research work was supported by Foundation of University of Science and Technology of Guangdong.

References

1. Dongguan Cultural Radio and Television Tourism Sports Conference. http://www.timedg.com/2019-03/14/20818306.shtml. Accessed 2 Nov 2019
2. De Mauro, A., Greco, M., Grimaldi, M.: A formal definition of Big Data based on its essential features. Libr. Rev. **65**(3), 122–135 (2016)
3. Lu, Y., Xu, X.: Cloud-based manufacturing equipment and big data analytics to enable on-demand manufacturing services. Robot. Comput. Integr. Manuf. **57**, 92–102 (2019)
4. Amadeo, M., et al.: Information-centric networking for the internet of things: challenges and opportunities. IEEE Network **30**(2), 92–100 (2016)
5. Zhao, X., Liu, L., Wang, H., Song, W.: Ontology construction of the field of tourism in Africa. In: 2015 8th International Symposium on Computational Intelligence and Design (ISCID), December 2015, vol. 1, pp. 47–50. IEEE (2015)
6. Chang, V., Wills, G.: A model to compare cloud and non-cloud storage of Big Data. Future Gener. Comput. Syst. **57**, 56–76 (2016)
7. Tang, B., et al.: Incorporating intelligence in fog computing for big data analysis in smart cities. IEEE Trans. Industr. Inf. **13**(5), 2140–2150 (2017)
8. Zuza, M., Pérez-Ilzarbe, I., Rivas, C., Rivas, A.: WCT tourism placemaking (poster). J. Transp. Health **7**, S4–S5 (2017)
9. Liao, K.C., et al.: An evaluation of coupling coordination between tourism and finance. Sustainability **10**(7), 2320 (2018)
10. Bujdosó, Z., et al.: Basis of heritagization and cultural tourism development. Procedia Soc. Behav. Sci. **188**, 307–315 (2015)
11. Brida, J.G., Lanzilotta, B., Pizzolon, F.: Dynamic relationship between tourism and economic growth in MERCOSUR countries: a nonlinear approach based on asymmetric time series models. Econ. Bull. **36**(2), 879–894 (2016)
12. Muhtaseb, B.M., Daoud, H.E.: Tourism and economic growth in Jordan: Evidence from linear and nonlinear frameworks. Int. J. Econ. Financ. Issues **7**(1), 214–223 (2017)
13. Sun, D.: Simulation analysis of peak period income estimation model in tourism scenic of minority areas. In: 2015 International Conference on Automation, Mechanical Control and Computational Engineering, April 2015. Atlantis Press (2015)

Virtualization and AI Applications

Fusion of Skin Color and Facial Movement for Facial Expression Recognition

Wanjuan Song[1,2](✉) and Wenyong Dong[1,3]

[1] Computer School, Wuhan University, Wuhan, China
{key_swj, dwy}@whu.edu.cn
[2] Hubei University of Education, Wuhan, China
[3] School of Software, Nanyang Institute of Technology, Nanyang, Henan, China

Abstract. In this paper, based on the combination of skin color and facial movement, the automatic expression recognition algorithm is studied. The image is transformed into YIQ color space by RGB, and image data is extracted in the first dimension of YIQ. Background and skin color are segmented from binary image. Pareto optimization algorithm is used to select facial expression features. This algorithm has less computation, simple structure and fast running speed. It can accurately detect and track the small-angle facial skin color, facial expression changes, facial rotation and occlusion. Experiments show that the proposed algorithm can be well adapted to face rotation at small angles. For the state of human eyes, the algorithm is not affected, and can adapt to the rich facial expression changes and different skin color better, and has a certain stability.

Keywords: Pareto optimization algorithm · Facial motion · Skin color · Genetic algorithm

1 Introduction

In recent years, with the increasing improvement of computer technology, real-time face recognition system has developed rapidly, and is widely used in supervision, retrieval and other related fields [1, 7, 10, 11]. However, at this stage, face recognition technology still has some shortcomings and is limited in practical application. In the face recognition system, face detection is an important part, so the detection part has a very crucial role. There are many detection algorithms in the face detection system, which can be divided into two categories. The first kind of detection algorithm is based on the pixel features, and the pixel features include contour, skin color, etc. The second kind of detection algorithm is based on the pixel features. It is based on biological features, in which the biological features include the micro-features among the pixels in the image, including the feature matrix and mean of the pixels, and the corresponding algorithms include neural networks, Ada Boost [2, 8, 9] and so on.

Face skin color and motion recognition refers to the process of determining the skin color information, size change and motion trajectory of a person's face in the image sequence. The method based on motion information and skin color information has faster implementation speed, but the requirement of skin color information is stricter in background color distribution, more misunderstandings and lower recognition rate.

K. Li et al. (Eds.): ISICA 2019, CCIS 1205, pp. 333–340, 2020.
https://doi.org/10.1007/978-981-15-5577-0_25

The tracking accuracy of face motion information needs to be further improved [4]. Therefore, the combination of detection and recognition methods based on motion information and skin color information can improve the accuracy of automatic expression recognition algorithm combined with face skin color and motion recognition, which can meet the real-time monitoring requirements in practical application [5, 12]. In this paper, based on the combination of skin color and facial movement, the automatic expression recognition algorithm is studied.

2 Extraction of Face Skin Color

Skin color is the key element of facial information. In fact, there are some differences on the surface of different people's skin color. But if the brightness and other factors are excluded, skin color shows a very high clustering, and its hue is basically the same [6]. In RGB color space, because there is no specific brightness bits information, skin color does not have good clustering, so it is difficult to segment skin color regions. In YIQ color space, skin color has high clustering, but only in the first dimension of the color space. This phenomenon shows that the algorithm of skin color segmentation is very simple and can meet the speed requirements of real-time monitoring system. Conversion based on RGB color space can be carried out to YIQ color space. The conversion formula is as follows (1):

$$\begin{bmatrix} Y \\ I \\ Q \end{bmatrix} = \begin{bmatrix} 0.228 & 0.586 & 0.113 \\ 0.595 & -0.273 & -0.321 \\ 0.210 & -0.522 & 0.311 \end{bmatrix} \times \begin{bmatrix} R \\ G \\ B \end{bmatrix} \quad (1)$$

The process of skin color extraction is divided into three steps: using formula 1 to transform the image from RGB color space to YIQ color space; extracting image data information in the first dimension of YIQ; setting appropriate threshold to segment the background and skin color of binary image.

3 Feature Selection of Facial Expression

3.1 Facial Expression Feature Selection Based on GA Algorithm

The main idea of genetic algorithm is Darwin's theory of biological evolution. By simulating the process of biological evolution in nature, the optimal value is searched iteratively. At present, the commonly used GA algorithm is simple genetic algorithm, namely SGA. Genetic algorithm is to search the optimal solution of a target problem by iteration method. In the process of population initialization for target problem, individuals can be obtained based on the gene coding of the population. Generally speaking, the individual population belongs to the entity, which has chromosomal characteristics. Chromosomes carry genetic material and are composed of many genes. Therefore, they are also expressed by a combination of genes. The external performance of individuals is determined by the combination of genes. In chromosomes, hair color is determined by a combination of genes. Thus, genes are encoded, i.e. genotypes are expressed by external mapping.

3.2 Face Expression Feature Selection Based on Improved Pareto Optimization Algorithm

Through the improved GA algorithm, the solution is obtained by optimization. The Pareto optimization algorithm is used to optimize the multi-objective of each population $F(S_k)(k = 1, 2, \cdots, n)$, and the optimal $\max F(S_k)$ is obtained. The solution s_k is obtained by formula (2).

$$subject\ to\ S_k = \begin{bmatrix} a_{11} & a_{12} & \ldots & a_{1m} \\ a_{21} & a_{22} & \ldots & a_{2m} \\ \ldots & \ldots & \ldots & \ldots \\ a_{m1} & a_{m2} & \ldots & a_{mm} \end{bmatrix} \tag{2}$$

When choosing the characteristics of Pareto optimization algorithm, two new optimization objective functions are established by using Fisher linear discriminant criterion. See formula (3).

$$\begin{cases} F_1(S_k) = \frac{1}{N}\sum_{i=1}^{m}\sum_{j=1}^{n}\left(S_{ij} - \frac{1}{N_w}\sum_{i=1}^{N_w} M_i\right)^2 \\ F_2(S_k) = \frac{1}{l}\sum\left(\frac{1}{N}\sum_{i=1}^{m}\sum_{j=1}^{n}\left(S_{ij} \quad \frac{1}{N_b}\sum_{j=1}^{N_b} M_j\right)^2\right) \end{cases} \tag{3}$$

In formula (3), the number of different kinds of expressions is expressed by l. M_i is the solution from one kind of expression by GA, and N_w is the number of solutions. M_j is the solution from different kinds by GA, and N_b is the number of corresponding solutions.

Through the objective function (3), we can see that $F_2(S_k)$ corresponds to enlarge the gap between classes, and $F_1(S_k)$ corresponds to narrow the gap between classes. Figure 1 is a face feature selection based on GA and pareto optimization algorithm.

3.3 Facial Expression Feature Classification Based on Stochastic Forest Method

After choosing the best features, we classify the expressions, including fear, anger, surprise, happiness and so on. According to the random forest classifier method, the accuracy of facial expression classification can be effectively improved. Random forest belongs to a combination classifier, which is essentially a set of tree classifiers. Among them, the base classifier is a decision tree constructed by the classification regression tree algorithm without pruning classification, and its output result is determined by the simple method of majority voting. Gini coefficient index is the splitting criterion of classified regression tree in Stochastic Forest algorithm. The calculation process is shown in formula (4), *mtry* represents the characteristic dimension of each node; P_i represents the probability of occurrence in the sample set S.

$$Gini(S) = 1 - \sum_{i=1}^{mtry} P_i^2 \tag{4}$$

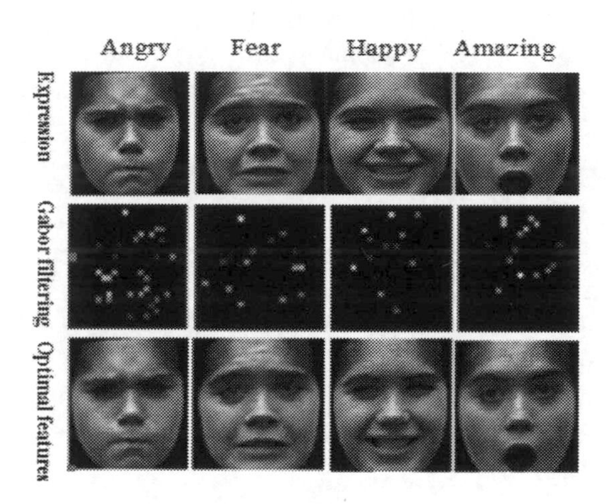

Fig. 1. Facial expression feature map

4 Face Expression Recognition Method

4.1 3-D Face Deformation Model

The 3D deformable model is based on a three-dimensional mesh face model. The face model is represented by vector $S \in R^{3N}$, which contains components on shape *x, y, z* and vector $T \in R^{3N}$ which contains RGB color model.

The number of vertices in a mesh is represented by N. In this 3D deformation model, there are two PCA models, one for color information and the other for shape. Formula (5) is used to represent any PCA model.

$$M := (\overline{v}, \sigma, V) \tag{5}$$

In the formula, the grid mean is represented by element $\overline{v} \in R^{3N}$, $\sigma \in R^{3N}$ by a standard deviation and $V = [v_1, \cdots, v_{n-1}] \in R^{3N \times n-1}$ by a vector. When building a model, the number of scans used is expressed in *n*.

4.2 Preprocessing of 3-D Face Posture Based on Eyes

The alignment algorithm SDM is used to align 67 alignment points in face based on binocular pose preprocessing. The position of nose, eye and mouth corner of face in SDM can be used as locating marker points. SDM defines an objective function, which is minimized by an algorithm. Solving this problem by a linear method is the core of SDM. See formula (6).

$$f(x_0 + \Delta x) = \|h(d(x_0 + \Delta x)) - \phi_*\|^2 \tag{6}$$

In the formula, the coordinates of p feature points are represented by $d(x) \in R^{p*1}$, the non-linear feature extraction function at each feature point is represented by h, and the feature in the three-dimensional face model is the feature of traditional 128-dimensional SIFT, i.e. $128\,h(d(x)) \in R^{128p*1}$, the optimal solution of manual calibration is represented by x_*, and the feature $\varphi_* = h(d(x_*))$ of SIFT is extracted at x_*. The solution of Δx can be regarded as the detection of facial feature points and the minimum objective function can be obtained.

After face alignment, there are six alignment points in each eye through SDM. The maximum x, y, minimum x and y values of the six alignment points are calculated. Through two coordinate points $(x_{\min}, y_{\min})$ and $(x_{\max}, y_{\max})$, in the interior of the rectangular frame, the eyes can be included. Through the intersection of four points of the rectangular line, the pupil position can be obtained. The two pupils are drawn together. The angle of two eyes is calculated by formula (7). In the formula, the left pupil coordinate is x_{left}, y_{left} a and the right pupil coordinate is x_{right}, y_{right}.

$$angle = \arctan\left(\frac{y_{right} - y_{left}}{x_{right} - x_{left}}\right) \tag{7}$$

The attitude of the face is adjusted by affine transformation from the acquired angle. Affine transformation is carried out by transforming coordinates and image transformation. After the image is input, its coordinates are mapped to the output picture coordinates. Figure 2 is a binocular-based face posture preprocessing.

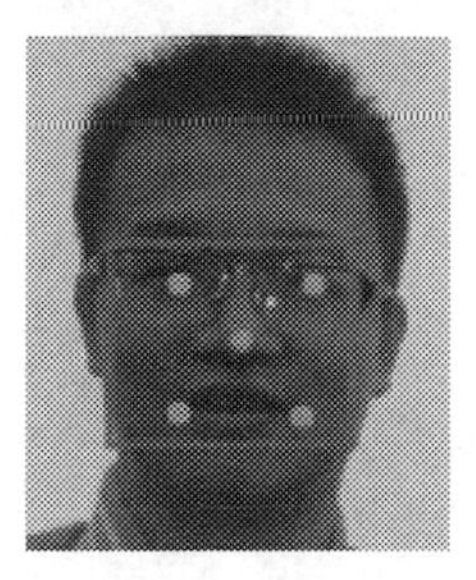

Fig. 2. Eye pose preprocessing based on binocular

5 Application of Support Vector Machine in Target Tracking

5.1 Training Process

The application of support vector machine to track and verify moving objects usually includes image classification and training process. In training, the sample images with known classification results are input and preprocessed. The preprocessed image features are extracted by feature extraction method and used as input data of SVM learner. By adjusting the kernels and parameters, optimizing the possible dimensions and deviations of the vector machine, optimizing the selection mode of the classifier, and finally realizing the classification of input training samples, the accuracy of the classifier is significantly improved.

5.2 Implementation of Dynamic Face Tracking

When studying k frames, the selected skin color model is mainly hybrid. The skin color of the image is input by segmenting. Based on the model, the skin color threshold is checked to get the candidate region of face. The optimized Sobel operator edge detection method is adopted to detect the candidate region and SNOW classifier is used to locate the candidate region. According to the previous results, face information is detected, and the rectangular area of face is finally obtained; the rectangular area of frame $k - 1$ is compared with the rectangular area of face, and the motion state of the rectangular area of face in the current frame is obtained by linear prediction; FWT algorithm and Haar wavelet transform are applied to enhance the rectangular area image. At the same time, SVM classifier is used to verify face information. If the corresponding face information can be detected, it can be labeled as a face area; if the corresponding face information can not be detected, it can be considered that the target is lost, tracking and detecting $k + 1$ frame within the allowable loss time range. If the lost time is beyond the allowable range, the tracking can be completed and a new round of face recognition, tracking and detection can be started at the same time.

6 Analysis of Experimental Results

On the basis of the algorithm flow given above, it realizes the accurate location and detection of face recognition and tracking, and realizes the automatic facial expression location combining face skin color and face movement. QC288 camera is used as image recognition tool for acquisition. The images are 640×480 resolution true color images. The circle labeled content is the face target in the image. Figure 3 is part of the result of face tracking. Through the automatic expression detection and tracking algorithm which combines the skin color of the face with the operation of the face in this paper, we can accurately detect the face. The skin color, facial expression change, face rotation and occlusion of face at small angle are measured and tracked.

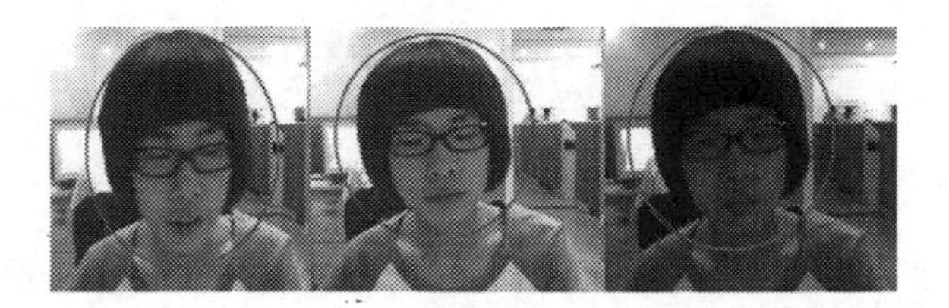

Fig. 3. Eye pose preprocessing based on binocular

7 Conclusion

In this paper, based on the combination of skin color and facial movement, the automatic expression recognition algorithm is studied. The image is transformed into YIQ color space by RGB, and image data is extracted in the first dimension of YIQ. Background and skin color are segmented from binary image. Pareto optimization algorithm is used to select facial expression features. This algorithm has less computation, simple structure and fast running speed. It can accurately detect and track the small-angle facial skin color, facial expression changes, facial rotation and occlusion. Experiments show that the proposed algorithm can be well adapted to face rotation at small angles. For the state of human eyes, the algorithm is not affected, and can adapt to the rich facial expression changes and different skin color better, and has a certain stability.

Acknowledgements. This research work is supported by the National Natural Science Foundation of China under Grant No. 61672024, 61170305 and 60873114, and National Key R&D Program of China (No. 2018YFB0904200 and 2018YFB2100500).

References

1. Gu, W., Liu, W., Zhu, Z., Xu, K.: A face detection method based on skin-color model and template matching. Microcomput. Appl. **30**(7), 13–16 (2014)
2. Cao, H.: Research on Face Recognition Algorithms Based on Geometric Features. QUFU Normal University (2015)
3. Ying, M., Guanzheng, T., Zhentao, L.: Review on facial expression recognition based on video image. J. Hunan Univ. Arts Sci. Technol. (2016)
4. Lian-Qiang, N., Zi-Tian, Z., Sheng-Nan, Z., et al.: Extraction method for facial expression features based on Gabor feature fusion and LBP histogram. J. Shenyang Univ. Technol. (2016)
5. Yucheng, L., Kun, L.: PCA and SVM face recognition method based on human face segmentation. Microcomput. Appl. **15**, 16 (2016)
6. Bo, S., Yongna, L., Jihong, L., et al.: Facial expression feature extraction based on tensor analysis. Comput. Eng. Appl. (2016)
7. Pantic, M.: Facial expression recognition. In: Li, S., Jain, A. (eds.) Encyclopedia of Biometrics. Springer, Boston (2014). https://doi.org/10.1007/978-3-642-27733-7
8. Shan, C., Gong, S., Mcowan, P.W.: Robust facial expression recognition using local binary patterns. In: IEEE International Conference on Image Processing, ICIP 2005. IEEE (2005)

9. Liu, P., Han, S., Meng, Z., et al.: Facial expression recognition via a boosted deep belief network. In: 2014 IEEE Conference on Computer Vision and Pattern Recognition (CVPR). IEEE Computer Society (2014)
10. Zhu, Y., Fan, H., Yuan, K.: Facial expression recognition research based on deep learning (2019)
11. Ashir, A.M., Eleyan, A., Akdemir, B.: Facial expression recognition with dynamic cascaded classifier. Neural Comput. Appl. **5**, 1–15 (2019)
12. Kuo, C.M., Lai, S.H., Sarkis, M.A.: Compact deep learning model for robust facial expression recognition. In: 2018 IEEE/CVF Conference on Computer Vision and Pattern Recognition Workshops (CVPRW) (2018)

Automatic Orange Fruit Disease Identification Using Visible Range Images

Vladimir Peter, Muhammad Asim Khan, and Huilan Luo(✉)

School of Information Engineering,
Jiangxi University of Science and Technology, 34100 Ganzhou, Jiangxi, China
vlad.edna@gmail.com, m.asimkhattak@gmail.com,
luohuilan@sina.com

Abstract. China is among the top producers of Navel Oranges in the world. However, the yield of these economically valuable fruits is affected by citrus diseases which cause a reduction in the amount and quality of the produced oranges. Traditionally, farmers rely on human experts to scout the plantations in order to spot infected fruits and identify the diseases. Scouting an entire plantation is a time-consuming task. Also farmers have to incur financial costs in order to pay the experts, who are not always readily available. This study proposes an automatic system for the identification of diseases from images of infected fruits taken by conventional digital cameras. The proposed approach has 5 general steps, background removal using Hough transform for orange shape detection, segmentation of symptoms via thresholding, selection, and extraction of features, and finally, training and classification by majority voting of 3 classifiers, namely, KNN, Random Forest and Multiple Support Vector Machine. The approach is evaluated on 3 diseases of the navel orange fruits namely Citrus canker, Citrus melanose, and Citrus black spot, achieving 93% accuracy using global color histogram, Local Binary Patterns, and Halarick texture features.

Keywords: Fruit symptoms · Background removal · Automatic disease identification · Color histograms · Texture classification

1 Introduction

China is among the top producers of the navel orange fruits worldwide. But citrus diseases such as citrus canker which affect navel oranges cause a reduction in the amount and quality of produced oranges. The symptoms of these diseases can be visually detected as they tend to affect the shape or color of the fruits, leaves, stems and other sections of the plants, and farmers have to detect citrus diseases early before they spread across the plantations. To do this, the conventional method is ground scouting, where human experts who can spot and identify citrus diseases scout the plantations in order to monitor the fruits, leaves, and trees. However, this process is labor-intensive and hence costly, both in terms of money and time [29] and is not entirely reliable because such experts are not readily available. Due to these challenges, researchers have worked to propose more advanced techniques.

K. Li et al. (Eds.): ISICA 2019, CCIS 1205, pp. 341–359, 2020.
https://doi.org/10.1007/978-981-15-5577-0_26

Technological advancements have resulted in the availability of consumer digital cameras as well as smartphone devices embedded with high-quality cameras at a relatively cheap price. This presents an opportunity for research on disease identification by analysis of conventional photographs of leaves and fruits in the visible range. This subject has gained the interest of many researchers (e.g. [2, 17, 26, 32]) in the field. The approach combines digital image processing techniques to analyze the images and machine learning techniques to classify, and hence identify the diseases. This approach generally consists of three steps; symptom segmentation, feature extraction and finally, training and classification. However, researchers have encountered many challenges in each of these steps.

The first and most challenging step is symptom segmentation. Given an image of a possibly infected fruit, this step aims to remove any background elements available as well as healthy regions of the fruit. This step is challenging due to mainly two reasons. First, background elements are complex [4] and vary from one image to another, making it hard to use automated techniques to separate the background from the region of interest (the fruit). For example, most fruits are green when unripe. An image of a fruit taken in the field will probably consist of leaves and other tree elements in the background, which are also green. In this case, using color to differentiate the background from the fruit will be very difficult. Thus, most researchers have opted to manually remove the background [9, 12]. Secondly, the disease symptoms tend to not have a well-defined boundary [4] making it hard to separate the healthy region from the infected areas. A common method for isolating the infected region from the healthy region is clustering based on color, using k-means clustering [12, 19, 28] or by fuzzy-c-means clustering [1, 5] techniques. These methods segment the image into different clusters, with each cluster having pixels that are more similar in terms of color. Researchers then analyze these clusters to identify which contains most of the infected symptoms depending on how the disease affects the fruits' color. Due to the challenges of this step, most techniques for disease identification are not fully automatic, requiring a manual step for removal of the background and healthy region.

The second step is feature extraction. Feature extraction prepares data for the training of machine learning classifiers. Since human experts look for color, shape and texture changes on fruits as symptoms manifest themselves in order to identify the disease, these are the same features researchers use. Thus, this step involves the extraction of color, shape as well as texture information encoded in the output images of the first step of symptom segmentation. The step also comes with its challenges. Mainly, most symptoms vary in appearance during different stages of the disease and secondly, symptoms of two or more diseases may have similarities in terms of how they affect color or shape or texture of the fruit.

Lastly, the extracted features are used to train machine learning classifiers or neural networks. Different researchers have used different classifiers such as Multiple Support Vector Machines (MSVM) [10], Random Forests [33] and K-Neural Networks (KNN] [19], and achieved different results because each classifier has its strengths and weaknesses depending on the application.

This study proposes an automatic system for the identification of diseases of the navel orange fruits. The proposed system enhances the general framework followed by previous researchers, by introducing a reliable method for symptom extraction using shape detection techniques and color space manipulation, followed by the determination and subsequent extraction of best features for a particular set of diseases, and lastly, training multiple classifiers, MSVM, KNN and Random Forests, in order to apply majority voting for classification. The approach is tested on 3 common diseases of navel orange fruits, Citrus canker, Citrus melanose, and Citrus black spot, highlighted next (Table 1).

Table 1. Three common citrus diseases affecting navel orange fruits.

Citrus canker	*Citrus black spot*	*Citrus melanose*

Citrus canker
Symptoms: Brown spots on fruits, often with an oily or water-soaked appearance. The spots (technically called lesions) are usually surrounded by a yellow halo.

Treatment: Control for citrus canker is better achieved through prevention. Once a lesion is spotted, it cannot be treated.

Citrus black spot
Symptoms: Can appear as any of 4 types of lesions on the fruit, varying in color from red to brown to black as well as shape.

Treatment: While control measures exist, black spot is not curable. Infected trees are normally removed.

Citrus melanose
Symptoms: Brown spots appear on the fruit, which grows together and starts to crack. As water from rain or other sources drips down the fruits, the spots also travel along it causing a stain.

Treatment: Citrus melanose is treatable with fungicides.

In addition to being common, these diseases were selected due to the similarities in how their symptoms manifest on the fruits. This means a more robust approach is required to efficiently classify and identify the diseases.

The next section offers a review of related literature. Section 3 follows next, presenting the proposed approach for the identification of the diseases, while Sect. 4 presents the dataset used and detailed analysis of the system's performance. This study concludes and highlights the limitations of the proposed approach in Sect. 5.

2 Literature Review

The emergence of new and advanced digital image processing and machine learning techniques, combined with the ubiquitous presence of digital cameras at affordable prices, has sparked many researchers to work on the problem of plant disease identification. The urgency of the problem is a result of the fact that these diseases result in lower yields, affecting the agriculture sector economically and the conventional method of scouting is time-consuming and costly [29]. Generally, the researchers use a 3 steps approach. First, segmentation is done to extract symptoms, then features such as color, texture or shape are extracted for the final step of training and classification using a machine learning model.

[9] used K-means clustering to remove symptoms from images of apple fruits, infected with blotch, rot, and scab diseases, and extracted complete local binary pattern features. They then used these features to train a Multiclass SVM (MSVM) classifier and achieved up to 93% classification accuracy. In their later work, [10] introduced an improved sum and difference histogram (ISADH) texture feature calculated from intensity values of neighboring pixels, which achieved near 99% accuracy when combined with the gradient filters proposed by [6, 14] used a similar approach to identify diseases of the pomegranate fruit, using K-means clustering for symptom segmentation and MSVM for classification. However, they achieved their best accuracy, 82%, by using shape features. In 2016, [11] combined color, texture and shape features the blotch, rot, and scab diseases of apple fruits and achieved 95% accuracy. They concluded that combining these features resulted in better performance than when the features were used separately. However, despite the good accuracy results, the proposed approach in all these studies required an extra manual step for background removal. Removal of background is an important preprocessing step required to ensure that noisy data is reduced for the remaining steps i.e. symptoms and feature extraction.

It is worth noting, that this 3 steps framework has also been applied to detect diseases in leaves of the plant [2, 31, 32]. For instance, [31]'s work focused on diseases affecting mango leaves. They used K means clustering technique to segment symptoms and used statistical features such as standard deviation and variance to train the MSVM classifier. These features gave good results for diseases that affected the coloring of the leaves, however, some diseases caused shape deformation, and different features were required. While the Open source computer vision library (OpenCV) they used was suitable for the statistical features, there was no implementation for the Elliptic Fourier analysis (EFA) technique used to describe leaf shape. They thus had to use the non-free Matlab library, which resulted in a slight increase in the cost of their system despite having better accuracy.

[4] reviewed the challenges in developing an automatic plant disease identification system using images taken in real conditions of the field. These challenges included the presence and complexity of background elements as well as illumination variations as a result of different capture conditions. Other challenges are a result of intrinsic factors, such as variations of symptoms of the same disease at the different stages of the disease, similarity of symptoms of different diseases and presence of more than one disease in the fruit. These challenges make it difficult to accurately remove the background and extract symptoms without manual intervention.

Due to these challenges, instead of manually removing the background, some researchers have used deep learning techniques for training and classification to improve the robustness of the system, instead of simple machine learning methods. However, deep learning techniques require a very large training dataset which is not always available and not easily obtainable. For instance, [23], required 13,689 images of diseased apple leaves to achieve 97.62% classification accuracy. These images were obtained by expanding a dataset of images taken from two provinces in China, through data augmentation techniques including image direction manipulation by rotation and adjusting the brightness, sharpness and contrast values of the original image.

[3] proposed an algorithm for separating infected regions in images of leaves from healthy regions. The algorithm manipulated the color histograms of the images taken from the h-channel and a-channel of the HSV and LAB color spaces. They observed that the symptoms of the diseases they investigated appeared brighter in a-channel and darker in the h-channel compared with the non-infected (healthy) regions, which were darker in the a-channel and brighter in the h-channel. The approach thus offered a semi-automatic way of extracting symptoms. The only manual step was the selection of which channel provided seemingly better results. Once the best channel was selected, a simple thresholding technique was applied to identify the healthy region. The value of the threshold was computed using a predetermined formula. This study uses the observation that symptoms of a disease may appear darker or brighter compared to the healthy regions in grayscale images extracted from color channels of HSV or Lab spaces to propose a fully automatic approach to disease identification of orange fruits.

3 The Proposed Approach

This study proposes a framework for the identification of orange fruit diseases as presented in Fig. 1 below. This framework is an extension of the general framework used by previous researchers.

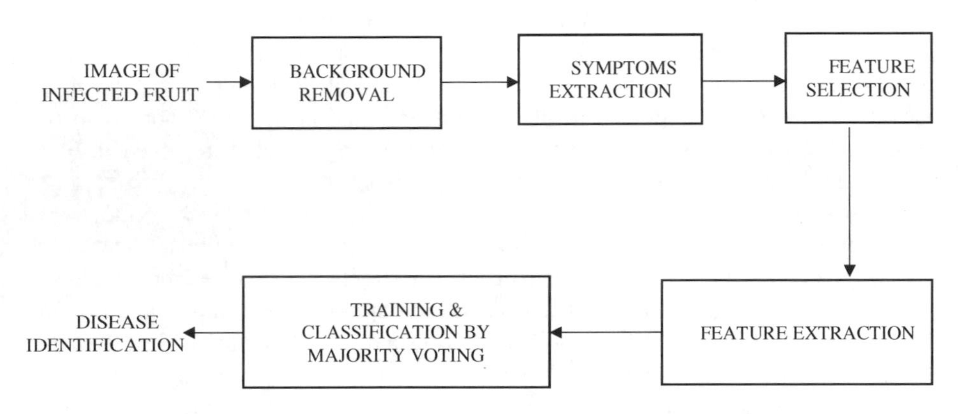

Fig. 1. Proposed framework

3.1 Background Removal

The proposed approach uses a key observation made on the collected images of infected orange fruits, including those not used in the dataset.

Observation 1: *The fruit tends to be the focus of the image and is thus located closer to the center of the image.*

Observation 2: *The shape of orange fruit is approximately circular in 2D space (as in images. It is spherical in 3D space).*

With these two observations, the algorithm used to remove the background elements from the images of infected fruits is developed and implemented in Python programming language using the Open-source Computer Vision (OpenCV) library. Below is the pseudo-code for background extraction and symptom extraction as these steps are performed subsequently in the code.

PseudoCode: Algorithm background removal and symptom extraction

```
program Background Removal & Symptoms extraction
  %input  = Image of infected fruit in BGR
  %output = Images of symptom-regions in BGR, Lab, gray and b* channel space

  import openCv as cv2,
         numpy as np,
         matplotlib as mpl
  mpl.use('TkAgg')

  begin
       const  og_bgr_img = (input)
       const  threshold_cut = 0.75;
       %resize the image
       og_rows, og_cols, dimension = np.shape(og_bgr_img)
       resized_rows = int((480 * og_rows) / og_cols)
       resized_img_bgr = cv2.resize(og_bgr_img, (480, resized_rows))
       % convert the resized image to lab, gray and extract b* channel
       lab_img = cv2.cvtColor(resized_img_bgr, cv2.COLOR_BGR2LAB)
       gray_img = cv2.cvtColor(resized_img_bgr, cv2.COLOR_BGR2GRAY)
       b_channel_img = lab_img[:, :, 2]
       % increase the contrast
        clahe = cv2.createCLAHE(clipLimit=4.0, tileGridSize=(8, 8))
        enhanced_img_b = clahe.apply(b_channel_img)
       % blur the image with bilateral filter to keep edges
        blurred_img_b = cv2.bilateralFilter(enhanced_img_b, 9, 75, 75)
       % find circles using HOUGH TRANSFORM
       circles = cv2.HoughCircles(blurred_img_b, cv2.HOUGH_GRADIENT, 1,
          minDist=120, param1=50, param2=5, minRadius=100, maxRadius=220)
        circles = np.uint16(np.around(circles))
       % get the best circle i.e the first circle
        i = circles[0, :][0]
        x, y, r = i[0], i[1], i[2]
       % create a mask of all zeros
        mask = np.zeros((resized_rows, 480), dtype=np.uint8)
       % change mask values that are within the detected circle to 1
        cv2.circle(mask, center=(x, y), radius=r, color=(1, 1, 1),
                   thickness=-1)
       % Background Removal STEP
       % get only pixels within the circle
       orange_pixels = []
       orange_pixels_pos_rows = []
       orange_pixels_pos_cols = []
       % do for all i.e $img = b_channel_img, lab_img, gray_img,…
       %… resized_bgr_img
       i = 0
       repeat
          i := i + 1;
          j = 0
            repeat
               j := j + 1;
                 if mask[i, j] == 1 % preserve $img[i,j] pixel color
                    orange_pixels.append(blurred_img_b[i, j])
                    orange_pixels_pos_rows.append(i)
                    orange_pixels_pos_cols.append(j)
                 else $img[i,j] := 0 % convert to 0 as background pixel
            until j = 480
       until i= resized_rows
     % get only pixels above threshold => Symptom Extraction
     % do for all i.e $img = b_channel_img, lab_img,…
```

```
        %… gray_img,resized_bgr_img
        md = np.median(orange_pixels)
        threshold = threshold_cut * md
          i = 0
          repeat
             i := i + 1;
             if orange_pixels[i] > threshold
                $img[i,j] := 0 % convert to 0 as background pixel
          until i= orange_pixels.size
        % return b_channel_img, lab_img, gray_img, resized_bgr_img
end
```

The width value of 480 and the parameters min-radius (100) and max radius (220), of the HoughCircles method, were selected as optimal after multiple experimentations with different values. Also, the use of the b channel of the Lab color space was selected after observing that the symptoms were much darker in the b channel, relative to the brighter healthy region (see Table 3). The use of Hough Transform to detect approximate circles in an edge-enhanced gray image was first introduced by [18] and has been used to detect fruit region in other research works [7, 15, 27, 30] due to its robustness

Table 2. Sample results of background removal step

Before background removal	*After background removal*
	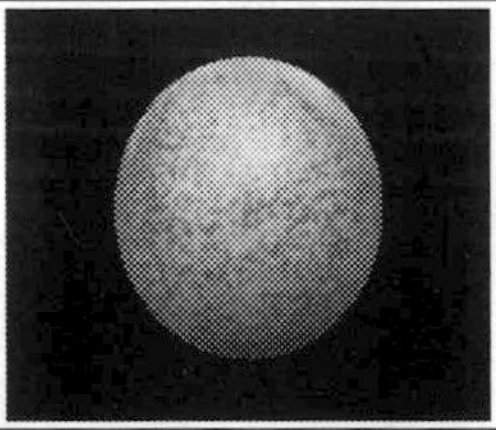
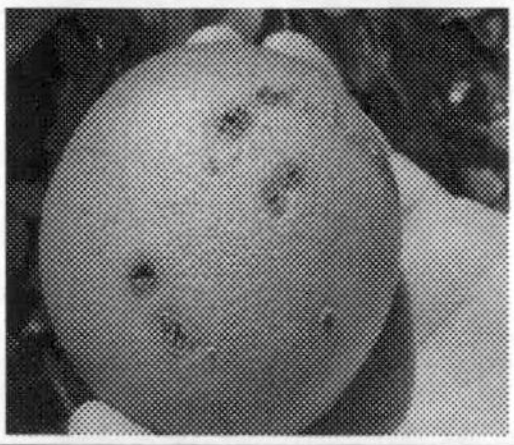	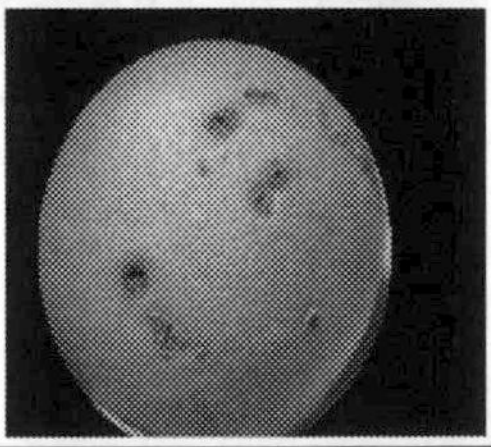

against noise and ability to detect even non-perfect circles [20]. The CLAHE algorithm, used to increase contrast, works by dividing the image into smaller regions called tiles before applying histogram equalization to each tile [27]. Table 2 shows sample outputs after the background removal algorithm is complete.

3.2 Symptom Extraction

Transforming the image to Lab was done first because the color information in Lab space is described using only two channels (a and b). L expresses light intensity, ranging from black to white. While a* expresses green to red color range and b*, blue to the yellow color range. Secondly, the Lab space is adapted to human color perception [16], which means that Lab was designed such that the numerical change in its values corresponds to approximately similar amount of visually perceived change.

On close observations of images in the a and b channel, however, it was found that the b channel is such that symptoms appear darker than the healthy region of the fruit. This difference, in how symptoms and healthy regions of fruits and leaves, has been observed by other researchers before [3].

Symptom extraction is hence possible by using a technique called thresholding. Thresholding is a simple technique whereby pixels which do not meet a pre-determined value are removed from the image. Since symptoms appear darker, their value is hence closer to 0 relative to brighter pixels. After experimentation with different thresholds to obtain the optimal threshold value, the threshold was obtained as;

$$Threshold = \frac{3}{4} \times median(b - channelpixels).$$

Pixels above this threshold were discarded as healthy regions and those that remained constitute the symptom region.

Table 3 below shows sample results of symptom segmentation and also highlights how symptoms of the citrus diseases appear darker in the b channel of Lab space while the healthy region appears brighter.

3.3 Feature Selection

Feature selection is a two-phase process. The first is done initially when determining which features to experiment with. The second phase is done after experimentation, where the best performing features with their corresponding classifiers are determined and coupled together to create the final system.

Different diseases manifest differently on the fruits (as well as other plant regions). It is crucial, therefore, to select features that best represent the diseases being identified. Citrus canker, citrus black spot and citrus melanose symptoms affect the color of the orange fruit as well as the texture but not the shape. Thus intuitively, shape features can be discarded in favor of color and texture features.

To extract color features, color histograms are used. The color histogram can be visualized as a graph or plot that keeps a count of pixel values in the image. It can be global, that is representing all pixels in a given image, or local, in which case only a part of the image is considered.

Table 3. Sample results of symptom segmentation

Infected fruit image in RGB	*Infected fruit Image in b channel*	*Extracted Symptoms*
	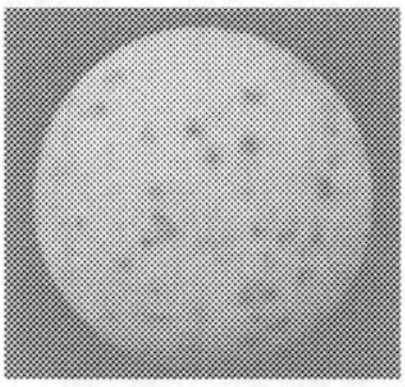	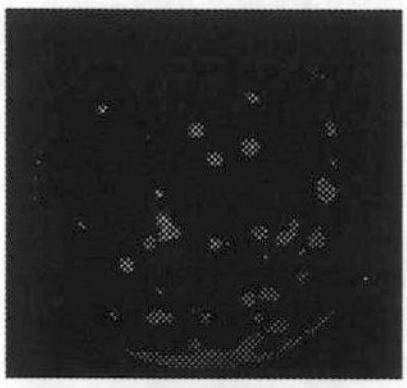
	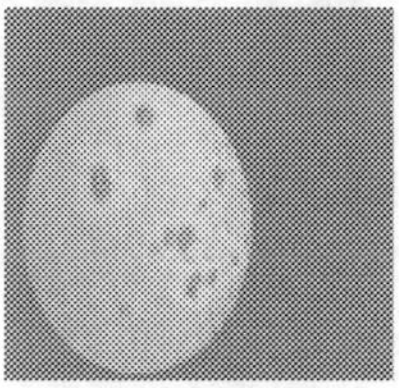	
	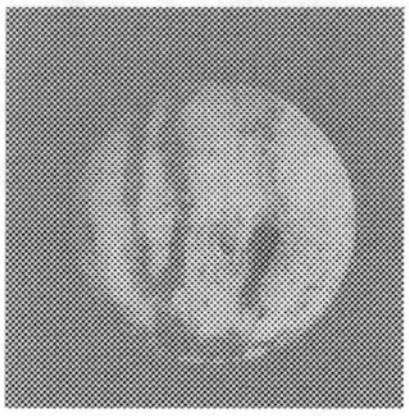	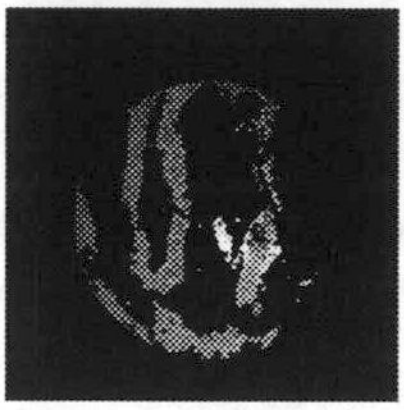

Local binary pattern (LBP) and Haralick features are used for texture representation. Haralick texture features provide a global representation of texture by computing a gray level co-occurrence matrix while LBP provides a local representation by comparing each pixel with its neighbors.

The python implementation of these features is available in the open-source libraries, OpenCV and Mahotas, and due to their popularity, their implementations are continuously tested and updated to make them more robust and error-proof.

3.4 Feature Extraction

During symptom segmentation, background pixels were encoded as black pixels (with values 0,0,0 in RGB space), while black pixels within the fruit region were encoded by adding 1 to their values (1,1,1 for RGB space) so that they are not to be confused with the background pixels in this next step, i.e. feature extraction.

In this step, global color histogram, local binary pattern and Halarick texture features are extracted from segmented images of the symptoms. For comparison, Scale Invariant Feature Transform (SIFT) features are also extracted, as they have been used by other plant disease detection researchers (e.g. [8, 11, 34]). SIFT features find key points, local regions that can be detected even if the scale, illumination or viewpoint of the image is changed, and provides descriptors that characterize these key points.

The OpenCV library provided methods for extraction of color histograms and SIFT features, while the Mahotas library was used to extract the texture features because its methods allowed the specification of black-pixels as background to be ignored. Both of these libraries are free to use.

The pseudocode for Halarick, LBP, and GCH features extraction algorithm is provided below.

```
program Feature Extraction
  %input := segmented symptoms image in grayscale, rgb, lab and b-channel
  %output := Halarick, LBP and GCH features and encoded label
  %used libraries
      Mahotas
      OpenCV as cv2
      from sklearn.preprocessing,
               LabelEncoder
               MinMaxScaler
  begin
      symptoms_img = cv2.imread(INPUT_IMG)
      img_variant = symptoms_img_color_variant
      img_label = disease_name

      %get lbp features for the 2D images
      if img_variant == grayscale OR img_variant == b-channel
           lbp_features =  mahotas.features.lbp(symptoms_img, radius=1,
                                        points=8, ignore_zeros=True)

      %get other features
      Halarick_features = mahotas.features.haralick(symptoms_img,
                                ignore_zeros=True, return_mean=True)
      %color histogram requires a mask to ignore-zeros as background
           ...values
      mask[i,j] = 1 if symptoms_img[i,j] > 0 else mask[i,j] = 0
      % compute the gray scale histogram
      bins = 16
      if img_variant == grayscale OR img_variant == b-channel
           GCH = cv2.calcHist(symptoms_img, [0, 1, 2], mask, [bins, bins,
                                bins],[0, max_channel_0_value,
                                0,max_channel_1_value,
                                0, max_channel_2_value])
      else
           GCH = cv2.calcHist(symptoms_img, [0], mask, [bins],
                                                            0, max_channel_0_value)
      % normalize the histogram
      cv2.normalize(GCH, GCH)
      GCH = GCH.flatten()
      % encode the labels
      encoded_labels = LabelEncoder.fit_transform(img_label)
      % normalize the feature vector in the range (0-1)
      scaler = MinMaxScaler(feature_range=(0, 1))
      Using GCH, Haralick, LBP as features
              rescaled_features = scaler.fit_transform(features)
      return rescaled_features and encoded_labels
end
```

The extraction of SIFT features requires a different approach. This utilizes a bag-of-words model that treats the features as words in a vocabulary, after which a histogram is created to represent the occurrence of each word in a particular image. This algorithm is provided below:

Algorithm for SIFT feature extraction using OpenCV.

Step 1. Using OpenCV, extract sift features and descriptors for the dataset images.
Step 2. Using 500 as the number of words, use OpenCV's Bag-of-Words K-Means trainer to cluster the feature descriptors.
Step 3. Set a vocabulary from the extracted features.
Step 4. For each symptoms image, extract sift features and descriptors for that image.
Step 5. Match the feature descriptors with the vocabulary created in step 4.

Step 6. Use the matching bag-of-words as bins to create a histogram

3.5 Training and Classification

The last step is training and classification, and this study used three classifiers, Random Forests, MSVM, and KNN that have been used in many related works (e.g. [27, 28]).

A Random Forest uses 'n' number of decision trees to fit the training set, such that each tree in the forest votes to categorize a randomly selected sub-sample from the set to the most probable class. Because the sub-samples to train each tree are selected randomly, this classifier has the advantages of being more robust to noise as well as less prone to overfitting. The majority vote from the trees is taken as the predicted class thus making the output of random forest classifier more accurate. Increasing the number of trees used improves the stability and accuracy of the prediction but also increases the processing time. Therefore, tuning this parameter to find an optimal value is a necessary and vital step during experimentation.

Support Vector Machines (SVM), represent the training samples as points in space, mapped such that there is a clear separating gap, called a hyperplane, between members of two classes based on the closeness in distance between the points. MSVM extends the binary SVM to a multi-class problem such that given 'n' number of classes;

$$m = \frac{n \times (n-1)}{2} \ \textit{classifiers}.$$

Where 'm' is the number of binary SVM classifiers to be trained using a 'one-vs-one' or 'one-vs-rest' approach. In a 'one-vs-all' approach, one class is treated as a positive class and all the others are treated as the negative class. The problem with this approach is that it is inefficient due to an imbalance in the number of training samples for each class. The 'one-vs-one' approach instead trains each of the 'm' classifiers on a unique set of binary classes. SVM is effective in high dimensional spaces, as is the case with features extracted from images, even when the number of dimensions is greater than the available samples.

KNN predicts a class by grouping K number of samples within the input data that are closest to each other, distance-wise. The number of groups and hence classes depend on the value of K provided. The algorithm is popular due to its simplicity and robustness. However, the performance of the system largely depends on the value of K provided as well as nature of the data points. The larger the value of K, the more stable and accurate the predictions, to a point, after which, the predictions become erroneous.

The scikit-learn python library implements these classification algorithms and was used in this study. In order to improve the accuracy of the classification results, majority voting by classifiers was done. Majority voting works by training three or more different classifiers and then taking the prediction that the majority of the classifiers give.

4 Experimental Results

4.1 Dataset Preparation

The dataset used to validate the proposed approach consisted of images of orange fruits infected with either of 3 diseases: citrus canker (100 images), citrus black spot (96 images) and citrus melanose (81 images). Some of these images were obtained online while others were taken from orange plantations in Ganzhou. The total number N of images used was hence 277. To make the sample more realistic, the images used had complex and varying background elements. The Fig. 2 below shows a few of the images used.

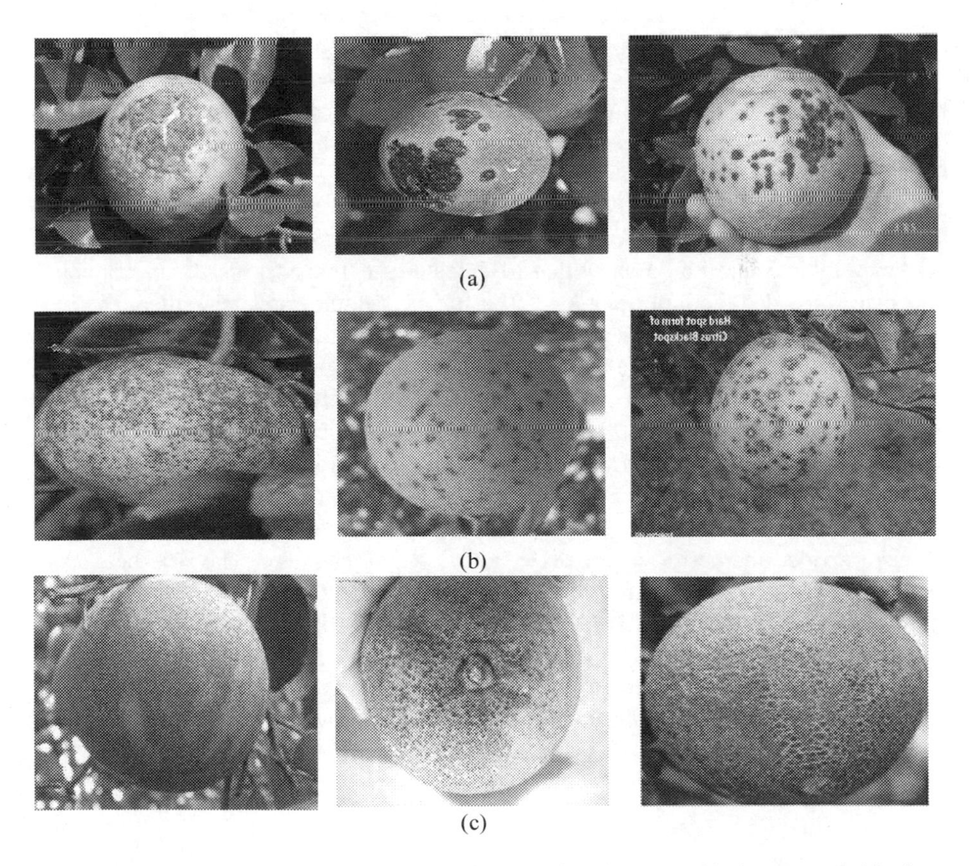

Fig. 2. Sample images in the dataset, (a) citrus canker (b) citrus black spot and (c) citrus melanose.

4.2 Analysis of Results

The dataset was split into training samples and testing samples, using 228 images for training (76 images per class) and 49 images for testing (18% of all the images in the dataset).

The symptom images which were initially in RGB form were also converted to 1 extra 3-dimension color space, Lab as well as two 2-dimension color spaces, gray and b (the b-channel in Lab).

Then 2 kinds of features were extracted from the RGB and Lab images, namely Haralick texture features and global color histograms (color features).

It was possible to extract 2 extra kinds of features for the gray and b images, namely Local binary pattern and SIFT features because these features supported only 2-dimension images.

The extracted features were then used to train 3 classifiers, Random Forests (with 10 trees), MSVM and KNN (with k = 50).

The performance results for each of the classifiers are presented in figures below, given that the system accuracy was defined as;

$$Accuracy(\%) = \frac{Total\,number\,of\,images\,correctly\,classified}{Total\,number\,of\,images\,used\,for\,testing} \times 100.$$

The Tables 5, 6 and 7 below present the performance of the approach for Random Forest, MSVM and KNN respectively. Each table providing a summary of the performance results is accompanied by a figure that presents a graphical view of the results in order to quickly highlight the best and lowest accuracies along with the extracted features and color channels used.

First, Table 5 and Fig. 3 show that LBP features extracted from the b-channel of Lab space provided the best accuracy, 67%, for the random forest classifier. The next best accuracy, 60%, was obtained from color features of Lab images, while the lowest accuracy, 13.5%, was from the color features extracted from images in grayscale.

Then, Table 6 and Fig. 4 show that color features extracted from gray images provided the best accuracy, 80%, for the MSVM classifier. The next best accuracy, 73.11%, was obtained from Haralick texture features from Lab images, while the lowest accuracy, 33%, was from the Haralick texture features of RGB images.

Lastly, Table 7 and Fig. 5 show that Haralick texture features extracted from Lab images provided the best accuracy, 81%, for the KNN classifier. The next best accuracy, 73.33%, was obtained from the Haralick texture features of gray images, while the lowest accuracy, 33%, was from the SIFT features of both gray and b-channel images.

Overall, KNN gave the best performance followed by MSVM, both achieving over 80% accuracy. It is also clear that different color spaces provided different classification accuracy. Features wise, SIFT features gave poor performance overall (Table 4).

Table 4. Performance of the system using random forest classifier

Image color type	Extracted features			
	Haralick	Global color histogram	SIFT	Local binary pattern
RGB	46.4	53.3		
L*a*b*	18.67	60		
Gray	53.33	13.5	33.11	53.33
b* (of Lab)	40	46.56	53.33	67

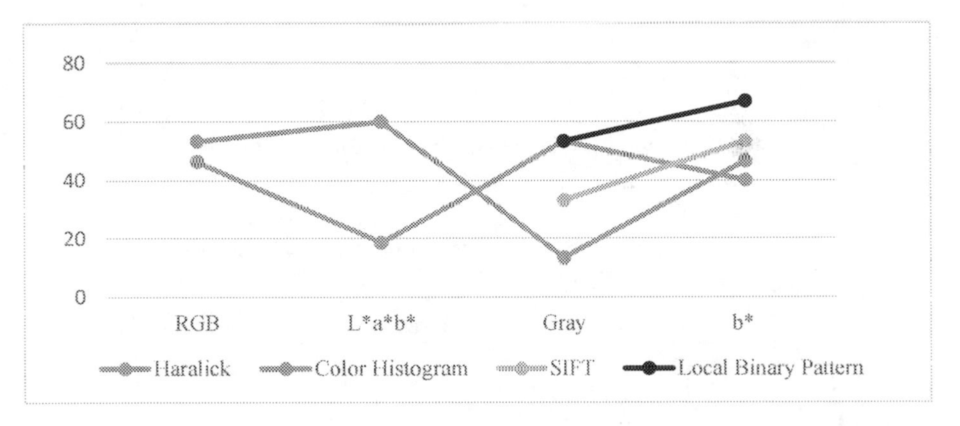

Fig. 3. Performance of the system using random forest classifier

Table 5. Performance of the system using MSVM classifier

Image color type	Extracted features			
	Haralick	Global color histogram	SIFT	Local binary pattern
RGB	33.33	53.3		
L*a*b*	73.11	33.45		
Gray	66.67	80	53.33	40
b* (of Lab)	53.33	40	46.78	33.33

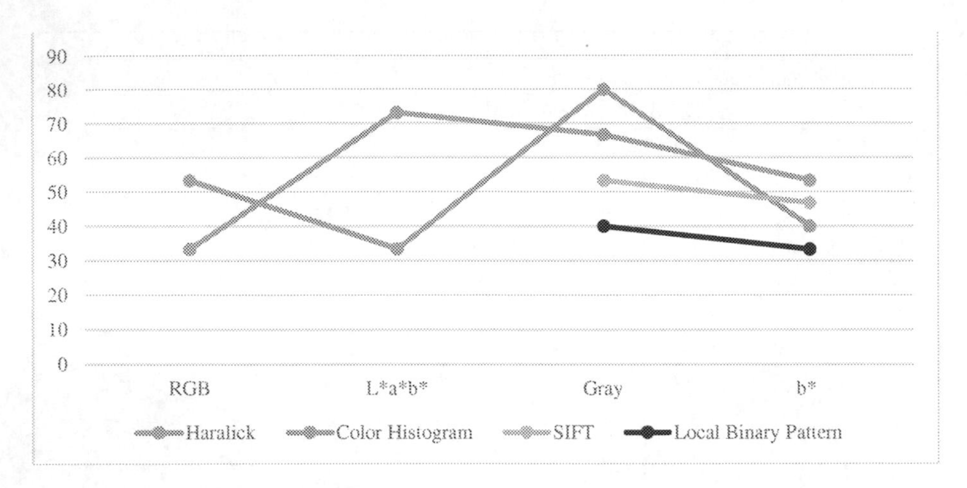

Fig. 4. Performance of the system using MSVM classifier

Table 6. Performance of the system using the KNN classifier.

Image Color Type	Extracted features			
	Haralick	Global Color Histogram	SIFT	Local Binary Pattern
RGB	46.66	60		
L*a*b*	81	33.45		
Gray	73.33	53.33	33.33	53.33
b* (of Lab)	56.6	53.33	33.33	60

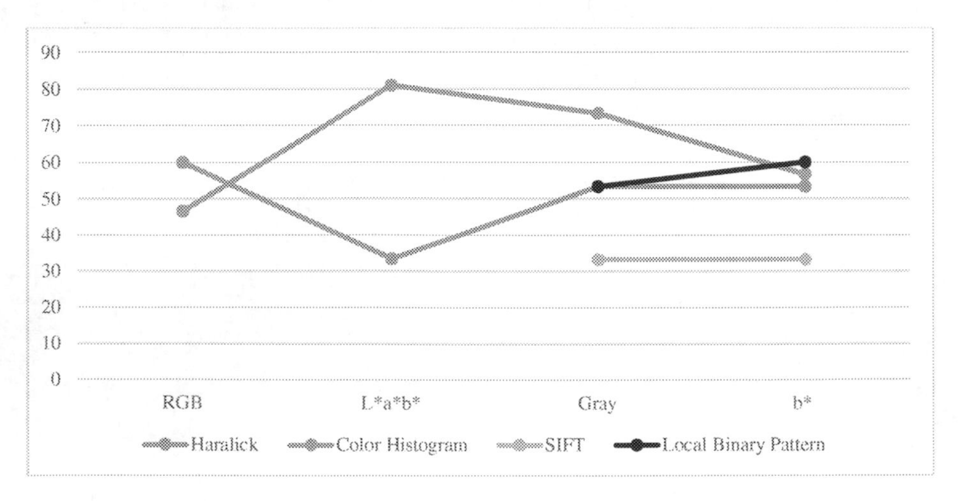

Fig. 5. Performance of the system using the KNN classifier.

Table 7. Best performance observed for each classifier

Classifier	Features extracted	Image color type	Accuracy (%)
KNN classifier (K = 50)	Haralick features	Lab	81
Random Forest Classifier (10 trees)	Local binary pattern	b from Lab	67
MSVM	Color histogram	Gray	80

The following table shows the features that provided the best accuracies for each of the classifiers.

To increase the performance of the system, the majority of voting was applied. The algorithm used was as follows:

Algorithm for classifier voting.

Step 1. Convert symptom images to gray and Lab color spaces.
Step 2. Get b channel of the Lab color space.
Step 3. Extract Haralick features from Lab images, Local binary pattern features from b images and color histograms from the gray images.
Step 4. Train the KNN classifier using Haralick features, Random Forest classifier using local binary pattern features and MSVM using color histograms.
Step 5. Using respective features for each trained classifier, use the classifiers to predict given test data.
Step 6. Take the prediction provided by the majority (at least 2) of the classifiers. In case they all disagree; the prediction of the Random Forest classifier was taken. This is because the random forest classifier is the one that deviates more from predictions of KNN and SVM.

Through majority voting, classification accuracy was increased to 93.33%.

5 Conclusion and Future Work

An automatic system for the identification of citrus diseases affecting the navel orange fruit has been proposed. The steps begin with background removal using circles detection by Hough transform to capture the orange region, followed by symptoms segmentation using a simple thresholding technique. In the third step, two texture features, Haralick and Local binary pattern, are extracted along with color histograms. The fourth and final step uses these features to train and classify 3 classifies, Random Forests, MSVM, and KNN. 3 diseases affecting the navel orange fruit, citrus canker, citrus black spot, and citrus melanose were used to test and validate the proposed system. The approach achieved 80% accuracy using a single classifier, which was improved to 93% through majority voting of three classifiers.

The limitations of this approach are that it works best when the orange fruit being checked is located closer to the center of the image and when the image is taken at a close distance to the fruit. Also, the approach works for a pre-selected set of diseases.

This study also proposes further work to be done. First, this approach was verified using only 3 diseases of the navel orange fruits. Other works could test it on other fruits and/or other diseases. Also, more images of infected fruits should be collected and made available as an open-source dataset for researchers, in order to validate and improve the proposed approach.

References

1. Ali, I., Ahmed, A.: Segmentation of different fruits using image processing based on fuzzy C-means method. In: 2018 7th International Conference on Reliability, Infocom Technologies and Optimization (Trends and Future Directions) (ICRITO), August 2018, pp. 441–447. IEEE (2018)
2. Barbedo, J.G.A.: An automatic method to detect and measure leaf disease symptoms using digital image processing. Plant Dis. **98**(12), 1709–1716 (2014)
3. Barbedo, J.G.A.: A novel algorithm for semi-automatic segmentation of plant leaf disease symptoms using digital image processing. Trop. Plant Pathol. **41**(4), 210–224 (2016)
4. Barbedo, J.G.A.: A review on the main challenges in automatic plant disease identification based on visible range images. Biosys. Eng. **144**, 52–60 (2016)
5. Behera, S.K., Jena, L., Rath, A.K., Sethy, P.K.: Disease classification and grading of orange using machine learning and fuzzy logic. In: 2018 International Conference on Communication and Signal Processing (ICCSP), April 2018, pp. 0678–0682. IEEE (2018)
6. Bhange, M., Hingoliwala, H.A.: Smart farming: Pomegranate disease detection using image processing. Procedia Comput. Sci. **58**, 280–288 (2015)
7. Changyi, X., Lihua, Z., Minzan, L., Yuan, C., Chunyan, M.: Apple detection from apple tree image based on BP neural network and Hough transform. Int. J. Agric. Biol. Eng. **8**(6), 46–53 (2015)
8. Dandawate, Y., Kokare, R.: An automated approach for classification of plant diseases towards development of futuristic decision support system in Indian perspective. In: 2015 International conference on advances in computing, communications and informatics (ICACCI), August 2015, pp. 794–799. IEEE (2015)
9. Dubey, S.R., Jalal, A.S.: Adapted approach for fruit disease identification using images. Int. J. Comput. Vis. Image Process. (IJCVIP) **2**(3), 44–58 (2012)
10. Dubey, S.R., Jalal, A.S.: Fruit disease recognition using improved sum and difference histogram from images. Int. J. Appl. Pattern Recognit. **1**(2), 199–220 (2014)
11. Dubey, S.R., Jalal, A.S.: Apple disease classification using color, texture and shape features from images. SIViP **10**(5), 819–826 (2016)
12. Dubey, S.R., Dixit, P., Singh, N., Gupta, J.P.: Infected fruit part detection using K-means clustering segmentation technique. Ijimai **2**(2), 65–72 (2013)
13. Dubey, S.R., Jalal, A.S.: Detection and classification of apple fruit diseases using complete local binary patterns. In: 2012 Third International Conference on Computer and Communication Technology, pp. 346–351. IEEE (2012)
14. James, A.P.: One-sample face recognition with local similarity decisions. Int. J. Appl. Pattern Recognit. **1**(1), 61–80 (2013)
15. Kadir, M.F.A., Yusri, N.A.N., Rizon, M., bin Mamat, A.R., Makhtar, M., Jamal, A.A.: Automatic mango detection using texture analysis and randomised hough transform. Appl. Math. Sci. **9**(129), 6427–6436 (2015)
16. Kasson, J.M., Plouffe, W.: An analysis of selected computer interchange color spaces. ACM Trans. Graph. (TOG) **11**(4), 373–405 (1992)
17. Khadke, L., Jadhav, S., Jadhav, K., Kasbe, A., Gudadhe, S.: A Survey on Identification of Plant Diseases. Image **4**(8) (2017)
18. Kimme, C., Ballard, D., Sklansky, J.: Finding circles by an array of accumulators. Commun. ACM **18**(2), 120–122 (1975)
19. Lamani, S.B.: Pomegranate Fruits Disease Classification with K Means Clustering (2018)

20. Lestriandoko, N.H., Sadikin, R.: Circle detection based on hough transform and Mexican Hat filter. In: 2016 International Conference on Computer, Control, Informatics and its Applications (IC3INA), October 2016, pp. 153–157. IEEE (2016)
21. Li, H., Lee, W.S., Wang, R., Ehsani, R., Yang, C.: Spectral angle mapper (SAM) based citrus greening disease detection using airborne hyperspectral imaging. In: Proceedings of 11th International Conference on Precision Agriculture. Monticello, Ill: International Society of Precision Agriculture, October 2012
22. Li, J., Rao, X., Ying, Y.: Development of algorithms for detecting citrus canker based on hyperspectral reflectance imaging. J. Sci. Food Agric. **92**(1), 125–134 (2012)
23. Liu, B., Zhang, Y., He, D., Li, Y.: Identification of apple leaf diseases based on deep convolutional neural networks. Symmetry **10**(1), 11 (2017)
24. Long, X., et al.: Flavonoids composition and antioxidant potential assessment of extracts from Gannanzao Navel Orange (Citrus sinensis Osbeck Cv. Gannanzao) peel. Natural product research, 1–5 (2019)
25. Lu, Z.J., et al.: The effects of inarching Citrus reticulata Blanco var. tangerine on the tree vigor, nutrient status and fruit quality of Citrus sinensis Osbeck 'Newhall'trees that have Poncirus trifoliata (L.) Raf. as rootstocks. Sci. Hortic. **256**, 108600 (2019)
26. Masood, R., Khan, S.A., Khan, M.N.A.: Plants disease segmentation using image processing. Int. J. Mod. Educ. Comput. Sci. **8**(1), 24 (2016)
27. Murillo-Bracamontes, E.A., Martinez-Rosas, M.E., Miranda-Velasco, M.M., Martinez-Reyes, H.L., Martinez-Sandoval, J.R., Cervantes-de-Avila, H.: Implementation of Hough transform for fruit image segmentation. Procedia Eng. **35**, 230–239 (2012)
28. Ranjit, K.N., Chethan, H.K., Naveena, C.: Identification and classification of fruit diseases. Int. J. Eng. Res. Appl. (IJERA) **6**(7), 11–14 (2016)
29. Sankaran, S., Mishra, A., Ehsani, R., Davis, C.: A review of advanced techniques for detecting plant diseases. Comput. Electron. Agric. **72**(1), 1–13 (2010)
30. Sengupta, S., Lee, W.S.: Identification and determination of the number of green citrus fruit under different ambient light conditions. In: International Conference of Agricultural Engineering CIGR-AgEng2012, July 2012
31. Sethupathy, J., Veni, S.: Opencv based disease identification of mango leaves. Int. J. Eng. Technol. **8**(5), 1990–1998 (2016)
32. Sujatha, S., Saravanan, N., Sona, R.: Disease identification in mango leaf using image processing. Adv. Nat. Appl. Sci. **11**(6 S), 1–8 (2017)
33. U.S. Department of Agriculture: Citrus: World Markets and Trade (2017). http://apps.fas.usda.gov/psdonline/circulars/citrus.pdf. Accessed 8 Nov 2017
34. Zawbaa, H.M., Hazman, M., Abbass, M., Hassanien, A.E.: Automatic fruit classification using random forest algorithm. In: 2014 14th International Conference on Hybrid Intelligent Systems, December 2014, pp. 164–168. IEEE (2014)

Orange Leaf Diseases Identification Using Digital Image Processing

Irene Anney Joseph, Muhammad Asim Khan, and Huilan Luo(✉)

School of Information Engineering, Jiangxi University of Science and Technology, Ganzhou 34100, Jiangxi, China
gissahirene@gmail.com, m.asimkhattak@gmail.com, luohuilan@sina.com

Abstract. A new framework is proposed to identify disease symptoms in navel oranges' leaves using images obtained in real conditions of the field. The approach proposed consists of the following steps, the first step is leaf symptoms segmentation using Grabcut algorithm to remove background elements and thresholding to remove healthy region, then color, texture and SIFT features were extracted and used to train KNN, Random Forest and MSVM classifiers.

The approach is then tested on identification of two common diseases affecting the navel oranges of Jiangxi; Citrus Greening and Citrus Canker images, and the accuracy achieved was 91.1% on Haralick features using images in Lab color space by Random Forest and K-Neural Network classifiers.

Keywords: Orange leaf diseases · Symptoms segmentation · Feature extraction · Digital image classification

1 Introduction

Agriculture plays a crucial role in the economic growth of many countries worldwide. Jiangxi, has traditionally been the top orange producing province in China, which ranks third globally in terms of annual orange production, after U.S.A and Brazil. However, Jiangxi has experienced stagnant production over the past five years. This is largely due to the negative impact of the citrus greening disease. The disease which has no cure so far, caused millions of greening disease-infected orange trees to be removed in Jiangxi province.

Before the development of technology in identifying plants diseases, farmers have been relying on experts to search through their plantations to identify diseases. This is an expensive and time method. Not to mention it is subject to bias and hence not reliably accurate. However, with the advancement of image processing techniques and machine learning technologies, there have been several proposed methods for identification of plant diseases that simplify the process, aimed at saving time and money while offering reliable accuracy. One area of research that has gained much interest recently, is in disease identification by analysis of conventional photographs of leaves and fruits in the visible range [3, 11, 14, 16, 29, 35]. This is the least expensive and most accessible technology, as the prices of digital cameras continue to drop, and most mobile devices include cameras that provide images with acceptable quality. The main

K. Li et al. (Eds.): ISICA 2019, CCIS 1205, pp. 360–378, 2020.
https://doi.org/10.1007/978-981-15-5577-0_27

framework used by researchers follows three main steps; defect segmentation, feature extraction, and training & classification.

The task of defect segmentation proves the most challenging and most advances have mostly been limited to cases in which the conditions, both in terms of disease manifestation and image capture, are tightly controlled, in-depth [5]. An interesting observation pointed out by [4] is to consider that in most images taken under real field conditions, the leaf will be more focused and centered than the background items. Researchers have also shown that different color spaces provide different accuracies in capturing the features of the infected region [5, 8, 15]. The extraction of features is also challenging due to the variations in the way disease symptoms manifest themselves. For instance, 8 the citrus greening disease causes asymmetrical blotchy mottling of leaves while citrus canker disease appears as raised spongy lesions on leaves, twigs, and fruit, which gradually increase in size to 5–10 mm over several months, eventually collapsing to form a crater-like surrounded by characteristic yellow halos. Thus, the feature extraction method must account for color changes, texture, shape, and size. For this, other researchers proposed using the improved sum and difference histogram from an image, which can be enhanced with gradient filters achieving nearly 99.9% classification accuracy in their experiments. The SVM Random Forest and KNN algorithms were used for training and classification [15].

The method proposed in this paper aims to provide an automatic orange leaf diseases recognition system that is applicable under the real condition of the fields. The method adapts the general framework used by other researchers and proposing fully automatic and improved procedures going through all the challenges observed from other researchers works for every step as shown in the framework. This method will help the farmers identify the orange diseases early rather than relying on a few experts available to scout through the large plantation to identify the diseases. This method will avoid the bias and inaccuracy when human try to identify the diseases on Orange Orchards through naked eye observation of seemingly the infected regions.

2 Related Work

The first step in the general framework of the research is background removal, for this step several researchers have opted to manually remove the background from the images [13, 22]. Others have avoided this issue altogether, by ensuring that the leaf is isolated prior to taking the photograph [24]. Placing a panel behind the leaf removing the leaves and placing them in Petri dishes [29] as well as using specially designed devices for the capture [6, 11] are just some of the methods of achieving this.

However, some researchers used some methods for background removal for instance Graph Cuts by developing a robust algorithm for "border matting" to estimate simultaneously the alpha-matte around an object boundary and the colors of foreground pixels [14]. Guiding Active contour for Tree Leaves Segmentation and Identification designed a light polygonal leaf model and later used to guide evolution of an active contour to combining global shape descriptors given by the polygonal model with local curvature-based features to classify the leaves.

For segmentation of the infected region from the leaf, the manual extraction of diseased parts is subject to bias and hence error-prone, since it is difficult to accurately pinpoint where the healthy part end and where the diseased part begins [16]. A computer algorithm was proposed to differentiate the signs and symptoms of plant disease from asymptomatic tissues in plant leaves. The simple algorithm manipulates the histograms of the H (from HSV color space) and a (from the L*a*b* color space) color channels. However, the segmentation was semi-automatic, requiring manual removal of the background as well as a manual final step of selecting the color space that captured the infected region more accurately [5].

Experts through naked eye observation of the leaves have traditionally identified plant leaf diseases. These experts look for changes in color and texture, as well as the shape and size of the seemingly diseased region. Researchers have done the same, trying to capture the color, texture, shape and size information during the feature extraction step. Others used color histograms [5], DCT features [36] and SIFT features for disease recognition [16].

Some researchers compared their results with the results of global color histograms (GCHs), color coherence vectors (CCV), and the descriptor concluded that the ISADH was far better.

They also concluded that using the SVM classifier [35] gave better results than using the KNN classifier [31]. SVM classifier was also used for automatic detection of plant leaf and diseases [16]. More sophisticated techniques have also been proposed, such as Markov Random Fields [13]. Graph Theory [6]. Deep Learning [15] Mean Shift [9] and Large Margin Nearest Neighbor (LMNN) classification [37].

3 Research Framework

The algorithm has four main stages, background removal, symptoms segmentation, feature extraction and finally training and classification. Figure 1 below presents the framework used in the proposed approach. In the initial step, the RGB images of all the leaf samples were collected. Some real samples of those Orange Leaf diseases are shown in Fig. 2 and 3. The next step is background removal. This step is followed by Symptoms segmentation whereby the healthy part of the leaf is removed. Feature extraction is the next step whereby color, textural and local features were extracted and finally training and classification which was done by SVM, Random Forest, and K-Neural Network Classifiers.

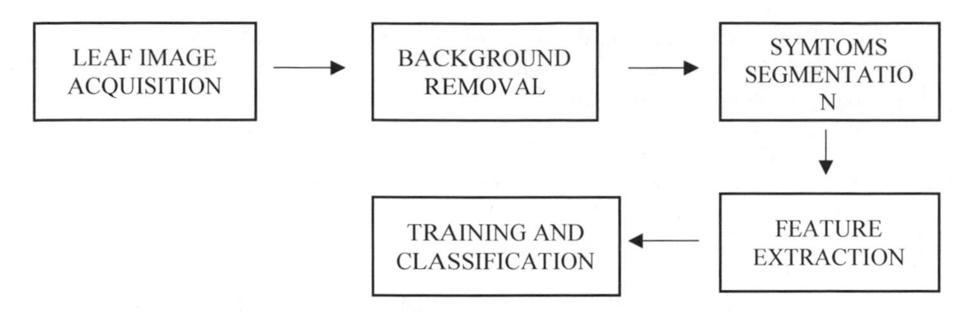

Fig. 1. Proposed research framework.

3.1 Leaf Image Acquisition

To identify the orange leaf diseases, a large collection of orange leaves is required. Most of the images were obtained by the smartphone camera taken in Orange Orchards located in Ganzhou, Jiangxi. Some of the images were downloaded in online-based databases. All the images were stored in RGB format. To prepare a dataset for training and classification all the images were resized to a width of 480 and a height of 360.

3.2 Background Removal

This is a process that eliminates the unwanted part of the image. Background removal was achieved by performing Grabcut algorithm [9]. This method starts with creating the specified masks around the leaf image, the algorithm estimates the color distributions of the leaf image and that of the background using a Gaussian mixture model. This is used to construct a random field over the pixel labels and running a graph-cut based optimization to deduce the values. This step repeats until convergence.

To achieve this the following observations were made, the orange leaf was the main focus of the image and it was located at the center of the image to get a uniform radius for the masks.

In this research's case, the following three masks were created:

1. GC_BGD that defines an obvious foreground, this was achieved by creating a circle of 50-pixel radius from the center of the image.
2. GC_PR_BGD that defines possible background pixels, the marking of all possible background was done by initializing each pixel value to 2.
3. GC_PR_FGD that defines possible foreground pixels. Marking a possible background was done by creating a circle with a radius of 90 pixels and initializing all pixel value to 1.

Several experiments were performed to obtain suitable radiuses for the masks.
Algorithm for background removal

1. Read input image and resize the image to 480 width which is the optimal width identified for leaf region detection.
2. Identify leaf section by using Grabcut technique thus creating masks.
3. Remove background pixels from the image.

Table 1 shows the orange leaf diseases before (original images) and after removing the background. The background of the image was turned to back color to differentiate it from the foreground thus only the leaf.

Table 1. Background removed leaf images.

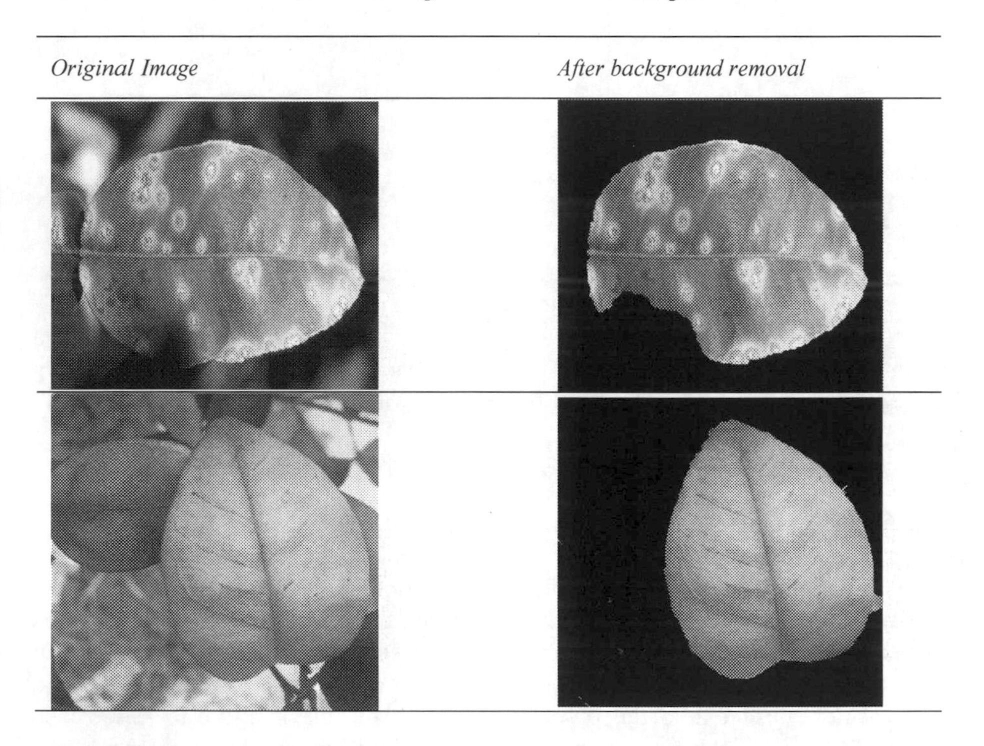

3.3 Symptoms Segmentation

This is a final stage of symptoms segmentation whereby the healthy part of an image has to be removed to leave the only diseased part for training. To achieve the symptoms segmentation various experiments with color space were performed [3]. The images were changed to Lab color space and channel a* and b* of Lab color space were obtained. The disease symptoms appeared to be brighter in channels a* and b* of Lab color space and the healthy parts appeared to be brighter. Through several experiments in obtaining the symptoms part of the leaf image, the threshold value was obtained. For channel A, 20% of brightest pixels were obtained and the darker part of the leaf which is the healthy part was removed and for channel B 5% of brightest values were obtained and the darker part of the leaf thus healthy part was removed. Lastly, erosion was done to remove the pixels that were not required. All the images were converted to four color spaces, i.e. RGB, grayscale, Lab, and b* color spaces. All the images were resized to a width of 480 and a height of 360.

$$threshold\ a^{*} = max - 0.2 \times max \tag{1}$$

$$threshold\ b^{*} = max - 0.5 \times max \tag{2}$$

Table 2 shows the results of experiments conducted when changing the color space from RGB color space to Lab and obtaining only a* channel, which shows the symptom part appear brighter than a healthy part of the leaf.

Table 2. Leaf images in a* color channel.

Infected fruit image in RGB	*After background removal*	*Infected fruit Image in a* channel*

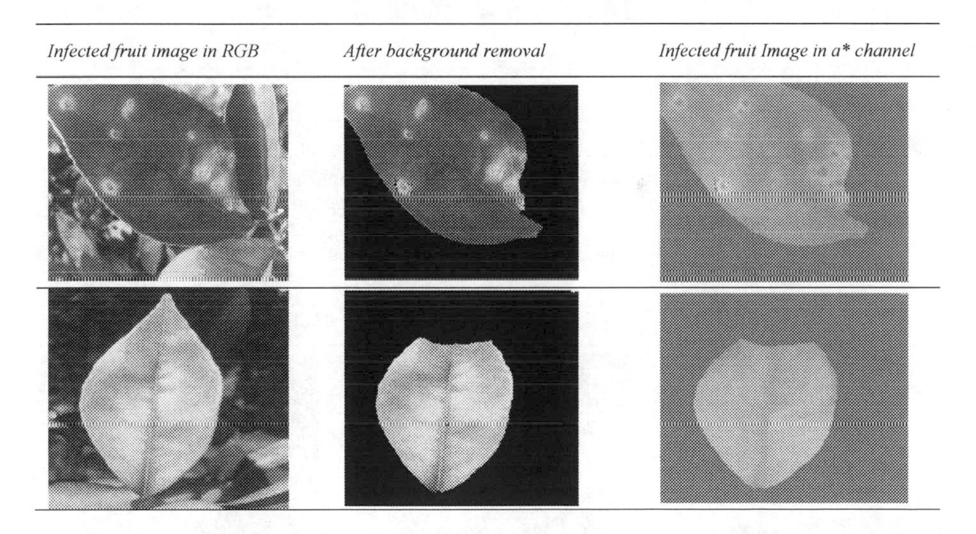

Table 3 shows the experiments conducted when changing the color space from RGB color space to Lab and obtaining only b* channel, which shows the symptoms part appear brighter than the healthy part of the leaf.

Table 3. Leaf images in b* color channel.

Infected fruit image in RGB	*After background removal*	*Infected fruit Image in b* channel*
		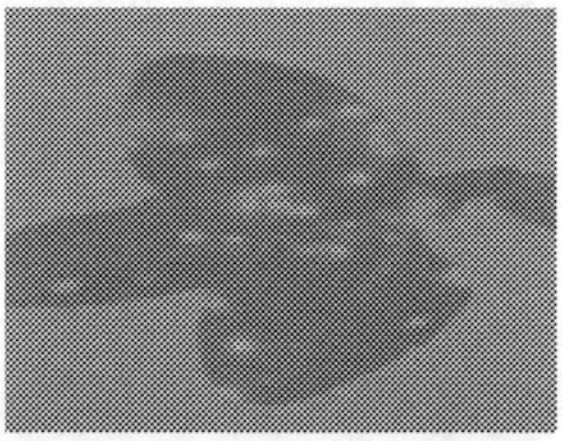
	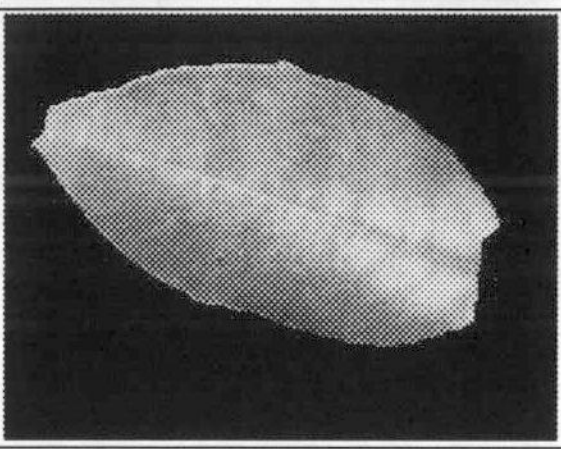	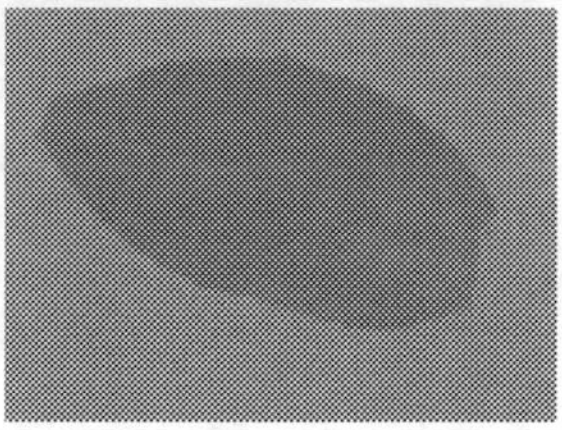

Table 4 presents the result after applying the thresholding by removing pixels under 20% of the brightest value for a* color channel and 5% for b* color channel. After thresholding, the disease symptom segmentations were obtained.

Table 4. Segmented leaf images.

Infected fruit image in RGB	*After background removal*	*symptom segmentations*
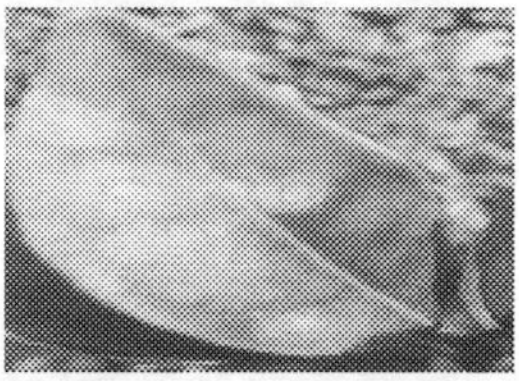	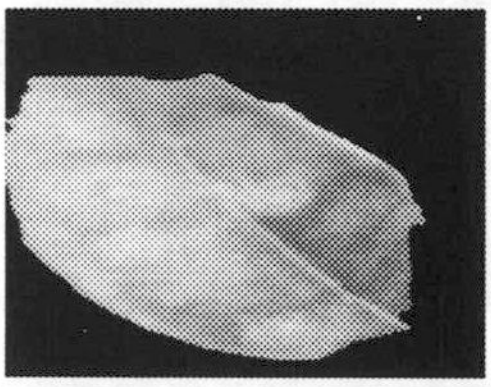	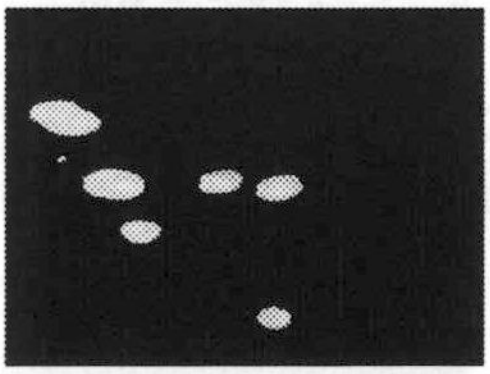
		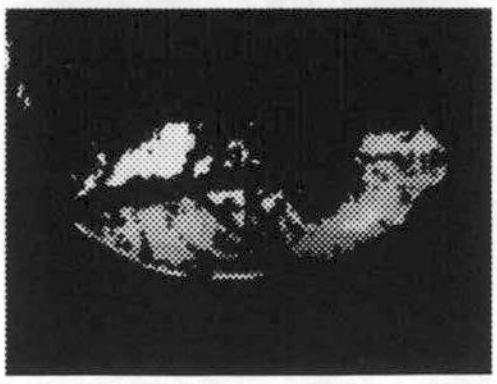

3.4 Feature Extraction

Feature extraction involves reducing the number of resources required to describe a large set of data. Feature extraction plays an important role in the classification of an image. In identifying orange leaf diseases human experts use color and texture features of a leaf to identify the diseases [19]. Considering this the same features were extracted to acquire the best results.

Both global and local features were used. Global feature represents an image as the whole object while local features describe the specific key-points of an image. For Global feature color and textural features were obtained using Color Histogram, Haralick Features and Local Binary Pattern (LBP) which is computed by comparing its neighbors [31, 16].

$$LPB_{N,R} = \sum\nolimits_{n=0}^{n-1} S(v_n - v_c)2^n, s(x) = \begin{cases} 1, x \geq 0 \\ 0, x < 0 \end{cases} \quad (3)$$

Where, N is the total number of neighbors, R is radius of the neighborhood, v_c is the value of the central pixel and v_n is the value of its neighbors.

Local feature extractor used was Scale-Invariant Feature Transform (SIFT). The feature extraction procedure was done for all four-color spaces, RGB, Lab, grayscale and b* from Lab color spaces. Due to the nature of the features, haralick features and color histogram were extracted using all the color spaces and for Local Binary pattern (LPB) and SIFT the features were extracted only in Grayscale and b* color space because the nature of the feature extracted only can work with 2-Dimension images.

3.5 Training and Classification

In this step, the extracted features were used to train machine learning classifiers, specifically, Support Vector Machine (SVM), Random Forest and K-Neural Network (KNN).

SVM is the training algorithm whose goal is to design a hyperplane that classifies all training vectors into two classes. The hyper plane selection depends on which will leave maximum margin for both classes. This method is used for both problems of classification and regression, however, SVM can be used from binary problems to multi-classification problems [16]. In this paper's this technique is used to train and classify the orange leaf disease. In Decision space information is separate in two classes as explained in the following figure.

Figure 2 shows the hyper plane and margin. Let set of n training data of two separable classes {(x1, y1), (x2, y2), …, (xn, yn)}, i = 1, 2, …, n, where xiϵ Rn is an n dimensional space and yi = ±1. Given a weight vector w and bias weight b, using the Eq. 4 and Eq. 5 presents the difference of hyper plane among two classes. This separation is called linear separation. Hyper plane can be defined by the Eq. 6

$$(wxi - b) \geq 1, \; if\, yi = 1 \quad (4)$$

$$(wxi -) \leq -1, \; if\, yi = -1 \quad (5)$$

$$w.xi - b = 0, \tag{6}$$

SVM tries to maximize the margin between these two classes by minimizing ½.||w||2. Quadratic optimization algorithms can identify which training points xi are support vectors with non-zero Lagrangian multipliers αi. The optimal solution problem is solved by the below equation.

$$L_d = \sum_{i=1}^{n} \alpha_i - \frac{1}{2} \sum_{i=1}^{n} \sum_{j=1}^{n} \alpha_i \alpha_j y_i y_j x_i x_j \tag{7}$$

To find out the decision function supporting vector of the above equation is used, and all other data are redundant. Generally, the real-world problems are not linear in nature. On-linear classification are necessary.

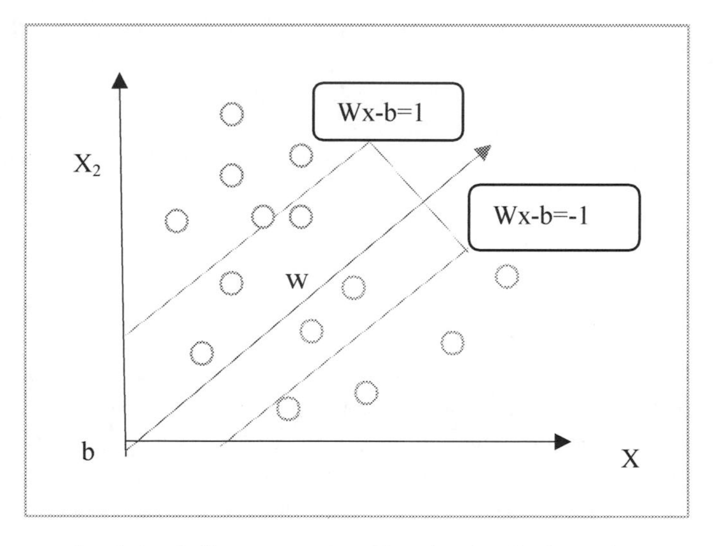

Fig. 2. An example of simple linear vector machine showing the hyper lane, sets of data and margin

Random Forest works as a large collection of decorrelated decision trees and use them to classify objects based on votes of all trees that is an object is given to a class that has most votes from all the trees. This method Random forest handles features by essentially binning them to the tree learners. Given a training dataset $D = d_1$ …., d_n with responses $E = e_1$ ……, e_n bagging repeatedly selects a random sampling replacement of the training set and fits the tree to these samples.

For b = 1, …, B, Sample, with replacement, n training examples from D *and* E call these D_b *and* E_b and train the classification or regression tree t_b on D_b *and* E_b. After the training process, predictions for unseen samples x' is created by calculating the average of the predictions from all the individual regression trees on x'

$$\hat{t} = \frac{1}{B}\sum\nolimits_{b=1}^{B} t_b(x') \tag{8}$$

It is also indifferent to non-linear features, can handle binary features, categorical features, and numerical features as well as it is fast and can work with high dimension since it splits data into subsets of data [8]. After several experiments performed 10 trees were found to be ideal and assigned to the images to classify the data

K-nearest neighbor (KNN) is one among supervised classifier. It stores all the data and classifies new data bases on the similarity measure. It suggest that if a data is similar to the neigbour then they are the same. Basically it classifies data according to how their neigbours are classified.

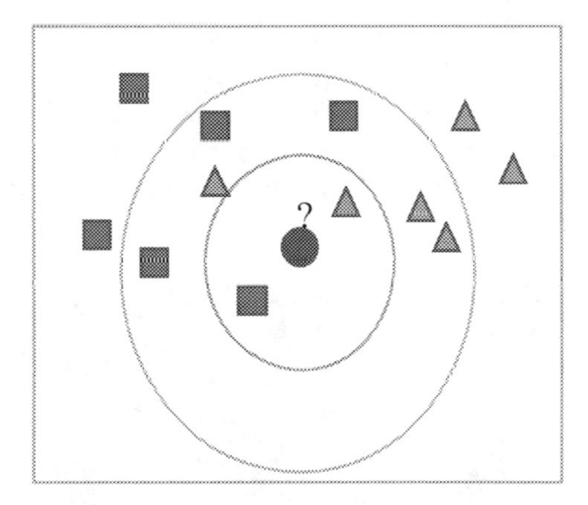

Fig. 3. Example of KNN classification. The test samples in the red circle should be classified to either green triangles or blue squares depending of the value of k which is user dependent value.

The training phase the algorithm contains storing features and class label of the training samples and in classification phase k (the user specified constant) and unlabeled vector is classified by assigning labels to the most frequent k training sample nearest to the query point.

Advantage of KNN is that it is well matched for multi-modal classes because its classification decision depends on a small neighborhood of similar objects. Good level of accuracy can be achieved only if the objective class is multi-modal. The negative aspect of the similarity measure being used within KNN is that it makes use of every one of features equally in calculating similarities. It can direct towards reduced likeness measures as well as classification faults when merely small division of the features is helpful for the purpose of classification. In this paper some experiments were performed to obtain the suitable k value and 50 was used as a k value.

4 Analysis of Results

4.1 Experimental Data

The image dataset was divided into two sets: training images and testing images, using 88% of the image dataset which makes 390 images for training (130 images per disease) and 51 images for testing. The segmentation was done in when images were in RGB (3-Dimension) format and later were converted to three more formats thus Lab (3-Dimension), b* from Lab (2-Dimension) as well as Gray Scale (2-Dimension), making four types of datasets with different color spaces. For Haralick features and Color Histogram, the features were extracted using all the color spaces and for Local Binary pattern (LPB) and SIFT the features were extracted only in Grayscale and b* color space because the nature of the feature extracted only can work with 2-Dimension images. The extracted features were then used to train 3 classifiers, Random Forests (with 10 trees), SVM and KNN (with k = 50).

This paper focuses on the following two common diseases of the navel oranges leaves. One is citrus canker. Citrus canker is a bacterial disease of citrus that causes premature leaf and fruit drop. Citrus canker is caused by the bacteria Xanthomonas citri. Citrus canker appears as raised spongy lesions on leaves, twigs, and fruit, which gradually increase in size to 5–10 mm over several months. Eventually, the lesions collapse forming a crater-like appearance. The lesions become surrounded by characteristic yellow halos, and the raised edges of the lesion may appear slimy (Fig. 4).

Fig. 4. Sample image of Citrus Canker infected leaf orange taken by a smartphone camera in Ganzhou Orange Orchard in Jiangxi

The other is citrus greening, also known as Huanglongbing (HLB) or yellow dragon disease, citrus greening infected trees produce fruits that are green, misshapen and bitter, unsuitable for sale as fresh fruit or juice. It has no known cure. The causative agents of citrus greening are motile bacteria, Candidatus Liberibacter spp. The disease is vectored and transmitted by the Asian citrus psyllid, Diaphorina citri, and the African citrus psyllid, Trioza erytreae which causes asymmetrical blotchy mottling of leaves (Fig. 5).

Fig. 5. Sample image of Citrus greening infected orange leaf taken by a smartphone camera in Ganzhou Orange Orchard in Jiangxi

The image dataset includes a total of 445 images, 151 for Citrus Canker, 147 for Citrus Greening Disease, and 147 images for random orange leaf diseases, healthy and other diseases. As mentioned earlier most of the images were captured by a smartphone camera in Ganzhou Orange Orchards and few were obtained from online based database. Of all the images 88% of the images were used in training and 12% was used in testing.

4.2 Performance Results

The results obtained from training the data in 4 different color spaces using 3 classifiers as explained above, the RGB color space images which were trained using haralick features and color histogram performed the best in haralick features obtaining 86.7% accuracy in all the classifiers and the last was color histogram using KNN classifier obtaining 64% accuracy. The Grayscale color space images trained using four features obtained the best accuracy using haralick features trained by KNN and color histogram trained by Random Forest both obtaining an accuracy of 80%, the least performance was SIFT features obtaining 33% accuracy by both SVM and KNN classifiers. Lab color space images achieved the best accuracy using haralick features obtaining 91.1%

in all the classifiers and color histogram using Random Forest, the last performance was color histogram using KNN which obtained 75% accuracy. And finally, b* color spaces obtained 86.7% accuracy using color histogram features trained by Random Forest classifier and the least result was 33% by SIFT features using the SVM classifier (Table 5).

Table 5. Performance and accuracy results.

	Haralick features			LPB			Color histogram			SIFT		
	KNN	Random forest	SVM	KNN	Random forest	SVM	KNN	Random forest	SVM	KNN	Random forest	SVM
RGB	86.7%	86.7%	86.7%				64%	86.7%	75%			
Gray Scale	80%	71%	71%	66%	75%	71%	68%	80%	71%	33%	46%	33%
Lab	91.1%	91.1%	91.1%				75%	91.1%	88%			
b*	77.7	73%	73%	77%	84%	71%	75%	86.7%	73%	35%	35%	33%

The accuracy was calculated by taking a total number of images correctly classified (tc) dividing by total number of images used for testing(tt) and multiplying to 100

$$Accuracy\ (\%) = \frac{tc}{tt} * 100 \tag{9}$$

The overall best classifier was Random Forest obtaining 91.1% for two features haralick features and color histogram as well as KNN obtaining 91.1% for one feature, haralick features. SVM came last obtaining 88% accuracy using color histogram (Table 6).

Table 6. The best performance observed for each classifier.

Classifier	Features extracted	Image color type	Accuracy (%)
KNN classifier (K = 50)	Haralick features	Lab	91.1%
Random Forest Classifier (10 trees)	Haralick features and Color histogram	Lab	91.1%
SVM	Color histogram	Lab	88%

4.3 Performance Results According to Color Space

The images were converted to 4 color spaces RGB, Grayscale, Lab and b* as mentioned earlier. The best color space for training the images was Lab color spaces since it obtained the best accuracy result in all the classifiers, and the least was Gray scale color space obtaining the least accuracy results in almost all the classifiers (Figs. 6, 7, 8 and 9).

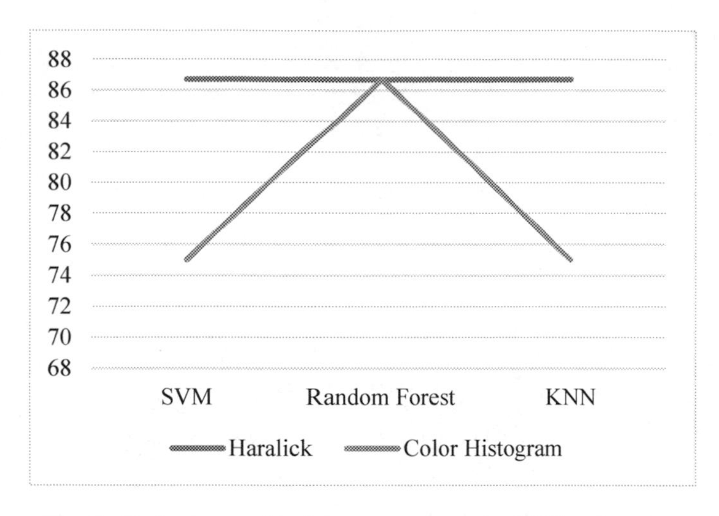

Fig. 6. Performance of training programs using RGB color space

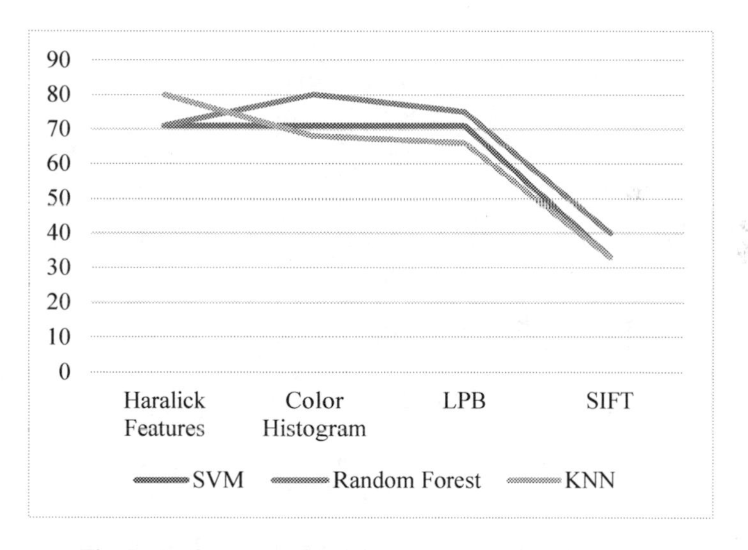

Fig. 7. Performance of training programs using Gray space

To summarize the overall performance, Random forest was the best by 91.1% accuracy using Haralick features and Color histogram, followed by KNN and SVM achieving 91.1% using Lab color space. Lab appears to be the best color space for training with Haralick feature since all the programs achieved the best results. For features, Haralick Features and Color Histogram were the best features to train and classify the diseases since they brought out the best results in all the training programs while SIFT features gave out the poor performance in all the training programs.

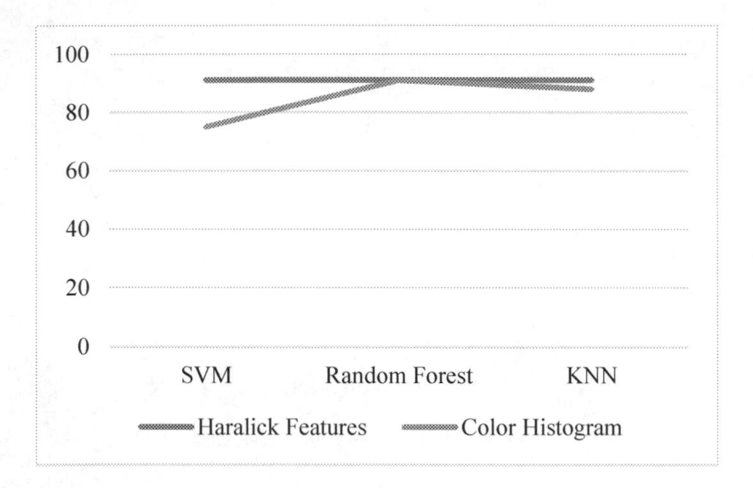

Fig. 8. Performance of training programs using Lab color space

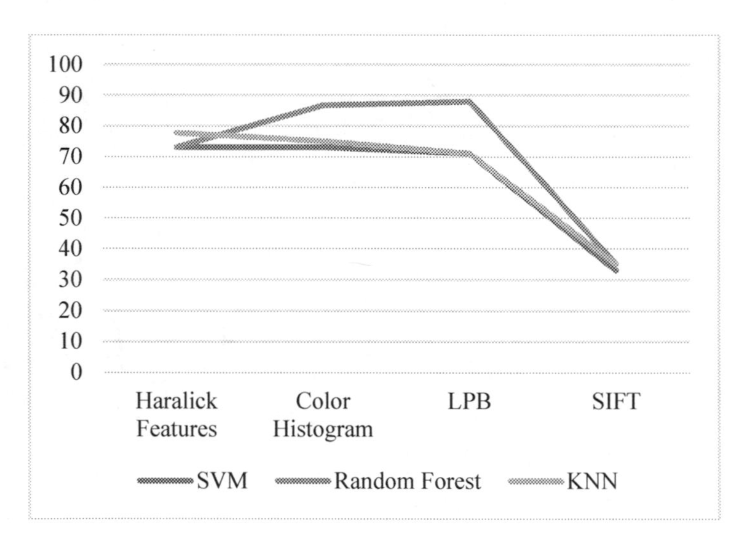

Fig. 9. Performance of training programs using b* color space

5 Comparative Study of Other Research Works

Dhabeed Al Bashish et al. [1] with their paper "A framework for detection and classification of plant leaf and stem diseases" proposed an image processing framework for classification of plant leaf and stem on early scorch, cotton mold, ash mold, late scorch and tiny whiteness diseases, using images taken with a digital camera they segmented plant images using independent color transformation structure and K means clustering. Feature extracted were color and texture features and lastly achieving an accuracy of 93% after training 32 samples per class by using back propagation neural networks.

Kruse et al. [22] in the publication Pixels classification method for identifying and quantifying leaf surface injury from digital images tested and compared four different approaches for classifying individual leaf pixels directly as healthy or injured from clover leaf RGB images with different degrees of ozone-induced visible injuries with the aim of determining which combination of feature vector and classifier provided the best all- around classifier. They used clover plants images taken by digital camera. The images were segmented by converting from RGB color Space to CIE L*A*B* color space and unfold of images, features used were color features of the pixels and its neighboring pixels controlled by a quadratic neighborhood windows with sides w of an odd number of pixels usually in the range 3–49 pixels and then classified using four different approaches namely Fit to a Pattern Multivariate Image Analysis (FPM-T2) acquiring an accuracy of 86%, Residual Sum of Squares (FPM-RSS) with an accuracy of 85%, linear discriminant analysis (LDA) getting accuracy of 95% and lastly K-means clustering and getting an accuracy of 93%.

Kim et al. [21] in Detection of Citrus Greening Using Microscopic Imaging using orange fruits introduced a detection method of citrus greening diseases with the images taken by microscopic imaging system, segmented the images using user-oriented ROI, feature used were color and texture features and classified using discriminating analysis getting and accuracy of 86.56% using only 60 samples each class.

Mohamed El Helly et al. [25] in the paper Integrating Diagnostic Expert System with Image Processing Via Loosely Coupled Technique acquired cucumber leaf images with powdery mildew, leaf miner, downey mildew diseases using high resolution camera with illuminati on light and segmented using Fuzzy C- Means Segmentation techniques and extracted mean of the red, green and blue channel, length of principal axes, eccentricity, diameter, compactness and Euler's number and then classified using Artificial Neural Network with number of training images 250 and testing 50 images for each disease getting an accuracy of 88.8%

Kulkarni and Patil [2] - Applying image processing to detect plants diseases, using alterneria anthracnose and bacterial blight pomegranate infected leaf images taken by a digital camera proposed an image processing method of diseases detection, they segmented the images using Conversion from RGB color Space to CIE L*A*B* color space and later extracted feature using Gabor filter technique and got 91% accuracy after classifying by back propagation neural network classifier.

Vipinad as et al. [36] in Detection and Grading of diseases in Banana leaves using Machine Learning paper used banana leaves infected with black sigatoka and panama wilt diseases to grade a leaf spot based disease segmenting the images by converting them from RGB color Space to YCBCR color space, extracting different features based on color shape and texture and used Support Vector Machine and Artificial Neural Network Fuzzy Interference System and getting 100% accuracy in ANFIS and 92% for SVM classifier.

Weizheng et al. [38] in Grading Method of Leaf Spot Disease Based on Image Processing using gray spots disease found in soybean plant leaves. The images were taken by a digital camera and later segmented by Otsu Thresholding and Sobel operator in hue component then diseased leaf area divided by total leaf area. Finally, plant diseases are graded by calculating the quotient of disease spot and leaf areas.

6 Conclusion

This paper has introduced an automatic method for identifying orange two orange leaf diseases for images obtained from real field condition Grabcut algorithm was applied for background removal, after which color manipulation in the lab color space was done to segment symptoms. Feature extracted were color features using Color Histogram, texture features using Haralick Features and Local Binary Pattern as well as local features using SIFT (Scale-Invariant Feature Transform). Finally, training and classification was done to classify the diseases using SVM (Support Vector Machine), Random Forest and K-Neural Network. The approach is then tested on identification of two common diseases affecting the navel oranges of Jiangxi; Citrus Greening and Citrus Canker images, and the accuracy achieved was 91.1% by Random Forest and K-Neural Network Lab color space appeared to be the best for training and classification since all the best results were obtained by using it.

References

1. Al Bashish, D., Braik, M., Bani-Ahmad, S.: A framework for detection and classification of plant leaf and stem diseases. In: 2010 International Conference on Signal and Image Processing, pp. 113–118. IEEE, Chennai (2010)
2. Anand, H.K., Aswin Patil, R.K.: Applying image processing technique to detect plant diseases. Int. J. Mod. Eng. Res. **2**(5), 3661–3664 (2012)
3. Barbedo, J.G.A.: An automatic method to detect and measure leaf disease symptoms using digital image processing. Plant Dis. **98**, 1709–1716 (2014)
4. Barbedo, J.G.A.: Digital image processing techniques for detecting, quantifying and classifying plant diseases. SpringerPlus **2**(1), 660 (2013)
5. Barbedo, J.G.A.: A review on the main challenges in automatic plant disease identification based on visible range images. Biosyst. Eng. **144**, 52–60 (2016)
6. Boese, B.L., Clinton, P.J., Dennis, D., Golden, R.C., Kim, B.: Digital image analysis of Zostera marina leaf injury. Aquat. Bot. **88**, 87–90 (2008)
7. Bondy, A., Murty, U.S.R.: Graph theory, 1st edn. Springer-Verlag, London (2008)
8. Breiman, L.: Mach. Learn. **45**, 5. https://doi.org/10.1023/A:1010933404324
9. Rother, C., Kolmogorov, V., Blake, A.: GrabCut: interactive foreground extraction using iterated graph cuts. ACM Trans. Graph. **23**, 309–314 (2004)
10. Camargo, A., Smith, J.S.: An image-processing based algorithm to automatically identify plant disease visual symptoms. Biosyst. Eng. **102**, 9–21 (2009)
11. Cheng, Y.: Mean shift, mode seeking, and clustering. IEEE Trans. Pattern Anal. Mach. Intell. **17**(8), 790–799 (1995)
12. Clement, A., Verfaille, T., Lormel, C., Jaloux, B.: A new color vision system to quantify automatically foliar discoloration caused by insect pests feeding on leaf cells. Biosyst. Eng. **133**, 128–140 (2015)
13. Cover, T., Hart, P.: Nearest neighbor pattern classification. IEEE Trans. Inf. Theor. **13**(1), 21–27 (1967)
14. Cui, D., Zhang, Q., Li, M., Zhao, Y., Hartman, G.L.: Detection of soybean rust using a multispectral image sensor. Sens. Instrum. Food Qual. Saf. **3**, 49–56 (2009)
15. Deng, L., Yu, D.: Deep learning: methods and applications. J. Found. Trends Sign. Process. **7**(3–4), 197–387 (2014). Electron. Agricult. **110**, 221–232

16. Dubey, S.R., Jalal, A.S.: Fruit disease recognition using improved sum and difference histogram from images. Int. J. Appl. Pattern Recogn. **1**(2), 199–220 (2014)
17. Fukunaga, K., Hummels, D.M.: Bayes error estimation using Parzen and k-NN procedures. IEEE Trans. Pattern Anal. Mach. Intell. **9**(5), 634–643 (1987)
18. James, A.P., Dimitrijev, S.: Inter-image outliers and their application to image classification. Pattern Recogn. **43**(12), 4101–4112 (2010)
19. Xiuhua, J., Li, M.: An Improved Algorithm Based On Color Feature Extraction For Image Retrieval. IEEE (2016). ISBN 978-1-5090-0768-4(2016)
20. Salhi, K., Jaara, E.M., Alaoui, M.T.: Texture image segmentation approach based on neural networks. iJES **6**(1), 2018 (2018)
21. Kim, D.G., Burks, T.F., Schumann, A.W., Zekri, M., Zhao, X., Qin, J.: Detection of citrus greening using microscopic imaging. Agric. Eng. Int.: CIGR Ejournal. Manuscript 1194 **XI** (2009)
22. Kruse, O.M.O., Prats-Montalban, J.M., Indahl, U.G., Kvaal, K., Ferrer, A., Futsaether, C.M.: Pixel classification methods for identifying and quantifying leaf surface injury from digital images. Comput. Electron. Agric. **108**, 155–165 (2014)
23. Li, S.Z.: Markov Random Field Modeling in Image Analysis, 3rd edn. Springer, London (2009). https://doi.org/10.1007/978-1-84800-279-1
24. Vipinadas, M.J., Thamizharasi, A.: Detection and grading of diseases in banana leaves using machine learning. Int. J. Sci. Eng. Res. **7**(7), 916–924 (2016)
25. El–Helly, M., Rafea, A., Ei–Gamal, S., Ei Whab, R.A.: Integrating Diagnostic Expert System With Image Processing Via Loosely Coupled Technique, Central Laboratory for Agricultural Expert System (CLAES) (2004)
26. Moya, E.A., Barrales, L.R., Apablaza, G.E.: Assessment of the disease severity of squash powdery mildew through visual analysis, digital image analysis and validation of these methodologies. Crop Protect. **24**(9), 785–789 (2005). https://doi.org/10.1016/j.cropro.2005.01.003
27. Bagri, N., Johari, K.: A Comparative Study on Feature Extraction using Texture and Shape for Content Based Image Retrieval (2015). ISSN 2005-4238 IJAST Vol.80(2015)
28. Puviarasan, N., Bhavani, R., Vasnthi, A.: Image retrieval using combination of texture and shape features. Int. J. Adv. Res. Comput. Commun. Eng. **3**, 5873–5877 (2014). ISSN 2319-5940
29. Olmstead, J.W., Lang, G.A., Grove, G.G.: Assessment of severity of powdery mildew infection of sweet cherry leaves by digital image analysis. HortScience **36**, 107–111 (2001)
30. Phadikar, S., Sil, J.: Rice Disease Identification Using Pattern Recognition Techniques, pp. 420–423. IEEE, Khulna (2008)
31. Pourreza, A., Lee, W.S., Ehsani, R., Schueller, J.K., Raveh, E.: An optimum method for real-time in-field detection of Huanglongbing disease using a vision sensor. Comput. Electron. Agric. **110**, 221–232 (2015)
32. Stehling, R., Nascimento, M., Falcao, A.: On shapes of colors for content-based image retrieval. In: Proceedings of the ACM Workshops on Multimedia Conference, Los Angeles, Calif, USA, November 2000
33. Arivazhagan, S., Newlin Shebiah, R., Ananthi, S., Vishnu Varthini, S.: Detection of unhealthy region of plant leaves and classification of plant leaf diseases using texture features. Agric. Eng. Int: CIGR J. **15**(1), 211–217 (2013)
34. Skaloudova, B., Krvan, V., Zemek, R.: Computer-assisted estimation of leaf damage caused by spider mites. Comput. Electron. Agric. **53**(2), 81–91 (2006). https://doi.org/10.1016/j.compag.2006.04.002
35. Unser, M.: Sum and difference histograms for texture classification. IEEE Trans. Pattern Anal. Mach. Intell. **8**(1), 118–125 (1986)

36. Varshney, P., Farooq, O., Upadhyaya, P.: Hindi Viseme recognition using subspace DCT features. Int. J. Appl. Pattern Recogn. (2014). Under Press
37. Weinberger, K.Q., Blitzer, J., Saul, L.K.: Distance metric learning for large margin nearest neighbor classification. J. Mach. Learn. Res. **10**, 207–244 (2009)
38. Weizheng, S., Yachun, W., Zhanliang, C., Hongda, W.: Grading method of leaf spot disease based on image processing. In: Proceedings of the 2008 international Conference on Computer Science and Softwarez Engineering, CSSE, 12–14 December 2008, vol. 06, pp. 491–494. IEEE Computer Society, Washington, DC (2008) http://dx.doi.org/10.1109/CSSE.2008.1649
39. Mahdy, Y., Shaaban, K., Abd El-Rahim, A.: Image retrieval based on content. J. Graph. Vis. Image Process. **6**(1), 55–60 (2006)
40. Zhou, R., Kaneko, S., Tanaka, F., Kayamori, M., Shimizu, M.: Disease detection of Cercospora Leaf Spot in sugar beet by robust template matching. Comput. Electron. Agric. **108**, 58–70 (2014)

A Lightweight Convolutional Neural Network for License Plate Character Recognition

Xingzhen Tao(✉), Lin Li, and Lei Lu

College of Information Engineering, Jiangxi College of Applied Technology, Ganzhou 341000, Jiangxi, China
348627805@qq.com

Abstract. License plate recognition is a very important component of intelligent parking lot management systems. The accuracy and speed of license plate recognition directly affect the speed of vehicles entering and leaving a parking lot. According to the characteristics of the character composition of a license plate, based on the convolutional neural network model LeNet-5, the license plate character recognition network model is designed. After training to the model, the recognition accuracy of each character in the license plate is higher than 99.97%, and the recognition duration of a single license plate is as fast as 1.31 ms. In this study, the detailed algorithm description of license plate recognition model is given; meanwhile, the optimization method for the neural network model is summarized.

Keywords: License plate character recognition · Convolutional neural network model LeNet-5 · Character recognition

1 Introduction

With urban development, the phenomenon of "Parking difficulties and chaotic parking" on the ground is increasingly prominent, which severely affects the urban environment and traffic order. The development of intelligent underground garage is an inevitable trend. Intelligent parking lot management system integrates advanced intelligent recognition technology and high-speed video image storage comparison, through the computer image processing and automatic identification, it facilitates the comprehensive management of charge, security and parking guidance and so on. In the system, license plate recognition is a key technology, of which, recognition accuracy and recognition duration of license plates are important indications of evaluation. The major issues of license plate recognition technology include license plate positioning and character recognition. The complete license plate recognition algorithm includes image capturing, license plate positioning, character segmentation and character recognition [1, 2]. Because the images used by the license plate recognition system is obtained from the actual environment, due to the influence of lighting, slope angle and other factors in the underground parking lot, the quality of the captured images is deteriorated, which makes the segmented characters of license plate have the defects of fractures, adhesion, blurs and coarseness [3]. How to accurately identify license plate characters has become a difficult assignment.

K. Li et al. (Eds.): ISICA 2019, CCIS 1205, pp. 379–387, 2020.
https://doi.org/10.1007/978-981-15-5577-0_28

2 License Plate Recognition

2.1 Characteristics of License Plates

There are many kinds of license plates in China. In this study, the civil license plates commonly seen in the underground parking lot are focused for the research. This kind of license plate is single row type, on which, the initial character is a Chinese character, the second character is an upper case English letter, and the third to seventh characters are composed of capital letters and numbers. The initial character is the abbreviation of the Chinese characters of the provinces, autonomous regions and municipalities directly under the central government in the mainland of China, with 31 characters available in total; the capital letters include 24 (except the letters I and O, which are easy to be confused with the numbers 1 and 0); the numbers are 10 Arabic numerals, with 65 characters in total [4].

2.2 Character Recognition Algorithm

License plate character recognition technology refers to the automatic extraction of vehicle license plate information from the graphic data of a license plate image, including Chinese characters, capital letters, Arabic numerals and license plate color, and the identification of information [5]. At present, the traditional license plate character recognition methods mainly include character template matching, character feature statistics and machine learning methods [2].

1) Character recognition based on template matching

Template matching is one of the most typical and fundamental methods in the field of image recognition [6], which is to match several feature vectors extracted from the image or image region I(i, j) to be recognized with the feature vectors corresponding to the template T(i, j) established in advance. The basic process of character recognition based on template matching is as follows: firstly, in the normalization process, convert the characters to be recognized into the same size as the images in the template library, then match them with all templates, and finally select the best matched ones as the result. The matching coefficient is calculated by using the following formula:

$$R(x,y) = \frac{\sum_{i,j}(T(i,j)I(x+i,y+j))}{\sqrt{\sum_{i,j}T(i,j)^2\sum_{i,j}I(x+i,y+j)^2}} \tag{1}$$

Where threshold of R (x, y) is taken with the value of [0, 1]. Nevertheless, the template matching method still has deficiencies, which can be listed as follows:

a) In the process of normalization, when the target characters are converted into images of the same size as the template images, it is easy to cause partial distortion of the characters.
b) When there are multiple recognition categories, the template library is high in volume, while the matching speed will be slower and the recognition duration will be prolonged.
c) There are many similar characters in a license plate, for those characters the recognition rate of template matching is not high.

2) Character recognition based on feature statistics
Image recognition based on feature statistics first requires the establishment of the feature vector of each character, and according to the different features, definition of classifiers is performed [7, 8]. In the recognition stage, character features are extracted from the segmented binarized character images. Common features include structural features and statistical features (outer contour features, internal structure features, stroke variation characteristics, etc.). After gathering a large number of statistics, the features being used are obtained, and then matching with the feature sets in the character library to obtain recognition results of the input characters. There are two major shortcomings of character recognition based on feature statistics:

a) There are higher requirements for license plate segmentation and noise processing in the initial stages. If the characters in the images are fuzzy and slanted, which leads to inaccurate segmentation and loss of features, the subsequent recognition rate will decrease.
b) The strokes of Chinese characters in license plates are complex, and it is necessary to decompose the characters into a collection of one or several structural features for the lines of horizontal, vertical, left-falling, right-falling, turning and circle. In order to obtained the features used in the characters, enormous statistics have to be gathered, resulting in a prolonged duration of recognition.

3) Machine learning based character recognition
The character recognition approach based on machine learning primarily includes the procedures of image pre-processing, feature extraction and recognition [9]. In feature extraction, both manually operated extraction or convolutional neural networks could be used to extract image features. Currently, convolutional neural network is the most suitable neural network model for character recognition [10], with local receptive field and weight sharing characteristics, it greatly reduces computational complexity, for simple classification or specific object classification, by borrowing machine learning methods, only a relatively small number of data sets are needed to achieve a better effect. The most basic architecture of convolutional neural networks including convolutional layers, pooling layers, and full connection layers [11].

a) *Convolutional layers.* The purpose of convolutional computation is to extract the different characteristics of the input image [12]. As the most important layer of a convolutional neural network, the key lies in the local association and window sliding operation. The convolutional layer is the operation of the matrix, which would perform padding operation when the stride is not sufficient. The convolutional layer is characterized by weight sharing, a neuron focuses on a feature of the image, and the entire layer is equivalent to a feature extractor. On the premise of not reducing feature extraction effect, convolution operation can greatly reduce the number of parameters, effectively preventing the problem of overfitting. The filters for the convolutional layer are defined as follows:
   (1) Observation window size F is generally odd (1×1, 3×3 or 5×5)
   (2) The stride S of window movement is the pixel size of each movement, usually one pixel.

(3) The calculation formula: input volume size is $H_1 \times W_1 \times D_1$, four parameter filters number K, Filter size F, stride S, zero padding size P, output volume size is $H_2 \times W_2 \times D_2$.

$$H_2 = (H_1 - F + 2P)/S + 1$$

$$W_2 = (W_1 - F + 2P)/S + 1$$

$$D_2 = K$$

b) *Pooling layers.* The process of pooling and convolution are somewhat similar, that they both adopt the filter approach, but the calculation methods are different [13]. In the pooling layer, the maximum pooling and average pooling operations are used, that is, within a filter size to obtain the maximum or average values of the matrix, by removing the unimportant samples in the feature map, the amount of parameters could be further reduced. In the pooling layer, the size of the feature map matrix can be effectively reduced, but the depth generally does not change, thus effectively reducing the parameters in the fully connected neural networks, in order to speed up the computation speed and prevent the occurrence of overfitting problems.

c) *Full connection layers.* The previous convolutional layer and the pooling layer are feature-extracted [14], the subsequent full connection layer is feature-weighted, and the final full connection layer acts as a "classifier" in the whole convolutional neural network, which outputs the classification results.

3 LeNet-5 Based License Plate Character Recognition Algorithm

3.1 Convolutional Neural Network LeNet-5

The LeNet-5 model was proposed by Professor Yann LeCun [15] in 1998 in his paper "Gradient-based learning applied to document recognition", which is the first convolutional neural network successfully applied to digital recognition issues. The LeNet-5 model has a total of seven layers, and Fig. 1 shows the architecture of the LeNet-5 model:

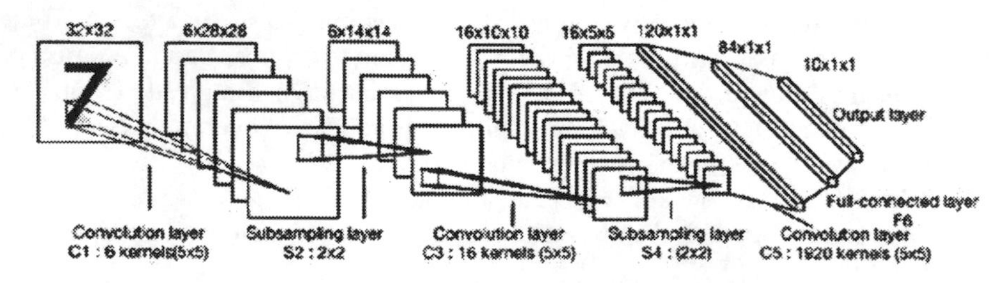

Fig. 1. The architecture of the LeNet-5 model.

(1) After inputting the image, convolution operation is performed through 6 trainable convolution kernels, then 6 feature vector maps each in the size of 28×28 are produced on the C1 layer;
(2) After each feature response graph has been performed pooling operation, the S2 layer produces 6 feature vector maps of 14×14;
(3) Performing convolution operation to 16 trainable convolution kernels, then the C3 layer with 16 feature vector maps each in the size of 10×10 is obtained;
(4) After pooling operation, the S4 layer produces 16 feature vector maps each in the size of 5×5;
(5) Vectorize all the feature maps of the S4 layer, input them into a fully connected neural network, then output 120 neurons;
(6) After going through two full connection layers, it reaches the output layer, which has 10 neurons, each represents an Arabic numeral. By obtaining a row matrix of 10 in length, it is finally determined which number it belongs to.

3.2 Vehicle License Plate Character Recognition Model of LeNet-5

LeNet-5 was originally used to recognize handwritten numerals. The number of categories entered was 10. Compared with numerals, license plates need to be classified into many more categories. In addition to 10 Arabic numerals, it also includes 31 Chinese characters and 24 English letters. In view of the difference between the composition of license plate characters and handwritten numbers, the traditional LeNet-5 is reorganized to design a license plate character recognition model by making full use of the characteristics of license plate itself. Through the observation of the captured image of the license plate as shown in the Fig. 2 below, it can be seen that characters of different categories on the license plate have distinct features, and the characters of the same category have little variability. Meanwhile, the background color is blue, which is easy to extract and segment character contour. These provide the premise for us to design a simplified and efficient deep learning network.

Fig. 2. The captured image of the license plate.

Because the background color of the license plate is blue and the characters are white, that by using the OpenCV method, we can easily segment and binarize the characters. The processed results are shown in the Fig. 3. After the processing, the effective information contained in the picture becomes more distinct and the interference information is greatly reduced. Thus, the difficulty of deep learning network recognition is reduced. For input data like this, through deep learning theory, it can be seen that only by a few data training, a simple network can complete the classification task, without overfitting and underfitting phenomenon.

Fig. 3. The processed results of segment and binarize.

With the foundation of pre-processed data, we can choose LeNet-5, the simplest network structure in deep learning, as our classification network. The structure of the license plate character recognition network is shown as below:

(1) The first is the input layer. Based on the pre-processed results of the license plate mentioned earlier, the input layer here is set to 28 × 28 single-channel monochrome license plate character image data for input.
(2) The first layer contains six 5 × 5 convolution kernels, and the slide stride length is 1. Which is mainly used for extracting low-dimensional features.
(3) In the second layer, the output of the previous layer (the convolutional layer) is performed with maximum value pooling.
(4) The third layer contains 16 convolution kernels each in the size of 5 × 5, and the slide stride length is 1. It is mainly used for extracting low-dimensional features.
(5) In the fourth layer, the output of the previous layer (the convolutional layer) is performed with maximum value pooling.
(6) The fifth and sixth layers are fully connected layers, which are mainly used to extract high-dimensional features.
(7) The seventh layer is the classification output layer, which outputs the classification of license plate characters by full connection. Each character category occupies one output channel as a possible probability value.

4 Experiment

4.1 Experimental Environment

The experimental operating environment is Windows 10 (x64) operating system, the CPU is Intel i5, of which the main frequency is 3.2 GHz, and the memory is 4 GB. TensorFlow program framework is adopted for program editing. The data set was collected from the Internet, with a total of 5,771 images, each consists of 1 character.

4.2 Conduct of the Experiment

The neural network work operation is divided into three steps: first to establish the neural network; second, feed the data training network; and finally, enter the data prediction results. Specific experimental procedures are as follows:

Step 1: Sort the data into training and validation data sets. The amount of images in the training dataset is 5539, and the amount of images in the validation data sets is 232.

Step 2: Read the data. First, the path of the image files and the list of corresponding labels are established in the program. Then, the official interface of TensorFlow is used to read the images in a random sequence according to the list data.
Step 3: Preprocess the data. Two steps of processing are required for the picture data. Firstly, by re-sizing the dimension to $32 \times 32 \times 1$, so that the data can be entered into the model. Then conduct enhanced data processing, the main operation is to randomly crop character pictures. This is mainly aim to prevent the inadequate amount of data, which is necessary for network training.
Step 4: Train the model. The processed data is entered into the model, the gradient computation is initiated, and the weight parameters are updated. By observing the variation result of the correct rate, the hyper-parameters are adjusted, and in the final debugging result, set the hyper-parameter value in the experiment as follows: the amount of training rounds is set to 256. The learning rate is set to 0.001.
Step 5: Verify the effect of the model. Enter the validation data into the trained model to get the correct rate, if the accuracy rate is similar to the training rate, it is indicated that the network has been successfully trained.
Step 6: Test the model. Read the license plate, pre-process to get a binary single character image, and then enter into the LeNet-5 model to get the prediction results.

4.3 Results of the Experiment

By using the license plate character recognition network proposed in this study, recognition of the characters in the license plate mentioned above is conducted, it turns out the recognition rate is higher than 99.97%, and the recognition duration is 1.31 ms. The detailed results are shown in Table 1, comparison with other recognition algorithms are shown in Table 2. The results suggest that the license plate character recognition algorithm designed for underground parking lot in this study has high recognition rate, short-time recognition duration, which can facilities smooth entering and leaving for the vehicles.

Table 1. The detailed results of recognition

Character	京	A	F	0	2	3	6
Recognition rate	100%	100%	100%	100%	100%	99.97%	100%

Table 2. The comparison with other recognition algorithms

Recognition model	Template matching	BP network	LeNet-5 [10]	Ours
Recognition rate	82%	93.3%	98.68%	99.97%

5 Conclusions

License plate character recognition is a key component of intelligent parking management system, which requires high recognition rate and short recognition duration. Based on the LeNet-5 model, its structure is modified in this study, through preprocessing the license plate character images and training the network in the existing data set, eventually the correct rate of character recognition of the network proposed in this study is higher than 99.97%, while the duration of single license plate recognition is only 1.31 ms. Through this experiment, several approaches to optimize convolutional neural networks are summarized. According to the theory of deep learning, the only approach to improve the ability of network feature extraction is to increase the network node, and after deepening the network according to this requirement, it will encounter the issues of gradient attenuation, gradient explosion, increased parameters, inadequate data for training, overfitting and so on, then by the following methods, these can be solved.

(1) Gradient attenuation: use Relu function as the activation function and the residual connection method.
(2) Gradient explosion: normalize/standardize the output of each layer.
(3) Increased parameters: use convolution instead of full connection, use pooling as the transition of convolution.
(4) Inadequate data: use Dropout method and batch input approach.

Acknowledgement. This work was supported by the Key Research and Development Project of Ganzhou, the name is "Research and Application of Key Technologies of License Plate Recognition and Parking Space Guidance in Intelligent Parking Lot".

References

1. Wang, W.: License plate recognition algorithm based on radial basis function neural networks. In: International Symposium on Intelligent Ubiquitous Computing & Education (2009)
2. Saha, S.: A review on automatic license plate recognition system (2019)
3. Li, L., Feng, G.: The license plate recognition system based on fuzzy theory and BP neural network. In: Fourth International Conference on Intelligent Computation Technology & Automation (2011)
4. Yao, Z., Yi, W.: Bionic vision system and its application in license plate recognition. Nat. Comput. **3**, 1–11 (2019)
5. Jokic, A., Vukovic, N.: License plate recognition with compressive sensing based feature extraction (2019)
6. Iwasaki, A.: Independent randomness tests based on the orthogonalized non-overlapping template matching test (2019)
7. Raj, M.A.R., Abirami, S.: Structural representation-based off-line tamil handwritten character recognition. Soft Comput. **24**(2), 1447–1472 (2019). https://doi.org/10.1007/s00500-019-03978-5
8. Redmon, J., et al.: You only look once: unified, real-time object detection (2015)

9. Wang, N., Zhu, X., Jian, Z.: License plate segmentation and recognition of chinese vehicle based on BPNN. In: International Conference on Computational Intelligence & Security. (2017)
10. Zhao, Z., Yang, S., Ma, Z.: License plate character recognition based on convolutional neural network LeNet-5. J. System Simul. **22**(3), 638–641 (2010)
11. Melnyk, P., You, Z., Li, K.: A high-performance CNN method for offline handwritten Chinese character recognition and visualization. Soft Comput. (2019). https://doi.org/10.1007/s00500-019-04083-3
12. Skaria, S., Al-Hourani, A., Lech, M., et al.: Hand-gesture recognition using two-antenna doppler radar with deep convolutional neural networks. IEEE Sens. J. **19**(8), 3041–3048 (2019)
13. Quan, Z., Yang, W., Gao, G., et al.: Multi-scale deep context convolutional neural networks for semantic segmentation. World Wide Web **22**(2), 555–570 (2019)
14. Jin, R., Lu, L., Lee, J., et al.: Multi-representational convolutional neural networks for text classification. Comput. Intell. **35**, 599–609 (2019)
15. Lecun, Y., Bottou, L., Bengio, Y., et al.: Gradient-based learning applied to document recognition. Proc. IEEE **86**(11), 2278–2324 (1998)

A Robust Green Grape Image Segmentation Algorithm Against Varying Illumination Conditions

Haojie Huang, Qinghua Lu(✉), Lufeng Luo, Zhuangzhuang Zhou, and Zongjie Lin

Department of Mechatronics, Foshan University, Foshan 528000, Guangdong Province, China
qhlu@fosu.edu.cn

Abstract. This paper proposes a green grape target segmentation algorithm against an unstructured environment. The algorithm includes three main step: (1) the coordinates of the green grape in the images were obtained and marked of the regression box by using an improved Faster-R-CNN target detection algorithm; (2) Crop the green grapes area through the regression box to reduce the influence of complex background on the image segmentation process, extract the green grape contour based on the HSV color space transformation and edge extraction algorithm; (3) The extracted green grape contour is used as a mask and synthesized onto the original image to complete the segmentation. Ten pairs of green grape images under sunny and cloudy days were segment using the proposed method respectively. The experiments show that the recognition accuracy for unoccluded and partially occluded of green grape were 92% and 82% on sunny day. On the cloud day, the recognition accuracy for unoccluded and partially occluded were 85% and 83%, respectively. The result show the accuracy and feasibility of this method for green grapes recognition during varying illumination conditions. This research provides technical support of visual localisation technology for green grapes picking robots.

1 Introduction

To promote the intelligent robot to the wild environment, one of the biggest challenges is the identification and positioning of the target. Because the illumination changes unevenly in the real natural environment, the illumination of each time period may not be fixed. In addition, due to the color of green grapes is similar to the foliage, which is difficult to distinguish by conventional image segmentation methods that increases the difficulty of intelligent robots in identifying targets. To this end, the predecessors have done a lot of research work. Among them, the color-based classification is the most basic classification method for fruits in images. In order to find the color index that is most favorable for image segmentation, some works extract the color components of YCbCr, HIS, YIQ, HSV, L*a*b [1–3]. For some fruits with relatively regular shapes such as apples,

K. Li et al. (Eds.): ISICA 2019, CCIS 1205, pp. 388–398, 2020.
https://doi.org/10.1007/978-981-15-5577-0_29

citruses, litchis, [4–7] extract fruits by Hough circle transformation or random ring method (RRM). [8] used the texture difference between fruit and foliage to segment the green apples. Some works use the K-means clustering algorithm to segment black grapes, litchis, and citruses [9,11]. With the excellent performance of deep learning in classification tasks, researchers have also tried to use neural networks to classify green grapes [10].

The contribution of this paper is to propose a robust green grape image segmentation algorithm against varying illumination conditions. Firstly, using the improved Faster-R-CNN target detection algorithm detected the green grape area in the image, and then the contour extraction of the green grape region is carried out by using HSV color space and edge extraction algorithm. Finally, the extracted green grape outline is synthesized as a mask onto the original image to complete the segmentation.

2 Related Work

In the wild environment, the change of illumination conditions is a factor of uncertainty which it is easy to form spots on fruits, besides, the surface color of the green grape is similar to the color of the foliage. Therefore, it is difficult to extract green grapes using a segmentation method based on fruit color analysis. In order to solve the above problems, the green grape segmentation algorithm proposed in this paper is mainly divided into three steps, which as shown in the following Fig. 1.

Step 1: Firstly, in order to reduce the complexity of image segmentation, the improved Faster-R-CNN target detection algorithm is used to detect image and identify and locate green grapes in the image. Use the regression box to mark the position of the green grape and output the classification label and confidence of the green grape, and return the coordinates of the upper left corner of the regression box and its height and width.

Step 2: Cut the green grapes in the regression box from the original picture as the green grape area. The HSV color space conversion is performed on the green grape region, which from an image of the S component was separated. Edge extraction is performed on the separated S component image to obtain the edge contour of the green grape.

Step 3: A mask region was generated by the green grape edge contour according to the size of the previously extracted green grape region, and the mask region is synthesized onto the original image by using the coordinate parameters of the regression box outputted in the previous step 1. Images segmentation was completed after the pixels inside the outline of the green grape are all converted into blue pixels.

2.1 The Improved Faster-R-CNN Target Detection Algorithm

With the popularity of deep learning and the excellent performance of convolutional neural networks in classification tasks, many researchers have begun to

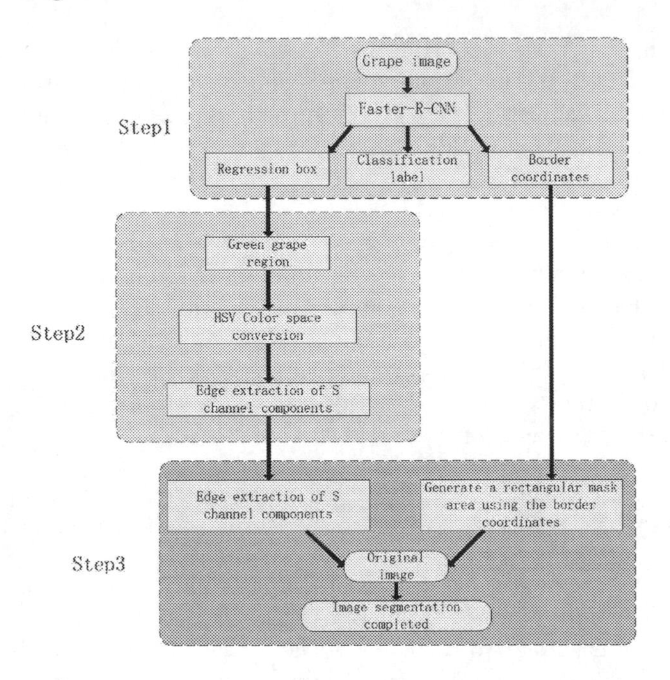

Fig. 1. Algorithm flowchart

apply convolutional neural networks to the task of location and recognition of targets. Among them, many excellent target detection algorithms have emerged, such as the early R-CNN [12], Fast-R-CNN [13], and later Faster-R-CNN [14], YOLOv3 [15] etc. According to the characteristics of the green grapes, we finally chose Faster-R-CNN as the basic framework for the detection of green grapes.

First of all, it is the preparation of the green grape training set. In order to enable the trained model to accurately identify green grapes in an unstructured environment such as uneven illumination, occlusion or overlap, we searched for 200 green grapes from Google. The selected green grapes are completely random, including various time periods (e.g., morning, noon, dusk), various weather conditions (e.g., sunny, cloudy, rainy days), various lighting conditions (e.g., overexposure, underexposure), and a single string of green grapes in a simple background. At the same time, in order to increase the diversity of the training samples, we have horizontally flipped the 200 training samples, rotated a small angle (3°, 6°), and expanded the image size by 1.5 times, increasing the number of training samples to 1000.

We divide the 1000 training samples into training and verification sets, training sets, verification sets, and test sets. The criteria for division follow the following formula:

$$\begin{cases} T_v = \frac{1}{2}W_t \\ T_e = \frac{1}{2}W_t \\ T_v \bigcup T_e = W_t \\ T_v \bigcap T_e = \phi \\ T_r = \frac{1}{2}T_v \\ V_a = \frac{1}{2}T_v \\ T_r \bigcup V_a = T_v \\ T_r \bigcap V_a = \phi \end{cases}$$

where W_t is the total set of these 1000 training samples, T_v is the total set of training and verification sets, T_e is test set, T_a is training set and V_a is verification set.

We manually label the green grape targets in these training samples which were considered as a actual coordinate reference to correct the prediction frame of the training output in a linear regression manner. The correction of the prediction box follows the following formula [13]:

$$\begin{cases} x^* = \frac{(x-x_a)}{w_a} \\ y^* = \frac{(y-y_a)}{y_a} \\ w^* = log(\frac{w}{w_a}) \\ h^* = log(\frac{h}{h_a}) \end{cases}$$

Where x, y, w, and h denote the boxs upper left center coordinates and its width and height, and x_a, y_a, w_a, h_a denote the upper left corner and its width and height of the prediction box output by the training. x^*, y^*, w^*, h^* denote the optimal offset between the prediction frame and ground truth respectively. After completing the training, Faster-R-CNN has the ability to identify the best offset between the prediction box and ground truth.

Second, it is an improvement of the Faster-R-CNN network structure. By modifying the network structure of Faster-R-CNN, it can achieve the target detection task suitable for green grapes. We show the structure of the improved Faster-R-CNN through the following Fig. 2. There are two feature extractors used by Faster-R-CNN, namely VGG16 Net[16] and ResNet[17]. Since the depth of the layer of the neural network affects the processing speed of the entire algorithm, the green grape detection task is relatively simple, and only needs to distinguish the green grape from the background. Therefore, compared to the 101-layer ResNet, the 21-layer VGG16 Net is more suitable as a target Detection for green grapes. The parameters and structure of VGG16 Net used in this paper are shown in Fig. 3. We used ImageNet's pre-training weights to initialize the weights and biases of the convolutional neural network, thus greatly reducing the number of iterations of the model convergence. At the same time, we set two sets of learning rates to prevent over-fitting during training. The learning rate of the initial training setting is 0.001. When the number of training reaches one-third of the maximum number of iterations, the learning rate is multiplied

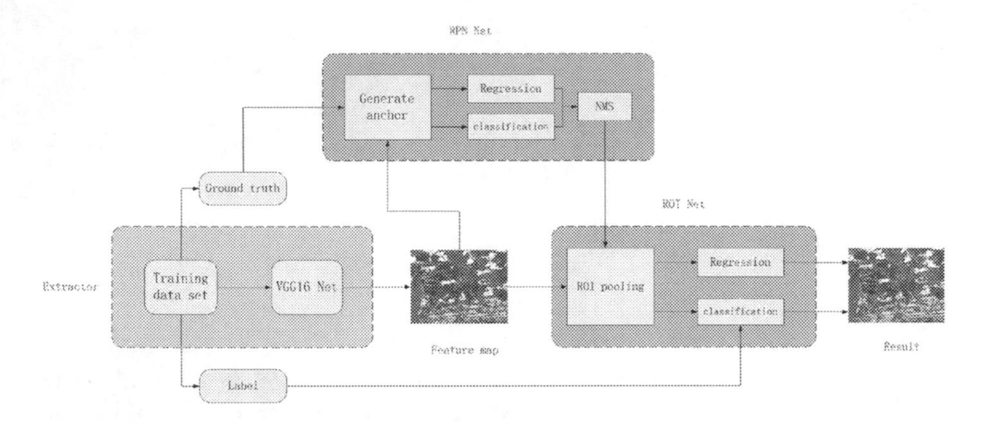

Fig. 2. The structure of improved Faster-R-CNN

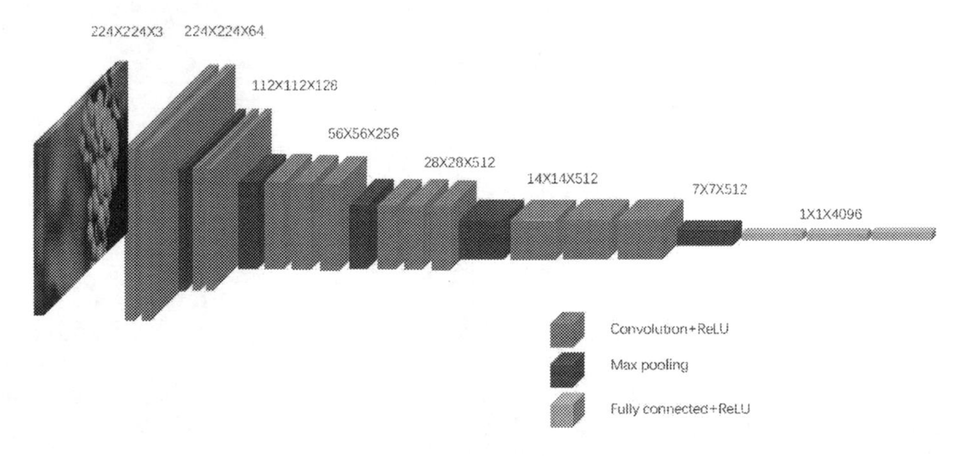

Fig. 3. The parameters and structure of VGG16 Net

by a fixed operator u to reduce the learning rate during training. After empirical debugging, the value of u is set to 0.1. In addition, in order to remove some unnecessary interference, the computational efficiency of the algorithm is improved at the same time. We changed the classification identification number in RPN Net and ROI Net from the original 21 classification to the 2 classification, and only retained the classification labels of green grapes and background. The classifier in the RPN module classifies the regions extracted by the extractor to determine whether the green grapes are included and divided into positive and negative samples. On the contrary, it is considered a negative sample (background) when below the threshold. Areas between the two thresholds are discarded. The division of positive and negative samples follows the following rules:

$$\begin{cases} IOU > 0.7, positive \\ IOU < 0.3, negative \end{cases}$$

After the positive and negative samples were Classified the non-maximum suppression algorithm (NMS) is used to suppress those redundant positive samples, and the threshold of the NMS is set to 0.7. Eventually the total number of samples entered into the ROI pooling module is 256. The ROI pooling module unifies the sizes of samples of different sizes from the NMS. At the same time, the classifier in the ROI pooling module will perform an accurate classification of 256 samples with correcting the position and coordinates of the prediction box. Finally, using the classification label and position coordinates output in the ROI pooling module, the detection results of the green grapes are displayed on the original image.

2.2 Edge Extraction of Green Grapes

After the detection by Faster-R-CNN, the position of the green grape is also detected from the image while the regression box is output. We regard the green grapes in the regression box as the green grape area, and cut the height and width of the box as the size of the green grape area from the images. By cropping to minimize the effects of background (e.g., foliage, sky, grass, etc.) on the image segmentation of green grapes. The cropped green grape area retains only a small amount of background information, and then we perform color analysis on the green grape area. In order to find the color component that is most conducive to image segmentation, we converted the green grape region to the HIS, HSV, L*a*b, and YCbCr color spaces, and found that the S component of the green grape region in the HSV color space can highlight the green grape, which is in sharp contrast to the background (Fig. 4(b)). Therefore, we use the S channel component as the main color indicator for the green grape image segmentation.

Since the green grape image were captured from in an unstructured environment, there was a substantial amount of noise usually exists in the images, which may adversely affect image segmentation. To this end, a Gaussian filter with a kernel size of 5×5 was applied to the green grape region during preprocessing. Then, we make a threshold processing to the green grape area and converted the green grape area image into a binary image (Fig. 4(c)). We perform edge contour extraction on Fig. 4(c), which is based on the border following algorithm [18] used to track the outer border of the binary image (Fig. 4(d)).

(a) Original image (b) S component image (c) Binary image

(d) Contour image (e) Mask image (f) Result image

Fig. 4. The operation of the green grape region

In the meanwhile, we designed a screening algorithm to remove the local contour of the interference. The algorithm follows the following rules:

$$C = \begin{cases} 0 & 0 < A_c < A_e \\ A_c & A_c > A_e \end{cases} \tag{1}$$

Where A_c represents the contour area, and A_e represents the average value of the contour area.

2.3 Green Grape Target Segmentation

In Sect. 2.2, we have obtained the edge contour of the green grape area, and it will be masked in next step. Before the masking operation, in order to make the extracted green grape contour closer to the actual situation, we carried out the subsequent processing of the green grape contour. Firstly, we will solidify the contour of the green grape and fill the green grape area. Then we dilate and smooth the edge contour of green grapes (Fig. 4(e)). The green grape region was finally masked and fused with the original image using the predicted box coordinates which export in Sect. 2.1. In the end, the image segmentation of green grapes in an unstructured environment was achieved (Fig. 4(f)).

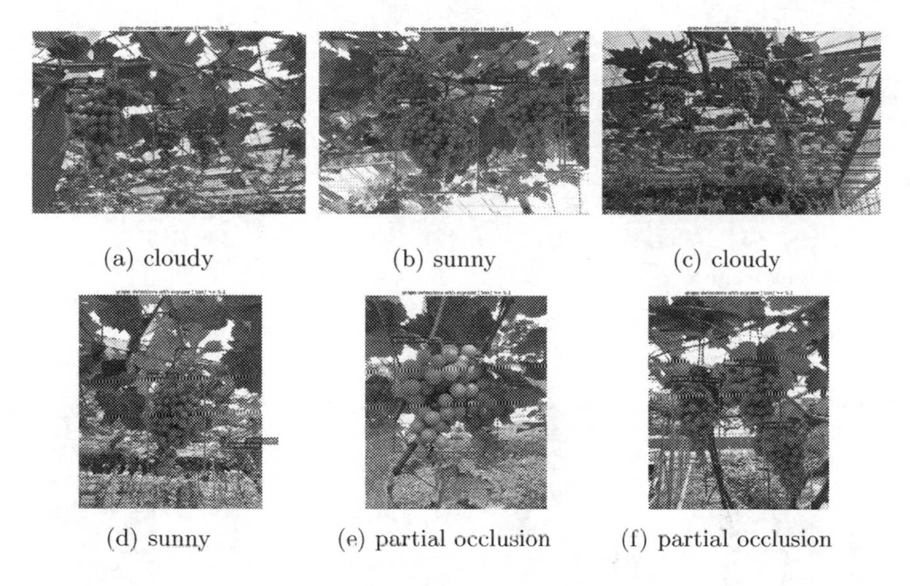

(a) cloudy (b) sunny (c) cloudy

(d) sunny (e) partial occlusion (f) partial occlusion

Fig. 5. The green grape image detection results in Grape Science Park

3 Experiments and Results

The purpose of this paper is to propose a algorithm to image segmentation of captured green grape images in an unstructured environment. Green grape images were usually captured in the wild or in a greenhouse, where the illumination conditions weren't certain. Moreover, the disturbance of the wind in the wild environment is inevitable, and the disturbance of the wind may cause the foliage to block the green grapes, which will affect the final detection results. In order to enable Faster-R-CNN to accurately detect green grapes in complex environments, we have adjusted the parameters of Faster-R-CNN accordingly. The image of the grape photographed in the Grape Science Park was tested as shown in the Fig. 5.

As the detection results shown in the Fig. 5, the Fig. 5(b,d) were captured in the morning with plenty of light, and the green grapes showed a bright color. The Fig. 5(a,c) were captured in the afternoon with dim light, and there were shadows on the surface of the green grape. In the Fig. 5(e,f), the grape is partially obscured due to the disturbance of the wind. Below we show the detection results of green grapes through a table. It can be seen from the results in the Table 1 that the detection of green grapes is robust under illumination conversion conditions. The average detection time per figure is 292 ms.

Table 1. Detection results of green grape under four conditions

Condition	Actual amount	Detected amount	Undetected amount	Success rate	Failure rate
Sunny and non-occlusion	14	13	1	92.85%	7.15%
Sunny and occlusion	11	9	2	81.81%	18.19%
Cloudy and non-occlusion	21	18	3	85.71%	14.29%
Cloudy and occlusion	6	5	1	83.33%	16.67%
Total	52	45	7	86.53%	13.47%

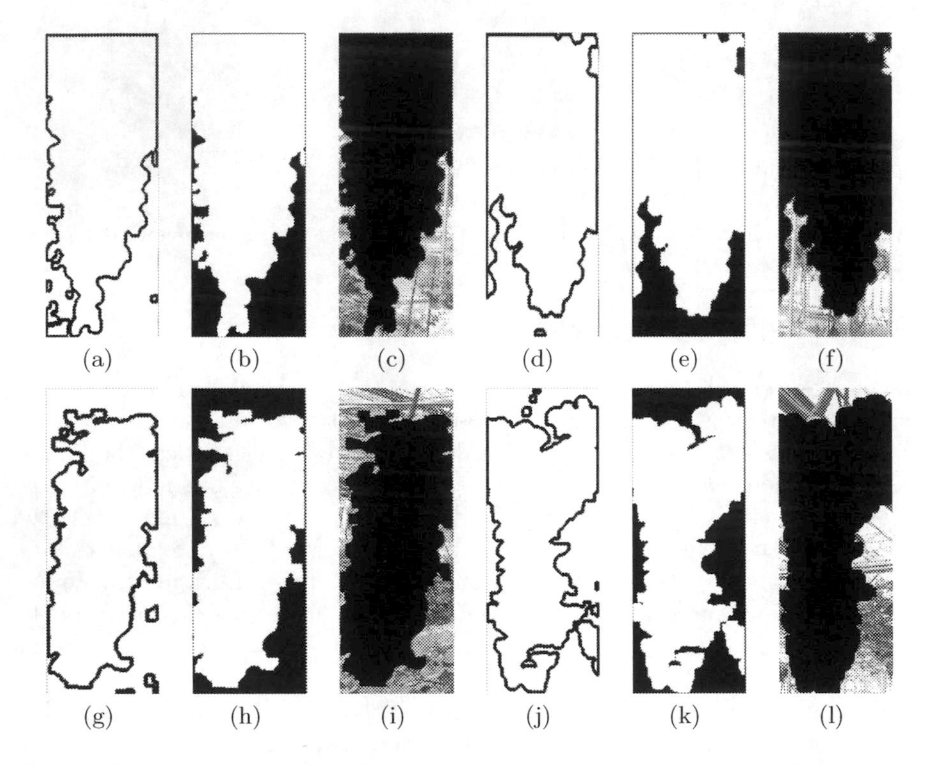

Fig. 6. Edge extraction results of green grapes. a, d, g, j are the results of edge extraction; b, e, h, k are the results after the largest area screening; c, f, i, l are the results of displaying the image segmentation on the original image.

We extracted the contours of the green grape region detected by Faster-R-CNN, and locally synthesized the extracted green grape contour mask with the original image. As shown in the Fig. 6, the results of the local synthesis demonstrate that the extracted green grape outline conforms to the actual shape of the grape. Figure 7 shows the final result of image segmentation. The results of image segmentation are affected by the results of Faster-R-CNN.

(a) Detection result (b) Segmentation result

(c) Detection result (d) Segmentation result

Fig. 7. Image segmentation results

4 Conclusions

This paper proposes a robust green grape image segmentation algorithm against varying illumination conditions. Combining deep learning with traditional image segmentation algorithms, the advantages of neural networks in detecting tasks are used to reduce the difficulty of image segmentation. It solves the problem that it is difficult for the harvesting robot to identify the target of the green grape in an unstructured environment. The experimental results show that the proposed algorithm is robust which can accurately detect green grapes under the condition of illumination varied, and the accuracy rate is 86.53%. Moreover, the average time of detection was 292 ms, and the average time of image segmentation was 53 ms, which has good real-time performance and can meet the needs of real-time control of fruit harvesting robots.

Acknowledgments. This work was supported by grants from the National Key R&D Program of China (2018YFB1308000), National Natural Science Foundation of China (51705365) and Research Projects of Universities Guangdong Province (2018KZDXM074).

References

1. Luo, L., Tang, Y., Zou, X., Ye, M., Feng, X., Li, G.: Vision-based extraction of spatial information in grape clusters for harvesting robots. Biosyst. Eng. **151**, 90–104 (2016)
2. Zhao, Y., Gong, L., Zhou, B., et al.: Detecting tomatoes in greenhouse scenes by combining AdaBoost classifier and colour analysis. Biosyst. Eng. **148**(8), 127–137 (2016)
3. Luo, L., Tang, Y., Lu, Q., et al.: A vision methodology for harvesting robot to detect cutting points on peduncles of double overlapping grape clusters in a vineyard. Comput. Ind. **99**, 130–139 (2018)
4. Si, Y., Liu, G., Feng, J.: Location of apples in trees using stereoscopic vision. Comput. Electron. Agric. **112**, 68–74 (2015)
5. Mehta, S.S., Burks, T.F.: Vision-based control of robotic manipulator for citrus harvesting. Comput. Electron. Agric. **102**, 146–158 (2014)
6. Wang, C., Zou, X., Tang, Y., Luo, L., Feng, W.: Localisation of litchi in an unstructured environment using binocular stereo vision. Biosyst. Eng. **145**, 39–51 (2016)
7. Xiong, J., He, Z., Lin, R., Liu, Z., et al.: Visual positioning technology of picking robots for dynamic litchi clusters with disturbance. Comput. Electron. Agric. **151**, 226–237 (2018)
8. Linker, R., Cohen, O., Naor, A.: Determination of the number of green apples in RGB images recorded in orchards. Comput. Electron. Agric. **81**, 45–57 (2012)
9. Nuske, S., Achar, S., Bates, T., Narasimhan, S., Singh, S.: Yield estimation in vineyards by visual grape detection. In: Proceedings of the IEEE/RSJ International Conference on Intelligent Robots and Systems, San Francisco, CA, USA, 25-30 September 2011, pp. 2352–2358
10. Behroozi-Khazaei, N., Maleki, M.R.: A robust algorithm based on color features for grape cluster segmentation. Comput. Electron. Agric. **142**(Part A), 41–49 (2017)
11. Wang, C., Tang, Y., Zou, X., et al.: A robust fruit image segmentation algorithm against varying illumination for vision system of fruit harvesting robot. Optik - Int. J. Light Electron Opt. **131**, 626–631 (2017)
12. Girshick, R., Donahue, J., Darrell, T., Malik, J.: Rich feature hierarchies for accurate object detection and semantic segmentation. In: The IEEE Conference on Computer Vision and Pattern Recognition (CVPR), pp. 580–587 (2014)
13. Girshick, R.: Fast R-CNN. In: 2015 IEEE International Conference on Computer Vision (ICCV), pp. 1440–1448 (2015)
14. Ren, S., He, K., Girshick, R., et al.: Faster R-CNN: towards real-time object detection with region proposal networks. International Conference on Neural Information Processing Systems (2015)
15. Redmon, J., Farhadi, A.: YOLOv3: an incremental improvement (2018)
16. Long, J., Shelhamer, E., Darrell, T.: Fully convolutional networks for semantic segmentation. IEEE Trans. Pattern Anal. Mach. Intell. **39**(4), 640–651 (2014)
17. He, K., Zhang, X., Ren, S., et al.: Deep residual learning for image recognition (2015)
18. Suzuki, S., Be, K.: Topological structural analysis of digitized binary images by border following. Comput. Vis. Graph. Image Process. **30**(1), 32–46 (1985)

A Convolutional Neural Network Model of Image Denoising in Real Scenes

Shumin Xie(✉), Xingzhen Tao, and Lei Deng

College of Information Engineering, Jiangxi College of Applied Technology, Ganzhou, Jiangxi 341000, China
459955101@qq.com

Abstract. For dealing with the noise images in real scenes, this paper proposed a deep convolutional neural network (CNN) based on real noise scene, which is closer to real life. By taking advantage of CNN multiple convolutional layers, the model learns the data characteristics of noised images in real scenes, and further constantly adjusts its own parameters. The experiment results show that the CNN network model proposed in this paper has better denoising performance for noised images in real scenes. The image processed by the method proposed in this paper is clearer, with better visual effect. At the same time, the edge details in the image are better maintained.

Keywords: Convolutional neural network · Image denoising · Noised image · Real scene

1 Introduction

As one of the most efficient information carriers, images carry a large amount of information, which play an important role in our daily life. However, affected by factors like noise environment and imaging device, images are often disturbed by noise to varying degrees. In order to effectively utilize the information contained in images, it is usually necessary to conduct image denoising. Therefore, image denoising, one of the most basic steps in image processing, has become one of the most popular topics in current research [1–3].

Image denoising is a comprehensive technology involving optics, electronic technology and mathematical analysis [4]. In recent years, as digital image technology is widely applied in many fields and machine vision have made major breakthrough, denoising methods based on deep learning have been widely studied [5–7]. Zhang et al. proposed a super-resolution network for multiple degradations (SRMD) [8], in this paper, image denoising is regarded as a process of mapping a noise image to an ideal image, by using the multi-layer perceptron (MLP) in the neural network to fit the image, it can effectively reduce the noise, and achieving a better denoising results. [9] proposed a denoising convolutional neural networks (DnCNN) by combining residual learning and batch normalization, it can attain better denoising results without processed data.

However, in reality, many factors can contribute to image noise, such as dark current noise, short noise and thermal noise, which are also processed by the camera's system, such as denoising, image grayscale correction, and compression [10]. Because of this,

K. Li et al. (Eds.): ISICA 2019, CCIS 1205, pp. 399–404, 2020.
https://doi.org/10.1007/978-981-15-5577-0_30

the real image noise is different from additive white Gaussian noise (AWGN). Therefore, SRMD and DnCNN's denoising results of the noise images in real scenes will be greatly reduced. To solve this problem, this paper proposed a network model based on real scenes, and trains the model according to the noise images which is closer to the real life scene. Compared with the traditional network model, the proposed network model can achieve better denoising results for the noised image in real scene, moreover, the image after denoising is more clear, and the edge details in the image are well preserved.

2 System Model

2.1 Noise Model

Compared with AWGN, the noise in real scenes is usually more complicated, which generally includes thermal noise and dark current noise. The noise produced by a photon sensor during imaging generally obeys a Poisson distribution, while the other interferences are usually considered as Gaussian noise [11]. Therefore, the noise of imaging sensor can be considered to obey the Poisson-Gaussian distribution, i.e., $n(I) \sim \mathcal{N}(0, \sigma^2(I))$, where I represents the original pixel image.

In addition, the generation of images in real scenes usually goes through in-camera processing (ISP), which further increases the complexity of the real image noise. ISP generally has two main steps, denoising and image grayscale correction. Therefore, the noised image in real scene can be further expressed as:

$$y = f(De(I + n(I))) \tag{1}$$

Where $f(\bullet)$ represents the camera response function (CRF), and $De(\bullet)$ represents the denoising function in ISP.

2.2 Network Model

In order to recover the ideal image from the noise image, the CNN is used to establish the denoising model so as to achieve image denoising in the real scene. The design of the network model proposed in this paper is based on the noise model in Sect. 2.1, and the process is as shown in the Fig. 1.

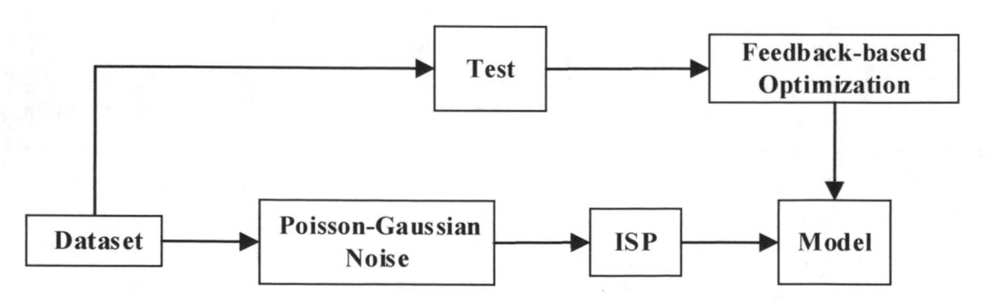

Fig. 1. The process of the proposed network model.

The goal of the model is to minimize the error between the image denoised by the network model and the ideal image. In order to achieve denoising performance, the target formulation of CNN can be expressed as:

$$c = \arg\min \sum_{i=1}^{n} \|c(y_i) - I_i\|_2^2 \tag{2}$$

Where I_i denotes an image randomly selected from the dataset, y_i is the noised image corresponding to I_i, and $c(\bullet)$ represents the convolutional neural network. Through multiple convolution operation of the convolutional neural network, the network model can continuously learn the data characteristics of the image and adjust the parameters of the model, further improving its performance. In the network model, the output of each convolutional layer can be used as the input of the next layer, then the output data feature F_j of the j-th convolutional layer can be expressed as:

$$F_j = \Gamma(Ke_j * F_{j-1} + \eta_j) \tag{3}$$

Where $\Gamma(\bullet)$ is an activation function, Ke_j denotes the convolution kernel between the j-th layer and the $(j-1)$-th layer, and η_j represents the offset parameter of the j-th layer. Furthermore, the loss function generated during convolution operation of each layer can be expressed as:

$$L = \frac{1}{n} \sum_{j=1}^{n} \|\psi(I_j) - I_j\|^2 \tag{4}$$

Where $\psi(I_j)$ represents the forward convolution output of I_j.

Consequently, in the process of establishing the CNN network model, the input image data is alternately operated by a plurality of convolution layers, the data of output image features can be continuously obtained. At the same time, according to the criterion of minimizing the loss function, parameters of the network model are continuously adjusted, and finally the proposed CNN network model of image denoising based on real scene is obtained.

3 Experiments

For training the proposed network model, we select 300 images from the BSD500 dataset [12] and 1500 images from the Waterloo dataset [13]. In addition, the noise model in Sect. 2.1 was used to obtain the noised image, which is taken as the data of training the network model. In the experiments of image denoising, this paper adopts the commonly used Set5 as test data and compares it with the existing DnCNN and SRMD network models.

The denoising results of various network models is shown in the Fig. 2, from which it can be seen that for the image in real scenes, the denoising results of the DnCNN and SRMD network models is not very well as it still remain a lot of noise. The network

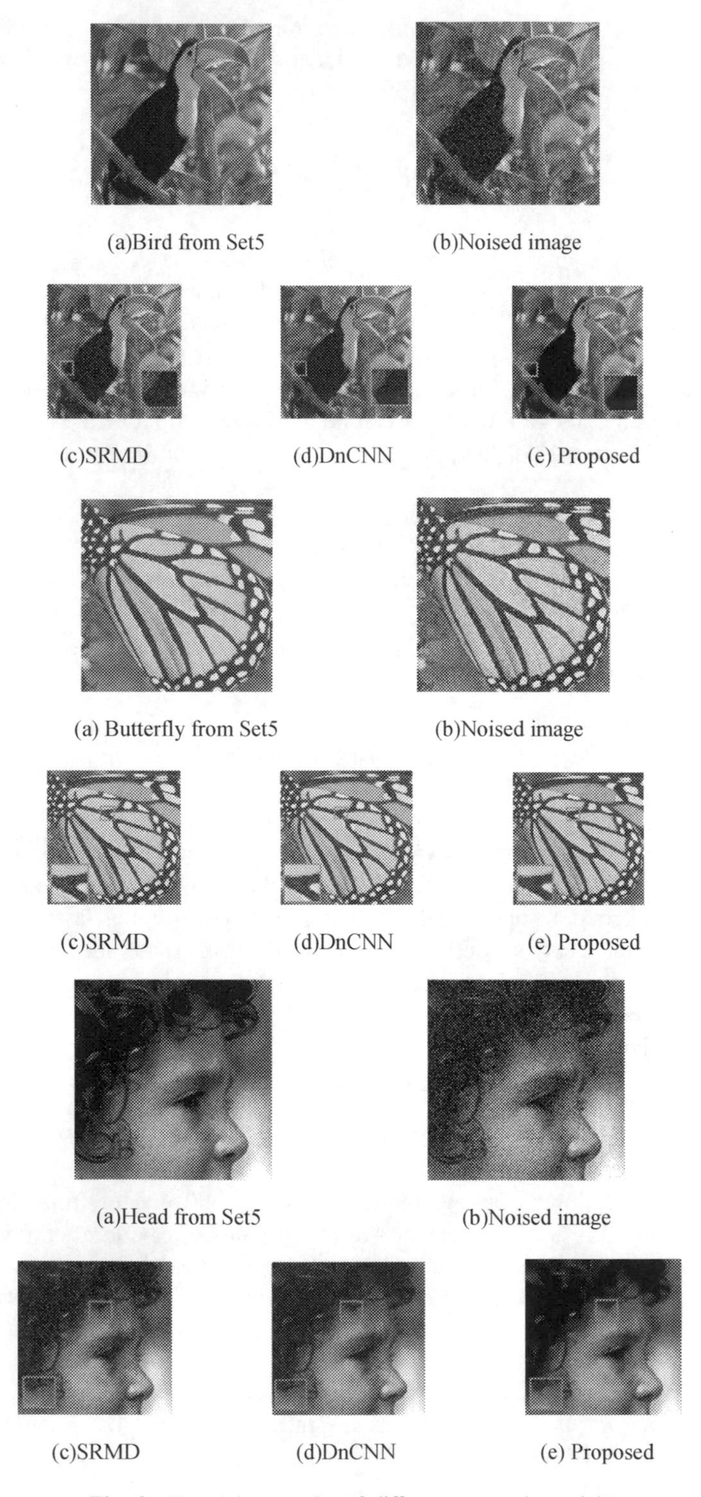

Fig. 2. Denoising results of different network models

model proposed in this paper has better denoising performance for noised images in real scenes. The image after denoising is clearer, with better visual effect, and the edge details in the image are better preserved.

The Table 1 gives the peak signal to noise ratio (PSNR) and structure similarity (SSIM) values of the SRMD, DnCNN network model, and the network model proposed in this paper. From the Table 1, it can be seen that for the butterfly image, the PSNR values of the proposed network model are higher than those of DnCNN and SRMD, the SSIM value of the proposed model is greater than that of the SRMD model and is very close to that of DnCNN. In terms of the bird and head images, the PSNR and SSIM values of the proposed model are greater than that of the DnCNN and SRMD models.

Table 1. The PSNR and SSIM of different models.

	Noised image		SRMD		DnCNN		Proposed	
	PSNR	SSIM	PSNR	SSIM	PSNR	SSIM	PSNR	SSIM
Bird	23.866	0.506	26.046	0.644	27.634	0.764	29.627	0.821
Butterfly	24.721	0.722	27.555	0.864	29.294	0.922	29.530	0.912
Head	23.544	0.429	25.498	0.513	27.427	0.602	28.495	0.663

4 Conclusion

For dealing with the noised images in real scenes, this paper proposed a CNN network model based on real noise scenes, and the noise model which is closer to the real noise was trained through the dataset. The experiment results show that the CNN network model proposed in this paper has better denoising performance for noised images in real life scenes. The image after denoising is clearer, with better visual effect, and the edge details in the image are better preserved.

Acknowledgement. This work was jointly supported by the Key Research and Development Project of Gan-zhou, the name is “Research and Application of Key Technologies of License Plate Recognition and Parking Space Guidance in Intelligent Parking Lot”, the Education Department of Jiangxi Province of China Science and Technology research projects with the Grant No. GJJ181265.

References

1. Tian, C., Xu, Y., Fei, L., et al.: Deep Learning for image denoising: a survey. arXiv: computer vision and pattern recognition (2018)
2. Zha, Z., Zhang, X., Wang, Q., et al.: Group sparsity residual constraint for image denoising. arXiv: computer vision and pattern recognition (2017)
3. Lefkimmiatis, S.: Non-local color image denoising with convolutional neural networks. In: Computer Vision and Pattern Recognition, pp. 5882–5891 (2017)

4. Preethi, S., Narmadha, D.: A survey on image denoising techniques. Int. J. Comput. Appl. **58**(6), 27–30 (2012)
5. Hou, Q., Cheng, M., Hu, X., et al.: Deeply supervised salient object detection with short connections. IEEE Trans. Pattern Anal. Mach. Intell. **41**(4), 815–828 (2019)
6. Krizhevsky, A., Sutskever, I., Hinton, G.E., et al.: ImageNet classification with deep convolutional neural networks. Commun. ACM **60**(6), 84–90 (2017)
7. Lefkimmiatis, S.: Universal denoising networks: a novel CNN architecture for image denoising. arXiv: computer vision and pattern recognition (2017)
8. Zhang, K., Zuo, W., Zhang, L.: Learning a single convolutional super-resolution network for multiple degradations. In: Proceedings of the IEEE Computer Society Conference on Computer Vision and Pattern Recognition, CVPR (2017)
9. Zhang, K., Zuo, W., Chen, Y., et al.: Beyond a Gaussian denoiser: residual learning of deep CNN for image denoising. IEEE Trans. Image Process. **26**, 3142–3155 (2017)
10. Heide, F., Steinberger, M., Tsai, Y., et al.: FlexISP: a flexible camera image processing framework. ACM Trans. Graph. **33**(6), 1–13 (2014). International Conference on Computer Graphics and Interactive Techniques
11. Foi, A., Trimeche, M., Katkovnik, V., et al.: Practical Poissonian-Gaussian noise modeling and fitting for single-image raw-data. IEEE Trans. Image Process. **17**(10), 1737–1754 (2008)
12. Martin, D., Fowlkes, C.C., Tal, D., et al.: A database of human segmented natural images and its application to evaluating segmentation algorithms and measuring ecological statistics. In: International Conference on Computer Vision, pp. 416–423 (2001)
13. Ma, K., Duanmu, Z., Wu, Q., et al.: Waterloo exploration database: new challenges for image quality assessment models. IEEE Trans. Image Process. **26**(2), 1004–1016 (2017)

Multilevel Image Thresholding Based on Renyi Entropy Using Cuckoo Search Algorithm

Zhijun Liang and Yi Wang(✉)

Guangdong University of Science and Technology, Dongguan 532000, Guangzhou, China
171651963@qq.com

Abstract. In this paper, optimal thresholds for multi-level thresholding in image segmentation are gained by maximizing the Renyi entropy using cuckoo search algorithm. The proposed method is tested on standard set of images. Besides, the control parameter of the Renyi entropy is discussed. Experiment results show that the effect of segmented image is not significantly affected by varying the parameter value when the thresholds to 2 then to 5.

Keywords: Image segmentation · Multi-level thresholding · Renyi entropy · Cuckoo search algorithm

1 Introduction

Image segmentation has play an important role as a preprocessing step in image processing. Thresholding is useful in image segmentation [1]. Note that Image segmentation deals with subdividing the image into objects of meaningful information, which is widely used in computer vision [2], biomedical imaging [3], pattern recognition [4]. Recent years, many techniques for image segmentation have been developed and proposed in the literature [5]. Due to the fact that image segmentation based on thresholding is very simple and efficient. Thresholding is considered the most preferred technique out of all the existing methods used for image segmentation [6]. Various methods for global thresholding are available in the literature to segment images and extract meaningful patterns of interest. Bi-level global thresholding is used to divide the image into object and background [7, 8]. However, for real life images bi-level thresholding does not acquire appropriate results. Hence, there is a strong need for multi-level thresholding which divides the histogram of the image into number of classes of homogenous gray levels such that some criterion is optimized [9]. Many such criteria are employed to achieve multi-level thresholding. Otsu's criteria maximize the sum of between-class variances for separating the classes. However, it is difficult to identify the optimal thresholds effectively for multi-level thresholding. The computational complexity increases as the number of thresholds increases. Obviously the best solution is to adopt evolutionary computational techniques. In this paper, various thresholding algorithms are proposed which use different kinds of heuristic method such as PSO [10], ABC [11], GA [12], CS [13] and hybrid algorithms [14, 15].

K. Li et al. (Eds.): ISICA 2019, CCIS 1205, pp. 405–413, 2020.
https://doi.org/10.1007/978-981-15-5577-0_31

This has motivated us to introduce a new method for finding the optimal thresholds effectively for multi-level thresholding. Here we consider optimal thresholding as a constrained optimization problem. The desired stability yields appropriate constraints for the maximization problem. Interesting and stable solutions are obtained through the convergence of a new meta-heuristic algorithm called cuckoo search. Renyi entropy also called non-extensive entropy has been studied in image segmentation.

The paper is organized as follows: Sect. 2 presents concepts of Renyi entropy. Section 3 describes concepts of Cuckoo search algorithm. Experiment results are presented in Sect. 4. Conclusions are drawn in Sect. 5.

2 Formulation of the Multilevel Image Thresholding

Entropy is basically a thermodynamic concept associated with the order of irreversible processes from a traditional point of view. For an image with L gray levels, probability distribution of the gray levels, P_i can be measured by

$$N = \sum_{i=1}^{L} h(i),\ P_i = h(i)/N \tag{1}$$

especially, $h(i)$ denotes the occurrence of gray-level i.

Suppose that a gray image can be divided into two classes, one for the object (class A), and the other for the background (class B). The probability distributions of the object and background class A, and B, are given by

$$\begin{aligned} P_A&: \frac{P_1}{P^A}, \frac{P_2}{P^A}, \ldots, \frac{P_t}{P^A} \\ P_B&: \frac{P_{t+1}}{P^B}, \frac{P_{t+2}}{P^B}, \ldots, \frac{P_L}{P^B} \end{aligned} \tag{2}$$

Where, $P^A = \sum_{i=1}^{t} P_i,\ P^B = \sum_{i=t+1}^{L} P_i,\ and\ P^A + P^B = 1.$

Shannon entropy has been used in a variety of applications. Extensions of Shannon original work have resulted in many alternative measures of information. Renyi proposed a one parameter generalization of the Shannon entropy as

$$R_\alpha(A) = \frac{1}{1-\alpha} \ln \sum_{i=1}^{L} p_i^\alpha \tag{3}$$

Where, the $\alpha > 0$, is a non-extensive control parameter which depend on the Accordingly for each class, Kapur's entropy can be expressed as:

$$R_\alpha^{A_1}(t) = \frac{1}{1-\alpha} \ln \sum_{i=1}^{t} \left(P_i/P^{A_1}\right)^\alpha,\quad R_\alpha^{A_2}(t) = \frac{1}{1-\alpha} \ln \sum_{i=t+1}^{L} \left(P_i/P^{A_2}\right)^\alpha \tag{4}$$

Where, $\alpha > 0$ is a parameter. The different from Tsallis entropy, the Renyi entropy will meet the extensive property

$$R_\alpha(A_1 + A_2) = R_\alpha(A_1) + R_\alpha(A_2) \tag{5}$$

Similarity, according to the additive rule, the extensive Renyi entropy can be computed by

$$\begin{aligned} &R_\alpha^{A_1}(t) = \frac{1}{1-\alpha}\ln\sum_{i=1}^{t_1}\left(P_i/P^{A_1}\right)^\alpha, \quad R_\alpha^{A_2}(t) = \frac{1}{1-\alpha}\ln\sum_{i=t_1+1}^{t_2}\left(P_i/P^{A_2}\right)^\alpha, \\ &\cdots, \; R_\alpha^{A_{m+1}}(t) = \frac{1}{1-\alpha}\ln\sum_{i=t_m+1}^{L}\left(P_i/P^{A_{m+1}}\right)^\alpha \end{aligned} \tag{6}$$

The total Renyi entropy can be gained by

$$R_\alpha(A_1 + A_2 + \cdots + A_{m+1}) = \sum_{i=1}^{m+1} R_\alpha(A_i) \tag{7}$$

Then, Renyi entropy based thresholding approach obtains the optimal threshold can be described:

$$R_\alpha^{opt} = \arg\max[R_\alpha(A_1 + A_2 + \cdots + A_{m+1})] \tag{8}$$

3 Cuckoo Search Algorithm

Recently, the Cuckoo Search (CS) algorithm is proposed by Yang and Deb. CS is also a heuristic global search algorithm which is wildly used to solve different optimization problems. We can use CS for finding a global optimal solution. CS algorithm is inspired by the obligate brood parasitism of some cuckoo species by laying their eggs in the nests of other host birds of other species found in different places. While taking into account the complexity space where the optimal solution is sought, the CS algorithm offer us efficient means for maximizing or minimum the objective function are proved. Note that the quality or fitness of a solution can simply be proportional to the value of the objective function as it is the case with other search algorithms. CS is different from other evolution algorithms in the sense that it is a meta-heuristic search algorithm. Since the CS has fewer parameters compare to other algorithm, Therefore the parameter is easy to set, so CS may be useful for nonlinear problems and multi-objective optimizations. Here in a CS algorithm, a pattern corresponds to a nest and each individual attribute of the pattern corresponds to a Cuckoo-egg.

The following principles of CS algorithm are:

1. Interestingly, each cuckoo bird lays one egg at a time, and dumps its egg in a randomly chosen nest of another bird from other species.

2. Usually the best nests containing high quality eggs are carried over to the next generations.
3. The number of available host nests is fixed. And the egg laid by a cuckoo bird is discovered by the host bird with a probability $p_\alpha \in [0,\ 1]$. Note that the worst nests are discovered and dropped from further calculations.

The choice of control parameters in CS algorithm is simple and required for implementing the algorithm. Note that the control parameters of the CS algorithm are the scale factor (β) and the mutation probability value (p_α). While generating new solution $x^{(t+1)}$, for a cuckoo i, a Levy flight is performed:

$$x_i^{t+1} = x_i^t + \alpha \oplus Levy(\lambda) \quad (9)$$

where $\alpha > 0$, is the step size. Here we choose $\alpha = 1$. Levy flights provide a random walk while their random steps are drawn from a Levy distribution for large steps defined by:

$$\text{Levy} \sim u = t^{-\lambda}, \quad (1 < \lambda \leq 3) \quad (10)$$

which has an infinite variance and infinite mean.

Based on above three principles, the basic steps of the CS algorithm are summarized below (presented as a Pseudo Code in Table 1).

Table 1. Pseudo code for cuckoo search algorithm

Cockoo Search algorithm
Begin
Objective function $f(x)$, $x = (x_1, \ldots, x_d)^T$;
Initialize a population of n host nests, $x_i (i = 1,2,\ldots,n)$;
While (t < MaxIterations) or (f_{min} > tol);
Get a cuckoo (say *m*) randomly by Levy flights, $Levy \sim u = t^{-\lambda}$;
Generate new nests, $x_i^{t+1} = x_i^t + \alpha \oplus Levy(\lambda)$;
Evaluate its quality/fitness F_m;
Choose a nest among n (say *n*) randomly;
if ($F_m > F_n$),
Replace *j* by the new solution;
end
Abandon a fraction (p_a) of worst nests
[and build new ones at new locations via Levy flights];
Keep the best solutions (or nests with quality solutions);
Rank the solutions and find the current best;
end while
end

4 Experiments Results

4.1 Parameters Setting

The proposed method has been tested under a set of benchmark images. Some of these images are widely used in the multilevel image segmentation literature to test different methods (Lena, Barbara, Goldhill and Peppers), as shown in Fig. 1. The CS implemented in Matlab2017b on a personal computer with Intel Corei5 CPU, 4G RAM running window 10 system. The designed programs are revised from ones given by the homepage of cuckoo search algorithm. The parameters of the CS are population size is 20, number of iterations is 200, threshold from 2 to 5, lower bound is 1, upper bound is 256. The popular performance indicator, peak signal to noise ratio (PSNR), is used to compare the segmentation results by using the multilevel image threshold techniques. For the sake of completeness we define PSNR, measured in decibel (dB) as

$$PNSR - 20\log_{10}\left(\frac{255}{RMSE}\right),\ (dB) \tag{11}$$

where RMSE is the root mean-squared error, defined as:

$$RMSE = \sqrt{\frac{\sum_{i=1}^{M}\sum_{j=1}^{N}\left(I(i,j) - \widehat{I}(i,j)\right)^2}{M * N}} \tag{12}$$

Here I and $\widehat{I}$ are original and segmented images of size M^*N, respectively.

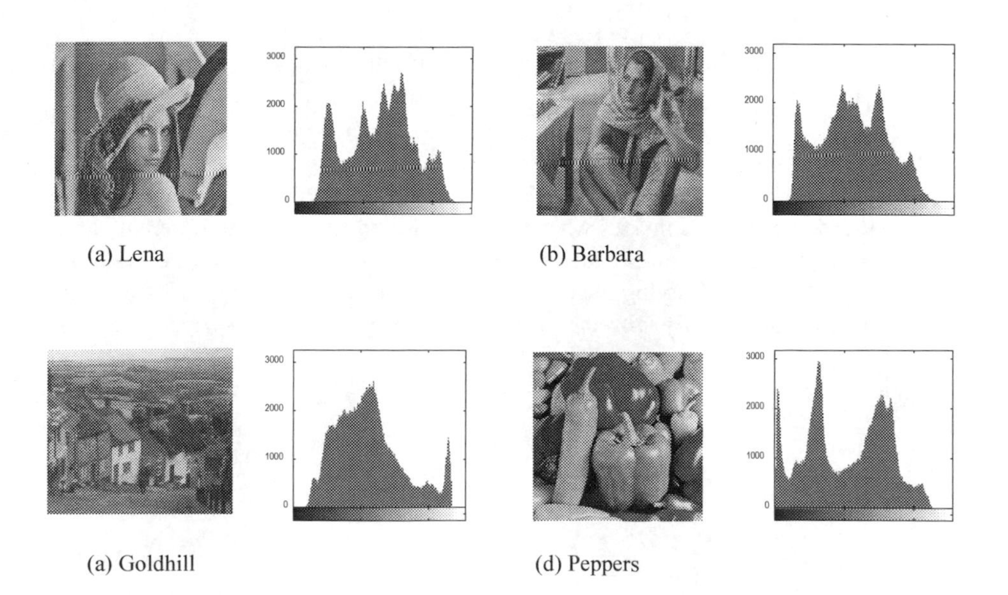

(a) Lena (b) Barbara

(a) Goldhill (d) Peppers

Fig. 1. The original images and their histograms

4.2 The Image Segmentation Result Based on Renyi Entropy with Different Control Parameters Using CS

For evaluating the performance of the method, we have implemented this method on the four test images, when the typical parameter α is 0.3. The performance metrics for checking the effectiveness of the method are chosen as the PSNR and RMSE. At first, we focus on the image segmentation quality sensitivity to the number of thresholds k. Table 2 shows the selected thresholds, RMSE value and PSNR value with different thresholds. Analysis from the table shows that the number of thresholds increase, the RMSE are decrease and the PSNR is enlarge.

Table 2. Experiment results based on Renyi entropy with typical parameter $\alpha = 0.3$

Images	k	Thresholds	RMSE	PSNR
Lena	2	96 164	0.9549	21.2301
	3	79 128 178	0.6975	24.0058
	4	69 109 148 187	0.6082	25.2711
	5	65 101 139 177 191	0.5849	25.6394
Barbara	2	90 160	1.0267	20.7698
	3	74 125 176	0.7091	23.9719
	4	68 112 156 198	0.6136	25.2521
	5	58 93 129 166 183	0.5515	26.1555
Goldhill	2	90 160	0.9951	21.0574
	3	64 117 173	0.7644	23.3453
	4	62 105 148 191	0.6118	25.2749
	5	56 95 134 172 191	0.5431	26.3047
Peppers	2	74 147	1.2134	19.2759
	3	60 120 176	0.8188	22.7348
	4	44 87 132 178	0.5882	25.5982
	5	44 86 131 177 191	0.5491	26.2049

In the following, we will set the number of thresholds to 2 then to 5. The Rényi entropy is compared for different values of α ranging from 0.1 to 0.9. Table 3 shows the PSNR index over the 4 images when using the Rényi entropy with different values of α considering 2 and 5 thresholds. The interesting result is that the PSNR is not significantly affected by varying the parameter value for Renyi entropy. However, the PSNR increase considerably with the number of thresholds (Table 4).

Table 3. PSNR based on Renyi entropy with α = 0.2, 0.4, 0.6, 0.8

Images	k	PSNR			
		$\alpha = 0.2$	$\alpha = 0.4$	$\alpha = 0.6$	$\alpha = 0.8$
Lena	2	21.2943	21.2301	21.2458	21.2458
	3	23.8828	23.8828	23.9023	23.8828
	4	25.1399	25.2296	22.2987	24.8357
	5	25.5661	25.5133	25.3810	25.2764
Barbara	2	20.8036	20.7698	20.7849	20.7833
	3	23.8185	22.2966	23.3758	23.9386
	4	25.1878	25.3180	25.2203	25.2729
	5	26.4308	25.8531	25.3966	25.2074
Goldhill	2	21.0971	21.0574	21.0621	21.0509
	3	23.3795	23.2559	22.5393	23.3795
	4	25.3137	25.3147	25.3339	25.2615
	5	26.4712	26.0739	25.2618	25.6426
Peppers	2	19.3609	19.1168	20.4461	20.4442
	3	22.5443	22.5659	22.7514	22.5504
	4	25.6175	25.2507	25.5750	25.5600
	5	24.6362	26.0512	25.5605	25.6053

Table 4. PSNR based on Renyi entropy with α = 0.1, 0.5, 0.7, 0.9

Images	k	PSNR			
		$\alpha = 0.1$	$\alpha = 0.5$	$\alpha = 0.7$	$\alpha = 0.9$
Lena	2	21.2683	21.2458	21.2458	21.2458
	3	23.9793	24.0659	23.8828	23.8828
	4	25.0233	25.3821	25.3416	24.8555
	5	25.8790	25.4773	25.4776	25.0317
Barbara	2	20.7806	20.7849	20.7883	20.7883
	3	23.9218	23.8729	23.9219	23.9796
	4	25.1296	25.2599	25.3326	25.2729
	5	26.5427	25.2238	25.3659	25.3402
Goldhill	2	21.1264	21.0504	21.0621	21.0509
	3	23.3606	23.2975	23.3591	23.2611
	4	25.3076	25.3564	25.2578	25.2578
	5	26.7377	25.5331	25.4513	23.9996
Peppers	2	19.4281	20.4049	20.4461	20.5319
	3	22.8686	22.5504	22.7348	22.5659
	4	25.5650	25.5637	25.5476	25.5661
	5	27.1134	25.6716	23.5852	25.6798

5 Conclusion

An extensive study on the application of Cuckoo search algorithm for multilevel thresholding for image segmentation is made. As seen from the experimental results, non-extensive entropy based image thresholding using Cuckoo search algorithm is useful for image segmentation. An interesting feature of the proposed method is that Renyi entropy uses global and objective property of the image histogram and is easily implemented. The parameter α of Renyi entropy be used as a tuning parameter for improvising image thresholding results.

Acknowledgments. This work was supported by the Guangdong Youth Characteristic project under Grant No. 2017KQNCX227.

References

1. Felzenszwalb, P.F., Huttenlocher, D.P.: Efficient graph-based image segmentation. Int. J. Comput. Vision **59**(2), 167–181 (2004). https://doi.org/10.1023/B:VISI.0000022288.19776.77
2. Pham, D.L., Xu, C., Prince, J.L.: Current methods in medical image segmentation. Annu. Rev. Biomed. Eng. **2**(1), 315–337 (2000)
3. Fu, K.S., Mui, J.K.: A survey on image segmentation. Pattern Recogn. **13**(1), 3–16 (1981)
4. Haralick, R.M., Shapiro, L.G.: Image segmentation techniques. Comput. Vis. Graph. Image Process. **29**(1), 100–132 (1985)
5. Al-Amri, S.S., Kalyankar, N.V.: Image segmentation by using threshold techniques. arXiv preprint arXiv:1005.4020 (2010)
6. Xiao, C., Zhu, W.: Threshold selection algorithm for image segmentation based on Otsu rule and image entropy. Jisuanji Gongcheng/Comput. Eng. **33**(14), 188–189 (2007)
7. Haque, J.: Dynamically adjusting and predicting image segmentation threshold: U.S. Patent 7,653,242, 26 January 2010
8. Arora, S., Acharya, J., Verma, A., Shi, J., Malik, J.: Normalized cuts and image segmentation. Departmental Papers (CIS), p. 107 (2000)
9. Arora, S., Acharya, J., Verma, A., Panigrahi, P.K.: Multilevel thresholding for image segmentation through a fast statistical recursive algorithm. Pattern Recogn. Lett. **29**(2), 119–125 (2008)
10. Maitra, M., Chatterjee, A.: A hybrid cooperative–comprehensive learning based PSO algorithm for image segmentation using multilevel thresholding. Expert Syst. Appl. **34**(2), 1341–1350 (2008)
11. Horng, M.H.: Multilevel thresholding selection based on the artificial bee colony algorithm for image segmentation. Expert Syst. Appl. **38**(11), 13785–13791 (2011)
12. Hammouche, K., Diaf, M., Siarry, P.: A multilevel automatic thresholding method based on a genetic algorithm for a fast image segmentation. Comput. Vis. Image Underst. **109**(2), 163–175 (2008)

13. Bhandari, A.K., Singh, V.K., Kumar, A., et al.: Cuckoo search algorithm and wind driven optimization based study of satellite image segmentation for multilevel thresholding using Kapur's entropy. Expert Syst. Appl. **41**(7), 3538–3560 (2014)
14. Sathya, P.D., Kayalvizhi, R.: Modified bacterial foraging algorithm based multilevel thresholding for image segmentation. Eng. Appl. Artif. Intell. **24**(4), 595–615 (2011)
15. Bhandari, A.K., Kumar, A., Singh, G.K.: Modified artificial bee colony based computationally efficient multilevel thresholding for satellite image segmentation using Kapur's, Otsu and Tsallis functions. Expert Syst. Appl. **42**(3), 1573–1601 (2015)

Multilevel Image Thresholding Using Bat Algorithm Based on Otsu

Suping Liu and Yi Wang(✉)

College of Software Engineering, Guangdong University of Science and Technology, Dongguan 532000, Guangdong, China
1471836256@qq.com

Abstract. In this paper, optimal thresholds for multi-level thresholding in image segmentation are gained by maximizing Otsu's between-class variance using bat algorithm (BA). The performances of the proposed algorithm are demonstrated by considering four benchmark images. The performance assessment is carried using peak-to-signal ratio (PSNR) and root mean square error (RMSE). The experiment results show that the more threshold, the better the segmentation effect.

Keywords: Bat algorithm · Multilevel thresholding · Image segmentation · Otsu

1 Introduction

Image segmentation plays a important role in image preprocessing, it is wildly used in computer vision, pattern recognition and damage detection [1, 2]. Over the years, several techniques for segmentation have been proposed and implemented in the literature [3], but, Thresholding is considered the most desired procedure out of all the existing procedures used for image segmentation, because of its simplicity, robustness, accuracy, and competence [4, 5]. If the image can be divided into two classes, such as the background and the object of interest, it is called bi-level thresholding. Further more bi-level thresholding also can be extended to multilevel thresholding to obtain more than two classes [6]. In the non-parametric approaches, the thresholds are determined by maximize some criteria, such as between-class variance or entropy measures [7].

Traditional methods work well for a bi-level thresholding problem, when the number of threshold value increases, computing complexity of the thresholding problem also will increase [8–10]. Therefore, the traditional method is different to get the desirable result. Hence, in recent years, kinds of heuristic search algorithms are applied to solve for the multilevel threshold such as BFO [11], ABC [12], CS [13] and so on.In this work, the BA is adopted for solving multilevel thresholding image segmentation problem using Otsu's between-class variance method [14].

K. Li et al. (Eds.): ISICA 2019, CCIS 1205, pp. 414–420, 2020.
https://doi.org/10.1007/978-981-15-5577-0_32

The paper is organized as follows. Section 2 presents the Otsu based multilevel thresholding problem. Section 3 presents the overview of the BA. Experimental results are evaluated and discussed in Sect. 4. Conclusion of the present research work is given in Sect. 5.

2 Methodology

Multilevel Thresholding image segmentation method are employed to find the best possible threshold in the segmented histogram by satisfying some guiding parameters. Otsu based image thresholding is initially proposed in 1979 [15]. This method obtains the optimal solution by maximizing the objective function. In the present work, Otsu's non-parametric segmentation method known as between-class variance is considered. A detailed description of the between-class variance method could be found in bi-level thresholding (for $m = 2$), input image is divided into two classes such as A_0 and A_1 (background and objects) by a threshold at a level "t." The class A_0 encloses the gray levels in the range 0 to $t-1$ and class A_1 encloses the gray levels from t to $L-1$. The probability distributions for the gray levels A_0 and A_1 can be expressed as

$$A_0 = \frac{p_0}{\omega_0(t)} \cdots \frac{p_{t-1}}{\omega_0(t)}, \quad A_1 = \frac{p_t}{\omega_1(t)} \cdots \frac{p_{L-1}}{\omega_1(t)} \tag{1}$$

Where, p_i is the gray level probability, $\omega_0(t) = \sum_{i=0}^{t-1} p_i$, $\omega_1(t) = \sum_{i=t}^{L-1} p_i$, *and* $L = 256$. And, the mean levels μ_0 and μ_1 for C_0 and C_1 can be measured by

$$\mu_0 = \sum_{i=0}^{t-1} \frac{ip_i}{\omega_0(t)}, \quad \mu_1 = \sum_{i=t}^{L-1} \frac{ip_i}{\omega_1(t)} \tag{2}$$

The mean intensity (μ_T) of the entire image can be described as

$$u_T = \omega_0\mu_0 + \omega_1 u_1, \qquad \omega_0 + \omega_1 = 1 \tag{3}$$

The objective function for the bi-level thresholding problem can be defined as

$$F_t^{opt} = \arg\max(\delta_0 + \delta_1) \tag{4}$$

Where, $\sigma_0 = \omega_0(u_0 - u_T)^2$ and $\sigma_1 = \omega_1(u_1 - u_T)^2$.

The bi-level thresholding can be extended to multilevel thresholding problem by increase the various "m" values as follows. Let us consider that there are "m" thresholds (t_1, t_2,…,t_m), which divide the image into "m" classes: A_0 with gray levels in the range 0 to t-1, A_1 with enclosed gray levels in the range t_1 to t_2-1,…, and A_m with gray levels from t_m to L-1.The objective function for the multilevel thresholding problem can be expressed as

$$F_t^{opt} = \arg\max(\delta_0 + \delta_1 + \ldots + \sigma_m) \quad (5)$$

Where, $\sigma_0 = \omega_0(u_0 - u_T)^2, \sigma_1 = \omega_1(u_1 - u_T)^2, \ldots, \sigma_m = \omega_m(u_m - u_T)^2$.

3 Ba Algorithm

The BA algorithm is new population based metaheuristic approach proposed by Yang [16]. The algorithm exploits the so-called echolocation of the bats. The bat use sonar echoes to detect and avoid obstacles. It's generally known that sound pulses are transformed into a frequency which reflects from obstacles. The bats navigate by using the time delay from emission to reflection. The pulse rate is usually defined as 10 to 20 times per second, and it only lasts up about 8 to 10 ms. After hitting and reflecting, the bats transform their own pulse into useful information to explore how far away the prey is. The bats are using wavelength λ that vary in the range from 0.7 to 17 mm or inbound frequencies f of 20–500 kHz. Hence, we can also vary f while fixing λ, because λ and f are related due to the fact λf is constant. The pulse rate can be simply determined in the range from 0 to 1, where 0 means that there is no emission and 1 means that the bat's emitting is their maximum. The bat behaviour can be used to formulate a new BAT. Yang used three generalized rules when implementing the BA algorithms:

1. All bats use echolocation to sense distance, and they also 'know' the difference between food/prey and background barriers in some magical way;
2. Bats fly randomly with velocity v_i at position x_i with a fixed frequency f_{min}, varying wavelength λ and loudness A_0 to search for prey. They can automatically adjust the wavelength of their emitted pulses and adjust the rate of pulse emission $r \in [0, 1]$, depending on the proximity of their target;
3. Although the loudness can vary in many ways, we assume that the loudness varies from a large (positive) A_0 to a minimum constant value A_{min}.

In BA algorithm, initialization of the bat population is performed randomly. In our simulations, we use virtual bats naturally. Namely, generating new solutions is performed by moving virtual bats according to the following equations:

$$f_i = f_{\min} + (f_{\max} - f_{\min})\beta \quad (6)$$

$$v_i^t = v_i^{t-1} + (x_i^t - x*)f_i \quad (7)$$

$$x_i^t = x_i^{t-1} + v_i^t \quad (8)$$

where $\beta \in [0, 1]$ is a random vector drawn from a uniform distribution. Here x_* is the current global best location (solution) which is located after comparing all the solutions among all the bats. In our implementation, we will use $f_{min} = 0$ and $f_{\max} = 1$, depending the domain size of the problem of interest. Initially, each bat is randomly assigned a frequency which is drawn uniformly from $[f_{min}, f_{\max}]$. A random walk with direct exploitation is used for the local search that modifies the current best solution according the equation:

$$x_{new} = x_{old} + \partial A^t \tag{9}$$

where $\partial \in [-1,1]$ is a random number, while A^t is the average loudness of all the best at this time step. The local search is launched with the proximity depending on the rate r_i of pulse emission. A_s the loudness usually decreases once a bat has found its pray, while the rate of pulse emission increases, the loudness can be chosen as any value of convenience. Hence, both characteristics imitate the natural bats. Mathematically, these characteristics are captured with the following equations:

$$A_i^{t+1} = \alpha A_i^t,\ r_i^t = r_i^0[1 - \exp(-\gamma t)] \tag{10}$$

where α and γ are constants. Actually, α parameter plays a similar role as the cooling factor of a cooling schedule in the simulated annealing.

4 Experiments Results and Analysis

4.1 Parameters Setting

The proposed method has been tested under a set of benchmark images. Some of these images are widely used in the multilevel image segmentation literature to test different methods (Monkey, Couple, Boats and Butterfly), as shown in Fig. 1. The BA implemented in Matlab2018b on a personal computer with Intel Corei7 CPU, 8G RAM running window 10 system. The parameters of the BA are population size is 25, number of iterations is 200, threshold value range from 2 to 5. The popular performance indicator, peak signal to noise ratio (PSNR) is used to compare the segmentation results by using the multilevel image threshold method [17]. The PSNR is measured in decibel (dB) as

$$PNSR = 20\log_{10}\left(\frac{255}{RMSE}\right),\ (dB) \tag{11}$$

where RMSE is the root mean-squared error, defined as:

$$RMSE = \sqrt{\frac{\sum_{i=1}^{M}\sum_{j=1}^{N}\left(I(i,j) - \widehat{I}(i,j)\right)^2}{M * N}} \tag{12}$$

Here I and $\widehat{I}$ are original and segmented images of size $M * N$, respectively.

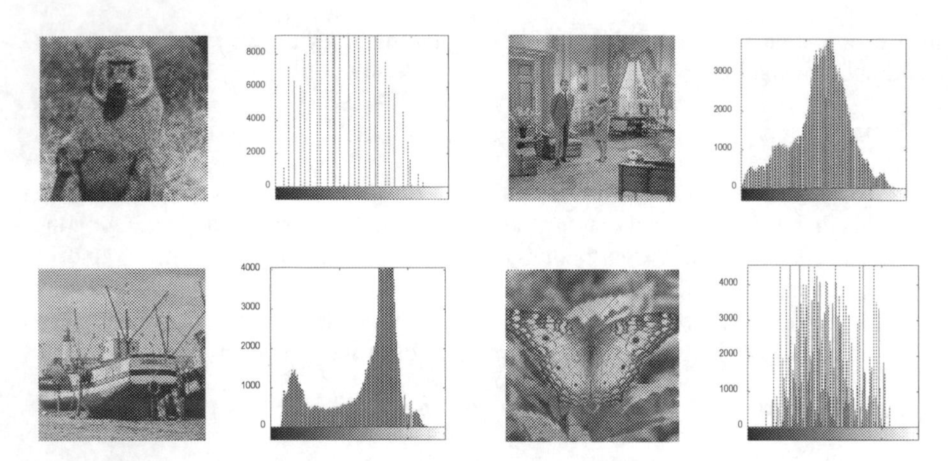

Fig. 1. The test images and their histograms

4.2 The Image Segmentation Results Based BA

In order to verify the performance of the method, Four test images are used to implement this experiment. The performance metrics for checking the effectiveness of the method are chosen as the PSNR and RMSE. At first, we focus on the image segmentation quality sensitivity to the number of thresholds k. Table 1 shows the selected thresholds, RMSE value and PSNR value with different thresholds, and Fig. 2

Table 1. Experiment results based on Otsu

Images	k	Thresholds	RMSE	PSNR
Monkey	2	70 127	1.2060	19.3369
	3	67 121 175	0.7745	23.2412
	4	56 99 143 186	0.5744	25.8146
	5	50 86 124 155 186	0.5248	26.6019
Couple	2	87 145	1.1164	20.0113
	3	75 124 162	0.7942	22.9971
	4	59 102 137 171	0.6101	25.2291
	5	55 93 125 150 182	0.4944	27.0670
Boats	2	82 146	1.2463	19.1150
	3	67 123 164	1.0378	20.6803
	4	51 101 143 171	0.8680	22.1598
	5	44 84 125 155 177	0.6371	24.7347
Butterfly	2	97 149	1.0920	20.2549
	3	80 118 161	0.8567	22.3471
	4	71 97 126 161	0.7482	23.5061
	5	60 89 119 150 177	0.5749	25.7807

show the segmented images when the thresholding is 2, 3, 4 and 5. From the Table 1 shows that the number of thresholds increase, the RMSE is decrease and the PSNR is enlarge, and the threshold is evenly distributed. Corresponding the images segmentation effect showed in Fig. 2 is consistent with the experiment results in Table 1.

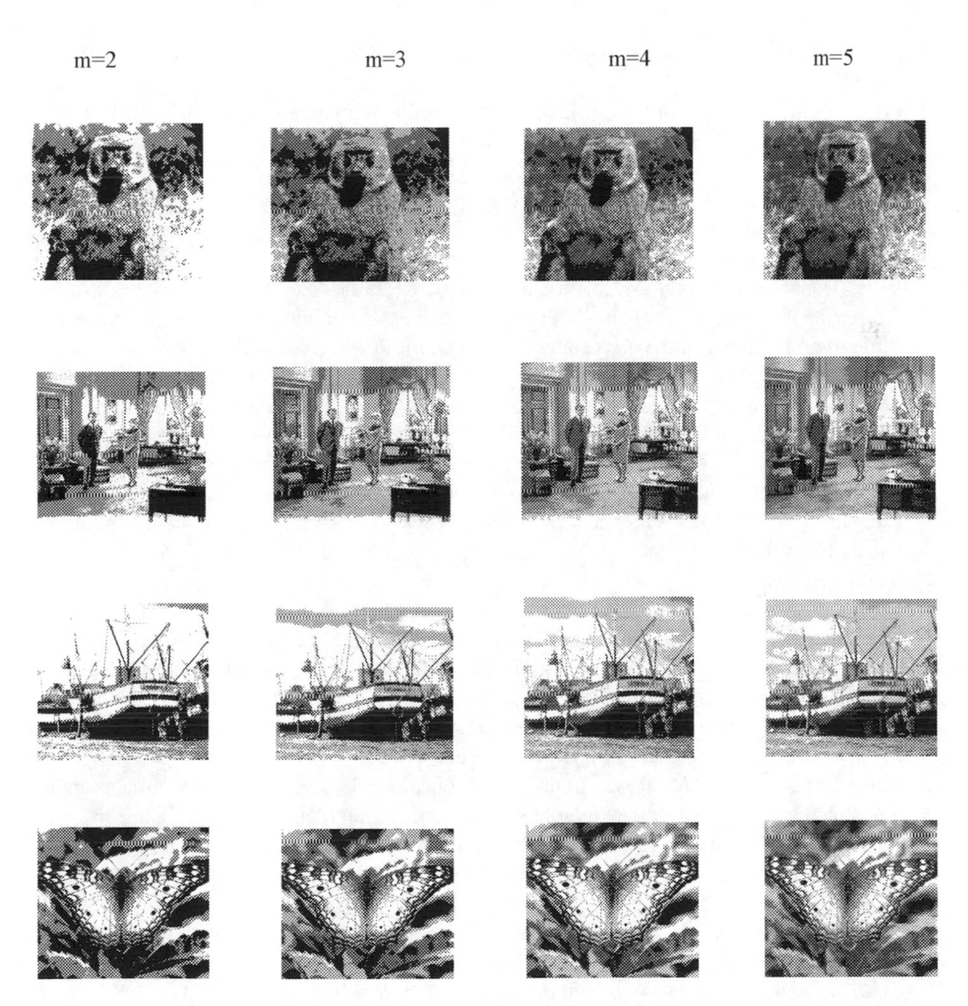

Fig. 2. The segmented images when the thresholding from 2 to 5

5 Conclusion

In this paper, optimal multilevel image thresholding problem is addressed using Otsu guided BA. The proposed histogram based bounded search technique helps in reducing the computation time. The PSNR and RMSE are adopted to evaluate the performance of the algorithm, the simulation results show that as the number of threshold increases, the segmentation effect is better. It is an effective method in image segmentation.

Acknowledgements. This work was supported by the Guangdong Youth Characteristic project under Grant No. 2017KQNCX227.

References

1. Shi, J., Malik, J.: Normalized cuts and image segmentation. Departmental Papers (CIS), 107 (2000)
2. Fu, K.S., Mui, J.K.: A survey on image segmentation. Pattern Recogn. **13**(1), 3–16 (1981)
3. Pham, D.L., Xu, C., Prince, J.L.: Current methods in medical image segmentation. Annu. Rev. Biomed. Eng. **2**(1), 315–337 (2000)
4. Pal, N.R., Pal, S.K.: A review on image segmentation techniques. Pattern Recogn. **26**(9), 1277–1294 (1993)
5. Mardia, K.V., Hainsworth, T.J.: A spatial thresholding method for image segmentation. IEEE Trans. Pattern Anal. Mach. Intell. **10**(6), 919–927 (1988)
6. Arora, S., Acharya, J., Verma, A., et al.: Multilevel thresholding for image segmentation through a fast statistical recursive algorithm. Pattern Recogn. Lett. **29**(2), 119–125 (2008)
7. Bhandari, A.K., Kumar, A., Singh, G.K.: Modified artificial bee colony based computationally efficient multilevel thresholding for satellite image segmentation using Kapur's, Otsu and Tsallis functions. Expert Syst. Appl. **42**(3), 1573–1601 (2015)
8. Khairuzzaman, A.K.M., Chaudhury, S.: Multilevel thresholding using grey wolf optimizer for image segmentation. Expert Syst. Appl. **86**, 64–76 (2017)
9. Liu, Y., Mu, C., Kou, W., et al.: Modified particle swarm optimization-based multilevel thresholding for image segmentation. Soft. Comput. **19**(5), 1311–1327 (2015)
10. Sarkar, S., Das, S., Chaudhuri, S.S.: A multilevel color image thresholding scheme based on minimum cross entropy and differential evolution. Pattern Recogn. Lett. **54**, 27–35 (2015)
11. Sathya, P.D., Kayalvizhi, R.: Image segmentation using minimum cross entropy and bacterial foraging optimization algorithm. In: 2011 International Conference on Emerging Trends in Electrical and Computer Technology, pp. 500–506. IEEE (2011)
12. Horng, M.H.: Multilevel thresholding selection based on the artificial bee colony algorithm for image segmentation. Expert Syst. Appl. **38**(11), 13785–13791 (2011)
13. Rajinikanth, V., Sri Madhava Raja, N., Satapathy, S.C.: Robust color image multi-thresholding using between-class variance and cuckoo search algorithm. In: Satapathy, S.C., Mandal, J.K., Udgata, S.K., Bhateja, V. (eds.) Information Systems Design and Intelligent Applications. AISC, vol. 433, pp. 379–386. Springer, New Delhi (2016). https://doi.org/10.1007/978-81-322-2755-7_40
14. Vala, H.J., Baxi, A.: A review on Otsu image segmentation algorithm. Int. J. Adv. Res. Comput. Eng. Technol. (IJARCET) **2**(2), 387–389 (2013)
15. Otsu, N.: A threshold selection method from gray-level histograms. IEEE Trans. Syst. Man Cybern. **9**(1), 62–66 (1979)
16. Yang, X.S.: A new metaheuristic bat-inspired algorithm. In: González, J.R., Pelta, D.A., Cruz, C., Terrazas, G., Krasnogor, N. (eds.) Nature Inspired Cooperative Strategies for Optimization (NICSO 2010). Studies in Computational Intelligence, vol. 284, pp. 65–74. Springer, Heidelberg (2010). https://doi.org/10.1007/978-3-642-12538-6_6
17. Huynh-Thu, Q., Ghanbari, M.: Scope of validity of PSNR in image/video quality assessment. Electron. Lett. **44**(13), 800–801 (2008)

A Color-Filling Algorithm for Dialect Atlas

Jiaqi Yuan[1] and Huaxiang Cai[2(✉)]

[1] Nantong University, Nantong 226019, JiangSu, China
[2] Jiangnan University, Wuxi 214122, Jiangsu, China
caihx@jiangnan.edu.cn

Abstract. This paper introduces the concept of *Atlas of Dialects*, showing the comprehensive difference in selected themes of specific dialects on the maps. A new solution is used to show the difference in the way of rendering the regions with various colors instead of traditional complicated symbols. This method comes with a problem that if the similar colors are used in neighboring regions, it will be inconvenient to distinguish them on the map. Instead of choosing the colors randomly, a color-filling algorithm is proposed to generate colors from the H(hue)S(saturation)V(value) color space which will make the colors easily be distinguished especially in the neighboring regions. The difference between the colors is quantified by the method of calculating weighed Euclidean distance.

Keywords: Atlas of Dialects · HSV color space · Color-filling algorithm

1 Introduction

In many countries with a long history it is meaningful to do some research work about the dialect. Many scholars have done a lot of jobs to invest the difference between the statements about a certain thing or parlance in selected regions. The conventional process is to establish a database contains representative vocabulary proposed by linguists. Then the ways of saying a certain thing will be recorded as the materials to learn the culture and history [1]. To ensure the accuracy of information, those volunteers should be old enough so that they could know some old statements in their hometown. The volunteers should also meet the requirement that they have not left their hometown for more than three months to avoid the influence of different accents. The relationship between regions may be found after studying the pitch changes, syllables and intonation. In the past years, various symbols [2] are used to represent different statements such as a pentagon stands for a certain statement and a triangle stands for another [3]. The same symbol in different regions means that people in these regions have a same saying about a certain topic. Every single map with different symbols corresponds to one topic. This kind of method used to be the most common way to show the difference on a map. Until recent years location symbol-maps can be found in almost all modern atlases. To represent the linguistic data, geometrical signs such as lines, circles, triangles, etc., are put on the particular location [4]. However, after all these symbols have been marked on the map, it is difficult to find the relationships because of the difficulty to recognize the same symbol on a map especially when a single map contains dozens

K. Li et al. (Eds.): ISICA 2019, CCIS 1205, pp. 421–430, 2020.
https://doi.org/10.1007/978-981-15-5577-0_33

of symbols. On the other hand, the symbols sometimes will be mixed with the labels showing names of the towns which makes it more difficult to recognize the same symbol. Since thousands of hours have been spent to collect the information, a new method should be used to show the data more straightforward so that it will be more easily to do the subsequent research. A new solution which is inspired by the classic four-color map theorem [5] could provide a new way to show the difference among the regions. Various colors can be used to represent different regions with different styles, of course, the number of colors could be much more than four. Since the output form will be the maps, the GIS software will be used to fill the regions with different colors by joining the geographic database and the language-style database which is summed by linguists to distinguish the dialect styles of different towns. So, the color information and spatial topological information [6] will be used to render the maps. The coming problem is that software of GIS could paint different regions in various colors depending on its style but the colors are completely random. That means two or more neighboring areas may be painted with similar colors. This is obviously another obstacle to distinguish the regions which may have different styles but they were painted with similar colors. So, the algorithm will focus on generating very different colors to render neighboring regions with different styles in order to distinguish them more conveniently.

2 Data Preprocessing

To generate the Linguistic Atlas which could consequently enable a better understanding of the nature of particular speech [7], the geographic database from official Surveying and Mapping Institute will be essential. Basic database will be ready after deleting the areas which are not intend to be investigated. The investigation has lasted dozens of months because of the reasons I mentioned above. Linguistic Scholars have done many works to collect and organize the dialect materials. Finally, these areas will be categorized into various styles in different themes. The more detailed the study, the more themes it will have. Every theme will refer to an independent map on which different areas will be painted with different colors. It is not necessary to create so many GIS databases to record all the conclusions. A more convenient method is to create excel-tables. The number of the tables is the number of the themes intended to be investigated. It means that every single table refer to a specific theme. Every excel-table contains two columns, the first one is the names of the regions which is exactly correspondent to the names in GIS basic database and the other one is the styles of the regions. The style can be recorded in the form of Arabic numerals which can be handled easily in the later processing. In fact, an Arabic number stands for a specific style in the current theme and their relationship should be recorded. At last, these numbers will also be used to represent different colors. The most famous software to handle the GIS database called ArcGIS has provided a powerful tool "Spatial join [8]" to find the neighboring towns of every town in the basic database and output them in a DBF file. The algorithm is intent to determine which color should be rendered on each town in the selected area so that the relationship can be shown clearly. Both of the two kinds of files will be very important because the Excel files created mappings between the towns and their styles in different themes, as well as the DBF file recorded adjacent relations of the towns. Corresponding to each Excel file, the dialect map can be generated to match the theme while the DBF file will be invariant if the investigate area is not changed.

3 Color-Filling Algorithm

The purpose of the algorithm is to set the colors of different regions. Geographical location relations are stored in the DBF file and the styles of the dialect are stores in Excel tables, both of them are needed to be considered while choosing the color. Python can be used to process the files by importing necessary packages. The relationship of the towns with their dialect styles can be described as follows:

(1) Create a Dictionary in Python to record the neighboring information about every town by reading the DBF file. Every KEY in the Dictionary is the name of a town and the VALUE is a List of its adjacent towns.
(2) Create another Dictionary to record the location information about the "Styles" of towns. The KEY is the style number of the specific town and the correspondent VALUE is a list of the style numbers stands for its adjacent towns. The relationship comes from the first Dictionary and the Excel file. If a KEY has existed in the Dictionary, it means the style number has been recorded so the new list relates to it should be merged into the VALUE of the existing KEY. Abandon the first Dictionary after finishing the new Dictionary.
(3) Check the elements in the Dictionary and delete the repeat style numbers in the lists.
(4) Resort the Dictionary by the length of the list. The key with longest list will be arranged at the top of the Dictionary.

The Table 1 shows the relationship of the statements about father in Rugao Dialect after being processed in the data structure of Dictionary.

Table 1. Dictionary of statements about *father*.

KEY (Style number)	Value (List of adjacent style numbers)
3	1,2,4,5,6,7,8,9,10
5	1,3,6,7,8,9,10
1	2,3,4,5,6,9
6	1,3,5,9
7	2,3,5,10
9	1,3,5,6
2	3,4,7
4	1,2,3
10	3,5,7
8	3,5

The numbers 1 to 10 represent that there are ten different statements about father in Rugao Dialect and each number refer to one of them. Every specific meaning of each statement should be recorded and studied by the linguists. We only need to consider about adjacent relations of these numbers in order to draw the Linguistic Atlas with colors which are easy to be distinguished from each other.

The Dictionary has been established to research the problem of Color-Filling. The numbers recorded in the Dictionary could be considered as several row vectors. Since every style number will be linked to a color, an appropriate color spaces should be chosen to match the Dictionary. Among the common color spaces, the H.S.V color space provide a color by Hue, Saturation and Value (see Fig. 1).

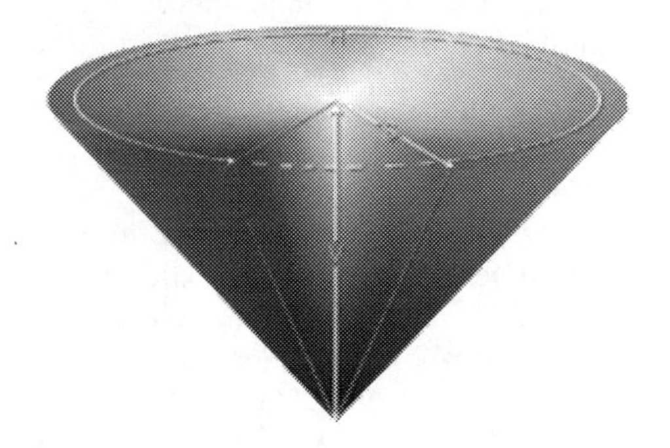

Fig. 1. HSV color space (Color figure online)

It looks like an inverted cone. In the HSV color space Hue stands for the quality of a color as determined by its dominant wavelength and it changes from 0° to 360°. For instance, 0° means red and 120° means green. Saturation changes from 0.0 to 1.0 and it represents the ratio between the purity of the selected color and the maximum purity of the color, when it turns into 0 means the gray-scale color space. Value indicates the brightness of the color and it also changes from 0.0 to 1.0. When the Value is 0.0, the color will come to total black which is the vertex of the cone in Fig. 1. HSV color space will be a circle of various colors when it is viewed from the top. At this moment the Value and the Saturation should be 1.0. So, it will be easy to choose different colors from the edge of the circle. RGB color space which is used in computer system is a linear space coordinate system. The maximum and minimum in linear space coordinate system will have huge difference (just like the white and black). So, to pick colors in RGB color space is not so intuitive as in HSV color space [9] which approximates the human vision [10]. For example, nine different colors should be applied around the style number 3 in the first-row vector in Table 1. Style number 3 will be set to the color when Hue is 0° and its adjacent style numbers will be evenly distributed around the circle (see Fig. 2).

In this way, different colors can be got by dividing 360° equally. Notice two adjacent colors could be very similar, we can change the Saturation when generate a new color every time. The next row vector will be handled in another colored circular palette by changing the Value. The process will be finish when all the style numbers

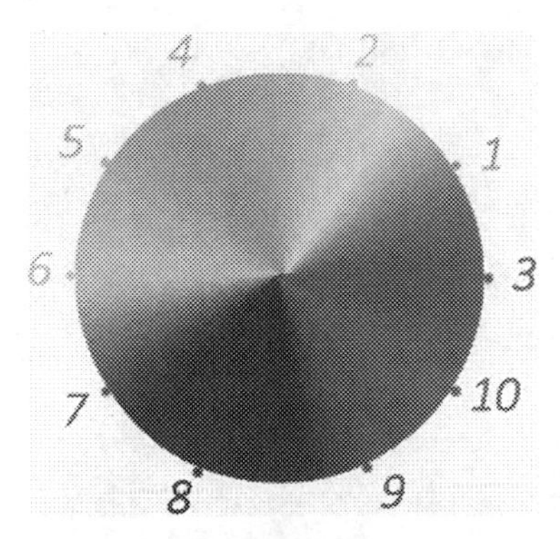

Fig. 2. Evenly distributed colors (Color figure online)

have been mapped to a color. In this example the job will be done after finishing the first-row vector because all the style numbers are contained in the vector. In other cases, if a style number has been set to a color in a previous row vector it is not needed to be changed.

The steps can be described as follow:

(1) Generate a color in HSV Color Space. Initialize H as 0 or a random angle, S as 1.0, V as 1.0, K as 0.
(2) Read the Key-Value pair and give the key number the color with current HSV.
(3) Consider the length of the value as Φ, if Φ is an even number then set changeS as 0.5, otherwise set changeS as 0.33.
(4) Set n as the serial number of the elements in Value. n loops from 1 to Φ.
If the element has been given a color then ignored it without processing else:

$$H = \left(H + \frac{360^\circ}{\Phi}\right) \bmod 360^\circ$$

S = S – changeS: if S < 0.3 then S = 1.0
the color with HSV is given to the style number correspond to the n
(5) If all the style numbers have been paired with colors then exit the function otherwise:

$$K+=1$$

$$V = (1-\alpha)^K (0 < \alpha < 1)$$

H = random(0, 360°)
then read the next Key-Value pair and go to Step (2).

The different colors can be got from the HSV color space; however, the color space is not provided in common GIS systems. As a result, these colors should be changed into RGB mode. The conversion formula is as follow:

When $0 \leq H < 360$, $0 \leq S \leq 1$ and $0 \leq V \leq 1$:

$$C = V \times S$$

$$X = C \times \left(1 - \left|\left(\frac{H}{60^\circ}\right) mod\ 2 - 1\right|\right)$$

$$m = V - C$$

$$(R', G', B') = \begin{cases} (C, X, 0), & 0^\circ \leq H < 60^\circ \\ (X, C, 0), & 60^\circ \leq H < 120^\circ \\ (0, C, X), & 120^\circ \leq H < 180^\circ \\ (0, X, C), & 180^\circ \leq H < 240^\circ \\ (X, 0, C), & 240^\circ \leq H < 300^\circ \\ (C, 0, X), & 300^\circ \leq H < 360^\circ \end{cases}$$

$$(R, G, B) = ((R' + m) \times 255, (G' + m) \times 255, (B' + m) \times 255)$$

The colors of the style numbers in Table 1. are paired with colors in Table 2.

Table 2. Color table of statements about father

Style number	HSV	RGB
1	(36,0.5,2)	(255, 204, 127)
2	(72, 1, 1)	(204, 255, 0)
3	(0, 1, 1)	(255, 0, 0)
4	(108, 0.5, 1)	(153, 255, 127)
5	(144, 1, 1)	(0, 255, 101)
6	(180, 0.5, 1)	(127, 255, 255)
7	(216, 1, 1)	(0, 101, 255)
8	(252, 0.5, 1)	(153, 127, 255)
9	(288, 1, 1)	(203, 0, 255)
10	(324, 0.5, 1)	(255, 127, 203)

With these generated colors paired to the style numbers, the map can be rendered in the ArcGIS.

4 Chromatism Analysis

In order to analyze the performance of the color-filling algorithm, we should quantify the difference between the colors of neighboring regions. A widely commended method which has been used in commercial products is a combination both weighted Euclidean distance functions, where the weight factors depend on how big the "red" component of the color is [11]. First one calculates the mean level of "red" and then weights the $\Delta R'$ and $\Delta B'$ signals as a function of the mean red lever. The distance between colors C_1 and C_2 is:

$$\bar{r} = \frac{C_1, R + C_2, R}{2}$$

$$\Delta R = C_{1,}R - C_2, R$$

$$\Delta G = C_1, G - C_2, G$$

$$\Delta B = C_{1,}B - C_2 B$$

$$\Delta C = \sqrt{\left(2 + \frac{\bar{r}}{256}\right) \times \Delta R^2 + 4 \times \Delta G^2 + \left(2 + \frac{255 - \bar{r}}{256}\right) \times \Delta B^2}$$

We calculated all the ΔC between the color of Key number and its adjacent colors in Table 2. The results are showed in the matrix in Table 3.

Table 3. Quantitative value of the difference of the generated colors

	1	2	3	4	5	6	7	8	9	10
3	446	517		570	669	672	605	486	380	383
5	418		669			319	407	428	649	504
1		228	446	199	418	304			457	
6	304		672		319				525	
7		603	605		407					415
9	457		380		649	525				
2			517	210			603			
4	199	210	570							
10			383		504		415			
8			486		428					

We also chose the colors in random mode which means the colors are distributed randomly by the system. The results are showed in Table 4.

By comparing the Quantitative value of the chromatism, we can conclude (see in Fig. 3) that the colors generated from the algorithm are more different than the random colors. The X axis represents the style numbers and de Y axis is the quantitative difference of two colors. Each point stands for a result of two neighboring regions. We can see most of the blue points are above the orange ones which means the colors generated from the color-filling algorithm looked more different than the random colors.

Table 4. Quantitative value of the difference of the random colors

	1	2	3	4	5	6	7	8	9	10
3	146	82		100	202	216	234	266	323	407
5	347		202			37	72	122	202	307
1		68	146	246	347	360			445	
6	360		216		37				170	
7		316	234		72					255
9	445		322		202	170				
2			82	181			316			
4	246	181	100							
10			407		307		255			
8			266		122					

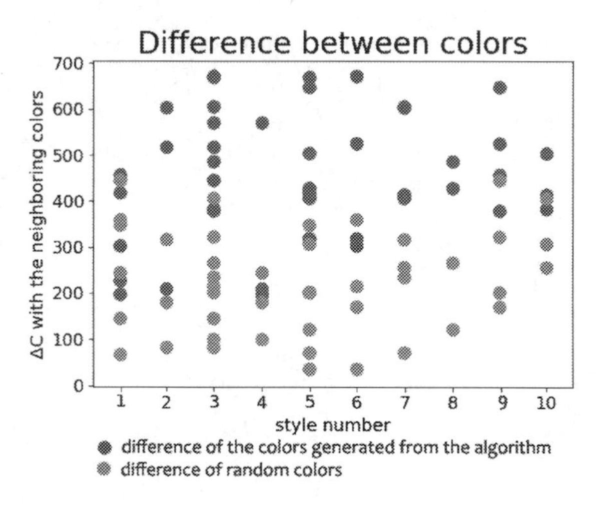

Fig. 3. Scatter points about chromatism (Color figure online)

5 Conclusions

In this study, an algorithm is proposed to generate colors in HSV color space intent to make them more different and easier to be distinguished in the map. The adjacent relations of the regions are also considered to make sure the similar colors will not appear in adjacent regions. The chromatism of adjacent colors is quantified by calculating weighted Euclidean distance and the results showed that it has a better performance than choosing colors by random. It can be widely used in the linguistic atlas because language can always be categorized into various categories in different themes. The example used in this paper is the statement about *father* in *Rugao* dialect and after the processing with the color-filling algorithm, it can be painted as Fig. 4. The conclusion drawn in this paper had been used to render most of the maps in *The Linguistic Atlas of the Rugao Dialect.* Although choosing the colors from the HSV color space is similar to the way humans perceive colors, some generated colors are not suitable for the publication as they are too dark or colorful. So, more parameters should be taken into consideration in the algorithm to make sure the generated colors could meet some certain conditions. However, how to define the concept of colorful is still a difficult and little understood problem and we should create more mathematical models to quantify it in the color space.

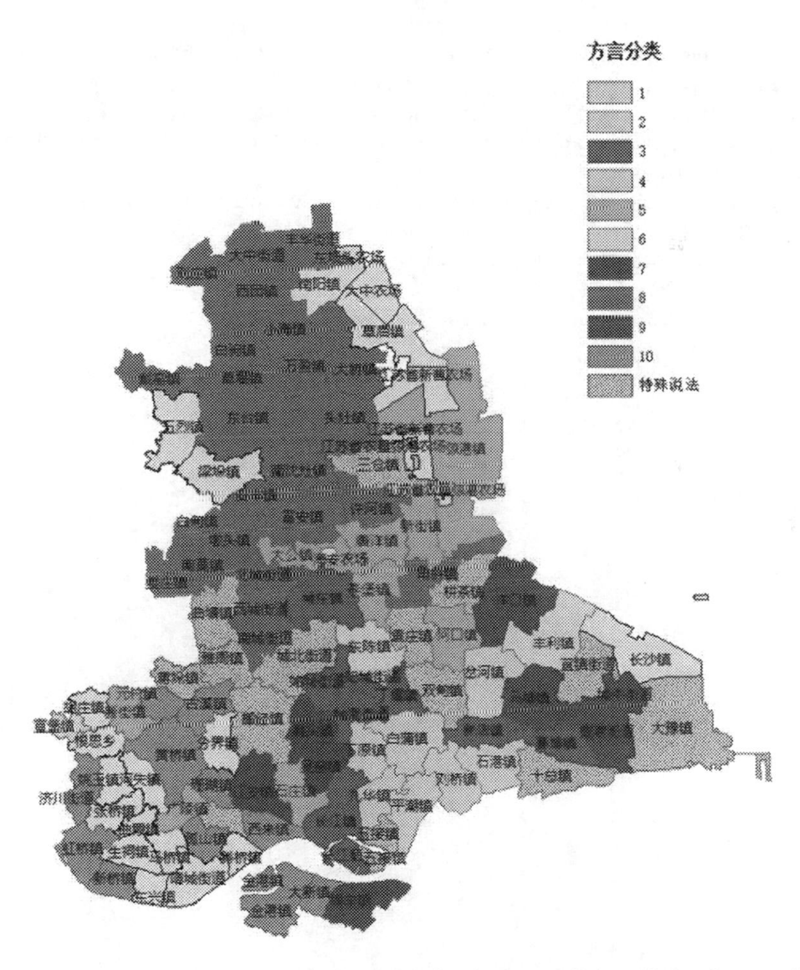

Fig. 4. A linguistic map about statement of *father* in *Rugao* dialect (Color figure online)

References

1. Cao, Z.: Linguistic atlas of Chinese dialects: preface. Lang. Teach. Linguist. Stud. **2**, 1–8 (2008)

2. Zhou, D., Tong, X., Li, L.: Production of dialect thematic maps base on ArcGIS and CorelDraw. Popular Sci. Technol. **3**(17), 205–207 (2015)
3. Qin, L., Gan, Y.: Analysis of symbol design in dialect map. J. Guangdong Polytech. Normal Univ. **8**, 1–7 (2014)
4. Veith, W.H.: Encyclopedia of Language & Linguistics, 2nd edn, pp. 517–528. Elsevier, Amsterdam (2006)
5. Appel, K., Haken, W.: The solution of the four-color-map problem. Sci. Am. **10**, 108–121 (1977)
6. Wu, R.-Q.: Extracting contour lines from topographic maps based on cartography and graphics knowledge. JCS&T **9**(2), 58–64 (2009)
7. Podobnikar, T., Škofic, J., Horvat, M.: Mapping and analysing the local language areas for Slovenian linguistic atlas. In: Gartner, G., Ortag, F. (eds.) Cartography in Central and Eastern Europe. Lecture Notes in Geoinformation and Cartography, vol. 199089, pp. 361–382. Springer, Heidelberg (2009). https://doi.org/10.1007/978-3-642-03294-3_23
8. https://resources.arcgis.com/en/help/main/10.1/index.html#//00080000001600000. Accessed 30 Sept 2013
9. Forsyth, D.A., Ponce, J.: Computer Vision a Modern Approach. Tsinghua University Press, Beijing (2004)
10. Ibraheem, N.A., Hasan, M.M.: Understanding color models: a review. ARPN J. Sci. Technol. **2**(3), 265–275 (2012)
11. http://compuphase.com/cmetric.htm. Accessed 31 May 2019

Person Re-identification Based on Spatially Constraints and Kernel Consensus PCA

Bin Hu[1(✉)], Yanjing Cai[1,2], Shi Cheng[1], and Zelin Wang[1]

[1] School of Information Science and Technology, Nantong University, Nantong, Jiangsu, China
hubin@ntu.edu.cn
[2] Jiangsu Vocational College of Business, Nantong, Jiangsu, China

Abstract. Person re-identification (Re-ID) aims at pedestrian matching and ranking across non-overlapping field-of-view. However, Re-ID is a challenge work due to the variations in resolution, pose, illumination and camera viewpoint. We learn a measurement score, which consists of multiple sub-similarity measurements with each taking in charge of a sub-region. In particular, the measurement function combines the similarity function and the distance function, so it exploits the complementary strength of the two functions. Dense color and texture features are extracted to describe the person image, and consensus PCA, which captures a consensus projection between multi-block data, is employed to deal with the features in a linear space. Furthermore, kernel consensus PCA is proposed to settle camera transition and dimensionality reduction in the non-linear space. The experimental results on four benchmark show significant and consistent improvements over the state-of-the-art methods.

Keywords: Person re-identification · Kernel consensus PCA · Metric learning · Spatially constraints

1 Introduction

Person re-identification (Re-ID) aims at finding the same pedestrian across a network of camera with non-overlapping field-of-view. It plays an important part in video surveillance by saving lots of human efforts on searching for a person from massive videos. However, it is a very challenging task due to these cameras have very little or no overlapping field of view. First, a camera observe many pedestrians in one day, and some of them may have similar appearance. Second, the same individual across a network of cameras may have a dissimilar appearances due to the variations in resolution, pose, camera settings, illumination, occlusions, background and camera viewpoint [1]. Most state-of-the-art approaches adopt supervised learning [2–5] and require lots of labeled image pairs. Though Semi-supervised and unsupervised methods do not need labeled image pairs, they are uncompetitive with supervised techniques and have received little attention in the research community [6, 7]. Some studies attempt to exploit person structure information to improve the performance of Re-ID methods. They believe that person structure information, such as body parts, human poses, person attributes, and background context information, can help Re-ID methods

K. Li et al. (Eds.): ISICA 2019, CCIS 1205, pp. 431–441, 2020.
https://doi.org/10.1007/978-981-15-5577-0_34

capture discriminative local visual features [8]. Deep learning architectures perform better than the approaches above, however training a deep convolution neural networks (DCNN) model requires thousands labeled images or more [9, 10].

In this paper, we propose an approach based on supervised metric learning. We employ the similarity metric learning model, which conveniently exploit the complementary strength of similarity and distance function, to measure the image pair. And we argue that the similarity between the image pair should obey certain spatial constraints, for example, the region containing the body of a pedestrian should be compared with the region containing the body rather than the region containing the head. With the constraint, each region has its own measurement score. For the high dimension of person description, consensus principal component analysis is used to reduce the dimension. We evaluate our approach by operating in-depth experiments in four benchmarks, and the results outperform the state-of-art.

2 Related Work

Most state-of-the-art approaches are based on metric learning and feature extraction. Most methods have been considered for learning a robust metric and developing distinctive feature representations. The pedestrian's face cannot be well captured in surveillance cameras, so the face cues cannot be used. Gray et al. proposed EFL by fusing 8 color channels with 19 texture channels [11], while Farenzena et al. considered the symmetric and asymmetric prior of human body part and employed the weighted color histograms, Maximally Stable Color Regions (MSCR) and Recurrent High-Structured Patches (RHSP) to describe the images [12]. Many researchers extracted dense local features and concatenated them to capture different image properties [2, 5, 13, 14]. Recent years, many works proposed various DCNN based methods which can be very effective, but extremely large amounts of images are required to pre-train a DCNN model and hundreds or thousands image pairs are need to fine-tune the model [8–10]. Above all, lots of different distinctive feature representations are proposed to describe the image, and the representations usually have a very high dimension. Most works adopt PCA [21], White PCA [2, 15] or kernel PCA [14] to reduce the feature dimension. However, probe and gallery images are captured by distinct cameras, PCA is a challenging scenario to the application in person Re-ID problem. Using multi-block multivariate models [16] to learn a common subspace where the direct comparison between probe and gallery images results in a high matching performance is a better solution to tackles the person Re-ID problem. These models have been employed when additional information is available for grouping variables in a meaningful blocks such as different camera views. Consensus PCA (CPCA) [17] is a multi-block extensions of PCA which seeks for a consensus direction among all the blocks and is useful to compare the descriptors of the same object. We adopt CPCA to learn a subspace where we match the probe and gallery images. Furthermore, a nonlinear extension of CPCA (Kernel CPCA) is proposed since CPCA is a linear model.

Metric learning has gradually shown its powerful strength in Re-ID problem [2, 3, 13–15, 18, 21], and among all of them, Mahalanobis distance functions have received the most attention in the Re-ID community. LDFA learned the metric matric by maximizing the inter-class separability [19], and LADF combined the distance metric with a locally adaptive thresholding rule for each pair of sample images [20]. Based on a likelihood-ratio test of two Gaussian distributions modeling positive and negative pairwise differences between intra-class and inter-class, a very effective and efficient metric learning named KISSME is proposed [2]. XQDA improved KISSME by learning a more discriminative distance metric and a low-dimensional subspace simultaneously [5].

Besides, some works formulated the Re-ID problem as a ranking problem and tried to learn a ranking function to order the relevant pedestrian pairs before irrelevant ones. Ranking methods based on SVM learning received the most attention. For instance, Prosser et al. [22] combined a set of weak RankSVMs into a stronger ranker using a boosting principle. Instead of computing a fixed weight vector for all pedestrians, Zhang et al. [23] learned specific weight parameters so that the ranking function is highly tuned to the person's appearance.

3 Proposed Method

The flowchart of proposed method illustrated in Fig. 1. A person image is divided into R regions in the horizon direction, and the similarity score combines the global score of the whole image and local scores for associated those regions. We explain it with more details as follows, including feature extraction, dimension reduce method CPCA and its kernel extension and the metric learning model that exploit the complementary strength of similarity and distance function.

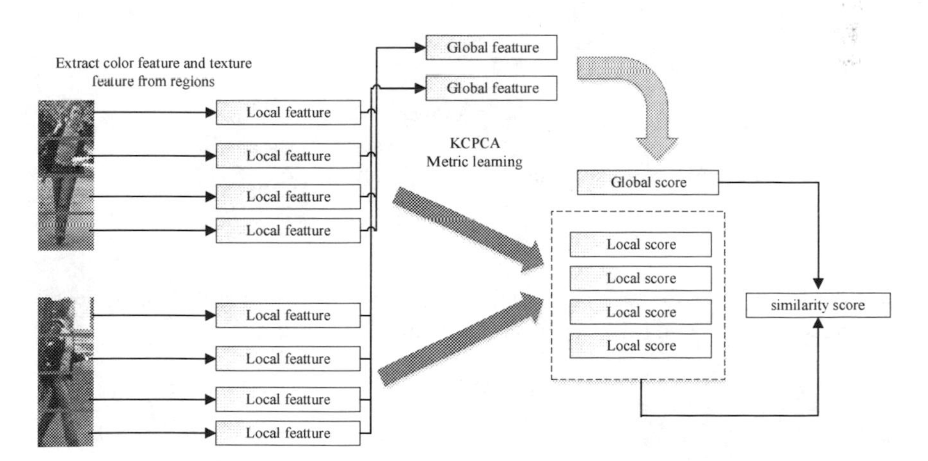

Fig. 1. The flowchart of proposed method. A person image is partitioned into R non-overlap horizontal stripe regions. Color feature and texture feature are extracted from each region and kernel CPCA is employed for dimension reduction. The final similarity score combines local scores and global score.

Feature extraction. A person image is divided into R non-overlap regions in the horizon direction, and each region is partitioned into a collection of overlapped patches. From each patch, color and texture histograms are extracted, and the feature of each patch that belongs to a same stripe region are concatenated together as x^r, where $r \in \{1, \ldots, R\}$, and the whole image can be described as $\{x^1, \ldots, x^R\}$.

Dimension reduce. CPCA has been widely employed in chemo-metrics and bio-chemical process monitoring [17], it is useful when there is a meaningful division of data into blocks. A typical example corresponds to multiple measures the images captured by multiple cameras of the same object.

Let $\mathbf{X}_a$ and $\mathbf{X}_b$ be two blocks of data, the algorithm begins with a super score $\mathbf{t} \in \mathbb{R}^n$, which is an initial vector of consensus and can be a column of these blocks. Then $\mathbf{t}$ is used to compute the loadings $\mathbf{w}_a, \mathbf{w}_b \in \mathbb{R}^m$:

$$\mathbf{w}_a = \mathbf{X}_a^{\mathrm{T}}\mathbf{t}/\mathbf{t}^{\mathrm{T}}\mathbf{t}$$

$$\mathbf{w}_b = \mathbf{X}_b^{\mathrm{T}}\mathbf{t}/\mathbf{t}^{\mathrm{T}}\mathbf{t}$$

$$\mathbf{w}_a = \frac{\mathbf{w}_a}{\|\mathbf{w}_a\|}$$

$$\mathbf{w}_b = \frac{\mathbf{w}_b}{\|\mathbf{w}_b\|}$$

Then the normalized loadings are used to compute the respective block scores $\mathbf{s}_a, \mathbf{s}_b \in \mathbb{R}^n$:

$$\mathbf{s}_a = \mathbf{X}_a\mathbf{w}_a$$

$$\mathbf{s}_b = \mathbf{X}_b\mathbf{w}_b$$

$\mathbf{s}_a$ and $\mathbf{s}_b$ are combined as a super block score $\mathbf{S}$ defined as:

$$\mathbf{S} = [\mathbf{s}_a, \mathbf{s}_b]$$

At last the super score $\mathbf{t}$ is regressed on the super block $\mathbf{S}$ to obtain the super weight $\mathbf{w}$ and $\mathbf{t}$ update by projecting the super block $\mathbf{S}$ onto super weight $\mathbf{w}$:

$$\mathbf{w} = \mathbf{S}^{\mathrm{T}}\mathbf{t}/\mathbf{t}^{\mathrm{T}}\mathbf{t}$$

$$\mathbf{w} = \frac{\mathbf{w}}{\|\mathbf{w}\|}$$

$$\mathbf{t} = \mathbf{S}\mathbf{w}$$

This process repeats until the convergence of super score $\mathbf{t}$ to a predefined precision, and block variables $\mathbf{X}_a$ and $\mathbf{X}_b$ are deflated to obtain a new score vector.

$$\mathbf{X}_a = \mathbf{X}_a - \mathbf{t}\mathbf{w}_a^{\mathrm{T}}$$

$$\mathbf{X}_b = \mathbf{X}_b - \mathbf{t}\mathbf{w}_b^{\mathrm{T}}$$

Let f be the number of factors, repeats the process f times, a f - dimension vector is got. $\mathbf{W}_a \in \mathbb{R}^{m \times f}$ and $\mathbf{T} \in \mathbb{R}^{n \times f}$ are constructed storing the loading vector $\mathbf{w}_a$ and $\mathbf{t}$. Due to the normalization, $\mathbf{w}_a$ is orthonormal, $\mathbf{T}$ can be obtained as a closed-form:

$$\begin{aligned} \mathbf{X}_a &= \mathbf{T}\mathbf{W}_a^{\mathrm{T}} \\ \mathbf{X}_a\mathbf{W}_a &= \mathbf{T}\mathbf{W}_a^{\mathrm{T}}\mathbf{W}_a \\ \mathbf{T} &= \mathbf{X}_a\mathbf{W}_a \end{aligned} \tag{1}$$

The common latent subspace of a test vector $\hat{\boldsymbol{x}}_a$ can be computed as:

$$\hat{\boldsymbol{t}}_a = \hat{\boldsymbol{x}}_a\mathbf{W}_a$$

Similarly, the projection vector of test sample $\hat{\boldsymbol{x}}_b$ is:

$$\hat{\boldsymbol{t}}_b = \hat{\boldsymbol{x}}_b\mathbf{W}_b$$

Above is the lincar CPCA model, a nonlinear model can reach improved results [13, 14], next the proposed kernel CPCA method is described.

The kernel matrices $\mathbf{K}_a$ and $\mathbf{K}_b$, represent the application of a kernel function using samples from blocks A and B, are high-dimensional space substituting the cross-product by kernel space $\mathbf{\Phi}$:

$$\mathbf{K} = \mathbf{\Phi}\mathbf{\Phi}^{\mathrm{T}}$$

From CPCA algorithm, we obtain that:

$$\mathbf{s}_a = \mathbf{X}_a\mathbf{X}_a^{\mathrm{T}}\mathbf{t}/\mathbf{t}^{\mathrm{T}}\mathbf{t}$$

$$\mathbf{s}_a = \mathbf{\Phi}\mathbf{\Phi}^{\mathrm{T}}\mathbf{X}_a^{\mathrm{T}}\mathbf{t}/\mathbf{t}^{\mathrm{T}}\mathbf{t}$$

$$\mathbf{s}_a = \mathbf{K}_a\mathbf{t}/\mathbf{t}^{\mathrm{T}}\mathbf{t}$$

Likewise,

$$\mathbf{s}_b = \mathbf{K}_b\mathbf{t}/\mathbf{t}^{\mathrm{T}}\mathbf{t}$$

According to Eq. (1), an approximation of $\mathbf{K}_a$ is derived:

$$\mathbf{X}_a = \mathbf{T}\mathbf{W}_a^{\mathrm{T}}$$

$$\mathbf{X}_a^{\mathrm{T}}\mathbf{X}_a = \mathbf{X}_a^{\mathrm{T}}\mathbf{W}_a\mathbf{W}_a^{\mathrm{T}}$$

$$\mathbf{K}_a = \left(\mathbf{W}_a \mathbf{W}_a^{\mathrm{T}}\right)^{\mathrm{T}} \mathbf{K}_a$$

$$\mathbf{K}_a \approx \mathbf{w}_a^{\mathrm{T}} \mathbf{w}_a \mathbf{K}_a$$

Similarly,

$$\mathbf{K}_b \approx \mathbf{w}_b^{\mathrm{T}} \mathbf{w}_b \mathbf{K}_b$$

Therefore, the rank-one deflation of kernel matrices $\mathbf{K}_a$ and $\mathbf{K}_b$ are, respectively,

$$\mathbf{K}_a = \mathbf{K}_a - \mathbf{w}_a^{\mathrm{T}} \mathbf{w}_a \mathbf{K}_a$$

$$\mathbf{K}_b = \mathbf{K}_b - \mathbf{w}_b^{\mathrm{T}} \mathbf{w}_b \mathbf{K}_b$$

Metric learning. Given the image pair, we adopt similarity metric score [15] to measure the similarity of the image pair:

$$f_{(M,G)}\left(x_i, x_j\right) = s_G\left(x_i, x_j\right) - d_M\left(x_i, x_j\right)$$

Where d_M is the distance measurement and s_G is the similarity function. The score of r-th region between image pair is defined:

$$s^r\left(x_i, x_j\right) = f^r_{(M,G)}\left(x_i, x_j\right)$$

We employ a linear function to combine the score of each region to exploit the complementary strengths of local regions:

$$s^{local}\left(x_i, x_j\right) = \sum_{i=1}^{R} s^r\left(x_i, x_j\right)$$

The score between local regions of image pair cannot describe the matching of large patterns across the regions. To compensate the insufficiency of local score, we also make use of the polynomial feature map by concatenating the feature of each region for the whole image, yielding global score:

$$s^{global}\left(x_i, x_j\right) = f_{(M,G)}\left(x_i, x_j\right)$$

The global score and local score are linearly combined, and the overall similarity score of the image pair is given by:

$$\mathrm{s}\left(x_i, x_j\right) = s^{local}\left(x_i, x_j\right) + s^{global}\left(x_i, x_j\right)$$

4 Experiments

We evaluate our approach by operating in-depth experiments in four benchmarks, VIPeR [24], PRID450S [25], iLIDS [26] and CUHK01 [27]. The results are reported using Cumulative Match Characteristic (CMC) performance curves, which corresponds to match performance at the top-r positions.

4.1 Evaluation Protocols

We fixed the evaluation protocol across all datasets. All of the images are scaled to be 128×48 pixels and partitioned into 8 regions. For each region, the patch size is 4×8 pixels, and the step is 2×4 pixels. The color features are extracted from RGB, norm-RGB, HSV, LUV and L1L2L3 color space, norm-RGB is defined as:

$$R = R/(R + G + B + \xi)$$

$$G = G/(R + G + B + \xi)$$

$$B = B/(R + G + B + \xi)$$

We concatenate the color histograms, each with 16 bins per channel, from the five color space above, into 240-dimensional color feature vector. And we extract HOG [28] and SILTP as the texture feature. For evaluation, we generated probe/gallery images accordingly to the common setting: VIPeR: 316/316; PRID: 100/649; iLIDS: 60/60 and CUHK01: 486/486.

4.2 Comparison to State-of-the-Art Approaches

The comparison results of four benchmark are demonstrated from Table 1 to Table 4. From the experimental results, our method performs better than the others. The framework is not sensitive to local occlusions owing to the measurement score combines the local scores and global score. The metric function collaborate similarity function as well as distance function to exploit their complementary strength, so our method is better than the methods use only similarity function or distance function. And the multiblock PCA we adopt is more effective than PCA in analyzing the images captured by multiple cameras of the same object. Though CNN based methods outperform the methods below, they require extreme large number images to pre-train the model and thousands of image pairs to fine-tune, and most datasets are too small to fine-tune the CNN model.

Table 1 shows the results of the state-of-the-art method on VIPeR dataset, OSML [29] only reports its rank 1 accuracy, and our method outperforms others.

PRID450S is a new dataset, and only a few methods report the results on it. We compare our method with these algorithms in Table 2, and our method outperforms them. Our result is similar to LOMO + XQDA [5], but a little better at rank 1.

Table 3 shows the results of the state-of-the-art method on iLIDS dataset, and our result performs much better than the others at rank 1.

Table 1. Top-n matching rates (%) of different methods on the VIPeR dataset

Method	Rank1 (%)	Rank5 (%)	Rank10 (%)	Rank20 (%)
LFDA [19]	19.7	46.7	62.1	77.0
SVMML [20]	27.0	60.9	75.4	87.3
KISSME [2]	23.8	54.8	71.0	85.3
Rpcca [14]	22.0	54.8	71.0	85.3
OSML [29]	34.3	–	–	–
Klfda [14]	32.3	65.8	79.7	90.9
MFA [14]	32.2	66.0	79.7	90.6
RDC [13]	15.7	38.42	53.86	70.1
LOMO [5]	40.0	68.13	80.51	91.1
LSSCDL [23]	42.7	–	84.3	91.9
Ours	44.5	74.3	85.5	93.1

Table 2. Top-n matching rates (%) of different methods on the PRID450S dataset

Method	Rank1 (%)	Rank5 (%)	Rank10 (%)	Rank20 (%)
SCNCD [30]	41.6	68.9	79.4	87.8
KISSME [2]	33.0	–	71.0	79.0
LOMO [5]	61.2	84.8	90.9	95.1
LSSCDL [23]	60.5	–	88.5	93.6
EIML [31]	35.0	–	68.0	77.0
Ours	61.4	84.9	90.8	95.6

Table 3. Top-n matching rates (%) of different methods on the iLIDS dataset

Method	Rank1 (%)	Rank5 (%)	Rank10 (%)	Rank20 (%)
KISSME [2]	28.0	54.2	67.9	81.6
LFDA [19]	32.2	56.0	68.7	81.6
SVMML [20]	20.8	49.1	65.4	81.7
rPCCA [14]	28.0	56.5	71.8	85.9
kLFDA [14]	36.9	65.3	78.3	89.4
MFA [14]	32.1	58.8	72.2	85.9
LOMO [5]	43.0	66.8	78.2	88.2
Ours	49.7	69.8	82.6	91.4

On the CUHK01 dataset, our method perform as well as LOMO + XQDA [5], and the two methods both outperform the others.

Table 4. Top-n matching rates (%) of different methods on the CUHK01 dataset

Method	Rank1 (%)	Rank5 (%)	Rank10 (%)	Rank20 (%)
KISSME [2]	12.5	31.5	42.5	54.9
LFDA [19]	13.3	31.1	42.2	54.3
SVMML [20]	18.0	42.3	55.4	68.8
Rpcca [14]	21.6	47.4	59.8	72.6
kLFDA [14]	29.1	55.2	66.4	77.3
MFA [14]	29.6	55.8	66.4	77.3
LOMO [5]	63.2	83.9	90.0	94.4
OSML [29]	45.6	–	–	–
Ours	63.2	84.1	90.7	95.0

5 Conclusions

This paper presents a person Re-ID method based on spatially constrained feature extraction and similarity function. The framework combines local similarities with global similarity, and collaborate similarity function as well as distance function to exploit their complementary strength. The kernel CPCA shows its power in comparing the descriptors of the same object by projecting the feature to a subspace. The experimental results in four benchmarks outperform the state-of-art.

Acknowledgements. This work was partially supported by National Natural Science Foundation of China (NO. 61976120), Research on Teaching Reform at Nantong University (NO.2018B43) and Project Topics of Nantong Science and Technology Plan (Guidance) in 2018 (NO. MSZ18080).

References

1. Chen, Y.C., Zhu, X., Zheng, W.S., et al.: Person re-identification by camera correlation aware feature augmentation. IEEE Trans. Pattern Anal. Mach. Intell. **40**(2), 392–408 (2018)
2. Koestinger, M., Hirzer, M., Wohlhart, P., et al.: Large scale metric learning from equivalence constraints. In: CVPR 2012, pp. 2288–2295 (2012)
3. Cheng, D., Gong, Y., Shi, W., et al.: Person re-identification by the asymmetric triplet and identification loss function. Multimedia Tools Appl. **77**(3), 3533–3550 (2018)
4. Matsukawa, T., Okabe, T., Suzuki, E., et al.: Hierarchical Gaussian descriptor for person re-identification. In: CVPR 2016, pp. 1363–1372 (2016)
5. Liao, S., Hu, Y., Zhu, X., et al.: Person re-identification by local maximal occurrence representation and metric learning. In: CVPR 2015, pp. 2197–2206 (2015)
6. Lv, J., Chen, W., Li, Q., et al.: Unsupervised cross-dataset person re-identification by transfer learning of spatial-temporal patterns. In: CVPR 2018, pp. 7948–7956 (2018)
7. Peng, P., Xiang, T., Wang, Y., et al.: Unsupervised cross-dataset transfer learning for person re-identification. In: CVPR 2016, pp. 1306–1315 (2016)

8. Song, C., Huang, Y., Ouyang, W., et al.: Mask-guided contrastive attention model for person re-identification. In: CVPR 2016, pp. 1179–1188 (2016)
9. Wu, Q., Dai, P., Chen, P., et al.: Deep adversarial data augmentation with attribute guided for person re-identification. SIViP **5**, 1–8 (2019). https://doi.org/10.1007/s11760-019-01523-3
10. Shen, Y., Li, H., Yi, S., Chen, D., Wang, X.: Person re-identification with deep similarity-guided graph neural network. In: Ferrari, V., Hebert, M., Sminchisescu, C., Weiss, Y. (eds.) ECCV 2018. LNCS, vol. 11219, pp. 508–526. Springer, Cham (2018). https://doi.org/10.1007/978-3-030-01267-0_30
11. Gray, D., Tao, H.: Viewpoint invariant pedestrian recognition with an ensemble of localized features. In: Forsyth, D., Torr, P., Zisserman, A. (eds.) ECCV 2008. LNCS, vol. 5302, pp. 262–275. Springer, Heidelberg (2008). https://doi.org/10.1007/978-3-540-88682-2_21
12. Farenzena, M., Bazzani, L., Perina, A., et al.: Person re-identification by symmetry-driven accumulation of local features. In: CVPR 2010, pp. 2360–2367 (2010)
13. Zheng, W.S., Gong, S., Xiang, T.: Reidentification by relative distance comparison. IEEE Trans. Pattern Anal. Mach. Intell. **35**(3), 653 (2013)
14. Xiong, F., Gou, M., Camps, O., Sznaier, M.: Person re-identification using kernel-based metric learning methods. In: Fleet, D., Pajdla, T., Schiele, B., Tuytelaars, T. (eds.) ECCV 2014. LNCS, vol. 8695, pp. 1–16. Springer, Cham (2014). https://doi.org/10.1007/978-3-319-10584-0_1
15. Cao, Q., Ying, Y., Li, P.: Similarity metric learning for face recognition. In: ICCV 2013, pp. 2408–2415 (2013)
16. Westerhuis, J.A., Kourti, T., Macgregor, J.F.: Analysis of multiblock and hierarchical PCA and PLS models. J. Chemom. **12**(5), 301–321 (1998)
17. Wold, M.S., Hellberg, H.S., et al.: PLS modeling with latent variables in two or more dimensions. In: PLS Model Building: Theory and Applications (1987)
18. Davis, J.V., Kulis, B., Jain, P., et al.: Information-theoretic metric learning. In: International Conference on Machine Learning, pp. 209–216. ACM (2007)
19. Pedagadi, S., Orwell, J., Velastin, S., et al.: Local fisher discriminant analysis for pedestrian re-identification. In: CVPR 2013, pp. 3318–3325 (2013)
20. Li, Z., Chang, S., Liang, F., et al.: Learning locally-adaptive decision functions for person verification. CVPR 2013, pp. 3610–3617 (2013)
21. Sun, C., Wang, D., Lu, H.: Person Re-identification via distance metric learning with latent variables. IEEE Trans. Image Process. **26**(1), 23–34 (2017)
22. Engel, C., Baumgartner, P., Holzmann, M., et al.: Person re-identification by support vector ranking. In: BMVC 2010, pp. 1–11 (2010)
23. Zhang, Y., Li, B., Lu, H., et al.: Sample-specific SVM learning for person re-identification. In: CVPR 2016, pp. 1278–1287 (2016)
24. Gray, D., Brennan, S., Tao, H.: Evaluating appearance models for recognition, reacquisition, and tracking. In: PETS 2007 (2007)
25. Roth, Peter M., Hirzer, M., Köstinger, M., Beleznai, C., Bischof, H.: Mahalanobis distance learning for person re-identification. In: Gong, S., Cristani, M., Yan, S., Loy, C.C. (eds.) Person Re-Identification. ACVPR, pp. 247–267. Springer, London (2014). https://doi.org/10.1007/978-1-4471-6296-4_12
26. Zheng, W.S., Gong, S., Xiang, T.: Associating groups of people. In: BMVC 2009 (2009)
27. Li, W., Zhao, R., Wang, X.: Human reidentification with transferred metric learning. In: Lee, K.M., Matsushita, Y., Rehg, James M., Hu, Z. (eds.) ACCV 2012. LNCS, vol. 7724, pp. 31–44. Springer, Heidelberg (2013). https://doi.org/10.1007/978-3-642-37331-2_3

28. Dalal, N., Triggs, B.: Histograms of oriented gradients for human detection. In: CVPR 2005, pp. 886–893 (2005)
29. Bak, S., Carr, P.: One-shot metric learning for person re-identification. In: CVPR 2017 (2017)
30. Yang, Y., et al.: Salient color names for person re-identification. In: Fleet, D., Pajdla, T., Schiele, B., Tuytelaars, T. (eds.) ECCV 2014. LNCS, vol. 8689, pp. 536–551. Springer, Cham (2014). https://doi.org/10.1007/978-3-319-10590-1_35
31. Hirzer, M., Roth, P.M., Bischof, H.: Person re-identification by efficient impostor-based metric learning. In: IEEE International Conference on Advanced Video and Signal-Based Surveillance, pp. 203–208 (2012)

Three-Dimensional Reconstruction and Monitoring of Large-Scale Structures via Real-Time Multi-vision System

Yunchao Tang[1,4](✉), Mingyou Chen[2,4], Xiangguo Wu[3,4](✉), Kuangyu Huang[2,4], Fengyun Wu[2,4], Xiangjun Zou[2,4], and Yuxin He[2,4]

[1] College of Urban and Rural Construction, Zhongkai University of Agriculture and Engineering, Guangzhou 510225, China
ryan.twain@gmail.com

[2] Key Laboratory of Key Technology on Agricultural Machine and Equipment, College of Engineering, South China Agricultural University, Guangzhou 510642, China

[3] Key Lab of Structures Dynamic Behavior and Control of the Ministry of Education and School of Civil Engineering, Harbin Institute of Technology, Harbin 150001, China

[4] Key Lab of Smart Prevention and Mitigation of Civil Engineering Disasters of the Ministry of Industry and Information Technology, Harbin Institute of Technology, Harbin 150090, China

Abstract. A four-ocular vision system is proposed for the three-dimensional (3D) reconstruction of large-scale concrete-filled steel tube (CFST) materials under complex testing conditions. These measurements are vitally important for evaluating the seismic performance and 3D deformation of large-scale specimens. A four-ocular vision system is constructed to sample the large-scale CFST, then point cloud acquisition, filtering, and stitching algorithms are applied to obtain 3D point cloud of the specimen surface. Novel point cloud correction algorithms based on geometric features and deep learning are proposed to correct the coordinates of the stitched point cloud. The proposed algorithms center on the stitching error of the multi-view point cloud and the geometric and spatial characteristics of the targets for error compensation, which makes them highly adaptive and efficient. A high-accuracy multi-view 3D model for the purposes of real-time complex surface monitoring can be obtained via this method. Performance indicators of the two algorithms were evaluated on actual tasks. The cross-section diameters at specific heights in the reconstructed models were calculated and compared against laser range finder data to test the performance of the proposed method. A visual tracking test on a CFST under cyclic loading shows that the reconstructed output well reflects the complex 3D surface after point cloud correction and meets the requirements for dynamic monitoring. The proposed method is applicable to complex environments featuring dynamic movement, mechanical vibration, and continuously changing features.

Y. Tang and M. Chen—Authors contributed equally to this work.

K. Li et al. (Eds.): ISICA 2019, CCIS 1205, pp. 442–457, 2020.
https://doi.org/10.1007/978-981-15-5577-0_35

Keywords: Multi-vision · Point cloud correction · Visual tracking · Real-time detection · Structural monitoring

1 Introduction

Three-dimensional (3D) visual information is the most intuitive data available to an intelligent machine as it attempts to sense the external world [1–3]. Vision 3D reconstruction technology can be utilized to acquire the spatial information of target objects for efficient and accurate non-contact measurement [4, 5]. It is an effective approach to tasks such as real-time target tracking, quality monitoring, and surface data acquisition; further, it is the key to realize automatic, intelligent, and safe machine operations [6–15].

In the field of civil engineering, researchers struggle to reveal the failure mechanisms of certain materials or structures in seeking the exact properties of composite materials. Traditional contact measurement methods rely on strain gauges, displacement meters, or other technology which may be inconvenient and inefficient. A vision sensor can comprehensively reveal the optical information of the target surface, allow the user to develop a highly targeted measurement scheme for different targets, and achieve high precision and non-contact measurement.

Researchers and developers in the computer vision field additionally struggle to effectively track and measure dynamic surface-deformed objects with stereo vision. The 3D reconstruction of curved surfaces under large fields of view (FOVs) is particularly challenging in terms of full field dynamic tracking [16]. The core algorithm is the key component of any tracking system. Problematic core algorithms restrict the application of 3D visual reconstruction technology including omnidirectional sampling and high-quality point cloud stitching.

Extant target surface monitoring techniques based on 3D reconstruction include monocular vision, binocular vision, and multi-vision methods [15, 17–19]. The inherently narrow FOV of monocular vision and binocular vision systems does not allow the user to sample large-scale information [20, 21]. The construction of a multi-vision system, supported by model solutions and error analysis methodology under coordinate correlation theory, is the key to successful omni-directional sampling. Candau [22], for example, correlated two independent binocular vision systems with a calibration object while applying a spray on an elastic target object as a random marker for the dynamic mechanical analysis of an elastomer. Zhou [23] binary-encoded a computer-generated standard sinusoidal fringe pattern. Shen [24] conducted 3D profilometric reconstruction via flexible sensing integral imaging with object recognition and automatic occlusion removal. Liu [21] automatically reconstructed a real, 3D human body in motion as-captured by multiple RGB-D (depth) cameras in the form of a polygonal mesh; this method could, in practice, help users to navigate virtual worlds or even collaborative immersive environments. Malesa [25] used two strategies for the spatial stitching of data obtained by multi-camera digital image correlation (DIC) systems for engineering failure analysis: one with overlapping FOVs of 3D DIC setups and another with distributed 3D DIC setups that have not-necessarily-overlapping FOVs.

The above point cloud stitching applications transform the point cloud into a uniform coordinate system by coordinate correlation. In demanding situations, the precision of the stitched point cloud may be decisive while the raw output of the coordinate correlation is ineffective. Persistent issues with dynamic deformation, illumination, vibration, and characteristic changes as well as visual measurement error caused by equipment and instrument deviations yet restrict the efficacy of high-precision 3D reconstruction applications. To this effect, there is demand for new techniques to analyze point cloud stitching error and for designing novel correction methods.

In addition to classical geometric methods and optical methods, deep neural networks have also received increasing attention in the 3D vision field due to their robustness. Ayan [26] obtained the topology and structure of 3D shape by means of coding, so that convolutional neural network (CNN) could be directly used to learn 3D shapes and therefore perform 3D reconstruction; Li [27] combined structured light techniques and deep learning to calculate the depth of targets. These methods alone outperform traditional methods on occlusion and un-textured areas; they can also be used as a complement to traditional methods. Zhang [28] proposed a method for measuring the distance of a given target using deep learning and binocular vision methods, where target detection network methodology and geometric measurement theory were combined to obtain 3D target information. Sun [29] designed a CNN and established a multi-view system for high-precision attitude determination, which effectively mitigates the lack of reliable visual features in visual measurement. Yang [30] established a binocular stereo vision system combined with online deep learning technology for semantic segmentation and ultimately generated an outdoor large-scale 3D dense semantic map.

Our research team has conducted several studies combining machine vision and 3D reconstruction in various attempts to address the above problems [31–41]. In the present study, we focus on stitching error and the recovery of multi-view point clouds for the high-accuracy monitoring of large-scale CFST specimens. Our goal is to examine the structure and material properties in a non-contact manner. We ran a construction and error analysis of a multi-vision model followed by improvements to the high-quality point cloud correction algorithms to achieve real-time correction and reconstruction of large-scale CFST structures. The point cloud correction algorithms, which center on the stitching error of the multi-view point cloud, are implemented via geometry-based algorithm and deep-learning-based algorithm. Since the proposed point cloud correction algorithms are constructed based on the geometric and spatial characteristics of the target, they are adaptive and efficient under complex conditions featuring dynamic movement, mechanical vibration, and continuously changing features. We hope that the observations discussed below will provide theoretical and technical support in improving the monitoring performance of current visual methods for CFSTs and other large-scale objects.

2 Methodology

The algorithms and seismic test operations used in this study for obtaining the dynamic CFST surfaces by four-ocular vision system are described in this section. The dynamic specimen surfaces were obtained and the relevant geometric parameters were extracted by means of stereo rectification, point cloud acquisition, point cloud filtering, point cloud stitching, and point cloud correction algorithms. A flow chart of this process is shown in Fig. 1.

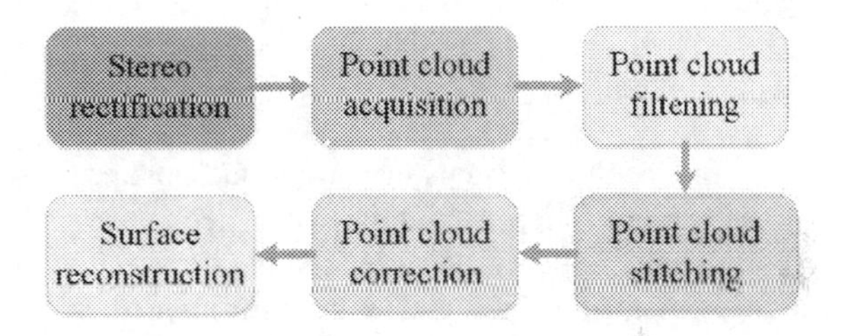

Fig. 1. Algorithm flow overview.

2.1 Stereo Rectification

A 2D circle grid with 0.3 mm manufacturing error was placed in 20 different poses to provide independent optical information for calibration. We applied camera calibration based on Zhang's method [42] to determine the matrixes of each individual camera.

$$Z_C \begin{bmatrix} u \\ v \\ 1 \end{bmatrix} = \mathrm{M}[\mathbf{R} \vdots \mathbf{T}] \begin{bmatrix} X_w \\ Y_w \\ Z_w \\ 1 \end{bmatrix} \tag{1}$$

where $[u\ v\ 1]^T$ is the image coordinate of the circle center, $[X_w\ Y_w\ Z_w]^T$ is the world coordinate of the circle center, Zc is the depth of the circle center, and M and [R T] are the camera matrix and extrinsic matrix of the single camera, respectively. Structural parameters $[R_T\ T_T]$ of the binocular cameras were calculated based on the extrinsic parameters to realize stereo calibration.

After stereo calibration, we used the classical Bouguet's method [43] to implement stereo rectification. Corresponding points are placed from left and right images into the same row in the images.

2.2 Point Cloud Acquisition

The camera in our setup samples images after stereo rectification. It is necessary to obtain as much target surface information as possible to perform 3D reconstruction and obtain accurate geometric parameters of the target, so we sought to generate as dense a 3D point cloud as possible. The triangulation-based calculation for dense 3D points is as follows:

$$X_C = \frac{x_l T_x}{d},\ Y_C = \frac{y_l T_x}{d},\ Z_C = \frac{f T_x}{d} \tag{2}$$

where $[X_C, Y_C, Z_C]^T$ is the camera coordinate of the target, $[x_l, y_l]^T$ is the left imaging plane coordinate of the target, T_x is the length of the baseline, f is the focal length of the camera, and $d = x_l - x_r$ is the disparity. Here, we used a classical 3D stereo matching algorithm [44] to generate a dense disparity map from each pixel in the images.

2.3 Point Cloud Filtering

Point cloud filtering is one of the key steps in point cloud post processing. It involves denoising and structural optimization according to the geometric features and spatial distribution of the point cloud, which yields a relatively compact and streamlined point cloud structure for further post processing. The pass-through filter, bilateral filter and radius outlier removal (ROR) filter is utilized in the research.

The pass-through filter [45] defines inner and outer points according to a cuboid with edges parallel to three coordinate axes. Points that fall outside of the cuboid are deleted. The cuboid is expressed as follows:

$$\{(X_C, Y_C, Z_C)|X_{\min} \le X_C \le X_{\max}, Y_{\min} \le Y_C \le Y_{\max}, Z_{\min} \le Z_C \le Z_{\max}\} \tag{3}$$

where $\begin{bmatrix} X_C & Y_C & Z_C \end{bmatrix}^T$ is the inlier, $X_{\min}$, $Y_{\min}$, $Z_{\min}$, $X_{\max}$, $Y_{\max}$ and $Z_{\max}$ are the boundaries of the cuboid, respectively.

The pass-through filter can be used to roughly obtain the main point cloud structure. The relative position of the specimen and the optical system can be determined in advance so that the threshold in each direction can be determined accurately. It is worth noting that although the specimen discussed here is cylindrical, it had a large amplitude swing during the test; the cuboid area has better adaptability than the cylindrical area in the filtering task for this reason.

The bilateral filter [46] can remove small-scale noise, i.e., anomalous 3D points that are intermingled with the surface of the point cloud. Bilateral filtering smooths the point cloud surface while retaining the necessary edge information. Noise moves along the following direction:

$$p_i^{'} = p_i + \alpha \cdot n \tag{4}$$

where $p_i^{'}$ is the filtered point, p_i is the original point, α is the weight of bilateral filtering, and n is the normal vector of p_i.

The ROR filter [47] distinguishes inliers from outliers based on the number of neighbors of each element in the point cloud. For any point $\boldsymbol{P}_i$ in point cloud $\{\boldsymbol{P}\}$, a sphere is constructed with a radius of r centered on it. If the number of elements in $\{S_i\}$ is less than N, the ROR filter regards $\boldsymbol{P}_i$ as noise.

2.4 Point Cloud Stitching

Multiple cameras can be employed in the sampling task to gather as much information as possible about the target. As shown in Fig. 2, four cameras constitute two pairs of binocular cameras in our experimental setup. The two left camera coordinate systems are denoted "CCS-A" and "CCS-B", respectively, and the world coordinate system as "WCS". The principle of point cloud stitching is to solve the coordinate transformation matrixes ${}^{CA}_{W}T$ and ${}^{CB}_{W}T$ between WCS to CCS-A and CCS-B, then transfer the point cloud from CCS-A and CCS-B to WCS to complete the coordinate correlation. The WCS can be established via a high-precision calibration board with known parameters.

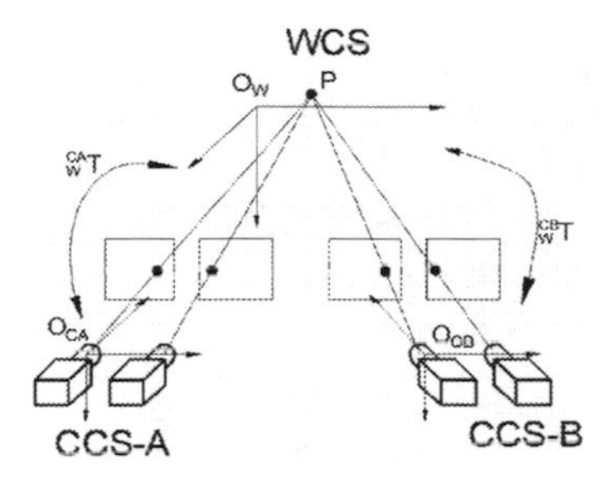

Fig. 2. Coordinate transformation of multi-camera system

Our calibration board has 99 circular patterns. The coordinates of all the centers on CCS-A, CCS-B, and WCS were combined column-by-column to obtain coordinate data matrixes M_{CA} M_{CB}, and M_{W} (sized 3*99):

$$M_{CA} = {}^{CA}_{W}TM_{W} \tag{5}$$

$$M_{CB} = {}^{CB}_{W}TM_{W} \tag{6}$$

The following is obtained through matrix transformation according to the principal of least squares:

$${}^{\mathrm{CA}}_{W}\boldsymbol{T} = M_{\mathrm{CA}}M_{W}^{\mathrm{T}}(M_{W}M_{W}^{\mathrm{T}})^{-1} \tag{7}$$

$${}^{\mathrm{CB}}_{W}\boldsymbol{T} = M_{\mathrm{CB}}M_{W}^{\mathrm{T}}(M_{W}M_{W}^{\mathrm{T}})^{-1} \tag{8}$$

The coordinate data matrixes Q_{CA} and Q_{CB} of the target relative to CCS-A and CCS-B were calculated according to Formula (2) and then converted into the WCS as follows:

$$Q_{WA} = {}^{CA}_{W}TQ_{CA} \tag{9}$$

$$Q_{WB} = {}^{CB}_{W}TQ_{CB} \tag{10}$$

2.5 Point Cloud Correction

The stress state of the devices in this setup is very complex due to the combination of large axial pressure, cyclic tension, and fastening force. Random, difficult-to-measure shocks and vibrations often occur during such tests which cause the camera frames to move slightly on the ground, although they are fixed in advance and far from the specimen (about 1 m). The effect of the slight movement of the camera frames may cause the stitched point clouds to appear staggered, as shown in Fig. 3, which can be destructive in demanding monitoring missions. Assume the translation vector and rotation matrix of two staggered point clouds caused by the complex loading conditions are P_M and R_M.

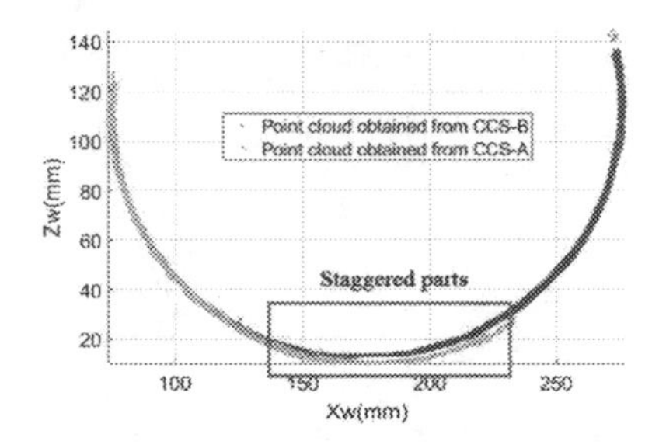

Fig. 3. Visualization of point cloud stitching problem

We designed geometry-based and deep-learning-based algorithms in this study to correct two staggered point clouds.

In geometry-based algorithm, we assume that the optical axis of the cameras remain parallel to the (horizontal) ground surface after they move, so the moving direction of the point cloud is perpendicular to the Y_W axis and $p_y = 0$, and there is no rotation between the two point clouds caused by stitching error:

$$P_M = \begin{bmatrix} p_x & 0 & p_z \end{bmatrix}^T, \ R_M = E$$

where E refers to the unit matrix. The translation vector mentioned above was acquired by nonlinear least squares, then the staggered point cloud was translated accordingly. Deformation of the upper part of the specimen was not severe at any point in the test, so it is close to an ideal cylinder even the bottom presents obvious deformation. We set a plane $Y_W = h_0^{'}$ parallel to the $X_W O_W Z_W$ plane to intercept the upper part of the specimen and obtain a circular arc. The expression of the arc in 3D space is:

$$\begin{cases} F_{circle}(X_W, Z_W) = (X_W - a)^2 + (Z_W - b)^2 - r^2 = 0 \\ Y_W = h_0^{'} \end{cases} \quad (11)$$

where $[a \quad b]^T$ is the projection of the circle center on the $X_W O_W Z_W$ plane and r is the radius of the circle. The optimization target of nonlinear least squares is:

$$\min S_{circle}(a, b, r) = \sum_{i=1}^{n} F_{circle}^2(X_W^{(i)}, Z_W^{(i)}) \tag{12}$$

where $S_{circle}(a, b, r)$ is the objective function of nonlinear least squares and $\begin{bmatrix} X_W^{(i)} & Z_W^{(i)} \end{bmatrix}^T$ is the i-th sample point in the section. The optimal values $\hat{a}$, $\hat{b}$, and $\hat{r}$ were iteratively obtained by the Levenberg-Marquardt (LM) algorithm [48].

The respective estimated values $c_A = [\hat{a}_A, \hat{b}_A]^T$ and $c_B = [\hat{a}_B, \hat{b}_B]^T$ for the height h_0' can be obtained. $\mathbf{P}_M$ can be calculated as shown in Fig. 4.

$$\mathbf{P}_M = c_A - c_B \tag{13}$$

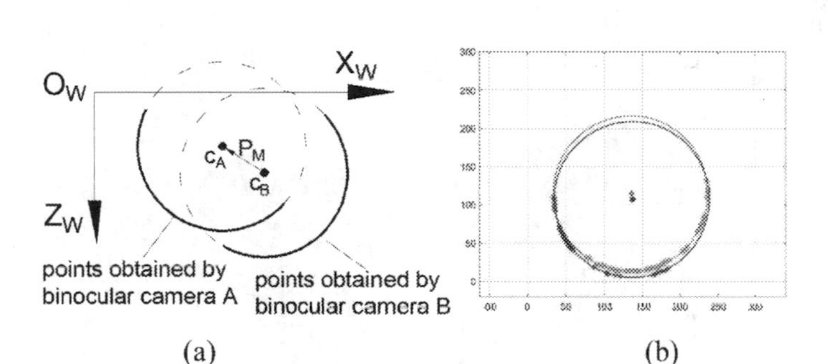

Fig. 4. (a) Schematic diagram of splicing correction; (b) Circle fitting example

Next, we used PointNet++ [49], a robust 3D semantic network, to extract the common parts of the point cloud and performed ICP registration [50] on them to correct the relative positions of the two clouds. This method has no stronger assumptions than the geometry-based method.

$$P_M = [p_x \quad p_y \quad p_z]^T, \; R_M = \begin{bmatrix} r_{11} & r_{12} & r_{13} \\ r_{21} & r_{22} & r_{23} \\ r_{31} & r_{32} & r_{33} \end{bmatrix}$$

where P_y and R_M are not necessarily equal to 0 and unit matrix.

As shown in Fig. 5, we manually marked the common parts of two staggered point clouds in a large number of samples and fed them to the network for training. After the training was complete, we could use the network to identify and extract the common parts of the two given point clouds.

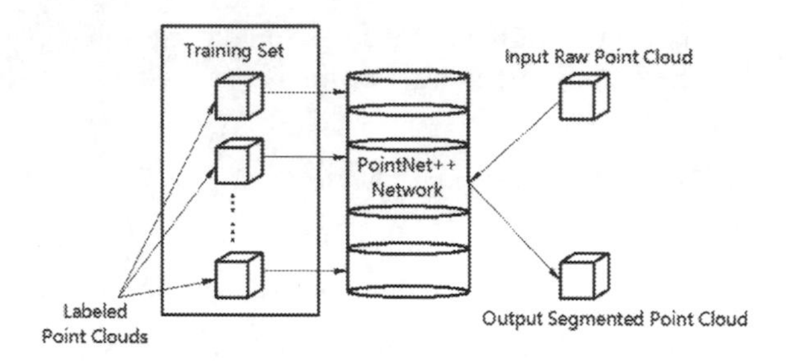

Fig. 5. PointNet++ network training and application

The blue shaded area in Fig. 6 represents the common parts extracted by the PointNet ++ network which were used to implement ICP registration for point cloud correction.

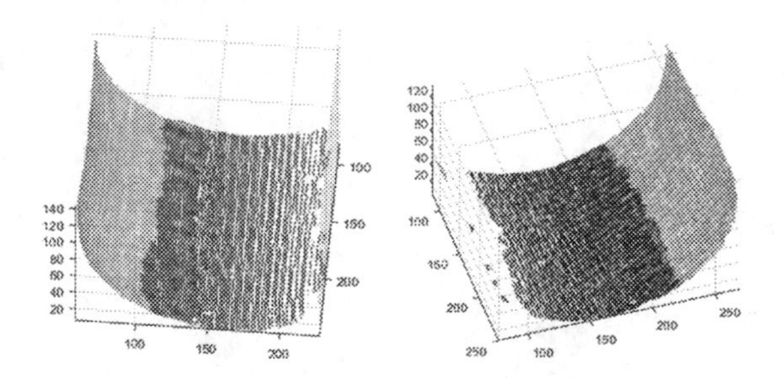

Fig. 6. Common parts extracted by PointNet++

In the point cloud stitching step, there is always a small angle θ between the plane of the calibration board and the axis of the specimen. Since the vision measurement technique is based on the WCS, so it is difficult to determine the actual measuring position in this setup. To address this problem, we transformed the points from the WCS to a new specimen coordinate system fixed to the specimen, so as to guaranteed that the section in the Y_F direction actually corresponds to the real cross-section of the specimen.

The specimen coordinate system was established as discussed below. The equation for an arbitrarily posed cylinder in the WCS is:

$$[A(Z_W - z_0) - C(X_W - x_0)]^2 + [A(Y_W - y_0) - B(X_W - x_0)]^2 + [B(Z_W - z_0) - C(Y_W - y_0)]^2 - R^2 = 0 \quad (14)$$

where $s = [A \quad B \quad C]^T$ is the unit vector of the axis of the cylinder, $O_F = [x_0 \quad y_0 \quad z_0]^T$ is a 3D point on the axis of the cylinder, and R is the radius of the cylinder.

Similar to the circle-fitting process discussed above, the upper half of the cylinder was fitted using nonlinear least squares via LM algorithm, and the optimal values $s = \begin{bmatrix} \hat{A} & \hat{B} & \hat{C} \end{bmatrix}^T$ and $O_F = \begin{bmatrix} \hat{x}_0 & \hat{y}_0 & \hat{z}_0 \end{bmatrix}^T$ can be iteratively obtained. A unique point $Y_1 = [0 \quad y_1 \quad 0]^T$ for $\overrightarrow{O_F Y_1} \bullet s = 0$ can be determined on the Y_W axis:

$$y_1 = \frac{\hat{A}\hat{x}_0 + \hat{C}\hat{z}_0}{\hat{B}} + \hat{y}_0 \tag{15}$$

The base of the new specimen system is available after unitization.

$$\begin{bmatrix} e_{X_F} \\ e_{Y_F} \\ e_{Z_F} \end{bmatrix} = \begin{bmatrix} \left\| \overrightarrow{O_F Y_1} \right\|_2^{-1} \bullet \overrightarrow{O_F Y_1} \\ s \\ e_{X_F} \times e_{Y_F} \end{bmatrix} \tag{16}$$

Next, the rotation matrix and translation vector of the specimen coordinate system to the WCS can be calculated as follows:

$$^W_F R = [e_{X_F} \quad e_{Y_F} \quad e_{Z_F}], \; ^W_F P = O_F \tag{17}$$

The point cloud from the WCS is transformed to the specimen coordinate system as follows:

$$^F P = {}^W_F R^{-1} ({}^W P - {}^W_F P) \tag{18}$$

After the correction is accomplished, poison reconstruction [51] is implement to obtain 3D surface.

3 Experiment

A four-ocular vision system was used to perform 3D surface tracking experiments on a CSFT column under axial and cyclic radial loads. We focused on the accuracy of two selected sample points on the 3D model, which approximately reflects the precision of the reconstructed 3D surface. Diameters of specimen cross-sections were measured by laser rangefinders as standard values, then the visual measurements were compared against the standard values. Finally, indicators for error evaluation were calculated to validate the proposed method. Three CFST specimens were selected for this experiment.

During the test, an axial load was applied to the top of the specimen at a constant 10 kN. A cyclic radial load was applied along the horizontal direction with 20 kN added in each cycle. After the specimen yielded, the horizontal displacement of the press increased in each cycle; the increment is an integral multiple of the yield displacement of the specimen. Each run of the experiment ended once significant deformation or damage was observed in the bottom of the specimen.

As shown in Fig. 7(a), the laser rangefinders and vision system collected dynamic measurements d_i^L and d_i^V once per minute. The distance of the laser point to the base h_0 was determined in advance as a standard measurement. In the reconstructed surface model, a cross-section of h_0 height was taken to obtain a visual measurement as shown in Fig. 7(b). To determine this section on the 3D model, the point with the smallest Y_W value in the model was targeted, then the section with height h0 above it was taken as the target section. The visual measurement was taken according to the distance between the leftmost and rightmost points in this section.

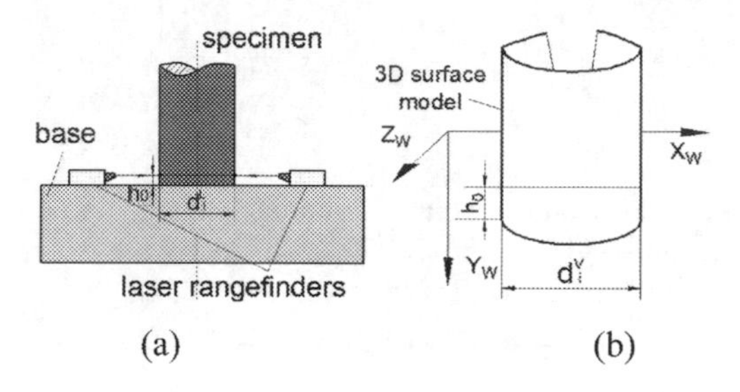

Fig. 7. (a) Acquisition of laser measurements; (b) Acquisition of vision measurements

The initial diameters of the specimens were about 203 mm and the tolerance was between IT12 and IT16. Line graphs of the measured values of each specimen are shown in Fig. 8(a)–(c). We used a personal computer (i7-8700 K CPU, Nvidia GTX 1080Ti GPU, and 16 GB RAM) to accomplish the calculation.

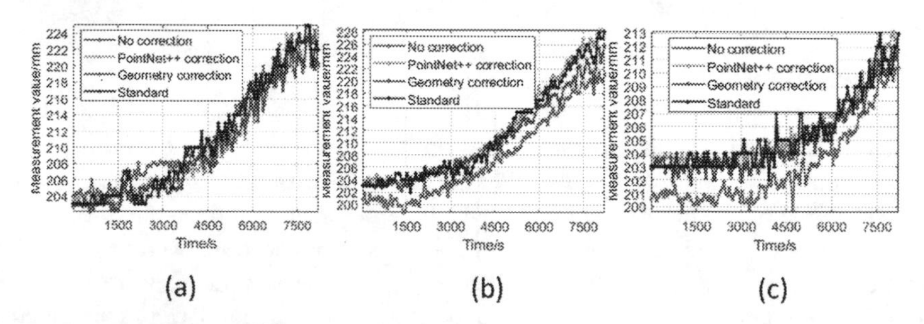

Fig. 8. (a) Real-time measurement of specimen 1; (b) Real-time measurement of specimen 2; (c) Real-time measurement of specimen 3

We collected the calculation times for all sample points as shown in Table 1. Each method took under 2.5 s; their respective averages were 1.75 s, 2.28 s, and 1.87 s. The time described here includes the time necessary to calculate a 3D point cloud from a 2D image.

Table 1. Average 3D reconstruction calculation time

	No correction(s)	With PointNet++ based correction(s)	With geometry based correction(s)
Specimen 1	1.73	2.35	1.84
Specimen 2	1.73	2.22	1.97
Specimen 3	1.80	2.27	1.81
Average	1.75	2.28	1.87

Figure 8 show that the point cloud correction algorithm effectively reduces visual measurement error; it can compensate for any error caused by vibration of the press and inaccurate establishment of the WCS to provide accurate visual measurements of the 3D model.

In order to evaluate the performance of the vision system and the correction algorithm, the maximum absolute error (M), mean absolute error (MAE), mean relative error (MRE), and root mean square error (RMSE) were used to evaluate the dynamic measurement error. For the deep-learning-based algorithm, the average M of each specimen is 3.00 mm, the average MAE is 1.11 mm, the average MRE is 0.52%, and the average RMSE is 1.84 mm. For the geometry-based correction algorithm, the average M of each specimen is 3.21 mm, the average MAE is 1.23 mm, the average MRE is 0.58%, and the average RMSE is 2.34 mm. These values altogether satisfy the requirements for high-accuracy measurement.

Both of the algorithms we developed in this study can effectively correct the spatial position of point clouds. Their effects do not significantly differ.

4 Conclusion

In this study, we focused on a series of unfavorable factors that degrade the accuracy of surface reconstruction tasks. A point cloud correction algorithm was proposed to manage the unexpected shocks and vibration which occur under actual testing conditions and to correct the stitched point cloud obtained by multi-vision systems. The essential geometric parameters of the reconstructed surface were measured, then stereo rectification, point cloud acquisition, point cloud filtering, and point cloud stitching were applied to obtain a 3D model of a complex dynamic surface. In this process, a deep-learning-based and geometry-based algorithm were deployed to compensate for the stitching error of multi-view point clouds and secure high-accuracy 3D structures of the target objects.

Geometric analysis and coordinate transformation were applied to design the geometry-based point cloud correction algorithm. This method is based on strong mathematical assumptions, so it has fast calculation speed with satisfactory correction accuracy. By contrast, the deep-learning-based algorithm relies on a large number of training samples and the forward propagation of the network is more computationally complicated than the geometry-based algorithm; it takes a longer time to accomplish point cloud correction. However, since the applicable object of the network is determined by the type of objects in the training set, it does not rely on manually designed geometric assumptions and is thus much more generalizable to different types of 3D objects.

The proposed point cloud correction algorithms make full use of the geometric and spatial characteristics of targets for error compensation, so both are more adaptive and efficient than standard-marker-based correction frameworks. They effectively enhance the accuracy of point cloud stitching over traditional methods and their effects do not significantly differ. The deep-learning-based algorithm is highly versatile, while the geometry-based algorithm is more computationally effective for cylindrical objects. They may serve as a reference for improving the accuracy of multi-view, high-accuracy, and dynamic 3D reconstructions for CFSTs and other large-scale structures under complex conditions. They are also workable as-is for completing tasks such as structural monitoring and data collection.

Acknowledgments. This work was supported by the National Natural Science Foundation of China (51578162), the Key-area Research and Development Program of Guangdong Province (2019B020223003), and the Scientific and Technological Research Project of Guangdong Province (2016B090912005).

References

1. Pan, B.: Thermal error analysis and compensation for digital image/volume correlation. Opt. Lasers Eng. **101**, 1–15 (2018)
2. Genovese, K., Chi, Y., Pan, B.: Stereo-camera calibration for large-scale DIC measurements with active phase targets and planar mirrors. Opt. Express **27**, 9040–9053 (2019)
3. Dong, Y., Pan, B.: In-situ 3D shape and recession measurements of ablative materials in an arc-heated wind tunnel by UV stereo-digital image correlation. Opt. Lasers Eng. **116**, 75–81 (2019)
4. Fathi, H., Dai, F., Lourakis, M.: Automated as-built 3D reconstruction of civil infrastructure using computer vision: Achievements, opportunities, and challenges. Adv. Eng. Inform. **29**, 149–161 (2015)
5. Kim, H., Leutenegger, S., Davison, Andrew J.: Real-Time 3D reconstruction and 6-DoF tracking with an event camera. In: Leibe, B., Matas, J., Sebe, N., Welling, M. (eds.) ECCV 2016. LNCS, vol. 9910, pp. 349–364. Springer, Cham (2016). https://doi.org/10.1007/978-3-319-46466-4_21
6. Munda, G., Reinbacher, C., Pock, T.: Real-time intensity-image reconstruction for event cameras using manifold regularisation. Int. J. Comput. Vision **126**, 1381–1393 (2018)
7. Feng, D.-M., Feng, M.Q.: Computer vision for SHM of civil infrastructure: From dynamic response measurement to damage detection–a review. Eng. Struct. **156**, 105–117 (2018)
8. Feng, D.-M., Feng, M.Q.: Vision-based multipoint displacement measurement for structural health monitoring. Struct. Control Health Monit. **23**, 876–890 (2016)
9. Cai, Z., Liu, X., Li, A., Tang, Q., Peng, X., Gao, B.Z.: Phase-3D mapping method developed from back-projection stereovision model for fringe projection profilometry. Opt. Express **25**, 1262–1277 (2017)

10. Hyun, J.S., Chiu, G.T., Zhang, S.: High-speed and high-accuracy 3D surface measurement using a mechanical projector. Opt. Express **26**, 1474 (2018)
11. Zhen, L., Li, X., Li, F., Zhang, G.: Flexible dynamic measurement method of three-dimensional surface profilometry based on multiple vision sensors. Opt. Express **23**, 384–400 (2015)
12. Wu, Q., Zhang, B., Huang, J., Wu, Z., Zeng, Z.: Flexible 3D reconstruction method based on phase-matching in multi-sensor system. Opt. Express **24**, 7299–7318 (2016)
13. Yang, X., Li, H., Yu, Y., Luo, X., Huang, T., Yang, X.: Automatic pixel-level crack detection and measurement using fully convolutional network. Comput. Aided Civ. Infrastruct. Eng. **33**, 1090–1109 (2018)
14. Huňady, R., Hagara, M.: A new procedure of modal parameter estimation for high-speed digital image correlation. Mech. Syst. Signal Process. **93**, 66–79 (2017)
15. Huňady, R., Pavelka, P., Lengvarský, P.: Vibration and modal analysis of a rotating disc using high-speed 3D digital image correlation. Mech. Syst. Signal Process. **121**, 201–214 (2019)
16. Tang, Y., et al.: Real-time detection of surface deformation and strain in recycled aggregate concrete-filled steel tubular columns via four-ocular vision. Robot. Comput.-Integr. Manuf. **59**, 36–46 (2019)
17. Ma, Z.-L., Liu, S.-L.: A review of 3D reconstruction techniques in civil engineering and their applications. Adv. Eng. Inform. **37**, 163–174 (2018)
18. Kim, H., Kim, H.: 3D reconstruction of a concrete mixer truck for training object detectors. Autom. Constr. **88**, 23–30 (2018)
19. Sun, L., Abolhasannejad, V., Gao, L., Li, Y.-W.: Non-contact optical sensing of asphalt mixture deformation using 3D stereo vision. Measurement **85**, 100–117 (2016)
20. Liu, Y., Yang, J.-C., Meng, Q.-G., Lv, Z.-H., Song, Z.-J., Gao, Z.-Q.: Stereoscopic image quality assessment method based on binocular combination saliency model. Sig. Process. **125**, 237–248 (2016)
21. Liu, Z., et al.: 3D real human reconstruction via multiple low-cost depth cameras. Sig. Process. **112**, 162–179 (2015)
22. Candau, N., Pradille, C., Bouvard, J.-L., Billon, N.: On the use of a four-cameras stereovision system to characterize large 3D deformation in elastomers. Polym. Testing **56**, 314–320 (2016)
23. Zhou, P., et al.: Experimental study of temporal-spatial binary pattern projection for 3D shape acquisition. Appl. Opt. **56**, 2995–3003 (2017)
24. Shen, X., Markman, A., Javidi, B.: Three-dimensional profilometric reconstruction using flexible sensing integral imaging and occlusion removal. Appl. Opt. **56**, D151–D157 (2017)
25. Malesa, M., et al.: Non-destructive testing of industrial structures with the use of multi-camera Digital Image Correlation method. Eng. Fail. Anal. **69**, 122–134 (2016)
26. Sinha, A., Bai, J., Ramani, K.: Deep learning 3D shape surfaces using geometry images. In: Leibe, B., Matas, J., Sebe, N., Welling, M. (eds.) ECCV 2016. LNCS, vol. 9910, pp. 223–240. Springer, Cham (2016). https://doi.org/10.1007/978-3-319-46466-4_14
27. Li, F., et al.: Depth acquisition with the combination of structured light and deep learning stereo matching. Signal Process. Image Commun. **75**, 111–117 (2019)
28. Zhang, J., Hu, S., Shi, H.: Deep learning based object distance measurement method for binocular stereo vision blind area. Methods **9** (2018)

29. Sun, S., Liu, R., Pan, Y., Du, Q., Sun, S., Su, H.: Pose determination from multi-view image using deep learning. In: 2019 15th International Wireless Communications & Mobile Computing Conference (IWCMC), pp. 1494–1498. IEEE, June 2019
30. Yang, Y., Qiu, F., Li, H., Zhang, L., Wang, M.-L., Fu, M.-Y.: Large-scale 3D semantic mapping using stereo vision. Int. J. Autom. Comput. **15**(2), 194–206 (2018). https://doi.org/10.1007/s11633-018-1118-y
31. Zou, X., Zou, H., Lu, J.: Virtual manipulator-based binocular stereo vision positioning system and errors modelling. Mach. Vis. Appl. **23**, 43–63 (2012). https://doi.org/10.1007/s00138-010-0291-y
32. Lin, G., Tang, Y., Zou, X., Xiong, J., Fang, Y.: Color-, depth-, and shape-based 3D fruit detection. Precision Agric. **21**(1), 1–17 (2019). https://doi.org/10.1007/s11119-019-09654-w
33. Lin, G., Tang, Y., Zou, X., Cheng, J., Xiong, J.: Fruit detection in natural environment using partial shape matching and probabilistic Hough transform. Precision Agric. **21**(1), 160–177 (2019). https://doi.org/10.1007/s11119-019-09662-w
34. Lin, G., Tang, Y., Zou, X., Xiong, J., Li, J.: Guava detection and pose estimation using a low-cost RGB-D sensor in the field. Sensors **19**, 428 (2019)
35. Luo, L., Tang, Y., Zou, X., Wang, C., Zhang, P., Feng, W.: Robust grape cluster detection in a vineyard by combining the AdaBoost framework and multiple color components. Sensors **16**, 2098 (2016)
36. Luo, L., Tang, Y., Zou, X., Ye, M., Feng, W., Li, G.: Vision-based extraction of spatial information in grape clusters for harvesting robots. Biosys. Eng. **151**, 90–104 (2016)
37. Tang, Y., Li, L., Feng, W., Liu, F., Zou, X., Chen, M.: Binocular vision measurement and its application in full-field convex deformation of concrete-filled steel tubular columns. Measurement **130**, 372–383 (2018)
38. Wang, C., Tang, Y., Zou, X., Luo, L., Chen, X.: Recognition and matching of clustered mature litchi fruits using binocular charge-coupled device (CCD) color cameras. Sensors **17**, 2564 (2017)
39. Wang, C., Tang, Y., Zou, X., SiTu, W., Feng, W.: A robust fruit image segmentation algorithm against varying illumination for vision system of fruit harvesting robot. Optik-Int. J. Light Electron Opt. **131**, 626–631 (2017)
40. Wang, C., Zou, X., Tang, Y., Luo, L., Feng, W.: Localisation of litchi in an unstructured environment using binocular stereo vision. Biosys. Eng. **145**, 39–51 (2016)
41. Song, S., Duan, J., Yang, Z., Zou, X., Fu, L., Ou, Z.: A three-dimensional reconstruction algorithm for extracting parameters of the banana pseudo-stem. Optik **185**, 486–496 (2019)
42. Zhang, Z.: A flexible new technique for camera calibration. IEEE Trans. Pattern Anal. Mach. Intell. **22**, 1330–1334 (2000)
43. Sereewattana, M., Ruchanurucks, M., Siddhichai, S.: Depth estimation of markers for UAV automatic landing control using stereo vision with a single camera. In: International Conference on Information and Communication Technology for Embedded System (2014)
44. Hirschmuller, H.: Accurate and efficient stereo processing by semi-global matching and mutual information. In: IEEE Computer Society Conference on Computer Vision and Pattern Recognition (IEEE 2005), vol. 2, pp. 807–814 (2005)
45. Zeineldin, R.A., El-Fishawy, N.A.: Fast and accurate ground plane detection for the visually impaired from 3D organized point clouds. In: 2016 SAI Computing Conference (SAI), (IEEE 2016), pp. 373–379 (2016)
46. Tomasi, C., Manduchi, R.: Bilateral filtering for gray and color images. In: ICCV, p. 2 (1998)

47. Skinner, B., Vidal-Calleja, T., Miro, J.V., De Bruijn, F., Falque, R.: 3D point cloud upsampling for accurate reconstruction of dense 2.5 D thickness maps. In: Australasian Conference on Robotics and Automation, ACRA (2014)
48. Marquardt, D.W.: An algorithm for least-squares estimation of nonlinear parameters. J. Soc. Ind. Appl. Math. **11**, 431–441 (1963)
49. Qi, C.R., Yi, L., Su, H., et al.: Pointnet++: deep hierarchical feature learning on point sets in a metric space. In: Advances in Neural Information Processing Systems, pp. 5099–5108 (2017)
50. Besl, P.J., McKay, N.D.: Method for registration of 3-D shapes. In: Sensor Fusion IV Control Paradigms Data Structure, vol. 1611, pp. 586–607. International Society for Optics and Photonics (1992)
51. Kazhdan, M., Bolitho, M., Hoppe, H.: Poisson surface reconstruction. In: Proceedings of the Fourth Eurographics Symposium on Geometry Processing (2006)

Facial Expression Recognition Adopting Combined Geometric and Texture-Based Features

Yujiao Gong(✉) and Yongbo Yuan

Chengdu Neusoft University, Dujiangyan, Chengdu, China
gongyujiao@nsu.edu.cn

Abstract. In recent facial expression recognition competitions, top approaches were using either geometric relationships that best captured facial dynamics or an accurate registration technique to develop texture features. These two methods capture two different types of facial information that is similar to how the human visual system divides information when perceiving faces. This paper discusses a framework of a fully automated comprehensive facial expression detection and classification. We study the capture of facial expressions through geometric and texture-based features, and demonstrate that a simple concatenation of these features can lead to significant improvement in facial expression classification. Each type of expression has individual differences in the commonality of facial expression features due to differences in appearance and other factors. The geometric feature tends to emphasize the facial parts that are changed from the neutral and peak expressions, which can represent the common features of the expression, thus reducing the influence of the difference in appearance and effectively eliminating the individual differences. Meanwhile, the consolidation of gradient-level normalized cross correlation and Gabor wavelet is utilized to present the texture features. We perform experiments using the well-known extended Cohn-Kanade (CK+) database, compared to the other state of the art algorithms, the proposed method achieved provide better performance with an average accuracy of 95.3%.

Keywords: Facial expression recognition · Geometry feature · Texture feature

1 Introduction

Facial expression recognition is a challenging technique in numerous researches, and impacts important applications like medical, lie detection, cognitive activity, robotics interaction, forensic section, automated training systems, security, intellectual state identification, music for mood, operator fatigue detection, etc. So much progress has been made, all the methods that have been used for facial expression recognition have similar steps. After face detection and segmentation, the next step is feature extraction and feature selection. The final step is feature classification. The differences between methods are based on the variety of features selected, the feature extraction method and the feature classification method.

K. Li et al. (Eds.): ISICA 2019, CCIS 1205, pp. 458–469, 2020.
https://doi.org/10.1007/978-981-15-5577-0_36

In recent years, facial expression classification approaches mainly include two categories: (a) geometric feature-based and (b) texture feature-based. The geometric feature focus on where action units (AU) were detected tracking changes in permanent and transient facial features through accurate geometric modeling. The texture feature-based methods involve the use of local (face parts) or global (entire facial image) descriptors which intend at describing facial appearance.

Deepthi, Archana et al. [1] implemented FER using a 2D-DCT for feature extraction and a neural network is used as a classifier by using the JAFFE database. Punitha, Geetha et al. [2] extracted the mouth intensity code value (MICV) difference between the first and the greatest facial expression intensity frame, a Hidden Markov Model (HMM) is used as a classifier, with the own created dataset and achieved 94% accuracy. Zhang, Liu et al. [3] used the Gabor LBP feature and Gabor LPQ features as an input of Multiclass SVM classifier in JAFFE database. The accuracy obtained is 98%. Owusu and Zhan [4] fed selected Gabor features into a support vector machine (SVM) classifier and obtain an average recognition rate of 97.57%. Shah, Khanna et al. [5] implemented FER for Color Images using Gabor, Log Gabor Filters and PCA, and the Euclidian distance is used to classify the reduced features. The self-database is used for testing with an accuracy of 86.7%. Lajevardi, Husain et al. [6] presented the FER system based on hybrid face regions (HFR). Using Log Gabor filter features are extracted based on the whole face image and face region Naïve Bayes is used as a classifier. JAFFE database and the Cohn-Kanade database are used for testing with an accuracy of 97% and 91% respectively.

ELLaban, Ewees, Elsaeed et al. [7] tracked the Facial Expression Recognition using Support Vector Machines and k-Nearest Neighbor Classifier. In their work, Gabor, PCA are used for feature extraction. SVM and KNN classifiers are used for classification of the features extracted. Accuracy they achieved by testing the self-database using SVM is 90% and SVM outperformed than KNN. Lee, Uddin and Kim et al. [8] presented FER Using Fisher Independent Component Analysis and Hidden Markov Model from the Cohn-Kanade database. The FICA Fisher Linear Discriminant (FLD) is used for feature extraction based on a class-specific learning algorithm. Sumathi, Santhanam and Mahadevi et al. [9] investigated using Facial Action Coding System (FACS) action units and the methods which recognize the action units parameter using facial expression data that are extracted, various kinds of human facial expressions are recognized based on their geometric facial appearance, and hybrid features. In [10], a pose-invariant spatial-temporal textural descriptor is proposed, which is used to achieve 94.48% average accuracy using SVMs on the CK+ database. Some other researches in classifying six or seven expressions based on CK+ database can be found in [11–13].

There are numerous methodologies that have been proposed, they give us an overall idea about how effective the geometric and texture-based features are, but recognizing facial expression with a high accuracy remains a difficult problem due to the complexity and variety of facial expressions.

In this work, we propose an automated approach for recognizing facial expressions with a combination of geometric and texture-based features. The task of an automated FER method can be split into a sequence of processing stages such as preprocessing, feature extraction, and classification. Initially, the preprocessing is done before extracting features, to make texture invariant to translation, rotation, and scaling. Then, we combined geometric features and texture features for feature extraction. Geometric features are described by facial feature point displacements, slope and angle difference between normalized neutral and peak expression images, and texture features are represented by gradient-level normalized cross correlation and Gabor wavelet. Finally, the feature vector comprises of features, which are fed to SVM classifier for recognizing facial expressions from CK+ database.

The remaining section of the paper is organized as follows: Sect. 2 describes the preprocessing procedure. Feature extraction is introduced in Sect. 3. We evaluate the performance of the proposed method in Sect. 4 and finally, the paper concludes in Sect. 5.

2 Preprocessing

Preprocessing is a vital step in facial expression recognition, including face detection, face alignment, illumination processing, etc. For a given image, we first localize the centers of eyes with Adaboost learning algorithm [14] for reference points, followed by an image rotation to live up to eye coordinates (see Fig. 1).

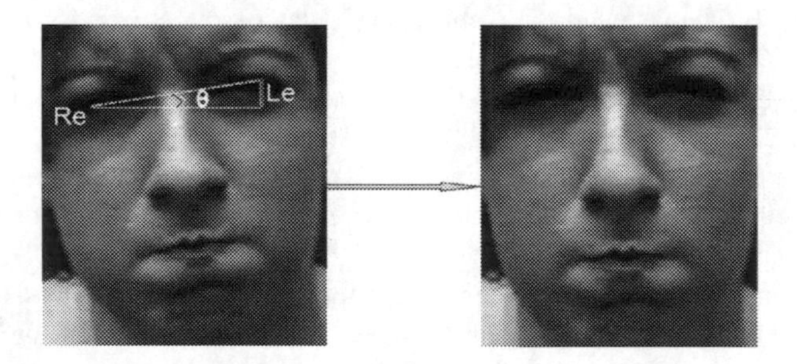

Fig. 1. Normalized face acquisition.

Where *Le* is the center position of the left eye and *Re* is the center position of the right eye, θ is the angle between the direction of the two eyes and the horizontal direction.

3 Feature Extraction

3.1 Geometric Feature Representation

In this work, we use the difference-ASM (DASM) features describe the facial shape difference between the neutral and peak expressions. The coordinates of salient points can be described as the change of facial landmark positions. On the other hand, the difference in the values in the corresponding facial angles and slopes between the neutral and peak expressions serves as the 24 geometric features for the face which help capture the coarse (large) distortions in the facial geometry across expression classes.

First of all, we use Adaboost algorithm and the extended Active Shape Model (ASM) [15, 16] to detect face and locate facial feature point respectively, the Stasm implementation [17] of the ASM module which is a C++ library built to run on top of OpenCV has been used as part of this work. Figure 2 shows the facial feature points and the detail face region obtained from the ASM algorithms. Then the displacements between x and y coordinates of 21 feature points on the neutral and expressive faces are calculated as a 21 × 2 dimensional feature vector.

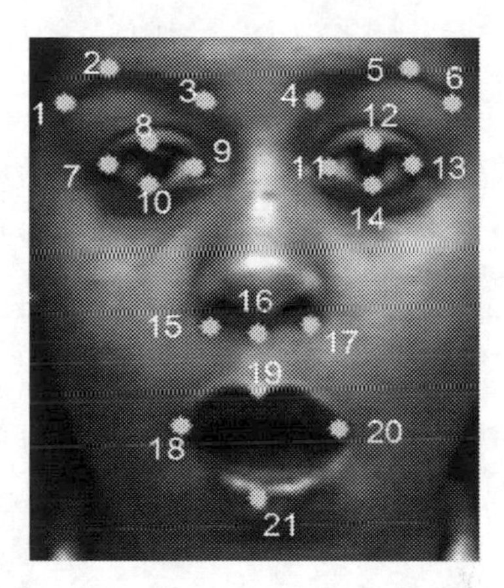

Fig. 2. ASM-based face region acquisition.

Furthermore, geometric features such as slope and angle of the eye or mouth from the ASM landmark information are extracted. All the facial landmarks should be normalized to the same scale due to the different coordinates and the different scales. The normalization is described in the following 4 steps.

The 21 × 2 feature vertices can be described as follows:

$$\alpha_i = (x, y) \quad i = 1, 2, \cdots 21. \tag{1}$$

The bottom of the nose is assigned as the 16th feature point shown in Fig. 2, and it has been defined as follows:

$$bas = \alpha_{16} = \left(x_{16,} y_{16}\right) \tag{2}$$

To transform into a common coordinate system, the coordinate values of each feature point subtract the coordinate values of bas as follows:

$$\beta_i = \alpha_i - bas \text{ i} = 1, 2, \cdots 21. \tag{3}$$

Meanwhile, each should also be normalized to make all the facial models with the same scale:

$$N_i = \beta_i/(\beta_{11} - \beta_9)\ \mathrm{i} = 1, 2, \cdots 21. \tag{4}$$

Where β11 and β9 denote two inner eyes vertices. Based on the normalized feature points slope features are obtained as follows:

Eye slope features:

$$\begin{gathered} S_1 = (N_9 - N_8);\ S_2 = (N_9 - N_{10});\ S_3 = (N_9 - N_7); \\ S_4 = (N_8 - N_7);\ S_5 = (N_{10} - N_7) \end{gathered} \tag{5}$$

Mouth slope features:

$$\begin{gathered} S_6 = (N_{20} - N_{19});\ S_7 = (N_{20} - N_{21});\ S_8 = (N_{20} - N_{18}); \\ S_9 = (N_{19} - N_{18}); S_{10} = (N_{21} - N_{18}) \end{gathered} \tag{6}$$

Also, angle features can be generated as follows:

Eye angle features:

$$\begin{aligned} a_1 &= \cos^{-1}((s_1 \times s_3)/(\|s_1\| \times \|s_3\|)); \\ a_2 &= cos^{-1}((s_2 \times s_3)/(\|s_2\| \times \|s_3\|)); \\ a_3 &= cos^{-1}((s_4 \times s_3)/(\|s_4\| \times \|s_3\|)); \\ a_4 &= cos^{-1}((s_3 \times s_5)/(\|s_3\| \times \|s_5\|)); \\ a_5 &= cos^{-1}((s_1 \times s_4)/(\|s_1\| \times \|s_4\|)); \\ a_6 &= cos^{-1}((s_2 \times s_5)/(\|s_2\| \times \|s_5\|)); \end{aligned} \tag{7}$$

Mouth angle features, where ||*|| is the norm operator:

$$\begin{aligned} a_7 &= \cos^{-1}((\mathrm{s}_6 \times \mathrm{s}_8)/(\|\mathrm{s}_6\| \times \|\mathrm{s}_8\|)); \\ a_8 &= \cos^{-1}((s_7 \times s_8)/(\|s_7\| \times \|s_8\|)); \\ a_9 &= \cos^{-1}((s_9 \times s_8)/(\|s_9\| \times \|s_8\|)); \\ a_{10} &= \cos^{-1}((s_{10} \times s_8)/(\|s_{10}\| \times \|s_8\|)); \\ a_{11} &= \cos^{-1}((s_6 \times s_9)/(\|s_6\| \times \|s_9\|)); \\ a_{12} &= \cos^{-1}((s_7 \times s_{10})/(\|s_7\| \times \|s_{10}\|)); \end{aligned} \tag{8}$$

Here, including a 21 × 2 dimensional displacement feature vector, a 12-dimensional slope feature vector and a 12-dimensional angle feature vector, the difference-ASM features are generated for classification.

3.2 Texture Feature Representation

In this paper, gradient-level normalized cross correlation (GNCC) and Gabor wavelet are utilized to extract texture features of facial faces, with the gradient-level matrix in the spatial space that has a better processing outcome on the local region of images and

Gabor operator have been selected due to their simplicity, intuitiveness and computational efficiency.

The texture difference measure based on the relations between gradient vectors is robust with respect to noise and illumination changes [18], Ahmed and Hossain [19] investigated using gradient-based ternary texture for facial expression recognition operate faster than using gray level. Thus, the gradient value is a good measure to describe how the gray level changes within a neighborhood and it can be used to derive the local texture difference measure. Let p = {x,y}, $g_n(p)$, $g_e(p)$ and $g'_n(p)$, $g_e'(p)$ be a feature point, the neutral and peak expression frames and the neutral frame gradient vector ($g_n^x(p)$, $g_n^y(p)$) with $g_n^x(p) = \Delta x g_n(p)$ and $g_n^y(p) = \Delta y g_n(p)$, the peak expression frames gradient vector ($g_e^x(p)$, $g_e^y(p)$) with $g_e^x(p) = \Delta x g_e(p)$ and $g_e^y(p) = \Delta y g_e(p)$, respectively. After that, the normalized cross correlation is calculated to evaluate the gradient difference in a feature point's neighborhood between the normalized neutral and peak expression frames, defined as follows:

$$NC = \frac{\sum_{P \in M}\left(g_n'(p) - \overline{g_n'(P)}\right)\left(g_e'(p) - \overline{g_e'(P)}\right)}{\sqrt{\sum_{P \in M}\left(g_n'(p) - \overline{g_n'(P)}\right)^2 \sum_{P \in M}\left(g_e'(p) - \overline{g_e'(P)}\right)^2}} \quad (9)$$

Where $\overline{g_n'(P)}$ and $\overline{g_e'(P)}$ are the averages of $g_n'(p)$ and $g_e'(p)$, M is the 10 × 10 neighborhood centered at point p. Texture features are calculated in the neighborhoods centered at 21 feature points (see Fig. 3).

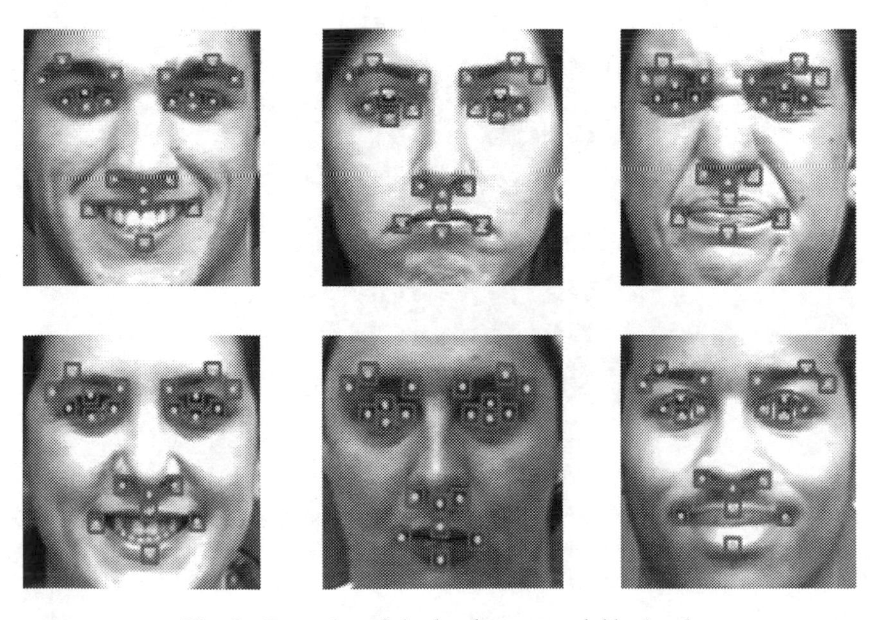

Fig. 3. Examples of the local texture neighborhood.

In the frequency domain, Gabor wavelet [20, 21] has outstanding performance in obtaining multi-scale, multi-direction local information. Initially, the Gabor function is estimated for the image and it is computed as follows:

$$\varphi_{u,v}(z) = \frac{\left\|k_{u,v}\right\|}{\sigma^2} e^{\left(-\|ku,v\|^2\|z\|^2/\|2\sigma\|^2\right)}\left[e^{ik_{u,v}z} - e^{-\sigma\frac{2}{2}}\right] \tag{10}$$

z = (x, y) gives the pixel position in the spatial domain, and frequency vector *ku, v* is defined as follows:

$$k_v = \frac{k_{max}}{f_v} e^{i\phi_u} \qquad v = 0, 1, \cdots, 4; u = 0, 1, \ldots, 7 \tag{11}$$

Where $\phi_u = u_\pi/u_{max}$, $\phi_u \in (0, \pi)$, and u and v denote the orientation and scale factors of Gabor filters respectively. In our system, we adopt the Gabor filters of five scales and eight orientations, with $\sigma = 2\pi$, $k_{max} = \pi/2$. For each pixel of the peak expression frame, totally 40 Gabor features are obtained when 40 Gabor filters are used.

In practice, inhomogeneous sampling is applied to extract distinctive expression information mainly located at eyes, mouth, and nose. Considering the high correlation of adjacent pixels, and that the Gabor filter is not sensitive to the position of the gray value, we extract the Gabor features of some fixed geometric positions of the facial landmarks [22] as shown in Fig. 4, the Gabor features of each sampling point form feature vector with dimension 8 × 8 × 40 = 2560. Before fusing the Gabor and other features, we reduce their dimensionality to remove some of the redundant information with PCA [23, 24]. PCA is a useful technique used to reduce the dimensionality of a feature vector and has been successfully used in face analysis.

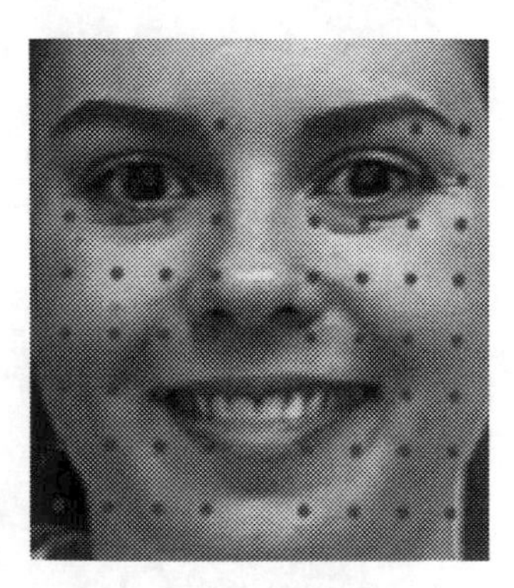

Fig. 4. The distribution of the sampling points on the face.

In the classification phase, we attempt to combine the advantages of geometric features and texture features to attain better FER performance. As mentioned in this section, in a facial expression image sequence, we work with the neutral and peak expression frames for DASM and GNCC features, with only the peak expression frames for Gabor features. We consider DASM as $f1$, GNCC as $f2$, Gabor as $f3$, in order to remove the unfavorable effect resulting from unequal dimensions, $f1, f2$, and $f3$, are all be normalized as $V1, V2, V3$ given by:

$$V1 = \left[\frac{f1}{||f1||}\right], V2 = \left[\frac{f2}{||f2||}\right], V3 = \left[\frac{f3}{||f3||}\right] \tag{12}$$

The fusion feature F can be defined as:

$$F = \frac{|V1V2V3|}{||[V1V2V3]||} \tag{13}$$

Where $||*||$ is the norm operator. Here, we consider the simple concatenation of geometric and textural features is generated for classification.

4 Experiment

4.1 Dataset

The extended Cohn-Kanade (CK+) database [25] has been used in all our experiments. The database consists of image sequences across 123 subjects displaying 7 expressions: happy, sad, surprise, fear, anger, disgust, and contempt. The image sequences vary in duration from neutral faces to apexes, the peak information of the prototypical facial expressions. Since our work focuses only on the 6 basic expressions: happy, sad, surprise, fear, anger and disgust, the image sequences corresponding to the 'contempt' expression class have been left out from consideration. The total number of labeled image sequences therefore considered for our work turned out to be 309. Table 1 presents the detailed statistics for the portion of the dataset that is used.

Table 1. Overview of the dataset.

Expression	AN	DI	FE	HA	SA	SU	Total
No. of image	45	59	25	69	28	83	309

AN = Anger, DI = Disgust, FE = Fear, HA = Happiness, SA = Sadness, SU = Surprise.

4.2 Experiment Results

Support Vector Machines (SVMs) [26] are primarily binary classifiers that find the best separating hyperplane by maximizing the margins from the support vectors. Support vectors are defined as the data points that lie closest to the decision boundary. LIBSVM [27] has been used as part of this work. To recognize the facial expressions, a one-versus all multi-class SVM is used, and all the results presented in this section have been 5-fold cross validated. The dataset is randomly divided into 5 groups of roughly equal numbers of subjects and each prototypical expression.

In order to demonstrate the efficiency of the concatenated feature, the findings on the individual geometry and texture-based features and on the concatenated feature are established through classification experiments. The experiment results are shown in Table 2 and Fig. 5. The proposed method achieves an average recognition accuracy of 95.3%. From Table 2, one can see that the results of every class of facial expression, among which, the recognition rate of happy and surprise are highest. It is an intuitive result as these expressions cause many more facial movements mainly located around the mouth and thus are relatively easy to recognize. But fear and sadness are similar sometimes and attain lower classification accuracies due to the lack of facial deformation and training samples, so they are not easily identified. The confusion matrix for

Table 2. Recognition rate(%) Recognition rates of different features.

Feature\Expression	Anger	Disgust	Fear	Happy	Sadness	Surprise	Average
DASM	91.2	91.4	89.3	92.8	87.6	94.3	91.1
GNCC	90.5	89.8	88.71	81.6	87.4	85.8	87.3
Gabor	84.1	85.2	83.4	83.6	84.2	85.3	84.3
DASM + GNCC + Gabor	93.2	94.8	92.5	97.7	95.1	98.5	95.3

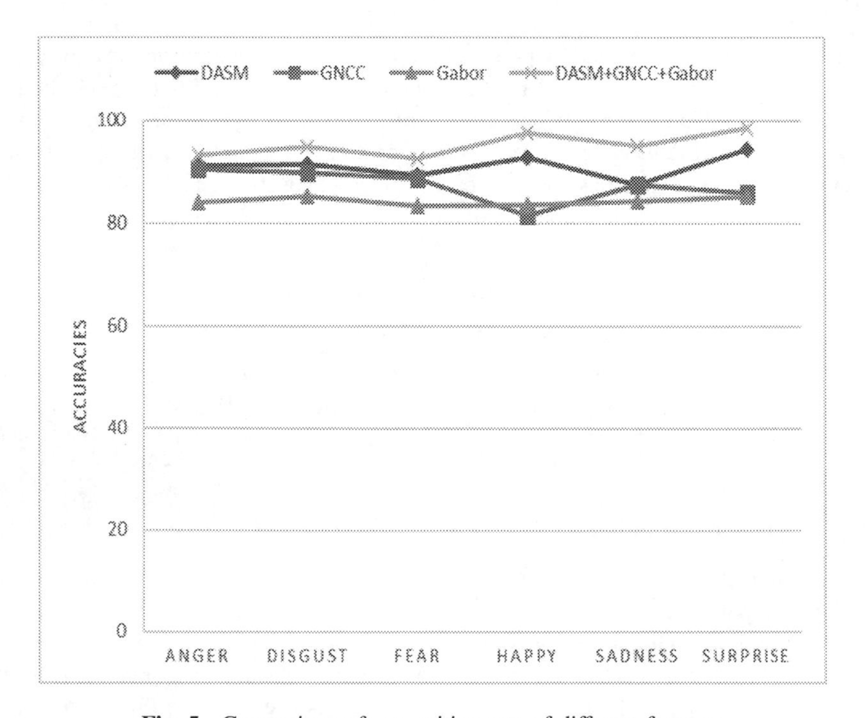

Fig. 5. Comparison of recognition rate of different features.

6-class classification in Fig. 5 shows that the concatenated feature performs better than the individual features.

To further evaluate our proposed method, we consider the statistical comparison between performances of the proposed approach and the other state of the art algorithms considered in Sect. 1, employing the CK+ database. The experiment results are shown in Table 3. From Table 3, the comparison between our method and previously published approaches shows that our method outperforms other methods.

Table 3. A comparison of the proposed method with some examples from the related ones.

Author	Expressions	Feature extraction	Expression classification	Accuracy (%)
Sho [10]	6	WH, UWH	Extreme Sparse Learning (ESL)	94.48
Lit [11]	7	Gabor	SVM	90.1
Raz [12]	6	FIT	ANN	84
Kar [13]	7	Gabor, ADA	PCA	94.72
We	6	DASM, GNCC, Gabor	SVM	95.3

5 Conclusion

In this paper, we proposed a framework of fusing geometric and appearance features of the difference between the neutral and peak expressive facial expression images to recognize facial expressions. The difference tends to emphasize the facial parts that are changed from the neutral to expressive face and eliminate in that way the identity of the facial image. The feature fusion method fully utilizes the local geometric feature and texture information to extract expression features. Based on the combination of DASM, GNCC and Gabor wavelet extraction method, a SVM classification method is used to recognize six facial expressions from CK+ database, namely Happiness, Sadness, Anger, Fear, Disgust and Surprise, plus the Contempt. Thus, we have obtained a suitable classification system to work with the six basic emotions. Extensive experiments show that the proposed method achieves more reliable results comparing to DASM, GNCC and Gabor descriptor alone and outperforms several other methods on the CK+ database. It indicates that the proposed method has a strong potential as an alternative method for building a facial expression recognition system.

In the future, this work will be extended in three aspects. Firstly, more feature descriptors will be used to give more comprehensive facial representations. Secondly, the fusion strategy will be improved to increase the recognition rate. Finally, we expect to extend this framework to analyze people's emotional state.

Acknowledgments. This work was supported by the Scientific Research Fund of Sichuan Provincial Education Department under Grant No. 18ZB0013, the Science and Technology Project of Dujiangyan under Grant No. 2018FW01.

References

1. Deepthi, S., Archana, G.S., JagathyRaj, V.P.: Facial expression recognition using artificial neural networks. OSR J. Comput. Eng. (IOSR-JCE) **8**(4), 01–06 (2013). ISSN 2278–0661, ISBN2278-8727
2. Punitha, A., Geetha, M.K.: HMM based real time facial expression recognition. Int. J. Emerg. Technol. Adv. Eng. **3**(1), 180–185 (2013)
3. Zhang, B., Liu, G.: Facial expression recognition using LBP and LPQ based on Gabor wavelet transform based on Gabor face image. In: IEEE International Conference on Computer and Communications (2016)
4. Owusu, E., Zhan, Y., Mao, Q.R.: An SVM-AdaBoost facial expression recognition system. Appl. Intell. **40**(3), 536–545 (2014)
5. Shah, S.K., Khanna, V.: Facial expression recognition for color images using Gabor, log Gabor filters and PCA. Int. J. Comput. Appl. **113**(4), 42–46 (2015)
6. Lajevardi, S.M., Hussain, Z.M.: Feature extraction for facial expression recognition based on hybrid face regions. Adv. Electr. Comput. Eng. **9**(3), 63–67 (2009)
7. ELLaban, H.A., Ewees, A.A., Elsaeed, A.E.: A real-time system for facial expression recognition using support vector machines and k-nearest neighbor classifier. Int. J. Comput. Appl. **159**(8), 0975–8887 (2017)
8. Lee, J.J, Uddin, M.Z., Kim, T.S.: Spatiotemporal human facial expression recognition using fisher independent component analysis and hidden markov model. In: 2008 30th Annual International Conference of the IEEE Engineering in Medicine and Biology Society, EMBS 2008, pp 2546–2549. IEEE (2008)
9. Sumathi, C.P., Santhanam, T., Mahadevi, M.: Automatic facial expression analysis a survey. IEEE Int. J. Comput. Sci. Eng. Surv. **3**(6), 47 (2012)
10. Shojaeilangari, S., Yau, W.Y., Nandakumar, K., Li, J., Teoh, E.K.: Robust representation and recognition of facial emotions using extreme sparse learning. IEEE Trans. Image Process. **24**, 2140–2152 (2015)
11. Littlewort, G., et al.: The computer expression recognition toolbox (CERT). In: IEEE International Conference on Automatic Face & Gesture Recognition and Workshops (FG 2011) (2011). https://doi.org/10.1109/fg.2011.5771414
12. Razuri, J.G., Sundgren, D., Rahmani, R., Cardenas, A.M.: Automatic emotion recognition through facial expression analysis in merged images based on an artificial neural network. In: 12th Mexican International Conference on Artificial Intelligence, pp. 85–96 (2013). https://doi.org/10.1109/micai.2013.16
13. Kar, A., Mukerjee, A.: Facial expression classification using visual cues and language. In: IIT (2011). http://www.cs.berkeley.edu/*akar/se367/project/report.pdf
14. Viola, P.A., Jones, M.J.: Rapid object detection using a boosted cascade of simple features. In: Proceedings of the 2001 IEEE Computer Society Conference on Computer Vision and Pattern Recognition, CVPR 2001. IEEE (2001)
15. Jabid, T., Kabir, M.H., Chae, O.: Robust facial expression recognition based on local directional pattern. ETRI J. **32**(5), 784–794 (2010)
16. Milborrow, S., Nicolls, F.: Locating facial features with an extended active shape model. In: Forsyth, D., Torr, P., Zisserman, A. (eds.) ECCV 2008. LNCS, vol. 5305, pp. 504–513. Springer, Heidelberg (2008). https://doi.org/10.1007/978-3-540-88693-8_37
17. Milborrow, S., Nicolls, F.: Active shape models with SIFT descriptors and MARS. In: Proceedings of the International Conference on Computer Vision Theory and Applications (VISAPP), vol. 2, pp. 380–387 (2014)

18. Li, L., Leung, M.K.H.: Integrating intensity and texture differences for robust change detection. IEEE Trans. Image Process. **2002**, 105–112 (2002)
19. Faisal, A., Emam, H.: Automated facial expression recognition using gradient-based ternary texture patterns. Chin. J. Eng. **2013**, 8 (2013). Article ID 831747
20. Donato, G., Bartlett, M.S., Hager, J.C., Ekman, P., Sejnowski, T.J.: Classifying facial actions. IEEE Trans. Pattern Anal. Mach. Intell. **21**(10), 974–989 (1999)
21. Liu, C., Wechsler, H.: Gabor feature based classification using the enhanced fisher linear discriminant model for face recognition. IEEE Trans. Image Process. **11**(4), 467–476 (2002)
22. Zhu, J.X., Su, G.D., Li, Y.E.: Facial expression recognition based on Gabor feature and Adaboost. J. Optoelectron. Laser **17**, 993–998 (2006)
23. Turk, M.A., Pentland, A.P.: Face recognition using eigenfaces. In: IEEE Computer Society Conference on Computer Vision and Pattern Recognition, pp. 586–591 (1991)
24. Belhumeur, P.N., Hespanha, J.P., Kriegman, D.J.: Eigenfaces vs fisherfaces: recognition using class specific linear projection. IEEE. Trans. Pattern Anal. Mach. Intell. **19**(7), 711–720 (1997)
25. Chang, C.C., Lin, C.J.: LIBSVM: a library for support vector machines. ACM Trans. Intell. Syst. Technol. **27**(2), 1–27 (2011)
26. Vapnik, V.: Statistical Learning Theory. Wiley, New York (1998)
27. Lucey, P., Cohn, J.F., Kanade, T., Saragih, J., Ambadar, Z., Matthews, I.: The extended Cohn-Kanade dataset (CK+): a complete facial expression dataset for action unit and emotion-specified expression. In: Proceedings 2010 IEEE Computer Society Conference on Computer Vision and Pattern Recognition Workshops, pp. 94–101 (2010)

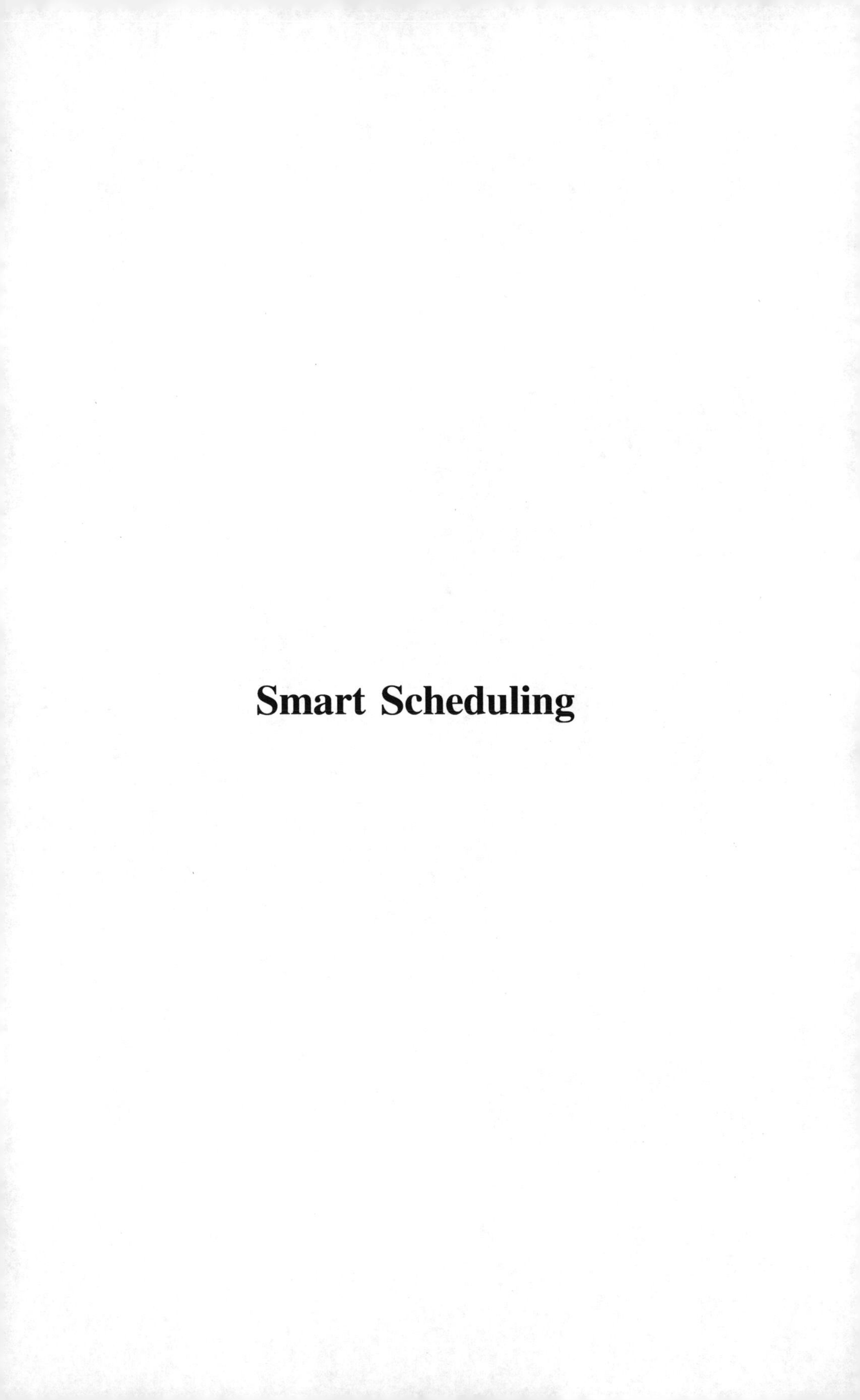

Smart Scheduling

Nested Simulated Annealing Algorithm to Solve Large-Scale TSP Problem

Lei Yang[1(✉)], Xin Hu[1], Kangshun Li[1], Weijia Ji[1], Qiongdan Hu[1], Rui Xu[1], and Dongya Wang[2]

[1] College of Mathematics and Informatics, South China Agricultural University, Guangzhou 510642, China
yanglei_s@scau.edu.cn

[2] University of Exeter, College of Engineering, Mathematics and Physical Sciences, Exeter EX4 4QF, UK

Abstract. Traveling Salesman Problem is one of the most valuable combination optimization NP-hard in mathematics and computer science. Simulated Annealing Algorithm is one of the better algorithms to solve the TSP problem. However, with the increasing scale of the TSP problem, this algorithm also shows some limitations. That is, the solution performance is poor, and the efficiency is low. This paper proposes a new Nested Simulated Annealing Algorithm (NSA), which divides the megacity group into small and medium-sized urban groups through recursive diffusion and setting the threshold method. In this way, the large-scale TSP problem can be transformed into small and medium-sized TSP problem at first. And then optimize the small and medium-sized TSP problem by using the Simulated Annealing Algorithm. Finally, synthesize a whole to achieve the final optimization effect. In this paper, we use three large-scale TSP problems in the TSPLIB database to test the proposed algorithm and compare it with the traditional Simulated Annealing Algorithm. The results show that the Nested Simulated Annealing Algorithm proposed in this paper has better efficiency and effect than the original algorithm in solving large-scale TSP problems.

Keywords: Large-scale TSP Problem · Nested Simulated Annealing Algorithm · Recursive diffusion · K-means

1 Introduction

In the era of information explosion, the amount of information may be too large to imagine. New requirements are necessary to put forward for the processing of massive data. However, many non-linear and uncertain factors make it difficult to establish accurate mathematical models for many problems. And the combinatorial optimization is just one of them, it is still an important issue when the computer capacity cannot be broken, and the TSP problem is one of the NPC problems that are difficult to solve completely through computer operations. The TSP is described as follows: Given a series of cities and their coordinates, the traveler is required to start from a certain city, then visit all the cities one by one,

K. Li et al. (Eds.): ISICA 2019, CCIS 1205, pp. 473–487, 2020.
https://doi.org/10.1007/978-981-15-5577-0_37

and each city only visits once, and finally returns to the starting city. The goal is to find the path that is as small as possible. For a TSP problem with n cities, the number of solutions is $(n-1)!/2$ [10], if n is a small number, such as 10 or 100, under the great computing power of the computer, we can exhaustively find the optimal path. With the increase of n, the scale of the solution has exploded, the approximate infinity of the solution can be said to be the biggest problem of large-scale TSP problem. For the number of approximate infinite solutions, it is also difficult for computer algorithms to obtain quickly better solutions and evaluate the solution.

With the increasing scale of the TSP problem, the exponential growth rate of the solution scale makes the solution extremely complicated. Branch-and-bound algorithm, branch-and-cut algorithm, cutting plane algorithm, plane discovery algorithm, or branch and price are known as the most effective precise algorithms. As an algorithm for solving large-scale TSP problem, the branch-and-bound algorithm requires one or more loosely structured decisions. It can be used to process TSP branches and cutting algorithms with 40 to 60 cities. Ralph Gomory proposed cutting plane algorithm in the 1950s, which can solve integer programming and mixed-integer programming problems. The cutting plane algorithm based on linear programming proposed by George Dantzig discovered the exact solution of the 15112 German towns in TSPLIB in 2001, whose computation time is equivalent to 22.6 years. In the branch and price algorithm, the model of the original problem usually contains a large number of decision variables [1]. Many experts and scholars at home and abroad have studied the solution of large-scale TSP problem. However, there is still no perfect solution to this problem, and no algorithm can solve all possible examples in polynomial time [9]. The approximation algorithm achieves a good solution compared to the precise algorithm but does not guarantee that the optimal solution will be found. Therefore, if a small deviation can be accepted, an approximation algorithm can be used to solve the large-scale TSP problems. It is necessary to use heuristic algorithms (or approximation) to provide the best possible, but not necessarily the best feasible solution. The Simulated Annealing Algorithm is derived from the principle of solid annealing, starting from a certain higher initial temperature, with the continuous decline of temperature parameters. The algorithm combines the probability jump feature to randomly find the global optimal solution of the objective function in the solution space, i.e., the local optimal solution can jump out probabilistically and eventually tend to be globally optimal. This algorithm is a probability-based algorithm [14], which regards the initially disordered state as high temperature, and gradually finds the best path by looking for stable low-temperature ordered state. It has been proved to be an excellent and efficient algorithm with mild and fast convergence rate and exceptional efficiency in solving TSP problems [8,15,16]. However, with the increase of city scale, the Simulated Annealing Algorithm has low efficiency and weak optimization effect when solving large-scale TSP problems.

Given the characteristics of large-scale TSP problems, this paper improves the Simulated Annealing Algorithm by incorporating recursive diffusion strategies, which makes it more adaptable to large-scale TSP problems. Also, the speed and efficiency of finding the optimal solution are improved. Then we test the proposed algorithm using three large-scale TSP problems in the TSPLIB database and compare it with the traditional Simulated Annealing Algorithm. The experimental results verify the usefulness and effectiveness of the new algorithm.

The remainder of this study is organized as follows. Section 2 presents a discussion of the basic Simulated Annealing Algorithm. Our new Nested Simulated Annealing Algorithm(NSA) is presented in Sect. 3. The experimental results and discussion are shown in Sect. 4. Finally, Sect. 5 concludes this study and gives future research directions.

2 Background

Because of the characteristics of large-scale TSP problems, this paper improves the Simulated Annealing Algorithm. Both Diffusion-and-Clustering Strategies and K-means clustering algorithms are used in this paper. So this section, we will introduce these two basic algorithms.

2.1 Simulated Annealing Algorithm

The idea of the Simulated Annealing Algorithm was first put forward by N. Metropolis et al., and it was introduced into the field of combinatorial optimization by S. Kirkpatrick. The Simulated Annealing Algorithm is a stochastic optimization algorithm based on the Monte-Carlo iterative solution strategy, which utilizes the similarity between the general combinatorial optimization problem and the annealing process of solid matter in physics [4]. The Simulated Annealing Algorithm is derived from the principle of solid annealing. The solid is warmed to a sufficiently high temperature and then allowed to cool slowly. When heating, the internal particles of the solid become disordered with the increase of heat, and its internal energy also increase. While slowly cooling, the particles become more and more orderly and reaching an equilibrium state at each temperature, finally reaching the ground state at normal temperature, and its internal energy is also reduced to a minimum [6].

The Simulated Annealing Algorithm can be decomposed into three parts: solution space, objective function, and initial solution. Simulated Annealing Algorithm has nothing to do with the initial value. The solution obtained by the algorithm is also independent of the initial solution state (i.e., the starting point of algorithm iteration). The Simulated Annealing Algorithm is asymptotically convergent and has been theoretically proved to be a global optimization algorithm converging to the global optimal solution with probability 1. The Simulated Annealing Algorithm has parallelism [11].

Nested Simulated Annealing Algorithm to Solve Large-Scale TSP Problem proposed in this paper calls the Simulated Annealing Algorithm twice. It is used to calculate the shortest loop and the shortest path of the urban agglomerations, and also the shortest path within the urban agglomeration.

2.2 K-means Algorithm

Give a set of n data points, and the K-means algorithm is to divide them into k groups in the d-dimensional Euclidean space r (i.e., clustering) [5].

The K-means algorithm first randomly selects K objects as the initial clustering center, then calculates the distance between each object and each cluster center, and assign each object to the cluster center closest to it. The cluster centers and objects assigned to them represent a cluster. Once all objects are allocated, the center of each cluster will be recalculated according to existing objects in the cluster. This process will be repeated until a stopping criterion is met [3]. The stopping criterion may be no objects are reassigned to different clusters. No cluster centers are re-changed, or the Square sum of errors reaches local minimum local [2].

The computational complexity of the K-means algorithm is $O(Nkn)$, where N is the number of iterations, n is the number of cities, k is the number of clusters. The criterion of the K-means algorithm is that the distance between data points in each cluster is as close as possible, and each cluster is as far as possible.

If $X = \{x_1, x_2, \ldots, x_n\}$, where $X_1(l = 1, 2, \ldots, k)$ is k partitions given by K-means and $E = \{e_1, e_2, \ldots, e_k\}$ represent the center of the k partitions, then the objective function of the K-means algorithm is:

$$F = \min \sum_{l=1}^{k} \sum_{x_i \in x_l} \|x_i - e_1\|^2 \tag{1}$$

The K-means algorithm has the advantages of simple implementation and high efficiency, so it is suitable for clustering large-scale data sets.

3 The Nested Simulated Annealing Algorithm (NSA)

This section will describe the new algorithm as a whole. The general idea of the new algorithm is to cluster the urban coordinate points of the large-scale TSP problem into a cluster firstly, then to optimize the cluster path to form the inter-cluster path. And then, intra-cluster optimization is performed to find intra-cluster paths. Finally, an overall optimal path for large-scale TSP problems will be found. The new algorithm is introduced from step 3.1 to step 3.4. And step 3.5 will describe the new algorithm as a whole.

3.1 Clustering to Large-Scale Urban Agglomeration

The first step of the Nested Simulated Annealing Algorithm is to cluster. After receiving the urban coordinate data from the total algorithm, they are clustered by the clustering algorithm, then the cluster grouping is labeled after each coordinate, and the central coordinate system E is generated to record the central coordinates of each cluster at the same time. Two clustering algorithms are designed in the Nested Simulated Annealing Algorithm (NSA) proposed in this paper. The first one is the improved K-means algorithm. Some appropriate improvements are made to K-means to adapt to the total algorithm better. The second one is the Diffusion Clustering Algorithm proposed in this paper.

The Improved K-Means Algorithm. K-means algorithm has the advantage of simplicity and efficiency. Still, its disadvantage is that the initial center position is random, and the number of clusters needs to be set manually. In view of the above shortcomings and the characteristics of large-scale TSP problems, we have made appropriate improvements to make its position no longer random and does not need to set the parameters manually, but automatically generated. Thus, a cluster grid map with the average grid line distribution can be obtained. For avoiding the over-concentration of initial points, each cluster located intersection between horizontal and vertical lines of the grid (except for the boundary).

Diffusion Clustering Algorithm. To transform the city into an urban agglomeration, we assume the biggest difference between them is the distance. Specifically, it is that an obvious isolation zone should be formed around each urban agglomeration, which is separated from other urban agglomerations. Therefore, the urban agglomeration is more likely to be regarded as a point to calculate the shortest circuit between urban agglomerations. So there is the idea of diffusion clustering. The idea of diffusion clustering is as follows: given a distance (the distance from the shortest isolation belt), choose a city arbitrarily firstly, then find other cities to determine whether there is a city with a distance less than the isolation zone. If there is, it will be incorporated into the urban agglomeration. Then it will be recursively diffused until every city in the urban agglomeration confirming that there is no other city in the distance. The algorithm will ensure that distance between any point within the urban agglomeration and other urban agglomerations will be greater than a fixed value d_max, while there is at least one point within it that is less than d_max. Thus isolation between the urban agglomeration and the urban agglomeration will be formed.

3.2 Determine the Order of Large-Scale Urban Agglomeration

After the clustering of megacities, several clusters are formed. Each cluster can be regarded as a point. The central coordinates of the clusters replace the coordinates of this point. The Simulated Annealing Algorithm is used to find out the best inter-cluster loop, that is, the order of urban agglomerations, to identify the inter-cluster path of megacities.

As clustering, the center of each urban agglomeration has been set:

$$E = \{e_1, e_2, \ldots, e_k\} \tag{2}$$

Using this point as the coordinates of urban agglomeration, we use the Simulated Annealing Algorithm to find out the optimal Hamiltonian circuit.

3.3 The Choice of Border City in Urban Agglomerations

From the loop of large-scale urban agglomerations, the location of each urban agglomeration in the loop can be determined. Next, we should determine the path within the urban agglomeration. Before deciding the groups' path, We should define the boundary city in the urban agglomeration. It can deem as a connection point with other urban agglomerations. Generally speaking, the urban agglomeration needs to connect two cities from top to bottom, so it needs two boundary cities at the beginning and the end. The inner part of the urban cluster is not a loop but a single line. As a result, it only needs to establish the head city, while the rear city is the head city of the next urban agglomeration.

3.4 Description of the Whole Algorithm

For large-scale TSP problems, the city size is too large, the number of solutions for TSP problems with n cities is $(n-1)!/2$ [7,10]. If $n > 10000$, then the number of solutions is huge. The time it takes to find the optimal path is hard to imagine or even impossible. To find a better path in as short a time as possible, the new algorithm use cluster firstly, then the Simulated Annealing Algorithm [12,13] is applied to urban agglomerations formed by clustering and cities in urban agglomeration respectively to find their optimal path. Finally, they are connected as a whole to reduce time.

The overall description of the algorithm is as follows:

Step 1, Divide TSP problem with tens of thousands of cities into several urban agglomerations, that is, clustering;

Step 2, Consider the urban agglomeration as a point and use the Simulated Annealing Algorithm to find the optimal Hamiltonian circuit;

Step 3, Identify a boundary city inside each urban agglomeration, i.e., the city connected to other urban agglomerations. Thus a complete Hamiltonian circuit can be formed afterward;

Step 4, After determining the boundary cities, the shortest path within the urban agglomeration is determined by taking the boundary city as the head and tail. Because it is an improvement of the Simulated Annealing Algorithm, we still choose the Simulated Annealing Algorithm. The core of this algorithm is double Simulated Annealing Algorithm and Clustering Algorithm;

Finally, we determined the order of the urban agglomeration and the order of the inner cities in the urban agglomeration, and linked them into rings, i.e., the path we need.

Algorithm 1 shows the whole pseudo-code of the Nested Simulated Annealing Algorithm.

4 Experiments and Comparisons

4.1 Experimental Explanation

TSP standard database TSPLIB for experimental is used in this paper. The Nested Simulated Annealing Algorithm is written in the C# language on the

Algorithm 1. Nested Simulated Annealing Algorithm

Require: coordinates[ ][ ]
Ensure: len, bestPath[ ] // centrePoint: central coordinates of each cluster and the coordinates possessed by the cluster // centreList: optimal path loop of clusters // cluster: algorithms for clustering // deterBound: algorithm for determining the boundary city // saa: Simulated Annealing Algorithm
1: cluster(coordinates)
2: **then** get(centrePoint[ ][ ])
3: saa(centrePoint[ ][ ])
4: **then** get(centrePoint[ ][ ])
5: **while** i <centreList **do**
6: T = get(centreList[i])
7: deterBound(centrePoint[T])
8: len+=saa(centrePoint[T][ ])
9: Write(centrePoint[T][ ], bestPath)
10: **end while**
11: return A

Visual Studio software. It runs on the platform of AMD Ryzen 5 2500U in memory 8G (software allocated memory with dynamic distribution, about 100M-2G). The original Simulated Annealing Algorithm is abbreviated as SA (Simulated Annealing Algorithm). The Nested Simulated Annealing Algorithms has two algorithm models; one is the K-means clustering improved algorithm, which is abbreviated as SA-KS (Simulated annealing algorithm-K-means). And the other is the improved algorithm using diffusion clustering, which is abbreviated as SA-SP (Simulated annealing algorithm-Spread).

Experiments compare the advantages and disadvantages of the two algorithms. The comparison method is to find out the shortest path calculated at the same time. The experimental interval is 5, 10, 15, 20, 25, 30, 40, 50, 60, the unit is in seconds, and termination time is determined by whether the optimal result is obvious or not. Since the simulated annealing algorithm is calculated by continuous iteration, it is only necessary to set the number of iterations to be close to infinity, and then insert the time for judgment. As long as the time arrives, the distance is output to get the shortest path at that time.

The Nested Simulated Annealing Algorithm divides the data set into several urban agglomerations for calculation. Therefore, it is necessary to divide the total time into time segments and then distribute them to the urban agglomeration. A uniform strategy is adopted at the beginning of the algorithm. And the result shows that the time to find out the shortest path of SA-SP is often several times that of other algorithms. It is finally found that the use of recursive diffusion clustering makes it easy to form a large scale urban agglomeration and some small scale urban agglomerations. Therefore, we use the principle of proportionality for the time fragment of SA-SP. That is, the time segment of large-scale urban agglomeration is large, while the time segment of small-scale urban agglomeration is small. Due to the improvement of the previous algorithm, SA-KS can form an urban agglomeration with about 100 cities. The error is about plus or

Table 1. Experimental results of d15112

Time/min	SA	SA-KS	SA-SP
5	66143147.6653943	66172344.3532432	99950532.283926
10	63625653.1995809	62488074.5813857	49847499.652352
15	55867564.3039949	40427092.2009234	40481877.233455
20	48155297.5289634	27777578.3184985	36235223.2523424
25	40250735.6080993	26539623.3342345	34252344.5223444
30	38130347.4113312	25439246.6989969	32077533.035223

minus three orders of magnitude during the experimental process. To avoiding complicated calculations, the segments are divided evenly (Table 1).

The data sets selected in this paper are d15112, pla33810, rl5934. Among them, the number of cities in d15112 is 15112, the number of cities in pla33810 is 33810, the number of cities in rl5934 is l5934. The data comes from https://www.proxy.iwr.uni-heidelberg.de/groups/comopt/software/TSPLIB95/TSP/.

4.2 Experimental Results and Analysis

(1) d15112

Number of cities: 15112

Number of city groups formed by SA-KS: 151

Number of city groups formed by SA-SP: 153, Distance between boundary cities: 250

From the chart, it can be intuitively felt that the shortest path obtained by the new algorithm is better than the original algorithm. However, the result of the new algorithm is inferior to the original algorithm at the beginning of 5 min, which shows that the new algorithm is not superior to the original algorithm at the beginning of the operation. With the increase of time, the convergence speed of the new algorithm is greatly accelerated, while the original algorithm that is relatively stable. The convergence speed of the new algorithm has a great potential of opening and closing, which makes it nearly complete in 20 min. However, the original algorithm is nearly complete in 30 min. That is to say, the new algorithm not only achieves a shorter path but also takes a shorter time. It shows that the efficiency of the new algorithm is better. Compared with the two new algorithms, the final result of SA-KS is better than that of SA-SP (Figs. 1, 2, 3 and 4).

(2) pla33810

Number of cities: 33810

Number of city groups formed by SA-KS: 338

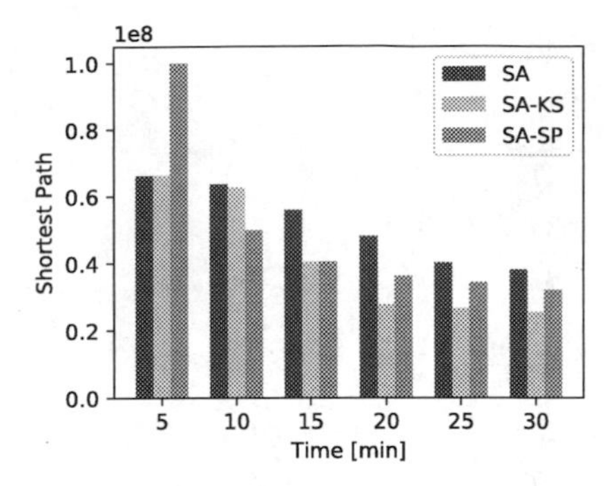

Fig. 1. d15112 Compare the shortest path of the three algorithms

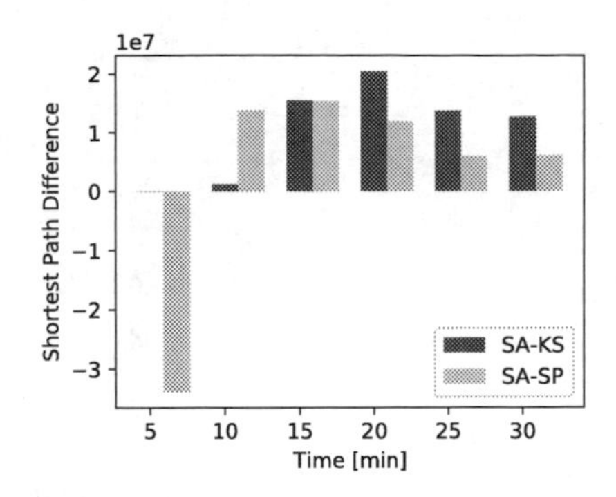

Fig. 2. d15112 The shortest path difference between the new and the original algorithms

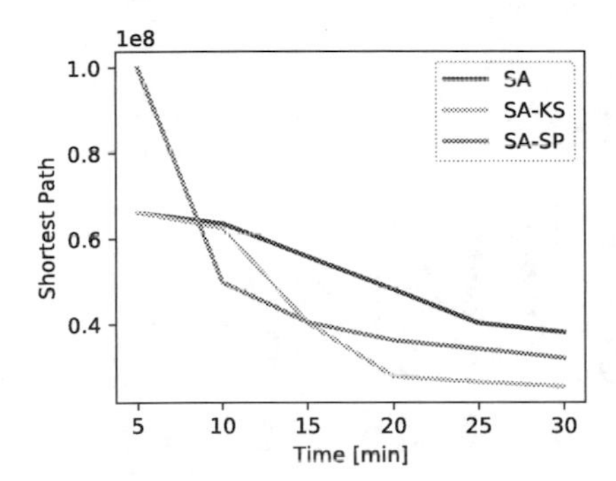

Fig. 3. d15112 Compare the shortest path of the three algorithms

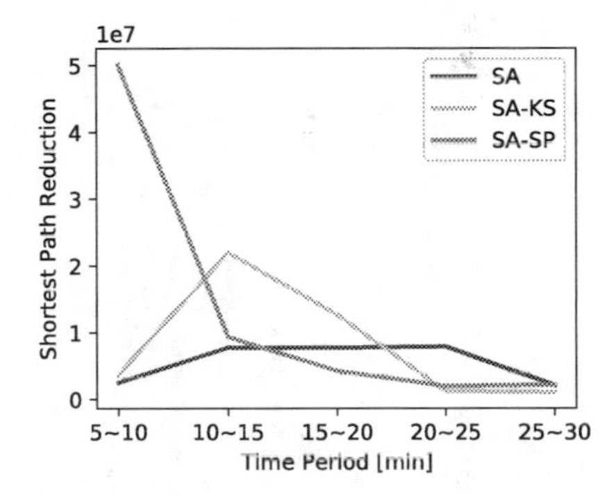

Fig. 4. d15112 The shortest path difference between the new and the original algorithms

Number of city groups formed by SA-SP: 548, Distance between boundary cities: 2500

With the largest data volume in TSPLIB, Pla33810 has symbolic significance. As can be seen from the graph set clearly, the limitation for large-scale TSP problems of the original algorithm is that the calculation time is considerable. From Fig. 7, the shortest distance calculated at 60 min for the original algorithm is more than SA-KS, or even three times that of SA-SP.

Although it can be seen from Fig. 8 that the original algorithm still has the space to decrease, the shortest path of SA-KS is half of it at 60 min. And SA-SP is even one-fifth of it. The original algorithm only optimizes about one-third of the shortest path in 60 min. It is inconceivable to spend very little time to get the shortest path of the original algorithm. For this data set, the new algorithm is undoubtedly superior to the original algorithm in both efficiency and effect (Table 2).

Table 2. Experimental results of pla33810

Time/min	SA	SA-KS	SA-SP
5	6538118097.42389	3246265731.9917	1089159002
10	5815506732.03497	2450163420.09839	843617237
15	5432568869.1673	2049669371.06635	762447365
20	5093680637.42291	1970049125.26245	735888407
25	4882292916.0636	1969127618.50726	702342344
30	4729375010.53876	1952342342.32423	686448051
40	4466574081.10608	1939423422.32423	643463234
50	4252562812.41895	1934814233.78528	609214147
60	4082714139.03558	1933134134.23422	609135146

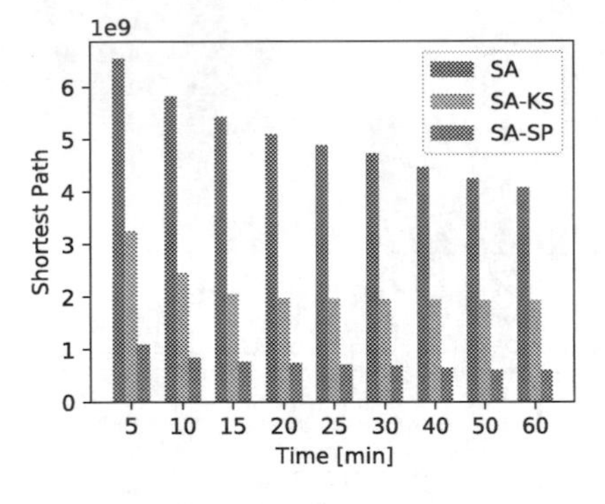

Fig. 5. pla33810 Compare the shortest path of the three algorithms

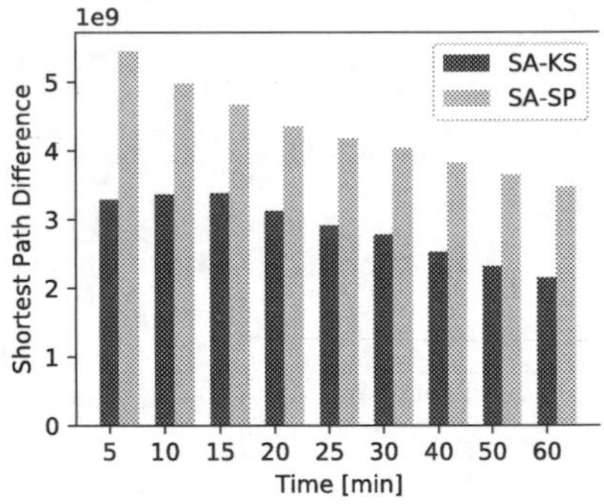

Fig. 6. pla33810 The shortest path difference between the new and the original algorithms

Compared with the two new algorithms, SA-SP is undoubtedly better. Figure 5 shows that the shortest path of SA-SP tends to be stable five minutes ahead of SA-KS. And the value is only half of that (Fig. 6).

(3) rl5934

Number of cities: 5934

Number of city groups formed by SA-KS: 59

Number of city groups formed by SA-SP: 352, Distance between boundary cities: 25

The gap between the new algorithm and the old one both reaches the maximum in 20 min. After that, the shortest distance gap of algorithms begins to decrease. The shortest distance of the SA-KS algorithm decreases greatly in the first 20 min, while the original algorithm decreases uniformly in 60 min. And finally, the original one reaches the same level as the SA-KS algorithm. That is, the SA-KS algorithm achieves the effect of the original algorithm at 60 min

Table 3. Experimental results of rl5934

Time/min	SA	SA-KS	SA-SP
5	15932897.715384	12717264.5666647	1094287.34256053
10	15128362.9296885	3730546.88289547	1037325.64975929
15	14893925.9242811	3546742.95794344	927265.894996285
20	13831427.2653251	3022189.93628502	899818.836472754
25	12224572.3554425	2904052.15571928	871517.595030069
30	9418625.5698123	2885636.25743437	650327.521313071
40	6119607.55737066	2842333.03874254	334137.65235281
50	4483642.47985458	2897860.72196817	302343.235232734
60	3787122.49793386	2854062.5260334	283429.610421419

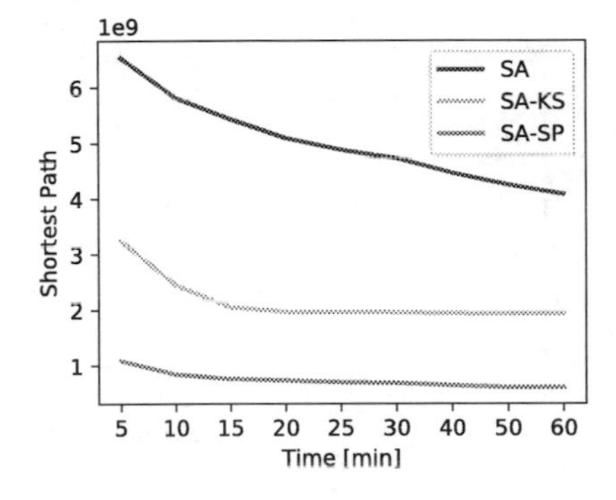

Fig. 7. pla33810 Compare the shortest path of the three algorithms

Fig. 8. pla33810 The shortest path difference between the new and the original algorithms

in 20 min. Although it is not like pla3810, which gets the effect with five minutes, it can also be regarded as a three-fold increase in efficiency. For SA-SP, its advantages are distinct. For its efficiency, it reaches the optimal result in 5 min. The shortest path obtained by SA-SP is also much better than the other two algorithms (Table 3).

4.3 Analysis and Evaluation

Firstly, the difference between the NSA algorithms and the original algorithm is compared. According to the above experimental results for the large-scale TSP problem, the NSA algorithms are superior to the original algorithm in the efficient and final result. The NSA algorithms obtain the optimal path within 25 min, while the original algorithm generally takes 60 min. For most of the data, the NSA algorithm can get an optimal path far beyond the original algorithm within 5 min, or even just a fraction of it. After 20 min, the shortest path of the NSA algorithm

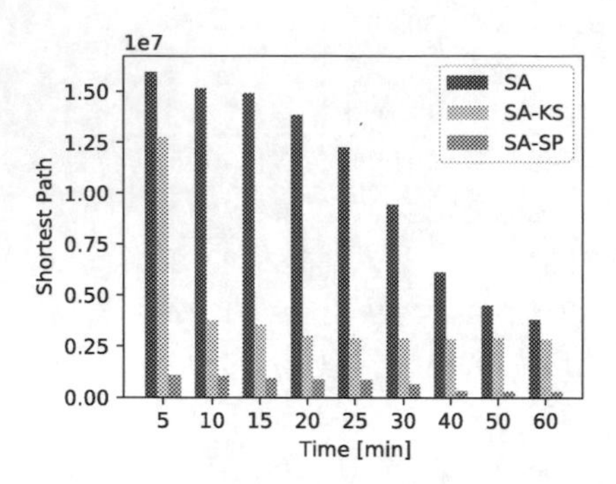

Fig. 9. rl5934 Compare the shortest path of the three algorithms

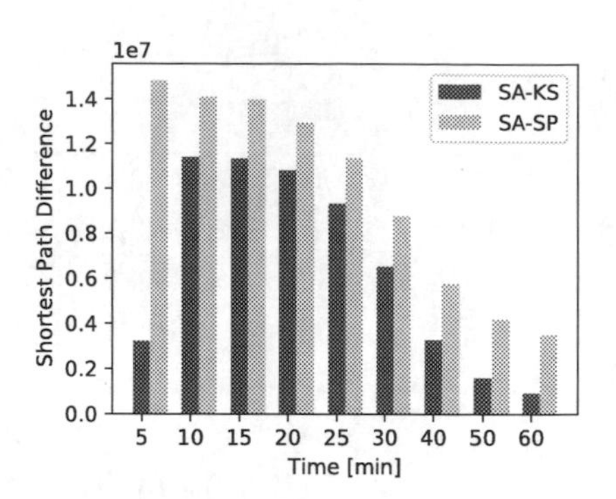

Fig. 10. rl5934 The shortest path difference between the new and the original algorithms

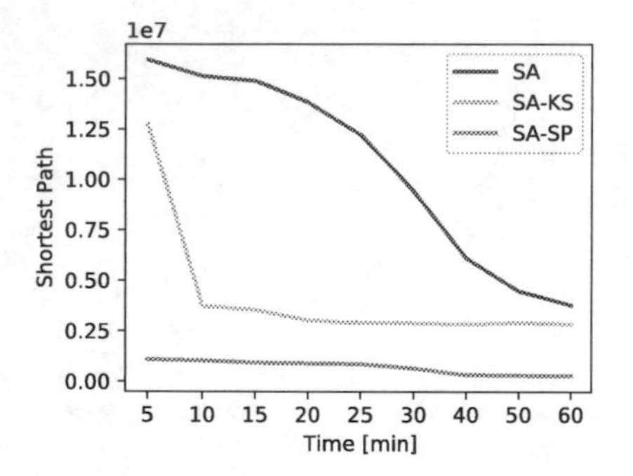

Fig. 11. rl5934 Compare the shortest path of the three algorithms

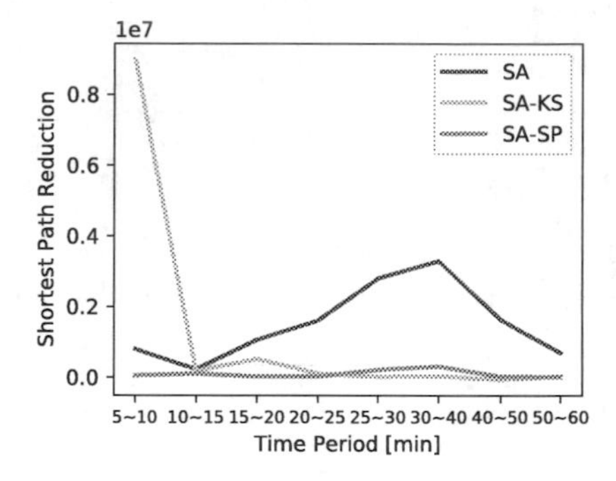

Fig. 12. rl5934 The shortest path difference between the new and the original algorithms

is basically no longer updated. The original algorithm can also achieve the same shortest path that the NSA algorithm can make on some data sets, but it takes several times the time of the new algorithm. The original algorithm gets the shortest path smoothly and gradually, while the NSA algorithm, especially SA-KS, gets the shortest path more vigorously and rapidly. So the improvement of the NSA algorithms is successful. Through the NSA algorithm's basic ideas, we can know that it first calculates the shortest path between urban agglomerations, laying a general framework. The shortest path will not be too bad under this framework. This is why the NSA algorithms can quickly get the shortest path. Then there is the refinement calculation within the urban agglomeration. This process is completed within 20 min, which is also the reason why SA-KS is difficult to make further progress after 20 min. Because internal refinement has been achieved, it is challenging to continue to optimize. According to this principle, the original algorithm can theoretically obtain a shorter optimal path than the NSA algorithm because the shortest path of the new algorithm is limited to the urban group. Going to an urban group then returns to the original city. It is maybe the shortest distance. But it takes a lot of time to get this result. From the above experimental result, we can know it did not appear within 60 min. The original algorithm is stable at 60 min, so it is impossible to achieve the shortest path produced by the new algorithm (Figs. 9, 10, 11 and 12).

Then compare the difference between the two NSA algorithms SA-KS and SA-SP. From the experimental results, there are advantages and disadvantages. For larger urban agglomerations, SA-KS has better results, while SA-SP is more adaptable. The SA-SP has an excellent solution to address large-scale and small-scale TSP problems. The reason is that they use different clustering algorithms. SA-KS uses the improved K-means clustering algorithm to get urban agglomerations. The SA-SP uses a recursive diffusion algorithm to easily form an oversized urban cluster and several small-scale urban agglomerations. The experimental time is proportionally distributed so that large-scale urban agglomerations can calculate thoroughly.

During the experiment, it is easy to see the difference between the two kinds of clustering. When obtaining the shortest path, SA-KS often has a sharp decline in a certain period. This is because the size of urban agglomeration is similar, and the allocation time is the same. With increases in time, the internal operation get deepens in each urban agglomeration, and this is easy to form a superposition effect. That is, every little reduction of each urban agglomeration is a lot of distance. SA-SP can be regarded as a combination of SA-KS and SA because of its megacities. From the experimental chart, it can be seen that the SA-SP descent process is relatively smooth, and the descent time is longer, too. SA-KS can hardly be further explored after the stabilization of each urban agglomeration (it is generally found in the experiment that this time is 20 min).

In summary, the two NSA algorithms have improved to a certain extent, and SA-SP has better adaptability. SA-KS is better at processing some specific data sets, especially large-scale data sets.

5 Conclusion

Although this experiment proves the feasibility of the Nested Simulated Annealing Algorithm, there is still room for further improvement. In the growth of K-means in this paper, it automatically generates urban agglomerations with about 100 cities, ignoring the total number of cities. For example, for 1000 cities, only ten urban agglomerations are created, which is fatal for finding the shortest path. This may be one of the reasons for the gap between SA-KS and the original algorithm on small-scale TSP problems. As for SA-SP, firstly, it is a relatively primitive method to input the boundary distance manually. It can try to make a secondary improvement by determining the boundary distance through the number of cities and the maximum value of city coordinates. Also, we can automatically adjust the boundary distance and reclassify when the number of city clusters is too large or too small.

In the future, we will further study and improve the algorithm, then try to investigate whether the NSA algorithm can be extended for other classical combinatorial optimization problems.

Acknowledgments. This work was partially supported by the National Natural Science Foundation of China (Grant Nos. 61573157 and 61703170), Science and Technology Project of Guangdong Province of China (Grant Nos. 2018A0124 and 2017A020224004), Science and Technology Project of Tianhe District of Guangzhou City(Grant No. 201702YG061), Science and technology innovation project for College Students(Grant No. 201910564127). The authors also gratefully acknowledge the reviewers for their helpful comments and suggestions that helped to improve the presentation.

References

1. Gao, J.: Hierarchical Solving Method for Large Scale TSP Problems. Master's thesis, Inner Mongolia University For Nationlities (2014)
2. Hu, X., Zhang, J., Qi, P., Zhang, B.: Modeling response properties of v2 neurons using a hierarchical k-means model. Neurocomputing **134**, 198–205 (2014). https://doi.org/10.1016/j.neucom.2013.07.052
3. Huang, Q.: Based on Fusion on Feature-Level Image Technique Study Intelligent. Master's thesis, China University of Petroleum (2012)
4. Khan, N.A., Shaikh, A.: A smart amalgamation of spectral neural algorithm for nonlinear lane-emden equations with simulated annealing. J. Artif. Intell. Soft Comput. Res. **7**(3), 215–224 (2017). https://doi.org/10.1515/jaiscr-2017-0015
5. Lam, Y.K., Tsang, P.W.: eXploratory k-means: A new simple and efficient algorithm for gene clustering. Appl. Soft Comput. **12**(3), 1149–1157 (2012). https://doi.org/10.1016/j.asoc.2011.11.008
6. Li, J.: Color separation research and implementation based on printing and dyeing CAD system. Master's thesis, Zhejiang University (2004)
7. Liu, X.: The Application and Research of an Improved Genetic Algoruthm on the TSP Problem. Master's thesis, Harbin University of Science and Technology (2011)
8. Pan, J., Cheng, Z.G., Lv, J.Y.: Study on the TSP problem based on SA algorithm. In: Applied Mechanics and Materials, vol. 687–691, pp. 1316–1319 (2014). https://doi.org/10.4028/www.scientific.net/amm.687-691.1316
9. Shi, Z.: The single machine parallel-batching scheduling problem with family-jobs. Master's thesis, Zhengzhou University (2008)
10. Wang, P.: Research on Parallel Ant Colony Algorithm and Application. Master's thesis, Southwest Jiaotong University (2008)
11. Xiaoping, Z., Shengbowen, L.: Fetal electrocardiogram extraction based on independent component analysis and particle swarm optimizer. Inf. Electron. Eng. **6** (2009)
12. Xie, Y.: A summary on the simulated annealing algorithm. Control Autom. **5**, 66–68 (1995)
13. Yang, M., Gao, X., Li, L.: Improved bidirectional CABOSFV based on multi-adjustment clustering and simulated annealing. Cybern. Inf. Technol. **16**(6), 27–42 (2016). https://doi.org/10.1515/cait-2016-0075
14. Lu, Y.T., Lin, Y.Y., Peng, Q.Z., Wang, Y.Z.: A review of improvement and research on parameters of simulated annealing algorithm. Coll. Math. **31**(6), 96–103 (2015)

15. Zhang, Z.C., Han, W., Mao, B.: Adaptive discrete cuckoo algorithm based on simulated annealing for solving tsp. Tien Tzu Hsueh Pao/Acta Electronica Sinica **46**, 1849–1857 (2018). https://doi.org/10.3969/j.issn.0372-2112.2018.08.008
16. Zhi, J.Z., Yu, G.B., Deng, S.J., Chen, Z.L., Bai, W.Y.: Modeling and simulation about tsp based on simulated annealing algorithm. In: Applied Mechanics and Materials, vol. 380, pp. 1109–1112 (2013). Trans Tech Publ

Modeling and Scheduling for the Clean Operation of Semiconductor Manufacturing

Ya-Chih Tsai[1], Jihong Pang[2], and Fuh-Der Chou[2](✉)

[1] Department of Hotel Management, Vanung University, Tao Yuan, Taiwan, Republic of China
amywang@mail.vnu.edu.tw

[2] The College of Mechanical and Electrical Engineering, Wenzhou University, Wenzhou 325035, Zhejiang, China
pangjihong@163.com, fdchou@tpts7.seed.net.tw

Abstract. Motivated by the clean operation in the semiconductor manufacturing, this paper model it as a non-identical parallel machine scheduling problem with machine flexible periodical maintenance, in which the machines must to be stopped for changing cleaning agent periodically to avoid that too much the dirt residue in the machine damages the wafer. The objective is to minimize the makespan. For the problem, we proposed a mixed integer programming (MIP) model to find all optimal solutions for small problems, additionally, an efficient particle swarm optimization (PSO) algorithm is develop to obtain near-optimal solutions. Computational results show that the proposed PSO algorithm is quite successful on both solution accuracy and efficiency to solve the considered problem.

Keywords: Non-identical parallel machine · Makespan · Flexible maintenance

1 Introduction

An integrated model where job scheduling and maintenance activities are considered simultaneously has attracted a lot of attention recently. The motivation comes from the manufacturing industry, for example, a machine may be designed to be cleaned, refueled or maintained after working for a certain period, such as a steel strip production in a steel plant [1, 2] or clean operation in a semiconductor manufacturing [3, 4]. Since the machines are not available all the time which causes machine maintenance become one of the critical factors that can significantly affect job scheduling performance, therefore, it is an important issue to consider job scheduling and maintenance activities simultaneously to obtain a better system performance.

This paper models the clean operation of semiconductor manufacturing as a non-identical parallel machine scheduling problem with flexible maintenance activities. In the clean operation, the dirt such as particle, organic material, and metal-salts, etc., on a wafer will left in the machine, and due to the dirt allowance, the machines must be stopped and change cleaning agent periodically in order to process a series of jobs later. This operational behavior is similar to batching problems as shown in Fig. 1.

K. Li et al. (Eds.): ISICA 2019, CCIS 1205, pp. 488–496, 2020.
https://doi.org/10.1007/978-981-15-5577-0_38

Our considered problem can be stated as follows: There are n jobs to be processed on m non-identical parallel machines. Each job J_j has a positive processing time p_{ij}, a release time r_j, and dirt d_{ij}, wherein p_{ij} and d_{ij} are determined by the assigned machine M_i. In addition, each machine M_i has a positive clean time w_i and a maximum dirt allowance T_i. Given a feasible schedule of S of jobs, the accumulation of dirt on each machine must not exceed its dirt allowance. The objective is to find a schedule to minimize the makespan. According to the standard machine scheduling classification [5], this paper is denoted as $R_m|r_j, d_{ij} \leq T_i, fpa|C_{max}$, where fpa in the second field means the maintenance activity is flexible and occurs periodically [6].

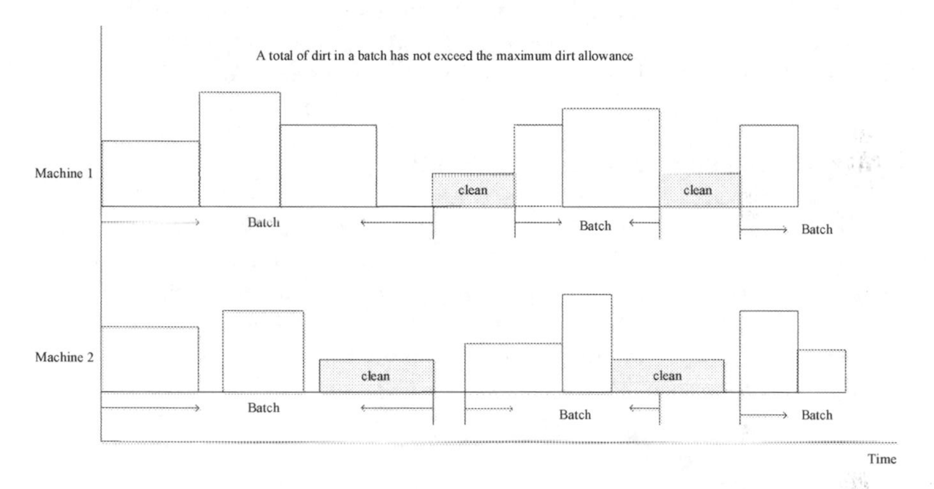

Fig. 1. The operational behavior of clean operation

In the literature, scheduling problems with machine maintenance activity can be classified into "fixed" and "flexible" model. The first model assumes that the starting and completion time of maintenance activity are fixed and given in advance. There are lots of researches about the first model, and Schmidt [7] and Ma [8] provided a relative comprehensive overview. The second model integrates job's production with maintenance activity together, that is, the scheduler has to determine when to process each job and when to conduct each maintenance activity simultaneously. Qi et al. [9] were the first to address this type of scheduling problems, in which the objective was to minimize total completion time of jobs. Since then, scheduling problem with flexible maintenance model have received considerable attention [6, 10–14].

To the best of our knowledge, Su and Wang [3] is the first to address the clean operation of semiconductor manufacturing. They consider a single-machine scheduling problem and proposed a mixed integer programming model and a heuristic based on DP method and V-rule of Kanet [15] to minimize the total absolute deviation of the job completion time. Pang et al. [16] extent the problem of Su and Wang [3] to dynamic case, i.e., the job's release time is involved, additionally, they considered bi-objective functions of minimizing total weighted tardiness and total completion time, for the problem, they developed a scatter simulated annealing (SSA) algorithm and compared with the well-known meta-heuristic algorithms including SMOSA [17] and NSGA-II [18] in the computational experiment, the results showed that the SSA algorithm performed well. Su et al. [4] considered unrelated parallel machine problem with periodical maintenance activities, the objective is to minimize the number of tardy jobs. They developed improved heuristic algorithms based on forward/backward insert mechanisms to obtain high-quality solutions. In this paper, we consider a non-identical parallel machines problem with flexible maintenance activities, where jobs have different release time, and dirt left in any machine are also different. The objective is to minimize the makespan. For the problem, we develop a mixed integer programming (MIP) model to obtain optimal solutions for small-size problems since there are no benchmarks for the considered problem so far. Additionally, a particle swarm optimization (PSO) algorithm is also developed for solving large-size problems.

2 Mixed Integer Programming Model

In this section, a mixed integer programming (MIP) model is constructed to describe the characteristics of the considered problems in which a set of jobs $J = \{J_1, J_2,\ldots, J_n\}$ is to be scheduled on m non-identical parallel machines. Preemption is not allowed, and machines are not available all the time because they must be stopped to clean dirt for next job processing. The parameters and variables are listed as follow. Our objective is to minimize the makespan, i.e., C_{max}.

- Parameters

n: number of jobs
m: number of machines
i: index of machines, i = 1,…,m
j: index of jobs, j = 1,…,n
k: index of processing sequence, k = 1,…,n
M_i: machine i
J_j: job j
r_j: the release time for job j
p_{ij}: the processing time of job j on machine i
d_{ij}: the left dirt for job j while processed on machine i
T_i: the maximum dirt allowance for machine i
w_i: the clean time for machine i
BM: a very large positive integer; $BM = \sum_{i=1}^{m} \sum_{j=1}^{n} p_{ij}$

- Decision variables

X_{ijk}: 1 if job j is processed on machine i at position k, 0 otherwise
Y_{ik}: 1 if the maintenance is implemented at position k *on* machine i, 0 otherwise
ST_{ik}: the start time for machine i at position k for processing jobs
PT_{ik}: the processing time for machine i at position k

CT_{ik}: the completion time for machine i at position k
Q_{ik}: the accumulated dirt of machine i at position k
C_j: the completion time of job j

$$C_{max} = \max C_j$$

- Model

$$Min. \quad C_{max} \tag{1}$$

s.t.

$$\sum\nolimits_{i=1}^{m} \sum\nolimits_{k=1}^{n} X_{ijk} = 1 \;\; i = 1, \ldots, m \tag{2}$$

$$\sum\nolimits_{j=1}^{n} X_{ijk} \leq 1 \qquad i = 1, \ldots, m; k = 1, \ldots, n \tag{3}$$

$$ST_k \geq \sum\nolimits_{j=1}^{n} r_j \cdot X_{ijk} \qquad i = 1, \ldots, m; k = 1, \ldots, n \tag{4}$$

$$PT_{ik} = \sum\nolimits_{j=1}^{n} p_{ij} X_{ijk} \quad i = 1, \ldots, m; k = 1, \ldots, n \tag{5}$$

$$ST_{ik} \geq CT_{i,k-1} + w_i \times Y_{i,k-1} \qquad i = 1, \ldots,; \;\; k = 2, \ldots, n \tag{6}$$

$$CT_{ik} \geq ST_{i,k} + PT_{ik} \quad i = 1, \ldots, m; k = 1, \ldots, n \tag{7}$$

$$Q_{i1} = \sum\nolimits_{j=1}^{n} (d_{ij} \times X_{ij1)} \qquad i = 1, \ldots, m \tag{8}$$

$$Q_{i,k-1} + \sum\nolimits_{j=1}^{n} (d_{ij} \times X_{ijk}) \leq Q_{ik} + BM \times Y_{i,k-1} \; i = 1, \ldots, m; k = 2, \ldots, n \tag{9}$$

$$\sum\nolimits_{j=1}^{n} (d_{ij} \times X_{ijk}) \leq Q_{ik} + BM\left(1 - Y_{i,k-1}\right) \; i = 1, \ldots, m; k = 2, \ldots, n \tag{10}$$

$$Q_{ik} \leq T_i \quad i = 1, \ldots, m; k = 1, \ldots, n \tag{11}$$

$$\sum\nolimits_{j=1}^{n} X_{ijk} \leq \sum\nolimits_{j=1}^{n} X_{ij,k-1} \; i = 1, \ldots, m; k = 2, \ldots, n \tag{12}$$

$$C_j \geq CT_{ik} + BM\left(X_{ijk} - 1\right) \; i = 1, \ldots, m; k = 2, \ldots, n \tag{13}$$

$$C_{max} \geq C_j \quad j = 1, \ldots, n \tag{14}$$

The objective function (1) is to minimize the makespan. Constraint (2) ensures that each job is assigned to only one position for on a machine. Constraint (3) ensures that not more than one job is assigned to any position in the sequence for any machine.

Constraint (4) define the available time at the kth position on machine i for processing jobs. Constraint (5) specifies that the processing time at the kth position on machine i for processing jobs. Constraint (6) ensure that the start time at the kth position on machine i should be greater than or equal to the completion time of the previous position. Constraint (7) prevent that a job from processing before its ready time. Constraint (8) defines that the total of dirt left at the first position on machine i. Constraints (9) and (10) define that the total of dirt left at the kth position on machine, excluding the first position. Constraint (11) ensures that the total of dirt has not be greater than the maximum dirt allowance for each machine. Constraint (12) forces that at least one job has to be processed between maintenance activities. Constraint (13) define the completion time of job j. Constraint (14) defines that the makespan.

3 Particle Swarm Optimization

PSO is one of population-based meta-heuristic algorithms, which mimics the behavior of flying birds and their communication mechanism to search optimal solutions [19]. In PSO, each particle with its position and velocity is represented as a solution in the problem solution space, and each particle adjusts their position by its own direction and flock direction. The original PSO is used to solve continuous optimization problems, for discrete optimization problems such as scheduling problems, thus, the particle position representation, velocity, and particle movement for each particle in our PSO need to modify.

Permutation list is one of popular encoding schemes for meta-heuristic algorithms, aiming to make a solution recognizable for an algorithm. For example, the permutation list for 5-job-2-parallel machine problem as follow.

Index	1	2	3	4	5	6
Random number	0.23	0.98	0.13	0.44	0.53	0.67
permutation	J_3	J_1	J_4	J_5	–	J_2

where four jobs $(J_3\ J_1, J_4, J_5)$ are assigned to machine M_1, one job (J_2) is assigned to machine M_2. In our PSO, the position of particle l is denoted as X^l represented by an $1 \times (n+m-1)$ matrix for an n-job m-parallel machine problem. As mentioned above, the position of particle l can be transferred to

$$X^l = [0.23\,0.98\,0.13\,0.44\,0.53\,0.67]$$

Initially, the iteration counter $k = 0$, and the position and velocity of particle i is denoted as $\mathrm{X}_i^0 = \left[x_{i1}^0 x_{i2}^0 \cdots x_{i,n+m-1}^0\right]$ and $\mathrm{V}_i^0 = \left[v_{i1}^0 v_{i2}^0 \cdots v_{i,n+m-1}^0\right]$, respectively, and the initial value of x_{ij}^0 and v_{ij}^0 are randomly generated as follows:

$$x_{ij}^0 = x_{\min} + (x_{\max} - x_{\min}) \times rand$$

$$v_{ij}^{0} = v_{\min} + (v_{\max} - v_{\min}) \times rand$$

where $x_{\max} = 5.0$, $x_{\min} = -5.0$, $v_{\max} = 5.0$, $v_{\min} = -5.0$, and *rand* denotes a random number uniformly distributed in [0, 1].

At each iteration, the velocity of each particle is updated by the *gbest* and the *pbest* solutions, where *gbest* and *pbest* mean the overall best position visited by its companions and the best position the particle visited so far. Then the position of each particle is updated using its updated velocity per iteration. Specifically, the value x_i^{k+1} of the position X_i^{k+1} is updated for the next iteration using the following relationship.

$$x_{ij}^{k+1} = x_{ij}^{k} + v_{ij}^{k+1},$$

where

$$v_{ij}^{k+1} = v_{ij}^{k} + r_1 c_1 \left(pbest_{ij}^{k} - x_{ij}^{k}\right) + r_2 c_2 \left(gbest_{gj}^{k} - x_{ij}^{k}\right).$$

The parameters r_1 and r_2 are uniformly distributed random number in [0, 1]. c_1 and c_2 are the cognitive and social parameters respectively, which are used to determine whether particles prefer to move closer to the *pbest* or *gbest* positions, while r_1 and r_2 are uniform random numbers between 0 and 1.

This process of updating the velocities V_i^k, position X_i^k, $pbest^k$ and the $gbest^k$ is repeated until a user-defined stopping condition is met.

At each iteration, the *pbest* and *gbest* are chosen depending on particle's fitness value. In this paper, the fitness value of a particle is defined by the makespan. To obtain the fitness value of particle *i,* first the position of particle *l* is transferred to permutation list. Second, based on the permutation list, jobs are assigned to the corresponding machine as mentioned above. Finally, a feasible schedule considering dirt constraint is obtained according to the full-batch policy. An example with 5-job and 2-machine instance is used to illustrate the procedure, where the clean time of (w_1, w_2) is (3, 4) and dirt allowance of (T_1, T_2) is (10, 8), other relative information is shown in Table 1. The procedures to obtain the fitness value are shown in Table 2.

4 Experiments

A numerical experiment is conducted to investigate the effectiveness of the constructed MIP model and the proposed PSO against the optimal solution obtained by the model. Since the considered problem is NP-hard, the MIP model could not solve large-sized instances in a reasonable computational time. Thus, in this section, a set of small-sized instances is generated randomly. There are six numbers of jobs: n = 5, 6, 7, 8, 9, and 10 processed on two machines. The release times, processing times, and dirt were generated uniformly on the interval [0, 20], [3, 15], and [1, 9], respectively. The clean times (w_1, w_2) of machines are (3, 5), (5, 10), and (10, 20), and the maximum dirt allowances (T_1, T_2) for the two machines are (10, 12), (10, 15), and (15, 25). For a pair of clean time and maximum dirt allowances, 10 instances are generated and solved, thus, a total of instances are 90 for each job.

The proposed MIP model was processed by ILOG CPLEX Optimization Studio Version 12.7.1, and the proposed PSO is coded in C++. The experiment is conducted on a PC Xeon E5-1620 CPU with 3.6 GHz and 12 GB RAM.

In Table 3, the columns under "Max. PD", "Min. PD", and "APD" give the maximum, minimum and average percentage deviation of the proposed PSO algorithm from the MIP over 90 instances. The PD of the PSO algorithm is defined as follow:

$PD = \frac{(Sol_{PSO} - Sol_{MIP})}{Sol_{MIP}} \times 100$ where Sol_{PSO}, and Sol_{MIP} is the solution obtained by the PSO and MIP, respectively. Since the size of test problems is small, all optimal solutions could be obtained by the MIP method in this experiment. The other columns under "#best" denote the number of instances each method finds the best solution.

From Table 3, it is obvious that the difference between the average (APD) and worst-case (Max. PD) performance are large, it could be caused by the decoding method. In this paper, we applied the full-batch policy to obtain a feasible solution based on the permutation list, the search space may not cover in such a way to obtain optimal solutions as possible. Fortunately, the average performance (APD) of the PSO is from 0.84 to 1.3, this value suggests that the PSO algorithm averagely finds 99.16% to 98.70% of optimality in the experiments. Additionally, the average CPU time consumed by the PSO is much lower than that of MIP and does not increase greatly as the problem size increases. From the result, the PSO could not all optimal solutions for the small problems, it is because that the permutation list and full-batch policy neglect the effect of job release time. However, when the number of jobs increase, the impact of job release time will decrease. Overall, the PSO algorithm would obtain the best solution efficiently, on average. Thus, the proposed PSO algorithm could be applied in a real-world semiconductor manufacturing setting.

Table 1. A 5-job and 2-machine instance

	r_j	p_{1j}	p_{2j}	d_{1j}	d_{2j}
J_1	0	3	4	5	4
J_2	3	7	8	3	3
J_3	5	4	5	4	3
J_4	2	3	4	6	5
J_5	6	5	6	6	4

Table 2. Obtaining fitness value for a particle i

X_i^k	[0.11, 0.76, 0.37, 0.89, 0.41, 0.45]
Permutation list	(1, 3, 5, 6, 2, 4)
Schedule	M_1: $J_1 \rightarrow J_3$, clean, $\rightarrow J_5$ M_2: $J_2 \rightarrow J_4$
Fitness value (C_{max})	17

Table 3. Comparing the two proposed methods

n	Max. PD	Min. PD	APD	#Best	Average CPU time	
					MIP	PSO
5	21.43	0.00	1.06	80	0.16859	0.01000
6	18.18	0.00	1.30	76	0.16669	0.01200
7	27.59	0.00	1.10	74	0.30120	0.01400
8	12.12	0.00	0.84	79	0.73441	0.01600
9	13.21	0.00	0.99	69	2.30020	0.01800
10	11.32	0.00	1.20	66	5.89620	0.02000

5 Conclusions

Motivated by clean operation of semiconductor manufacturing, this paper modeled it as a non-identical parallel machine scheduling problem with flexible maintenance activities, and the objective is to minimize the makespan. For the problem, a MIP model and a PSO algorithm are developed. For small-size problems, all optimal solutions could be obtained by the MIP model and they could also be treated as benchmarks for evaluating the performances of heuristic algorithms including the proposed PSO algorithm. Based on our computational experiment, the proposed PSO algorithm has significant performance for small-size problems, moreover, the PSO algorithm would obtain the best solution efficiently, on average. Therefore, the proposed PSO algorithm could be applied in a real-world semiconductor manufacturing setting. The research can be extended to deal with multiple objectives or develop other meta-heuristic algorithm in the future.

Acknowledgment. The authors are grateful to the editor and the anonymous referees whose constructive comments have led to a substantial improvement in the presentation of the paper. This work was supported by the Natural Science Foundation of Zhejiang Province (Grant No. LY18G010012) and the National Natural Science Foundation of China (No.71671130).

References

1. Ying, K.-C., Lu, C.-C., Chen, J.-C.: Exact algorithms for single-machine scheduling problems with a variable maintenance. Comput. Ind. Eng. **98**, 427–433 (2016)
2. Luo, W., Cheng, T.C.E., Ji, M.: Single-machine scheduling with a variable maintenance activity. Comput. Ind. Eng. **79**, 168–174 (2015)
3. Su, L.-H., Wang, H.-M.: Minimizing total absolute deviation of job completion times on a singel machine with cleaning activities. Comput. Ind. Eng. **103**, 242–249 (2017)
4. Su, L.H., Hsiao, M.-C., Zhou, H., Chou, F.-D.: Minimizing the number of tardy jobs on unrelated parallel machines with dirt consideration. J. Ind. Prod. Eng. **35**, 383–393 (2018)
5. Pinedo, M.: Scheduling: Theory, Algorithms, and Systems. Prentice-Hall, New Jersey (2002)

6. Chen, J.S.: Scheduling of nonresumable jobs and flexible maintenance activities on a single machine to minimize makespan. Eur. J. Oper. Res. **190**, 90–102 (2008)
7. Schmidt, G.: Scheduling with limited machine availability". Eur. J. Oper. Res. **121**, 1–15 (2000)
8. Ma, Y., Chu, C., Zuo, C.: A survey of scheduling with deterministic machine availability constraints. Comput. Ind. Eng. **58**, 199–211 (2010)
9. Qi, X., Chen, T., Tu, F.: Scheduling the maintenance on a single machine. J. Oper. Res. Soc. **50**, 1071–1078 (1999)
10. Cui, W.W., Lu, Z.: Minimizing the makespan on a single machine with flexible maintenances and jobs release dates. Comput. Oper. Res. **80**, 11–22 (2017)
11. Sbihi, M., Varnier, C.: Single-machine scheduling with periodic and flexible periodic maintenance to minimize maximum tardiness. Comput. Ind. Eng. **55**, 830–840 (2008)
12. Chen, J.S.: Optimization models for the machine scheduling problem with a single flexible maintenance activity. Eng. Optim. **38**, 53–71 (2006)
13. Yang, S.L., Ma, Y., Xu, D.L., Yang, J.B.: Minimizing total completion time on a single machine with a flexible maintenance activity. Comput. Oper. Res. **38**, 755–770 (2011)
14. Chen, J.S.: Single machine scheduling with flexible and periodic maintenance. J. Oper. Res. Soc. **57**, 703–710 (2006)
15. Kanet, J.J.: Minimizing variation of flow time in single machine systems. Manag. Sci. **27**, 1453–1459 (1981)
16. Pang, J., Zhou, H., Tsai, Y.-C., Chou, F.-D.: A scatter simulated annealing algorithm for the bi-objective scheduling problem for the wet station of semiconductor manufacturing. Comput. Ind. Eng. **123**, 54–66 (2018)
17. Suppapitnarm, A., Seffen, K.A., Parks, G.T., Clarkson, P.J.: Simulated annealing: An alternative approach to true multi-objective optimization. Eng. Optim. **33**, 59–85 (2000)
18. Deb, K., Pratap, A., Sgarwal, S., Meyarivan, T.: A fast and elitist multiobjective genetic algorithm: NSGA-II. IEEE Trans. Evolut. Comput. **6**, 182–197 (2002)
19. Kennedy, J., Eberhart, R.C.: Particle swarm optimization. In: Proceedings of the 1995 IEEE International Conference on Neural Networks, pp. 1942–1948. IEEE press, New Jersey 1995

Research on IRP of Perishable Products Based on Improved Differential Evolution Algorithm

Zelin Wang and Jiansheng Pan(✉)

School of Information Science and Technology,
Nantong University, Nantong 226009, China
whwzl@whu.edu.cn, ntwzl@ntu.edu.cn

Abstract. Inventory routing problem is a NP-hard problem, and its research has always been a hot issue. The special operation of perishable goods puts forward higher requirements for inventory and transportation. In order to reduce the quantity of deteriorated goods, improve the storage efficiency of perishable goods, and further reduce the operating cost of enterprises, the inventory path problem is studied on the basis of inventory path problem. In order to rationally arrange the distribution time, quantity and route of each customer point, a mathematical model is established on the basis of some assumptions, taking inventory and vehicles as constraints, and aiming at optimizing the total cost of the system. In view of the particularity of perishable goods inventory routing problem, the proposed algorithm (IDE) improves the differential evolution algorithm from two perspectives. The grid is used to initialize the population and the greedy local optimization algorithm is combined with the differential evolution algorithm to improve the convergence speed of the algorithm. The accuracy of the algorithm is improved by adaptive scaling factor, two evolutionary modes and changing the constraints of the problem. Then the improved algorithm is used to solve the inventory routing problem. The numerical results show that the algorithm is effective and feasible, and can improve the accuracy of the algorithm and accelerate the convergence speed of the algorithm.

Keywords: Differential evolution algorithm · Inventory routing problem · Accuracy of the algorithm · Convergence speed of the algorithm

1 Introduction

Inventory-routing problem is to determine the inventory strategy and distribution strategy. The purpose of inventory strategy is to determine the distribution objects and quantity of goods in each planning period. Distribution strategy is to determine the distribution route of goods. IRP seems to minimize the sum of inventory costs and distribution costs. IRP is the integration of inventory problem and distribution problem, which needs to be solved on one platform at the same time. Because these two problems are contrary to each other, in the pursuit of the minimum inventory cost, it will inevitably bring the maximum distribution cost; on the contrary, if the pursuit of the minimum distribution cost, it will inevitably bring the maximum inventory cost. But at the same time, solving these two problems is a very difficult event. Both of them

K. Li et al. (Eds.): ISICA 2019, CCIS 1205, pp. 497–513, 2020.
https://doi.org/10.1007/978-981-15-5577-0_39

are N-P difficult problems. Especially when the number of customers in distribution is large and the demand of each customer is random, the optimal strategy of IRP problem is often very complex, and the solution of the problem often makes the quantity of distribution and the distribution room. Distribution routes lack stability. The value and freshness of perishable products are highly correlated, which is a special population of IRP. Because of the decrease of freshness and the increase of deterioration, the goods are damaged seriously. It is more urgent and realistic to study a special population of IRP of perishable products. In this paper, an improved differential evolution algorithm is used to find the approximate optimal solution of the IRP problem for perishable products.

In the available literature, Yu Li, Shuhua Zhang, Jingwen Han [1] studies an inventory level model that depends on demand and dynamic pricing Chinese style. In [2], Samira Mirzaei, Abbas Seifi established an inventory path optimization model based on freight cost, inventory cost and sales loss cost. Combined with simulated annealing and tabu search, he designed a meta heuristic algorithm. Bhattacharjee et al. [3] established a multi-period ordering and pricing model, and solved a deteriorating product with a fixed period by using two different heuristic algorithms. Levin Y [4] and others assume that the demand of perishable products obeys Poisson distribution, and that the retailer's dynamic pricing strategy under the condition that demand will be affected by price. Goyal and Giri [5] firstly reviewed the inventory status of perishable goods. During this period, the research on supply chain theory reached a new height.

Literature [6] through theoretical analysis, it is proved that the sum of inventory cost and distribution cost under the optimal strategy of partitioning customers is 98.5% optimal. Zoning can effectively simplify the problem and reduce the difficulty of the problem, so this paper also uses the idea of fixed zoning, and does not need to spend a lot of cost for low probability events. Document [7] is the earliest introducer of IRP partitioning idea, but the author decomposes the needs of individual customers and allows no more than one vehicle to serve them in a distribution period, which means the actual operation of the problem more difficult. Document [8] improves document [7], which restricts that each customer can only be distributed by one vehicle in a delivery period, but cannot be separately distributed. However, because the strategy is to seem the optimal solution of the problem, the scale of solving the problem is relatively small. Literature [9] uses classical Lagrange relaxation algorithm to solve IRP problem. The idea is complex and difficult to implement. Moreover, with the increase of the scale of the problem, the complexity of the algorithm increases exponentially.

In solving IRP problems, because of the complexity of the problem itself, when the scale of the problem itself is large, it is a very difficult event to find the optimal solution. Document [10] tries to solve IRP problems by using heuristic variable neighbor search intelligent method. IRP is solved in two stages. Firstly, heuristic variable neighbor search is adopted. In this stage, the inventory cost is not considered. The purpose is to obtain a feasible initial solution. Then the second stage iteratively optimizes the initial solution and achieves very good results. Literature [11] tries to use heuristic tabu search intelligent algorithm to solve the shortest inventory path problem, and compares the operation effect of the algorithm with that of the original Lagrange relaxation algorithm, which proves that this method has obvious superiority.

From the literature [12–16], it can be seen that in recent years, differential evolution algorithm has achieved remarkable results in solving large-scale combinatorial optimization problems because of its global convergence and robustness. In this paper, an improved differential evolution algorithm is proposed to deal with the IRP problem of perishable products.

2 Problem Model

2.1 Study on Freshness of Perishable Products

The freshness of perishable commodities is the Mey point of this paper. In the study of freshness loss of perishable products, scholars mainly focus on the change of freshness of perishable products with time, while the research on the relationship between freshness of fresh products and transportation distance is relatively small.

In the study of the freshness of perishable products changing with time, freshness is generally regarded as a continuous decreasing function with time. The function is generally expressed as follows: $\theta(t) = \frac{1}{1+\alpha t^2}$, where a > 0 indicates the sensitivity of freshness of perishable products to time. If we derive the freshness function of perishable products, we can get $\frac{d\theta(t)}{dt} = \frac{2\alpha t}{(1+\alpha t^2)^2} \leq 0$. From the first derivative, we can see that freshness is a continuous decreasing function of time, that is, with the passage of time, the freshness of perishable products will gradually decline. The freshness of perishable products in this paper is only related to the distance. The longer the transportation distance is, the freshness decreases. That is, freshness is also a continuous decreasing function of the distribution route. We assume that there is a certain functional relationship between freshness loss L and distance. After statistical analysis and functional fitting of the relevant data, we get that the function in the relevant definition domain is an exponential function, that is, $\Psi = \alpha\theta^{kd}, \frac{d\Psi}{d_d} = \alpha k\theta^{kd}$. Among them, a and k are constants greater than 0, $0 <= \Psi <= 1$, d denotes the distance of perishable products from suppliers to retailers. Then Ψ is the freshness of perishable products.

2.2 Model Construction

2.2.1 Model Hypothesis

Before the model is finally established, several assumptions need to be made about the model:

(1) Retailers' demand for goods is related to freshness and price, the demand of retailers is random and independent, and obeys normal distribution.
(2) Within a delivery period, each vehicle can provide distribution services to multiple retailers, but only one vehicle can serve one retailer, and each vehicle only travels once for delivery.
(3) For inventory, the retailer's inventory cost is considered only, without considering the supplier's inventory cost and inventory quantity.
(4) The supplier carries on the transportation of a perishable product to the retailer by only road style.

(5) The supplier managed inventory (VMI) model is adopted, and the supplier can control the retailer's inventory through monitoring.
(6) Retailers adopt (S, R, T) inventory strategy, and there is no replenishment lead time problem.
(7) The supplier has enough delivery vehicles and the distribution vehicles are identical, and the vehicle load is limited.
(8) There is a supplier, many retailers, and the location coordinates of these businesses are determined. In order to reflect the problem more clearly and intuitively, the transportation distance only considers the straight line distance between the points.
(9) No consideration is given to the service time of the vehicle at the customer's office, i.e. instantaneous completion;
(10) During transportation, the freshness of perishable products will change, but it is only related to the transportation distance.

2.2.2 Relevant Parameters and Symbolic Descriptions

(1) Relevant parameters

Q_i: Retailer i's maximum inventory.
t: cycle time.
M: Number of distribution vehicles.
N: Number of retailers.
W: Load limits for distribution vehicles.
U_i: Retailer i's inventory upper limit.
L_i: Retailer i's inventory lower limit.
D_{ij}: the distance between retailer i and retailer j, $i = 0$ means supplier;
Ψ_i: Freshness loss of perishable product i;
Pv: the original price per unit product;
P_i: unit out-of-stock cost of retailer i;
C_0: Fixed start-up costs incurred by suppliers per delivery;
C_1: unit transportation cost of vehicle transportation;
h_i: Retailer i unit products do not consider freshness reduction in storage costs;
a_i: the level of service that retailer i should provide to consumers is related to the minimum inventory level.

(2) Variables

X_{ti}: The demand of retailer i in the t-cycle conforms to the normal distribution $N(\mu, \sigma^2)$, and the distribution function is $F_i(x)$.
S_{ti}: Initial inventory of retailer i in t cycle;
Q_{ti}: Supplier's replenishment to retailer i at the end of the t-cycle;

$$x_{Mtij} = \begin{cases} 1, & \text{at the end of cycle } t, \text{ vehicle } k \text{ service for retailer } i \\ 0, & \text{otherwise} \end{cases}$$

$$Y_{Mti} = \begin{cases} 1, & \text{at the end of cycle } t, \text{ vehicle } k \text{ drives from retaile } i \text{ to retailer } j \\ 0, & \text{otherwise} \end{cases}$$

2.3 Cost Analysis

The secondary supply chain system established in this paper is a typical R-System. R-System is based on the assumption that all the demanders in the supply chain system should be managed by a central organizer or decision maker, regardless of the order cost of the supplier and the inventory management cost of the supplier, or it is a centralized decision-making that the supplier does not set up a distribution center. The task of the central decision-maker is to determine how the goods should be allocated among different demand points and how to distribute the goods to the demand point. The purpose of decision makers is to minimize the total cost of the whole secondary supply chain system, because the standard supply chain system is a typical R-System. Therefore, in the total cost, the supplier only considers the transportation cost in the process of distribution to the downstream retailer, while the retailer's cost includes the retailer's order cost, the retailer's inventory cost and the out-of-stock cost.

2.3.1 Cost Analysis of Retailers

(1) Inventory analysis of retailers

1) Inventory Limit Analysis of Retailer i

For retailers, if the inventory carrying capacity is greater, the frequency of suppliers replenishing goods to downstream retailers will decrease correspondingly in a certain period of time. If so, the cost of suppliers will undoubtedly be lower, but the problem is that retailers hold more goods because of the large inventory carrying capacity. In contrast, if the retailer's inventory cap is smaller, it will increase the frequency of suppliers replenishing retailers in a certain period of time, so that the retailer's inventory cost will be lower, but because replenishments to retailers are more frequent, suppliers will also replenish. Transportation costs are rising. Therefore, the inventory cap should be considered in the whole supply chain system, instead of making the inventory cap infinite.

2) Inventory Lower Limit Analysis of Retailer i

Because retailers adopt (t, R, S) inventory strategy, which corresponds to (t, L_i, U_i) inventory strategy established in this paper. In this inventory strategy, the key is to see the relationship between retailer's inventory level and L_i. If the inventory level of retailer i is larger than that of retailer L_i at the end of the t cycle, then suppliers do not have to replenish the retailer.

Another way to understand this is that at the beginning of the $t + 1$ cycle, retailer i's inventory should have all the lowest levels of $L_i + 1$. If we require the retailer's customer service level not to be lower than d_i, that is, $P\{x_{(t+1)i} = L_i + 1\} \geq d_i$, and because X_i obeys the normal distribution $N(\mu_i, \sigma_i^2)$, then there is a minimum value for L_i.

$$\varnothing\left(\frac{L_i + 1 - \mu_i}{\sigma_i}\right) = d_i \tag{1}$$

For all retailers, assuming that their inventory level at the end of this cycle is greater than the minimum inventory level L_i, the initial inventory level of the retailer at the beginning of the next cycle is equal to the inventory level at the end of this cycle; on the contrary, assuming that their inventory level at the end of this cycle is less than the minimum inventory level L_i, then it is zero. The initial inventory of the seller in the next cycle will become the maximum after supplier replenishment. The formula can be expressed more directly as follows:

$$S_{(t+1)i} = \begin{cases} Ui, & S_{ti} - x_{ti} \leq L_i \\ S_{ti} - x_{ti}, & S_{ti} - x_{ti} > L_i \end{cases} \tag{2}$$

(2) Retailer Cost Analysis

Assuming that under the premise of no replenishment, the retailer's main expenses mainly include three, namely, the ordering cost, the inventory holding cost and the shortage cost.

1) Order cost

We set the order cost as *CO*. The order cost borne by the retailer mainly depends on the quantity of goods delivered by the supplier to the retailer and the instant price of perishable products when they arrive at the retailer. Because retailers adopt (*s, R, T*) inventory strategy, the distribution volume of perishable products from suppliers to retailers varies according to the retailer's inventory. When the retailer is out of stock at the end of the period, the perishable products that the supplier distributes to the retailer are the upper limit of the retailer's inventory; when the retailer's final inventory is less than the lower limit of the inventory, the quantity of perishable products that the supplier distributes to the retailer is the difference between the upper limit of the retailer's inventory and the retailer's final inventory; when the retailer's final When the inventory is larger than its lower limit, the supplier does not need to distribute the goods to the retailer. The mathematical expression is as follows: the supplier's distribution to retailer I at the end of the t-cycle is as follows:

$$Q_{ti} = \begin{cases} U_i, & S_{ti} - x_{ti} \leq 0 \\ U_i - (S_{ti} - x_{ti}), & 0 < S_{ti} - x_{ti} < L_i \\ 0, & S_{ti} - x_{ti} > L_i \end{cases} \tag{3}$$

As for the freshness loss of perishable products, the freshness of perishable products is related to the transportation distance. The larger the transportation distance the greater the freshness loss. This will further affect the price of the product. Therefore, the focus of this paper is how to determine the loss of freshness. Assuming that the distribution route of perishable products is *i*, then it needs to be analyzed first. Suppose that the last retailer is retailer g before the delivery vehicle arrives at retailer I and the last retailer on Retailer G is retailer *j*. In the *t*-cycle, zero-i vendor I is distributed by vehicle M_0, supplier-retailer *i* is distributed by l_{ti} and supplier-retailer J is distributed by l_{tj}.

$$l_{ti} = l_{tj} + X_{M_0tjg}d_{jg} + X_{M_0tgi}d_{gi} + X_{M_0tji}d_{ji} \quad where\ (X_{M_0tjg} + X_{M_0tgi})X_{M_0tij} = 0$$

Similarly, the same analogy is applied to the distribution distance of other retailers before the retailer *j*, and the final value of l_{tj} can be obtained. At this point, the price of perishable products to retailer I can be determined by $\Psi_{ti} = \alpha\left(e^{bl_{ti}} - 1\right)$. It can be determined that the price of perishable products to retailer *i* is

$$P_{ti} = \mathrm{Pv} \cdot (1 - \Psi_{ti}) \tag{4}$$

Then, retailer I's t-cycle order cost is:

$$CO_{ti} = P_{ti} \cdot Q_{ti} = \begin{cases} \mathrm{Pv} \cdot (1 - \Psi_{ti})U_i, & S_{ti} - x_{ti} \leq 0, \\ \mathrm{Pv} \cdot (1 - \Psi_{ti})(U_i - (S_{ti} - x_{ti})), & 0 < S_{ti} - x_{ti} < L_i \\ 0, & S_{ti} - x_{ti} > L_i \end{cases} \tag{5}$$

Therefore, the order cost of all retailers is

$$\mathrm{CO} = \sum\nolimits_{t=1}^{m} \sum\nolimits_{i=1}^{n} CO_{ti} \tag{6}$$

2) Inventory Cost of Retailer I in the Third Cycle

Inventory cost of retailer is *IC*. In order to express it more directly and clearly, this paper will take the average inventory of each cycle as the research object, that is, to sum and average the initial inventory level and the end inventory level of a certain cycle.

When there is $s_{ti} - x_{ti} <= 0$ and the inventory at the end of the period is zero, the average inventory in the T period is $s_{ti}/2$.

When there is $0 < s_{ti} - x_{ti} <=$ Li and the inventory at the end of the period is not zero, the average inventory in the T period is $(2\ s_{ti} - x_{ti})/2$.

When there is $s_{ti} - x_{ti} >$ Li, that is, when there is no need to replenish the retailer, the average inventory in the T period is $(2\ s_{ti}\ x_{ti})/2$.

Then retailer i's t-cycle inventory cost is:

$$IC_{ti} = \begin{cases} ln(h_i/(1+m\mu)) \cdot S_{ti}/2, & S_{ti} - x_{ti} \leq 0 \\ ln(h_i/(1+m\mu)) \cdot (2S_{ti} - x_{ti})/2, & 0 < S_{ti} - x_{ti} < L_i \\ ln(h_i/(1+m\mu)) \cdot (2S_{ti} - x_{ti})/2, & S_{ti} - x_{ti} > L_i \end{cases} \quad (7)$$

Therefore, the inventory cost of all retailers is IC $= \sum_{t=1}^{m} \sum_{i=1}^{n} IC_{ti}$.

3) Shortage Cost of Retailer I in the Third Cycle

Suppose the cost of shortage of retailers is SC. We Know that if the demand of consumers is greater than the inventory level of perishable products of retailers at the beginning of this cycle, there will be a shortage of supply and demand, that is, shortage. At this time, the shortage of retailer I in the t-cycle is $x_{ti} - s_{ti}$, and the cost of shortage is

$$SC_{ti} = p \cdot (X_{ti} - S_{ti}) \quad (8)$$

Therefore, the shortage cost of all retailers is SC $= \sum_{t=1}^{m} \sum_{i=1}^{n} SC_{ti}$.

When the retailer's inventory at the end of the period is larger than its inventory limit, $0 <= x_{ti} < s_{ti} - L_i$, the supplier is not required to distribute the goods to the retailer and there is no shortage of the goods, so the retailer's cost at this time only includes the inventory cost.

$$C_1 = \sum_{t=1}^{m} \sum_{i=1}^{n} SC_{ti} = \sum_{t=1}^{m} \sum_{i=1}^{n} \ln(h_i \cdot (2S_{ti} - x_{ti})/2) \quad (9)$$

When the retailer's final inventory is greater than zero but less than the retailer's lower inventory limit, i.e. $s_{ti} - L_i <= x_{ti} < s_{ti}$, the supplier is required to replenish the inventory. At this time, the retailer's cost includes the order cost and inventory cost.

$$C_1 = \sum_{t=1}^{m} \sum_{i=1}^{n} OC_{ti} + \sum_{t=1}^{m} \sum_{i=1}^{n} SC_{ti} = \sum_{t=1}^{m} \sum_{i=1}^{n} \mathrm{Pv} \cdot (1 - \Psi_{ti})(\mu_i - (S_{ti} - x_{ti})) + \sum_{t=1}^{m} \sum_{i=1}^{n} h_i \cdot \frac{(2S_{ti} - x_{ti})}{2} \quad (10)$$

When the retailer's inventory at the end of the period is out of stock, that is, $x_{ti} > s_{ti}$, the retailer's cost includes not only the order cost and inventory cost, but also the out-of-stock cost.

$$\begin{aligned} C_1 &= \sum_{t=1}^{m} \sum_{i=1}^{n} (OC_{ti} + IC_{ti} + SC_{ti} \\ &= \sum_{t=1}^{m} \sum_{i=1}^{n} \mathrm{Pv} \cdot (1 - \gamma_{ti})\mu_i + \sum_{t=1}^{m} \sum_{i=1}^{n} h_i \cdot S_{ti}/2 + \sum_{t=1}^{m} \sum_{i=1}^{n} p(x_{ti} - S_{ti}) \end{aligned} \quad (11)$$

2.3.2 Supplier Cost Analysis

Because this paper does not consider the inventory cost of suppliers, the cost of suppliers is mainly in the process of distribution of goods.

Transportation costs incurred. In this paper, the transportation cost is mainly divided into two components: the fixed starting cost and the fixed starting cost of the vehicle.

The total transportation cost is:

$$C_2 = mc_0 + c_1 \cdot \sum_{k=1}^{K} \sum_{t=1}^{m} \sum_{i=0}^{n} \sum_{j=0}^{n} X_{ktij} d_{ij} \tag{12}$$

2.4 Model Establishment

Through the above analysis, the total cost of the system is divided into four parts: retailer's order cost, retailer's inventory holding cost, retailer's shortage cost and supplier's transportation cost. However, under different demands, there will be different total cost calculation.

When $x_i >= \&\& \; x_i < s_{ti} - L_i$:

$$\begin{aligned} \mathrm{TC} &= C_1 + C_2 \\ &= \sum_{t=1}^{m} \sum_{i=1}^{n} h_i \cdot (2S_{ti} - x_{ti})/2 + mc_0 + c_1 \cdot \sum_{M=1}^{M} \sum_{t=1}^{m} \sum_{i=0}^{n} \sum_{j=0}^{n} X_{Mtij} d_{ij} \end{aligned} \tag{13}$$

When $x_i <= s_{ti} \; \&\& \; x_i >= s_{ti} - L_i$

$$\begin{aligned} \mathrm{TC} = C_1 + C_2 = & \sum_{t=1}^{m} \sum_{i=1}^{n} \mathrm{Pv} \cdot (1 - \Psi_{ti})(\mu_i - (S_{ti} - x_{ti})) \\ & + \sum_{t=1}^{m} \sum_{i=1}^{n} h_i \cdot (2S_{ti} - x_{ti})/2 + mc_0 + c_1 \cdot \sum_{k=1}^{K} \sum_{t=1}^{m} \sum_{i=0}^{n} \sum_{j=0}^{n} X_{ktij} d_{ij} \end{aligned} \tag{14}$$

When $x_i \rangle \; s_{ti}$

$$\begin{aligned} \mathrm{TC} = C_1 + C_2 = & \sum_{t=1}^{m} \sum_{i=1}^{n} \mathrm{Pv} \cdot (1 - \gamma_{ti})\mu_i + \sum_{t=1}^{m} \sum_{i=1}^{n} h_i \cdot (S_{ti})/2 \\ & + \sum_{t=1}^{m} \sum_{i=1}^{n} p(x_{ti} - S_{ti}) + mc_0 + c_1 \cdot \sum_{M=1}^{M} \sum_{t=1}^{m} \sum_{i=0}^{n} \sum_{j=0}^{n} X_{Mtij} d_{ij} \end{aligned} \tag{15}$$

2.5 Model Conversion

In practical problems, because the demand of perishable products X_{ti} obeys the random variable of normal distribution $N\ (\mu_i,\ \sigma_i^2)$, the whole function TC contains random variable, and then the objective function also contains random variable, which makes it impossible to find the minimum value of the desired objective function and can not be directly transported through the model. Calculate and solve. We use the stochastic expectation model to determine the stochastic demand, transform the stochastic variable into the deterministic variable, and then further solve the objective function. So the ultimate model to minimize the total cost of the system is:

$$
\begin{aligned}
& TC' = C_1 + C_2 = \\
& \sum_{t=1}^{m} \sum_{i=1}^{n} \int_{0}^{S_{ti}-L_i} lnh_i(2S_{ti} - x_{ti})/2dF(x) \\
& + \sum_{t=1}^{m} \sum_{i=1}^{n} \int_{S_{ti}-L_{ti}}^{S_{ti}} \mathrm{Pv} \cdot (1 - \Psi_{ti})\,(\mu_i - (S_{ti} - x_{ti}))dF(x) \\
& + \sum_{t=1}^{m} \sum_{i=1}^{n} \int_{S_{ti}-L_{ti}}^{S_{ti}} h_i \cdot \frac{(2S_{ti} - x_{ti})}{2} dF(x) + \sum_{t=1}^{m} \sum_{i=1}^{n} \int_{S_{ti}}^{+\infty} \mathrm{Pv} \cdot (1 - \gamma_{ti})\,\mu_i dF(x) \\
& + \sum_{t=1}^{m} \sum_{i=1}^{n} h_i S_{ti}/2dF(x) + \sum_{t=1}^{m} \sum_{i=1}^{n} p(x_{ti} - S_{ti}) + mc_0 \\
& + c_1 \cdot \sum_{M=1}^{m} \sum_{t=1}^{s} \sum_{i=0}^{n} \sum_{j=0}^{n} X_{Mtij} d_{ij}
\end{aligned}
\tag{16}
$$

$$\text{s.t.} \sum_{i=1}^{n} Q_{ti} Y_{kti} \le W, \forall k \in K,\ t \in T \tag{17}$$

$$0 \le L_i \le U_i \le Q_i', \forall i \in N \tag{18}$$

$$\sum_{M=1}^{m} Y_{kti} = 1, \quad \forall t \in T, i \in N \tag{19}$$

$$\sum_{i=0}^{N} x_{ktoi} = \sum_{j=0}^{N} x_{ktoj} \le 1, \quad \forall k \in K\ t \in T \tag{20}$$

$$\sum_{k=1}^{K} \sum_{\substack{i=0 \\ i \ne j}}^{n} X_{ktij} = 1 \quad \forall i \in N, j \in N, t \in T \tag{21}$$

$$\sum_{i=0}^{n} X_{Mtij} = Y_{Mtj}, \quad \forall j \in N, M \in m, t \in T \tag{22}$$

$$\sum_{j=0}^{n} X_{ktij} = Y_{kti}, \quad \forall i \in N, M \in m, t \in T \tag{23}$$

$$X_{Mtij} \in \{0, 1\}, \forall i, j \in N, M \in m, t \in T \tag{24}$$

$$Y_{Mti} \in \{0, 1\}, \forall i \in N, M \in m, t \in T \tag{25}$$

In the model constructed above, constraint (17) indicates that the quantity of goods transported by a vehicle does not exceed its maximum load capacity; constraint (18) indicates that the retailer's inventory limit and inventory limit should be between 0 and the maximum inventory; constraint (19) ensures that each demand point has one and only one vehicle. Vehicle serves it, each demand point is only on one path and does not go through a loop; constraint (20) means that the starting point and final destination of each vehicle are suppliers; constraint (21) means that when two retailers appear on a distribution path, they must replenish their goods; (22) and (23) means that, for example, when two retailers appear on a distribution path, they must replenish their goods. If the retailer needs to distribute, then when the vehicle M passes through the retailer, it must serve the retailer; (24), (25) the variable is 0–1.

3 Algorithmic Design

3.1 Basic Ideas of Differential Evolution Algorithms

Assuming that the problem to be solved is a minimization problem, the mathematical model of the problem is min $f(x_1, x_2, \ldots, x_n)$, where $x_j \in [L_j, U_j]$, and 1 <= j <= n. $X(0)$ is the initial population, so $X_i(t) = (x_{i1}(t), x_{i2}(t), \ldots, x_{in}(t))$ is the first individual i n the t-generation population, the population is n-dimensional spatial structure, and the population size is NP. Let X1 and X2 be two different individuals, then the difference vector formed by them is $X_1 - X_2$.

The differential evolution algorithm is described as follows:

Step 1 randomly generates the initial population:

$$P(0) = \{X_i(0)|x_{ij}(0) = \text{rand}(0,1) * (U_j - L_j) + L_j, 1 \le i \le s \cap 1 \le j \le n\}, \tag{26}$$

The Rand (0, 1) function is to find the random number on (0, 1) and set the initial value of the evolution time t = 0.

Step 2: Perform mutation operation: Randomly select three different individuals $X_a(t)$, $X_b(t)$, $X_c(t)$, a $\neq$ b $\neq$ c, a_b_c from the current population P(t). Calculate individual differences $D_i(t+1) = (d_{i1}(t+1), d_{i2}(t+1), \ldots, d_{in}(t+1))$:

$$d_{ij}(t+1) = x_{aj}(t) + F * (x_{bj}(t) - x_{cj}(t)), \tag{27}$$

Where $1 < i < s$, $1 < j < n$;

Step 3 performs the crossover operation: randomly generate the random decimal $R_1 < (0, 1)$ and the random integer $R_2 < [1, n]$, and calculate the temporary individual $E_i(t+1) = (e_{i1}(t+1), e_{i2}(t+1), \ldots, e_{in}(t+1))$:

$$e_{ij}(t+1) = \begin{cases} d_{ij}(t+1) & r_1 \le CR \text{ or } r_2 = j \\ x_{ij}(t) & \text{otherwise} \end{cases} \tag{28}$$

Where $1 < i < s$ and $1 < j < n$;

Step 4 performs the selection operation:

$$e_{ij}(t+1) = \begin{cases} d_{ij}(t+1) & r_1 \le CR \text{ or } r_2 = j \\ x_{ij}(t) & \text{otherwise} \end{cases} \tag{29}$$

Where $1 < i < s$;

Step 5 calculates $X_{best}(t+1)$ of the individuals with the smallest fitness in population $P(t+1)$.

Step 6 If the termination condition is not satisfied, then t = t + 1 and step 2; otherwise output X_{best} and $f(X_{best})$, and end.

3.2 Improvement of Differential Evolution Algorithms

As the number of retailers increases, the running time of the algorithm will increase exponentially, and the accuracy of the algorithm will decrease. In order to accelerate the convergence speed and improve the accuracy of the algorithm, the classical differential evolution algorithm must be improved.

In order to accelerate the convergence speed of the algorithm, the following measures are taken:

When the initial population is initialized, the traditional idea of random generation is changed, and the grid thinking is adopted to distribute the individual of the initial population evenly into the solution space. For customer I inventory, [Lj, Uj] can be divided into n intervals.

$$P(0) = \{X_{\rm i}(0)|x_{\rm ij}(0) = i * \left(U_j - L_j\right)/{\rm n} + L_j, \{1 < {\rm i} < {\rm s}, 1 < {\rm j} < {\rm n}\} \tag{30}$$

On the basis of step 4 of Sect. 3.1, greedy algorithm is introduced to find the optimal solution based on Xi(t). The algorithm is as follows:

```
While nm < Nm
    If f(xi)>f(xi+sin(pi/2*nm)) then xi=xi+sin((pi/2)*nm);
        Nm = nm + 1;
    End while
```

In order to improve the accuracy of the algorithm, some methods are adopted to avoid falling into local optimum. The specific measures are as follows:

The scaling factor F of the mutation operation of the classical differential evolution algorithm is adaptively and dynamically changed. The value of the scaling factor decreases as the evolution algebra increases, $F = 1 + e^{-n}$.

Secondly, two evolutionary models, DE/rand/1/bin and DE/rand-to-best/1/bin, were chosen randomly, or two populations were used to adopt different evolutionary models. Because DE/rand/1/bin evolutionary model has high precision, but the convergence speed of the algorithm is relatively slow, while DE/rand-to-best/1/bin evolutionary model has the opposite speed, but it is easy to fall into local optimum.

For the condition of selection operation, aiming at the IRP problem, all constraints are neglected selectively, and the *f'value* is obtained. That is to change the fitness function appropriately. The choice here is conditional. We use *f* as the fitness value on the premise that the population tends to concentrate and is in danger of falling into the local optimum. Here we use the variance of the population to judge the degree of population aggregation. When the variance is less than a threshold, we use the cancellation of all constraints to calculate the fitness value.

4 Simulation Experiment

Suppose there is a one-to-many two-stage perishable product distribution system, in which there are one supplier and 20 retailers. The system plan period is 7 days, and every day counts as a cycle. Delivery of perishable products is apple, freshness loss function $\Psi = ae^{kl} - a$, which is determined by analysis that a = 0.05, B = 0.003. The fixed start-up cost of suppliers delivering goods to retailers is 150 yuan per day, the transportation cost of vehicles is 2 yuan/Km, the maximum load of transport vehicles is 900 pieces, the unit inventory cost of retailers is 1 yuan/piece per day, the delivery price of perishable products from suppliers is 40 yuan/piece, and the unit out-of-stock cost of retailers is 4 yuan/piece. The retailer's maximum inventory is 300 pieces. Retailers have the same level of service. The numbers of suppliers and retailers and the corresponding left-hand parameters are shown in Table 1.

Table 1. Relevant parameters of suppliers and retailers

Retailer number	Coordinate	Initial inventory	(μ, σ^2)
0	50,50		
1	23,80	100	50,400
2	31,23	50	50,900
3	22,85	0	70,400
4	28,66	33	80,400
5	3,98	258	60,1600
6	34,61	25	80,4900
7	4,84	0	50,900
8	48,38	159	60,400
9	5,24	68	90,900
10	65,63	0	70,400
11	7,72	239	50,900
12	72,28	87	60,900
13	53,59	150	50,400
14	32,41	123	60,1600
15	78,32	76	80,900
16	90,12	68	50,400
17	96,16	135	70,100
18	65,61	226	90,1600
19	48,33	90	70,2500
20	83,3	86	50,1600

This paper establishes a joint optimization model for inventory and transportation of perishable products under Stochastic Demand Based on vendor managed inventory (VMI) with seven-day planning period. The objective function is the total cost of the system, which includes four parts: retailer's order cost, retailer's inventory holding cost, retailer's shortage cost and supplier's transportation cost. In this paper, the improved differential evolution algorithm is used to solve the mathematical model, and the output results are obtained (Figs. 1, 2 and 3).

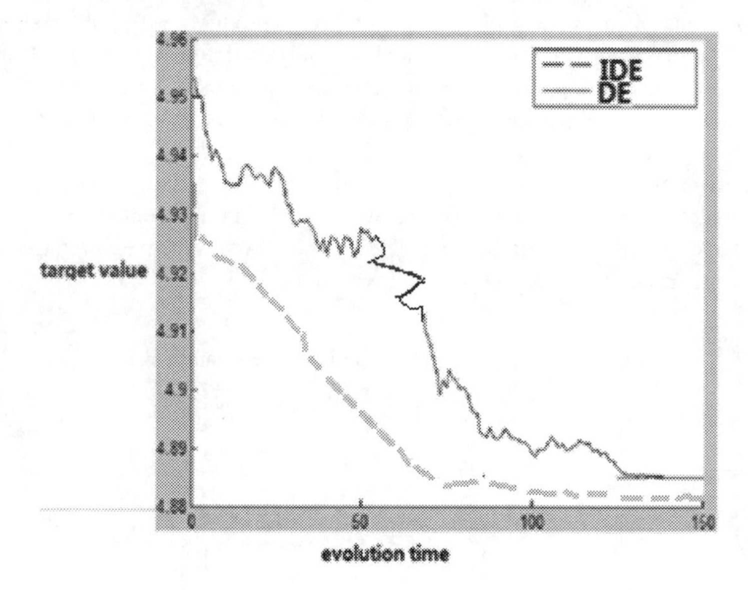

Fig. 1. Inventory routing optimization total cost

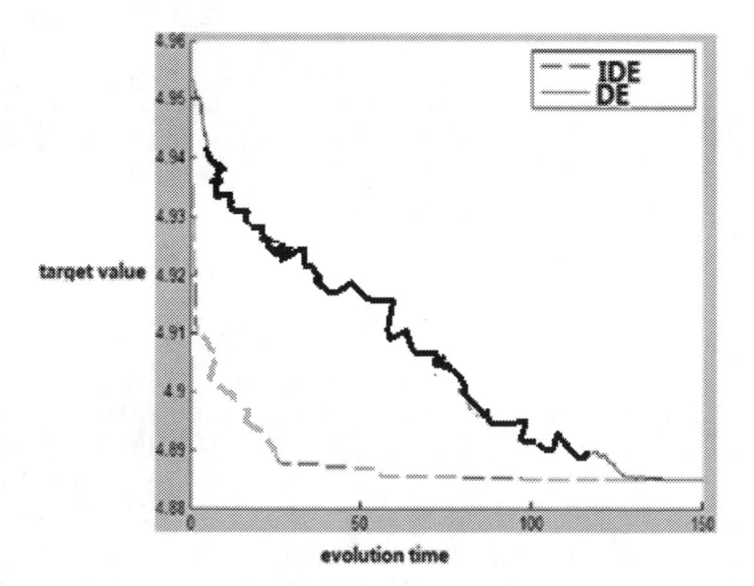

Fig. 2. Order value

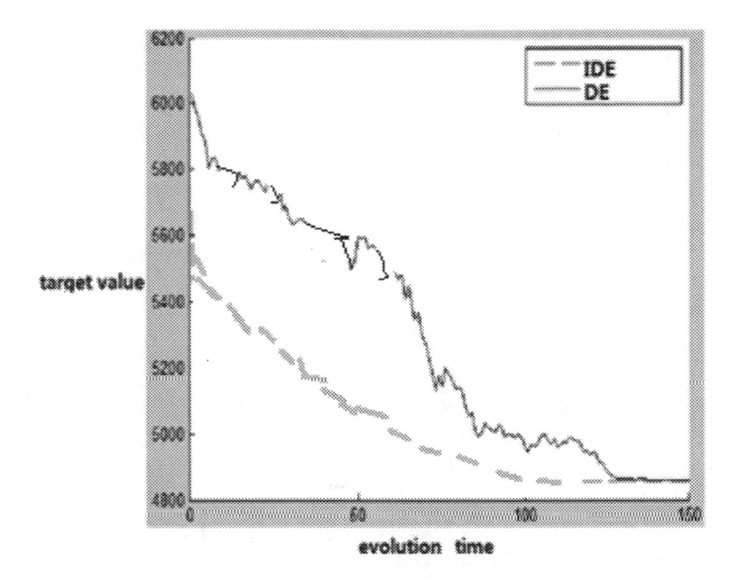

Fig. 3. Routing cost

The whole distribution route of IDE algorithm is:
First day:
0-10(276)-3(278)-7(288)-0-6(245)-4(230)-2(227)-0-19(201)-14(247)-9(211)-0-12 (203)-16(231)-20(274)-0-15(209)-17(235)-0
The second day:
0-8(261)-18(254)-1(142)-5(222)-0
The third day:
0-3(270)-19(180)-9(180)-0
The forth day:
0-3(261)-11(180)-7(210)-5(210)-0-6(180)-4(240)-14(210)-15(240)-0-8(284)-10 (290)-0
The fifth day:
0-19(180)-9(180)-0-12(240)-17(240)-16(280)-20(240)-0
The sixth day:
0-5(180)-1(180)-14(240)-8(240)-0-18(253)-0
The seventh day:
0-6(240)-4(240)-7(240)-3(210)-0-9(180)-0-14(240)-19(240)-10(180)-15(210)-0
The whole distribution route of DE algorithm is:
The firth day:
0-3(292)-7(270)-10(254)-0-6(255)-4(242)-0-2(200)-19(211)-14(187)-9(207)-0-20 (211)-15(239)-17(256)-0-12(201)-16(274)-0
The second day:
0-8(272)-18(209)-1(176)-5(203)-0
The third day:
0-3(270)-19(180)-9(180)-0

The forth day:
0-11(180)-7(210)-5(210)-0-6(180)-4(240)-14(210)-15(240)-0-8(284)-10(290)-0
The fifth day:
0-3(230)-19(180)-9(180)-0-12(240)-17(240)-16(280)-20(240)-0
The sixth day:
0-8(240)-14(180)-1(240)-5(180)-0-18(250)-0
The seventh day:
0-3(240)-7(240)-4(240)-6(210)-0-9(180)-0-15(240)-18(240)-19(180)-14(211)-0

After IDE algorithm optimization, the total cost of the secondary supply chain is 483577 yuan, of which the retailer's order cost is 455240 yuan, the retailer's inventory cost is 23207 yuan, the retailer's out-of-stock cost is 365 yuan, and the supplier's transportation cost is 4765 yuan. The transportation distance is 2,031.72 Milometers.

After optimization by DE algorithm, the total cost of the secondary supply chain is 493 139 yuan, of which the retailer's order cost is 463 018 yuan, the retailer's inventory cost is 24582 yuan, the retailer's out-of-stock cost is 407 yuan, and the supplier's transportation cost is 5132 yuan. The transportation distance is 2,408.03 Milometers.

The numerical experiments show that the IDE algorithm proposed in this paper is not only feasible in solving the IRP problem of perishable products, but also has great superiority over DE algorithm.

5 Summary

Inventory and distribution of perishable products is a very difficult problem, because it is an IRP problem and a N-P problem. In addition, perishable products have their own particularities. With the extension of distribution time and inventory time, their value will gradually decrease. In order to arrange the distribution time, quantity and route of each customer point reasonably, on the basis of some assumptions, taking inventory and vehicle as constraints, and aiming at optimizing the total cost of the system, a mathematical model is established. Aiming at the particularity of perishable goods inventory routing problem, the differential evolution algorithm is presented from two perspectives. The degree of convergence is improved by grid initialization and greedy local optimization algorithm combined with differential evolution algorithm. The accuracy of the algorithm is improved by adaptive scaling factor, two evolution modes and changing the constraints of the problem. Then the improved algorithm is used to solve the inventory routing problem. The numerical results show that the algorithm is effective and feasible, and the performance of the algorithm is greatly improved compared with the DE algorithm.

References

1. Li, Y., Zhang, S., Han, J.: Dynamic pricing and periodic ordering for a stochastic inventory system with deteriorating items. Automatica **5**, 200–213 (2017)
2. Mirzaei, S., Seifi, A.: Considering lost sale in inventory routing problems for perishable goods. Comput. Ind. Eng. **47**, 213–227 (2015)
3. Bhattacharjee, S., Ramesh, R.: A multi-period profit maximizing model for retail supply chain management: an integration of demand and supply-side mechanisms. Eur. J. Oper. Res. **122**(3), 584–601 (2007)
4. Levin, Y., McGill, J., Nediak, M.: Risk in revenue management and dynamic pricing. Oper. Res. **56**(2), 326–343 (2008)
5. Goyal, S.M., Giri, B.C.: Recent trends in modeling deteriorating inventory. Eur. J. Oper. Res. **134**, 1–16 (2001)
6. Donselaar, M.V., Woensel, T.V., Broekmeulen, R.: Inventory control of perishables in supermarkets. Int. J. Prod. Econ. **104**(2), 462–472 (2006)
7. Anily, S., Bramel, J.: An asymptotic 98.5% effective lower bound on fixed partition policies for the inventory-routing problem. Discret. Appl. Math. **145**(1), 22–39 (2004)
8. Anily, S., Federgruen, A.: One warehouse multiple retailer systems with vehicle routing costs. Manag. Sci. **36**(1), 92–114 (1990)
9. Chan, L., Federgruen, A., Simchi-Levi, D.: Probabilistic analyses and practical algorithms for inventory-routing models. Oper. Res. Int. J. **46**(1), 96–106 (1998)
10. Rafie-Majd, Z., Pasandideh, S.H.R., Naderi, B.: Modelling and solving the integrated inventory-location-routing problem in a multi-period and multi-perishable product supply chain with uncertainty: lagrangian relaxation algorithm. Comput. Chem. Eng. **109**, 9–22 (2017)
11. Mjirda, A., Jarboui, B., Macedo, R., Hanafi, S., Mladenovic, N.: A two phase variable neighborhood search for the multi-product inventory routing problem. Comput. Oper. Res. **52**, 291–299 (2014)
12. Li, K., Chen, B., Sivakumar, A.I., Wu, Y.: An inventory routing problem with the objective of travel time minimization. Eur. J. Oper. Res. **236**, 936–945 (2014)
13. Tang, R.: Decentralizing and coevolving differential evolution for large-scale global optimization problems. Appl. Intell. **4**, 1–16 (2017)
14. GhasemishabanMareh, B., Li, X., Ozlen, M.: Cooperative coevolutionary differential evolution with improved augmented Lagrangian to solve constrained optimisation problems. Inf. Sci. **369**, 441–456 (2016)
15. Salman, A.A., Ahmad, I., Omran, M.G.H.: A metaheuristic algorithm to solve satellite broadcast scheduling problem. Inf. Sci. **322**(C), 72–91 (2015)
16. Wang, Z., Wu, Z., Zhang, B.: Packet matching algorithm based on improving differential evolution. Wuhan Univ. J. Nat. Sci. **17**(5), 447–453 (2012)

An Improved Hybrid Particle Swarm Optimization for Travel Salesman Problem

Bo Wei[1], Ying Xing[1], Xuewen Xia[2(✉)], and Ling Gui[2]

[1] School of Software, East China Jiaotong University, Nanchang 330013, China
weibo@whu.edu.cn

[2] College of Physics and Information Engineering, Minnan Normal University, Zhangzhou 363000, China
xwxia@whu.edu.cn

Abstract. Travel salesman problem (TSP) is a typical NP-complete problem, which is very hard to be optimized by traditional methods. In this paper, an improved hybrid particle swarm optimization (IHPSO) is proposed to solve the travel salesman problem (TSP). In IHPSO, there are four novel strategies proposed for improving its comprehensive performance. Firstly, a probability initialization is used to add prior knowledge into the initialization, so that much computing resources can save during the evolution of the algorithm. Furthermore, in order to improve the algorithm's convergence accuracy and population diversity, two kinds of crossover are proposed to make better use of **Gbest** and **Pbest**. Lastly, a directional mutation is applied to overcome the randomness of the traditional mutation operator. Comparison results among IHPSO and other EAs, including PSO, genetic algorithm (GA), Tabu Search (TB), and simulated annealing algorithm (SA) on the standard TSPLIB format manifest that IHPSO is more superior to other methods, especially on large scale TSP.

Keywords: Travel salesman problem · Particle swarm optimization · Genetic operators · Probabilistic and deterministic

1 Introduction

The travel salesman problem (TSP) firstly formulated as a mathematical problem in 1930 is a typical NP-complete problem. In recent years, it attracts much attention in mathematics and computation sciences communities and has been one of the most intensively studied problems in optimization [1]. The purpose of the TSP is to find the shortest journey, which each city can be visited only once and returns to the starting city. Concretely, it means to search for an arrangement $\pi(\mathbf{C}) = (c_1, c_2 \ldots c_p)$ of natural subsets $\mathbf{C} = (1, 2, \ldots p)$ (the elements of C are the serial number of cities). The TSP can be described as following:

$$\mathbf{T}_d = \sum_{d=1}^{n-1} d_{c_d c_{d+1}} + d_{c_n c_1} \tag{1}$$

where $d_{c_d c_{d+1}}$ represents the distance between the city c_d and city c_{d+1}.

K. Li et al. (Eds.): ISICA 2019, CCIS 1205, pp. 514–525, 2020.
https://doi.org/10.1007/978-981-15-5577-0_40

Since it is very hard to use classical methods to optimize the TSP, many heuristic methods are applied to deal with it in recent years, such as simulated annealing (SA) [2], tabu search (TB) [3], neural networks (NN) [4], and genetic algorithm (GA) [5]. Particle swarm optimization (PSO) proposed by Kennedy and Eberhart [6] in 1995 is an effective approach to solve continuous problems. In recent years, PSO has been successfully applied to many optimization problems. However, the conventional PSO cannot directly apply in TSP which is a kind of discrete problem rather than a continuous problem. So the PSO method was proposed to solve TSP firstly by Clerc [7] in 2004. Then Hendtlass [8] used PSO to solve small-size TSP by adding a memory capacity for each particle. Pang et al. [9] combined the characteristics of PSO with the concept of fuzzy theory. Based on the previous research, an improved hybrid PSO (IHPSO) is proposed in this work, in which some new features are introduced to improve its comprehensive performance.

The rest of this paper is organized as following. Section 2 describes the framework of the canonical PSO. The description of TSP is shown at Sect. 3. The hybrid particle swarm optimization is used for discrete problems is mentioned in Sect. 4. Details of IHPSO are described in Sect. 4. And the experimental results between IHPSO and other 4 test functions in 9 benchmark problems in TSPLIB [10] are detailed in Sect. 5. Finally, conclusions are given in Sect. 6.

2 PSO

2.1 Canonical PSO

PSO is typical swarm intelligence algorithms inspired by birds' predatory behaviors. At the beginning of PSO, a group of particles is initiated in a feasible space, and each particle represents a penitential solution of an optimization problem. During the searching process in hyperspace, the *i*th particle is associated with two vectors, i.e., a position vector $X_i = [x_{i1}, x_{i2}, \ldots, x_{iD}]$ and a velocity vector $V_i = [v_{i1}, v_{i2}, \ldots, v_{iD}]$, where D represents the dimensionality of an objective problem. The vector X_i can be regarded as a candidate solution to the problem while the vector V_i denotes the search direction and step size of the ith particle.

In each generation, the *i*th particle updates its velocity and position relying on the global best solution $Gbest = [gb_1, gb_2, \ldots, gb_D]$ and its personal best solution $Pbest = [pb_{i1}, pb_{i2}, \ldots, pb_{iD}]$. The update rules are defined as following.

$$v_{ij}^{t+1} = \omega \cdot v_{ij}^{t} + c_1 \cdot r_1 \cdot (pb_{ij} - x_{ij}^{t}) + c_2 \cdot r_1 \cdot (gb_j - x_{ij}^{t}) \tag{2}$$

$$x_{ij}^{t+1} = x_{ij}^{t} + v_{ij}^{t+1} \tag{3}$$

where ω is an inertia weight that determines the influence of the previous velocity; c_1 and c_2 are two numbers randomly distributed in the range of [0,1]; x_{ij}^{t} and v_{ij}^{t} represent the *i*th particle's position and velocity in the *j*th dimension at generation t, respectively.

2.2 Hybrid PSO for TSP

The algorithm described above is regarded as the traditional PSO, which is very easy applied in continuous problems. However, for discrete problems, the traditional PSO doesn't work without modifications. Therefore, to solve the discrete problem, there have been attempts to modify the PSO algorithm.

Clerc proposed a hybrid PSO (HPSO) method for the discrete problem of TSP firstly [7]. In the research, some modified operators for update of the position and velocity in PSO are proposed to meet requirements of TSP, such as the opposite of velocity, the addition of position and velocity (movement), the subtraction of two positions, the subtraction and addition of two velocities, and the multiplication of velocity and constant. Nevertheless, size of TSP problems involved in the work is all smaller than 17. Hendtlass [8] use PSO to solve small-size TSP by adding a memory capacity for each particle. The purpose of adding a memory capacity is to use the specific alternate target points instead of the current local best position. In this way, the efficiency and diversity of the algorithm are improved. Pang et al. [9] combined the characteristics of PSO with the concept of fuzzy theory, and each particle represents a fuzzy matrix. Fuzzy matrices are used to represent the position and velocity of the particles in PSO, meanwhile, the operators in the original PSO are redefined. Wang et al. [11] redefined operators of PSO by introducing the concepts of "swap operator" and "swap sequence". Therefore, TSP could be solved by the PSO in another way [12]. Kennedy and Eberhart proposed a discrete binary version of PSO by defining particles' trajectories and velocities in terms of changes of probabilities that a bit will be in one state or the other proposed an improved version of PSO [13], which includes two phases. The first phase includes fuzzy *C*-means clustering, a rule-based route permutation, a random swap strategy, and a cluster merger procedure. Meanwhile, a genetic-based PSO procedure is used in the second phase [14]. Hybrid PSO [7] abandons the function of tracking the extreme value to update the particle's position in the traditional PSO. In fact, hybrid PSO includes two crossover operators and a mutation operator into PSO. Owing to the advantages of HPSO, it has been widely used to solve discrete optimization problems, such as TSP.

3 IHPSO for TSP

The HPSO introduced above is a simple superposition of PSO and genetic operator. With the modification, the PSO variant can be directly used in discrete problems. However, its convergence speed is not high enough. In other words, the original hybrid PSO is easy to fall into a locally optimal solution. In order to improve the algorithm performance in TSP, we propose an IHPSO, in which there are four novel strategies. Firstly, a probability initialization is used to reduce the complexity as well as to achieve better performance on a large scale in TSP. Relying on the strategy, the initial population can obtain guidance of prior knowledge. Secondly, two different matrices are introduced to enhance two crossover operators. Through two crossover operators the algorithm can take full advantages of the global best solution **Gbest** and the personal best solution **Pbest**. Thirdly, a directional mutation operator is proposed to avoid the excellent genes broken, thus to improve its solution accuracy. The details of new strategies and the flowchart of IHPSO are presented as follows one by one.

3.1 Probability Initialization

Although a random initialization method, which is very popular in the PSO community, can increase the diversity of the initial population, it ignores much helpful guidance information that can be extracted from cities' coordinates. Hence, the aimless initialization method has lower efficiency, especially in large scale TSP. In this research, a new initialization mode is proposed to overcome the above-mentioned shortcomings.

To takes advantage of some prior knowledge based on the cities' positions, an adjacency matrix is constructed as (4).

So that individuals in the initial population have very high fitness.

$$\mathbf{citydist} = \begin{bmatrix} d_{11}, d_{12}, \ldots, d_{1n} \\ d_{21}, d_{22}, \ldots d_{2n} \\ \ldots\ldots\ldots\ldots\ldots \\ d_{n1}, d_{n2}, \ldots, d_{nn} \end{bmatrix} \tag{4}$$

where d_{ij} denotes the distance between the *i*th city and *j*th city.

Based on the matrix **citydist**, a heuristic initialization parameter matrix can be defined as (5).

$$\mathbf{hini} = \begin{bmatrix} 1/d_{11}, 1/d_{12}, \ldots, 1/d_{1n} \\ 1/d_{21}, 1/d_{22}, \ldots, 1/d_{2n} \\ \ldots\ldots\ldots\ldots\ldots\ldots\ldots \\ 1/d_{n1}, 1/d_{n2}, \ldots, 1/d_{nn} \end{bmatrix} \tag{5}$$

The key idea of the probability initialization is that the initialization of population is based on a transfer probability matrix, an element in which indicates the probability that from the current city to the next city. Its formula is as (6):

$$\mathbf{P}(v) = \mathbf{hini}(i, v)^{\alpha} \tag{6}$$

$$\text{Or } \mathbf{P} = (p_1, \ldots, p_v, \ldots, p_k) \tag{7}$$

Where $\alpha = 5$; k = length (**visiting**), $0 \leq k \leq (n-1)$ **visiting** is the visiting cities vector. **hini** is the heuristic initialization parameter matrix, and the size of **hini** is $n * n$ (n is the number of cities). Thereinto, i represents a serial number of the last one of visited cities, and j represents a serial number of all the visiting cities. In this way, $\mathbf{P}(v)$ represents the probability between the last one of visited cities i and the visiting cities v. With the initialization process going on, the scale of **visiting** is getting smaller and smaller. In other words, the size of **P** is decreasing from $(n-1)$ to 0 with the initialization process.

After the calculation of the transition probability, we need to select which city we should shift next. At this moment, each visiting city is given a ratio, and this ratio is the proportion of this city in entire visiting cities. As the algorithm progresses, the number of cities that need to be visited will gradually decrease, then the ratio will change. An additive approach is used for the final probability selection matrix. In the end a random number is generated as a criterion for choosing the next city to visit.

Based on the above description, the strategy of probability initialization can be detailed as Algorithm 1.

Algorithm 1. Probability initialization

```
01: For j=2:n     // n is the number of cities
02:   For i=1:ps   // psdenotes the population size
03:   Build visited cities vector visited, visiting cities vector visiting, and
        the selection probability of visiting cities Vector P;
04:     Update visiting according to the data in visited;
05:     For k=1:length(visiting) //calculate the probability of being
                    // selected for each candidate city
06:       P(k)=Ta(visited(end), visiting (k))^α ;
07:     End
08:     P=P/(sum(P));
09:     Pcum=cumsum(P);
10:     select=find(Pcum>=rand);
11:     to_visit=visiting(select(1));
12:     individual(i, j)=to_visit; // putting visited cities into individual;
    //The complete individual is initial population
13:   End
14: End
```

3.2 Probabilistic Crossover Operator

In PSO, **Pbest** of a particle retains some helpful personal experience. Thus, applying the **Pbest** to guide the particle has a positive effect during the optimization process. In order to efficiently utilize some favorable characteristics of the **Pbest**, we use the probabilistic crossover operator in this research.

The probabilistic crossover operator is on the basis of the probabilistic matrix **pcitylink**.

$$\mathbf{pcitylink} = \begin{bmatrix} a_{11}, a_{12}, \ldots, a_{1r} \\ a_{21}, a_{22}, \ldots, a_{2r} \\ \cdots\cdots\cdots\cdots \\ a_{n1}, a_{n2}, \ldots, a_{nr} \end{bmatrix} \tag{8}$$

The deterministic matrix **pcitylink** = ($\mathbf{A}_1, \mathbf{A}_2, \ldots, \mathbf{A}_i, \ldots, \mathbf{A}_n$) ($n$ is the number of cities, $i = 1,2,\ldots, n$) stores a serial number of cities, and the size of this matrix is $n * r$. And the $\mathbf{A}_i = (a_{i1}, a_{i2}, \ldots, a_{il}, \ldots, a_{ir})$ represent information about ith city. Furthermore, l denotes the lth nearest city for the ith city. For example, a_{il} represents the lth nearest city from ith city. 1 to r indicates the distance between two cities from near to far.

The probability is reflected in the selection of the gene's **cros** insertion position, which is collected from **Pbest**. Matrix **pcitydistlinkcum** represents the probability of the corresponding city transfer to the next one. The uniqueness of this matrix is that each element is the sum of itself and the element in front of him in **pcitydistlink**. **pcitydistlink** is a distance probability matrix. According to the last element of **cros**, the corresponding element *rcc* in **pcitydistlinkcum** can be selected. By comparing the numerical values of *rcc* and random number *rand*, a range of insertion positions **selectdd** can be got from **pcitylink**. The element of **pcitylink**is the ordering information from near to far, so that we choose the first element of **selected** as an insertion point of gene fragment **cros**. We put **cros** on the back of *dd*. There is a special case here. The same elements in current particle **aa** and **cros** are previously deleted, so if the *dd* in the element of **cros**, we cannot find *dd* in **aa**. In this case, gene **cros** will be putting at the last **aa**.

The above expression can be represented by Fig. 1 and Fig. 2 while the pseudo code of it is presented in Algorithm 2.

If *dd* is inside **cros**, the probabilistic crossover operator can be showed as Fig. 1

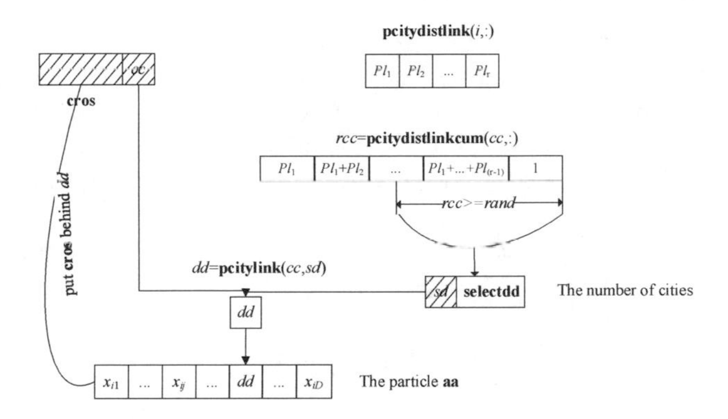

Fig. 1. Case 1 of probabilistic crossover operator

If *dd* is not inside **cros**, the probabilistic crossover operator can be showed as Fig. 2.

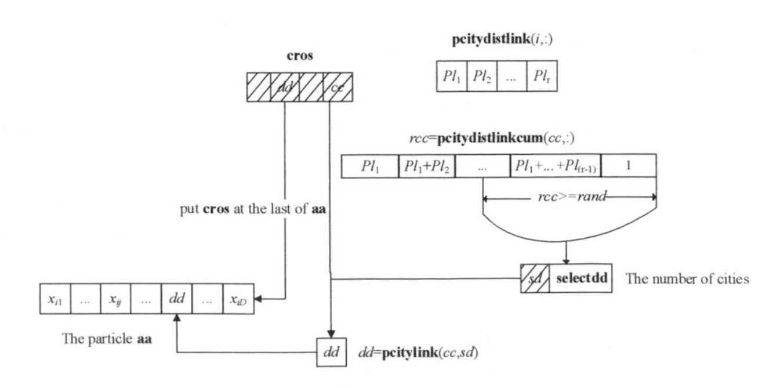

Fig. 2. Case 2 of probabilistic crossover operator

Algorithm 2. Probabilistic crossover operator

01: Generate two intersection bits c1 and c2 on **Pbest**;
02: Generate gene fragment to be crossed **cros**;
03: Deleting the same part between current particle **aa** and **cros**;
04: cc=**cros**(1);
05: rcc=**pcitydistlinkcum**(cc,:);
06: **selectdd**=find(rcc>=rand);
07: dd=**pcitylink**(cc, sd);
08: If dd does not equal to the element of **cros**
09:   Put **cros** on the back of **dd**;
10: Else
11:   Put cros at the last of current particle **aa**;
12: End

3.3 Deterministic Crossover Operator

In PSO, **Gbest** is the global best-so-far position, and there is an only **Gbest** during the evolution of the algorithm. Compared to other particles, **Gbest** contains much more promising information, which can provide more effective knowledge for other individuals. This means that **Gbest** has a strong guiding significance for population evolution. For the characteristics of **Gbest**, deterministic crossover operator is used to operate on **Gbest**.

The deterministic crossover operator is on the basis of the deterministic vector **dcitylink**.

$$\mathbf{dcitylink} = [m_{11}, \ldots, m_{i1}, \ldots, m_{n1}]^{\mathrm{T}} \tag{9}$$

n is the number of cities, $i = 1,2,\ldots, n$; m_{i1} represent the nearest city to city i. 1 to $(n-1)$ indicates the distance between two cities from near to far.

Deterministic crossover operator means the selection of the gene's **cros** insertion position is stationary. This kind of fixation doesn't mean the insertion position is changeless for each particle or in each generation. This means that the selection method is stationary. In simple terms, the gene **cros** is inserted behind the element which is the closest one to the first element of **cros**.

cc is the first element of **cros**, and *dd* is the closest element according to the vector **citylink**. Putting **cros** behind the *dd* means the distance between *dd* and *cc* is the shortest. Consequently, it is a purposeful insertion, and the better algorithm performance can be achieved. There is also a special case here. The same elements in current particle **aa** and **cros** are previously deleted, so if the *dd* in the element of **cros**, we cannot find *dd* in **aa**. In this case, gene fragment **cros** will be putting at the last **aa**.

The above process can be demonstrated by Fig. 3 and Fig. 4 while the pseudo code of it is presented in Algorithm 3.

If *dd* is inside **cros**, the deterministic crossover operator can be showed as Fig. 3.

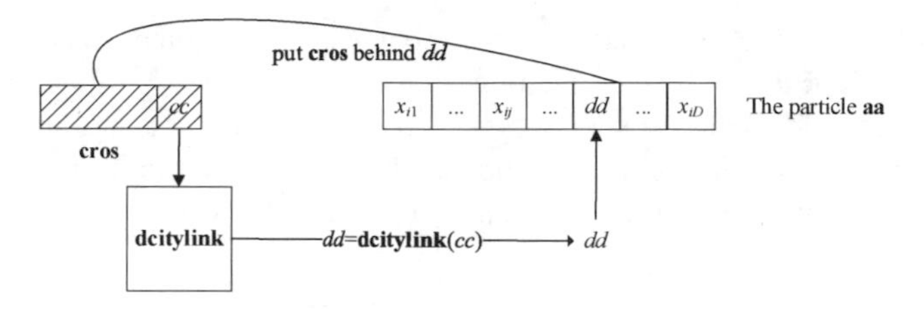

Fig. 3. Case 1 of deterministic crossover operator

If *dd* is not inside **cros**, the deterministic crossover operator can be showed as Fig. 4.

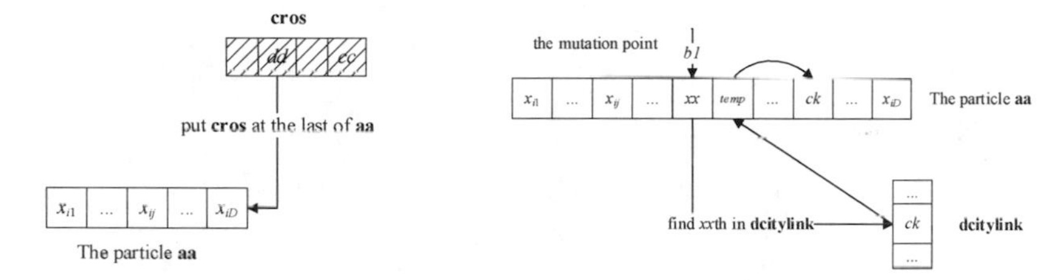

Fig. 4. Case 2 of deterministic crossover operator

Fig. 5. Directional mutation operator

Algorithm 3. Deterministic crossover operator
01: Generate two intersection bits c3 and c4 on **Gbest**;
02: Generate gene fragment to be crossed cros;
03: Delete the same part between current particle **aa** and **cros**;
04: cc=**cros**(1);
05: dd=**dcitylink**(cc);
06: If dd does not equal to the element of **cros**
07: Put cros on the back of **dd**;
08: Else
09: Put **cros** at the last of current particle **aa**;
10: End

3.4 Directional Mutation Operator

Mutation operator can add a little disturbance to an individual, and then improve population diversity. Although a random mutation operator can increase population diversity to some extent, it may damage the excellent genes of those promising particles. Thus, to overcome the shortcomings of the random mutation operator, a directional mutation operator is proposed in this work. The main idea of the directional mutation operator is that change the data of the point behind the mutation point, and this change is on the basis of the matrix **dcitylink**. Specifically, *xx* is the element of the mutation point in the current particle **aa**, and the closest element *ck* can be found according to **dcitylink**. So the combination of *xx* and *ck* is the best for element *xx*. At last, the position of *ck* in the **aa** should be found and replaced by *temp*. In this way, the algorithm achieves the directional mutation without any overlapping parts.

The above expression can be illustrated as Fig. 5 while the pseudocode of it is presented in Algorithm 4.

Algorithm 4. Directional mutation operator
01: Generate a mutation point randomly *b1*;
02: Take the element *xx* at the mutation point and *temp* behind the mutation operator from the current particle **aa**;
03: *ck*=**citylink**(*xx*)
04: **aa**(*b1*+1)=*ck*; // *i* repents the serial number of individual.
05: Find the position of *ck*, and taking place this position with *temp*

3.5 Algorithm Flow of IHPSO

Based on the components introduced above, IHPSO can be described as follows (Fig. 6):

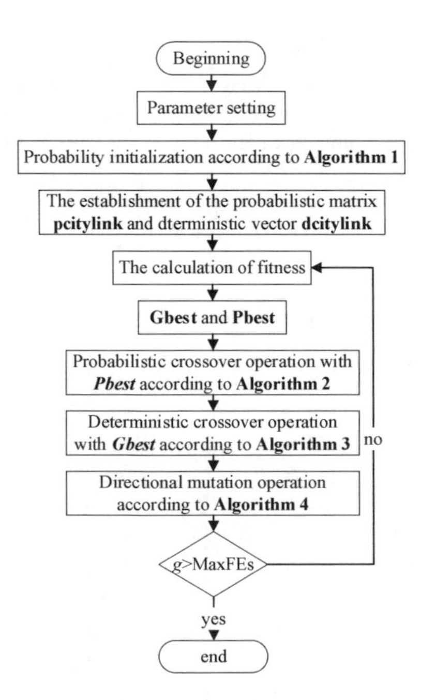

Fig. 6. IHPSO

4 Experiment Result and Analysis

The comparison results between IHPSO and other 4 contrast algorithms, in terms of the mean value (Mean) and standard deviation (Std.Dev) of 30 independent runs, are listed in Table 1. Moreover, the best tour paths for the 9 TSP achieved by IHPSO are demonstrated in Fig. 7.

From the results listed in Table 1, we can see that IHPSO displays the best performance on all the test problems. Although IHPSO and HPSO belong to the PSO community, IHPSO dominates HPSO on all the 9 problems. Furthermore, with the increase of the cities' number, the advantages of IHPSO are obvious and increase. Thus, we can obtain a preliminary conclusion that IHPSO can yield very favorable performance on large scale TSP, and has promised scalability.

Table 1. Comparison results between IHPSO and other 4 contrast algorithms

		IHPSO	HPSO	GA	SA	TB
eil51	Best	**428.87**	441.84	446.34	461.79	447.76
	Mean	**448.35**	463.52	467.37	503.01	476.90
	Std.Dev	9.23	20.74	8.97	28.27	**3.45**
berlin52	Best	**7544.4**	7716.7	7647.2	8645.3	8167.7
	Mean	**7938.49**	8302.01	8259.24	9276.67	9238.71
	Std.Dev	**179.43**	382.55	79.48	188.87	389.48
eil76	Best	**557.50**	597.0247	620.83	627.66	595.46
	Mean	**581.66**	650.95	687.31	703.76	654.20
	Std.Dev	**15.51**	30.9	47.60	26.01	30.30
rat99	Best	**1249.8**	1584.4	1850.9	1659.1	1516.4
	Mean	**1356.21**	1758.66	2439.19	1909.14	1704.74
	Std.Dev	**38.44**	67.68	331.22	115.41	120.83
eil101	Best	**677.08**	820.46	1349.8	799.22	753.88
	Mean	**706.12**	879.62	1502.71	888.44	821.606
	Std.Dev	**17.86**	35.36	81.54	48.58	41.08
lin105	Best	**14866**	18886	18835	22716	21742
	Mean	**15775.3**	21528.13	24850.57	26261.37	24559.77
	Std.Dev	**435.41**	1364.24	3183.79	1823.45	2115.15
bier127	Best	**124550**	165840	294650	18892	154490
	Mean	**133512.3**	181364.67	333406.67	168639.4	167539.67
	Std.Dev	**3491.55**	7036.82	14019.24	29928.47	7540.24
ch130	Best	**6372.3**	10029	21481	9339.2	8757.4
	Mean	**6936.59**	10959.5	24078.37	10414.54	9860.80
	Std.Dev	**200.00**	661.58	1063.32	719.059	529.52
ch150	Best	**7047.9**	12626	19775	10500	9875.6
	Mean	**7558.40**	13598.3	24356.17	11882.33	11085.08
	Std.Dev	**239.26**	672.58	3774.56	725.86	611.69

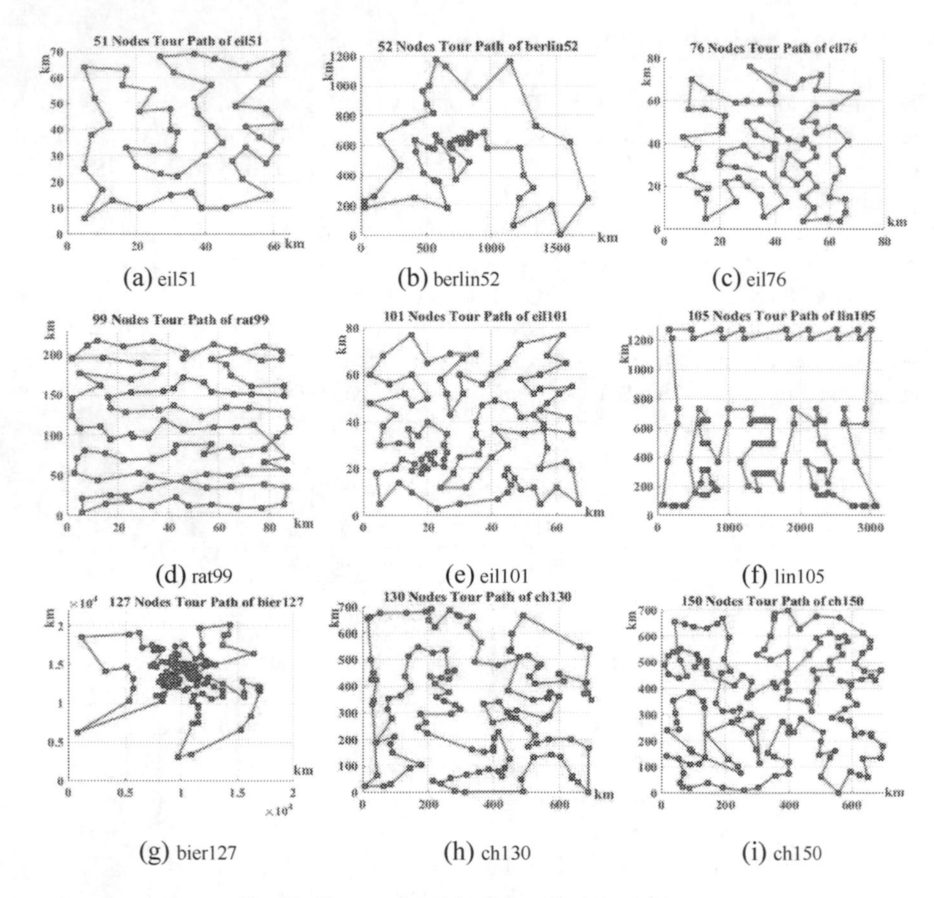

(a) eil51 (b) berlin52 (c) eil76

(d) rat99 (e) eil101 (f) lin105

(g) bier127 (h) ch130 (i) ch150

Fig. 7. Tour path of the 9 benchmark problems

5 Conclusion

In this paper, we proposed an IHPSO to deal with TSP. In IHPSO, there are four novel strategies are introduced to improve its performance. Firstly, a probability initialization method rather than the popular randomly initialization is introduced to improve the population's initial fitness. Secondly, a probabilistic crossover operator and a deterministic crossover operator are conducted between an individual and its **Pbest** and the **Gbest**. In this condition, useful information with **Pbest** and **Gbest** can be effectively utilized to balance the convergence speed and population diversity. Lastly, a directional mutation operator is adopted to add a little disturbance to improve population diversity. The comparison results between IHPSO and other four types of EAs on nine TSP test problems toverify the outstanding performance of IHPSO. In addition, compared to the canonical PSO, IHPSO displays more favorable performance on large scale TSP than on small size TSP.

Acknowledgement. This study was funded by the National Natural Science Foundation of China (Grant Nos.: 61806204, 61663009, 61602174), the National Natural Science Foundation of Jiangxi Province (Grant Nos.: 20171BAB202019, 20161BAB202064, 20161BAB212052, 20151BAB207022), the National Natural Science Foundation of Jiangxi Provincial Department of Education (Grant Nos.: GJJ160469), and the Research Project of Jiangxi Provincial Department of Communication and Transportation (No. 2017D0038).

References

1. Bello, R., Gomez, Y., Nowe, A., et al.: Two-step particle swarm optimization to solve the feature selection problem. In: 7th International Conference on Intelligent Systems Design and Applications, pp. 691–696. IEEE, Atlanta (2007)
2. Li, Y., Zhou, A., Zhang, G.: Simulated annealing with probabilistic neighborhood for traveling salesman problems. In: 7th International Conference on Natural Computation, pp. 1565–1569. IEEE, Shanghai (2011)
3. Lin, Y., Bian, Z., Liu, X.: Developing a dynamic neighborhood structure for an adaptive hybrid simulated annealing - tabu search algorithm to solve the symmetrical traveling salesman problem. Applied Soft Comput. **49**, 937–952 (2016)
4. Li, R., Qiao, J., Li, W.: A modified hopfield neural network for solving TSP problem. In: 12th World Congress on Intelligent Control and Automation, pp. 1775–1780. IEEE, Guilin (2016)
5. Pang, W., Wang, K.P., Zhou, C.G., et al: Fuzzy discrete particle swarm optimization for solving traveling salesman problem. In: The 4th International Conference on Computer and Information Technology, vol. 3, pp. 796–800. IEEE, Beijing (2004)
6. Kennedy, J., Eberhart, R.C.: Particle swarm optimization. In: Proceedings of IEEE International Conference on Neural Network, ICNN 1995, vol. 4, pp. 1942–1948. IEEE, Perth (1995)
7. Clerc, M.: Discrete particle swarm optimization, illustrated by the traveling salesman problem. In: New Optimization Techniques in Engineering. Studies in Fuzziness and Soft Computing, vol. 141. Springer, Heidelberg (2004). https://doi.org/10.1007/978-3-540-39930-8_8
8. Hendtlass, T.: Preserving diversity in particle swarm optimization. Lect. Notes Comput. Sci. **2718**, 4104–4108 (2003)
9. Reinelt, G.: TSPLIB—a traveling salesman problem library. Orsa J. Comput. **3**(4), 376–384 (1991)
10. Shi, X.H., Liang, Y.C., Lee, H.P., et al.: Particle swarm optimization-based algorithms for TSP and generalized TSP. Inf. Process. Lett. **103**(5), 169–176 (2007)
11. Wang, K.P., Huang, N.L., Zhou, C.G., et al.: Particle swarm optimization for traveling salesman problem. Acta Scientiarium Naturalium Universitatis Jilinensis **3**(4), 1583–1585 (2003)
12. Shaj, V., Akhil, P.M., Asharaf, S.: Edge-PSO: a recombination operator based PSO algorithm for solving TSP. In: 6th International Conference on Advances in Computing, Communications and Informatics, pp. 35–41. IEEE, Kochi (2016)
13. Chang, J.F.: A parallel particle swarm optimization algorithm with communication strategies. J. Inf. Sci. Eng. **21**(4), 809–818 (2005)
14. Liao, Y.F., Yau, D.H., Chen, C.L.: Evolutionary algorithm to traveling salesman problems. Comput. Math Appl. **64**(5), 788–797 (2012)

Application of Parametric Design in Urban Planning

Rongrong Gu[1] and Wuzhong Zhou[2(✉)]

[1] School of Design, Shanghai Jiao Tong University, Minhang District, Shanghai 200240, China
[2] Nantong Gangzha Constructing Designing Institute Co., Ltd, Nantong 226000, China
wzzhou@sju.edu.cn

Abstract. With the increase in computational power and the emergence of a large number of digital data, the role of new methods and technologies in urban planning is becoming more and more important. The characteristics of the parametric design method determine its practicality and development potential in dealing with the planning and design of complex systems such as cities. Based on the concept and innovative ideas on urban planning, this paper adopts the methods of parametric technology to define the main measures and technological route. Also, based on the research methods of parametric technology, the platform of combining Rhino and Grasshopper can be used for the analysis on terrain elevation, contour, slope, sunshine and surface runoff. With this chance, this paper hopes to provide a rational and scientific aid for the implementation of urban planning at a later period.

Keywords: Urban planning · Rhino · Grasshopper

1 Innovative Ideas for Urban Planning

Building sustainable, resilient and livable environments requires taking increasingly complex design decisions and integrating knowledge from various areas of expertise [1]. Urban planning is the work of planning urban economic structure, spatial structure, and social structure development, often including urban district planning.

As the preliminary work of urban comprehensive management and the leader of urban management, it has the important role of guiding and standardizing urban construction. The characteristics of the complex system of the city indicate that urban planning is a complex, continuous decision-making process with long-term adjustments, continuous revisions and improvements on the urban development and operation. In the process of urban planning, data modeling methods can be used to quantify, model, and visualize data. At the same time, environmental planning, information technology, geographic information technology, and communication technology are used to complete spatial planning and residential design.

K. Li et al. (Eds.): ISICA 2019, CCIS 1205, pp. 526–540, 2020.
https://doi.org/10.1007/978-981-15-5577-0_41

2 Research on Parametric Technology

Recently, parametric design works have been leading the new trend of innovation and technology in the field of architecture. More and more research and applications have gradually gained in other design fields such as urban planning. With parametric design extensively spreading out, the discussion on the influence of the nature of thinking for parametric design and computer numbers has become more and more realistic.

Traditional urban design work usually takes site visits as a start. During the visits, designers gather information from their subjective experience. Parametric design tools, which originated from the CAD systems, were first proposed by Hillyard and Braid [2]. Parametric design tools accept variable input data, establish mathematical relationships and produce further data, including geometric information [3]. Geometrical Operation Approach is the most common approach by designers as they begin exploring the possibilities of parametric modelling [4].

Program modeling is one of the widely used applications in the field of computer graphics technology, even in urban design. This paper studies the functions and characteristics of Rhino and Grasshopper, combined with computer programming technology, in order to discover the new development direction of program design and design methods in the urban design field.

Design approaches are informed by trial solutions [5] and more recent computing assisted ideas of 'design patterns' [6]. Parametric urban design bridges two realms by drawing upon existing datasets and gathering and analyzing one's own data both at the site and off-site [7].

2.1 The Features and Characteristics of Rhino

Based on the principles of nurbs curves and surfaces, Rhino has precise and powerful modeling capabilities. Nurbs is presented in a digital way, accurately defining all three-dimensional objects such as curves, surfaces, and solids through mathematical calculations. Compared with the traditional software of the triangular mesh modeling method, Rhino is not subject to any constraints according to the curvature and scale of the curve. The construction model can provide a great flexibility and freedom for the user to adjust the accuracy of the model according to the required adjustment. It also provides users with work efficiency.

The Rhino digital model is very detailed and the surface is very smooth. The Rhino and its plugins can create models in a variety of ways, most of which are three commonly used modeling methods: conventional building modeling, T-splines curved building model creation and Grasshopper parametric model creation.

2.2 The Features and Characteristics of Grasshopper

Grasshopper is a 3D model plugin that runs in the Rhinoceros environment. It uses program algorithms to generate operations that are different from traditional modeling tools, transforming data logic into visual geometric models. The biggest feature of Grasshopper is that it can program more advanced and more complex logic modeling commands to the computer through program algorithms, so that the computer automatically generates model results based on the determined data logic structure. By writing an operator with a logical modeling concept, mechanical repetitive operations can be done by the computer's automatic loop operations. Unlike the Rhino Scrip, GH does not require any knowledge of the programming language. It can produce the model which the designer wants through some intuitive and understandable logical calculation process [8].

For designers who are exploring new shapes using generative algorithms, Grasshopper is a graphical algorithm editor tightly integrated with Rhino's 3-D modeling tools. Unlike Rhino Script, Grasshopper requires no knowledge of programming or scripting, but still allows designers to build form generators from the simple to the awe-inspiring.

2.3 Application of Rhino and Grasshopper in Planning and Design

In the process of parametric design, we take the main factors affecting the design as parameters. When there is a basic design parameter relationship, we also need to find some kind of rule system (i.e., algorithm) to construct the parameter relationship to generate the shape and describe the rule system in computer language to form the software parameter model.

The School of Architecture in Tsinghua University and other well-known architectural institutions use parametric design methods such as "Digital Diagram" and "Algorithm Generating" and combining stadium design and application with the future development of the city [9].

2.4 Terrain Elevation Analysis

2.4.1 The Concept and Application of Terrain Elevation Analysis

The terrain elevation analysis is based on the CAD of the elevation point, after the overall terrain model is established. The elevation of the terrain is determined according to the position and height of the terrain elevation point; the relationship of the terrain elevation is displayed by color. Terrain elevation analysis allows designers to understand the geomorphic conditions of the site and the overall elevation. It can visually understand the elevation gap between the highest and lowest points of the site, allowing designers to think about the height difference of the site and increase the rationality of the design.

2.4.2 The Analysis Principle of Grasshopper Logic Construction

(See Figs. 1, 2, 3, 4 and 5).

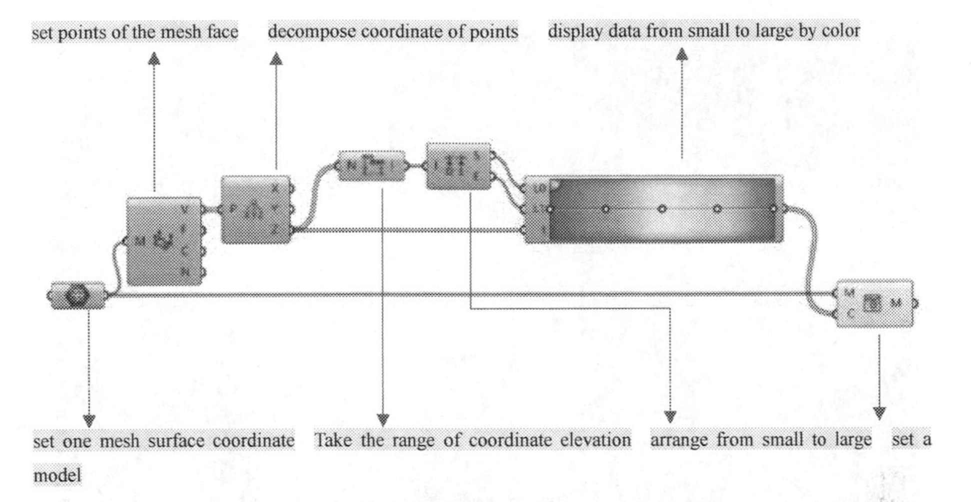

Fig. 1. The analysis principle of Grasshopper logic construction is based on the establishment of the terrain model, picking up the mesh surface of the terrain model, decomposing the elevation points on the mesh surface, obtaining the coordinates of all elevation points, and arranging the data on the Z axis from small to large. The display of different colors represents different elevation values. The terrain elevation analysis surface is formed according to the color of the terrain elevation corresponding to the displayed color.

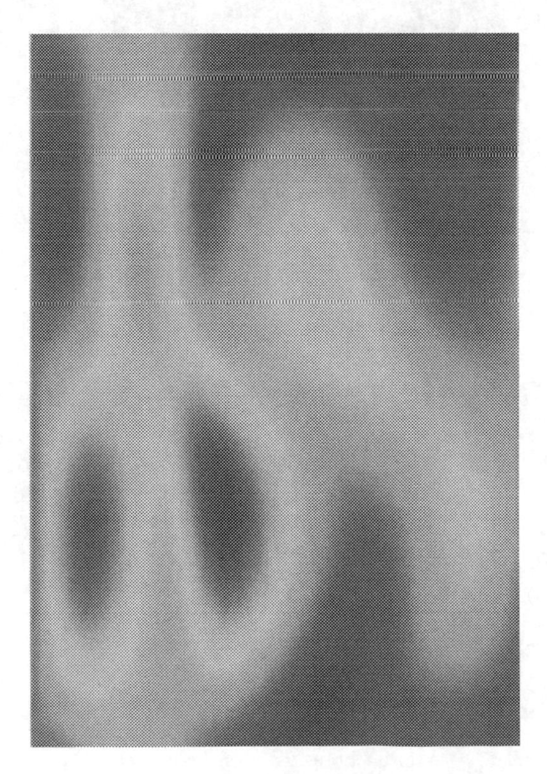

Fig. 2. Top view of the model

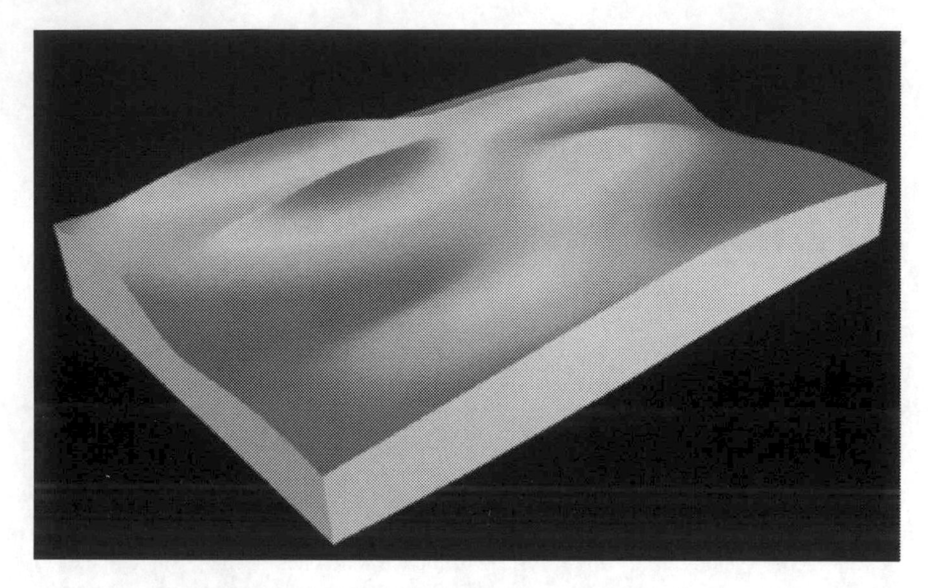

Fig. 3. Perspective view of the model

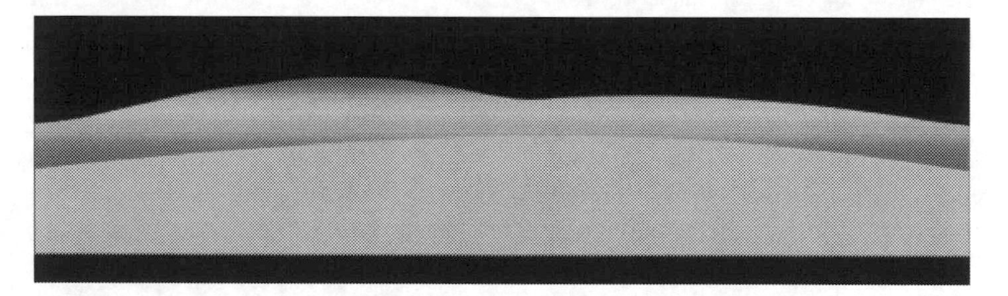

Fig. 4. Right view of the model

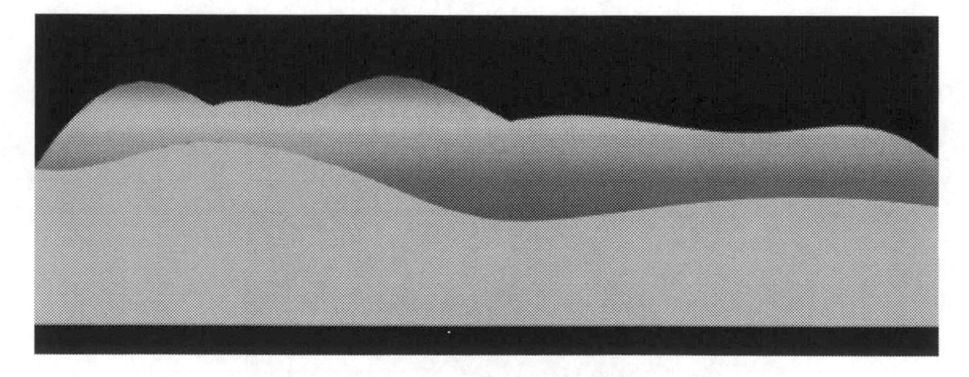

Fig. 5. Front view of the model

2.5 Contour Analysis

2.5.1 The Concept and Application of Contour Analysis

The contour analysis is performed by dividing the terrain elevation information into a plurality of groups at a certain interval and displaying them in a layered color. This can help planners to understand the contours and trends of the terrain. It has important guiding significance for planning designers to consider the multi-dimensional direction of planning route, landscape sequence, line of sight organization, building layout and site design.

2.5.2 The Analysis Principle of Grasshopper Logic Construction

The value of the button "distance" represents the height difference between the contour lines. The larger the value, the more uneven the contour line. In this experiment, the height difference of the contour line is 2 m. The experimental results obtained are as follows (Figs. 6, 7 and 8).

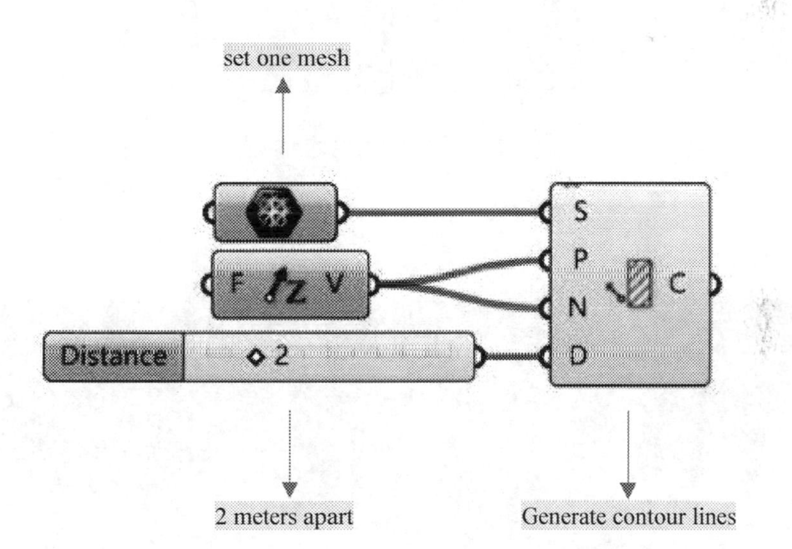

Fig. 6. The premise for analyzing the terrain contours is to obtain the contours of the terrain of the site. The contours generated by using the tif format image downloaded from the terrain data cloud to be imported into the Global Mapper may not fully reflect the terrain information of the site. Build terrain contours. The logic of Grasshopper is used to construct the contour line of the terrain mesh by inputting the spacing value. This method is suitable for the surface of the grid structure and the contours produced can be more realistic and smooth according to the optimization of the mesh surface.

Fig. 7. Top view of the model

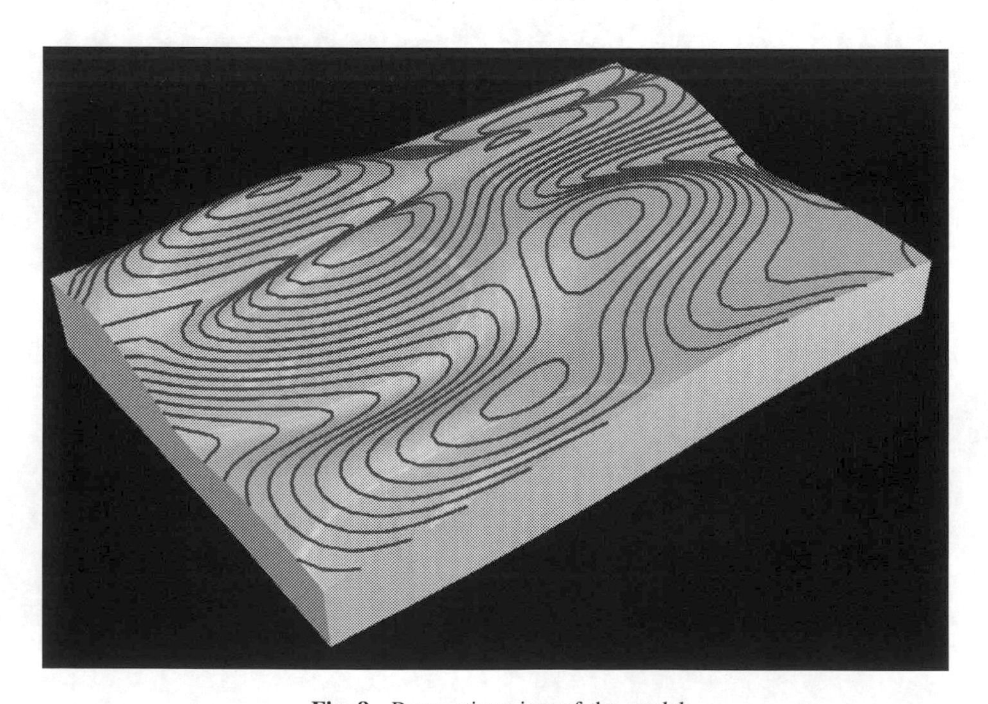

Fig. 8. Perspective view of the model

2.6 Slope Analysis

2.6.1 The Concept and Application of Slope Analysis

Slope has a decisive role in soil and water conservation planning and design, and is the first factor to be considered in the allocation of land use planning and governance measures. The terrain slope analysis is based on the terrain model. According to the angle between the tangent plane and the horizontal plane of the terrain surface, the slope of the terrain surface at this point is indicated.

2.6.2 The Analysis Principle of Grasshopper Logic Construction

Grasshopper translates the calculation principle of slope inclination angle into logic construction calculation. The author uses the angle between the z-axis direction of the point on the surface and its normal vector to calculate the slope. Pick the terrain mesh surface, decompose the vector of each point on the surface, and calculate the angle between the normal vector and the z-axis direction. The calculation of the angle in Grasshopper is based on radians, and the arc needs to be converted into an angle. Arrange according to the value of the angle, and the position where each angle is located is displayed by the layered color method (Figs. 9, 10, 11, 12 and 13).

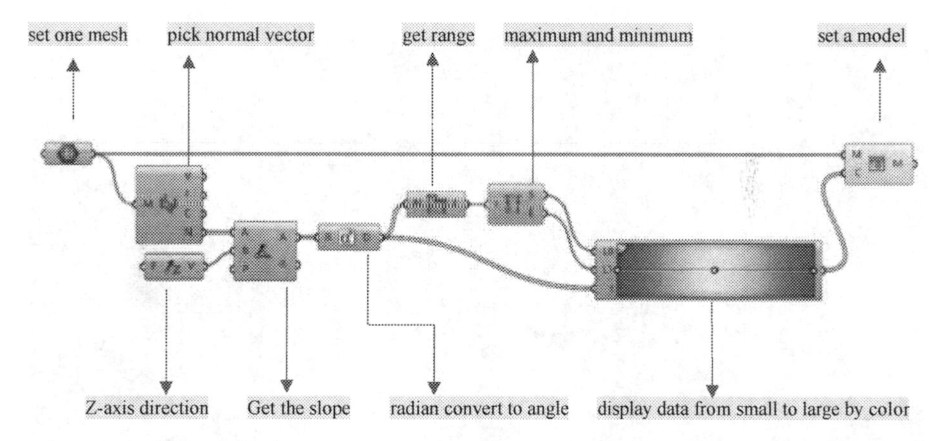

Fig. 9. The slope analysis is based on the calculation principle of the slope inclination angle and is converted into a logic construction. The point on the surface is decomposed into the tangent plane direction, the Y-axis direction, the Z-axis direction and the normal direction. According to the mathematical principle, the four directions are in the same plane, the Z-axis direction is perpendicular to the Y-axis direction, and the normal direction is perpendicular to the tangential plane direction. The relationship between the four is known from the figure, and the Z-axis and the normal direction are The angle c formed by the angle a and the y axis with the normal direction is equal to 90°, the angle b between the tangential plane direction and the y-axis direction, and the angle c formed by the normal direction and the y-axis direction are equal to 90°, based on a similar definition, so the angle a is equal to the angle b. According to the definition of the slope, the angle b indicates the degree of inclination of the point in the plane. The larger the b angle is, the greater the inclination is. Similarly, the larger the angle b, the larger the angle a, and the larger the slope.

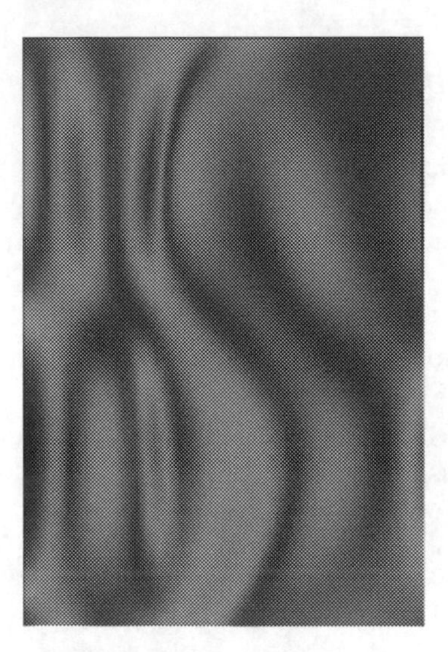

Fig. 10. Top view of the model

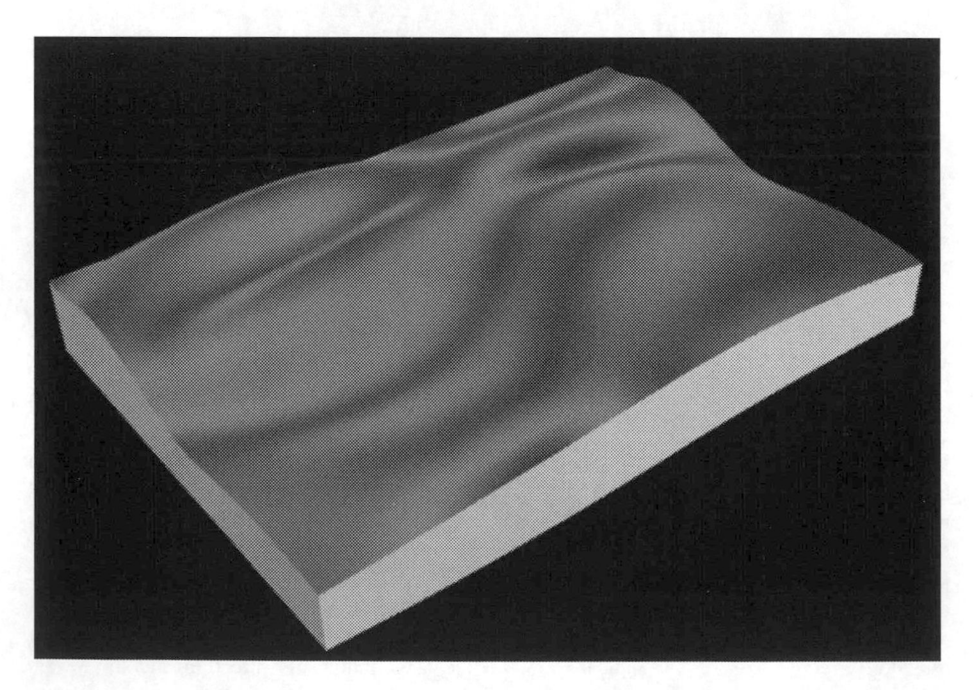

Fig. 11. Perspective view of the model

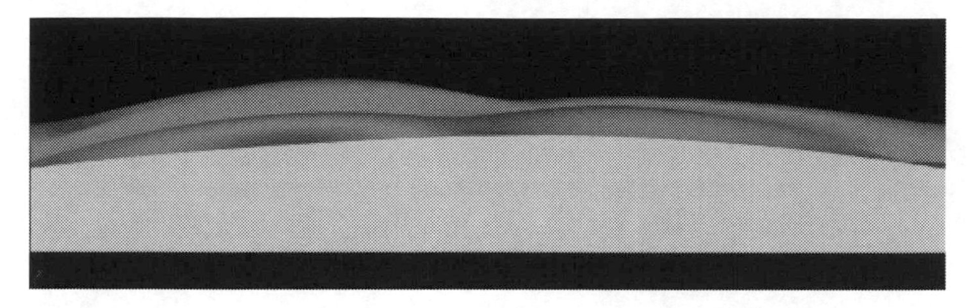

Fig. 12. Right view of the model

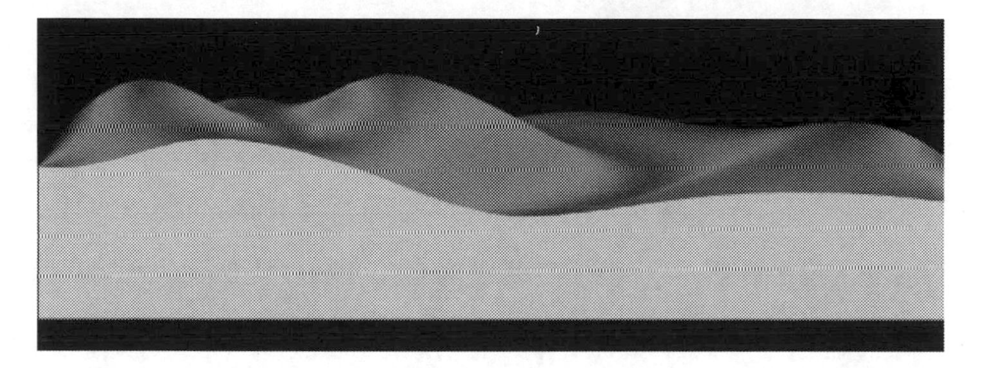

Fig. 13. Front view of the model

2.7 Sunshine Analysis

2.7.1 The Concept and Application of Sunshine Analysis

The sunshine analysis is based on the relationship between the position of the earth where the terrain is located and the time of the sun and the light. The relative value is calculated by logic to display the sunshine condition of the terrain. Sunshine analysis plays an important guiding role in the spatial layout of land use planning and design. For urban planning land, the factors affecting the results of the sunshine analysis include not only the buildings within the site, but also the interference of other buildings and terrain in the surrounding area.

2.7.2 The Analysis Principle of Grasshopper Logic Construction

Download a Rhino terrain on the cadmapper website as an analysis site to demonstrate the functionality of the Ladybug plugin. Open the terrain in Rhino and drag the surrounding buildings into the Grasshopper along with the buildings in the base. Download the epw file containing the complete weather data for the region on the Climate Consultant 6.0 software. Use the import epw command on the Grasshopper platform to analyze the imported epw file, then use the Sunpath operator to define the time we need to analyze (Great Cold Day or Winter Solstice Day), and finally use the plug-in's core operator sunlight hours analysis for sunshine analysis (Figs. 14 and 15).

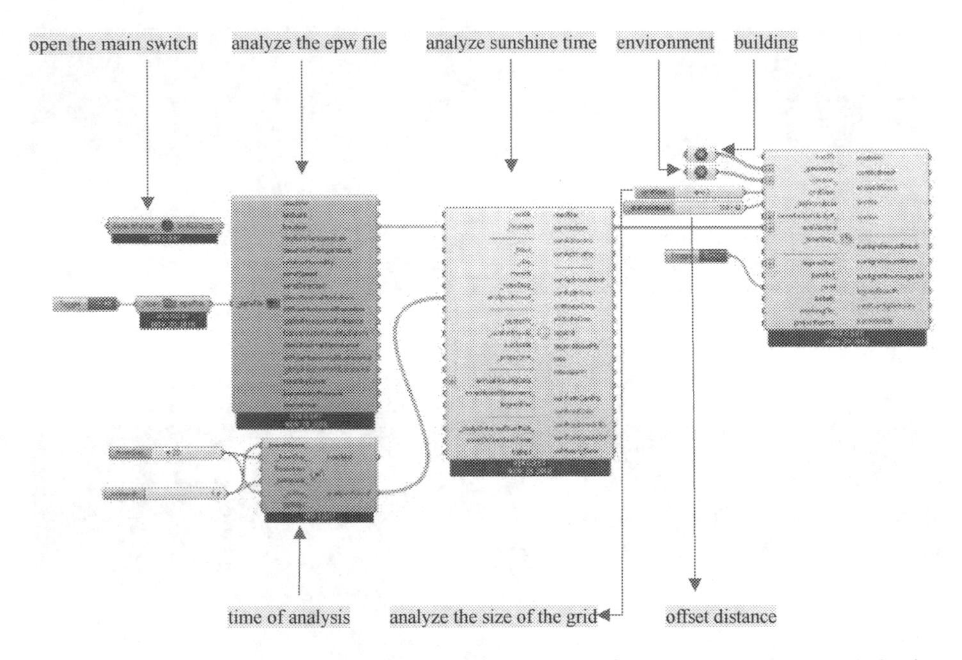

Fig. 14. Grasshopper's powerful Ladybug plugin allows for daylight analysis and daytime sunshine calculations. Enter and analyze the standard weather data in Grasshopper in Ladybug, draw the sun path, wind direction, radiation pattern, etc.; customize the chart in a variety of ways, perform radiation analysis, shadow analysis and view analysis. Ladybug follows the tree data structure and running method, and the input method is mainly based on slider and Boolean toggle. Ladybug's custom Legend parameter component allows users to customize the color, font and line shape of the drawing. The analysis results can be well matched with flat drawing software such as AI and PS.

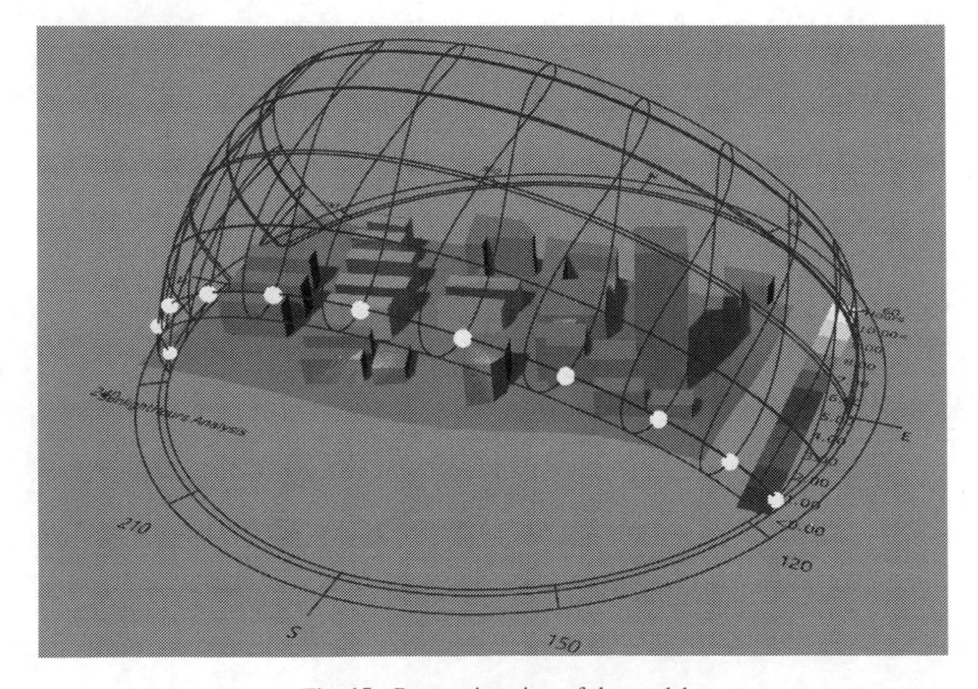

Fig. 15. Perspective view of the model

For many ladybug operators, most of their inputs use only the default values and do not require additional connections. This sunshine analysis experiment sets the day of the cold weather January 20 as the sunshine analysis calculation day. Assign at the input of the operator at the unlight hours analysis. The first group is geometry and context. They refer to objects that need to be analyzed for sunshine time and objects that may block sunlight. Then we connect the previous building and environment to them. Grid size refers to the size of the analysis grid. You can set a larger number for testing at the beginning, and then increase the subdivision to get more detailed analysis results when there is no problem. This experiment was set to 0.3. Disfrombase refers to the distance of the offset. Just set a small value here, and set the experiment to 0.01. Sunvector is the direction of the sun generated by the previous sunpath operator. Finally, the runit is connected to a TRUE Boolean value to start the calculation. The ensuing results are as follows.

2.8 Surface Runoff Analysis

2.8.1 The Concept and Application of Surface Runoff Analysis

Surface rainwater runoff refers to the process of establishing surface water flow models to study the source flow and convergence areas of surface water flows. With the rapid urbanization, more attention has been made on the relationship between rainwater treatment and urban areas. How to use surface rainwater resources reasonably and effectively in site planning becomes a hot thinking problem in current planning and design, as well as the surface rainwater resources and sites planning integration, improving the relationship between the site and rainwater, reducing the surface effluent of surface rainwater, and improving the utilization of surface stormwater runoff resources. The analysis of surface rainwater runoff allows designers to intuitively understand the source flow of rainwater, set relevant rainwater treatment measures through the characteristics of topography and geomorphology, optimize the spatial layout of the site, and achieve the goal of design and integration with nature.

2.8.2 The Analysis Principle of Grasshopper Logic Construction

When the rain falls, the direction is constantly changing. The concept of a loop is used to describe this process. The principle is that the input data for each calculation is based on the results of the previous calculation. When the set number of loops is 0, it means that no loop is performed. In this experiment, the number of cycles was set to 30 to simulate the experimental results. Pick a surface as an experiment object. Divide 30 divisions in the U/V direction on the surface to generate multiple control points and their tangent planes on the surface. Set the slider of 5.00 to adjust the length of the X vector arrow (Figs. 16, 17 and 18).

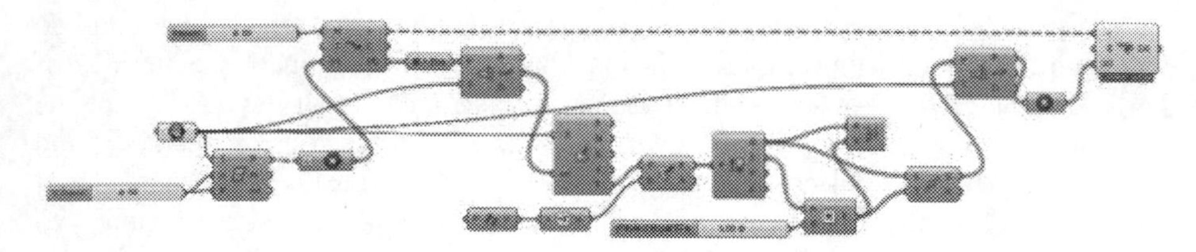

Fig. 16. The surface rainwater runoff analysis is different from other analyses. The logic construction system needs to pick up the terrain surface of the surface for analysis. If the terrain surface is a grid structure, it can be converted by Rhino's curtain surface. The principle of surface rainwater runoff analysis simulates the rainwater level and falls to the plane, flowing along the direction of the slope of the ground. The flow reaches the location where the slope is not enough to flow and becomes the place where the rainwater gathers. The multiple rainwater points circulate and flow into the surface rainwater. Based on this principle, the rainwater runoff process of the terrain of the site is simulated by the logic construction using the parameter setting, and the analysis result of the surface rainwater runoff is obtained.

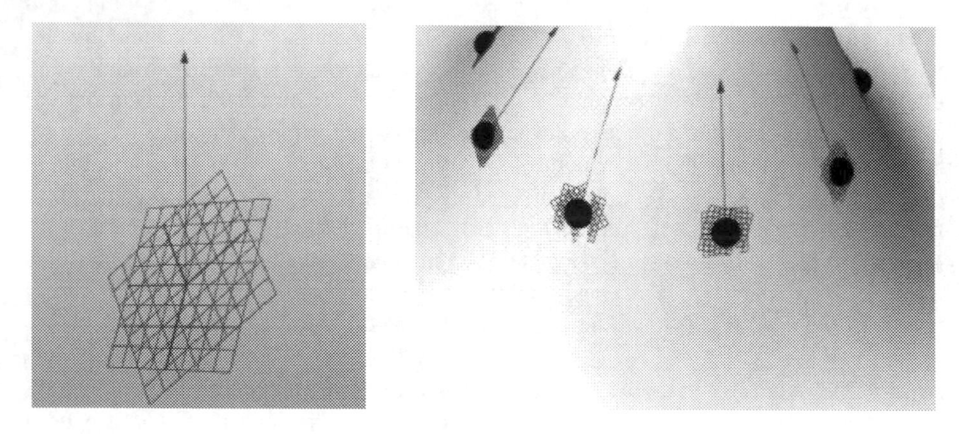

Fig. 17. Find the most upward direction on the surface of a space on a terrain surface with the smallest angle to the vertical

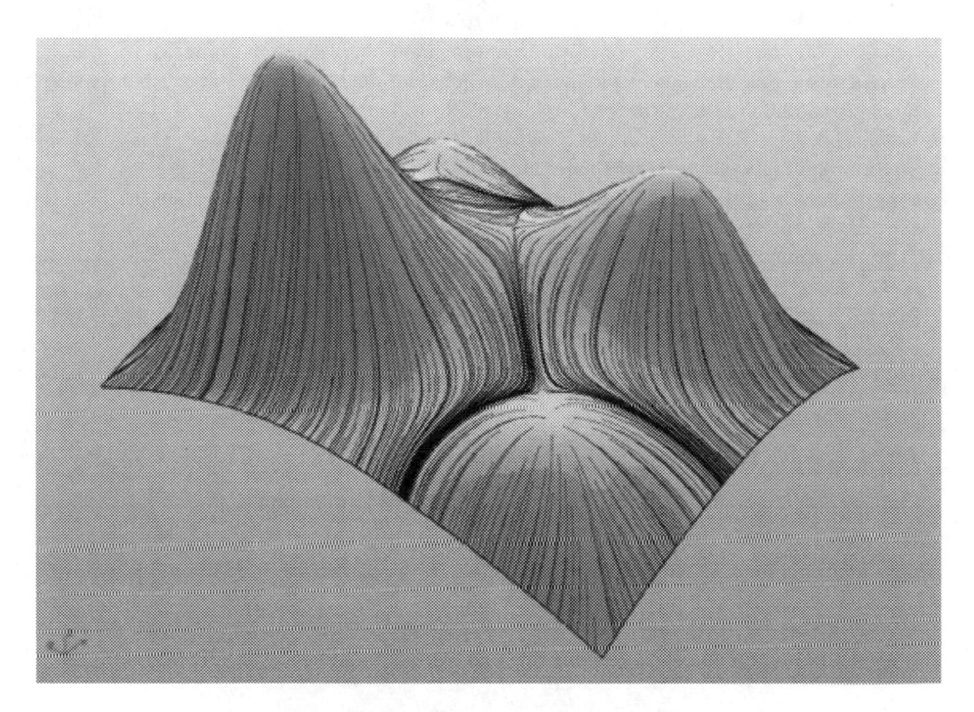

Fig. 18. Perspective view of the model

3 Conclusion

As the traditional urban planning and design assistance technology has been unable to meet the needs of the new era design, this paper takes the terrain analysis in urban planning as the entry point for research. With the help of Rhino and Grasshopper parameterization technology, it not only makes the terrain analysis of planning and design land scientific and standardized, but also ensures that the analysis results are accurate. They greatly improve the efficiency of pre-planning work, especially in the large-scale planning of complex and diverse terrain features. The analysis results obtained by parametric technology can reduce and avoid the subjectivity of planning designers to the site judgment. The results of the analysis are the basis for the scientific design of the design.

References

1. Batty, M., Longley, P.: Fractal Cities (2016)
2. Hillyard, R.C., Braid, I.C.: Analysis of dimensions and tolerances in computer-aided mechanical design. Comput. Des. **10**, 161–166 (1978)
3. Steinø, N., Veirum, N.: A parametric approach to urban design - tentative formulations of a methodology. In: Digital Design Methods - eCAADe Design, vol. 23, p. 679 (2010). Adv. Mater. Res. 1674 (2012)

4. Leung, T.M.: Parametric design modelling in urban art: approaches and future directions. In: Proceedings of the 2019 International Conference on Architecture: Heritage, Traditions and Innovations, AHTI 2019 (2019)
5. Lawson, B.: How Designers Think. Routledge, London (2006)
6. Woodbury, R.: Elements of Parametric Design. Routledge, London (2010)
7. Speranza, P.: Using parametric methods to understand place in urban design courses. J. Urban Des. **21**(5), 661–689 (2016)
8. Zeng, X.: Rhinoceros & Grasshopper Parametric Modeling. Huazhong University of Science & Technology Press, Wu Han (2011)
9. Zhang, P., Xu, W.: Parametric urban design: 2020 Tokyo Olympic city. Urbanism Archit. (04), 14–18 (2017)

Iterated Tabu Search Algorithm for the Multidemand Multidimensional Knapsack Problem

Dongni Luo, Xiangjing Lai(✉), and Qin Sun

Institute of Advanced Technology,
Nanjing University of Posts and Telecommunications, Nanjing 210023, China
xiangjinglai@qq.com

Abstract. The multidemand multidimensional knapsack problem (MDMKP) is a classic NP-hard combinatorial optimization problem with a number of real-world applications. In this paper, we propose an iterated tabu search (ITS) algorithm for solving this computationally intractable problem, by integrating two solution-based tabu search procedures aiming to locally improve the solutions and a perturbation operator aiming to jump out of local optimum traps. The performance of proposed algorithm was assessed on 54 benchmark instances commonly used in the literature, and the experimental results show that the proposed algorithm is very competitive compared to the state-of-the-art algorithms in the literature. In particular, the proposed ITS algorithm improved the best known results in the literature for 27 out of 54 instances.

Keywords: Multidemand multidimensional knapsack problem · Solution based tabu search · Perturbation operator · Heuristics

1 Introduction

The multidemand multidimensional knapsack problem (MDMKP) is an important variant of the classic multidimensional knapsack problem (MKP) [1]. Given a set $V = \{1, 2, \ldots, n\}$ of n items, a set $R = \{r_1, r_2, \ldots, r_m, r_{m+1}, \ldots, r_{m+q}\}$ of $m+q$ resources with a capacity lower or upper limit b_i for resource r_i $(1 \le i \le m+q)$. Each item of the set V is associated with a profit c_j which takes a positive, negative or zero value and consumes a given quantity a_{ij} for each resource $r_i(i \in \{1, 2, \ldots, m, m+1, \ldots, m+q\})$. The MDMKP consists in selecting a subset of items from V such that the resource consummation of selected items does not exceed the lower or upper limit of capacity for each resource in R. Formally, the multidemand multidimensional knapsack problem can be expressed as follows:

$$\text{Maximize } f(x) = \sum_{j=1}^{n} c_j x_j \tag{1}$$

K. Li et al. (Eds.): ISICA 2019, CCIS 1205, pp. 541–550, 2020.
https://doi.org/10.1007/978-981-15-5577-0_42

s.t

$$\sum_{j=1}^{n} a_{ij}x_j \leq b_i, \forall i \in \{1, 2, \ldots, m\} \quad (2)$$

$$\sum_{j=1}^{n} a_{ij}x_j \geq b_i, \forall i \in \{m+1, m+2, \ldots, m+q\} \quad (3)$$

$$x_j \in \{0, 1\}, \forall j \in \{1, 2, \ldots, n\} \quad (4)$$

where the following conditions are assumed:

$$b_i > 0, a_{ij} \geq 0, \forall i \in \{1, 2, \ldots, m+q\}, \forall j \in \{1, 2, \ldots, n\} \quad (5)$$

$$\sum_{j=1}^{n} a_{ij} \geq b_i, \forall i \in \{1, 2, \ldots, m+q\} \quad (6)$$

$$\max_j\{a_{ij}\} \leq b_i, \forall i \in \{1, 2, \ldots, m\} \quad (7)$$

$$\min_j\{a_{ij}\} < b_i, \forall i \in \{m+1, m+2, \ldots, m+q\} \quad (8)$$

The MDMKP has a variety of real-world applications [2], such as obnoxious and semiobnoxious facility location [3–5], capital-budgeting, and portfolio-selection [6], among others.

Due to its NP-hard feature, a number of heuristic algorithms have been proposed in the literature. For instance, in 2005, Cappanera and Trubian proposed a nested-tabu-search heuristic [2]. In 2006, Arntzen et al. proposed an adaptive memory search method called Almha in [7]. In 2010, Gortazar et al. introduced a black box scatter search method for general classes of binary optimization problems [8]. At the same year, Hvattum et al. proposed an alternating control tree (ACT) search framework for the MDMKP [9], which can lead to an exact algorithm or heuristic algorithm by choosing the routine of solving subproblems. In 2018, Lai et al. proposed a two-stage solution-based tabu search (TSTS) algorithm [10], and the computational results show that the TSTS algorithm is very competitive compared with the best performing algorithm in the literature, and can be viewed as the current state-of-the-art algorithm.

In this paper, to further enhance the performance of TSTS algorithm, we proposed an iterated tabu search (ITS) by integrating the main components of TSTS algorithm and a new perturbation operator. The experimental results show that the proposed ITS algorithm is very competitive compared to the state-of-the-art algorithm in the literature especially for the large scale instances.

The rest parts of paper are organized as follows. In Sect. 2, we describe briefly the proposed ITS algorithm. In Sect. 3, the performance of proposed ITS algorithm is assessed by reporting the computational results on 54 benchmark instances and making a comparison with the state-of-the-art algorithm in the literature. Finally, we summarize the present work in the last section.

2 Iterated Tabu Search (ITS) Algorithm for the MDMKP

The ITS algorithm proposed in this work is composed of three main components, including two tabu search procedures (i.e., $TabuSearch_1$ and $TabuSearch_2$) and a perturbation operator, where $TabuSearch_1$ aims to find a high-quality hyperplane in the search space, $TabuSearch_2$ aims to search for the improving solutions in the hyperplane located by $TabuSearch_1$, and the perturbation operator aims to jump out of the local optimum traps reached by $TabuSearch_2$. The general procedure and the components of the proposed algorithm are given in the following subsections.

2.1 Search Space and Solution Representation

The search space explored by our ITS algorithm is composed of all feasible and infeasible solutions, where each solution can be indicated by a n-dimensional 0-1 vector $(x_1, x_2, \ldots, x_n)$. Thus, the search space of the ITS algorithm can be expressed as follow:

$$\Omega = \{(x_1, x_2, \ldots, x_n) | x_i \in \{0, 1\}, 1 \leq i \leq n\} \tag{9}$$

2.2 General Procedure

```
Algorithm 1: General procedure of ITS algorithm for the MDMKP
1  Function ITS()
   Input: Instance I, time limit t_max, parameters K_step,K_max,K_min
   Output: The best solution S* found
2  begin
3      S ← InitialSol(I)
4      S1 ← TabuSearch1(S)
5      S2 ← TabuSearch2(S1)
6      S ← S2; S* ← S2; K ← K_min
7      while time < t_max do
8          S' ← Perturb(S, K);
9          S'' ← Tabusearch2(S');
10         if f(S'') > f(S*) then
11             S ← S''
12             S* ← S'';
13             K ← K_min;
14         end
15         else
16             K ← K + K_step;
17         end
18         if K > K_max then
19             K ← K_min;
20         end
21     end
22     return S*
23 end
```

Fig. 1. The general procedure of our ITS algorithm.

The ITS algorithm described in Fig. 1 starts from an initial solution obtained by assigning randomly 0 or 1 to each variable (line 3), and then performs $TabuSearch_1$ to find a high-quality hyperplane (line 4). Subsequently, the solution (S_1) returned by $TabuSearch_1$ is further improved by the $TabuSearch_2$. After that, the algorithm enters a 'while' loop in which $TabuSearch_2$ and the perturbation operator are iteratively performed until the time limit is reached. During this search process, only improving solution is accepted as the current solution. Finally, the best solution (S^*) found during the search process is returned as the result of ITS algorithm.

On the other hand, to enhance further the performance of algorithm, the strength K of the perturbation operator is adaptively adjusted as follows. First, the value of K is set to the minimum value K_{min}, then the value of K is increased by K_{step} if the current solution is not improved by the current $TabuSearch_2$ procedure, and is reset to K_{min} otherwise. In addition, the value of K is reset to K_{min} once its value reaches K_{max}. In this study, the default values of K_{min}, K_{step}, and K_{max} were set to 20, 5, and 40, respectively.

2.3 Tabu Search Methods

The proposed ITS algorithm employs two solution based tabu search methods proposed in the previous study [10]. For the sake of completeness, these two tabu search methods are briefly described in the following. For more details, the interested readers are referred to a pervious paper [10].

The First Tabu Search Method. The first tabu search $TabuSearch_1$ can be viewed as an iterative neighborhood search method whose main components include the tabu strategy to determine tabu status of neighbor solutions, neighborhood structure explored, and the evaluation function to measure the quality of a neighbor solution. Starting from the input solution, the $TabuSearch_1$ performs a number of iterations until the best solution has not been improved during the last α_1 iterations, where α_1 is called the tabu depth of tabu search.

The neighborhood N(S) explored by $TabuSearch_1$ procedure is the union of two basic neighborhoods, i.e., $N_1(S)$ and $N_2(S)$, where $N_1(S)$ is defined by 1-flip operator *Flip*() that consists in changing the value of one variable x_i to its complementary value $1\text{-}x_i$, and can be written as:

$$N_1(S) = \{S \oplus Flip(i) : 1 \leq i \leq n\} \tag{10}$$

The second basic neighborhood $N_2(S)$ is defined by the $Swap(u, v)$ operator. Given two items $u \in I^1$ and $v \in I^0$, the $Swap(u, v)$ operation exchanges the locations of items u and v, where I^1 denotes the set of the selected items and I^0 denotes the set of unselected items. Then, $N_2(S)$ is composed of all possible neighbors that can be obtained by applying *Swap* move to the current solution S and can be expressed as follows:

$$N_2(S) = \{S \oplus Swap(u, v) : u \in I^1, v \in I^0\} \tag{11}$$

Evaluation function employed by the $TabuSearch_1$ procedure is the weighted sum of the objective function of MDMKP and a penalty function $P(S)$ which measures the degree of constraint violation:

$$F(S) = \sum_{j=1}^{n} c_j x_j - \beta \times P(S) \tag{12}$$

where the parameter β is set to 10^2 in this work, and the penalty function $P(S)$ can be written as:

$$P(S) = \sum_{i=1}^{m} Max\left\{0, \sum\nolimits_{j=1}^{n} a_{ij}x_j - b_i\right\} + \sum_{i=q+1}^{q+m} Max\left\{0, b_i - \sum\nolimits_{j=1}^{n} a_{ij}x_j\right\} \tag{13}$$

In addition, to determine the tabu status of neighbor solutions, the $TabuSearch_1$ procedure employs a solution based tabu strategy that uses three hash vectors and three hash functions. For more details, the interested readers are referred to a previous paper [10].

The Second Tabu Search Method. The second tabu search algorithm $TabuSearch_2$ works on a fixed hyperplane $\Omega_{[t]}$ obtained by the first tabu search $TabuSearch_1$, which can be written as:

$$\Omega_{[t]} = \{x \in \{0,1\}^n \bigg| \sum_{i=1}^{n} x_i = t\} \tag{14}$$

In order to speed up the neighborhood search, the $TabuSearch_2$ procedure adopts a reduced swap neighborhood that can be written as follows:

$$N_3(S) = \{S \oplus Swap(u,v) : u \in I^1, v \in I^0; f(S \oplus Swap(u,v)) > f(S^*)\} \tag{15}$$

where S denotes the current solution, S^* denotes the best feasible solution found so far, and I^0 is the set of unselected items in S and I^1 is the set of selected items. On the other hand, the $TabuSearch_2$ procedure uses the same evaluation function and tabu strategy as in $TabuSearch_1$ procedure to evaluate the quality of the neighbor solutions and determine the tabu status of neighbor solutions, respectively.

2.4 Perturbation Operator

To enhance its diversification ability, our ITS algorithm employs a perturbation operator to jump out of local optimum traps reached by the local search procedure. The perturbation operator with a strength K is composed of K consecutively performed *Swap* operations, where each *Swap* operation exchanges the positions of two items randomly selected from I^0 and I^1, respectively, where I^0 is the set of unselected items in the current solution S and I^1 is the set of selected items. The pseudo-code of the perturbation operator is given in Fig. 2.

```
Algorithm 2: Perturbation Operator
1 Function Perturb(S,K)
  Input: Input solution S, strength of perturbation K
  Output: The perturbed solution S
2 for i ← 1 to K do
3 |  Randomly select an element v from the set I^1
4 |  Randomly select an element u from the set I^0
5 |  S ← S ⊕ Swap(v, u)
6 end
7 return S
```

Fig. 2. The pseudo-code of the perturbation operator.

3 Computational Experiments and Comparison

In this section, we assess the performance of the proposed ITS algorithm by carrying out extensive computational experiments on a number of benchmark instances widely used in the literature and by making a comparison with the state-of-the art algorithm in the literature.

3.1 Benchmark Instance

To assess the performance of our ITS algorithm, we used in this work 54 benchmark instances from the OR-library (http://people.brunel.ac.uk/~mastjjb/jeb/orlib/mdmkp-info.html) as our test bed. These instances can be divided into two sets. The first set is composed of 24 instances with $n = 250$, $m = 10$, and $q = 5$ or 10. The second set is composed of 30 large scale instances with $n = 500$, $m = 30$, and $q = 30$.

3.2 Parameter Settings and Experimental Protocol

Our ITS algorithm adopts several parameters whose settings and descriptions can be given as follows. The tabu depth α_1 of the first tabu search method (i.e., $TabuSearch_1$) was set to 10^4 and the tabu depth α_2 for the second tabu search method (i.e., $TabuSearch_2$) was set to 5×10^4. The minimum perturbation strength $K_{\min}$ was set to 20, the maximum perturbation strength $K_{\max}$ was set to 40, the incremental step size K_{step} of perturbation strength was set to 5. The timeout limits t_{max} of our ITS algorithm were set according to the sizes of instances. The values of t_{max} were set to 300 and 1000 s for the instances with n = 250 and n = 500, respectively.

In addition, our ITS algorithm was implemented in C ++ and complied by g++ complier with –O3 option. All the computational experiments were carried out on a computer with an Intel E5-2670 processor and 2 GB RAM, running Linux operating system. Due to the stochastic feature of algorithms, the ITS algorithm and the reference algorithm (i.e., the TSTS algorithm) both were run 30 runs for each tested instance.

3.3 Computational Results and Comparison

The first experiment aims to assess the performance of the ITS algorithm on the small instances with $n = 250$, and the computational results are summarized in Table 1. The first two column of table give the name of instances and the best known objective value reported in the literature. Columns 3-4 of the table give the best objective values (f_{best}) obtained over 30 independent runs respectively for the ITS algorithm and the TSTS algorithm that is the best performing algorithm in the literature, columns 5-6 give the average objective value (f_{avg}) for two compared algorithms, columns 7-8 give the worst objective values (f_{worst}), and the last two columns give the standard deviation (σ) of objective values obtained over 30 runs. The better results between two compared algorithms are indicated in bold, the worse results are indicated in italic, and the improved results are marked by '*'. The rows '#Better', '#Equal', '#Worse' of the table show the number of instances for which the associated algorithm obtained a better, equal, and worse result compared to other algorithm. The last row '#Improve' shows the number of instances for which the associated algorithm improves the best known result reported in the literature.

Table 1. The experimental results and comparison on the small instances with n = 250, m = 10, q <=10.

Instance	Best known	f_{best}		f_{avg}		f_{worst}		σ	
		ITS	TSTS	ITS	TSTS	ITS	TSTS	ITS	TSTS
250.10.5.0.0	56306	**56308***	*56255*	**56132.83**	*56092.67*	**55871**	*55859*	**84.76**	*114.32*
250.10.5.0.1	59619	59619	59619	**59590.90**	*59575.67*	**59506**	*59392*	**25.01**	*44.64*
250.10.5.0.2	54912	**54952***	*54950**	**54771.60**	*54765.97*	*54422*	**54633**	*103.89*	**85.14**
250.10.5.0.3	52399	*52331*	**52345**	**52223.33**	*52204.57*	*51926*	**52054**	*87.93*	**69.84**
250.10.5.0.4	58234	*58175*	**58234**	*57927.07*	**57999.00**	*57584*	**57610**	*191.42*	**142.46**
250.10.5.0.5	99682	*99616*	**99637**	*99398.17*	**99414.57**	**99074**	*98991*	**139.42**	*154.75*
250.10.5.1.0	26976	26976	26976	*26913.40*	**26933.10**	*26845*	**26872**	*32.33*	**27.63**
250.10.5.1.1	26658	26684*	26684*	**26589.67**	*26561.07*	**26492**	*26447*	**53.94**	*54.18*
250.10.5.1.2	25749	*25749*	**25791***	**25668.07**	*25655.23*	*25478*	**25532**	*74.52*	**57.69**
250.10.5.1.3	27181	**27181**	*27161*	**27127.50**	*27120.80*	**27071**	*27047*	**29.51**	*35.34*
250.10.5.1.4	26856	26856	26856	*26775.57*	**26786.67**	**26711**	*26692*	**54.33**	*56.84*
250.10.5.1.5	46244	46244	46244	**46170.40**	*46166.17*	**46126**	*46075*	**26.86**	*27.96*
250.10.10.0.0	52441	52442*	52442*	*52385.40*	**52392.47**	*52130*	**52171**	*52.53*	**50.34**
250.10.10.0.1	53745	53745	53745	**53685.07**	*53684.50*	*53631*	**53633**	**25.23**	*28.51*
250.10.10.0.2	46927	*46879*	**46927**	**46830.90**	*46799.60*	**46746**	*46684*	**33.69**	*57.81*
250.10.10.0.3	54856	*54816*	**54856**	*54774.93*	**54792.07**	**54751**	*54684*	**16.69**	*38.48*
250.10.10.0.4	49675	*49675*	**49699***	**49635.17**	*49628.00*	**49555**	*49349*	**28.09**	*60.33*
250.10.10.0.5	92989	*92989*	**93060***	**92936.47**	*92930.30*	**92788**	*92762*	*49.52*	**48.19**
250.10.10.1.0	26696	26696	26696	**26684.90**	*26681.53*	*26517*	**26606**	*34.38*	**28.17**
250.10.10.1.1	25893	*25876*	**25893**	**25815.87**	*25798.77*	**25763**	*25521*	**15.30**	*60.93*
250.10.10.1.2	26517	26517	26517	*26484.83*	**26500.33**	26438	26438	*36.34*	**29.16**

(*continued*)

Table 1. (*continued*)

Instance	Best known	f_{best}		f_{avg}		f_{worst}		σ	
		ITS	TSTS	ITS	TSTS	ITS	TSTS	ITS	TSTS
250.10.10.1.3	26684	26684	26684	**26652.73**	*26643.33*	**26513**	*26507*	**42.31**	*48.15*
250.10.10.1.4	26676	*26631*	**26676**	*26617.53*	**26630.10**	26592	26592	**12.60**	*18.16*
250.10.10.1.5	42629	42629	42629	**42534.57**	*42510.70*	**42299**	*42238*	**69.76**	*86.70*
#Better		3		16		14		15	
#Equal		11		0		2		0	
#Worse		10		8		8		9	
#Improve		4	6						

Table 1 shows that the ITS algorithm and the TSTS algorithm have a similar performance on the small instances with $n = 250$. Specifically, in terms of f_{best}, the ITS algorithm obtained a better, equal, and worse result for 3, 11 and 10 instances, respectively, compared to the TSTS algorithm. In terms of f_{avg}, the ITS algorithm obtained a better result on 16 instances and a worse result on the remaining 8 instances. In terms of f_{worst}, the ITS algorithm obtained a better, equal, and worse result for 14, 2, and 8 instances. When regarding the standard deviation (σ) of objective values obtained, one finds that ITS algorithm obtained a better result on 15 out of 24 instances, which means that the proposed ITS algorithm is more robust than the TSTS algorithm.

The second experiment aims to assess the ITS algorithm on the large instances with $n = 500$, and the experimental results are summarized in Table 2, where the same notations are used as in Table 1.

Table 2 discloses that the proposed ITS algorithm outperforms significantly the TSTS algorithm from the literature on the large instances with $n = 500$. Specifically, the ITS algorithm obtained a better result for 21 out of 30 instances than the TSTS algorithm in terms of f_{best}. In terms of f_{avg}, the ITS algorithm obtained a better and worse result for 25 and 5 instances, respectively. In addition, the ITS algorithm and the TSTS algorithm obtained a similar result in terms of both f_{worst} and the standard deviation (σ) of objective values. On the other hand, one observes that the proposed ITS algorithm improved the best known results for 23 out of 30 instances, which means a strong search ability of algorithm.

In summary, these experiments show that the proposed ITS algorithm is very competitive compared to the state-of-the-art algorithm in the literature especially for the large scale instances with $n = 500$.

Table 2. The experimental results and comparison on the large scale instances with n = 500, m = 30, q = 30.

Instance	Best known	f_{best}		f_{avg}		f_{worst}		σ	
		ITS	TSTS	ITS	TSTS	ITS	TSTS	ITS	TSTS
500.30.30.0.2.1	85188	85496*	85294*	85134.53	85085.60	84782	84718	151.35	116.77
500.30.30.0.2.2	82073	82226*	82178*	82017.50	82013.87	81606	81556	144.80	149.82
500.30.30.0.2.3	77393	77343	77212	77038.00	77031.00	76823	76827	126.98	95.29
500.30.30.0.2.4	82304	82420*	82398*	82206.33	82168.23	81991	81985	95.25	124.26
500.30.30.0.2.5	83525	83694*	83634*	83408.03	83342.57	83037	82990	142.94	154.72
500.30.30.0.2.6	145967	145877	145854	145703.57	145682.33	145531	145518	67.12	79.69
500.30.30.0.2.7	152246	152267*	152126	152045.67	152009.67	151829	151894	96.23	59.42
500.30.30.0.2.8	157687	157794*	157720*	157521.87	157451.13	157100	157137	124.72	126.40
500.30.30.0.2.9	153751	153912*	153903*	153674.27	153642.33	153348	153405	113.33	106.09
500.30.30.0.2.10	142173	142208*	142224*	142051.07	142010.27	141839	141817	95.84	96.39
500.30.30.0.2.11	185226	185329*	185137	185013.47	184992.93	184812	184822	99.37	73.62
500.30.30.0.2.12	194614	194660*	194566	194468.50	194449.40	194357	194334	76.60	58.86
500.30.30.0.2.13	208246	208275*	208275*	208139.43	208131.37	207992	207870	66.23	88.09
500.30.30.0.2.14	215849	215837	215817	215684.03	215681.37	215489	215500	94.02	76.71
500.30.30.0.2.15	194224	194311*	194215	194139.80	194115.07	194028	194020	62.90	50.39
500.30.30.1.5.1	51666	51689*	51770*	51601.63	51581.10	51497	51465	48.10	70.10
500.30.30.1.5.2	50101	50256*	50202*	49950.37	49922.23	49696	49693	104.75	112.74
500.30.30.1.5.3	51226	51178	51127	50994.70	50929.97	50734	50675	84.54	107.44
500.30.30.1.5.4	51637	51753*	51660*	51480.03	51452.50	51312	51318	92.34	75.72
500.30.30.1.5.5	52078	52128*	52016	51886.33	51892.00	51611	51804	111.99	66.71
500.30.30.1.5.6	84052	84070*	84079*	83877.57	83868.93	83683	83690	88.02	91.50
500.30.30.1.5.7	82850	82952*	82819	82702.17	82642.60	82473	82406	104.49	105.31
500.30.30.1.5.8	82722	82762*	82843*	82557.63	82609.27	82191	82460	107.47	98.38
500.30.30.1.5.9	82825	82712	82681	82538.23	82502.10	82283	82297	104.19	90.53
500.30.30.1.5.10	82845	82860*	82827	82622.80	82599.43	82305	82203	123.05	140.50
500.30.30.1.5.11	88887	88883	88855	88788.97	88730.50	88628	88571	53.40	73.26
500.30.30.1.5.12	87254	87304*	87365*	87173.53	87194.77	87066	87104	55.79	53.04
500.30.30.1.5.13	87315	87253	87315	87178.83	87189.80	87043	87078	51.15	60.45
500.30.30.1.5.14	87583	87655*	87711*	87542.90	87503.43	87302	87326	74.36	86.03
500.30.30.1.5.15	87956	87965*	88040*	87845.30	87865.07	87679	87721	57.03	58.97
#Better		21		25		15		17	
#Equal		1		0		0		0	
#Worse		8		5		15		13	
#Improve		23	16						

4 Conclusions

In this work, we propose an iterated tabu search algorithm (ITS) for the NP-hard multidemand multidimensional knapsack problem (MDMKP) with many real-world applications. The proposed ITS algorithm integrates two solution-based tabu search procedures and a perturbation operator to reach a suitable balance between the intensification and diversification of search. Extensive experiments were carried out to assess the performance of proposed ITS algorithm. Computational results on 54 benchmark instances showed that the proposed algorithm is competitive compared to the state-of-art results in the literature especially for the large scale instances.

This study shows that the perturbation operator is useful to jump out of local optimum traps reached by the solution based tabu search method for solving the large scale MDMKP instances. In addition, the idea of the proposed ITS algorithm is very general, and it is very interesting to check its effectiveness on other binary optimization problems.

Acknowledgements. This work was partially supported by the Natural Science Foundation of Jiangsu Province of China (Grant No. BK20170904), the National Natural Science Foundation of China (Grant No. 61703213), six talent peaks project in Jiangsu Province (Grant No. RJFW-011), and NUPTSF (Grant Nos. NY217154 and RK043YZZ18004).

References

1. Chu, P.C., Beasley, J.E.: A genetic algorithm for the multidimensional knapsack problem. J. Heuristics **4**, 63–86 (1998)
2. Cappanera, P., Trubian, M.: A local-search-based heuristic for the demand-constrained multidimensional knapsack problem. INFORMS J. Comput. **17**(1), 82–98 (2005)
3. Cappanera, P., Gallo, G., Maffioli, F.: Discrete facility location and routing of obnoxious activities. Discrete Appl. Math. **33**(1–3), 3–28 (2003)
4. Plastria, F.: Static competitive facility location: An overview of optimisation approaches. Eur. J. Oper. Res. **129**(3), 461–470 (2001)
5. Romero-Morales, D., Carrizosa, E., Conde, E.: Semi-obnoxious location models: a global optimization approach. Eur. J. Oper. Res. **102**(2), 295–301 (1997)
6. Beaujon, G.J., Marin, S.P., Mcdonald, G.C.: Balancing and optimizing a portfolio of R&D projects. Naval Res. Logistics (NRL) **48**(1), 18–40 (2001)
7. Arntzen, H., Hvattum, L.M., LKketangen, A.: Adaptive memory search for multidemand multidimensional knapsack problems. Comput. Oper. Res. **33**(9), 2508–2525 (2006)
8. Gortázar, F., Duarte, A., Laguna, M., Martí, R.: Black box scatter search for general classes of binary optimization problems. Comput. Oper. Res. **37**, 1977–1986 (2010)
9. Hvattum, L.M., Arntzen, H., Løkketangen, A., Glover, F.: Alternating control tree search for knapsack/covering problems. J. Heuristics **16**(3), 239–258 (2010)
10. Lai, X.J., Hao, J.K., Yue, D.: Two-stage solution-based tabu search for the multidemand multidimensional knapsack problem. Eur. J. Oper. Res. **274**(1), 35–48 (2018)

Research on CCE Allocation Algorithm in LTE

Yuechen Yang, Qiutong Li(✉), and Wenjuan Wei

Chengdu Neusoft University, Dujiangyan, Chengdu, Sichuan, China
Li-Qiutong@qq.com

Abstract. In order to solve CCE allocation fairness problem in LTE system, this paper proposes a CCE allocation algorithm to ensure the fairness by limiting the largest UE numbers of the CCE allocated in the uplink and downlink scheduling. At the same time, another new algorithm is proposed to calculate the maximum number of UEs dynamically by balancing the number of users participating in the dispatching and their channel quality. Finally, the simulations results of data flow of single UE and multiple UEs under fixed MCS show that the downlink resource allocation algorithm implemented in this paper successfully limits the number of UEs in the uplink allocation and solves the RB waste problem caused by the unfair allocation of CCE.

Keywords: LTE · CCE · RB

1 Background

CCE is the wireless resource which is used to carry down the control information-DCI, the resource occupied by PDCCH channel. The data transmitted in PDCCH channel is DCI, the wireless resource occupied by the data called CCE, and the minimum resource occupied by data is one CCE. Specific to each user, the number of CCE required for a DCI depends on its channel quality. The better the channel quality is, the smaller the number of CCE required. In the LTE physical layer protocol, it is specified that how to solve PDCCH blindly for each UE, so CCE allocation must be carried out according to this process when eNodeB allocates CCE resources. CCE allocation algorithm is very important. After introducing the CCE allocation process of eNodeB, this paper proposes a CCE allocation algorithm to solve the existing problems, and finally verifies the feasibility of the algorithm through simulation.

2 CCE Distribution

2.1 CCE Allocation Process

The process of receiving DCI by UE is dictated in the protocol, so the sender of DCI information must also follow this process to transmit DCI information. Otherwise the sender and UE cannot reach an agreement, which will lead to the failure of the communication.

Therefore, the process of assigning CCE for eNodeB must satisfy each process in the above section. While assigning CCE, the starting position of CCE should be

K. Li et al. (Eds.): ISICA 2019, CCIS 1205, pp. 551–559, 2020.
https://doi.org/10.1007/978-981-15-5577-0_43

calculated according to the C_RNTI of the UE and the current subframe number. Then eNodeB determines the amount of CCE required to transmit the DCI information according to the quality of downlink channel.

In accordance with the initial position and quantity relationship, eNodeB queries whether there is CCE can meets the requirements in the search space to be allocated to UE in the current cell. If yes, the allocation is successful, and it needs to be used at the mark position in the variable that records the CCE allocation in the current cell so that other UEs or DCI can no longer occupy the physical resources of this CCE. Also these CCE need to be combined with the DCI of that UE, otherwise the allocation fails. Since the schedule interval TTI in LTE system is 1 ms, the valid time of CCE is only 1 ms, that is to say, in each schedule, the CCE allocation will be proceed after the occupied flag bit of CCE being initialized to 0.

2.2 Existing Problems

CCE is a shared resource in uplink and downlink scheduling as well as a kind of shortage physical resources in LTE system. The allocation of CCE resources in scheduling includes both uplink scheduling and downlink scheduling, which will lead to the following problem: fairness of CCE allocation.

CCE allocation fairness problem is not the fairness between UE, but the fairness of DCI number of equity in a TTI hosted in the uplink and downlink scheduler. If there is no restriction on the allocation of CCE resources in the uplink and downlink scheduling, CCE resources will be exhausted after the uplink scheduling cause it runs very fast and needs to occupy a lot of CCE resources. When applying for CCE resources in downlink scheduling, the application fails or only a few UE can be successfully applied. Even RB in the PDSCH channel is not allocated when resources are allocated in downlink scheduling, so RB can only be wasted in current TTI because there is no CCE resource to carry its DCI information. In turn, the idle RB in PUSCH can only be wasted in TTI at the moment. In order to solve this problem, we can calculate in advance that how much UEs can be allocated of CCE resources for current TTI in both uplinks and downlink before dispatching. When the number of UE that have been allocated CCE resources of uplink is exceed the number we calculated, allocating process stops. The same is true for downlink, so that CCE resources can be used fairly by uplink and downlink scheduling. The waste of RB problem on the uplink or downlink will not happen due to the unfair of CCE resources.

3 Design to Ensure Fairness of CCE Allocation

The idea of this new algorithm is to count the number of users who need CCE and to explore the channel quality of downlink before allocating CCE in the uplink and downlink scheduling tasks. Then the average channel quality of users that needs to be allocated CCE resources in the uplink and downlink scheduling task should be calculated. According to the average channel quality of all users who need to allocate CCE, the total number of CCE in the current cell can be divided into PDCCH, that is, how much DCI information can be carried. Of course, this is just an average value.

Then, the maximum number of users that can be assigned in both uplink and downlink scheduling tasks at the current scheduling time is calculated according to the requirements of uplink and downlink scheduling tasks, t. For illustration purposes, define the variables in Table 1.

Table 1. Variables and their meanings in the algorithm.

Variable name	Meaning
max_ul	The maximum UE number of CCE can be allocated in uplink scheduling
max_dl	The maximum UE number that can be assigned to CCE in downlink scheduling
dl	The number of UE's that the downlink wants to dispatch
ul	The number of UE's that the uplink wants to dispatch
CCE_NUM	The number of CCE currently available for allocation
cce_all	The number of CCE used at the end of the dispatch
ue_dl	The UE number of CCE successfully allocated by downlink after dispatching is completed
ue_ul	UE number of uplink successfully allocated CCE after dispatching is completed
cqi_dl	Downlink channel quality of UE
cqi_ul	Uplink channel quality of UE
aver_dl	Downlink average channel quality
aver_ul	Uplink average channel quality
aver_all	Total average channel quality
pdcch_num	Under the current situation, CCE can be divided into PDCCH number according to *aver_all* in the cell

Suppose that during the scheduling process there are so many UEs in both the uplink and downlink that need to be allocated CCE resources, *max_ul* and *max_dl* need to be calculated before CCE being dispatched. Then, through its limitation, the relationship between *max_ul* and *aver_ul* in the distribution process is directly proportional. That is, when the uplink channel quality is generally good, *max_ul* will be larger than *max_dl*. Meanwhile, try to improve the utilization rate of CCE: $all\_cce : CCE\_NUM \rightarrow 100\%$

The specific process of the algorithm is as follows:

(1) *cqi_dl* and *cqi_ul* of UEs that need to be scheduled are obtained in the uplink and downlink scheduling, and then calculate *aver_dl* and *aver_ul*.
(2) Calculate *aver_all* according to simple average through *aver_dl* and *aver_ul*, then take *aver_all* as CQI, calculating L through the correspondence between channel qualities and *pdcch_num* is calculated according to formula (1):

$$pdcch\_num = \lceil CCE\_NUM/L \rceil \tag{1}$$

Through Eqs. (2) and (3), *max_ul* and *max_dl* are preliminarily calculated:

$$\max\_ul = \lceil pdcch\_num \times aver\_ul/(aver\_dl + aver\_ul) \rceil \tag{2}$$

$$\max\_dl = pdcch\_num - \max\_ul \tag{3}$$

(4) Adjust *max_ul* and *max_dl*: first, add *ad_ul* to *max_ul*, and *ad_dl* to *dl_dl* (if *ad_ul* and *ad_dl* are too small will lead to low utilization rate of CCE resources, and too large of the two will destroy the proportion relationship between *max_ul* and *max_dl*, generally suppose $ad\_ul = 3$, $ad\_dl = 4$). Then it is adjusted by Eqs. (4) and (5)

$$\max\_ul = (\max\_ul > ul)?ul : \max\_ul \tag{4}$$

$$\max\_dl = (\max\_dl > dl)?dl : \max\_dl \tag{5}$$

It should be noted that if one side executing first is known in the schedule process, we just need to limit the maximum number of users scheduled by that side, and the maximum number of users scheduled by the other side can be set as the number of users participating in the schedule. If not, both *max_dl* and *max_ul* values are obtained by this algorithm.

After these four steps, *max_ul* and *max_dl* can be determined used as the limit value of UE in uplink and downlink respectively when CCE is allocated. Once the number of UEs which has been successfully allocated a CCE in the dispatch is larger than the above parameter, the application fails and will no longer allocate anymore.

Through the calculation of step (3), it will be found that the algorithm actually guaranties a proportional fairness in uplink and downlink. When the downlink channel quality is generally bad, *aver_dl* will be relatively small, thereby *max_ul* will be larger, thus it ensures the CCE resources are reasonable used. It is a waste of CCE that to allocate more CCE to UE in downlink scheduling when downlink channel quality is generally poor. The first formula in step (4) is to add *max_ul* by 4 and *max_dl* by 3 to ensure the utilization of CCE. In reality, UE that is expected to be scheduled may not be scheduled in actual scheduling, such adjustment can improve the number of small CQI values in UE that participate in calculation. This adjustment improves the value of CQI more which is smaller in the UE involved in calculation but the CQI value of UE that actually applies for CCE inversion according to the resource allocation order is relatively large.

Without the limitation of this algorithm, when the downlink scheduling runs first, and there are so many UEs that required to be scheduled in the uplink and down link, CCE will be exhausted in the downlink scheduling, which will result of RB wasted in PUSCH. Because the method limits the maximum UE that it can assign a CCE to a downlink within the TTI, if the restriction of this method is adopted, the fairness of

UEs in the uplink and downlink in the CCE allocation process can be guaranteed, hence the waste of RB caused by the unfairness of CCE allocation can be avoided.

4 Algorithm Simulation Analysis

The eNodeB system in the laboratory simulation environment is shown in Fig. 1.

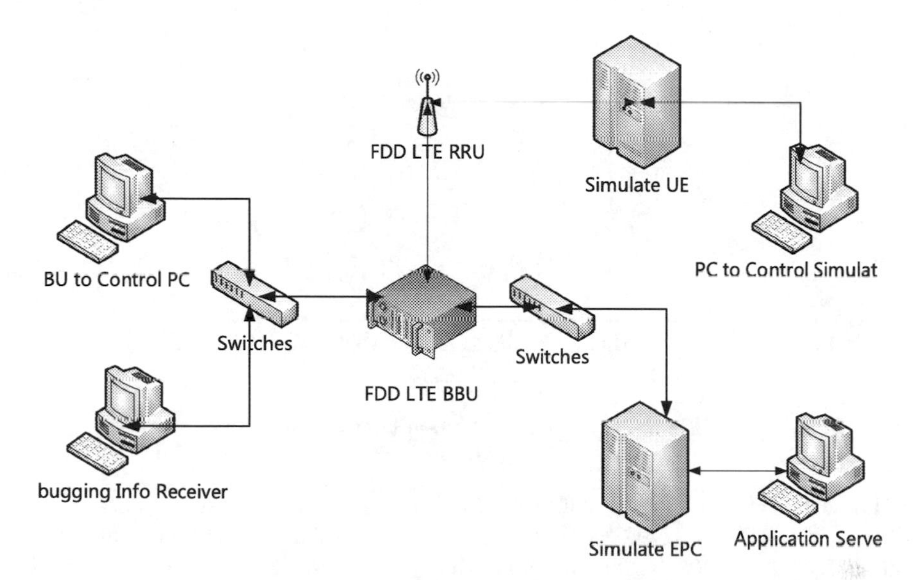

Fig. 1. Test environment schematic diagram

In this test environment, UE and EPC adopt analog equipment, and analog UE and base station adopt wired connection. A signal attenuator is added in the middle to attenuate both up and down signals. BBU and RRU are connected via optical fiber, while S1 link between base station and core network are connected via network cable. A simulated application server is used on the core network side to communicate with UE, sending and receiving data. In the test process, BBU first controls the PC to transmit all the codes of the whole base station to be tested to the hardware board of BBU through the network port, then start the base station to establish the community. After the establishment of the cell, it is connected to the cell by using the simulation UE through random access and attachment process, and then relevant tests can be conducted. The correctness of this paper is illustrated by the downflow of single UE and the downflow of multiple UE respectively.

4.1 Single UE Downlink Flow Rate

After the establishment of the cell, getting access 1 UE, make the simulated UE control PC to power the Jperf server on and to receive data with this port number and test downlink data transmission and downlink resource allocation. During this test the following AMC is closed, static MCS = 9 is adopted as the down-link modulation encoding method. The base station is configured as 1 cell; the up and down link bandwidth are 10 MHz, that is 50 RB; using 2 antennas to transmit; PDCCH occupies 3 OFDM symbols; Ng in the PHICH = 1/2. To see the rate more clearly, we converted the collected data into the rate diagram shown in Fig. 2 by using MATLAB:

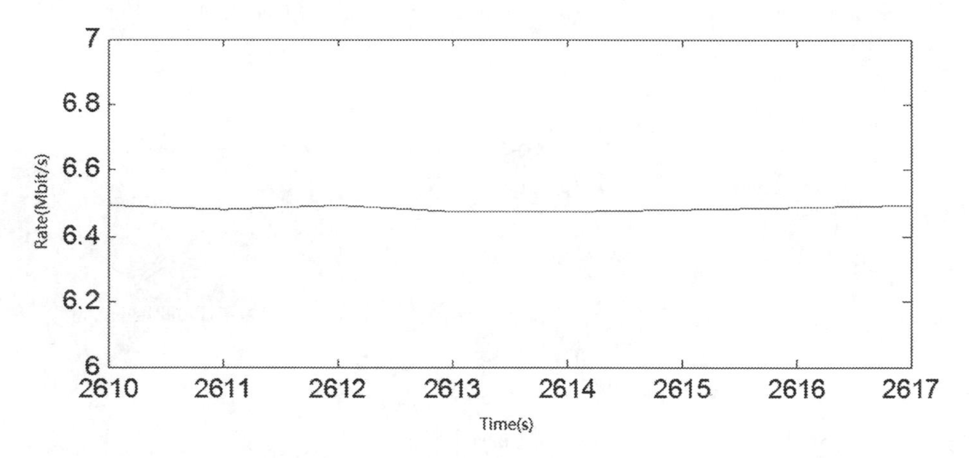

Fig. 2. Single UE application layer rate diagram

It can be seen from Fig. 2 that, when MCS is 9, the amount of data that 50 RB can transmit in one TTI is 7992 bit, so the amount of data that can be transmitted in one second is 7992 * 1000 = 7992000 bit/s. That is, when MCS is 9, the rate of 50 RB single code words rate is (7992000/1024/1024) Mbit/s, which is about 7.6 Mbit/s. In this paper, the algorithm of converting byte number to RB number in the process of number allocation uses the relation of $y = kx$ to calculate, where x represents the number of RB, k is the number of bytes that can be transmitted by one RB under the MCS, and y is the number of bytes that can be transmitted. The comparison and analysis between this paper and LTE protocol are drawn as Fig. 3:

As can be seen from Fig. 3, under the condition that MCS is 9, 7992 bit can be transmitted in the protocol, while only 6800 bit can be transmitted in the pre-allocation of this paper. Therefore, at the peak rate, although 50 RB were assigned to the UE by one TTI during MAC scheduling, only 6800 bit given by RLC was transmitted. It can actually transfer 7992 bits of data, so the remaining 1192 bits are useless data, which reduces the rate of the application layer. At the same time, the traffic also shows the correctness of the downlink resource dynamic allocation realized in this paper.

4.2 Multi UE Downlink Flow Rate

The cell configuration is the same as that in Sect. 4.1 but 4 UEs are connected. As can be seen from Fig. 4, the rate of 4 UEs is around 1.61 Mbit/s, while the application layer under single UE is 6.5 Mbit/s, which means that these 4 UEs share these data transmission opportunities equally.

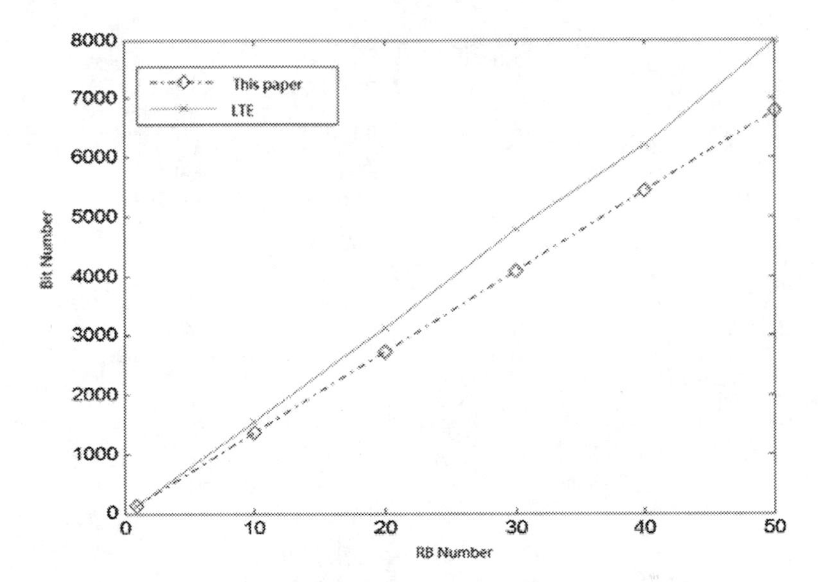

Fig. 3. Rate difference analysis diagram

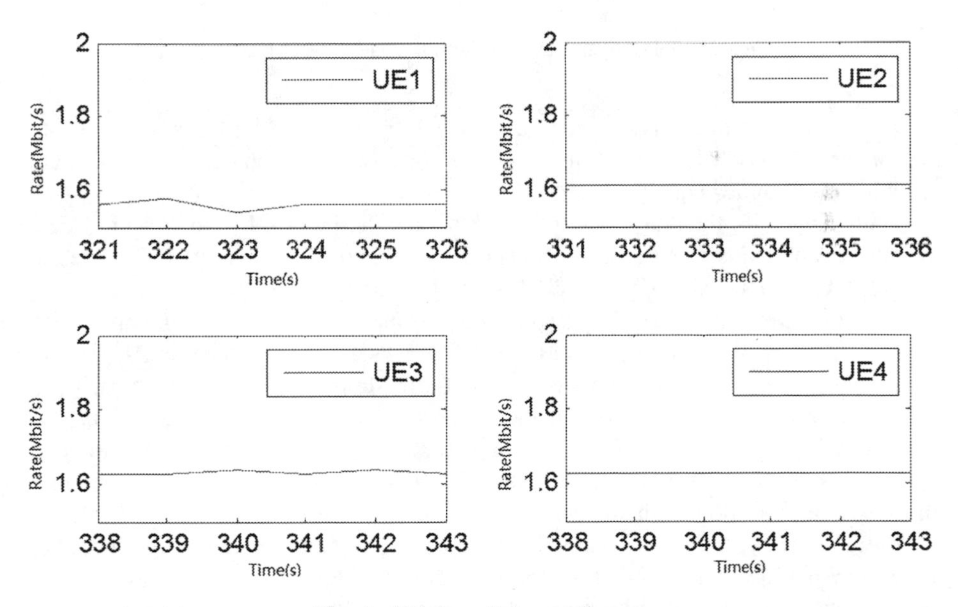

Fig. 4. UE downlink rate diagram

In order to display the rate of each UE more clearly, we draw the rate part with MATLAB as shown in Fig. 5:

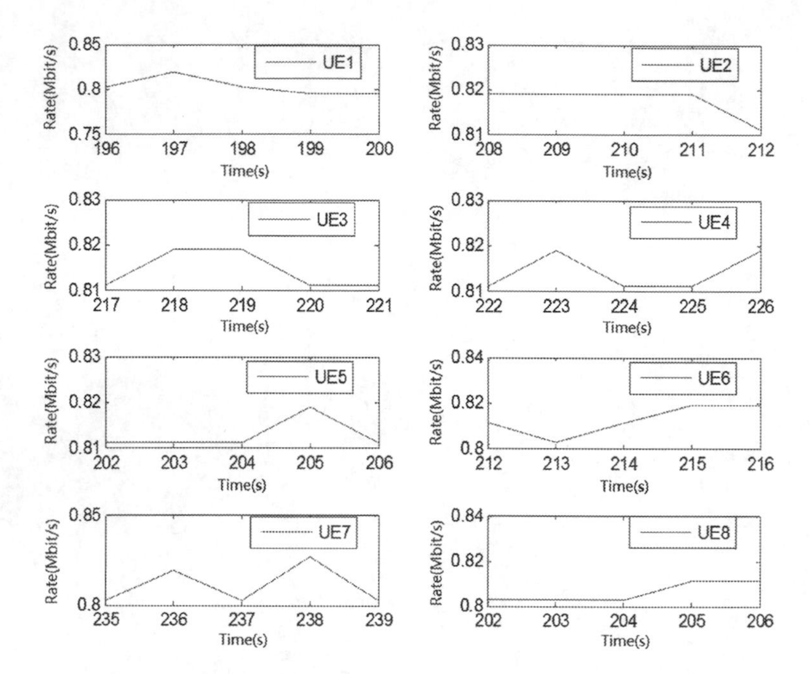

Fig. 5. Eight UE downlink flow rate chart when MCS is 9

As can be seen from Fig. 5, the rate of all UEs is about 0.81 Mbit/s, and the total rate of the whole 8 UEs are about 6.5 Mbit/s, that is, 8 UEs share these data transmission opportunities equally.

With single UE, the rate is about 6.5 Mbit/s and the jitter is 1.87 ms. With 4 UEs, the rate of each UE is about 1.62 Mbit/s and the jitter is 6 to 8 ms. With 8 UEs, the rate of each UE is about 0.81 Mbit/s with a jitter of 13 to 19 ms.

Based on the analysis of the test scenario, in the case of that the maximum data transfer rate of application layer is close to 6.5 Mbit/s case, while there are 4 UEs and each UE's downlink data transmission demand is 2 Mbit/s, the rate at which each user get should be around 1.6 Mbit/s. In the same way, with 8 UEs and downlink data transmission demand for each UE is 1 Mbit/s, the rate at which each user get should be 0.81 Mbit/s. That is to say, the rate results of Figs. 4 and 5 conform to the scheduling and resource allocation of this paper.

From the above analysis, it can be seen that the rate graph obtained by scheduling and resource allocation proposed in this paper conforms to the results achieved by the polling scheduling algorithm and downlink resource allocation in this paper. It also shows that the allocation of CCE and RB in this paper not only supports dynamic allocation of physical resources in single UE but also supports dynamic allocation of physical resources in multiple UEs.

5 Summary

Based on the analysis of the problems in the allocation of CCE resources by scheduling, this paper proposes a method to ensure the fairness of CCE allocation in LTE system, which ensure the balance between the number of UEs which can be allocated CCE resources and the UEs channel quality. It solves the RB waste problem caused by the unfair allocation of CCE. Finally, the system simulation tests of the downlink rate of single UE and multiple UEs verify the correctness and effectiveness of downlink physical resource allocation realized in this paper.

References

1. GPP TSG RAN WG1 Meeting #62bis R1-105115, Xian, China, 11th–15th October 2010
2. GPP TSG RAN WG1 Meeting #63 R1-105825, Jacksonville, USA, 15th–19th November 2010
3. GPP TSG RAN WG1 Meeting #63bis R1-110000, Dublin, Ireland, 17th-21th January 2011
4. Shi, S., Feng, C., Guo, C.: A resource scheduling algorithm based on user grouping for LTE-advanced system with carrier aggregation. Computer Network Multimedia Technology, CNMT2009, Wuhan, 18–20 January 2009
5. Yuan, G., Zhang, X., Wang, W., Yang, Y.: Carrier aggregation for LTE-advanced mobile communications systems. IEEE Commun. Mag. **48**, 88–93 (2010)
6. Huang, X.Z., Wu, H.C., Wu, Y.Y.: Novel pilot-free adaptive modulation for wireless OFDM system. IEEE Trans. Veh. Techol. **57**, 3863–3867 (2008)
7. GPP TS 36.321 V8.9.0, Evolved Universal Terrestrial Radio Access (E-UTRA). Medium Access Control (MAC) protocol specification (Release 8), pp. 10–40, June 2010
8. GPP TS 36.322 V8.8.0, Evolved Universal Terrestrial Radio Access (E-UTRA). Radio Link Control (RLC) protocol specification (Release 8), pp. 7–31, June 2010

Research on Tobacco Silk Making Scheduling Based on Improved DE

Qi Ji(✉), Wei Wang, Mingmeng Meng, Chengliang Yang, and Zhongmin Zhang

Zhejiang China Tobacco Industry Co., Ltd, Hangzhou, China
ntwzl@ntu.edu.cn

Abstract. An improved differential evolution algorithm was proposed to solve the shop scheduling problem in tobacco industry. Taking a cigarette manufacturing enterprise as the background of research and application, the process path and rules of tobacco exhaust production are analyzed, and a differential evolutionary production system model based on improved DE is constructed. The differential evolution algorithm and constraint theory are combined to optimize shop scheduling. In the process of optimization, production scheduling is firstly made to meet the bottleneck resource rules, and then, under this condition, the second production scheduling is optimized. By limiting the search space, the convergence of the algorithm is accelerated. Finally, the results of numerical experiments show that the proposed algorithm is feasible in tobacco production.

Keywords: Tobacco manufacturing · Differential evolution algorithm · Shop scheduling

1 The Introduction

China's tobacco industry is a major contributor to the country's tax revenue, which accounts for about 10% of the country's total annual revenue and can support the armed forces. Tobacco industry in China is an industry with great influence of policy, and the state implements the operation system of unified leadership and vertical management. However, due to the development of market economy and the urgent need of internationalization of domestic market, the competition among industries is increasingly fierce. At the same time, with the improvement of manufacturing automation information, tobacco manufacturing equipment information extraction and control decisions have reached a very high level, thus providing the necessary information foundation for enterprise organization and management. Therefore, the implementation of information production organization and management and optimization of decision-making, improve the production efficiency of tobacco enterprises has become an urgent need of the tobacco industry.

Tobacco shop scheduling is the problem of arranging specific process sequence on specific equipment, and shop scheduling optimization is how to rationalize the execution sequence, so as to achieve an optimal performance index (such as the highest machine utilization rate, the shortest processing cycle, etc.). It is extremely important to optimize shop scheduling in manufacturing enterprises, especially in small batch

K. Li et al. (Eds.): ISICA 2019, CCIS 1205, pp. 560–571, 2020.
https://doi.org/10.1007/978-981-15-5577-0_44

production workshops. Due to the variety of products and uncertainty of tasks, manual shop scheduling is very difficult, and an intelligent auxiliary shop scheduling software is urgently needed. Shop scheduling optimization problem has been proved to be NP Hard problem mathematically [1], that is, there is no polynomial solution method to obtain the optimal solution. In recent decades, a lot of achievements have been made in the research on this problem at home and abroad, and many algorithms have been developed, such as genetic algorithm, heuristic algorithm, branch and bound method, integer programming method, Lagrange relaxation method and so on [2]. Among them, genetic algorithm has attracted the attention of many scholars with its fast global search ability [3–6]. However, most literature studies on genetic scheduling algorithm are focused on scattered parts, and few studies on scheduling with assembly relationship. In particular, there are very few studies on tobacco shop scheduling in this particular industry. Therefore, this paper proposes an improved differential evolution algorithm based on shop scheduling to solve the problem of shop scheduling optimization. The experimental results show that the optimization effect is obvious.

2 Algorithm and Application of Tobacco Shop Scheduling

There are basically four types of production scheduling models in the manufacturing industry:

1. Flow type, mainly solve the problem of order optimization
2. Discrete, mainly to solve the problem of multi - process, multi - resource optimization
3. Process and discrete mix, mainly to solve the order and scheduling optimization
4. Project management, focus on key chain and cost and time minimization

For tobacoo shop scheduling is largely scattered type model, the main solving process logic and resource capacity constraints, materials and processes, optimization rule choice, calculate the earliest start time, the latest material redistribution and alternative, redistribution and alternative resources, flexible scheduling, cost constraints, many kinds of scheme optimization, to meet the demand of the customers accordingly.

As there are many potential rules in the production process of tobacco industry, so far, the automatic shop scheduling software of tobacco industry enterprises is only implemented by some simple shop scheduling algorithms. The following is a brief summary of the existing shop scheduling algorithms in tobacco industry.

2.1 Theory of Constraints

TOC (theory of constraints) is an Israeli physicist and business management consultant named OPT (optimized production technology), which proposes standardized methods for the definition and elimination of some constraints in manufacturing production and operation activities to support continuous improvement [8]. The essence of TOC is to identify the source of bottle neck of the system and make full use of the bottleneck resources of the system, reduce the restrictive effect of bottleneck process on production, and arrange the resource allocation of non-bottleneck process, so that it can keep pace with the productivity of bottleneck process and reduce wip to the lowest level. Therefore, on the basis of the master production plan, the production process is studied to determine the bottleneck factors, and then the operation plan based on the key processes is prepared to achieve the optimal production.

2.2 Branch and Bound Method

In essence, branch and bound method is a kind of enumeration method, because it can be a strong lower bound for partial path calculation, it has become a more effective method to solve small-scale workshop scheduling problem. But for large scale scheduling problem, branch and bound method needs more time.

2.3 Lagrange Relaxation Algorithm

Lagrange relaxation algorithm can obtain the suboptimal solution of complex programming problems well, and the quantitative evaluation of the suboptimal solution has become one of the important methods to solve complex production scheduling problems. Compared with branch and bound method, Lagrange relaxation algorithm takes longer time.

2.4 Dynamic Programming

Dynamic programming is an effective algorithm for solving small-scale production scheduling problems, but its computational complexity increases exponentially with the expansion of the problem scale, so it is not suitable for solving large-scale production scheduling problems.

2.5 Computational Intelligence Method

Computational intelligence method refers to the algorithm inspired by natural phenomena and intelligent methods. This algorithm starts from the initial scheduling and updates the solution in each iteration with little interference (which is provided by heuristic rules), so it is also a kind of heuristic method. Computational intelligence method is mainly divided into local search intelligence algorithm and swarm intelligence algorithm. The former mainly includes some serial algorithms, such as basic local search algorithm, tabu search algorithm and simulated annealing algorithm. The latter mainly includes some parallel algorithms, such as genetic algorithm, ant colony algorithm and particle swarm optimization.

In this paper, the constraint rules in tobacco production scheduling are hidden in the algorithm design, and the improved differential evolution algorithm is proposed to study tobacco production scheduling directly.

3 Description of Tobacco Production Process

Tobacco production process is different from general machinery products processing, in the process of tobacco production, because of its special process requirements, the production process there are many constraint rules, some of these constraint rules for must follow the rules, the others to try to meet the rules, in the production scheduling, must be considered.

3.1 Tobacco Processing Process Analysis

The algorithm research in this paper is based on the background of a tobacco production enterprise, in which the production process mainly involves two workshops: silk making workshop and rolling and wrapping workshop. Among them, the silk making workshop is mainly composed of 3 silk making wires, including a stalk wire, a leaf wire of 6000 kg/h, and a test wire of 3000 kg/h. The winding workshop is mainly composed of winding receiving machine, packing machine, packing machine and other equipment. The speed and number of winding receiving machine determine the output size of the production line. The silk making workshop and the rolling locomotive are connected by the cloth dispenser and the wire feeding machine. The actual processing process in tobacco enterprises is quite complex. In order to facilitate research, this paper makes a simplified modeling description of the tobacco production process. The specific processing process is shown in Fig. 1.

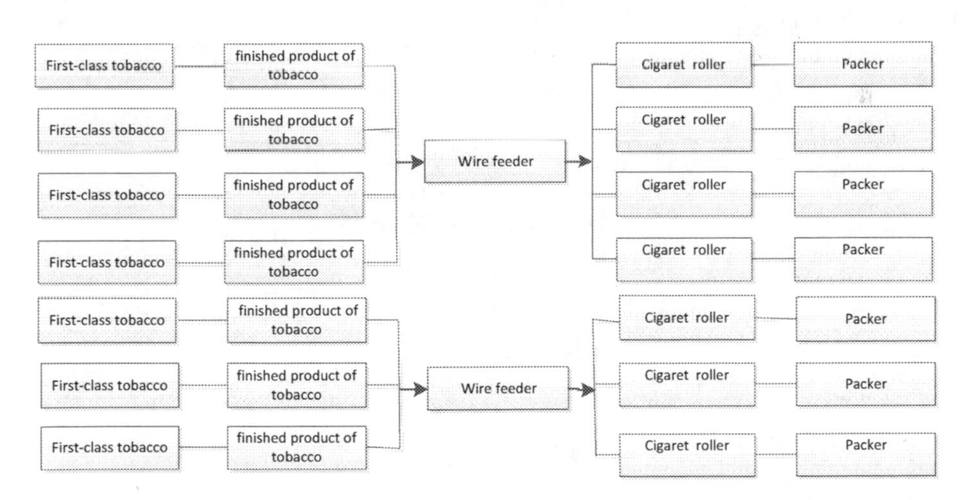

Fig. 1. Tobacco processing

Through the analysis of the tobacco processing process, it is not difficult to find that the tobacco manufacturing process belongs to the mixed pipeline processing mode. When each batch of brand tobacco is produced, the process path is determined unchanged, that is to say, the process sequence in the processing process is not interchangeable. At the same time, when the same process is processed, there is the choice and utilization of multiple devices, which constitutes the mixed pipeline scheduling model. However, in the process of tobacco processing, unlike ordinary mixed pipeline workshop scheduling, equipment can be arbitrarily selected in the process of tobacco production. In the process of tobacco production, there are fixed links between some equipment of different processes, that is, the discharge of the front-stage equipment can only be processed on the fixed off-stage equipment. In order to simplify the model, we can directly integrate and plan the equipment or process with fixed links as one process or one equipment.

3.2 Parameter Description

When using genetic algorithm for target scheduling, as with other algorithms, many quantitative parameter calculations are required as the basic data of scheduling. No matter what kind of target scheduling, time parameters are essential. The parameters needed by genetic algorithm in tobacco production are briefly introduced below.

1) a: speed of cigarette machine (unit: unit/minute);
2) m: single weight of cigarette (unit: g);
3) b: number of cigarette machines corresponding to each group of cabinets;
4) N: the number of cabinets contained in each group
5) Pi: the amount of silk stored in the no. I cabinet in each cabinet group (unit: Kg, where I is not greater than N);
6) Gi: quantity of tobacco required per cigarette brand;
7) M: silk consumption of cigarette machine;

$$M_{\mathrm{kg/m}} = \frac{m}{100} \times a \tag{1}$$

8) Z: unit time silk consumption of cabinet group

$$\mathrm{Zkg/min} = \mathrm{M} \cdot \mathrm{b} \tag{2}$$

9) Q: cabinet wire storage

$$\mathrm{Q} = \mathrm{p1} + \mathrm{p2} \ldots + \mathrm{pn} \tag{3}$$

10) T: processing time of different cigarette brands on different production lines

$$T_i = \frac{G_i}{Z_j} \tag{4}$$

I stands for cigarette brand and j for production line.

3.3 Introduction to Rules

In the process of tobacco production, there are many potential rules to be followed due to the special technological requirements of tobacco production and the fixed links between tobacco equipment. Here, we will just introduce some basic rules.

1) in the wire storage cabinet, only materials of the same brand of tobacco can be stored at the same time, and the storage time requirements can be met to complete the aging process.
2) in a single point of time, there is a one-to-one relationship between the wire storage cabinet and the wire feeding machine, and a one-to-many relationship between the wire feeding machine and the cigarette machine. That is to say, at the same time, the wire feeding machine can only obtain raw materials from a storage cabinet, and at the same time, it can feed materials to multiple winders on the assembly line.
3) on the same production line, the processed brands should be as high as possible from high to low. Brand from high to low, can be continuous production, on the contrary, production must stop after about 20 min. This is due to the brand from low to high production, in order to ensure quality problems, after the completion of the production of low-brand cigarettes, the equipment must be cleaned to ensure the purity requirements of high-brand tobacco.
4) ensure continuous feeding of the coiling unit. This is a principle that must be followed in tobacco production.

4 Application Research of Improved Differential Evolution Algorithm in Tobacco Shop Scheduling

4.1 Production Scheduling Model Construction

In the whole process of tobacco processing, there is a fixed link relationship among the finished product silk storage cabinet, silk feeding machine and winder. The equipment with fixed link relationship is regarded as a whole. At this time, because all the equipment before and after the link with these equipment are processed in a single pipeline, there is no selectivity problem in the equipment during process processing. Therefore, we regard the whole assembly line processing equipment with the same wire feeding machine as a whole, that is, genetic algorithm production scheduling is carried out based on the production line. For the material treatment, in order to facilitate the research and explanation of the problem, the principle of brand inseparability is implemented here. The whole production scheduling problem can also be treated as 0–1 backpack problem. The final production scheduling model is shown in Fig. 2.

4.2 Production Planning

1) implement the principle of batch indivisibility, that is, cigarettes of the same brand can only be processed continuously on the same assembly line. The algorithm is required to encode in batches and carry out decimal coding.

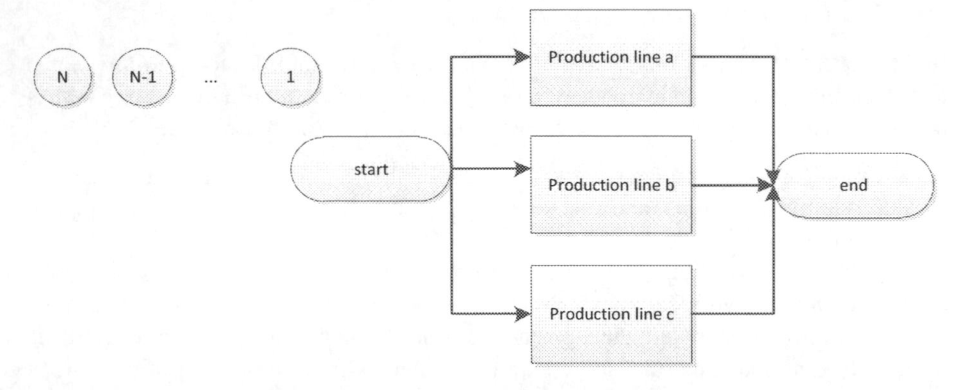

Fig. 2. Tobacco scheduling model

2) it is assumed that the equipment of each assembly line is reasonably configured and scheduled by the production line. The processing time is related to the production capacity of the assembly line and the output of this batch. The processing time of each batch at each assembly line is calculated by the given tobacco parameters. That is, the processing time in the input parameters of the algorithm is calculated.
3) the priority rule of brand from high to low is adopted on the same assembly line. That is, when two or more batches are arranged on an assembly line, the production sequence on this assembly line will be determined according to the priority of the brand, and the brand with high priority will be produced. For genetic algorithm, it is required to embed priority rule when decoding.

4.3 Improved Differential Evolution Algorithm and Practical Operation in Tobacco Emission

4.3.1 Basic Idea of Differential Evolution Algorithm

Suppose that the problem to be solved is a minimization problem, whose mathematical model is min $f(x_1, x_2,\ldots, x_n)$, where $x_j \in [L_j, U_j]$, and $1 \leq j \leq$ n. *X(0) is the initial population, let's say* $X_i(t) = (x_{i1}, x_{i2},\ldots, x_{in})$ is the ith individual in the population of the *t* generation, the individual of the population is the n-dimensional spatial structure, and the population size is NP. Set X_1 And X_2 is two different individuals, and the difference vector formed by them is $X_1 - X_2$ cars only.

The differential evolutionary algorithm is described as follows:

Step 1 randomly generate the initial population:

$$P(0) = \{X_i | x(0)_{ij}(0) = \text{rand}(0,1) * \left(U_j L_{\cdot j}\right) + L_j, \quad 1 \leq I \leq s \cap 1 \leq j \leq n\}, \quad (5)$$

Where the rand(0,1) function is to find the random number on (0,1), and set the initial value of the algebra t = 0;

Step 2 perform mutation operation: randomly select 3 different from X from the current population $P(t)_i(t)$ individual $X_a(t)$, $X_b(t)$, $X_c(t)$, a does not equal b does not equal c. Calculate the difference individual $D_i(t+1) = (d_{i1}(t+1), d_{i2}(t+1), \ldots, d_{in}(t+1))$:

$$d_{ij}(t+1) = \mathrm{x_{aj}(t)} + \mathrm{F} * \left(\mathrm{x}_{bj}(\mathrm{t}) - \mathrm{x}_{cj}(\mathrm{t})\right), \tag{6}$$

$1 \leq \mathrm{I} \leq \mathrm{s}, 1 \leq \mathrm{j} \leq \mathrm{n};$

Step 3 perform crossover operation: randomly generate the random decimal $r_1 \in (0,1)$ and random integer $r_2 \in [1, \mathrm{n}]$, calculate temporary individual E as follow $s_i(\mathrm{t}+1) = \left(\mathrm{e}_{i1}\mathrm{E}(\mathrm{t}+1)_{i2}(t+1), \ldots, \mathrm{e}_{in}(\mathrm{t}+1)\right)$:

$$e_{ij}(t+1) = \begin{cases} d_{ij}(t+1) & r_1 \leq CR \; or \; r_2 = j \\ x_{ij}(t) & \text{otherwise} \end{cases} \quad e_{ij}(\mathrm{t}+1) = \mathrm{d}_{ij}(\mathrm{t}+1), \tag{7}$$

Where $1 \leq \mathrm{I} \leq \mathrm{s}$ and $1 \leq \mathrm{j} \leq \mathrm{n}$;

Step 4 perform the selection operation:

$$X_i(t+1) = \begin{cases} E_i(t+1) & \text{if } f(E_i(t+1)) < f(X_i(t)) \\ X_i(t) & \text{otherwise} \end{cases} \tag{8}$$

Where $1 \leq \mathrm{I} \leq \mathrm{s}$;

Step 5 calculate the individual X with the minimum fitness in population $\mathrm{P(t + 1)}_{best}(\mathrm{t + 1})$;

Step 6: if the termination condition is not satisfied, then t = t + 1 and turn to step 2. Otherwise output X_{best} And f $(\mathrm{X})_{best})$, over.

4.3.2 Improvement of Differential Evolution Caused by Tobacco Emission

One of the characteristics of the differential evolution algorithm is its strong global search ability, for problem scheduling, generally can search for a better solution. However, as the solution size increases, the search space expands extremely, chromosomes useful for building blocks are more dispersed in the solution space, leading to an exponential increase in the cost of solving. In the actual scheduling process of tobacco production, the optimal production plan has certain regularity We find this rule and use it to guide the search direction of the genetic algorithm, which will accelerate the convergence rate of the algorithm.

Because of the variation factor F and interleaving CR Settings of differential evolution algorithm is global search ability and local search ability has a great influence, in the process of the operation of the algorithm, we hope global contraction ability more and more weak, strong local search ability is more and more, so that we can both convergence of the algorithm and avoid premature phenomenon. Obviously setting these two factors to fixed values does not weigh them. According to the analysis of the problem, the relationship of F factor should be reversed with the evolution algebra, that is, with the gradual advance of evolution, it gradually decreases. *CR* factor should be positively correlated with evolutionary algebra, that is, with the gradual advance of evolution, it will increase gradually.

In addition, X in the mutation operator formula (6) of the differential evolution $\mathrm{algorithm}_a(\mathrm{t})$. Random selection or purposeful selection can ensure the diversity of the population to the greatest extent, but it will affect the convergence of the algorithm.

However, if we choose purposefully, for example, we can choose the current optimal individual or one of several excellent individuals, which can accelerate the convergence of the algorithm, but may fall into local optimal. By analyzing the performance of the algorithm, it can be seen that the diversity of the population should be paid attention to in the early stage of the algorithm, and the convergence should be paid more attention to in the later stage of the algorithm.

To solve the IRP problem, the cauchy distribution is adopted in this paper to dynamically adjust F. The wide distribution of cauchy distribution on both wings is suitable for the study of IRP problem. In the process of algorithm evolution, the evolution range of the algorithm can be expanded to avoid falling into local optimal. Dynamic adjustment according to the following formula

$$F = \begin{cases} 1 - \left| \frac{fave - f}{fave - fbest} \right| & if\ f < fave \\ Cauchy(-2, 0.4) & \text{otherwise} \end{cases} \quad (9)$$

F is X_a(t) individual fitness value, f_{ave}Is the average fitness value of the current group, f_{best}Is the fitness value of the optimal individual in the current group. The probability density function of the cauchy distribution is

$$f(x) = \frac{1}{\pi}\left[\frac{0.4}{(x-2)^2 + 0.4}\right] \quad (10)$$

The dynamic adjustment of F is related to the adaptive value, and it is dynamically adjusted according to the adaptive value.

For the dynamic adjustment of crossover factor, this paper adjusts it according to algebra. In the initial stage of evolution, the value of crossover factor is reduced to maintain the diversity of the population. With the gradual advance of evolution, the value of crossover factor is gradually increased to accelerate the convergence speed of the algorithm. The crossover factor is adjusted according to the following formula.

$$CR = \begin{cases} \frac{g}{100-g} + 0.4 & CR < 0.9 \\ 0.9 & \text{otherwise} \end{cases} \quad (11)$$

Where g is the current evolutionary algebra. Let CR value be dynamically adjusted in the set [0.4, 0.9].

For X_a(t) individual selection. This paper adopts the random selection method at the early stage of evolution, the random selection principle of 1% optimal solution at the middle stage of evolution, and the optimal solution principle at the later stage of evolution.

In this paper, the standard difference evolution is improved from three aspects, so as to ensure the diversity of the population and the convergence of the algorithm.

5 Simulation and Analysis

This paper aims to minimize the cost of production for example verification.

5.1 Case Verification

5.1.1 Minimize the Quantitative Description of Production Cost

$$C_c^w = \sum\nolimits_{i=1}^{m} \left(\sum\nolimits_{i=1}^{n} \left(T_{ij} \times \left(Y_{ij \times} F_J^V \right) \right) \right) \tag{12}$$

Among them, C_c^w: production line processing cost; T_{ij}: processing time of brand I on production line j; F_J^V: processing rate on line j; Y_{ij}: brand I is 1 when it is processed on production line j; otherwise, it is 0.

5.1.2 Algorithm Input

In this case, there are altogether 8 brands of cigarettes and 3 processing lines. The tobacco demand of each brand is shown in Table 1 (brand name from high to low):

Table 1. Tobacco requirements for each brand

Brand	1	2	3	4	5	6	7	8
Wire amount	5000	6900	2900	4500	9800	3500	4500	1200

Note: the unit of tobacco quantity is kg

After calculation by formula (1), the processing time of each batch of tobacco brand in each production line is shown in Table 2.

Table 2. Production time of each brand of cigarettes in each production line (Ti unit: m) a: 6000 pieces/m: 0.9 g/piece b1 = 2 b2 = 3 b3 = 4

	1	2	3	4	5	6	7	8
Line a	398	635	278	423	915	316	422	102
Line b	321	427	184	278	609	208	278	73
Line c	241	321	141	212	448	164	213	52

Different production lines have different processing rates. The processing rates of the production line in this example are shown in Table 3.

Table 3. Processing rate of production line (Fjv unit $/m)

	Line a	Line b	Production line c
Rate	3.2	5.3	6.5

5.1.3 Results Presentation

Population size NP was set to 50, and evolution algebra g was set to 100.

After the completion of production scheduling, the gantt chart generated is shown in Fig. 3, and the iteration curve during the optimization process is shown in Fig. 4.

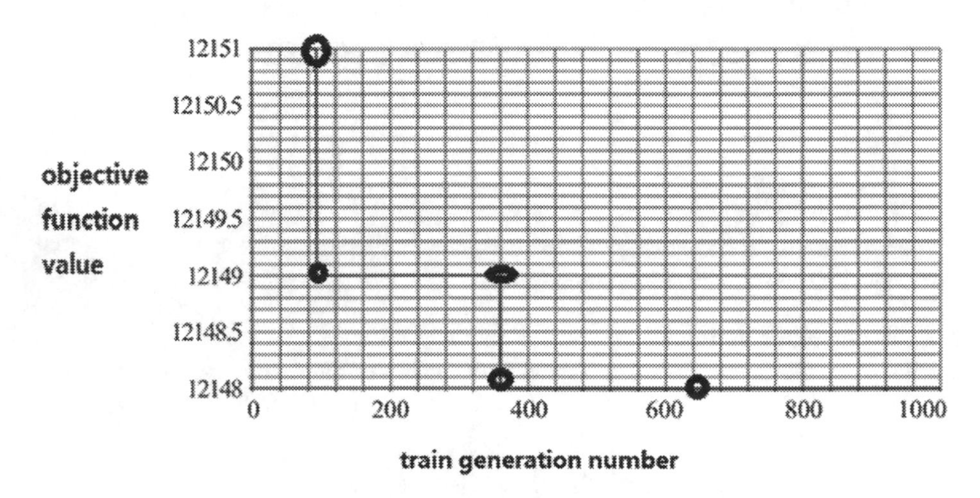

Fig. 3. The gantt chart

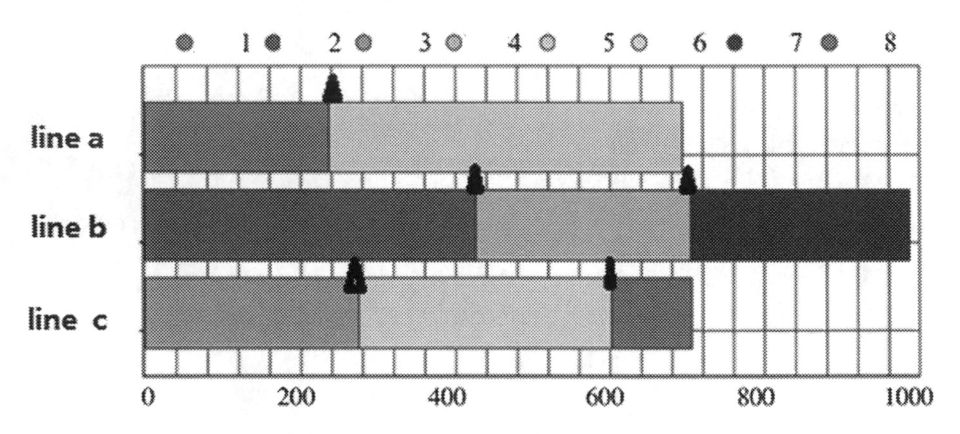

Fig. 4. Production cost iteration curve

5.2 Result Analysis

The above examples prove that the genetic algorithm has good application and good feasibility in tobacco production. However, this is just a simple production scheduling model. In the process of production scheduling, the same cigarette brand applies the principle of batch inseparable, just like the 0–1 knapsack problem. The utilization rate of resources is not very reasonable, and this algorithm needs to be further studied. In any case, this algorithm has laid a theoretical foundation for the targeted production scheduling of tobacco industry.

6 Conclusion

As there are many rules in the process of tobacco production, the production scheduling of the existing tobacco industry is mostly done by hand by planners, and automatic production scheduling also adopts some simple algorithms, such as forward scheduling and reverse scheduling. In this paper, the improved differential evolution algorithm is adopted to study tobacco scheduling and production scheduling, and various rules of production scheduling are automatically implied. Due to its parallel search ability, the differential evolution algorithm can well solve the problem of targeted production scheduling in the tobacco industry, laying a solid foundation for targeted production scheduling in the tobacco industry.

References

1. Watson, K.J., Blackstone, J.H., Gardiner, S.C.: The evolution of a management philosophy: the theory of constrains. J. Oper. Manag. **25**(2), 387–402 (2007)
2. Cox, J.F., Spencer, M.S.: The Constraints Management Handbook. St Lucie, Boca Raton (1997)
3. Zuo, Y., Gu, H.Y., Xi, Y.G.: Modified bottleneck-based heuristic for large-scale job-shop scheduling problems with a single bottleneck. J. Syst. Eng. Electron. **18**(3), 556–565 (2007)
4. Wang, Z.L., Wu, Z.J., Yin, L.: Hign-dimension large-scale packet matching algorithm in IPV6. Acta Eletronica Sinica **11**, 2181–2186 (2013)
5. Sreelaja, N.K., Pai, G.A.V.: Ant colony optimization based approach for efficient packet filtering in Firewall. Appl. Soft Comput. **10**(4), 1222–1236 (2010)
6. Wang, Z.L., Wu, Z.J., Yin, L.: Packet matching using self-adaptive chemical-reaction-inspired metaheuristic for optimization with probability distribution. Comput. Sci. **41**(5), 164–167 (2014)
7. Rafie-Majd, Z., Pasandideh, S.H.R., Naderi, B.: Modelling and solving the integrated inventory-location-routing problem in a multi-period and multi-perishable product supply chain with uncertainty: Lagrangian relaxation algorithm. Comput. Chem. Eng. **109**, 9–22 (2017)
8. Mjirda, A., Jarboui, B., Macedo, R., Hanafi, S., Mladenovic, N.: A two phase variable neighborhood search for the multi-product inventory routing problem. Comput. Oper. Res. **52**, 291–299 (2014)
9. Li, K.P., Chen, B., Sivakumar, A.I., Wu, Y.: An inventory routing problem with the objective of travel time minimization. Eur. J. Oper. Res. **236**, 936–945 (2014)
10. Tang, R.: Decentralizing and coevolving differential evolution for large-scale global optimization problems. Appl. Intell. **4**, 1–16 (2017)
11. Ghasemishabankareh, B., Li, X., Ozlen, M.: Cooperative coevolutionary differential evolution with improved augmented Lagrangian to solve constrained optimisation problems. Inf. Sci. **369**, 441–456 (2016)
12. Salman, A.A., Ahmad, I., Omran, M.G.H.: A metaheuristic algorithm to solve satellite broadcast scheduling problem. Inf. Sci. **322**(C), 72–91 (2015)

Intelligent Control

MODRL/D-AM: Multiobjective Deep Reinforcement Learning Algorithm Using Decomposition and Attention Model for Multiobjective Optimization

Hong Wu, Jiahai Wang(✉), and Zizhen Zhang

Department of Computer Science, Sun Yat-sen University, Guangzhou 510006, China
wangjiah@mail.sysu.edu.cn

Abstract. Recently, a deep reinforcement learning method is proposed to solve multiobjective optimization problem. In this method, the multiobjective optimization problem is decomposed to a number of single-objective optimization subproblems and all the subproblems are optimized in a collaborative manner. Each subproblem is modeled with a pointer network and the model is trained with reinforcement learning. However, when pointer network extracts the features of an instance, it ignores the underlying structure information of the input nodes. Thus, this paper proposes a multiobjective deep reinforcement learning method using decomposition and attention model to solve multiobjective optimization problem. In our method, each subproblem is solved by an attention model, which can exploit the structure features as well as node features of input nodes. The experiment results on multiobjective travelling salesman problem show the proposed algorithm achieves better performance compared with the previous method.

Keywords: Multiobjective optimization · Deep reinforcement learning · Attention model

1 Introduction

A multiobjective optimization problem (MOP) can be defined as follows:

$$\begin{aligned} \min f(\mathbf{x}) &= (f_1(\mathbf{x}),\ f_2(\mathbf{x}), \ldots, f_m(\mathbf{x})) \\ \text{subject to } & \mathbf{x} \in S, \end{aligned} \tag{1}$$

where S is the decision space, $f: S \to \mathbb{R}^m$ is composed of m real-valued objective functions where $\mathbb{R}^m$ is called the objective space, and $f_i(\mathbf{x})$ for $i \in \{1, 2, \ldots, m\}$ is the i-th objective of the MOP. Since different objectives in the MOP are usually conflicting, it is impossible to find one best solution that can optimize all objectives at the same time. Thus a trade-off is required among different objectives.

K. Li et al. (Eds.): ISICA 2019, CCIS 1205, pp. 575–589, 2020.
https://doi.org/10.1007/978-981-15-5577-0_45

Let $u, v \in \mathbb{R}^m$, u is said to dominate v if and only if $u_i \leq v_i$ for every $i \in \{1, 2, \ldots, m\}$ and $u_j < v_j$ for at least one index $j \in \{1, 2, \ldots, m\}$. A solution $x^* \in S$ is called a pareto optimal solution if there is no solution $x \in S$ such that $f(x)$ dominates $f(x^*)$ [1]. The set of all pareto optimal solutions is named as pareto set (PS), and the set $\{f(s)|s \in \text{PS}\}$ is called the pareto front (PF) [1].

Many MOPs are NP-hard, such as multiobjective travelling salesman problem (MOTSP), multiobjective vehicle routing problem, etc. It is often difficult to find the PF of a MOP using exact algorithms. There are mainly two categories of optimization algorithms for solving MOPs. The first category is heuristics, such as NSGA-II [2] and MOEA/D [1]. The second category is the learning heuristic based methods [3]. Heuristics are often used to solve MOPs [4–7], but there are several drawbacks for them. Firstly, it is time-consuming for heuristics to approximate the PF of a MOP. Secondly, once there is a slight change of the problem, the heuristic may need to re-perform again to compute the solutions [3]. As a problem-specific method, heuristics often need to be revised for different problems, even for the similar ones.

Recently, some researchers begin to focus on deep reinforcement learning (DRL) for single-objective optimization problem [8–12]. Instead of designing specific heuristics, DRL learn heuristics directly from data on end-to-end neural network. Taking travelling salesman problem (TSP) as an example, given n cities as input, the aim is to get a sequence of these cities with minimum tour length. DRL views the problem as a Markov decision problem. Then TSP can be formulated as follows: the state is defined by the features of the partial solution and unvisited cities, the action is represented by the selection of the next city, the reward is the negative path length of a solution, the policy is the heuristic that learning how to make decisions, which is parameterized by a neural network. The aim of DRL is to train the policy that maximizes the reward. Once a policy is trained, the solution can be generated directly from one feed forward pass of the trained neural network. Without repeatedly solving instances from the same distribution, DRL is more efficient and requires much less problem-specific expert knowledge than heuristics.

Inspired by MOEA/D and the DRL methods proposed recently, a deep reinforcement learning multiobjective optimization algorithm (DRL-MOA) [3] is proposed to learn heuristics for solving MOPs. In the DRL-MOA, MOTSP is decomposed to M single-objective optimization subproblems firstly. Then M modified pointer networks, each of them is similar to the pointer network in [13], are used to model these subproblems. Finally, these models are trained by Actor-Critic training algorithm [8] sequentially. The experiment results on MOTSP in [3] show that DRL-MOA achieves better performance than NSGA-II and MOEA/D.

As we know, MOTSP is defined on a graph, every node in the graph not only contains its own features, but also the graph structure features such as the distances from other nodes. In DRL-MOA, when the modified pointer network models the subproblems of MOTSP, it does not consider the graph structure features of the graph. Therefore, this paper proposes a multiobjective deep reinforcement learning algorithm using decomposition and attention model (MODRL/D-AM) to solve MOPs. Attention model can extract the node features as well as graph structure features of MOP instances, which is helpful in making decisions. To show the effectiveness of our method, MODRL/D-AM is compared with DRL-MOA for solving MOTSP, and a significant improvement is observed in the overall performance of convergence and diversity.

The remainder of our paper is organized as follows. In Sect. 2, DRL-MOA is described. MODRL/D-AM is introduced in Sect. 3. Experiment results and analysis are presented in Sect. 4. Finally, conclusions are given in Sect. 5.

2 Brief Review of DRL-MOA for MOTSP

2.1 Problem Formulation and Framework

We focus on MOTSP in this paper. Given n cities and m objective functions, and the j-th objective function of MOTSP is formulated as follows [14]:

$$f_j(\pi) = \sum_{i=1}^{n-1} c^j_{\pi(i),\pi(i+1)} + c^j_{\pi(n),\pi(1)}, \quad j \in \{1, 2, \dots, m\}, \tag{2}$$

where route π is a permutation of n cities and $c^j_{\pi(i),\pi(i+1)}$ is the j-th cost from city $\pi(i)$ to city $\pi(i+1)$. The goal of MOTSP is to find a set of routes that minimize the m objective functions simultaneously.

Just like MOEA/D, DRL-MOA decomposes MOTSP to M scalar optimization subproblems by the well-known weighted sum approach, which considers the combination of different objectives. Let $\lambda_i = (\lambda_{i1}, \lambda_{i2}, \dots, \lambda_{im})$, where $\lambda_{ij} \geq 0, j \in \{1, \dots, m\}$ and $\sum_{j=0}^{m} \lambda_{ij} = 1$, be a weight vector corresponding to the i-th scalar optimization subproblem of MOTSP, which is defined as follows:

$$g^{ws}(\pi|\lambda_i) = \sum_{j=1}^{m} \lambda_{ij} f_j(\pi) \tag{3}$$

The optimal solution of the scalar optimization problem above is a pareto optimal solution. Then, let $\{\lambda_1, \lambda_2, \dots, \lambda_M\}$ be a set of weight vectors, each weight vector corresponds to a scalar optimization subproblem. When $m = 2$, the weight vectors and corresponding subproblems are spread uniformly as in Fig. 1(a). The PS is made up of the non-dominated solutions of all subproblems.

After decomposing MOTSP to a set of scalar optimization subproblems, each subproblem can be modelled by a neural network and solved by DRL methods. However, training M models requires huge amount of time. Thus, to decrease the training time of the models, DRL-MOA adopts a neighborhood-based transfer strategy, which shows in Fig. 1(b). Each model corresponds to a subproblem. When one subproblem is solved, the parameters of the corresponding model will be transferred to the model of the neighborhood subproblem, then the neighborhood subproblem will be solved quickly. By making use of the neighborhood information among subprob-

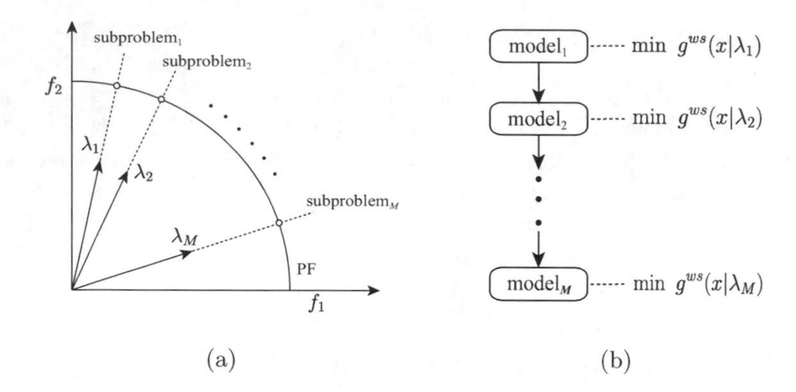

Fig. 1. (a) Decomposition strategy. (b) Neighborhood-based transfer strategy.

lems, all subproblems are tackled sequentially in a quick manner. The basic idea of DRL-MOA is shown in Algorithm 1. The subproblems are solved sequentially and M models are trained with Actor-Critic training algorithm by combining DRL and neighborhood-based transfer strategy. Finally, the PF can be approximated by a simple feed forward calculation of the M models.

Algorithm 1: Framework of DRL-MOA

Input: a well spread weight vectors $\{\lambda_1, \lambda_2, \ldots, \lambda_M\}$, the model of subproblems ω

Output: the optimal model ω^*

1 $\omega_{\lambda_1} \leftarrow$ initialization;
2 **for** $i = 1 : M$ **do**
3 **if** $i == 1$ **then**
4 $\omega^*_{\lambda_1} \leftarrow$ Actor-Critic$(\omega_{\lambda_1}, g^{ws}_{\lambda_1})$;
5 **else**
6 $\omega_{\lambda_i} \leftarrow \omega^*_{\lambda_{i-1}}$;
7 $\omega^*_{\lambda_i} \leftarrow$ Actor-Critic$(\omega_{\lambda_i}, g^{ws}_{\lambda_i})$;
8 **end**
9 **end**
10 **return** ω^*

2.2 Model of Subproblem: Pointer Network

A subproblem instance of MOTSP can be defined in a graph with n nodes, which is denoted by a set $X = \{x_1, x_2, \ldots, x_n\}$. Each node x_i has a feature vector $(x_{i1}, x_{i2}, \ldots, x_{im})$, which corresponds to the m different objectives of MOTSP. For example, a feature used widely is the 2-dimensional coordinate of Euclidean space. The solution denoted by $\pi = (\pi_1, \ldots, \pi_n)$ is a permutation of the graph nodes of MOTSP. The objective is minimizing the weighted sum of different objectives like Eq. (3). The process of generating a solution can be viewed as a sequential decision process, so each subproblem can be solved by an encoder-decoder model [15] parameterized by θ. Firstly, the encoder maps the node features to node embeddings in a high-dimensional vector space. Then the decoder generates the solution step by step. At each decoding step $t \in \{1, 2, 3, \dots, n\}$, one node π_t that has not been visited is selected. Hence, the probability of a solution can be modelled by the chain rule:

$$p_\theta(\pi|X) = \prod_{t=1}^{n} p_\theta(\pi_t|\pi_{1:t-1}, X). \quad (4)$$

In DRL-MOA, a modified pointer network is used to compute the probability in Eq. (4). The encoder of the modified pointer network transforms each node feature to an embedding in a high-dimensional vector space through a 1-dimensional (1-D) convolution layer. At each decoding time t, a gated recurrent unit (GRU) [15] and a variant of attention mechanism [16] are used to produce a probability distribution over the unvisited nodes, which is used to select the next node to visit. More details of the modified pointer network can be found in [3].

3 The Proposed Algorithm: MODRL/D-AM

3.1 Motivation

In DRL-MOA, a modified pointer network is used to model the subproblem of MOTSP. In the modified pointer network, an encoder extracts the node features using a simple 1-D convolutional layer. However, each subproblem of MOTSP is defined over a graph that is fully-connected (with self-connections). Such a simple encoder can not exploit the graph structure of a problem instance. At the decoding time t, the decoder uses a GRU to map a partial tour $\pi_{1:t-1}$ to a hidden state, which is used as a query vector to calculate the probability distribution of selecting the next node. However, the partial tour can not be changed and our goal is to construct a path from π_{t-1} to π_1 through all unvisited nodes. In other words, the selection of the next node is relevant only to the first and last node of the partial tour. Using a GRU in modified pointer network to map the total partial path to a hidden state may be not so helpful in selecting the next node, since there is much irrelevant information in the hidden state. Thus, this paper uses the attention model [11], instead of the modified pointer network, to model the subproblem.

3.2 Model of Subproblem: Attention Model

The attention model is also an encoder-decoder model. However, different from the modified pointer network, the encoder of attention model can be viewed as a graph attention network [17], which is used to compute the embedding of each

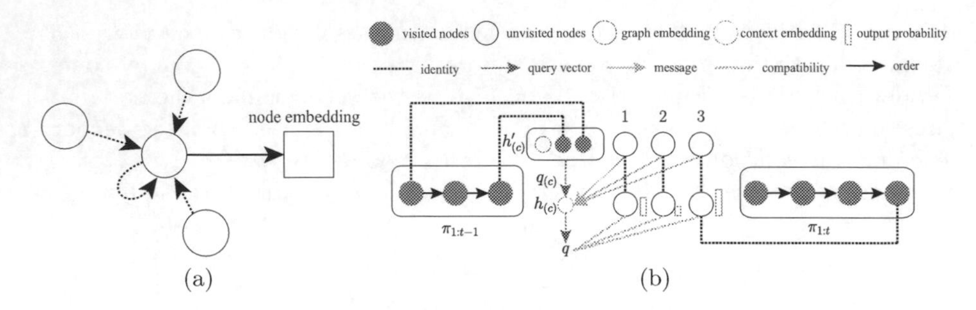

Fig. 2. (a) Encoder of attention model. (b) Decoding process at decoding step t.

node. As show in Fig. 2(a), by attending over other nodes, the embedding of each node contains the node features as well as the structure features. The decoder of attention model does not use a GRU to summarize the total partial path to a query vector, which is used to calculate the probability distribution of selecting the next node. Instead, the query vector is calculated by a transformation of a decoding context embedding consisting of the graph embedding, the first and last node embeddings of the partial tour, which is more useful in selecting the next node. The details of attention model are described below.

Encoder of Attention Model. The encoder of attention model transforms each node feature vector in the d_x-dimensional vector space to a node embedding in the d_h-dimensional vector space. The encoder is consisted of a linear transformation layer and N attention layers, which is similar to the encoder used in the Transformer architecture [18]. But the encoder of attention model does not use the positional encoding since the input order is not meaningful. For each node x_i, where $i \in \{1, \ldots, n\}$, the linear transformation layer with parameters $W \in \mathbb{R}^{d_h \times d_x}$ and $b \in \mathbb{R}^{d_h}$ transforms the node feature vector to the initial node embedding h_i^0:

$$h_i^0 = W x_i + b \tag{5}$$

Then the node embeddings $\{h_1^0, \ldots, h_n^0\}$ are fed into N attention layers. Each attention layer contains a multi-head attention sublayer and a feed-forward sublayer. For each sublayer, a batch normalization [19] layer and a skip connection [20] layer are used to accelerate the training process.

Multi-head Attention Sublayer. For each node x_i, this sublayer is used to aggregate different types of message from other nodes in the graph. Let the embedding of each node x_i in layer l be h_i^l, where $i \in \{1, \ldots, n\}$ and $l \in \{1, \ldots, N\}$. The output of multi-head attention sublayer $\hat{h}_i^l$ can be computed as follows:

$$\hat{h}_i^l = \mathrm{BN}^l(h_i^{l-1} + \mathrm{MHA}_i^l(h_1^{l-1}, \ldots, h_n^{l-1})), \tag{6}$$

where BN is the batch normalization layer and $\mathrm{MHA}_i^l(h_1^{l-1}, \ldots, h_n^{l-1})$ is the multi-head attention vector that contains different type of messages from other nodes. The number of heads is set to A. For each head a, the query vector $q_{ia}^l \in \mathbb{R}^{d_k}$, the key vector $k_{ia}^l \in \mathbb{R}^{d_k}$ and the value vector $v_{ia}^l \in \mathbb{R}^{d_v}$ is calculated by a transformation of the node embedding h_i^{l-1} for each node ($d_k = d_v = \frac{d_h}{A}$). Then the process of computing the multi-head attention vector is described as follows:

$$q_{ia}^l = W_{qa}^l h_i^{l-1}, k_{ia}^l = W_{ka}^l h_i^{l-1}, v_{ia}^l = W_{va}^l h_i^{l-1}, \tag{7}$$

$$u_{ija}^l = \frac{(q_{ia}^l)^T k_{ja}^l}{\sqrt{d_k}}, \quad w_{ija}^l = \frac{e^{u_{ija}^l}}{\sum_{j'=1}^n e^{u_{ij'a}^l}}, \tag{8}$$

$$h_{ia}^l = \sum_{j=1}^n w_{ija}^l v_{ja}^l, \quad \mathrm{MHA}_i^l(h_1^{l-1}, \ldots, h_n^{l-1}) = \sum_{a=1}^A W_{oa}^l h_{ia}^l, \tag{9}$$

where $W_{qa}^l \in \mathbb{R}^{d_k \times d_h}, W_{ka}^l \in \mathbb{R}^{d_k \times d_h}, W_{va}^l \in \mathbb{R}^{d_v \times d_h}, W_{oa}^l \in \mathbb{R}^{d_h \times d_v}$ are trainable attention weights of the l-th multi-head attention sublayer. $u_{ija}^l \in \mathbb{R}$ is the compatibility of the query vector q_{ia}^l of node x_i with the key vector k_{ja}^l of node x_j, the attention weight $w_{ija}^l \in [0, 1]$ is calculated using a softmax function. h_{ia}^l is the combination of messages from other nodes received by node x_i. The multi-head attention vector is computed with W_{oa}^l and h_{ia}^l.

Feed Forward Sublayer. In this sublayer, the node embedding of each node is updated by making use of the output of the multi-head attention layer. The feed forward sublayer (FF) is consisted of a fully-connected layer with ReLu activation function and another fully-connected layer. For each node x_i, the input of the feed forward sublayer is the output of the multi-head attention sublayer $\hat{h}_i^l$, the output is calculated as follows:

$$\mathrm{FF}^l(\hat{h}_i^l) = W_1^l \mathrm{ReLu}(W_0^l \hat{h}_i^l + b_0^l) + b_1^l, \tag{10}$$

$$h_i^l = \mathrm{BN}^l(\hat{h}_i^l + \mathrm{FF}^l(\hat{h}_i^l)), \tag{11}$$

where $W_0^l \in \mathbb{R}^{d_f \times d_h}, W_1^l \in \mathbb{R}^{d_h \times d_f}, b_0^l \in \mathbb{R}^{d_f}$ and $b_1^l \in \mathbb{R}^{d_h}$ are trainable parameters.

For each node x_i, the final node embedding h_i^N is calculated by N attention layers. Besides that, the graph embedding $\bar{h}^N$ is defined as follows:

$$\bar{h}^N = \frac{1}{n} \sum_{i=1}^n h_i^N, \tag{12}$$

both of the node embeddings and graph embedding will be passed to the decoder.

Decoder of Attention Model. At each decoding step $t \in \{1, \ldots, n\}$, the decoder needs to make a decision of π_t based on the partial tour $\pi_{1:t-1}$, the embeddings of each node and the total graph. Firstly, the initial context embedding $h'_{(c)} \in \mathbb{R}^{3d_h}$ is calculated by a concatenation of the graph embedding $\bar{h}^N$, the node embedding of the first node $h^N_{\pi_1}$ and the last node $h^N_{\pi_{t-1}}$. When $t=1$, $h^N_{\pi_1}, h^N_{\pi_{t-1}}$ are replaced by two trainable parameter vectors $v^1 \in \mathbb{R}^{d_h}, v^f \in \mathbb{R}^{d_h}$:

$$h'_{(c)} = \begin{cases} [\bar{h}^N, h^N_{\pi_1}, h^N_{\pi_{t-1}}] & t > 1 \\ [\bar{h}^N, v^1, v^f] & t = 1. \end{cases} \tag{13}$$

Then a new context embedding $h_{(c)}$ is computed with an A-head attention layer. The query vector $q_{(c)a} \in \mathbb{R}^{d_k}$ comes from the previous context embedding $h'_{(c)}$. For each node x_i, the key vector $k^{N+1}_{ia} \in \mathbb{R}^{d_k}$ and the value vector $v^{N+1}_{ia} \in \mathbb{R}^{d_v}$ are transformed from the node embedding h^N_i:

$$q_{(c)a} = W'_{qa} h'_{(c)}, k^{N+1}_{ia} = W^{N+1}_{ka} h^N_i, v^{N+1}_{ia} = W^{N+1}_{va} h^N_i, \tag{14}$$

where $W'_{qa} \in \mathbb{R}^{d_k \times 3d_h}, W^{N+1}_{ka} \in \mathbb{R}^{d_k \times d_h}$ and $W^{N+1}_{va} \in \mathbb{R}^{d_v \times d_h}$. Then the compatibilities of the query vector with all nodes are computed. Different from the encoder of attention model, the nodes that have been visited are masked when calculating the compatibilities:

$$u_{(c)ia} = \begin{cases} \frac{q^T_{(c)a} k^{N+1}_{ia}}{\sqrt{d_k}} & x_i \notin \pi_{1:t-1} \\ -\infty & \text{otherwise}. \end{cases} \tag{15}$$

Then the attention weights can be obtained by a softmax function and the new context embedding $h_{(c)}$ can be calculated as follows:

$$w_{(c)ia} = \frac{e^{u_{(c)ia}}}{\sum_{i'=1}^{n} e^{u_{(c)i'a}}}, \tag{16}$$

$$h_{(c)a} = \sum_{i=1}^{n} w_{(c)ia} v^{N+1}_{ia}, \quad h_{(c)} = \sum_{a=1}^{A} W^{N+1}_{oa} h_{(c)a}, \tag{17}$$

where $W^{N+1}_{oa} \in \mathbb{R}^{d_h \times d_v}$. Finally, based on the new context embedding $h_{(c)}$, the probability of selecting node x_i as the next node to visit $p_\theta(\pi_t = x_i | \pi_{1:t-1}, X)$ is calculated by a single-head attention layer:

$$q = W_q h_{(c)}, \quad k_i = W_k h^N_i, \tag{18}$$

$$u_i = \begin{cases} C \cdot \tanh(q^T k_i) & x_i \notin \pi_{1:t-1} \\ -\infty & \text{otherwise}, \end{cases} \tag{19}$$

$$p_\theta(\pi_t = x_i | \pi_{1:t-1}, X) = \frac{e^{u_i}}{\sum_{i'=1}^{n} e^{u_{i'}}}, \tag{20}$$

where $W_q \in \mathbb{R}^{d_h \times d_h}$ and $W_k \in \mathbb{R}^{d_h \times d_h}$ are trainable parameters. When we compute the compatibilities in Eq. (19), the result are limited in $[-C, C]$ ($C = 10$) by a tanh function.

The decoding process at decoding step t is shown in Fig. 2(b). Firstly, the context embedding is computed with a multi-head attention layer by making use of the partial solution and unvisited nodes. Then based on the context embedding and unvisited nodes, the probability distribution over unvisited nodes can be calculated by a single-head attention mechanism.

3.3 Framework and Training Method

The proposed algorithm uses the same MOEA/D framework as in DRL-MOA (Algorithm 1). The training method is briefly described as follows.

The Actor-Critic training algorithm [8], is used to train the model of the subproblem. For each subproblem, the training parameters ω_{λ_i} is composed of an actor network and a critic network. The actor network is the attention model, which is parameterized by θ. The critic network parameterized by ϕ has four 1-D convolutional layers to map the embeddings of a problem instance into a single value. The output of the critic network predicts an estimation of the objective function of the subproblem.

For the actor network, the training objective is the weighted sum of different objectives of solution π of a problem instance X. So the gradients of parameters θ can be defined as follows:

$$\nabla J(\theta|X) = E_{\pi \sim p_\theta(\cdot|X)}[(g^{ws}(\pi|\lambda_i; X) - b_\phi(X))\nabla_\theta log p_\theta(\pi|X)], \tag{21}$$

where $g^{ws}(\pi|\lambda_i; X)$ is the objective function of the i-th subproblem, which is the weighted sum of different objectives. λ_i is the corresponding weight vector. $b_\phi(X)$ is a value function calculated by the critic network, which predicts the expected objective value to estimate the policy learned by the actor network.

In the training process, the MOTSP instances are generated from distributions $(\Phi_1, \ldots, \Phi_m)$. Since for each node x_i of an instance X, different features $(x_{i1}, \ldots, x_{im})$ may come from different distributions $(\Phi_1, \ldots, \Phi_m)$. For example, x_{ij} can be a two-dimensional coordinate in Euclidean space and Φ_j can be a uniform distribution of $[0, 1] \times [0, 1]$. Then the gradients of parameters θ can be approximated by Monte Carlo sampling as follows:

$$\nabla J(\theta|X) \approx \frac{1}{B} \sum_{j=1}^{B} [(g^{ws}(\pi_j|\lambda_i; X_j) - b_\phi(X_j))\nabla_\theta log p_\theta(\pi_j|X_j)], \tag{22}$$

where B is the batch size, X_j is a problem instance sampled from $(\Phi_1, \ldots, \Phi_m)$ and π_j generated by the actor network is the solution of X_j.

Different from the actor network, the critic network aims to learn to estimate the expected objective value given an instance X. Hence, the objective function of the critic network can be a mean squared error function between the estimated objective value of the critic network $b_\phi(X)$ and the actual objective value of the solution generated by the actor network. The objective function of the critic network is formulated as follows:

$$\mathcal{L}_\phi = \frac{1}{B} \sum_{j=1}^{B} (b_\phi(X_j) - g^{ws}(\pi_j|\lambda_i; X_j))^2. \tag{23}$$

The training algorithm can be described in Algorithm 2.

Algorithm 2: Actor-Critic training Algorithm

Input: batch size B, dataset size D, number of epochs E, the parameters of actor network θ and the critic network ϕ

Output: the optimal parameters θ, ϕ

1 $\theta, \phi \leftarrow$ initialization from the parameters given in Algorithm 1
2 $T \leftarrow D/B$
3 **for** $epoch = 1 : E$ **do**
4   **for** $t = 1 : T$ **do**
5   **for** $j = 1 : B$ **do**
6       $X_j \leftarrow \text{SampleInstance}(\Phi_1, \ldots, \Phi_m)$
7       $\pi_j \leftarrow \text{SampleSolution}(p_\theta(\cdot|X_j))$
8       $b_j \leftarrow b_\phi(X_j)$
9   **end**
10     $d\theta \leftarrow \frac{1}{B}\sum_{j=1}^{B}[(g^{ws}(\pi_j|\lambda_i; X_j) - b_j)\nabla_\theta log p_\theta(\pi_j|X_j)]$
11     $\mathcal{L}_\phi \leftarrow \frac{1}{B}\sum_{j=1}^{B}(b_j - g^{ws}(\pi_j|\lambda_i; X_j))^2$
12     $\theta \leftarrow \text{ADAM}(\theta, d\theta)$
13     $\phi \leftarrow \text{ADAM}(\phi, \nabla_\phi \mathcal{L}_\phi)$
14   **end**
15 **end**
16 **return** θ, ϕ

4 Experiment

4.1 Problem Instances and Experimental Settings

MODRL/D-AM is tested on the Euclidean instances in [3]. In the Euclidean instances, the two node features are both sampled from $[0,1] \times [0,1]$ and both of the two cost functions between node i and node j are the Euclidean distance between them.

To train the models of MODRL/D-AM, problem instances with 20 and 40 nodes are used. After training, two models of MODRL/D-AM are obtained and the influence of different nodes in the training process can be discussed. To show the robustness of our method, the models are tested on problem instances with 20, 40, 70, 100, 150 and 200 nodes. Besides, kroAB100, kroAB150 and kroAB200 generated from TSPLIB [21] are used to test the performance of our method.

DRL-MOA is implemented and used as the baseline. Both our method and DRL-MOA are trained on datasets with 20 and 40 nodes, so there are four models in total: MODRL/D-AM (20), DRL-MOA (20), MODRL/D-AM (40), DRL-MOA (40). To make the result comparison more convincing, some parameters of our method and the baseline are set to the same value. The number of subproblems M is set to 100, the input dimension d_x is set to 4, the dimension of node embedding d_h is set to 128. In the training process, the batch size B is set to 200, the size of problem instances D is set to 500000, and the model of the first subproblem is trained for 5 epochs and each model of the remaining subproblems is trained for 1 epoch. Besides these parameters, the critic network is consisted of four 1-D convolutional layers. The input channels and output channels of the four convolutional layers are (4, 128), (128, 20), (20, 20) and (20, 1), where the first element of a tuple represents the input channel and the second element represents the output channel. For all convolutional layers, the kernel size and stride are set to 1.

In MODRL/D-AM, the number of attention layers N is set to 1, the number of heads A is set to 8, the dimension of the query vector d_k and the value vector d_v are both set to $\frac{d_h}{A} = 16$, and another dimension in the feed forward sublayer d_f is set to 512.

4.2 Results and Discussions

Hypervolume (HV) indicator is calculated to compare the performance of our method and DRL-MOA on tested instances. When computing the HV value, the objective values are normalized and the reference point is set to $(1.2, 1.2)$. The PFs obtained by MODRL/D-AM and DRL-MOA are also compared. Besides, the influence of different number of nodes in training process is also discussed. All test experiments are conducted by a GPU (GeForce RTX 2080Ti).

Table 1. The average of HV values and the calculation time obtained by MODRL/D-AM training with 20 and 40 nodes, and DRL-MOA training with 20 and 40 nodes. The test instances are random instances with 20, 40, 70, 100, 150, 200 nodes. The higher HV value is indicated in bold face.

#nodes		MODRL/D-AM (20)	DRL-MOA (20)	MODRL/D-AM (40)	DRL-MOA (40)
20	HV	**0.802**	0.796	**0.796**	0.785
	T(s)	4.6	2	4.6	1.9
40	HV	**0.813**	0.773	**0.821**	0.815
	T(s)	8.8	4.2	8.7	4.2
70	HV	**0.834**	0.803	**0.856**	0.842
	T(s)	15	6.7	14.7	6.5
100	HV	**0.846**	0.818	**0.872**	0.853
	T(s)	21.4	9.6	20.5	10.3
150	HV	**0.857**	0.838	**0.884**	0.864
	T(s)	29.5	15.9	30.7	15
200	HV	**0.866**	0.853	**0.894**	0.878
	T(s)	39.8	18.6	38.6	20.3

The HV values of random instances are shown in Table 1. For the random instances with 20, 40, 70, 100, 150 and 200 nodes, 10 instances are tested for each kind of random instances. The average of the HV values of each kind of random instances is calculated. In terms of the average of HV values, MODRL/D-AM

Table 2. The HV values obtained by MODRL/D-AM training with 20 and 40 nodes, and DRL-MOA training with 20 and 40 nodes. The test instances are kroAB100, kroAB150, kroAB200. The higher HV value is indicated in bold face.

#nodes		MODRL/D-AM (20)	DRL-MOA (20)	MODRL/D-AM (40)	DRL-MOA (40)
kroAB100	HV	**0.852**	0.832	**0.876**	**0.876**
	T(s)	19.8	9.2	19.2	9
kroAB150	HV	**0.874**	0.855	**0.891**	0.884
	T(s)	27.9	13.7	27.4	13.5
kroAB200	HV	**0.87**	0.85	**0.885**	0.882
	T(s)	37.3	18	38	17.7

(40) performs better than DRL-MOA (40) in all kinds of random instances. For kroAB100, kroAB150 and kroAB200, the HV values are computed in Table 2 and MODRL/D-AM (40) achieves a better performance than DRL-MOA (40). The calculation time of our method is longer than that of DRL-MOA. It is reasonable because the graph attention encoder of attention model requires more calculation resources than a single convolutional layer.

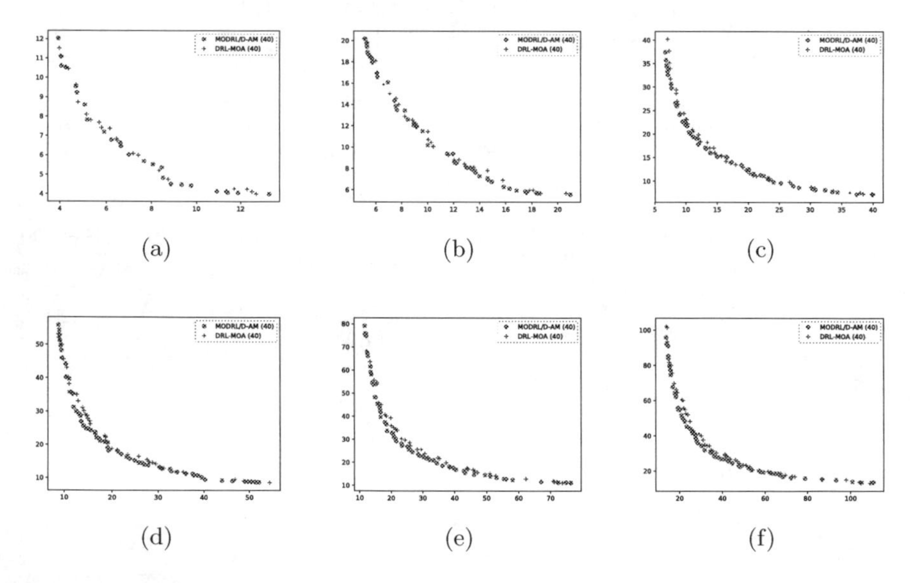

Fig. 3. The PFs obtained by MODRL/D-AM (40) and DRL-MOA (40) in solving random instances with (a) 20 nodes, (b) 40 nodes, (c) 70 nodes, (d) 100 nodes, (e) 150 nodes, (f) 200 nodes

The result of the tested instances with different nodes is shown in Fig. 3. By increasing the number of nodes, MODRL/D-AM (40) is able to get better performance in terms of convergence and diversity than that of DRL-MOA (40). Figure 4 shows the performance of MODRL/D-AM (40) and DRL-MOA (40)

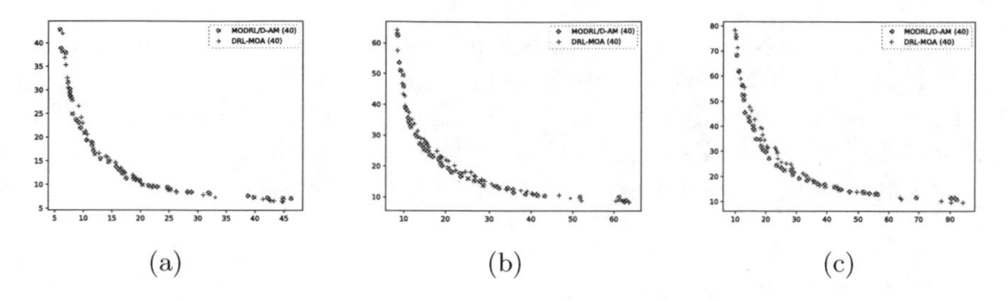

(a) (b) (c)

Fig. 4. The PFs obtained by MODRL/D-AM (40) and DRL-MOA (40) in solving instances of (a) kroAB100, (b) kroAB150, (c) kroAB200

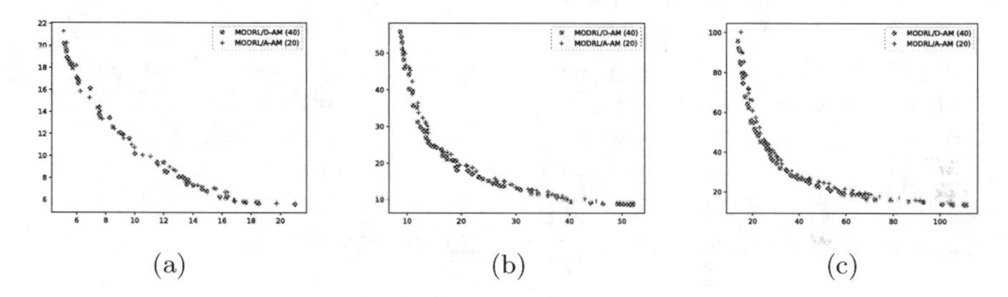

(a) (b) (c)

Fig. 5. The PFs obtained by MODRL/D-AM (40) and MODRL/D-AM (20) in solving random instances with (a) 40 nodes, (b) 100 nodes, (c) 200 nodes

on kroAB100, kroAB150 and kroAB200 instances. A significant improvement on convergence is observed for our method and the diversity achieved by our method is also slightly better.

Then, the performances of MODRL/D-AM (40) and MODRL/D-AM (20) are compared to investigate the influence of different number of nodes in training process. HV values in Table 1 show that MODRL/D-AM (40) performs better on random instances with 40, 70, 100, 150 and 200 nodes than MODRL/D-AM (20). For the random instances with 20 nodes, MODRL/D-AM (40) performs similar to MODRL/D-AM (20), while MODRL/D-AM (40) is slightly worse. From the PFs obtained by MODRL/D-AM (40) and MODRL/D-AM (20) in Fig. 5, a better performance is observed in terms of convergence and diversity. When training with instances with larger number of nodes, the model of MODRL/D-AM can learn to deal with more complex information about node features and structure features. Thus, a better model of MODRL/D-AM can be trained with more nodes.

From the experiment results above, it is observed that MODRL/D-AM has a good generalization performance in solving MOTSP. For MODRL/D-AM, the model trained with 40 nodes can be used to approximate the PF of problem instances with 200 nodes. In terms of convergence and diversity, MODRL/D-AM performs better than DRL-MOA.

The good performance of MODRL/D-AM indicates that the graph structure features are helpful in constructing solutions for MOTSP, and attention model can extract the structure information of a problem instance effectively. Thus, MODRL/D-AM can also be applied to other similar combinatorial optimization problems with graph structures such as multiobjective vehicle routing problem [4,5]. Finally, there is still an issue that the solutions of MOTSP instances are not distributed evenly in our experiment, which needs further research.

5 Conclusions

This paper proposes an multiobjective deep reinforcement learning algorithm using decomposition and attention model. MODRL/D-AM adopts an attention model to model the subproblems of MOPs. The attention model can extract structure features as well as node features of problem instances. Thus, more useful structure information is used to generate better solutions. MODRL/D-AM is tested on MOTSP instances, and compared with DRL-MOA which uses pointer network to model the subproblems of MOTSP. The results show MODRL/D-AM achieves better performance. A good generalization performance on different size of problem instances is also observed for MODRL/D-AM.

Acknowledgement. This work is supported by the National Key R&D Program of China (2018AAA0101203), and the National Natural Science Foundation of China (61673403, U1611262).

References

1. Zhang, Q., Li, H.: MOEA/D: a multiobjective evolutionary algorithm based on decomposition. IEEE Trans. Evol. Comput. **11**(6), 712–731 (2007)
2. Deb, K., Pratap, A., Agarwal, S., Meyarivan, T.: A fast and elitist multiobjective genetic algorithm: NSGA-II. IEEE Trans. Evol. Comput. **6**(2), 182–197 (2002)
3. Li, K., Zhang, T., Wang, R.: Deep reinforcement learning for multi-objective optimization. arXiv preprint arXiv:1906.02386 (2019)
4. Wang, J., Weng, T., Zhang, Q.: A two-stage multiobjective evolutionary algorithm for multiobjective multidepot vehicle routing problem with time windows. IEEE Trans. Cybern. **49**(7), 2467–2478 (2019)
5. Wang, J., Yuan, L., Zhang, Z., Gao, S., Sun, Y., Zhou, Y.: Multiobjective multiple neighborhood search algorithms for multiobjective fleet size and mix location-routing problem with time windows. IEEE Trans. Syst. Man Cybern. Syst. 1–15 (2019)
6. Cai, X., et al.: The collaborative local search based on dynamic-constrained decomposition with grids for combinatorial multiobjective optimization. IEEE Trans. Cybern. 1–12 (2019)
7. Yu, X., et al.: Set-based discrete particle swarm optimization based on decomposition for permutation-based multiobjective combinatorial optimization problems. IEEE Trans. Cybern. **48**(7), 2139–2153 (2017)
8. Bello, I., Pham, H., Le, Q.V., Norouzi, M., Bengio, S.: Neural combinatorial optimization with reinforcement learning. arXiv preprint arXiv:1611.09940 (2016)

9. Deudon, M., Cournut, P., Lacoste, A., Adulyasak, Y., Rousseau, L.-M.: Learning heuristics for the TSP by policy gradient. In: van Hoeve, W.-J. (ed.) CPAIOR 2018. LNCS, vol. 10848, pp. 170–181. Springer, Cham (2018). https://doi.org/10.1007/978-3-319-93031-2_12
10. Khalil, E., Dai, H., Zhang, Y., Dilkina, B., Song, L.: Learning combinatorial optimization algorithms over graphs. In: Advances in Neural Information Processing Systems, pp. 6348–6358 (2017)
11. Kool, W., van Hoof, H., Welling, M.: Attention, learn to solve routing problems! arXiv preprint arXiv:1803.08475 (2018)
12. Nazari, M., Oroojlooy, A., Snyder, L., Takác, M.: Reinforcement learning for solving the vehicle routing problem. In: Advances in Neural Information Processing Systems, pp. 9839–9849 (2018)
13. Vinyals, O., Fortunato, M., Jaitly, N.: Pointer networks. In: Advances in Neural Information Processing Systems, pp. 2692–2700 (2015)
14. Lust, T., Teghem, J.: The multiobjective traveling salesman problem: a survey and a new approach. In: Coello Coello, C.A., Dhaenens, C., Jourdan, L. (eds.) Advances in Multi-objective Nature Inspired Computing. Studies in Computational Intelligence, vol. 272, pp. 119–141. Springer, Heidelberg (2010). https://doi.org/10.1007/978-3-642-11218-8_6
15. Cho, K., et al.: Learning phrase representations using RNN encoder-decoder for statistical machine translation. arXiv preprint arXiv:1406.1078 (2014)
16. Bahdanau, D., Cho, K., Bengio, Y.: Neural machine translation by jointly learning to align and translate. arXiv preprint arXiv:1409.0473 (2014)
17. Veličković, P., Cucurull, G., Casanova, A., Romero, A., Lio, P., Bengio, Y.: Graph attention networks. arXiv preprint arXiv:1710.10903 (2017)
18. Vaswani, A., et al.: Attention is all you need. In: Advances in Neural Information Processing Systems, pp. 5998–6008 (2017)
19. Ioffe, S., Szegedy, C.: Batch normalization: accelerating deep network training by reducing internal covariate shift. arXiv preprint arXiv:1502.03167 (2015)
20. He, K., Zhang, X., Ren, S., Sun, J.: Deep residual learning for image recognition. In: Proceedings of the IEEE Conference on Computer Vision and Pattern Recognition, pp. 770–778 (2016)
21. Reinelt, G.: TSPLIB-traveling salesman problem library. ORSA J. Comput. **3**(4), 376–384 (1991)

Parameters Tuning of PID Based on Improved Particle Swarm Optimization

Wei Yu(✉) and Qingmei Zhao

Guangdong University of Science and Technology, Dongguan, China
58113129@qq.com

Abstract. Particle swarm optimization (PSO) is used to tune PID parameters because it is simple and efficient [1]. Particles learn from the global optimum value and the individual historical optimum value in PSO, but during the later stage of iteration it is easy to converge prematurely and fall into local optimum. In this article, an improved method is proposed, social learning in PSO (SL-PSO) is used to tune PID parameters. The fitness of all particles must be calculated and sorted from inferior to superior. The particles are allowed to learn from any individual better than themselves. In this way, the diversity of particles is increased greatly. At the end of the article, the PID parameters of quad-rotor aircraft confirm the validity of the algorithm.

Keywords: PID · Fitness · SL-PSO

1 Introduction

PID control is one of the most widely used control method in modern control field, and its effect largely depends on the selection of its parameters. So the tuning of parameters has always been the focus of PID control. Because of the increasing requirement of control accuracy and efficiency, Intelligent Cluster Algorithms and Optimal Algorithms are introduced to improve the controller. In this article, the parameters of quad-rotor aircraft are tuned by the improved particle swarm optimization algorithm (SL-PSO), which achieves the desired results.

2 PID Control

2.1 PID Control Theory

PID controller is one of the most widely used controllers in modern control field. Its principle is that the difference between the expected value and the actual value is taken as the input, and the output is obtained by summing up the proportional, integral and differential operations of the controller. As shown in Fig. 1.

K. Li et al. (Eds.): ISICA 2019, CCIS 1205, pp. 590–598, 2020.
https://doi.org/10.1007/978-981-15-5577-0_46

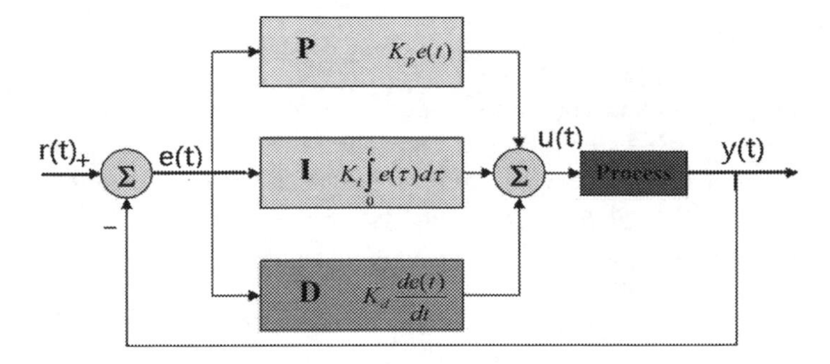

Fig. 1. PID control system block diagram

The input e(t) is expressed as:

$$e(t) = r(t) - y(t) \tag{1}$$

In Eq. 1, ***r(t)*** is the expected value and ***y(t)*** is the feedback, i.e. the output of the controlled object obtained from the previous control process.

The control equation is shown in Eq. 2 and the transfer function in Eq. 3.

$$u(t) = K_p \left[e(t) + \frac{1}{K_i} \int_0^t e(\tau)d\tau + K_d \frac{de(t)}{dt} \right] \tag{2}$$

$$G(s) = \frac{U(s)}{E(s)} = K_p \left(1 + \frac{1}{K_i s} + K_d s \right) \tag{3}$$

In the formulas, K_P is a proportional coefficient; K_i and K_d are integral and differential time constants respectively. The three parameters determine the performance of the PID controller.

2.2 The Methods of PID Parameters Tuning

The structure of the PID controller is simple, so long as the selected parameters are appropriate, it can achieve high-precision control. The common tuning methods of PID parameters are the Engineering tuning method which tuning cycle is too long and the effect is not ideal and the Intelligent PID parameter tuning method, in which by introducing continuous learning, evolution and iterations of intelligent optimization algorithm to adjust the parameters of PID, the ideal control effect can be achieved.

3 PSO Theory

PSO algorithm is a bionic algorithm, which is evolved from the process of birds looking for food. Each particle in this algorithm is a potential solution, which is expressed by its position and velocity [3]. Particles search for the optimal value by constantly replacing the individual and group optimal solutions. The fitness function is used to judge the quality of particles.

Assuming that a particle swarm of $\boldsymbol{m}$ moves in a D-dimensional space, $\boldsymbol{X_i} = (\boldsymbol{x_{i1}}, \boldsymbol{x_{i2}}, \ldots, \boldsymbol{x_{iD}})$ and $\boldsymbol{V_i} = (\boldsymbol{v_{i1}}, \boldsymbol{v_{i2}}, \ldots, \boldsymbol{v_{iD}})$ are the decomposition of the current position and velocity of particle $\boldsymbol{i}$ in each dimension respectively. The velocity of each particle varies dynamically with the position of other particles. $\boldsymbol{P_i} = (\boldsymbol{p_{i1}}, \boldsymbol{p_{i2}}, \ldots, \boldsymbol{p_{iD}})$ and $\boldsymbol{P_g} = (\boldsymbol{p_{g1}}, \boldsymbol{p_{g2}}, \ldots, \boldsymbol{p_{gD}})$ are the optimal position of the searched particle and the optimal position of the group respectively. The particles update their velocity and position according to Eq. 4.

$$\begin{cases} v_{id}^{k+1} = w * v_{id}^{k} + c_1 * r_1 * \left(p_{id} - x_{id}^{k}\right) + c_2 * r_2 * \left(p_{gd} - x_{id}^{k}\right) \\ x_{id}^{k+1} = x_{id}^{k} + v_{id}^{k+1} \end{cases} \tag{4}$$

In the formula, $\boldsymbol{d}$ is the component of the velocity of the particle $\boldsymbol{i}$ in the d dimension at the kth iteration. $\boldsymbol{p_{gd}}$ and $\boldsymbol{p_{id}}$ represent the component of the d-Dimension of the group optimal position and the particle optimal respectively.

$\boldsymbol{i} = 1, 2, \cdots, \boldsymbol{m}; \boldsymbol{d} = 1, 2, \cdots, \boldsymbol{D}$; k is the number of iterations; r_1 and r_2 are random numbers between 0 and 1; c_1 and c_2 are learning factors; w is inertia weight, used to adjust the local and global search ability.

In order to improve the efficiency and avoid blind search, the intervals of position and velocity ***[Xmin, Xmax], [Vmin, Vmax]*** need to be set in the algorithm [4].

In order to test the fitness of particles, Eq. 5 is used as fitness function [5]:

$$J = \int_{0}^{\infty} t|e(t)|dt \tag{5}$$

4 PSO-PID

Three parameters of PID controller are optimized and adjusted by PSO algorithm. The dimension of PSO algorithm is set to 3, and the values of PSO are assigned to $\mathbf{K_p}$, $\mathbf{K_i}$ and $\mathbf{K_d}$ in turn. After the controlled system is adjusted, calculate the fitness function and then check whether the fitness function satisfies the termination condition. If it does, the iteration ends; but if it does not, continue updating the particle swarm until the maximum number of iterations or the set threshold is reached. The flow chart is as follows in Fig. 2.

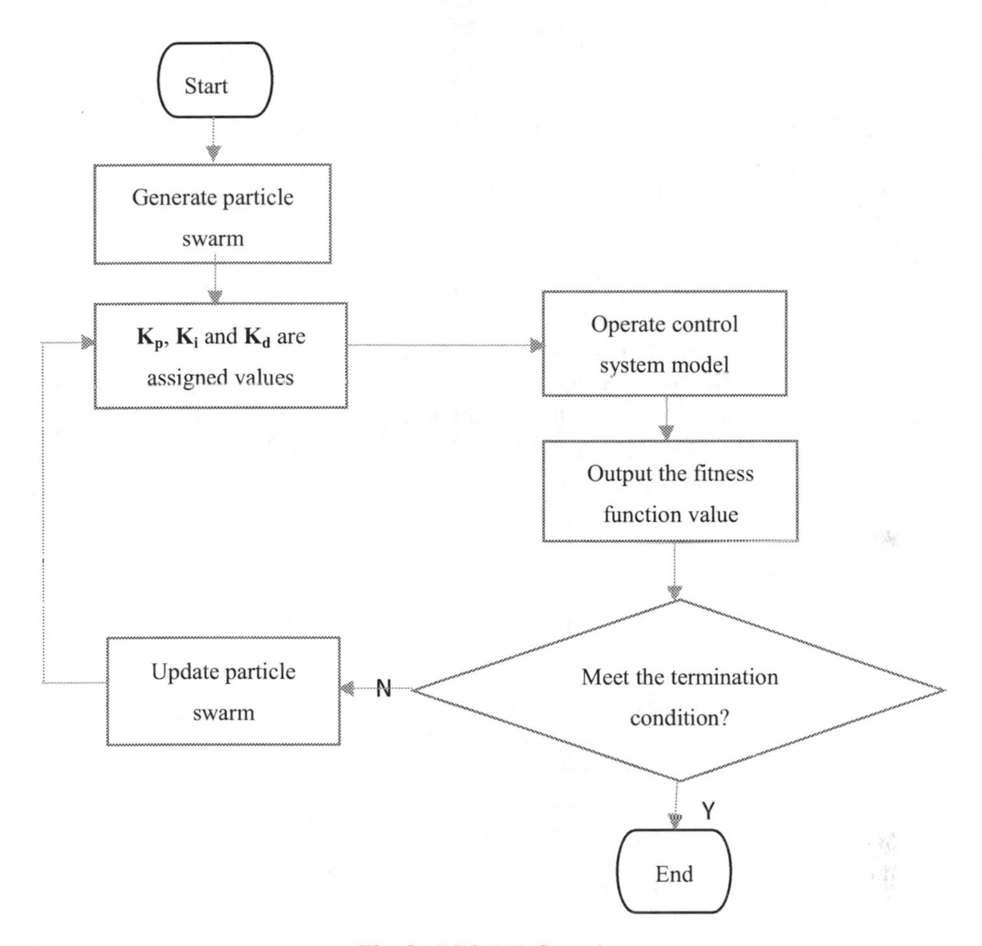

Fig. 2. PSO-PID flow chart

5 SLPSO-PID

In order to solve the problem that PSO algorithm tends to fall into local optimum in the later stage of iteration, which leads to premature convergence of the algorithm, an improved social learning particle swarm optimization (SLPSO) algorithm is introduced. In SLPSO, the particle does not learn from the global optimal value and the individual historical optimal value, but from the random particle which is better than itself. This method enhances the local information sharing ability of the algorithm, which greatly increases the diversity of particles. The design of the algorithm is as follows:

At the beginning, the number of particles in the group is necessary, which is proportional to the dimension of searching, and it is determined by Eq. 6, in which M is the minimum number of particles required by PSO, usually set to 100.

$$m = M + \lfloor D/10 \rfloor \tag{6}$$

Then according to Eq. 7, initialize the position and step size.

$$\begin{cases} x_i^0 = x_{min} + rand(x_{max} - x_{min}) \\ v_i^0 = v_{min} + rand(v_{max} - v_{min}) \end{cases} \tag{7}$$

In Eq. 7, $\boldsymbol{x_{max}}$, $\boldsymbol{x_{min}}$ and $\boldsymbol{v_{max}}$, $\boldsymbol{v_{min}}$ are the boundary value of position and velocity.

According to Eq. 5, calculate the fitness of all particles and sort the fitness from inferior to superior (from large to small). For each iteration, each particle needs to learn from any particle whose fitness is better than itself, including the global optimal particle. So the learner $\boldsymbol{i}$ needs to satisfy the condition: $\boldsymbol{1 \leq i \leq m}$, and the particle $\boldsymbol{l}$ who is learned from satisfies: $\boldsymbol{i < l \leq m}$, that is the fitness of latter is better than the former. As shown in Fig. 3.

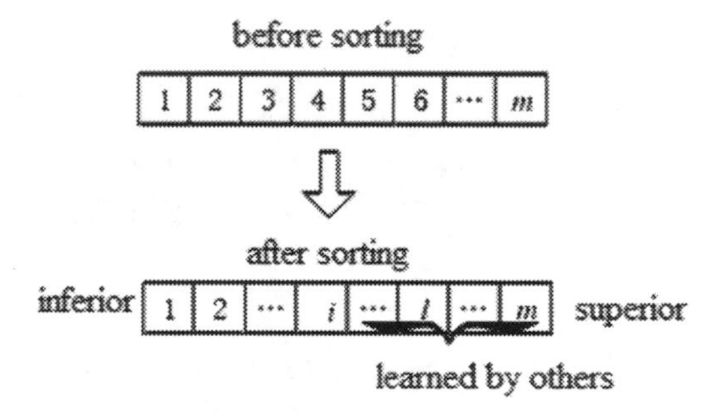

Fig. 3. Swarm sorting in SLPSO

After sorting, the average value of particles position is needed, which is given by Eq. 8.

$$\overline{x^k} = \sum_{i=1}^{m} x_i^k / m \tag{8}$$

In Eq. 8, $\boldsymbol{x_i^k}$ is the position of particle $\boldsymbol{i}$ at the **kth** iteration.

Then the proportion of learning from the average ε needs to be calculated, which is given by Eq. 9.

$$\varepsilon = \beta * \frac{D}{M} \tag{9}$$

The formula above shows that ε is proportional to the searching dimension $\boldsymbol{D}$. The parameter $\boldsymbol{\beta}$ is used to regulate the average which influences the algorithm. If $\boldsymbol{\beta}$ is excessive, the particle swarm will be premature. So it is usually set to a smaller value, in this article, let $\boldsymbol{\beta}$ equal 0.01.

After that, the learning scale from which fitness is better than the particle itself will be calculated, and also the learning scale from the average of the group. They are given in Eq. 10.

$$\begin{cases} I_i^k = x_l^k - x_i^k \\ C_i^k(t) = \overline{x^k} - x_i^k \end{cases} \tag{10}$$

Then the modified value of position is obtained according to Eq. 11.

$$\Delta x_i^{k+1} = r_1 \cdot \Delta x_i^k + r_2 \cdot I_i^k + r_3 \cdot \varepsilon \cdot C_i^k \tag{11}$$

In Eq. 11, both $\mathbf{r_2}$ and $\mathbf{r_3}$ are the arbitrary value between 0 and 1. $\mathbf{r_1}$ is the inertia weight, the value of which depends on the current fitness and the previous one. If the former is better than the latter, the value is between 0 and 1, otherwise, let $\mathbf{r_1}$ be 0 to eliminate errors.

The way of learning from any one of the group can increase the diversity of particle swarm, however, blindly adjusting particles without considering the searching dimension of the objective function will increase the running time of the algorithm. To solve the problem, learning probability is introduced, which is represented in Eq. 12.

$$P_i^L = \left(1 - \frac{i-1}{m}\right)^{\alpha * log\left(\frac{D}{M}\right)} \tag{12}$$

The base part of the formula shows that the learning probability is inversely proportional to the fitness, and the exponent shows that the learning probability is inversely proportional to the searching dimension of objective function. Usually the parameter α is set to a constant less than 1.

Assuming the learning probability is $\boldsymbol{P_i^K}$, next update the particle state according to Eq. 13.

$$x_i^{k+1} = \begin{cases} x_i^k + \Delta x_i^{k+1}, & if\ P_i^k \le P_i^L \\ x_i^k, & otherwise \end{cases} \tag{13}$$

Then calculate the fitness of the updated particle, and compare that with the previous, if it becomes better, the value of $\mathbf{r_1}$ is kept, otherwise, $\mathbf{r_1}$ is set to 0.

Next sort the fitness from inferior to superior once again and repeat iteratively until the threshold condition is satisfied or the maximum number of iterations is reached.

6 Experimental Results and Analysis

Three parameters of PID are tuned by SL-PSO algorithm, which are used to control the pitch angle of quad-rotor aircraft. To simplify the control, step signal is selected as the input, and the transfer function is given by Eq. 14 [5, 6].

$$G = \frac{57.95s + 4411}{s^3 + 107.5s^2 + 892.5s + 4435} \tag{14}$$

Figure 4 shows the proportional coefficient $\mathbf{K_p}$ and differential coefficient $\mathbf{K_d}$ decrease gradually with the increase of iteration times. The integral coefficient $\mathbf{K_i}$ has no obvious change in the whole process, and keeps the comparison all the time.

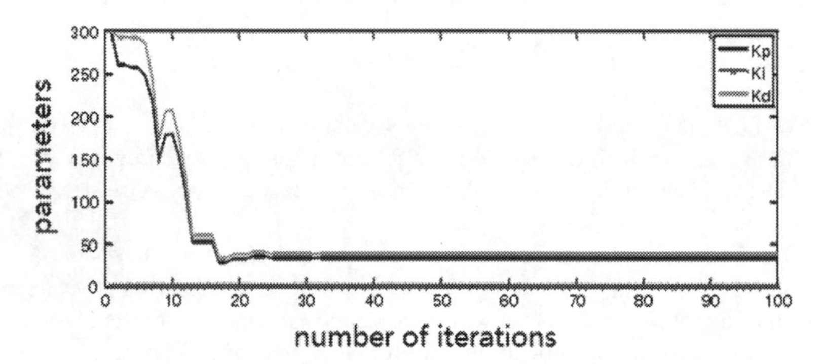

Fig. 4. Relationship between three parameters and iterations

Figure 5 and Fig. 6 show the relationship between the fitness and the iterations of PSO and improved SL-PSO respectively. The PSO-PID algorithm has a slow convergence speed, which is gradually stable after 57 iterations, but the SLPSO-PID algorithm only needs 17 iterations to become stable. In addition, the fitness of the optimal value obtained by the former is as high as 1.063, and the control accuracy is not high, while the latter is only 0.056, which is closer to the global optimal value.

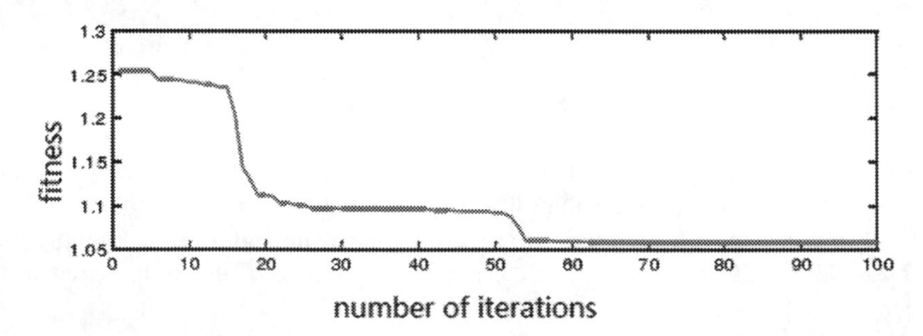

Fig. 5. Relationship between the fitness of PSO and the iterations

Also the performances of two algorithms are compared in Table 1, which show that the rising time and adjusting time of PID parameters tuned by SLPSO are shorter, the overshoot is smaller, and the control effect is better.

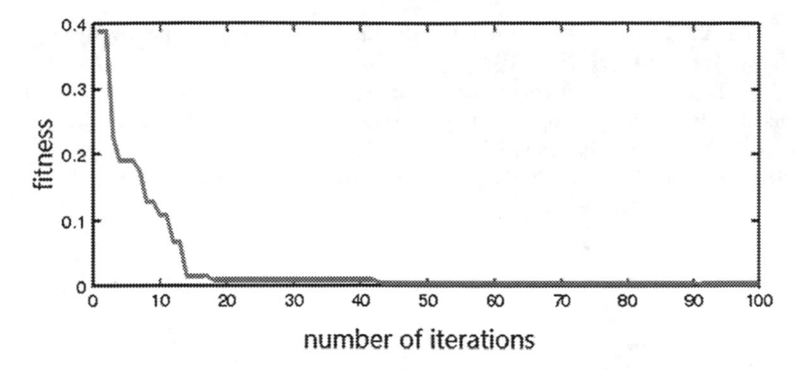

Fig. 6. Relationship between the fitness of SL-PSO and the iterations

Table 1. PID performance of the algorithms

	PSO-PID	SLPSO-PID
Parameters	$K_p = 29.9751$	$K_P = 34.9871$
	$K_i = 0.1521$	$K_i = 0.1321$
	$K_d = 32.4257$	$K_d = 35.2357$
Iterations	57	17
Fitness	1.063	0.056
Overshoot	0.102	0.004
Rising time	0.512	0.272
Adjusting time	1.318	0.335

7 Conclusion

In this paper, the PID parameters of quad-rotor aircraft are tuned by using the optimization characteristics of PSO algorithm. In order to avoid premature convergence in the late iteration and falling into local optimum, SL-PSO is introduced to improve the algorithm. The simulation results show that the SL-PSO algorithm can obviously improve the control efficiency and accuracy, and has great application value in actual flight control.

References

1. Oi, A., Nakazawa, C., Matsui, T., et al.: Development of PSO-based PID tuning method. In: International Conference on Control, Automation and Systems, pp. 1917–1920. IEEE (2008)
2. Tsoulos, I.G., Stavrakoudis, A.: Enhancing PSO methods for global optimization. Appl. Math. Comput. **216**(10), 2988–3001 (2010)
3. Kennedy, J, Eberhart, R.: Particle swarm optimization. In: Proceedings of IEEE International Conference on Neural Networks IV (1995)

4. Li, J., Liou, Y.C., Zhu, L.: Optimization of PID parameters with an improved simplex PSO. J. Inequal. Appl. **2015**(1), 1–5 (2015)
5. Kang, R., Ma, J., Jia, H.: Self-adapt scatter particle swarm optimization algorithm in the quad-rotor aircraft PID parameter optimization. Comput. Simul. **35**(3), 29–33 (2018)
6. Salih, A.L., et al.: Modelling and PID controller design for a quad-rotor unmanned air vehicle. In: 2010 IEEE International Conference on Automation Quality and Testing Robotics, pp. 1–5. IEEE (2010)

Design and Analysis of Knee-Joint Force Reduction Device and Fatigue Detection System

Caihua Qiu and Feng Ding(✉)

Computer Science Department, Guangdong University of Science and Technology, Dongguan, China
32320062@qq.com, 648489640@qq.com

Abstract. According to mechanics principle, force reduction device which is based on slider crank mechanism has been proposed. Our study developed the device's kinematics and dynamics model, and evaluated the value of reduced force by analyzing its dynamics and kinetics deeply. Moreover, by the means of changing conditions, factors which influence force reduction had been found. For fatigue detection, we came up with an algorithm which is based on surface electromyography digital transducer and is able to monitor fatigue degree in real-time.

Keywords: Knee-joint · Force reduction device · Fatigue detection

Knee-joint has complicated anatomy and high morbidity rate, while it plays a vital role in human's daily activity [1]. However, few protection device, which can protect people's knee-joint through force reduction, exists now. China has high morbidity rate of knee-joint illness and about 3% of it is osteoarthritis [2]. Four-linkage structure conforms to body engineering and possesses human knee-joint characters [3].

Existing research is mainly concerning about the analysis and designing of knee-joint's exoskeleton and optimization of exoskeleton's bionic performance. At present, there is less research fitting on protection of normal human's knee-joint force reduction during walking. In order to solve this problem, our study develops mathematics model which is based on slider-crank mechanism knee joint force reduction device, and verified simulation by MATLAB. In addition, optimization and improvement have been applied to the model.

1 Design of Knee-Joint Force Reduction

1.1 Analysis of Motion When Walking

One walking cycle can be divided into up-right phase and stride phase. Up-right phase consists of two period of time of double-foot support, which takes up to 0–12% and 50%–62%, respectively. In other stages, 12%–50% of them are (single leg supports 改为 supported by single leg) [4] (Fig. 1).

K. Li et al. (Eds.): ISICA 2019, CCIS 1205, pp. 599–608, 2020.
https://doi.org/10.1007/978-981-15-5577-0_47

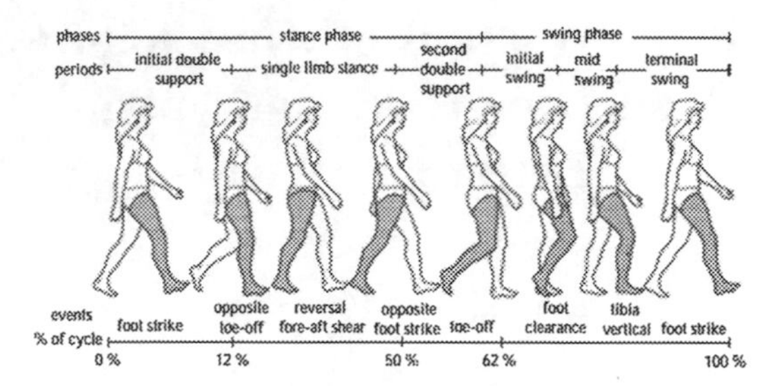

Fig. 1. Human walking cycle

1.2 Structure of Knee-Joint Force Reduction Device

When human walking, fast switching between up-right phase and stride phase causes impact vibration to knee-joint, and knee-joint force reduction device is able to decrease vibration and reduce part of its weight which would causes stress to knee-joint. Bionic and human knee joint operate simultaneously. Specifically, when bending over, bias force plank 5 uplifts with flexible kneecap 9 and shank, and joint board 2 cocks and departs from oil buffer 4; when standing up-right, bias force plank 5 protracts with flexible kneecap 9 and shank, and oil buffer 4 abuts on the first part of joint board 2 which plays a support role and decreases damage to knee joint.

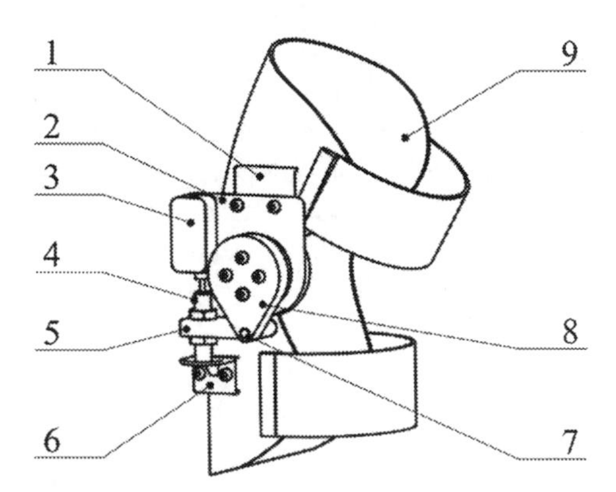

Fig. 2. Structure of knee-joint force reduction device. 1-thigh stator; 2-joint board; 3-controller box; 4-oil buffer; 5-bias force plank; 6-shank fixation plank; 7-coak; 8-limit block; 9-flexible kneecap

2 Kinematics and Dynamics Analysis of Knee-Joint Force Reduction Device

2.1 Mathematics Model of Knee Joint Force Reduction Device

According to organization diagram of Fig. 2, joint board 2 can be considered as body frame and fixed at thigh. Limit block 8 can be deemed as crank and is able to rotate around the joint board. Oil buffer 4's lower part fixed connected to bias force plank 5 which can be considered as connecting rod. The head part of oil buffer 4 is composed of vibration-absorptive material and is able to slide through joint board 2. Therefore it can be considered as sliding block. Based on the above analysis, mathematical model of knee-joint force reduction device can be built [5] (Fig. 3).

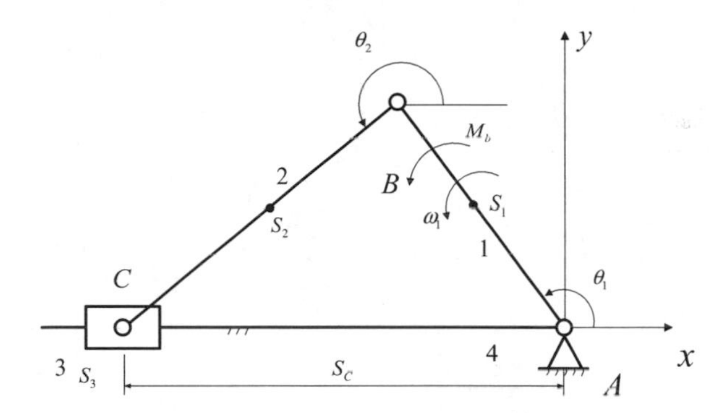

Fig. 3. Slider-crank mechanism model

Based on method of complex vector, developing kinematics model of slider-crank knee joint force reduction device, which including a closed loop named ABC. In practical application, oil buffer 4 is actuator.

$$\begin{cases} a_{S_1x} = -l_{AS_1}\omega_1^2 \cos\theta_1 \\ a_{S_2y} = -l_{AS_1}\omega_1^2 \sin\theta_1 \end{cases}$$

$$\begin{cases} a_{S_2x} = -l_1\omega_1^2 \cos\theta_1 - l_{BS_2}(\omega_2^2 \cos\theta_2 + \alpha_2 \sin\theta_2) \\ a_{S_2y} = -l_1\omega_1^2 \sin\theta_1 - l_{BS_2}(\omega_2^2 \sin\theta_2 - \alpha_2 \cos\theta_2) \end{cases}$$

$$\begin{cases} a_{S_3x} = -l_1\omega_1^2 \cos\theta_1 - l_{BS_2}(\omega_2^2 \cos\theta_2 + \alpha_2 \sin\theta_2) \\ a_{S_3y} = 0 \end{cases}$$

$$\begin{cases} F_{1x} = -m_1 a_{S_1 x} \\ F_{1y} = -m_1 a_{S_1 y} \\ F_{2x} = -m_2 a_{S_2 x} \\ F_{2y} = -m_2 a_{S_2 y} \\ F_{3x} = -m_3 a_{S_3 x} \\ F_{3y} = -m_3 a_{s_3 y} \\ M_1 = -J_{S_1 \alpha_1} \\ M_2 = -J_{S_2 \alpha_2} \end{cases}$$

$$\begin{cases} M_b - F_{R_{12}x}(y_{S_1} - y_B) - F_{R_{12}y}(x_B - x_{S_1}) - F_{R_{14}x}(y_{S_1} - y_A) - F_{R_{14}y}(x_A - x_{S_1}) = 0 \\ -F_{R_{12}x} - F_{R_{14}x} = -F_{1X} \\ -F_{R_{12}y} - F_{R_{14}y} = -F_{1y} \end{cases}$$

$$\begin{cases} F_{R_{12}x}(y_{S_2} - y_B) + F_{R_{12}y}(x_B - x_{S_1}) - F_{R_{23}x}(y_{S_2} - y_C) - F_{R_{23}y}(x_C - x_{S_2}) = -M_2 \\ F_{R_{12}x} - F_{R_{23}x} = -F_{2X} \\ F_{R_{12}y} - F_{R_{23}y} = -F_{2y} \end{cases}$$

$$\begin{cases} F_{R_{23}x} = -F_{3x} - F_r \\ F_{R_3 y} - F_{R_{34}y} = -F_{3y} \end{cases}$$

$$\begin{bmatrix} 1 & -(y_{S_1} - y_B) & -(x_B - x_{S_1}) & -(y_{S_1} - y_A) & -(x_A - x_{S_1}) & 0 & 0 & 0 \\ 0 & -1 & 0 & -1 & 0 & 0 & 0 & 0 \\ 0 & 0 & -1 & 0 & -1 & 0 & 0 & 0 \\ 0 & (y_{S_2} - y_B) & (x_B - x_{S_2}) & 0 & 0 & -(y_{S_2} - y_C) & -(x_C - x_{S_2}) & 0 \\ 0 & 1 & 0 & 0 & 0 & -1 & 0 & 0 \\ 0 & 0 & 1 & 0 & 0 & 0 & -1 & 0 \\ 0 & 0 & 0 & 0 & 0 & 1 & 0 & 0 \\ 0 & 0 & 0 & 0 & 0 & 0 & 1 & -1 \end{bmatrix} \begin{bmatrix} M_b \\ F_{R_{12}x} \\ F_{R_{12}y} \\ F_{R_{14}x} \\ F_{R_{14}y} \\ F_{R_{23}x} \\ F_{R_{23}y} \\ F_{R_{34}y} \end{bmatrix} = \begin{bmatrix} 0 \\ -F_{1x} \\ -F_{1y} \\ -M_2 \\ -F_{2x} \\ -F_{2y} \\ -F_{3x} - F_r \\ -F_{3y} \end{bmatrix}$$

2.2 Kinematics Solve of Knee-Joint Force Reduction Device

Human knee joint's range of motion is between −75°–0° [6]. So, we can set the crank nook angle 15°–90° in crank slide mechanism model (Fig. 4).

In up-right phase, crank nook angle is 90°. In order to make knee joint force reduction device accessible to human, length of limit block 8 is required to be the same as oil buffer 4, which assumed to be 35 mm. In the case of bias force plank equals 40 mm, I.e. crank length $l_1 = 35\,\text{mm}$, Connecting rod length $l_2 = \sqrt{40^2 + 35^2}\,\text{mm} = 53.15\,\text{mm}$. Sliding block's displacement, velocity and accelerated speed can be calculated through knee joint matrix data of reference [7] and calculation by MATLAB.

2.3 Kinematics Solve of Knee-Joint Force Reduction Device

In accordance with assumed size of different parts, physical index such as mass can be calculated, and therefore resolve of slide force can be concluded.

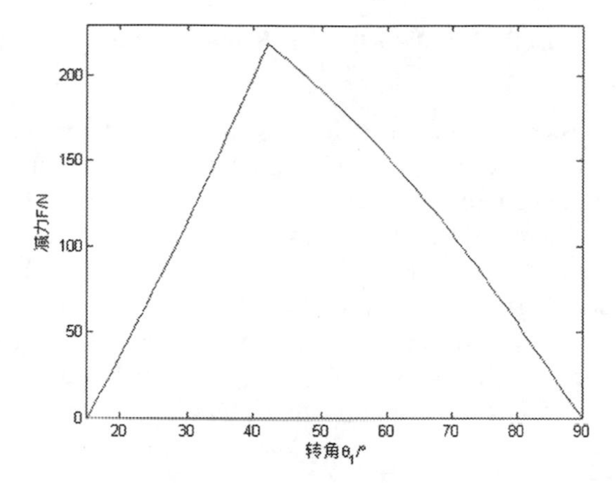

Fig. 4. Curve of knee joint force reduction device

3 Analysis and Optimization of Knee Joint Force Reduction Device

3.1 Effect Factor of Length of Crank to Knee Joint Force Reduction Device

According to the kinetics model, drawing value of force reduction device, change with length of crank, within the range of knee joint movement angle. Figure 5 presented curve of reduced force value under conditions when length of bias force plank equals 55 mm and length of crank equals 20 mm, 25 mm, 30 mm, 35 mm respectively. Through the analysis we can know that the length of crank has less influence in changing the value of reduced force. Under certain range, the increase of the length of crank would decrease the range of reduced force.

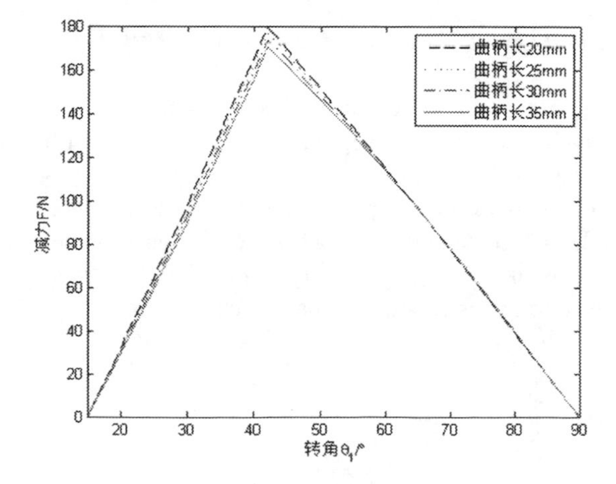

Fig. 5. Reduced force curve graph of changing crank length

3.2 Length of Connecting Rod's Effect to Knee Joint Force Reduction Device

According to established kinetics model, drawing value of force reduction device, change with length of connecting rod, within the range of knee joint movement angle Fig. 6 presented curve of reduced force value under conditions when length of bias force plank equals 35 mm and length of crank equals 40 mm, 45 mm, 50 mm, 55 mm respectively. Connecting rod's length is bias force plank and oil buffer's arithmetic square root of quadratic sum. The length of the force biasing plate increases and the length of the connecting rod increases. Through the analysis we can know that the length of connecting rod mainly affect the location of reduced force curve. Within certain range, the increase of the length of connecting rod would decrease the range of reduced force.

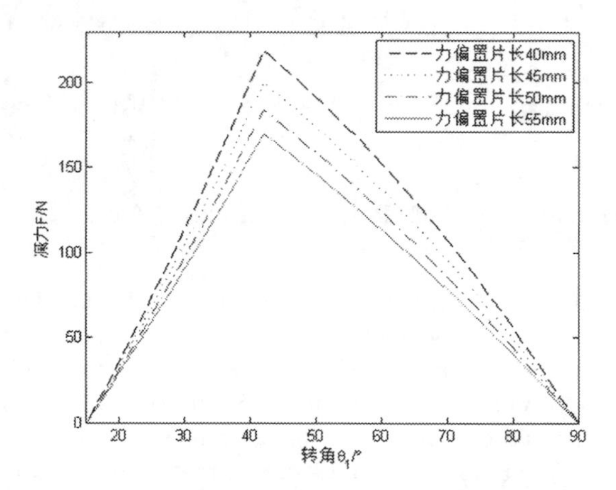

Fig. 6. Reduced force curve graph of changing bias force plank's length

4 Algorithm Design of Fatigue Detecting Model

This fatigue detecting model is based on analysis of information collected by surface electromyogram signal sensor, and quantitative fatigue value can be calculated.

Surface electromyogram signal sensor's place location is as Fig. 7 presented:

Assuming that human's walking speed and surface electromyogram signal sensor's range are independent of human fatigue's degree, uniformization algorithm of collected signal is required. t^* is time parameter in each cycle after uniformization, v^* is voltage value of surface electromyogram signal sensor after uniformization.

$$t^* = \frac{t - t_{\min}}{t_{\max} - t_{\min}} \tag{1}$$

$$v^* = \frac{v - v_{\min}}{v_{\max} - v_{\min}} \tag{2}$$

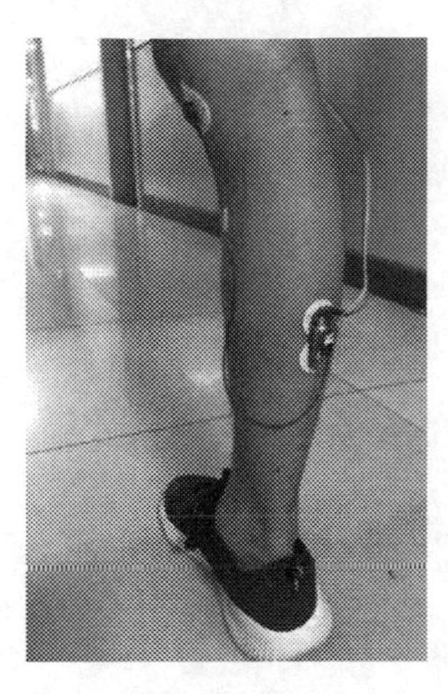

Fig. 7. Surface electromyogram signal sensor's place location

Several sets of experiments were carried out. After collecting results under fatigue and un-fatigue conditions of surface electromyogram signal sensor, choosing data from 5 walking cycle and drawing line chart as follows after uniformization (Figs. 8 and 9):

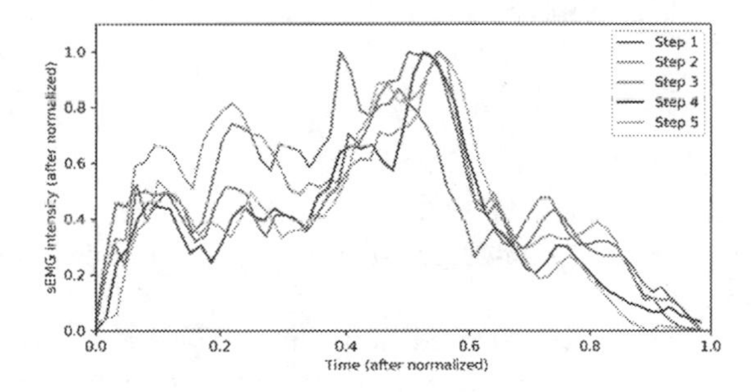

Fig. 8. Un-fatigue status walking electromyographic signal waveform

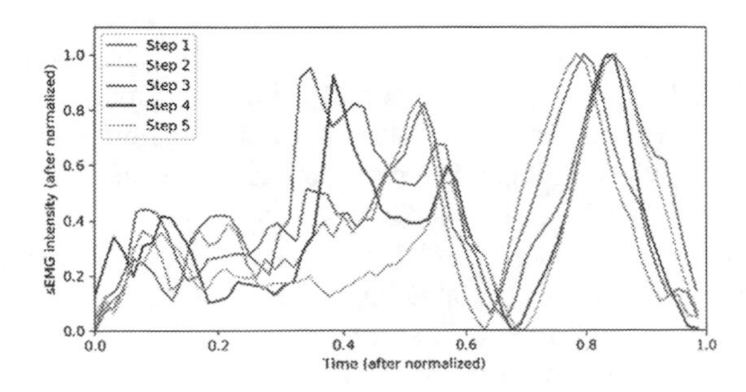

Fig. 9. Fatigue status walking electromyographic signal waveform

Appling fitting processing to data presented above and drawing fitted curve, making collected data point centralized at banding area, as Fig. 10 presented:

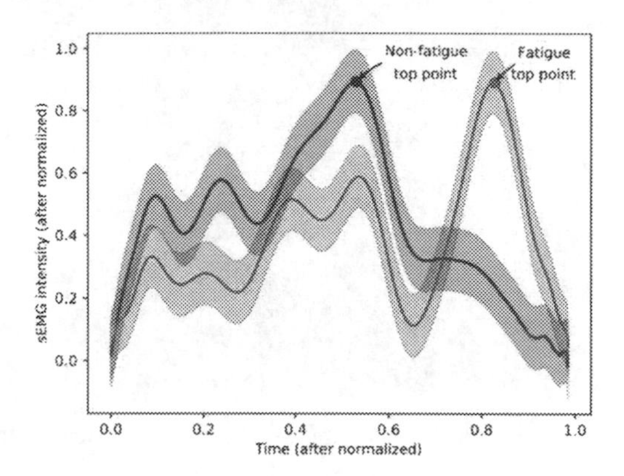

Fig. 10. Un-fatigue and fatigue status waveform

After analysis, the peak point of the waveform appears between 0.5 and 0.8 after normalization, and peak point of waveform's appear time is related to the level of human fatigue. Higher the level of human fatigue, later appear time of peak point, therefore, fatigue model can be concluded:

$$FL = \frac{t_p - t_{pn}}{t_{pf} - t_{pn}} \times 100\% \tag{3}$$

FL (Fatigue Level)—Fatigue Level, t_p—real time collected peak time point, t_{pn}—appear time of peak point under un-fatigue status, t_{pf}—appear time of peak point under fatigue status (Fig. 11).

Based on the above theory, Android Software for fatigue detection is designed, which can display the gait information and fatigue degree of human body in real time.

5 Conclusion

1) This paper proposes knee joint force reduction device which is based on crank slide mechanism. The basic principle is that crank slide under mechanism theory, which would provide force reduction effect during human changing from bending status to up-right status while walking and play buffering role during this process.
2) According to MATLAB's simulated curve, knee joint force reduction device's force reducing effect is closely related to crank and connecting rod's length. Crank's length has little effect in affecting force reducing effect while the increase of connecting rod would decrease the range of reducing force.

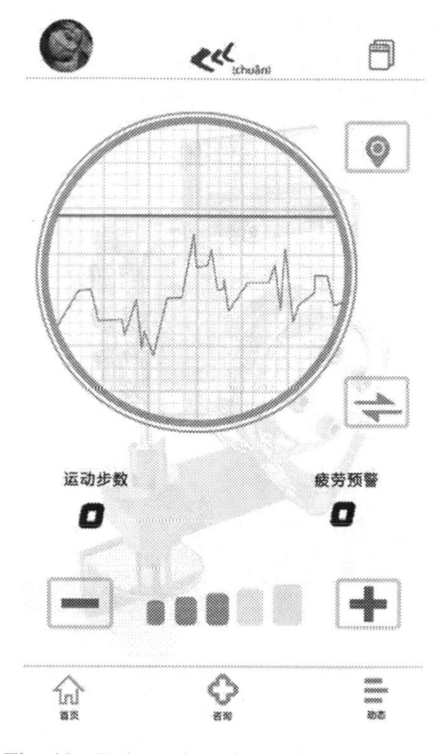

Fig. 11. Fatigue detection software interface

3) As a newly designed product, knee joint force reduction device still has room for improvement, such as increasing corresponding sensor which can display force reducing effect in real time and adopting structure optimization which can adjust crank and connecting rod's length to change force reducing effect at any time.
4) Fatigue model algorithm is able to detect and reflect the level of human fatigue through surface electromyogram signal's data in real time, and display the result through software to remind user to take a rest.

References

1. Wang, D.: Passive gravity supported flexible lower extremity exoskeleton system based on walking gait, p. 1. Zhejiang University (2016)
2. Yanhong, Lu, Shi, Xiaobing: Current situation and progress of epidemiology of knee osteoarthritis at home and abroad. Chin. J. Orthop. **20**(06), 81–84 (2012)
3. Zhang, Lijie, Wentao, Lu, Cao, Xuemin: Optimal design of four-bar bionic knee joint mechanism. Mach. Tool Hydraul. **43**(09), 67–70 (2015)

4. Abernethy, B., Mackinnon, L.T., Kippers, V., et al.: The Biophysical Foundations of Human Movement, pp. 10–100. Human Kinetics Publishers, Champaign (2004)
5. Li, B., Xu, C.: Matlab aided analysis of mechanical principles, vol. 8–13, pp. 44–51 (2011)
6. Wang, J.: Design and Simulation of the lower limb force increasing mechanism of exoskeleton robot, p. 25. Zhejiang University (2016)
7. Önen, Ü., Botsalı, F.M., Kalyoncu, M., et al.: Design and actuator selection of a lower extremity exoskeleton. IEEE/ASME Trans. Mechatron. **19**(2), 623–632 (2014)

Design and Implementation of Face Recognition Access Control System

Ling Peng(✉) and Yanchun Chen

Guangdong University of Science and Technology, Dongguan 523083, China
4829294@qq.com

Abstract. The convenience and security of traditional access control systems are poor. The face recognition system based on face recognition technology can effectively improve users' experience and security by controlling the door locks through the matching results of face recognition. In this design, face information is obtained through a camera, and an improved PCA algorithm is applied to obtain feature vectors, identify and match it. The controller operates an electromagnetic lock module according to the matching results, and displays the matching results on the OLED display screen.

Keywords: Raspberry Pi · PCA algorithm · Face recognition

1 Introduction

With the rapid development of information technology, face recognition technology is becoming more and more mature, and it is widely used in various fields, such as human tracking, commercial payment, mobile phone unlocking, and even railway stations have introduced face recognition devices to identify passengers [1]. The face recognition technology enriching our lives, and also brings us a better and convenient lifestyle. The traditional access control systems, or the card-type induction access control system is based on RFID technology, or the access control system is based on fingerprint recognition technology [2], and all of these are authenticated by contact and have proved poor convenience and security. However, the access control system based on face recognition technology can effectively improve the user experience and security.

2 Overall Design

This paper uses the Raspberry Pi 3B as the microprocessor and the Raspberry Pi camera to obtain face information, together with other components, such as photosensitive module, OLED display screen, electromagnetic lock module, etc. The overall framework of the system is as shown in Fig. 1. The Raspberry Pi camera collects face information and transmits it to the Raspberry Pi 3B controller, and the improved PCA algorithm is used for face recognition. The controller operates the electromagnetic lock module according to the matching results, and displays the matching results on the OLED display screen.

K. Li et al. (Eds.): ISICA 2019, CCIS 1205, pp. 609–615, 2020.
https://doi.org/10.1007/978-981-15-5577-0_48

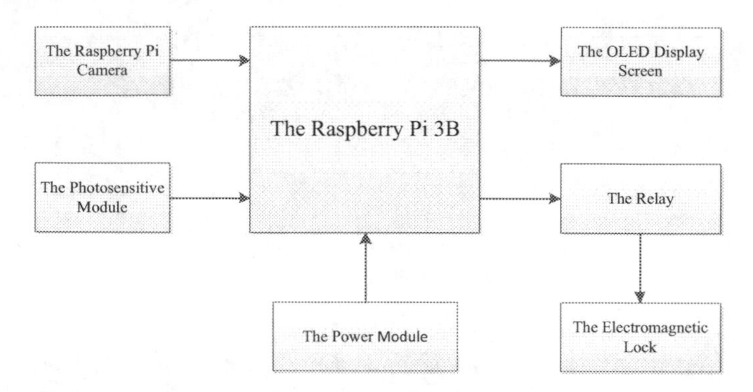

Fig. 1. The overall diagram of the system

3 Hardware Design

3.1 Microcontroller

The Raspberry Pi 3B is used as the master chip, based on ARM Cortex-A53 core Broadcom BCM2837 SOC, with built-in memory 1 GB (LPDDR2) and Micro SD card which is used as hard disk. The board can provide 4 USB interfaces and a 100M Ethernet interface. It has a TV output interface with video analog signals and a HDMI high-definition video output interface. It also has 40 general-purpose GPIO ports to provide pins for other peripherals, such as Camera interface, wireless WIFI interface, etc. [3].

3.2 Raspberry Pi Camera

The model of the Raspberry Pi camera is OV5647, which is a VGA CMOS image sensor produced by OV (OmniVision) [4]. The camera has 5 million pixels, aperture 2.8, and wide field of view. The camera can use CRIS (camera serial interface) interface to connect Raspberry Pi to obtain stable image data.

3.3 Photosensitive Module

In order to enhance the ability of the face recognition under dark light, this paper designs a photosensitive module. The photosensitive module uses sensitive photosensitive sensor. Because the intensity of the photodiode is different, and the current through the photodiode is different, the light intensity can be detected through the current. At the same time, this paper designs 4 LED fill lights. In the dark environment, 4 LED fill lights are automatically turned on so that the camera can also obtain clear face images.

3.4 Electromagnetic Lock Module

The electromagnetic lock module is a composite module, which is composed of a normally open relay and an external electromagnetic lock circuit. The two external control ports of the relay form a circuit with the external electromagnetic lock circuit. Under normal circumstances, the relay signal pin is low level. At this time, the relay is normally open, and forms a normally open circuit with the external electromagnetic, so the electromagnetic lock does not work (off state). When the relay signal pin goes high, the relay works and forms a closed loop with the external electromagnetic lock circuit, so the electromagnetic lock is working (on state) [5].

4 Software Design

4.1 Pretreatment

Affected by the environment, the face image collected by the camera may have problems such as illumination and contrast. Before face recognition, pre-processing is required. Preprocessing of face images includes operations such as grayscale transformation, geometry correction, filtering, and sharpening [6].

4.2 Extracting Image Features

PCA (Principal Component Analysis) can use the initial data to become a set of linearly independent representations of various dimensions through linear transformation. It is used to extract the main feature components of data, and is often used for dimensionality reduction of high dimensional data [7]. Applying the PCA algorithm in face recognition, the face image is processed to obtain the feature vector. In the process of face recognition, as the pixels of the camera become higher and higher, the matrix data to be processed is also getting larger. When the feature vector is extracted by the PCA algorithm, the amount of calculation is also larger. This paper improves the classical PCA algorithm. The first K column vectors are obtained according to the norm principle, and then the covariance matrix C is used QR decomposition to obtain a new eigenvector matrix, which can construct a low-dimensional matrix thereby. The specific implementation steps are as follows:

1. The sample matrix is entered a, assuming a size of m * n. It has m rows, each of which behaves an n-dimensional sample. Therefore the sample matrix can be expressed as $X_{m \times n}$.

$$X_{m \times n} = \begin{bmatrix} x_{11} & x_{12} & \cdots & x_{1n} \\ x_{21} & x_{22} & \cdots & x_{2n} \\ \cdots & \cdots & \cdots & \cdots \\ x_{m1} & x_{m2} & \cdots & x_{mn} \end{bmatrix} \tag{1}$$

2. According to the principle of PCA, this system really needs the first k with the main features in this matrix, instead of all eigenvalues and eigenvectors in this matrix. The principle of 2-norm is introduced, as shown in formula (2). The first k column vectors with the largest norm form the matrix $A_{m \times k}$.

$$||\mathrm{a}||_2 = \sqrt{x_1^2 + x_{12}^2 + \cdots x_n^2} = \sqrt{\sum_{i=1}^{n} |x_i|^2} \tag{2}$$

3. A is performed QR decomposition, A = QR, which can construct covariance matrix $\mathrm{C} = \mathrm{QR}R^T Q^T$.
4. R^T is performed singular value decomposition, $R^T = \mathrm{UD}V^T$, Owing $\mathrm{D} = \mathrm{diag}(\sigma_1, \sigma_2, \cdots, \sigma_t)$, then the covariance matrix can be expressed as the formula (3).

$$\mathrm{C} = \mathrm{QVD}U^T\mathrm{UD}V^T Q^T = QVD^2 V^T Q^T = QV\Lambda V^T Q^T \tag{3}$$

In the formula (3), $\Lambda = D^2$, owing $(QV)^T(\mathrm{QV}) = V^T Q^T QV = V^T\mathrm{V} = \mathrm{I}$, QV can diagonalize C. QV is the eigenvector matrix of C. Λ is the eigenvalue matrix of C.

5. P = QV is set the eigenvalue vector matrix, which can calculate the projection of the original data sample matrix X on the new eigenvector Y, as shown in formula (4).

$$\mathrm{Y} = \mathrm{X} * \mathrm{P} \tag{4}$$

The matrix Y is the data after dimension reduction.

4.3 Matching Recognition

The face matching process can be divided into training phase and testing phase. In the training phase, the Raspberry Pi camera collects face image. Through improved PCA algorithm, the Raspberry Pi 3B extracts the face image feature vector for forming the training set. In the testing phase, the Euclidean distance of the face image feature vector that will be recognized and the feature vector of the training set is calculated, as shown in formula 5. When the Euclidean distance is less than the threshold, it means that the face to be recognized is the same person with the k face in the training set. But when traversing all of the training set are greater than the threshold, it indicates that the face to be recognized is not in the training set.

$$\varepsilon_k = ||P - P_k^2|| \tag{5}$$

Since the PCA algorithm is sensitive to illumination, the paper proposes a weighted approach. The weighted approach firstly sorts the feature values from the largest to the smallest, and then the maximum four feature vector are used for weighting. The principle of weighting is to assign a weight less than 1 to the previous four vectors. While the other vectors remain unchanged, and use the weighted feature vector to calculate the Euclidean distance to determine whether the matching is successful. By the weighted approach, the size of the first four vectors can be reduced. Thereby reducing the impact on illumination but without reducing the effective information at the same time, it can also contain useful face image information.

5 Testing

5.1 Capture Camera Data

The system first builds the hardware peripherals of the Raspberry Pi 3B, and then enables the Raspberry Pi camera interface to collect face image. When face entry and recognition are performed, the light intensity is first judged. If the illumination is sufficient, we can directly enter and recognize the face image. And if it is in a dark environment, we can turn on the fill light and take a picture with fill light, as shown in Fig. 2.

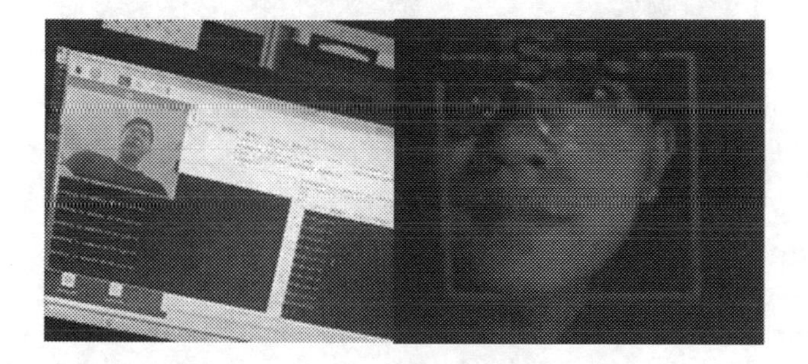

Fig. 2. The collecting face image diagram

5.2 Algorithm Comparison

Based on the experimental environment, this paper studies the classic PCA algorithm and the improved PCA algorithm, and mainly compares the recognition rate and recognition time. After repeated tests, the improved PCA algorithm has a great improvement in recognition rate and recognition time, especially when the weighting coefficient is 0.4, the recognition rate is more obvious. The experimental results are shown in Table 1.

Table 1. The algorithm comparison result table.

Algorithm type	Average recognition rate	Average recognition time
Classic PCA algorithm	78.45%	3.80 s
Improved PCA algorithm	81.30%	2.98 s

5.3 Electromagnetic Lock Control

According to the improved PCA algorithm, the electromagnetic lock is controlled by the recognition result. If face recognition does not match, the relay will not be triggered, the electromagnetic lock is in the default working state, that is, the off state; If face recognition matches, the relay is triggered, the green indicator light of the relay will light up, the electromagnetic lock works, and the state is open, as shown in the Fig. 3.

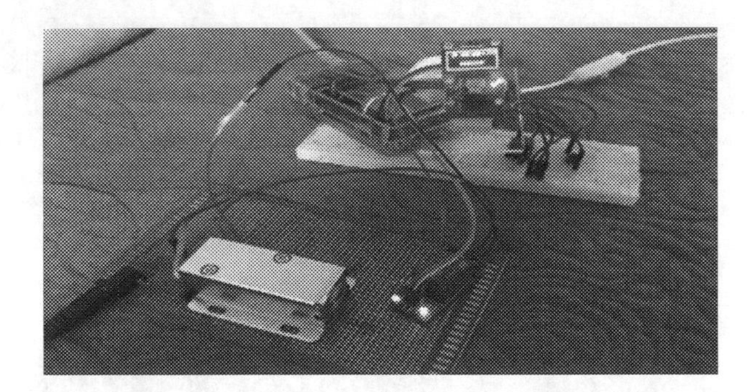

Fig. 3. The electromagnetic lock open state diagram

6 Summary

The system selects the raspberries pi as the micro controller with raspberry pi camera module, photosensitive module, OLED screen display module and relay control electromagnetic lock module. It not only can input image data when the light is enough, but it also can light up under the dim light, and the image information is processing. At the same time, using an improved PCA algorithm can improve the recognition rate and reduce data processing time so as to realize the function of entrance guard.

References

1. Zhong, G.: Research on face recognition system based on video sequence. China Sci. Technol., 5 (2016)
2. Wang, X.: The application of fingerprint recognition technology in access control system. Constr. Eng. Technol. Des. 14 (2017)
3. Zheng, W.: Design of Portable Plasmon-Enhanced Raman Fast Detector, p. 6. Xiamen University (2017)

4. Lei, H., Lan, H., Cai, W., Hu, Z.: Laser ranging system based on OV7725 monocular camera. J. Jianghan Univ. Nat. Sci. Ed. **45**(2), 120 (2017)
5. Liu, W.: Application of new safety door chain device, urban construction theory research (Electronic Edition), 20 (2013)
6. Zhang, D.: Research on the status and development trend of face recognition. Guangdong Commun. Technol. 6 (2018)
7. Wei, Y., et al: Data correlation study of SLE disease based on PCA mining. Software, 12 (2017)

Geohash Based Indoor Navigation

Yingshi Ye[1], Fuyi Wei[1](✉), Xianzi Cai[1], and Xiaofeng Hu[2]

[1] College of Mathematics and Information, South China Agricultural University, Guangzhou 510642, China
weifuyi@scau.edu.cn
[2] Shenzhen Wei Daan Computer Co., Ltd., Shenzhen 518109, China

Abstract. In order to reduce the dimension of coordinate information and locate quickly, the Geohash algorithm which was used for outdoor navigation originally, is extended to the indoor navigation. It divides the indoor space into many grids for finding optimal path and combines eight-direction maze for obtaining obstacle avoidance function. Through the technique of pretreatment, the new algorithm transforms the regional grids and inter-regional entry into graph theory model separately, uses the matrix to find the shortest path and compute the distance by designing the distance hash table and the inter-regional routing table to be a storage space. Experiment shows that this algorithm calculation speed is 82.4% higher than the A* algorithm approximately.

Keywords: Indoor navigation · Geohash algorithm · Routing table · Rasterize · A* algorithm

1 Introduction

With the rapid development of the economy, the scale of the city and the intensity of people's tourism have expanded rapidly. In the face of this change, it is urgent for us to have an automated navigation throughout every region. The development of outdoor navigation is earlier than the indoor navigation. So the research on outdoor navigation is more mature. A* algorithm and grid navigation are classic outdoor navigation algorithms. The Geohash algorithm [1] is the mainstream algorithm in grid navigation in recent years. This algorithm is a geo-data coding technology based on grid division, which can reduce the two-dimensional coordinate information to one-dimensional space, so that the coordinate information can be quickly located through various special data structures.

The indoor environment includes lots of obstacles, such as walls and columns, which makes the spatial information complicated. So the outdoor navigation technology cannot be directly applied to the indoor navigation. The indoor navigation can be roughly divided into three aspects [2]: positioning, model construction and path planning. Firstly, most of the indoor navigation technologies rely on the device's receipt of the signal to determine its position [3]. Then, in the model construction step, the most common processing methods uses geographic information to construct a road network model. In the existing research, Delaunay triangulation [4], complex plane vector grid coding [5], recursive partitioning

K. Li et al. (Eds.): ISICA 2019, CCIS 1205, pp. 616–627, 2020.
https://doi.org/10.1007/978-981-15-5577-0_49

regular octahedron [6], segmentation subspace [7] and graph semantics [8], have been used for constructing. Road network models can be used for model constructing. But it is difficult for these methods to model indoor obstacles. The grid model [9–12] can handle obstacle avoidance, but these methods cannot quickly locate the target coordinates to the grid points. In summary, urgent problems for indoor navigation to deal with are as follow: (1) efficiently positioning of the grid model; (2) avoiding the indoor obstacles; (3) improve the automation of modeling.

The indoor navigation algorithm in this article is based on Geohash method. In addition, the maze thought is involved to obtain obstacle avoidance function. Introducing area partitioning can solve the problem of memory information redundancy. Using pre-processing to store partial paths and distances make it fast-responding. The obvious optimizations make the new algorithm fast to position automatically, precision high and obstacle avoidance.

2 The Geohash Algorithm

The Geohash algorithm is a geographic data coding technology [1]. The algorithm is widely used in positioning query. It can reduce the redundancy generated by the traditional latitude and longitude representation method by converting the target latitude and longitude coordinates into string encoding, while ensuring the accuracy of the target position.

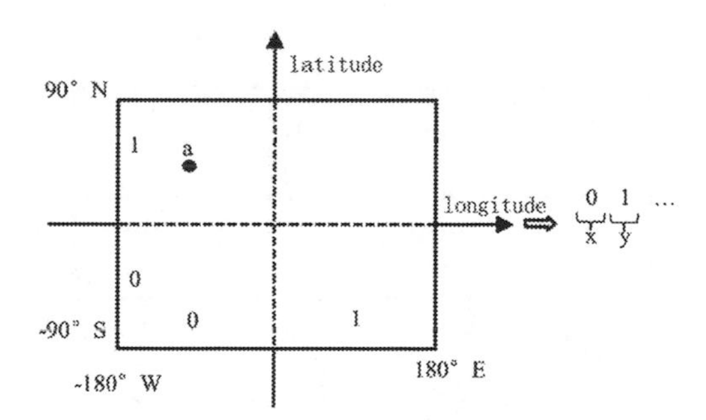

Fig. 1. Geohash coding diagram.

The Geohash algorithm can be described as a method which turns a two-dimensional coordinate into a one-dimensional sortable and comparable string. The implementation process is as follows: Firstly, the earth is regarded as a two-dimensional plane rectangle. Then the longitude(latitude) is divided equally into two left and right(up and down) intervals. When the coordinate point falls in the left (lower) interval, the code is 0. On the contrary, the code is 1. The interval where the coordinate points are located is divided continually until the set accuracy is met. A binary code corresponding to the coordinate point will be calculated. It is the geohash coding. For example, the point A in Fig. 1 is located on left interval of the x direction and the right interval of the y direction. So point A is coded as 01.

In view of the above characteristics of the Geohash algorithm, this algorithm is used for fine positioning of indoor environmental objects.

3 Indoor Geohash Algorithm

At present, the common navigation application software is based on real-time calculation. The indoor navigation system of industrial application has a large amount of data concurrency. So the practical application has a great requirement for calculation speed. Sacrificing the cost of space to optimize time performance is a common approach to program optimization. Preprocessing belongs to this optimization method.

3.1 Space Meshing Method

In this paper, the indoor space will be called as the space simply. The whole space is treated as a rectangular plane, which is modeled after the global positioning Geohash coding method.

Geohash algorithm is regional. In other words, the space can be evenly distributed into several same size rectangles by dichotomy. These rectangles will be referred as grids. And this dichotomy process will be referred as space meshing.

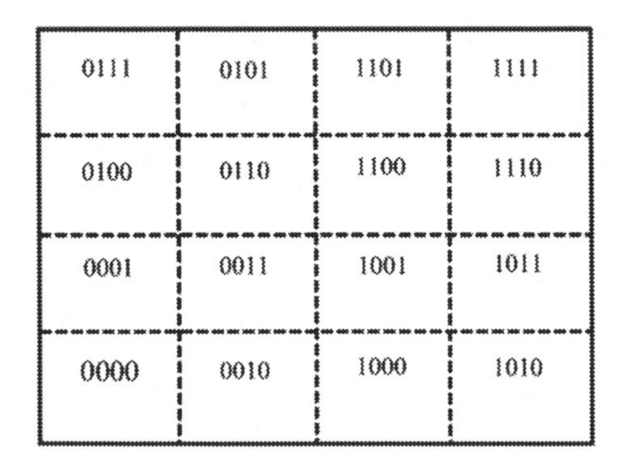

Fig. 2. Geohash dichotomy twice on the space

In Fig. 2, the space is divided into 16 same size grids. And each of them own a binary string. Once the space has been meshing, every grids owns an unique binary string. If we transform this binary string into decimal number, the number will be from zero and be continuous. This feature allows the encoding could be stored in an array. We need to define the following data structure to store raster information, which uses array to storage, as a raster information table:

3.2 Obstacle Information Processing

The indoor environment contains obstacles such as walls and load-bearing columns, which makes the spatial information very complicated. After the space meshing, some grids may become inaccessible due to these obstacles. Therefore, it is difficult to achieve accurate and efficient indoor positioning using only Geohash coding. How to use the obstacle information reasonably to determine the direction of motion of the path is the key of this section.

Navigation Direction Determination. Using the idea of maze [14], navigation direction is simply reduced to eight directions: east, south, west, north, southeast, northeast, southwest, and northwest. The direction of the grid can be stored in a one-dimensional array with eight elements. We call this array as the Direction array(D). The mapping relationship between direction and array subscript is shown in Fig. 3.

We define $D_i(n)$ as a value of the ith direction for a grid with Geohash decimal code n, in which $D_i(n) = 0/1(0 \leq i \leq 7)$. When $D_i(n) = 1$, the grid with n can pass to i^{th} direction. On the other hand, it cannot pass to i^{th} direction. While two adjacent grids can pass mutually, they can be regarded as being connected. Otherwise, the two grids are considered to be not connected. For example, in Fig. 4, grid a_1 and a_2 is adjaced. And a_2 is located in the south of a_1. So when $D_3(a_1) = D_1(a_2) = 1$, $a_1$and a_2 is connected.

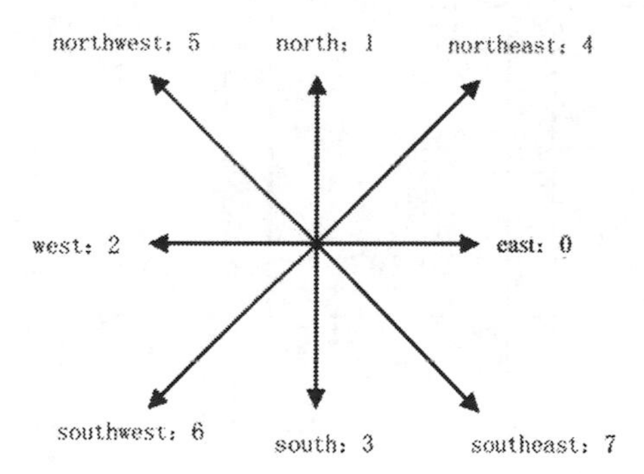

Fig. 3. The mapping of direction and subscript

According to the relationship between grid and obstacle items, the grid is divided into the following three classifications:

1. Free grid: There is no obstacle blocking in the grid. So it is free to pass this grid in eight directions, $D_i(n) = 1$ shown as a_1 and a_2 in Fig. 4.
2. Obstacle grid: There are obstacles in the grid which is blocked but not completely covered. This grid will not pass to some directions shown as the b_1 grid and b_2 in Fig. 4.

Firstly, we will discuss the judgment methods in one direction (i.e., east, south, west, and north). Based on the center point of the grid, when a single direction of the grid including central point is covered by an obstacle, it is called that the single direction is completely covered by the obstacle. When some direction is completely covered by an obstacle, this direction cannot pass and $D_i(n) = 0$. Figure shows that $D_0(b_1) = 0$ corresponding to the east of the b_1 in Fig. 4. Otherwise, it can pass like the west of the b_2 in Fig. 4, with $D_2(b_2) = 1$.
If all single direction in the composite direction (i.e., southeast, northeast, southwest, northwest) can pass, such as the east and south for the southeast, then the composite direction can pass, otherwise it cannot pass.

3. Blocked grid: The grid is completely blocked by obstacles, it can not pass in any direction. $D_i(n) = 0 (0 \leq i \leq 7)$. It shows in c grid in Fig. 4. $D_i(c) = 0 (0 \leq i \leq 7)$.

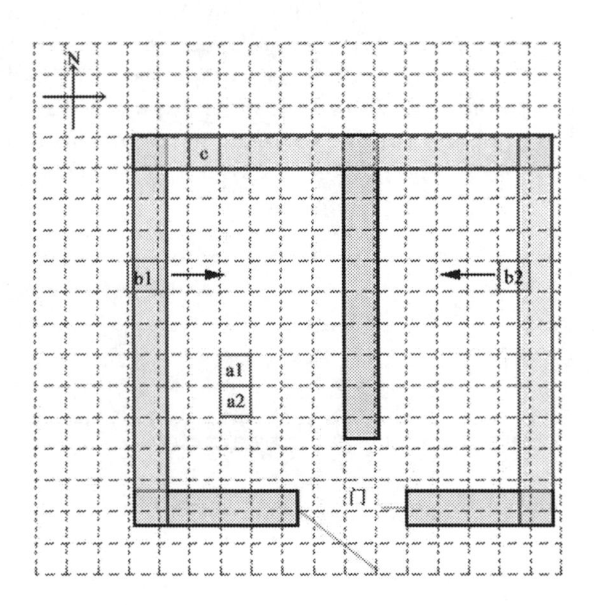

Fig. 4. Grid classification diagram

Grid Size Determination. With spatial rasterization, the space is divided into several rasters. In theory, it is possible to achieve arbitrarily small precision by continuous dichotomy. However, the higher the number of dichotomies is, the more grids the entire space are divided into, the more pre-stored information is. It will result in excessive space complexity in real-time systems.

The main difficulty in indoor navigation is to avoid obstacles such as walls and load-bearing pillars in the space and find the doors in the space. When the accuracy of the algorithm is too high, multiple grids will be divided into the wall. It will cause the grid precision to be lost. Conversely, if the grid is too large, the doorway will be covered, and the door passage information will be lost. It will result in misjudgment of the direction of the entrance and exit of the room. Therefore, the grid should be larger than the wall thickness, and smaller than the width of the door. It mean that:

$$max(wall\_size) \leq grid\_size \leq min(door\_size)$$

The number of dichotomies of the Geohash code is determined by the size interval of this. The relationship between the grid size and the number of dichotomies is as follows:

$$grid\_weight = \frac{x\_top - x\_low}{2^n}$$

$$grid\_height = \frac{y\_top - y\_low}{2^n}$$

The grid size will be gradually increased from 1 to n. n and will be determined when the grid size interval is reached for the first time.

3.3 Space Separated into Area

If the raster generated by spatial rasterization is used to calculate the distance directly, the memory space required is very large. In order to solve the problem of data redundancy, the space is divided into domains, and the distance information is processed.

The space is composed of rooms, corridors, etc., in which the rooms and corridors are connected by doors. The indoor space is hierarchical. Therefore, the rooms and the corridors are the basic units. The space is divided into several areas. The single room is labeled as a single entrance and exit area. The hallway and multi-door rooms are labeled as multiple entrance areas. We use $Area(n)$ to represent the number of area which the grid n belonged to.

3.4 Distance Pretreatment

The space is divided into M areas in 3.3. The distance can be calculated based on the relative position of the grid.

Undirected Graph Model in Area. The undirected graph models $G_m(V_m, E_m)(1 \leq m \leq M)$ are constructed in M areas. $V_m = v_0, ..., v_N$ indicates that area m has N grids. The grid is marked as a vertex. When $e \in E_m$, the v_i, v_j in area m is connected.

For fast search, a hashtable is designed for storing the path and distance. The data structure is as follows:

```
typedef struct Access{
    int target;/*target point Geohash decimal code*/
    int distance = 0;/*distance*/
    int path[PATH_NUM];/*path*/
```

```
    Access *next;/*address of next data point*/
};
Access accesses[MAX_ACCESS_NUM];
```

The hashnode structure is as follows:

```
typedef struct HashNode{
    int source;/* source point Geohash decimal code */
    Access *access;/* The first address of the reachable region
    array */
    Access *accessRear;/* The last of the reachable array */
    HashNode *next;/* Next node */
};
```

According to the undirected graph models, the adjacency matrix $A_m = (a_{ij})(i, j = 1, 2, ..., N)$ is structured. When $a_{ij} = 1$, the grids v_i, v_j in area m are connected with one step reachable. Conversely, when $a_{ij} = 0$, the grids v_i, v_j in area m are not connected without one step reachable. The power matrix$A_m^k = (a_{ij}^{(k)})$, is calculated. When $a_{ij}^{(k)} = 1$, the grids v_i, v_j can reach in k steps. Conversely, there is no path with k steps between the v_i, v_j grids. After this work, the distance and path between grids can be storaged in the hashtable.

Through the properties of the adjacency matrix described above, the distance and path between the grids in the are are calculated and stored in the intra-domain hash table. The specific steps are as follows:

(1) Initialization: Set $k = 2$ and the maximum number of steps $K = \max_{1 \leq i \leq M}(D_i)$. D_i is the diameter of the undirected graph. Turn (2);
(2) Calculate the k steps reachable matrix: $A_m^k = (a_{ij}^{(k)})$, $a_{ij}^{(k)} = \sum_{l=1}^{N} a_{il}^{(k-1)} \cdot a_{lj}^{(k-1)}$. Turn (3);
(3) Data cache: When $a_{ij}^{k-1} = 0$ and $a_{ij}^{k-1} \neq 0$, the grids v_i and v_j get the shortest path in k step. Due to $a_{ij}^{(k)} \neq 0$, it exists $0 \leq l \leq N$, so $a_{il}^{(k-1)} \cdot a_{lj}^{(k-1)} > 0$. It shows that the distance and path between the grids v_i and v_j can inquire in the hashtable. Turn (4)
(4) When $k \neq K$, turn (2). Otherwise, quit the program.

Routing Table Between Area. An inter-domain routing table between regions is generated by using a routing algorithm as follows. It can to achieve fast discovery and high timeliness. Due to the few of regions, an array is used as the storage structure, and the data structure of the inter-domain routing table is defined as.

```
typedef struct Route{
    int sourceArea;
    int targetArea;
    int sourceDoor;
```

```
    int targetDoor;
    int distance = 0;
    int path [PATHNUM];
};
Route routes[ROUTE_NUM];
```

Modeling the RIP algorithm. The following algorithm was designed.

(1) Constructed the undirected graph model $G^{'}(V, E, W)$ in area. The element in V represents the grid where all the areas in the space are located. When $e_{ij} \in E$, the grids v_i and v_j can be connected by multiple entrance area. Turn (2);
(2) Calculate the distance and path between each entry and exit through RIP, store it in the inter-domain routing table (this table is a static table), and exit.

3.5 System Initialization Process

The system initialization process are described as follows:

(1) Reading spatial information: Structural information (room, corridor, space size, etc.) and obstacle information (walls and doors) in the space;
(2) Accuracy determination: determined the accuracy and dichotomy by the obstacle information;
(3) Space rasterization: Use Geohash to divide the whole space into grids, and determine the intercommunication according to the positional relationship between the grids;
(4) Space division: The area information is stored in the grid information table by dividing the area into rooms and corridors in the space;
(5) In-domain hash table: establish an intra-domain undirected graph model in each region, and use the adjacency matrix of the graph to calculate the distance between each grids in the domain and find the shortest path, and store it in the inter-domain hash table;
(6) Inter-domain routing table: According to the regional entrance and exit, an inter-domain undirected graph model is established. The adjacency matrix of the graph is used to calculate the distance and find the shortest path respectively, and store it in the inter-domain routing table, and the algorithm initialization is completed.

After the algorithm is initialized, the raster information table, the inter-domain hash table, and the inter-domain routing table should be in the cache for future reference.

3.6 Distance and Shortest Path Acquisition

After the algorithm is initialized, it can accept the starting point and target point information at any time, to find the shortest path and calculate the distance.

Set $A(x_i, y_i)$ the starting point and $B(x_j, y_j)$ the ending point. The steps to find the shortest path for point A and point B and calculate the distance are designed as follows:

(1) Get two grids codes by Geohash encoding $(String_i, String_j)$.
(2) The two grids coded region codes are obtained from the raster information table $(Area_i, Area_j)$. If $Area_i = Area_j$, then two points are in the same area, and go to (3). Otherwise, go to (4).
(3) Use the intra-domain hash table to get the shortest distance $(d(A, B))$ and path $(l(A, B))$ for the two coordinate points, and exit the algorithm.
(4) Gets the set of all the entry and exit distances from the two coordinate points to their area (DA_i, DA_j). Then obtain the distance set between all the entrances and exits of the two areas through the inter-domain routing table (DR_{ij}). The shortest distance is:

$$d(A, B) = \min da_{i_0} + da_{j_0} + dr_{i_0 j_0} da_{i_0} \in DA_i, da_{j_0} \in DA_j, dr_{i_0 j_0} \in DR_{ij}$$

The shortest path is:

$$l(A, B) = l(A, entry_{i_0}, entry_{j_0}, B)$$

$entry_{i_0}(entry_{j_0})$ is the shortest path from point A(B) to B(A). Exit the algorithm.
Through the above algorithm, the shortest distance and the shortest path between the two coordinate points can be obtained.

4 Experiments

In order to verify the efficiency of the indoor navigation based on Geohash algorithm, two types of experiments are designed: simulated raster maps and complex geography.

The hardware environment of the experimental platform is Intel(R) Core(TM) which has four 3.6 GHz core CPUs. It supports eight threads in parallel on the hardware. The operating system is Windows 10 Professional (x64) with 8 GB memory. C language is adopted for programming.

4.1 Experiment in a Simple Simulation Map

To verify the effectiveness of the algorithm, the experiment is set on a simple simulation grid map[15]. This map is a grid map with 10 grids of length and width. In this environment, the path length can be calculated as the length of the grid side. The experiment graph is shown as Fig. 5.

The experimental results are shown as Table 1. In the same simulation environment, both algorithms can obtain the shortest path. The average time of the A* algorithm is 0.898 s, and the average time of the algorithm is 0.134 s. The algorithm in this paper is obviously better than the A* algorithm.

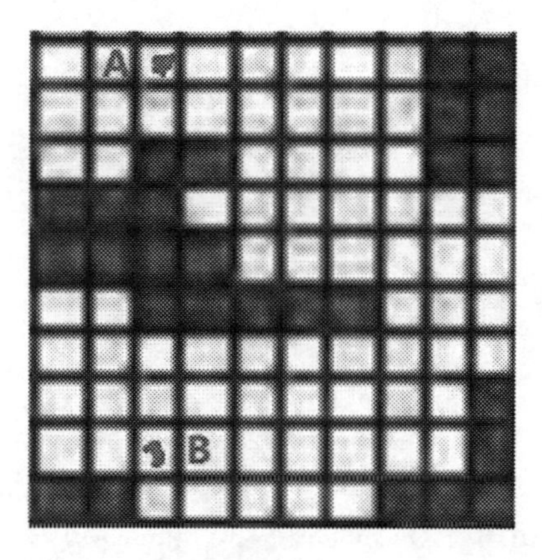

Fig. 5. A simple simulation map

Table 1. Two algorithm runtimes under a simple grid map

Algorithm	Time(s)	The length of path(m)
The A* algorithm	0.898	14.484
The algorithm in this paper	0.134	14.484

4.2 Experiment in a Complex Real Environment

The experimental environment is a place with 50 m long and 30 m wide (as shown in Fig. 6). There are 10 rooms in the experimental environment and the room has an internal wall block.

The dichotomy times is set as 12. In the initialization process, the space is divided into 4096 grids and 13 areas, which is shown in Fig. 7.

To verify the effectiveness of our algorithm, the traditional A* algorithm and the Geohash-based navigation algorithm designed in this paper are used to calculate the distances and path in the environment. This experiment will contain three pairs of points: the points in the same area(point a1, a2), the points cross single area (point b1, b2) and the points path cross multiple areas(point c1, c2). The shortest path is marked on Fig. 7. The process time and the distance are placed on Fig. 8.

The experimental results show that the calculating speed of the traditional A* algorithm decreases greatly as the distance between the source and target points increases. For the source and target points of the above four relationships the algorithm presented in this paper has well calculation speed. Its calculating speed for long distance calculations (crossing multiple regions) is also much better than the A* algorithm. The speed of the new algorithm increased 82.4% at most compared to the A* algorithm (across multiple regions). Multi-threading technology is adopted for multi-point computing. Experiments show that the speed of multi-point pair processing is better than that of A* algorithm.

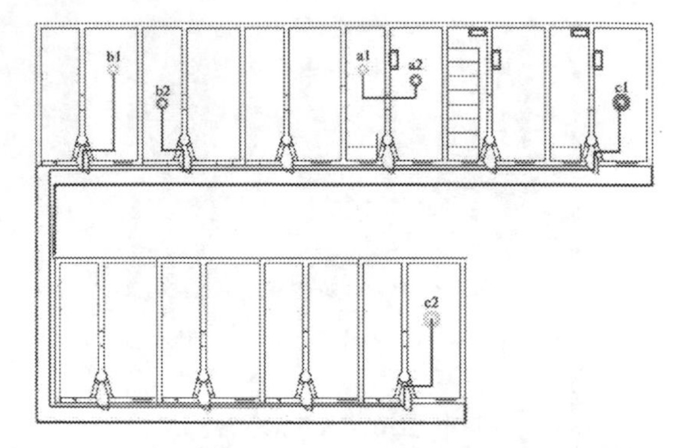

Fig. 6. Experimental environment graph

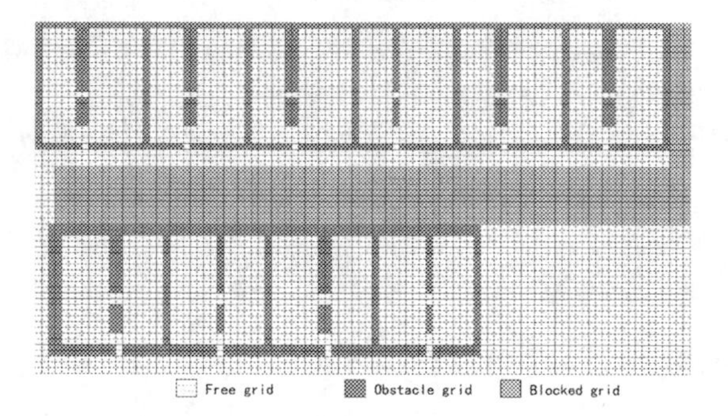

Fig. 7. Experimental environment grid figure

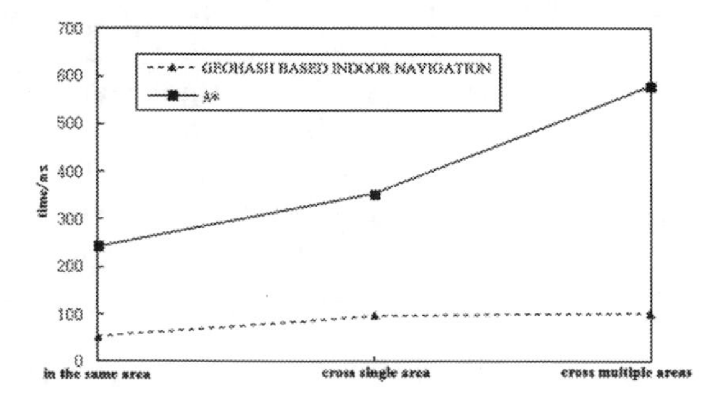

Fig. 8. Experimental result

5 Conclusion

This paper studies the method of obtaining the shortest path in the indoor environment. The pre-processing optimization strategy makes the real-time calculation more rapid. In terms of optimization and innovation, Geohash coding is used to reduce the dimensionality of the two-dimensional coordinate information, and the coordinate points are quickly located by this method. In view of the complex indoor environment and the need to avoid obstacles, methods such as obstacle handling and routing tables are introduced, so that the algorithm can quickly find the shortest path. Through the above optimization methods, the algorithm has the following advantages: (1) high automation and fast calculation response; (2) the capability of finding the shortest path and calculating its distance; (3) due to the use of pre-processing techniques, the higher average calculation speed for more coordinate points. Finally, the efficiency of the algorithm is verified by experiments.

References

1. Niemeyer, G.: Geohash Tips & Tricks (2013). http://geohash.org/site/tips.html
2. Yang, D.: Design and implementation of indoor precise positioning navigation system. Beijing University of Posts and Telecommunications (2011)
3. Gao, W., Hou, C., Xu, W., Chen, X.: Research progress and prospects of indoor navigation and positioning technology. J. Navig. Mapp. **7**(01), 10–17 (2019)
4. Lin, Y.: Indoor path planning algorithm introduced by navigation grid. J. Surv. Mapp. Sci. **41**(02), 39–43 (2016)
5. Vince, A., Zheng, X.: Arithmetic and Fourier transform for the PYXIS multi-resolution digital Earth model. Int. J. Digit. Earth **2**(1), 59–79 (2009)
6. Geng, J., Tong, X., Zhou, C., Zhang, K.: Generation algorithm of hexagonal discrete grid system of regular octahedron. J. Geo-Inf. Sci. **17**(07), 789–797 (2015)
7. Xu, Z., Zhong, S., Wang, Y.: An easy-to-update indoor navigation road network construction method. Comput. Simul. **32**(12), 267–271+275 (2015)
8. Lin, G., Song, G., Deng, C.: Research on the construction of semantic indoor navigation model based on graphs. Surv. Mapp. Eng. **24**(01), 48–52 (2015)
9. You, T., Wang, G., Lu, X., Sun, W., Zhang, W.: A traffic area model for indoor navigation and its automatic extraction algorithm. J. Wuhan Univ. (Information Science Edition) **44**(02), 177–184 (2019)
10. Jin, P., Wang, N., Zhang, X., Yue, L.: Mobile object data management for indoor space. Chin. J. Comput. **38**(09), 1777–1795 (2015)
11. Fu, M., Zhang, H., Wang, P., Wu, S., Lu, F.: Construction method of indoor navigation network based on moving object trajectory. J. Earth Inform. Sci. **21**(05), 631–640 (2019)
12. Zhang, A., Qu, Q., Deng, J., Pei, H., Chen, R.: Optimization of Dijkstra shortest path algorithm for indoor discrete grid space. J. Xiamen Univ. Technol. **26**(05), 36–43+67 (2018)
13. Liu, Y.: Comparison of rasterization of vector line elements by eight directions and full path method. Sci. Technol. Inform. **24**(10), 66 (2013)
14. Xie, R.: Computer Network, vol. 6, 6th edn, pp. 1–445. Electronic Industry Press, Beijing (2013)
15. Cheng, X., Yan, Y.: Indoor pointing path planning algorithm based on grid method. J. Chin. Inertial Technol. **26**(02), 236–240+267 (2018)

Application of NARX Dynamic Neural Network in Quantitative Investment Forecasting System

Jiao Peng(✉) and QingLong Tang

Guangdong University of Science and Technology, Dongguan, China
154711901@qq.com

Abstract. With the development of computer technology, quantitative investment based on data and rules has gradually emerged in China. Quantitative model has become a powerful tool to predict the market and guide investment. However, due to the complexity of the stock market, the traditional time series prediction technology has many shortcomings. The introduction of nonlinear autoregressive neural network NARX provides a basis for quantitative analysis of investment in stocks. The quantitative stock technical index data is taken as the training parameter and input parameter of NARX dynamic regression neural network, and the analysis results are obtained. We have collected Shanghai composite index data, tested the accuracy of quantitative analysis method, analyzed the prediction results of the model, compared the difference between the prediction results and the actual data so as to analyze the accuracy of the model.

Keywords: Neural network · NARX · Quantitative investment

1 Introduction

With the development of China's securities market and the maturity of technology, quantitative investment has become one of the important investment decision-making tools. As the automation of fundamental analysis of investment methods, quantitative investment can help investors to select substantial key information from more and more information and turn it into a investment decision. It is widely used in the field of stock investment, but involves a large amount of data, which requires computer software to assist analysis and processing. With the development of artificial intelligence technology, neural network model is gradually applied to the prediction of securities market. As a nonlinear autoregressive neural network, NARX dynamic neural network has better convergence and generalization ability, and is more widely used in the prediction of financial time series.

K. Li et al. (Eds.): ISICA 2019, CCIS 1205, pp. 628–635, 2020.
https://doi.org/10.1007/978-981-15-5577-0_50

2 Introduction to Quantitative Investment

2.1 Concept of Quantitative Investment

Quantitative investment mainly solves the problem of investment object and execution target. It refers to the investment method that uses mathematics, statistics and information technology to form mathematical models, uses computer to analyze the fundamental data of macro economy, industry and companies to generate investment strategies, and uses mathematical models to predict the future changes of investment portfolio. In short, quantitative investing is the automation of fundamental analysis. Quants can help investors choose the real key information with data mining technology, statistical technology and optimization technology and other scientific calculation methods for data processing, which has been the most optimal portfolio and the investment opportunity. The investors convert it into investment decision-making and widely use it in the field of stock investment. Since quantitative investment is an active investment strategy, the theoretical basis of active investment is the inefficiency or weak efficiency of the market. Through the analysis and research on the driving factors of price changes of individual stocks and industries, the investment portfolio can be constructed and optimized, so as to continuously beat the market and obtain returns beyond the market benchmark.

2.2 Characteristics of Quantitative Investment

The main characteristic of quantitative investment is to deduce the investment concept of qualitative research through mathematical model. With the help of the powerful information processing ability of the computer, the stocks that meet the "standard" in a full range are screened, the generation of any investment "blind spot" is avoided, and the "standard" investment object is captured to the maximum extent. It has the following characteristics:

Firstly, it has discipline. All investment decisions are based on models. Discipline is firstly shown in relying on and believing in the model. Every decision should be made according to the results of the model, which can overcome the weakness of human nature and cognitive deviation. In addition, every decision of traceable and quantitative investment is well grounded and supported by data.

Secondly, it is systematic. The main performance is "three multi": multi-level, multi-angle and multi-data. Multi-layer means that quantitative investment has models at different levels. Multi-angle means that investment prediction includes macro cycle, market structure, stubbornness, growth, profit quality, analysts' profit prediction, market sentiment and other perspectives. Multi-data refers to the processing of massive data.

Thirdly, it makes good use of the idea of arbitrage. Quantitative investment is to look for valuation depression, and catch the opportunities brought by misprizing and misevaluation through comprehensive and systematic scanning.

Fourthly, it wins by chance. It is manifested in two aspects: one is quantitative investment continuously excavates and USES historical laws which are expected to repeat in the future; The other is to rely on a group of stocks rather than one or several.

3 Introduction of Artificial Neural Network

3.1 Concept of Artificial Neural Network

Artificial neural network (or ANN, for short), hereinafter referred to as neural network, the abbreviation NN) or artificial neural network, is based on the basic principle of neural network in the biology, in understanding and abstract human brain structure and stimulation mechanism, the corresponding based on the theory of knowledge of network topology, simulation of the human brain neural system for complex information processing mechanism of a kind of mathematical model, used to estimate or approximate function.

Neural network is mainly composed of input layer, hidden layer and output layer. Each neuron in the network input layer represents a feature, the number of output layers represents the number of classification labels, and the number of hidden layers and hidden layer neurons are set manually. A basic three-layer neural network is shown in Fig. 1.

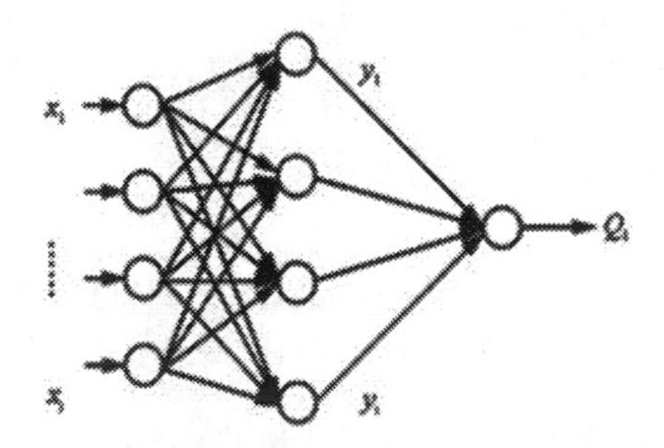

Fig. 1. Basic neural network diagram

Neural network is one of the important methods in data mining classification. Artificial neural network (ANN) is a mathematical model that uses structures similar to synaptic connections in the brain for information processing. In this model, a large number of nodes are connected to each other to form a network, or "neural network", achieving the purpose of processing information. Neural network usually needs training, and the training process is the learning process of network. Training changes the value of the connection weight of the network nodes to make them have the function of classification. After training, the network can be used for object identification. The advantage of neural network is that it can approximate any function with any accuracy. Neural network, a nonlinear model, can adapt to various complex data relations and has strong learning ability, which makes it better adapt to the change of data space than many classification algorithms. Drawing on the physical structure and mechanism of human brain, neural network can simulate some functions of human brain with characteristics of "intelligence".

3.2 Types of Neural Networks

Artificial neural network can be divided into static neural network and dynamic neural network. In a static neural network, the input semaphores are processed from the input layer to the front and back, and finally output the results through the neurons in the output layer. The process is relatively simple. When the output signal of the neural network in turn continues to train as the input signal, it is called dynamic neural network. Since the output signal is allowed to continue to enter the input as feedback to participate in the next iteration training, the dynamic neural network has the ability to remember the previous output results, so it has a great advantage in processing complex nonlinear systems. According to the different methods of dynamic neural network to realize system dynamics, it is mainly divided into regression neural network and neural network formed by neuron feedback. Among them, the regression neural network is a dynamic network composed of static neurons and output feedback of the network. The performance of NARX regression neural network is better than that of full regression neural network, and it can transform with full regression neural network, which is the most widely used neural network in nonlinear dynamic system.

3.3 NARX Dynamic Regression Neural Network

With feedback and based functions, the dynamic neural network can retain the data at the previous moment and add it to the calculation of the data at the next moment, making the network dynamic and more complete the information retained in the system. NARX dynamic regression neural network is one of the most widely used nonlinear dynamic neural networks. In time series models, NARX (Nonlinear autoregressive with external input) is a Nonlinear autoregressive model with external input, can learn to predict a given sequence, at the same time of the past value feedback input and another time series, this means that the current value of the model to time series with the same sequence of past values and drive (exogenous) sequence of the current values and past values.

The network model of NARX dynamic regression neural network is shown in Fig. 2, and its mathematical model can be expressed as follows:

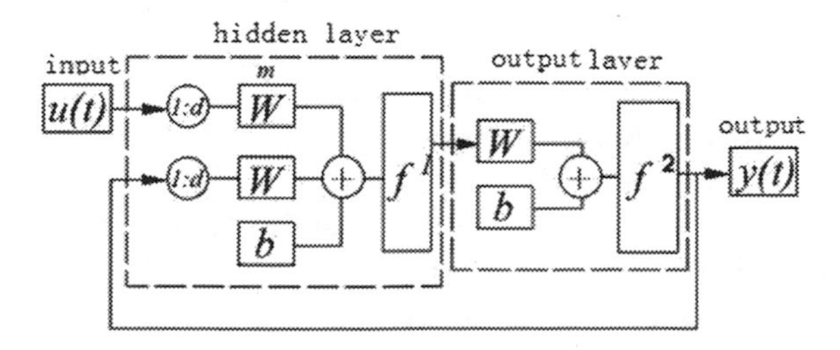

Fig. 2. NARX dynamic regression neural network structure

$$y(t) = f[y(t-1), y(t-2), \ldots\ldots, y(t-n_y), u(t-1), u(t-2), \ldots\ldots, u(t-n_y)] \quad (1)$$

Among of them, f() represents the nonlinear function realized by neural network, such as polynomial. F can be neural network, wavelet network, s-type network and so on. The output data y(t) is used as input in the feedback action, and the computational accuracy of the neural network is improved through open loop training or closed loop training.

In this scenario, information about u helps predict y, just like the previous y value itself. For example, y could be the expected number of stocks, and u could be a day of the year (the number of days of the year).

4 Application of NARX Dynamic Neural Network in Quantitative Investment Forecasting System

4.1 Research Significance of Neural Network in Quantitative Investment

Quantitative investment forecasting is to analyze and model these data based on the collection and processing of historical data and mathematical models such as mathematics, statistics and econometrics. After rigorous testing and simulation, it can be used to predict, judge and even trade the future securities market. The computer algorithm used in the conventional mathematical model processing is usually only able to deal with and solve the problems of linear system. With the development of artificial intelligence science and technology, the non-linear, fuzzy, self-learning and self-adaptive characteristics of neural network model are used to provide possibilities for the research and exploration of securities market investment prediction. For example, NARX dynamic neural network is used to predict the recent trend of rise and fall.

4.2 The Application of Neural Network in Quantitative Investment Forecasting System

(1) Obtain Shanghai index data

The closing price of Shanghai composite index refers to the price at the end of trading day, the most significant price at the observation, the closing price that often reflects the concern degree of market fund to some Shanghai composite index. It has the function that predicts the deductive direction of next trading day. The closing price is affected by the current opening, lowest and highest prices of the Shanghai composite index. Therefore, it is necessary to obtain the opening, lowest, highest, closing and trading volume of the Shanghai composite index to predict the closing price of the second day. Data of Shanghai composite index in 758 trading days from January 1, 2011 to December 31, 2013 were obtained. The data of the first 658 trading days is selected as the training set and the data of the next 100 trading days is selected as the test set.

(2) Data preprocessing

The change of the closing price of the Shanghai composite index is greatly affected by the highest, lowest and opening prices of the Shanghai composite index that day, which are therefore selected as input variables of NARX neural network to predict the closing price of the Shanghai composite index. In order to reduce the complexity of numerical calculation in the training process, the time series data of Shanghai composite index should be preprocessed before modeling.

In this system, MapMinMax is used to normalize the timing data of Shanghai composite index. The specific formula is as follows:

$$y = (y_{max} - y_{min}) * \frac{x - x_{min}}{x_{max} - x_{min}} + y_{min} \quad (2)$$

Among of them, x_{max} and x_{min} are the maximum and minimum values of the data sample, $y_{max} = 2$ and $y_{min} = 1$ are specified to normalize the original Shanghai composite index data to the interval of [1, 2], so as to accelerate the convergence of the neural network program (Figs. 3 and 4).

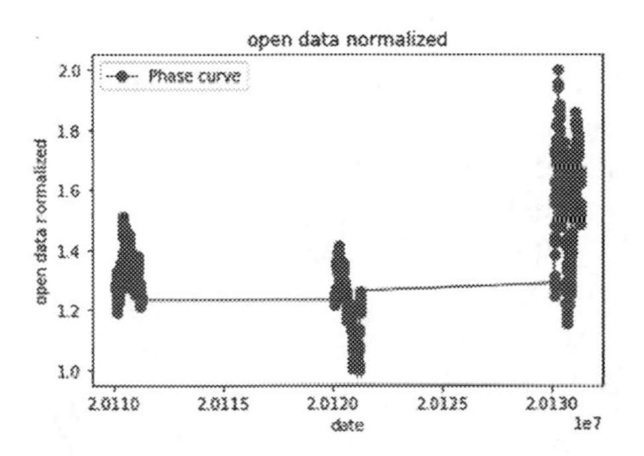

Fig. 3. Normalization of opening data

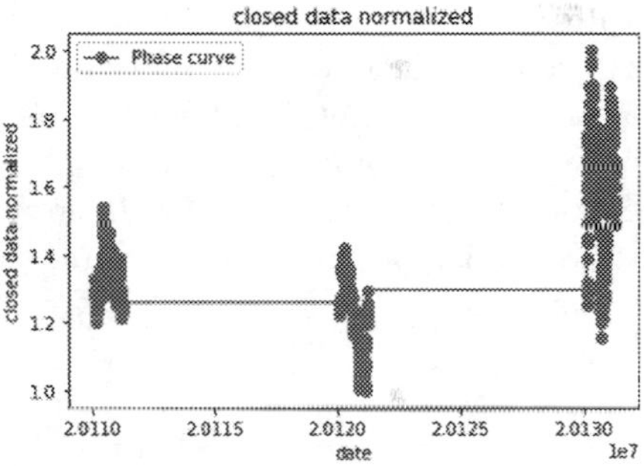

Fig. 4. Normalization of closing data

(3) Build the NARX model

The multivariable timing sequence of the normalized closing price, highest price, lowest price and opening price of Shanghai composite index is taken as the input variable of NARX neural network, which passes through the input layer containing input delay, hidden layer and output layer containing output delay. The hidden layer activation function of neural network is transfer function. Purelin transfer function is adopted to activate the output layer of neural network. In application, and the delay order of input and output as well as the number of hidden layer neurons should be determined in advance. The training, verification and test sets of NARX neural network are divided into 70%, 15% and 15%. In order to determine the number of hidden layer neurons in NARX, neural network model is set as 10 and the time delay order is set as 20 (Fig. 5).

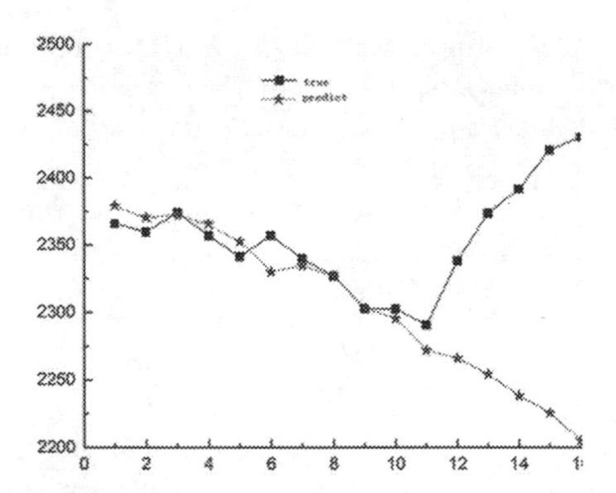

Fig. 5. Predicted value of closing data

(4) NARX model evaluation

RMSE (root-mean-square error) is used to evaluate the data model. The root mean square error is the square root of the deviation between the predicted value and the real value and the ratio of the number of observations n. In the actual measurement, the number of observations n is always limited, and the truth value can only be replaced by the most reliable (best) value. RMSE error is used to measure the deviation between the observed value y_(pre, I) and the true value y_(true, I). The specific formula is as follows:

$$\mathrm{RMSE} = \sqrt{\frac{\sum_{i=1}^{n}\left(y_{true,i} - y_{pre,i}\right)}{n}} \tag{3}$$

Among of them, n is the number of test samples (Fig. 6).

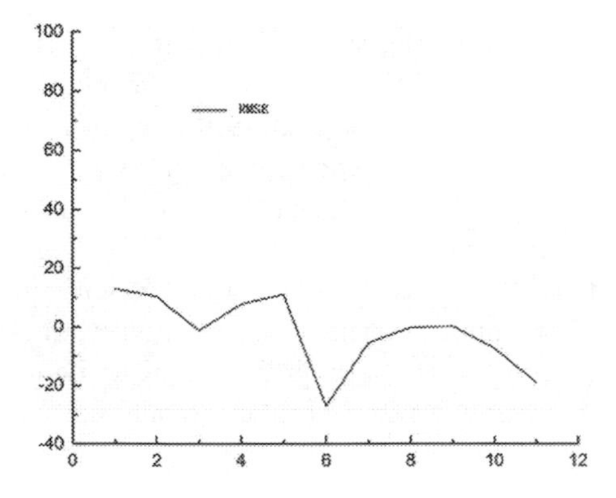

Fig. 6. Prediction error of closing data

In this prediction system, by using multiple variables such as the closing price, opening price, maximum price and minimum price of the Shanghai composite index, the prediction result is relatively close to the actual value, and the prediction accuracy decreases after the fifth day, with the deviation between the predicted value and the actual value being RMSE 12.13489.

5 Conclusion

The NARX dynamic neural network model has certain predictive ability for the problems of non-linear systems such as the security market. However, the data volume of the security market is very large, and various data information and indicators emerge endlessly, which are of practical significance and reflect certain market information. The policy and speculative characteristics of China's securities market are very obvious, and the stock price fluctuates a lot in a short time. These problems are difficult to be solved by neural network prediction alone, and need to be improved by combining with other prediction methods.

References

1. Hua, Y.: Research on stock timing based on BP neural network model. Shenyang University of Technology (2019)
2. Wu, Y.: Research on multi-factor stock selection strategy in a-share market based on neural network method. University of Electronic Science and Technology (2018)

A Deep Reinforcement Learning Algorithm Using Dynamic Attention Model for Vehicle Routing Problems

Bo Peng, Jiahai Wang(✉), and Zizhen Zhang

Department of Computer Science, Sun Yat-sen University, Guangzhou, China
wangjiah@mail.sysu.edu.cn

Abstract. Recent researches show that machine learning has the potential to learn better heuristics than the one designed by human for solving combinatorial optimization problems. The deep neural network is used to characterize the input instance for constructing a feasible solution incrementally. Recently, an attention model is proposed to solve routing problems. In this model, the state of an instance is represented by node features that are fixed over time. However, the fact is, the state of an instance is changed according to the decision that the model made at different construction steps, and the node features should be updated correspondingly. Therefore, this paper presents a dynamic attention model with dynamic encoder-decoder architecture, which enables the model to explore node features dynamically and exploit hidden structure information effectively at different construction steps. This paper focuses on a challenging NP-hard problem, vehicle routing problem. The experiments indicate that our model outperforms the previous methods and also shows a good generalization performance.

Keywords: Learning heuristics · Dynamic encoder-decoder architecture · Vehicle routing problem · Reinforcement learning · Neural network

1 Introduction

Vehicle routing problem (VRP) [1] is a well-known combinatorial optimization problem in which the objective is to find a set of routes with minimal total costs. For each route, the total demand cannot exceed the capacity of the vehicle. In literature, the algorithms for solving VRP can be divided into exact and heuristic algorithms. The exact algorithms provide optimal guaranteed solutions but are infeasible to tackle large-scale instances due to high computational complexity, while the heuristic algorithms are often fast but without theoretical guarantee. Considering the trade-off between optimality and computational costs, heuristic algorithms can find a suboptimal solution within an acceptable running time for large-scale instances. However, it is difficult to design a good heuristic algorithm, since it requires substantial problem-specific expert knowledge and hand-crafted features. Designing a heuristic algorithm is a tedious process, can we learn a heuristic automatically without human intervention?

K. Li et al. (Eds.): ISICA 2019, CCIS 1205, pp. 636–650, 2020.
https://doi.org/10.1007/978-981-15-5577-0_51

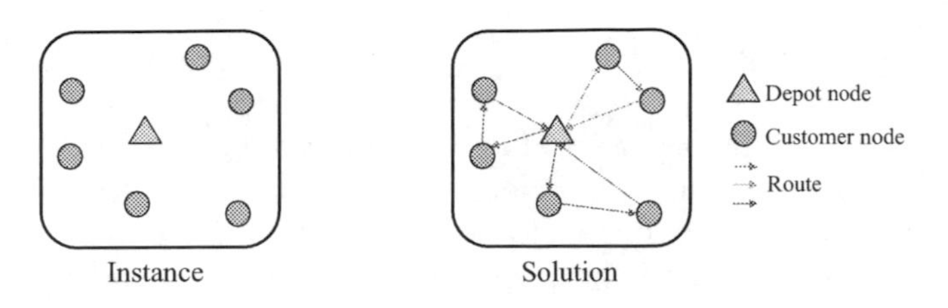

Fig. 1. A typical vehicle routing problem.

Motivated by recent advancements in machine learning, especially deep learning, there have been some works [2–7] on using end-to-end neural network to directly learn heuristics from data without any hand-engineered reasoning. Specifically, taking VRP for example, as shown in Fig. 1, the instance is a set of nodes, and the optimal solution is a permutation of these nodes, which can be seen as a sequence of decisions. Therefore, VRP can be viewed as a decision making problem that can be solved by reinforcement learning. From the perspective of reinforcement learning, typically, the state is viewed as the partial solution of instance and the features of each node, the action is the choice of next node to visit, the reward is the negative tour length, and the policy corresponds to heuristic strategy which is parameterized by a neural network. Then the policy is trained to make decisions for maximizing the reward. From the perspective of learning heuristics, given the instances from the distribution $\mathcal{S}$, a heuristics is learned to solve an unseen instance from the same distribution $\mathcal{S}$.

Recently, an attention model (AM) [7] is proposed to solve routing problems. In AM, an instance is viewed as a graph, and node features are extracted to represent such a complex graph structure, which captures the properties of a node in the context of its graph neighborhoods. Based on these node features, the solution is constructed incrementally. In AM, the node features are encoded as an embedding which is fixed over time. However, at different construction steps, the state of instance is changed according to the decision the model made, and the node features should be updated correspondingly.

This paper proposes a dynamic attention model (AM-D) with dynamic encoder-decoder architecture. The key of our improvement is to characterize each node dynamically in the context of the graph, which can explore and exploit hidden structure information effectively at different construction steps. To demonstrate the effectiveness of the proposed method, AM-D is applied to a challenging combinatorial optimization problem, vehicle routing problem. The numerical experiments indicate that AM-D performs significantly better than AM and obviously decreases the optimality gap.

This paper is structured as follows. Section 2 reviews related work. Section 3 describes original attention model for VRP. Section 4 and 5 present our dynamic attention model for VRP and the training method, respectively. Experimental results are given in Sect. 6. Section 7 concludes this paper.

2 Related Work

Learning based methods proposed in last several years can be divided into two categories in terms of types of problems solved. The first category focuses on solving permutation based combinatorial optimization problems, such as VRP and TSP. The second category solves 0–1 based combinatorial optimization problems, such as SAT and knapsack problem.

For the first category, the pointer network (PN) is introduced in [8], it takes combinatorial optimization problems as sequence to sequence problems where the input is a sequence of nodes and the output is a permutation of the input. PN overcomes the limitation that the output length depends on input by a “pointer”, which is a variant of attention mechanism [9]. This sequence to sequence model [10] is trained by the supervised manner and the label is given by an approximate solver.

However, PN is sensitive to the quality of labels and optimal solutions are expensive. In [3], the neural combinatorial optimization framework is proposed to solve combinatorial optimization problems, and the REINFORCE algorithm [11] is used to train a policy modeled by PN without supervised signals. In [4], the LSTM encoder of PN is replaced by element-wise projections which are invariant to the input order and will not introduce redundant sequential information.

In [5], combinatorial optimization is taken as a graph problem, and graph embedding [12] is used to capture combinatorial structure information between nodes. The model is trained by 1-step DQN [13] which is data-efficient, and the solution is constructed by the helper function.

In [6] and [7], graph attention network [14] is used to extract the features of each node in graph structure. In [6], an explicitly forgetting mechanism is introduced to construct a solution, which only requires the last three selected nodes per step. Then the constructed solution is improved by 2OPT local search [15]. In [7], a context vector is introduced to represent the decoding context, and the model is trained by the REINFORCE algorithm with a deterministic greedy rollout baseline.

For the second category, in [16], the graph convolutional network [17,18] is trained to estimate the likelihood, for each node in the instance, of whether this node is part of the optimal solution. In addition, the tree search is used to construct a large number of candidate solutions. In [19], GCOMB is proposed to solve combinatorial optimization problems over large graph based on graph convolutional network and Q-learning. In [20] and [21], the model is taken as a classifier. In [20], the message passing neural network [22] is trained to predict satisfiability on SAT problems. In [21], the graph neural network is used to solve decision TSP.

Since this study is targeted at solving VRP, [4,7] are the most related work with this paper. AM proposed recently in [7] for VRP is introduced as follows.

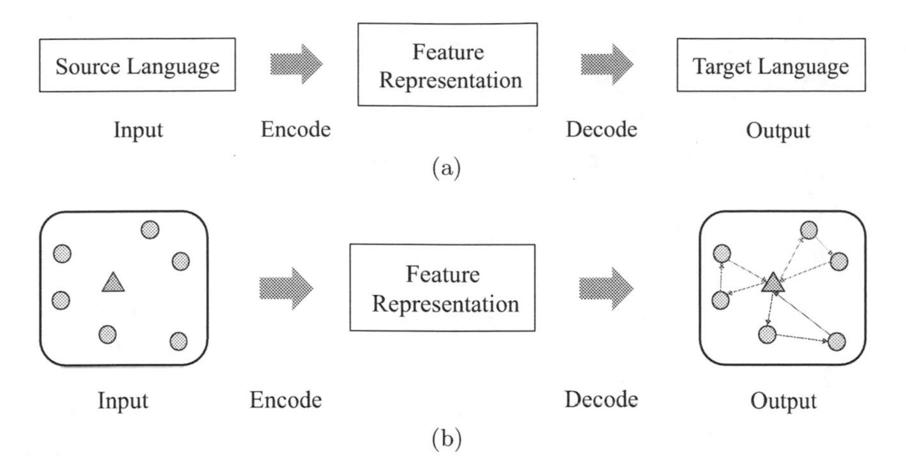

Fig. 2. (a) The encoder-decoder architecture for NMT. (b) The encoder-decoder architecture for VRP.

3 Attention Model for VRP

3.1 Problem Formulation and Preliminaries

This paper focuses on VRP. For the simplest form of the VRP, a single capacitated vehicle is responsible for delivering items to multiple customer nodes, and the vehicle must return to the depot to pick up additional items when it runs out of loads. The solution can be seen as a set of routes. In each route, it begins and ends at the depot.

Specifically, for VRP instance, the input $X = \{x_0, \ldots, x_n\}$ is a set of nodes and x_0 is the depot. Each node consists of two elements $x_i = (s_i, d_i)$, where s_i is a 2-dimensional coordinate of node i in euclidean space and d_i is its demand ($d_0 = 0$). The solution π is a sequence $\{\pi = (\pi_1, \ldots, \pi_T), \pi_t \in \{x_0, \ldots, x_n\}\}$, where each customer node is visited exactly once and the depot can be visited multiple times. T is the length of sequence π that may be varied from different solutions.

VRP can be viewed as a sequential decision making problem, and encoder-decoder architecture [10] is an effective framework for solving such kind of problems. Taking neural machine translation (NMT) for example, as shown in Fig. 2(a), the encoder extracts syntactic structure and semantic information from source language text. Then the decoder constructs target language text from the features given by encoder. Figure 2(b) shows that the encoder-decoder architecture can also be applied to solve VRP. Firstly, the structural features of the input instance are extracted by the encoder. Then the solution is constructed incrementally by the decoder. Specifically, at each construction step, the decoder predicts a distribution over nodes, then one node is selected and appended to the end of the partial solution. Hence, corresponding to the parameters θ and input instance X, the probability of solution $p_\theta(\pi|X)$ can be decomposed by chain rule as:

$$p_\theta(\pi|X) = \prod_{t=1}^{T} p_\theta(\pi_t|X, \pi_{1:t-1}). \tag{1}$$

3.2 Encoder

In encoder, graph attention network is used to encode node features to an embedding in context of graph. It is similar to the encoder in transformer architecture [23]. Firstly, for each d_x-dimensional (for VRP, $d_x = 3$, the coordinate and demand) input node x_i, the d_h-dimensional ($d_h = 128$) initial node embedding $h_i^{(0)}$ is computed through a linear transformation with learnable parameters $W \in \mathbb{R}^{d_h \times d_x}$ and $b \in \mathbb{R}^{d_h}$, separate parameters W_0 and b_0 are used for the depot:

$$h_i^{(0)} = \begin{cases} Wx_i + b & \text{if } i \neq 0 \\ W_0 x_i + b_0 & \text{if } i = 0. \end{cases} \tag{2}$$

These initial node embeddings are fed into the first layer of graph attention network and updated $N = 3$ times with N attention layers. For each layer, it consists of two sublayers: a multi-head attention (MHA) sublayer and a fully connected feed-forward (FF) sublayer.

Multi-head Attention Sublayer. As in [23], multi-head attention is used to extract different types of information. In the layer $\ell \in \{1, \ldots, N\}$, $h_i^{(\ell)}$ is denoted as the node embedding of each node i, and the output $\{h_0^{(\ell-1)}, \ldots, h_n^{(\ell-1)}\}$ of the layer $\ell - 1$ is the input of the layer ℓ. The multi-head attention vector $\mathrm{MHA}_i^{(\ell)}(h_0^{(\ell-1)}, \ldots, h_n^{(\ell-1)})$ of each node i can be computed as:

$$q_{im}^{(\ell)} = W_m^Q h_i^{(\ell-1)}, k_{im}^{(\ell)} = W_m^K h_i^{(\ell-1)}, v_{im}^{(\ell)} = W_m^V h_i^{(\ell-1)}, \tag{3}$$

$$u_{ijm}^{(\ell)} = (q_{im}^{(\ell)})^T k_{jm}^{(\ell)}, \tag{4}$$

$$a_{ijm}^{(\ell)} = \frac{e^{u_{ijm}^{(\ell)}}}{\sum_{j'=0}^{n} e^{u_{ij'm}^{(\ell)}}}, \tag{5}$$

$$h_{im}^{'(\ell)} = \sum_{j=0}^{n} a_{ijm}^{(\ell)} v_{jm}^{(\ell)}, \tag{6}$$

$$\mathrm{MHA}_i^{(\ell)}(h_0^{(\ell-1)}, \ldots, h_n^{(\ell-1)}) = \sum_{m=1}^{M} W_m^O h_{im}^{'(\ell)}. \tag{7}$$

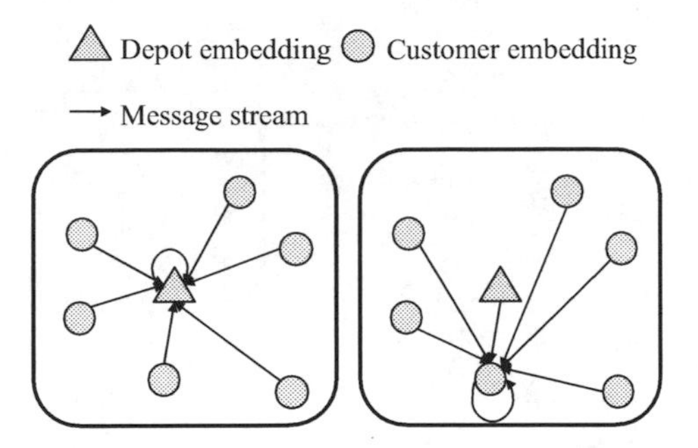

Fig. 3. The message stream in encoder. The embedding of each node is updated by aggregating the message of each node (including itself). On the left, the depot embedding is updated. On the right, the customer embedding is updated.

Here, the number of head is set $M = 8$, in each attention head $m \in \{1, \dots M\}$, the query vector $q_{im}^{(\ell)} \in \mathbb{R}^{d_k}$, key vector $k_{im}^{(\ell)} \in \mathbb{R}^{d_k}$ and value vector $v_{im}^{(\ell)} \in \mathbb{R}^{d_v}$ are computed with parameters $W_m^Q \in \mathbb{R}^{d_k \times d_h}$, $W_m^K \in \mathbb{R}^{d_k \times d_h}$ and $W_m^V \in \mathbb{R}^{d_v \times d_h}$ respectively. And the final vector is computed with $W_m^O \in \mathbb{R}^{d_h \times d_v}$ ($d_k = d_v = \frac{d_h}{M} = 16$).

Remark: The parameters W_m^Q, W_m^K and W_m^V do not share between each layer and the superscript ℓ is omitted for readability.

Feed-Forward Sublayer. In this sublayer, for each node i, based on multi-head attention vector, $h_i^{(\ell)}$ is computed by skip-connection and fully connected feed-forward (FF) network. For each node i:

$$\hat{h}_i^{(\ell)} = \tanh(h_i^{(\ell-1)} + \mathrm{MHA}_i^{(\ell)}(h_0^{(\ell-1)}, \dots, h_n^{(\ell-1)})), \tag{8}$$

$$\mathrm{FF}(\hat{h}_i^{(\ell)}) = W_1^F \mathrm{ReLu}(W_0^F \hat{h}_i^{(\ell)} + b_0^F) + b_1^F, \tag{9}$$

$$h_i^{(\ell)} = \tanh(\hat{h}_i^{(\ell)} + \mathrm{FF}(\hat{h}_i^{(\ell)})), \tag{10}$$

where $h_i^{(\ell)}$ is calculated with parameters $W_0^F \in \mathbb{R}^{d_F \times d_h}$, $W_1^F \in \mathbb{R}^{d_h \times d_F}$, $b_0^F \in \mathbb{R}^{d_F}$ and $b_1^F \in \mathbb{R}^{d_h} (d_F = 4 \times d_h)$.

After N attention layers, for each node i, the final node embedding h_i^N is calculated as:

$$h_i^N = \mathrm{ENCODE}_i^N(h_0^0, \dots, h_n^0). \tag{11}$$

$\mathrm{ENCODE}_i^N(h_0^0, \dots, h_n^0)$ is computed with Eqs. (3)–(10).

Figure 3 illustrates the stream of message between nodes. By aggregating the message of each node, the embedding of each node is updated according to the attention mechanism.

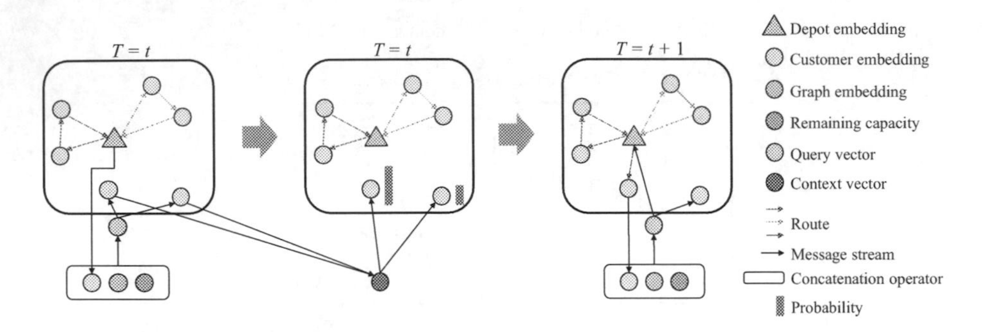

Fig. 4. The details of decoder at construction step t. At each construction step, according to the context vector and node embedding (except the nodes that violates the constraints), the decoder predicts a distribution over nodes and selects one to visit.

3.3 Decoder

In decoder, at each construction step $t \in \{1, \ldots, T\}$, one node is selected to visit based on the partial solution $\pi_{1:t-1}$ and the embedding of each node. As in [7], the context vector h_c is computed by M-head attention mechanism. Firstly, for VRP, a new vector $h_c^{'}$ is constructed as:

$$h_c' = \begin{cases} [\bar{h}_t; h_0^N; D_t] & \text{if } t = 1 \\ [\bar{h}_t; h_{\pi_{t-1}}^N; D_t] & \text{if } t > 1, \end{cases} \tag{12}$$

where [;] is concatenation operator, $h_{\pi_{t-1}}^N$ is the embedding of the node selected at construction step $t-1$, D_t is the remaining capacity of vehicle ($D_1 = D$), and $\bar{h}_t$ is the graph embedding, which is the mean vector of the embedding over nodes that have not been visited (including depot) at construction step t. Similar to the encoder, h_c is computed with a single M-head attention layer, and only a single query $q_{(c)}$ (per head) is computed (the parameters do not share with encoder):

$$q_{(c)m} = W_m^Q h_c^{'}, \quad k_{jm} = W_m^K h_j^N, \quad v_{jm} = W_m^V h_j^N, \tag{13}$$

$$u_{(c)jm} = \begin{cases} q_{(c)m}^T k_{jm} & \text{if } d_j <= D_t \text{ and } x_j \notin \pi_{1:t-1} \\ -\infty & \text{otherwise,} \end{cases} \tag{14}$$

$$a_{(c)jm} = \frac{e^{u_{(c)jm}}}{\sum_{j'=0}^{n} e^{u_{(c)j'm}}}, \tag{15}$$

$$h_{(c)m}^{'} = \sum_{j=0}^{n} a_{(c)jm} v_{jm}, \tag{16}$$

$$h_c = \sum_{m=1}^{M} W_m^O h_{(c)m}^{'}. \tag{17}$$

As shown in Eq. (14), in order to construct a feasible solution, the node that violates the constraints will be masked. For VRP, the following masking conditions are used. First, the customer node whose demand greater than the remaining capacity of the vehicle is masked. Second, the customer node that already been visited is masked.

Remark: The depot node can be visited multiple times and it will be masked only when $\pi_{t-1} = x_0$.

Finally, the probability $p_\theta(\pi_t|X, \pi_{1:t-1})$ is computed with a single-head attention layer:

$$q = W^Q h_c, \quad k_j = W^K h_j^N, \tag{18}$$

$$u_j = \begin{cases} C \cdot \tanh(q^T k_j) & \text{if } d_j <= D_t \text{ and } x_j \notin \pi_{1:t-1} \\ -\infty & \text{otherwise,} \end{cases} \tag{19}$$

$$p_\theta(\pi_t = x_j|X, \pi_{1:t-1}) = \frac{e^{u_j}}{\sum_{j'=0}^{n} e^{u_{j'}}}, \tag{20}$$

where C is used to clip the result within $[-C, C]$ $(C = 10)$. If node i is selected to visit at construction step t, the remaining capacity should be updated:

$$D_{t+1} = \begin{cases} D & \text{if } i = 0 \\ D_t - d_i & \text{otherwise.} \end{cases} \tag{21}$$

Figure 4 illustrates the details of decoder at construction step t. According to the partial solution and node embedding, the context vector is computed by the attention mechanism. Based on the context vector and the embedding of remaining nodes, the decoder predicts a distribution over these nodes and selects one to visit.

4 Dynamic Attention Model for VRP

As mentioned in Sect. 3, the solution is constructed incrementally by the decoder. At different construction steps, the state of the instance is changed, and the feature embedding of each node should be updated. As shown in Fig. 5, when the model constructed a partial solution, the remaining nodes, which do not be included in the partial solution yet, can be seen as a new instance. Constructing the remaining solution is equivalent to solve this new instance. Since some nodes have already been visited, the structure of this new instance is different from the original instance. Therefore, the structure information is changed and the node features should be updated accordingly. But in vanilla encoder-decoder architecture in AM for VRP, as shown in Fig. 6(a), the feature embedding of each node is computed only once, which corresponds to the initial state of instance.

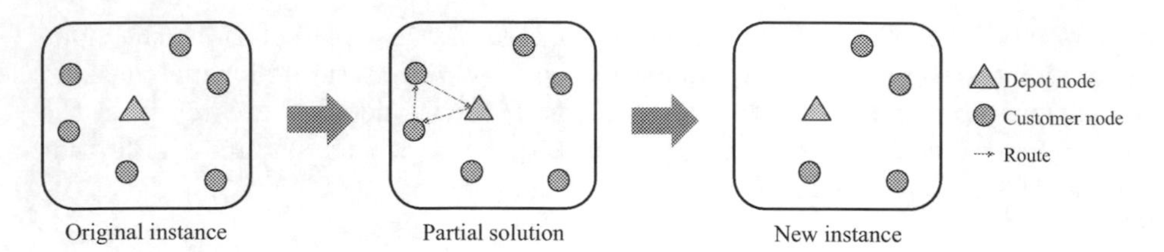

Fig. 5. The state of an instance is changed at different construction steps. When the model constructed a partial solution, the remaining nodes can be seen as a new instance. Since some nodes already been visited, the structure of this new instance is different from the original instance.

This paper proposes a dynamic encoder-decoder architecture to characterize the feature embedding of each node dynamically at different construction steps.

The dynamic encoder-decoder architecture, as shown in Fig. 6(b), is similar to vanilla encoder-decoder architecture. The key difference is that the embedding of each node will be immediately recomputed when the vehicle returns to the depot. Specifically, for each node i, the embedding can be updated at construction step t as:

$$h_i^t = \begin{cases} \text{ENCODE}_i^N(h_0^0, \dots, h_n^0) & \text{if } \pi_{t-1} = x_0 \\ h_i^{t-1} & \text{otherwise}, \end{cases} \tag{22}$$

where h_i^t is the embedding of node i at construction step t, and the layer number N is omitted. $\text{ENCODE}_i^N(h_0^0, \dots, h_n^0)$ is similar to Eq. (11) that is computed with N M-head attention layers. The only difference is that Eq. (4) is modified. In order to reflect that the structure of instance is changed, the nodes that have been visited are masked, and Eq. (4) is modified as:

$$u_{ijm}^{(\ell)} = \begin{cases} (q_{im}^{(\ell)})^T k_{jm}^{(\ell)} & \text{if } x_j \notin \pi_{1:t-1} \text{ or } j = 0 \\ -\infty & \text{otherwise}. \end{cases} \tag{23}$$

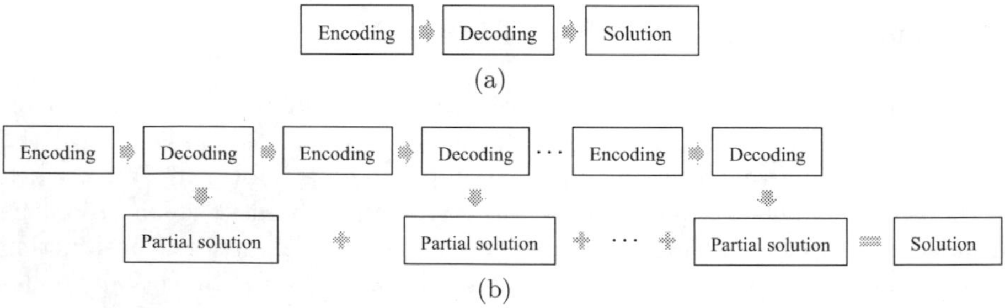

Encoding → Decoding → Solution

(a)

Encoding → Decoding → Encoding → Decoding ... Encoding → Decoding

Partial solution + Partial solution + ... + Partial solution = Solution

(b)

Fig. 6. The comparison between our dynamic architecture and the vanilla architecture. (a) In vanilla encoder-decoder architecture of AM for VRP, the encoder is used only once, the embedding of each node is fixed, which only can represent the initial state of the input instance. (b) In dynamic encoder-decoder architecture of AM-D, the encoder and decoder are used alternately to recode the embedding of each node and construct a partial solution.

During decoding, at each step t, the computation of Eqs. (13)–(18) is based on the latest embedding of each node $\{h_0^t, \ldots, h_n^t\}$ (the layer number N is omitted, and t is the construction step). As shown in Fig. 6(b), the entire architecture uses the encoder and decoder alternately to recode node embedding and construct a partial solution.

Given a distribution over nodes, there are two strategies to select the next node to visit. The one is sample rollout that selects a node using sampling. The other is greedy rollout that selects the node with maximum probability. The former is a stochastic policy and the latter is a deterministic policy.

5 Model Training

As in [3,4,6,7], solving combinatorial optimization problem is taken as Markov Decision Processes (MDP), and AM-D is trained by policy gradient using REINFORCE algorithm [11]. Given an instance X, our training objective is the tour length of solution π. Hence, based on instance X, the gradients of parameters θ are defined as:

$$\nabla_\theta J(\theta|X) = \mathbb{E}_{\pi \sim p_\theta(.|X)}[(L(\pi|X) - b(X))\nabla_\theta \log p_\theta(\pi|X)], \tag{24}$$

where $L(\pi|X)$ is the tour length of solution π, $b(X)$ is a baseline function for estimating the expected tour length $\mathbb{E}_{\pi \sim p_\theta(.|X)} L(\pi|X)$ of instance X which can reduce the variance of gradients and accelerate convergence effectively. In this paper, as in [7], the tour length of the greedy solution, which is constructed by greedy rollout, is taken as $b(X)$.

During training, the instances are drawn from the same distribution $\mathcal{S}$. The gradients of parameters θ are approximated by Monte Carlo sampling as:

$$\nabla_\theta J(\theta) \approx \frac{1}{B}\sum_{i=1}^{B}[(L(\pi_i^s|X_i) - L(\pi_i^g|X_i))\nabla_\theta \log p_\theta(\pi_i^s|X_i)], \tag{25}$$

where B is the batch size, π_i^s and π_i^g are the solutions of instance X_i constructed by sample rollout and greedy rollout respectively. The training algorithm is described in Algorithm 1.

Algorithm 1. REINFORCE algorithm

1: **Input:** number of epochs E, steps per epoch F, batch size B
2: Initialize parameters θ
3: **for** epoch $= 1, \ldots, E$ **do**
4:     **for** step $= 1, \ldots, F$ **do**
5:         $X_i \leftarrow$ RandomInstance() for $i \in \{1, \ldots, B\}$
6:         $\pi_i^s \leftarrow$ SampleRollout($p_\theta(.|X_i)$) for $i \in \{1, \ldots, B\}$
7:         $\pi_i^g \leftarrow$ GreedyRollout($p_\theta(.|X_i)$) for $i \in \{1, \ldots, B\}$
8:         $g_\theta \leftarrow \frac{1}{B}\sum_{i=1}^{B}(L(\pi_i^s) - L(\pi_i^g))\nabla_\theta \log p_\theta(\pi_i^s|X_i)$
9:         $\theta \leftarrow$ Adam(θ, g_θ)
10:     **end for**
11: **end for**

Table 1. Results on VRP. Len is the average length on test instance. Gap is the distance to state-of-the-art.

Method	VRP20, Cap30		VRP50, Cap40		VRP100, Cap50	
	Len	Gap	Len	Gap	Len	Gap
Gurobi	6.10	0.00%	–	–	–	–
LKH3	6.14	0.58%	10.38	0.00%	15.65	0.00%
RL (greedy) [4]	6.59	8.03%	11.39	9.78%	17.23	10.12%
AM (greedy) [7]	6.40	4.97%	10.98	5.86%	16.80	7.34%
AM-D (greedy)	**6.28**	**2.95%**	**10.78**	**3.85%**	**16.40**	**4.79%**
AM-D (2OPT)	6.25	2.46%	10.73	3.37%	16.27	3.96%
AM-D ($n = 20$)	–	–	11.00	5.97%	17.37	10.99%
AM-D ($n = 50$)	6.48	6.23%	–	–	16.55	5.75%
AM-D ($n = 100$)	6.65	9.02%	11.04	6.36%	–	–

6 Experiments

Experiments are conducted to investigate the performance of AM-D on VRP with node size $n = 20, 50, 100$. AM-D consists of two phases: training phase and testing phase. For each problem, in training phase, the model is trained with 30 epochs, and 10000 batches are processed in each epoch. In testing phase, the performance on 10000 test instances is reported, where the solution is constructed by greedy rollout, and the final results are the average length on all test instances.

6.1 Instances and Hyperparameters

As in [4] and [7], the instances are generated from a fixed distribution. For each node, the location are chosen randomly from the unit square $[0, 1] \times [0, 1]$, and the demand is a discrete number in $\{1, \ldots, 9\}$ chosen uniformly at random (the demand of depot is 0). The capacity of vehicle $D = 30$ for VRP with 20 customer nodes (denoted as VRP20), $D = 40$ for VRP50, $D = 50$ for VRP100, and the vehicle is located at the depot when $t = 1$. The batch size $B = 128$ and learning rate $\eta = 10^{-4}$ for both VRP20 and VRP50, $B = 108$ and $\eta = 5 \times 10^{-5}$ for VRP100. Finally, for each problem, the experiment is conducted by GPU (single 1080Ti for VRP20, VRP50, $3 \times$ 1080Ti for VRP100).

6.2 Results and Discussions

Comparison Results. Table 1 shows the results of VRP. Compared with AM, the performance of AM-D is notably improved for both VRP20 (2.02%), VRP50 (2.01%) and VRP100 (2.55%). AM-D significantly outperforms other baseline models as well.

The numerical experiments indicate that AM-D performs better than AM and other baseline methods. AM-D introduces the dynamic encoder-decoder architecture to explore structure features dynamically and exploit hidden structure information effectively at different construction steps. Hence, more hidden and useful structure information is taken into account, thereby leading to a better solution.

Generalization to Larger or Smaller Instances. How does the performance of the learned heuristics generalize to test instances with larger or smaller customer node size? Experiments are conducted to investigate the generalization performance of AM-D. Specifically, the model trained with instances with 20, 50 and 100 customer nodes are denoted as AM-D ($n = 20$), AM-D ($n = 50$) and AM-D ($n = 100$), respectively. AM-D ($n = 20$) is tested on instances with 50 and 100 customer nodes, AM-D ($n = 50$) is tested on instances with 20 and 100 customer nodes, and AM-D ($n = 100$) is tested on instances with 20 and 50 customer nodes, respectively.

The results are shown at the last three rows in Table 1. Specifically, on the one hand, the model trained with small instances ($n = 20$) has a good performance on large instances ($n = 50$, $n = 100$), and the results even better than some baseline methods. On the other hand, the model trained with large instance ($n = 100$) performs good on small instance ($n = 20$, $n = 50$) as well. The reason why AM-D has a good generalization performance may be as follows. AM-D constructs the solution incrementally, and this process can be divided into many stages. At each stage, only a partial solution is constructed, and thus the instance is transformed to a smaller one which is easier to solve.

Combination with Local Search. Local search is applied to further improve the results as in [6]. Firstly, for each instance, a solution is constructed by AM-D (greedy), then the 2OPT local search algorithm is applied to improve this solution. The resultant method is named AM-D (2OPT) and the results are shown in Table 1. The runtime of AM-D (greedy) and AM-D (2OPT) are also given in Table 2. The results indicate that the quality of the solution is improved by integrating local search, but the local search brings additional computational cost.

Table 2. Training time and testing time of AM-D. Testing time is average runtime on test instance.

	VRP20	VRP50	VRP100
Training time	14 h	58 h	250 h
Testing time (greedy)	0.29 ms	2.51 ms	15.92 ms
Testing time (2OPT)	0.05 s	0.34 s	2.21 s

Discussions. Machine learning and optimization are closely related, machine learning is often used as an assistant or helper component to improve the performance of solution or reduce computational costs in many optimization algorithms [24]. Totally different from these methods, AM-D is aiming to learn heuristics from data directly without human intervention. It means that knowledge or features can be extracted from the given problem instances automatically. Specifically, given an optimization problem and its instances generated from distribution $\mathcal{S}$, AM-D can learn an approximation or heuristic algorithm and solve the problem on unseen instances generated from distribution $\mathcal{S}$.

AM-D can be divided into training and testing phases like most of machine learning algorithms. The elapsed time of training and testing are shown in Table 2. Though the process of training is time-consuming, it is upfront, offline computation and can be seen as searching in algorithm space. Then, the trained model can be used directly to solve unseen instances without retraining from scratch, which is online even real-time computation process. Taking VRP20 for example, it takes 14 hours in training phase, but the process is one-time. Once the model has been trained, it only spends 0.29 milliseconds for solving each instance without retraining. Thus, AM-D is different from the classic heuristics, which search the solution iteratively in solution space from scratch for each instance.

The training phase of AM-D is time-consuming, thus it is trained only for problem instances with small and medium size due to the limitation of computing resources. It is promising to adopt existing parallel computing techniques to improve the computational efficiency for scaling to larger problem instances in the future.

7 Conclusion

This paper presents a dynamic attention model with dynamic encoder-decoder architecture for VRP. The key improvement is that the structure features of instances are explored dynamically, and hidden structure information is exploited effectively at different construction steps. Hence, more hidden and useful structure information is taken into account, for constructing a better solution. AM-D is tested by a challenging NP-hard problem, VRP. The results show that the performance of AM-D is better than AM and other baseline models for both problems. In addition, AM-D also shows a good generalization performance on different problem scales.

In the future, the proposed learning based methods, AM-D, can be extended to solve some real-world complex VRP variants [25–30] by hybridizing with operations research method, such as VRP with time windows, which will open a new era for combinatorial optimization algorithms [2].

Acknowledgment. This work is supported by the National Key R&D Program of China (2018AAA0101203), and the National Natural Science Foundation of China (61673403, U1611262).

References

1. Mańdziuk, J.: New shades of the vehicle routing problem: emerging problem formulations and computational intelligence solution methods. IEEE Trans. Emerg. Top. Comput. Intell. **3**(3), 230–244 (2019)
2. Bengio, Y., Lodi, A., Prouvost, A.: Machine learning for combinatorial optimization: a methodological tour d'Horizon. arXiv:1811.06128 (2018)
3. Bello, I., Pham, H., Le, Q.V., Norouzi, M., Bengio, S.: Neural combinatorial optimization with reinforcement learning. arXiv preprint arXiv:1611.09940 (2016)
4. Nazari, M., Oroojlooy, A., Snyder, L., Takác, M.: Reinforcement learning for solving the vehicle routing problem. In: Advances in Neural Information Processing Systems, pp. 9839–9849 (2018)
5. Khalil, E., Dai, H., Zhang, Y., Dilkina, B., Song, L.: Learning combinatorial optimization algorithms over graphs. In: Advances in Neural Information Processing Systems, pp. 6348–6358 (2017)
6. Deudon, M., Cournut, P., Lacoste, A., Adulyasak, Y., Rousseau, L.-M.: Learning heuristics for the TSP by policy gradient. In: van Hoeve, W.-J. (ed.) CPAIOR 2018. LNCS, vol. 10848, pp. 170–181. Springer, Cham (2018). https://doi.org/10.1007/978-3-319-93031-2_12
7. Kool, W., van Hoof, H., Welling, M.: Attention, learn to solve routing problems! In: International Conference on Learning Representations (2019)
8. Vinyals, O., Fortunato, M., Jaitly, N.: Pointer networks. In: Advances in Neural Information Processing Systems, pp. 2692–2700 (2015)
9. Bahdanau, D., Cho, K., Bengio, Y.: Neural machine translation by jointly learning to align and translate. In: International Conference on Learning Representations (2015)
10. Sutskever, I., Vinyals, O., Le, Q.V.: Sequence to sequence learning with neural networks. In: Advances in Neural Information Processing Systems, pp. 3104–3112 (2014)
11. Williams, R.J.: Simple statistical gradient-following algorithms for connectionist reinforcement learning. Mach. Learn. **8**(3–4), 229–256 (1992)
12. Dai, H., Dai, B., Song, L.: Discriminative embeddings of latent variable models for structured data. In: International Conference on Machine Learning, pp. 2702–2711 (2016)
13. Mnih, V., et al.: Human-level control through deep reinforcement learning. Nature **518**(7540), 529 (2015)
14. Veličković, P., Cucurull, G., Casanova, A., Romero, A., Lio, P., Bengio, Y.: Graph attention networks. In: International Conference on Learning Representations (2018)
15. Johnson, D.S.: Local optimization and the traveling salesman problem. In: Paterson, M.S. (ed.) ICALP 1990. LNCS, vol. 443, pp. 446–461. Springer, Heidelberg (1990). https://doi.org/10.1007/BFb0032050
16. Li, Z., Chen, Q., Koltun, V.: Combinatorial optimization with graph convolutional networks and guided tree search. In: Advances in Neural Information Processing Systems, pp. 539–548 (2018)
17. Scarselli, F., Gori, M., Tsoi, A.C., Hagenbuchner, M., Monfardini, G.: The graph neural network model. IEEE Trans. Neural Netw. **20**(1), 61–80 (2009)
18. Kipf, T.N., Welling, M.: Semi-supervised classification with graph convolutional networks. In: International Conference on Learning Representations (2017)

19. Mittal, A., Dhawan, A., Medya, S., Ranu, S., Singh, A.: Learning heuristics over large graphs via deep reinforcement learning. arXiv preprint arXiv:1903.03332 (2019)
20. Selsam, D., Lamm, M., Bünz, B., Liang, P., de Moura, L., Dill, D.L.: Learning a SAT solver from single-bit supervision. In: International Conference on Learning Representations (2019)
21. Prates, M.O., Avelar, P.H., Lemos, H., Lamb, L., Vardi, M.: Learning to solve NP-complete problems-a graph neural network for the decision TSP. arXiv preprint arXiv:1809.02721 (2018)
22. Gilmer, J., Schoenholz, S.S., Riley, P.F., Vinyals, O., Dahl, G.E.: Neural message passing for quantum chemistry. In: Proceedings of the 34th International Conference on Machine Learning-Volume 70, pp. 1263–1272 (2017). JMLR.org
23. Vaswani, A., et al.: Attention is all you need. In: Advances in Neural Information Processing Systems, pp. 5998–6008 (2017)
24. Song, H., Triguero, I., Özcan, E.: A review on the self and dual interactions between machine learning and optimisation. Prog. Artif. Intell. **8**(2), 143–165 (2019). https://doi.org/10.1007/s13748-019-00185-z
25. Yan, X., Huang, H., Hao, Z., Wang, J.: A graph-based fuzzy evolutionary algorithm for solving two-echelon vehicle routing problems. IEEE Trans. Evol. Comput. **24**(1), 129–141 (2020)
26. Wang, J., Yuan, L., Zhang, Z., Gao, S., Sun, Y., Zhou, Y.: Multiobjective multiple neighborhood search algorithms for multiobjective fleet size and mix location-routing problem with time windows. IEEE Trans. Syst. Man Cybern. Syst. 1–15 (2019, in press)
27. Wang, J., Ren, W., Zhang, Z., Huang, H., Zhou, Y.: A hybrid multiobjective memetic algorithm for multiobjective periodic vehicle routing problem with time windows. IEEE Trans. Syst. Man Cybern. Syst. 1–14 (2018, in press)
28. Wang, J., Weng, T., Zhang, Q.: A two-stage multiobjective evolutionary algorithm for multiobjective multidepot vehicle routing problem with time windows. IEEE Trans. Cybern. **49**(7), 2467–2478 (2019)
29. Wang, J., Zhou, Y., Wang, Y., Zhang, J., Chen, C.L.P., Zheng, Z.: Multiobjective vehicle routing problems with simultaneous delivery and pickup and time windows: formulation, instances, and algorithms. IEEE Trans. Cybern. **46**(3), 582–594 (2016)
30. Zhou, Y., Wang, J.: A local search-based multiobjective optimization algorithm for multiobjective vehicle routing problem with time windows. IEEE Syst. J. **9**(3), 1100–1113 (2015)

Elliptical Wide Slot Microstrip Patch Antenna Design by Using Dynamic Constrained Multiobjective Optimization Evolutionary Algorithm

Rangzhong Wu[1], Caie Hu[1(✉)], and Zhigao Zeng[2]

[1] School of Mechanical and Electronic Information, China University of Geosciences, Wuhan 430074, China
340450030@qq.com, caie_hu@cug.edu.cn
[2] School of Computer, Hunan University of Technology, Zhuzhou 412007, China
zzgzzg99@163.com

Abstract. This paper evolves an elliptical wide slot microstrip patch antenna by using Dynamic Constrained Multiobjective Optimization Evolutionary Algorithm (DCMOEA). The antenna structure is a square substrate with a size of 30 mm × 30 mm × 1.6 mm. The radiation patch in the front side, which is formed by subtracting two ellipses with same axial ratio R. The coupling element in the back side, where consists of difference between ground plate and an ellipse with an axial ratio R. To obtain optimum antenna structure with satisfying the design requirements, the antenna optimization problem is modeled as a constrained optimization problem (COP). The experimental results show the optimum antenna is obtained, as well as we find the radiation patch should be a whole ellipse instead of subtracting with two ellipses. Therefore, we draw a conclusion that the radiation patch should be complete rather than partial in the surface antennas design, which provides a valuable reference for follow-up study with antenna design in real engineering and scientific research.

Keywords: Microstrip patch antenna · Dynamic Constrained Multiobjective Optimization Evolutionary Algorithm · Design requirement · Constrained optimization problem · Optimum antenna

1 Introduction

Evolutionary algorithms (EAs) are global, parallel, search and optimization methods, founded on the principles of natural selection [1] and population genetics [2], and which are found a wide range of applications in various fields of science and engineering. EAs have been widely employed for solving antenna problems since the early 1990s [3,4]. Generally, antenna design problems are modeled as constrained optimization problems (COPs) [5–9]. Such as, Genetic Algorithm (GA) [10] is usually selected as an optimization tool for antenna design in [3,11–14]. Differential evolution (DE) [15] with an efficient global optimization

K. Li et al. (Eds.): ISICA 2019, CCIS 1205, pp. 651–661, 2020.
https://doi.org/10.1007/978-981-15-5577-0_52

is used frequently in the field of antenna design [9,16–18]. Evolution strategy (ES) [19] is very good in searching the accurate optimal solution in the antenna optimization [20,21].

Microstrip patch antennas with low profile, conformable to planar and non-planar surfaces, simple and inexpensive to manufacture using modern printed circuit technology, have been extensively researched in recent years. Elliptical microstrip patch antennas have been usually considered to design due to their geometries present greater potentials for low-profile antenna applications. The main advantage of elliptical microstrip patch antennas is that they give dual resonant frequency owing to two modes, even and odd mode [22,23]. Therefore, they are widely used in the aerospace field and the communication industry.

In antenna design, a well-designed practical antenna should meet the requirement for the performance parameters of the antenna, for instance, Gain, S-parameter, VSWR, etc. In addition, it also should meet the features such as small size, low profile, simple installation, and easy to manufacture so that it can be applied to real-world.

In this paper, the elliptical wide slot microstrip patch antenna is researched, and its structure is optimized by using Dynamic Constrained Multiobjective Optimization Evolutionary Algorithm (DCMOEA) [24]. This approach shows great potential in solving the complex COP.

After a brief introduction in Sect. 1, this paper is organized as follows: Sect. 2 introduces some related work. Section 3 gives the formation of the antenna design as a COP. Then the antenna design problem is solved by DCMOEA in Sect. 4. Finally, Sect. 5 summarizes the conclusion of this paper.

2 Related Work

In this section, we will introduce the basic concept for COP and DCMOEA. For antenna design, the antenna problem often is model as a COP, which usually is a non-linear problem. Traditional optimizers could not do well in solving the problem while EAs have potentiality in such complex problem. DCMOEA was proposed to solve COP with constraint difficulty and multimodal difficulty, which based on the idea that a COP is converted into an equivalent dynamic constrained multiobjective optimization problem (DCMOP) including three objectives. Therefore, it is employed in this paper.

Definition 1. ***(Constrained Optimization Problem (COP))***

A general COP consists of an objective function, a set of m constraints and n variables. The objective and constraints function are functions among the variables. A COP can be mathematically defined as:

$$\begin{array}{ll} min & y = f(\boldsymbol{x}) \\ st: & \overrightarrow{g}(\boldsymbol{x}) = (g_1(\boldsymbol{x}), g_2(\boldsymbol{x}), ..., g_m(\boldsymbol{x})) \leq \mathbf{0} \\ where & \boldsymbol{x} = (x_1, x_2, ..., x_n) \in \mathbf{X} \\ & \mathbf{X} = \{\boldsymbol{x} | \boldsymbol{l} \leq \boldsymbol{x} \leq \boldsymbol{u}\} \\ & \boldsymbol{l} = (l_1, l_2, ..., l_n), \boldsymbol{u} = (u_1, u_2, ..., u_n) \end{array} \tag{1}$$

where $f(\boldsymbol{x})$ is the objective function, $\vec{g}(\boldsymbol{x}) \leq \mathbf{0}$ is the constraint, $\mathbf{0}$ is the constrained boundary. $\boldsymbol{x}$ is the solution vector and $\mathbf{X}$ denotes the solution space, $\boldsymbol{l}$ and $\boldsymbol{u}$ are the lower bound and upper bound of the solution space.

Definition 2. ***(Feasible solution and Feasible set)***

A solution $\boldsymbol{x} = (x_1, x_2, ..., x_n) \in \mathbf{X}$ is said feasible, if $\vec{g}(\boldsymbol{x}) \leq \mathbf{0}$. The Feasible set of a COP is defined as:

$$S_F = \{\boldsymbol{x} : \boldsymbol{x} \in \mathbf{X}, \vec{g}(\boldsymbol{x}) \leq \mathbf{0}\} \tag{2}$$

Definition 3. ***(Constrained Violation)***

Given a solution $\boldsymbol{x}$, the Constrained Violation of a constraint in Eq. (1) is usually defined as:

$$G_i(\boldsymbol{x}) = \max\{g_i(\boldsymbol{x}), 0\}, i = 1, 2, ..., m. \tag{3}$$

Definition 4. ***(Solution Violation)***

Given a solution $\boldsymbol{x}$, the Solution Violation $\psi(\boldsymbol{x})$ is defined as:

$$\psi(\boldsymbol{x}) = \frac{1}{m}\sum_{i=1}^{m} \frac{G_i(\boldsymbol{x})}{\max_{\boldsymbol{x} \in P(0)}\{G_i(\boldsymbol{x})\}} \tag{4}$$

where $P(0)$ is the initial population of an EA, if $\max_{\boldsymbol{x} \in P(0)}\{G_i(\boldsymbol{x})\} < 1, i = 1, 2, ..., m$, we replace the $\max_{\boldsymbol{x} \in P(0)}\{G_i(\boldsymbol{x})\}$ with 1.

The order of two solutions is usually given as the Algorithm 1 , the comparison operator is to determine which one is better in comparison of two solutions [25].

Algorithm 1. Comparison of $\boldsymbol{x}_1$ and $\boldsymbol{x}_2$

Case1 Both are feasible, then the one with smaller objective f wins

Case2 One is feasible and the other is infeasible, then the feasible one wins

Case3 Both are infeasible, then the one with smaller violation Ψ wins.

DCMOEA with an efficient heuristic for global optimization. The motivation of the algorithm derived from converting a COP into DCMOP and to develop a general algorithm framework for solving DCMOPs by using multi-objective EAs (MOEAs). Therefore, the algorithm has lots of promising performances.

- A COP is converted into an equivalent DCMOP with three objectives, including an original objective, a constraint-violation objective and a niche-count objective.

- A method of gradually reducing the constraint boundary in order to handle the constraint difficulty.
- A method of gradually reducing the niche size in order to handle the multi-modal difficulty.

DCMOEA demonstrates superior performance in handing COPs, in particular to the kind of problems with highly constrained, multimodal and a large number of dimensions. Therefore, it is selected to solve the antenna design problem in this paper.

3 Formulating Antenna Design as COP

In this section, the design of the elliptical wide slot microstrip patch antenna is modeled as a COP.

3.1 Design Requirements of the Elliptical Wide Slot Microstrip Patch Antenna

Our purpose is to design an elliptical wide slot microstrip patch antenna with satisfying design requirements. The requirements are shown in Table 1:

Table 1. Antenna design requirements

Parameters	Requirements
Frequency	The range from 2.35 GHz to 2.55 GHz
Input impedance	50 Ω
VSWR	≤ 2
Gain	≥ 0 dB
Pattern range	$0° \leq \theta \leq 180°; \phi = 0°$
Size	30 mm × 30 mm × 1.6 mm

3.2 Geometric Structure of the Elliptical Wide Slot Microstrip Patch Antenna

The geometric structure of the antenna is shown in Fig. 1. It is formed in the following.

The square flame retardant 4 (FR4) substrate has a side length of 30 mm, the thickness of 1.6 mm, the relative permittivity of 4.4, and the input resistance is 50 Ω. Considering the ellipse smoother than the rectangle, the radiation patch on the front side, which is formed by subtracting two ellipses with the same axial ratio R. The coupling element in the back side, that is the difference between the ground plate and an ellipse with an axial ratio R. The feed point is located at bottom the side the antenna.

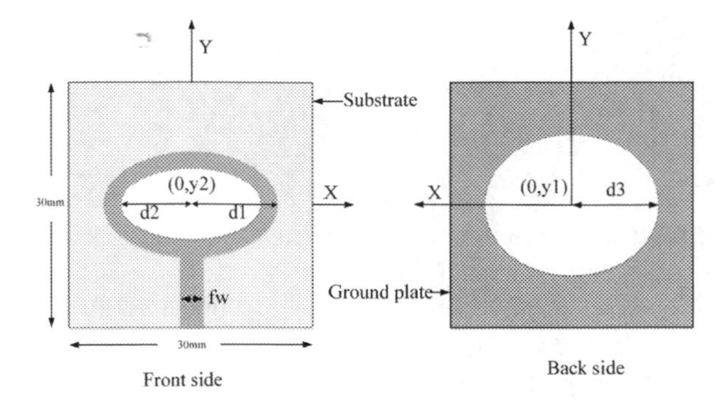

Fig. 1. Geometric structure of the elliptical wide slot microstrip patch antenna in vertical plane

3.3 Solution Vector and Solution Space

Both the location and the size of the ellipses and the microstrip feed line are changeable, as shown in Fig. 1. $d1, d2$ are the long axis of the frontal two ellipses with axial ratio R respectively, and $(0, y2)$ is the center of the ellipses. $d3$ is the long axis of the back ellipse with axial ratio R, and $(0, y1)$ is the center of the ellipse. fw is the width of the feed line.

A coordinate system is constructed in the patch antenna with the origin at the center of the patch antenna, as seen in Fig. 1. The geometric structure of the antenna is determined by seven variables. Six variables $d1, d2, d3, R, y1, y2$ are to determine the structure of the ellipses, where the ranges of $d1, d2, d3$ variables are $10\,\text{mm} \leq d1 \leq 14\,\text{mm}$, $0\,\text{mm} \leq d2 \leq 10\,\text{mm}$, $0\,\text{mm} \leq d3 \leq 14\,\text{mm}$ respectively, and the range of $R, y1, y2$ variables are $0.6 \leq R \leq 0.8$, $-2.8\,\text{mm} \leq y1, y2 \leq 2.8\,\text{mm}$ respectively. And the variable fw is to determine the size of the microstrip feed line, where the range of this variable is $1\,\text{mm} \leq fw \leq 3\,\text{mm}$.

All those seven variables $\boldsymbol{x} = (d1,\ d2,\ d3,\ R,\ fw,\ y1,\ y2)$ make up of the solution vector. And the solution space is the ranges of all the seven variables, see Eq. (5).

$$\begin{aligned}
&\mathbf{X} = \{\boldsymbol{x} | \boldsymbol{l} \leq \boldsymbol{x} \leq \boldsymbol{u}\} \\
&\boldsymbol{x} = (\text{d1, d2, d3, R, fw, y1, y2}) \\
&\boldsymbol{l} = (10,\ 0,\ 0,\ 0.6,\ 1,\ -2.8,\ -2.8) \\
&\boldsymbol{u} = (14,\ 10,\ 14,\ 0.8,\ 3,\ 2.8,\ 2.8)
\end{aligned} \tag{5}$$

3.4 Objective and Constraints

The objective and constraints are modeled according to the design requirement in Table 1. On top of that, the robustness of the antenna optimized also is taken into account in the paper. In practice, it is essential to concern about the robustness of antenna design, specially for real engineering. The gains are sampled in 5° increments over the region $0° \leq \theta \leq 180°$ and $\varphi = 0°$, and the frequencies are sampled in 5 MHz increments over the band from 2.35 GHz to 2.55 GHz.

In this paper, the objective $f(\boldsymbol{x})$ is defined based on the robustness will be taken into account, which is based on our previous work [9], see Eq. (6).

$$\begin{aligned} f(\boldsymbol{x}) = \textstyle\sum_{\theta}\sum_{\varphi}(&Gvariance_{(\theta,\varphi)} \\ &+ARvariance_{(\theta,\varphi)}) + VSWRvariance \end{aligned} \tag{6}$$

where (θ, φ) represents one direction in spherical coordinates. θ is the elevation angle with 5° increments over the range from 0° to 180° and φ is the azimuth angle with 0°. $Gvariance_{(\theta,\varphi)}$, $ARvariance_{(\theta,\varphi)}$ and $VSWRvariance$ are the variances of gain, axial ratio and VSWR respectively over the frequency band.

The details of the objective are shown as follows:

$$\begin{aligned} &Gvariance_{(\theta,\varphi)} \\ &\quad= \tfrac{1}{len(fr)}\textstyle\sum_{fr}(Gain_{(\theta,\varphi,fr)} - MeanG_{(\theta,\varphi)})^2 \\ &MeanG_{(\theta,\varphi)} \\ &\quad= \textstyle\sum_{fr}\frac{Gain_{(\theta,\varphi,fr)}}{len(fr)} \\ &ARvariance_{(\theta,\varphi)} \\ &\quad= \tfrac{1}{len(fr)}\textstyle\sum_{fr}(Axial_{(\theta,\varphi,fr)} - MeanAR_{(\theta,\varphi)})^2 \\ &MeanAR_{(\theta,\varphi)} \\ &\quad= \textstyle\sum_{fr}\frac{Axial_{(\theta,\varphi,fr)}}{len(fr)} \\ &VSWRvariance \\ &\quad= \tfrac{1}{len(fr)}\textstyle\sum_{fr}(VSWR_{fr} - MeanVSWR_{fr})^2 \\ &MeanVSWR_{fr} = \textstyle\sum_{fr}\frac{VSWR_{fr}}{len(fr)} \end{aligned} \tag{7}$$

where fr stands for a single frequency point, and $len_{(fr)}$ is the number of points over the frequency band. Therefore, $Gain_{(\theta,\varphi,fr)}$, $Axial_{(\theta,\varphi,fr)}$ and $VSWR_{(fr)}$ represent the gain, axial ratio and VSWR respectively in direction (θ, φ) and at frequency point fr.

fr the frequency, $fr = 2.35$ GHz, 2.355 GHz, $\cdots$, 2.545 GHz, 2.55 GHz;

The constraint on gains of the antenna design problem as shown in Eq. (8).

$$gGain_{(\theta,0^\circ,fr)}(\boldsymbol{x}) = -Gain_{(\theta,0^\circ,fr)} \le 0 \tag{8}$$

The constraint on VSWR are defined over the frequency band, see Eq. (9).

$$gVSWR_{fr}(\boldsymbol{x}) = VSWR_{fr} - 2 \le 0 \tag{9}$$

From the above, the antenna design is formed as a COP as follows:

$$\begin{aligned} min\ f(\boldsymbol{x}) &= VSWRvariance \\ &\quad+\textstyle\sum_{\theta=0^\circ}^{\theta=180^\circ}(Gvariance_{(\theta,0^\circ)} + ARvariance_{(\theta,0^\circ)}) \\ st:\ & gGain_{(\theta,0^\circ,fr)}(\boldsymbol{x}) = -Gain_{(\theta,0^\circ,fr)} \le 0 \\ & gVSWR_{fr}(\boldsymbol{x}) = VSWR_{fr} - 2 \le 0 \\ & \boldsymbol{x} = (d1, d2, d3, R, fw, y1, y2) \\ & \boldsymbol{l} = (10, 0, 0, 0.6, 1, -2.8, -2.8) \\ & \boldsymbol{u} = (14, 10, 14, 0.8, 3, 2.8, 2.8) \end{aligned} \tag{10}$$

In above COP, the objective $f(\boldsymbol{x})$ is minimization. As we all know, the smaller the variances are, the more robust the antenna is.

4 Solving Antenna Design by Using DCMOEA

In this section, the DCMOEA is used as search engine to optimize the structure of the elliptical wide slot microstrip antenna. The optimum antenna is obtained over the run of the DCMOEA, and the result is discussed.

4.1 Setting of DCMOEA

The parameters for the DCMOEA in solving the antenna are listed as follows:

1. Evolutionary generations $T = 500$.
2. Population size $NP = 30$.
3. The settings of the DE operator: crossover probability $CR = 0.9$, mutation probability $F = 0.5$.
4. The two constant parameters: precision requirement $\delta = 1.0e - 8$, control parameter of the decreasing $cp = 2$.
5. The number of the environmental changes: $S = 200$.

4.2 Results and Discussion

The electromagnetic simulation software Ansoft HFSS is adopted to simulate(evaluate) the antenna generated by the DCMOEA during optimization process. During optimization process, DCMOEA begins with a randomly initialized parent population of antennas with arbitrary shapes within the first generation. In each iteration during middle generations, an offspring population is generated via crossover and mutation of initialized parent population, and each individual in it is evaluated by Ansoft HFSS, after which the new parent population for the next iteration is selected from the union of the parent and offspring population according to DCMOEA. In the last generation, the optimum result is selected from the current population.

When the end of optimization, the optimum geometric structure of the antenna is obtained, as shown in Fig. 2. The variables value of the evolved antenna are ($d1 = 13.7$, $d2 = 1.7$, $d3 = 14$, $R = 0.7448$, $fw = 2.2$, $y1 = 0.3$, $y2 = -1.9$). We have fabricated this antenna by $3D$ metal printing. A photo of a prototype is shown in Fig. 3, the left side of the figure is the back of the evolved antenna, the right in front.

Performances of Evolved Antenna. VSWR of the evolved antenna by DCMOEA is less than 2 over the frequency from 2.35 MHz to 2.55 MHz in Fig. 4. When assuming that the antenna can work in a real environment when its VSWR is less than 2. Based on this assumption, the bandwidth of the evolved antenna covers the whole frequency range. This means that the bandwidth of the evolved antenna is wide with 200 MHz that satisfies the requirements table. Thus, an antenna with robust performance over the whole frequency range is achieved by the algorithm.

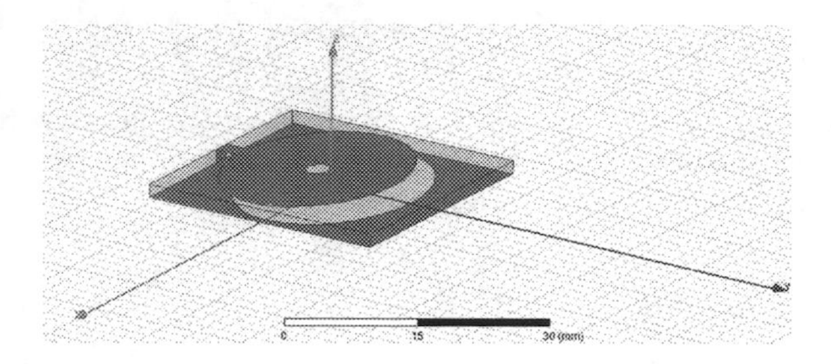

Fig. 2. The optimum geometric structure of the evolved antenna

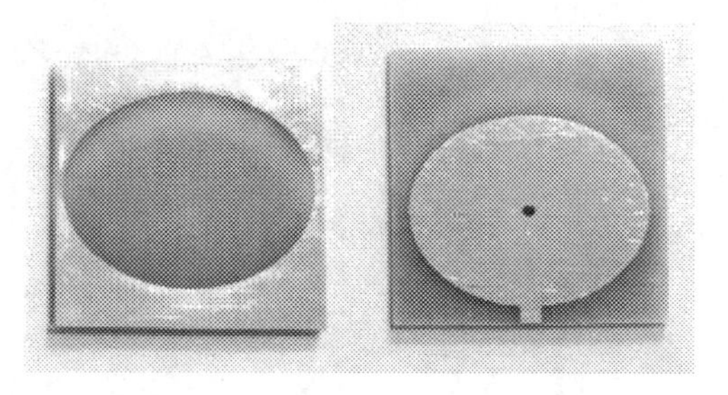

Fig. 3. The prototype of the evolved antenna by 3D printing

At the same time, the gains are greater than 0 over the range $0^\circ \leq \theta \leq 180^\circ$, $\varphi = 0^\circ$ are shown in Fig. 5. It indicates that the antenna evolved satisfies fully the design requirements. The three-dimensional radiation pattern over the range $0^\circ \leq \theta \leq 180^\circ$, $\varphi = 0^\circ$ in center frequency 2.45 GHz is shown in Fig. 6.

It is worth noting that analyzing the structure and results of the optimum elliptical wide slot microstrip patch antenna via the optimization of DCMOEA, we find that the radiation patch should be a whole ellipse instead of subtracting with two ellipses since $d2 = 1.7$ is largely less than 1/10 wavelength. In the paper, the elliptical wide slot microstrip patch antenna is one of the surface antennas. In the theory of the surface antennas, as it well known, the larger the effective aperture is, the better the performance of antenna is. Apparently, the optimum

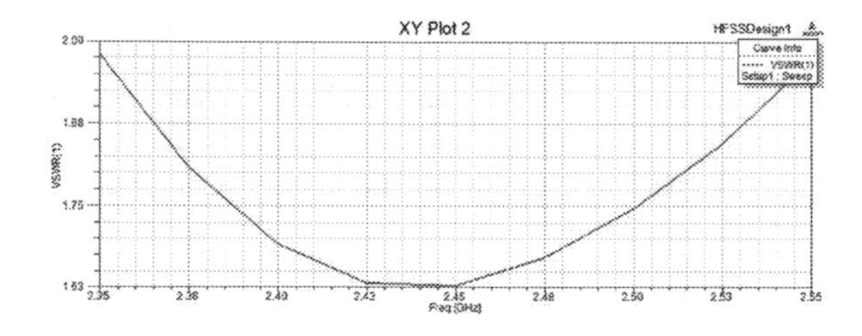

Fig. 4. VSWR of the evolved antenna

structure of the evolved antenna is all very well in the theory. Thus, we draw a conclusion that the radiation patch should be complete rather than partial, which provides a valuable reference for follow-up study with antenna design in real engineering and scientific research.

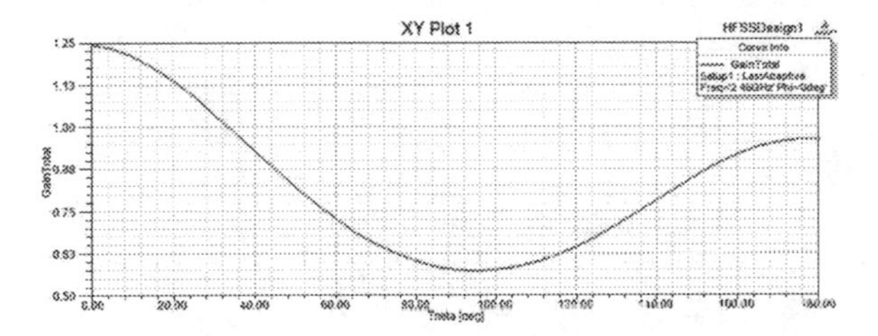

Fig. 5. Gains of the evolved antenna

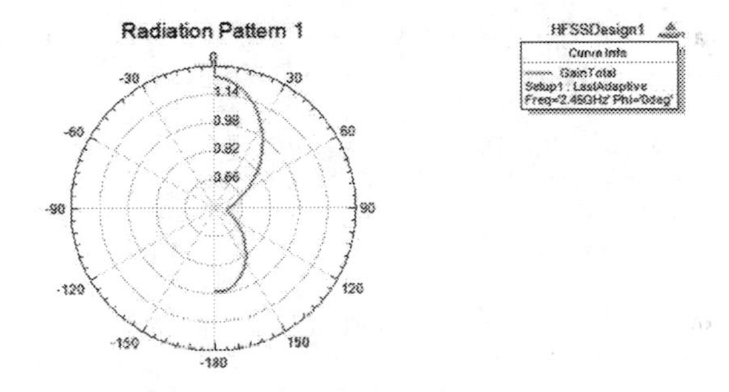

Fig. 6. Radiation patterns of the evolved antenna

4.3 Application of the Evolved Antenna

In Fig. 7, the evolved antenna is mounted to the router in the lab of our evolutionary antenna optimization group at China University of Geosciences, which is used to test whether the evolved antenna can work. The results did show the antenna works well.

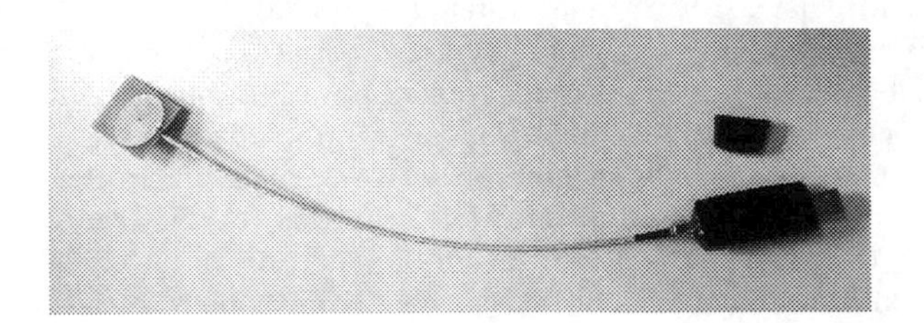

Fig. 7. Application of the evolved antenna

5 Conclusion

This paper shows that evolves an elliptical wide slot microstrip patch antenna by using DCMOEA. It has been shown that

1. Modeling for the design problem of the elliptical wide slot microstrip patch antenna as a constrained optimization problem (COP), and solving the problem by DCMOEA.
2. At the same time, the evolved antenna by the DCMOEA has obtained remarkable performance.
3. Otherwise, the evolved antenna is mounted to the router in the lab of our evolutionary antenna optimization group at China University of Geosciences, and it showed works well.
4. Last but not least, It is noteworthy that, we find that the radiation patch should be a whole ellipse instead of subtracting with two ellipses. It conforms to the theory of the surface antennas. Therefore, we draw a conclusion that the radiation patch should be complete rather than partial in the surface antennas design, which provides a valuable reference for follow-up study with antenna design in real engineering and scientific research.

We would like to pursue topic in the future:

1. To apply the expensive optimization algorithm for the elliptical wide slot microstrip patch antenna to reduce the computationally expensive.

Acknowledgments. This work was supported by Major Project for New Generation of AI Grant $No.$ 2018AAA0100400, the National Natural Science Foundation of China and other foundations (No.s: 61673355, 61271140, 61203306, 2012001202, 60871021), and high-performance computing platform of China University of Geosciences.

References

1. Darwin, C.: The origin of species, 1859–1959. Bios **30**(2), 67–72 (1959)
2. Fisher, R.A.: The genetical theory of natural selection. Bios (1930)
3. Linden, D.S., Altshuler, E.E.: Automating wire antenna design using genetic algorithms. Microwave J. **39**(3), 74–81 (1996)
4. Weile, D.S., Michielssen, E.: Genetic algorithm optimization applied to electromagnetics: a review. IEEE Trans. Antennas Propag. **45**(3), 343–353 (1997)
5. Jia, L., Zeng, S., Zhou, D., Zhou, A., Li, Z., Jing, H.: Dynamic multi-objective differential evolution for solving constrained optimization problem. In 2011 IEEE Congress of Evolutionary Computation (CEC), pp. 2649–2654 (2011)
6. Wang, Y., Cai, Z.: Combining multiobjective optimization with differential evolution to solve constrained optimization problems. IEEE Trans. Evol. Comput. **16**(1), 117–134 (2012)
7. Liu, Z., Zeng, S., Jiang, Y., Li, C., Ni, O.: Evolutionary design of a wide-band twisted dipole antenna for X-band application. In 2013 IEEE International Conference on Evolvable Systems (ICES), pp. 9–12 (2013)

8. Liu, B., Aliakbarian, H., Ma, Z., Vandenbosch, G.A.E., Gielen, G., Excell, P.: An efficient method for antenna design optimization based on evolutionary computation and machine learning techniques. IEEE Trans. Antennas Propag. **62**(1), 7–18 (2013)
9. Caie, H., Zeng, S., Jiang, Y., Sun, J., Sun, Y., Gao, S.: A robust technique without additional computational cost in evolutionary antenna optimization. IEEE Trans. Antennas Propag. **67**(4), 2252–2259 (2019)
10. Goldberg, D.E., Holland, J.H.: Genetic algorithms and machine learning. Mach. Learn. **3**(2), 95–99 (1988)
11. Altshuler, E.E., Linden, D.S.: Wire-antenna designs using genetic algorithms. IEEE Antennas Propag. Mag. **39**(2), 33–43 (1997)
12. Wen, Y.-Q., Wang, B.-Z., Ding, X.: A wide-angle scanning and low sidelobe level microstrip phased array based on genetic algorithm optimization. IEEE Trans. Antennas Propag. **64**(2), 805–810 (2015)
13. Makki, B., Ide, A., Svensson, T., Eriksson, T., Alouini, M.-S.: A genetic algorithm-based antenna selection approach for large-but-finite mimo networks. IEEE Trans. Veh. Technol. **66**(7), 6591–6595 (2016)
14. Chou, H.-T., Cheng, D.-Y.: Beam-pattern calibration in a realistic system of phased-array antennas via the implementation of a genetic algorithm with a measurement system. IEEE Trans. Antennas Propag. **65**(2), 593–601 (2016)
15. Storn, R., Price, K.: Differential evolution-a simple and efficient heuristic for global optimization over continuous spaces. J. Global Optim. **11**(4), 341–359 (1997). https://doi.org/10.1023/A:1008202821328
16. Deb, A., Roy, J.S., Gupta, B.: Performance comparison of differential evolution, particle swarm optimization and genetic algorithm in the design of circularly polarized microstrip antennas. IEEE Trans. Antennas Propag. **62**(8), 3920–3928 (2014)
17. Ding, Y., Jiao, Y.-C., Zhang, L., Li, B.: Solving port selection problem in multiple beam antenna satellite communication system by using differential evolution algorithm. IEEE Trans. Antennas Propag. **62**(10), 5357–5361 (2014)
18. Cui, C.-Y., Jiao, Y.-C., Zhang, L.: Synthesis of some low sidelobe linear arrays using hybrid differential evolution algorithm integrated with convex programming. IEEE Antennas Wirel. Propag. Lett. **16**, 2444–2448 (2017)
19. Rechenberg, I.: Evolutionsstrategie-optimierung technisher systeme nach prinzipien der biologischen evolution. frommann-holzboog, Stuttgart (1973)
20. Chen, Y., Yang, S., Nie, Z.: The application of a modified differential evolution strategy to some array pattern synthesis problems. IEEE Trans. Antennas Propag. **56**(7), 1919–1927 (2008)
21. Choi, K., Jang, D.-H., Kang, S.-I., Lee, J.-H., Chung, T.-K., Kim, H.-S.: Hybrid algorithm combing genetic algorithm with evolution strategy for antenna design. IEEE Trans. Magn. **52**(3), 1–4 (2015)
22. Agrawal, A., Vakula, D., Sarma, N.V.S.N.: Design of elliptical microstrip patch antenna using ANN. In: Piers Suzhou Progress in Electromagnetics Research Symposium, pp. 264–268 (2011)
23. Long, S., Shen, L., Schaubert, D., Farrar, F.: An experimental study of the circular-polarized elliptical printed-circuit antenna. IEEE Trans. Antennas Propag. **29**(1), 95–99 (1981)
24. Zeng, S., Jiao, R., Li, C., Li, X., Alkasassbeh, J.S.: A general framework of dynamic constrained multiobjective evolutionary algorithms for constrained optimization. IEEE Trans. Cybern. **47**(9), 2678–2688 (2017)
25. Deb, K.: An efficient constraint handling method for genetic algorithms. Comput. Methods Appl. Mech. Eng. **186**(2), 311–338 (2000)

Imputation Methods Used in Missing Traffic Data: A Literature Review

Pan Wu, Lunhui Xu(✉), and Zilin Huang

School of Civil Engineering and Transportation, South China University of Technology, Guangzhou, Guangdong, China
201810101746@mail.scut.edu.cn, lhxu@scut.edu.cn, scutjulian@163.com

Abstract. The missing traffic data has caused great obstacles and interference to further research, such as traffic flow prediction, which affects the traffic authorities' judgment for the real traffic operation state of road network and the new control strategies. It is very critical to select the imputation methods with good performance for maintaining the integrity and effectiveness of the traffic data. A large number of literatures have developed many methods to repair missing traffic data, yet lacking systematic comparison of these methods and an overview of the state-of-the-art development in imputation methods. In this paper, extensive research on imputation methods are sorted out and synthesized, the mechanism of missing traffic data is analyzed, and various algorithms in repairing missing data are systematically reviewed, highlighted some challenges and potential solutions. The purpose is to provide a structural diagram of the current recovery technology for missing traffic data, clearly pointing out the advantages and disadvantages of these methods, and helping researchers to conduct better exploration on the incomplete traffic data.

Keywords: Missing traffic data · Imputation method · Tensor decomposition

1 Introduction

Traffic missing data has long been a serious challenge for Intelligent Transportation Systems (ITS) [1, 2]. A crowd of traffic studies, such as traffic state identification, traffic management, have been limited by the incomplete data [3, 4]. Therefore, it is urgent to find an efficient imputation method to estimate missing data.

It is found that traffic authorities around the world have suffered from different degrees of information loss or anomaly caused by various factors, that is, traffic collection data (statistical traffic volume, speed, occupancy rate, etc.) appears missing, error and other phenomena, which can't reflect the actual traffic state of road network. For example, about 15% of the traffic data of highways in the United States are unavailable [5]. In Beijing, China, the missing rate of traffic data exceeds 10% [6], and the international famous traffic database PeMS has more than 5% of data appears missing [7], and so on. The missing data has brought serious consequences, especially for further research work such as traffic prediction [4]. Among them, data quality has been considered one of the main challenges for traffic forecasting [8]. If there is no

K. Li et al. (Eds.): ISICA 2019, CCIS 1205, pp. 662–677, 2020.
https://doi.org/10.1007/978-981-15-5577-0_53

suitable imputation method, the missing traffic data will be simply estimated or discarded, which will have a great impact on the subsequent analysis [7]. Generally, the missing data exists not only in the traffic field, but also in other disciplines [9]. In the past two decades, the imputation methods of missing data have been studied in many disciplines, such as statistics, sociology, medicine, etc. [10]. Filling the gaps with manually created data is considered the most straightforward method [11–13]. However, missing data remains a difficult problem to be solved in the traffic field [7].

In recent ten years, traffic missing data has aroused a hot topic in ITS. Experts and scholars in the traffic field all over the world have actively carried out research on this problem, and found that most methods to handle missing data in other fields are also suitable for traffic data [14]. In the early research, most of imputation methods were divided into three categories: prediction based, interpolation based and statistical learning based methods [1, 6]. However, many early methods proposed have not taken into account the spatial-temporal correlation of traffic flow or only considered the time series [15–18] or space dependence alone [19, 20], the implicit characteristics of traffic data have not been deeply excavated, and the accuracy of these methods needs to be further improved. Numerous methods proposed deeply mine the hidden temporal and spatial correlation of traffic flow, and complete the missing data computation by constructing the traffic data vector into a matrix model [21, 22]. The results show the existing matrix-based interpolation methods often fail when the data loss rate is large, especially when the data loss is in one day or a few days. On this basis, the researchers propose a tensor-based data interpolation method [1, 2, 23–25]. It is found that the imputation method based on tensor pattern can remain the diversity of traffic data. The method has good recovery effect on the missing data, especially in the case of large loss rate or bad weather, the interpolation performance is still maintained. However, the tensor-based method has not been widely used in missing traffic data, mainly because the method needs to capture the global structure of the data by higher-order decomposition, which is still a challenging problem.

To sum up, accurate and effective recovery of missing or abnormal traffic data still faces many challenges. Referring to a large number of literatures, it is found that the current literature review on imputation methods of missing traffic data is blank, lacking systematic comparative analysis of these methods. Therefore, this paper focuses on summary of imputation methods for missing traffic data to fill in the gaps in traffic field. Aiming at missing traffic data, this paper summarizes the current imputation methods, and systematically compares the principle, performance and application scope of these methods, as well as the limitations of various mainstream methods. The purposes are trifold. First, it makes readers have a clear understanding of the source and characteristics of traffic data. Second, readers can find out suitable method to deal with missing traffic data, having preliminary screening reference. Third, the paper can help readers to study imputation methods of other missing data, such as geographical location information. In this paper, the mode and the principle of loss data are briefly described, and the classification of imputation methods is summarized. It is found that probabilistic principal component analysis (PPCA) method is relatively common method, and the tensor-based methods have good performance. Regarding the principle and effect analysis of PPCA imputation methods, readers are recommended to refer to literature [7]. The tensor-based method is introduced well in literature [1] that has present the tensor method used in missing traffic data for the first time. On this basis, Chen et al. [25] has tried to recover the missing data through SVD combination tensor decomposition. Later Chen et al. [2] proposed Bayesian tensor decomposition method for repairing incomplete data. These four articles have introduced the most common imputation methods using in missing traffic data in detail.

The rest of the article is arranged as follows: the second chapter mainly analyzes the missing mechanism and classification of missing traffic data; the third chapter mainly introduces classification of the current common imputation methods of missing traffic data; the fourth chapter introduces the imputation methods based on temporal and spatial correlation, focusing on the tensor-based method to fill the missing data; the fifth chapter mainly gives index to evaluate the actual performance of these methods in dealing with missing traffic data; the sixth chapter describes the main conclusions, the limitations of these researches and the prospect of further studies.

2 The Traffic Missing Data

The qualitative characteristics of the traffic operational state are called traffic characteristics. The traffic volume, speed and density are three basic parameters that characterize the traffic characteristics.

2.1 Acquisition of Traffic Data

The collection methods of traffic data are generally divided into two types: mobile type and fixed type. The comparison of these methods is shown in Table 1 below.

Table 1. The comparison of acquisition techniques for traffic data

Types	Advantage	Disadvantage	Data acquisition means
Fixed type	The technology is mature, the detection accuracy is high, the acquisition data is convenient and large	The detection range is fixed, the equipment is vulnerable to damage and difficult to repair	Coil detector, microwave detector, ultrasonic detector, infrared detector, video detector, etc.
Mobile type	Large coverage, dynamic adjustment, more flexible	The manpower and material resources consumed are large, and the data collected is limited	Based on GPS data, the acquisition data based on identifying the automobile license plate, the electronic tag, etc.

At present, the traffic system mainly obtains the traffic data through the fixed collection method. The induction coil or the automatic traffic recorder (ATR) is common sources for traffic data [8].

2.2 Causes for Failure Data

The road traffic system is complex with high randomness and uncertainty, affected by the factors such as travel mode, travel habits, weather, environmental factors and traffic operation conditions. These factors are likely to cause the data acquired by the detector to not actually reflect the road traffic state.

In order to find the regular pattern of traffic data more effectively to complete subsequent analysis, it is necessary to identify and repair fault traffic data. The research work on the quality of traffic data abroad began in 1976. Payne et al. [26] took the induction coil on the highway as the research object, and first proposed the recognition method of wrong data based on single parameter threshold. Jacobsen et al. [27] used direct elimination method to deal with error data, but the remaining data are difficult to truly show the actual traffic operation state. Faulty data is usually divided into three categories: missing data, distorted data and abnormal data. This article focuses on the repair methods of missing data.

2.3 Classification of Missing Data

In most cases, the selection of suitable repair methods will consider the data missing pattern. In order to avoid erroneous statistical inference, it is important to understand the missing pattern and the missing mechanism. Rubin [28] indicated that the distribution of data was generally dependent on the observed missing pattern, and first proposed terms describing the missing patterns. Classification of data missing patterns is shown in Table 2 below.

Table 2. Classification of data missing patterns

Types	Missing pattern	Missing condition	Recovery methods
MAR	Missing at random	Only depends on the observed data	Common recovery methods can handle missing data in MAR/MACR, such as PPCA, BPCA, etc.
MCAR	Missing completely at random	Not depending on the input value	
MNAR	Missing is not at random	Depends on the missing variable	No available imputation method

Most studies deal with cases where missing data belongs to MAR or MCAR type [8, 29, 30].

3 Classification of Imputation Methods for Missing Traffic Data

A considerable number of researches have been done on repairing missing data. The main purpose is to improve the accuracy of data analysis in succession. At the beginning research of missing data, there are relatively few methods to recover missing data. The common methods are multiple interpolation and maximum likelihood estimation method based on EM algorithm [31], but the error of the two methods is large, and researchers have being exploring and developing new methods to repair the missing data. A large number of methods have been proposed for repairing missing traffic data. The classification of these methods is shown in Fig. 1.

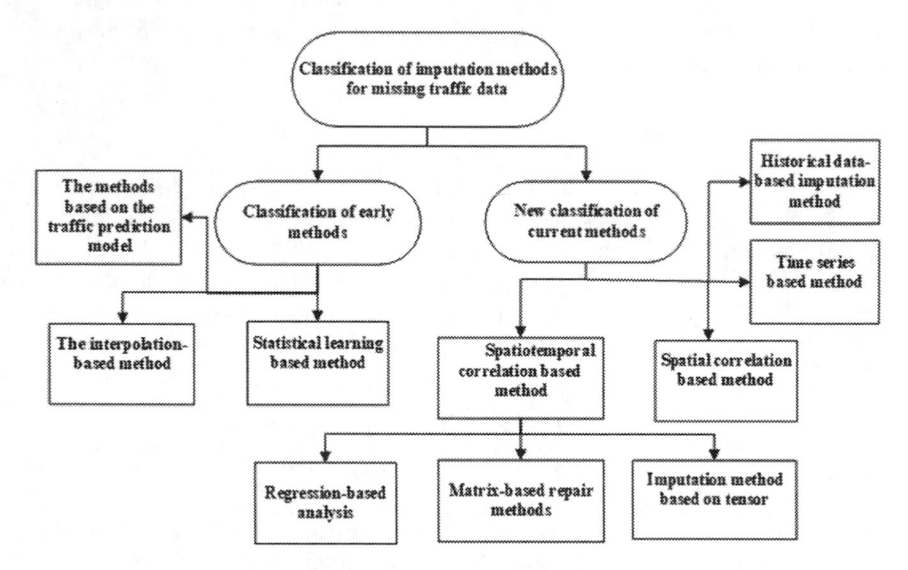

Fig. 1. Classification of imputation methods for missing traffic data, this scheme shows the different classification result in this work

3.1 Early Methods Classification of Repairing Missing Data

In the early days, a variety of methods have been proposed to recover missing traffic data. Most of them fall into three categories: prediction based, interpolation based and statistical learning based methods [1, 6]. The basic principles of methods are shown in Table 3 below.

Table 3. Characteristics of various methods

Classification	The mechanism of this type method	Capturing the random variation of traffic flow
Prediction based methods	The missing value is considered as a predicted value, according to the definite relationship between the past and the future data pairs in the historical database	Yes
Interpolation based methods	A weighted average of known data to fill the missing data which is time-adjacent or in an adjacent mode [32]	Yes
Statistical learning based methods	Traffic information obtained by emphasizing the traffic statistical characteristics, this method usually consists of two steps: firstly, assuming the probability distribution model of the data; secondly, iterating to estimate the parameters of the model, and estimating the missing data according to the observed data. The key of this method is to assume the special probability distribution followed by the observed data [6]	No

The Methods Based on the Traffic Prediction Model: The prediction models commonly used for repairing missing data mainly include the auto regressive integrated moving average (ARIMA) [5, 6, 33], exponential smoothing filtering method and time-series model, the nearest neighbor state parameter regression, non-parametric statistical model, spectral analysis technique, Gauss maximum likelihood method, Bayesian networks (BNS) [34], feed forward state parameter regression [35], etc. However, there are some limitations in these methods. The prediction model cannot fully utilize the data collected after the missing point, which may reduce the imputation performance. If a series of consecutive data is missing, the imputation methods may not work.

The Interpolation-Based Methods: The interpolation-based methods mainly include time interpolation method and pattern similarity method. That takes full advantage of the observation value in the space-time domain. Yin et al. [32] have shown that the time neighbor method is more effective than the spatial neighbor method. The time neighbor estimation method fills the missing point by averaging known data collected from the same detector at the same time on the adjacent day or on the same day at the same adjacent time, assuming that the daily traffic is very similar for several days in a row, which do not make full use of local daily traffic information. Pattern similarity method attempts to estimate the missing data by using the historical data of the same detector on different dates, which are very close to the data of the research day in the pattern. However, these interpolation methods assume that the traffic flow has strong similarity in the next few days, without explicitly describing the random change of the traffic flow, which may omit important information such as daily traffic distribution.

Statistical Learning Based Methods: The imputation methods based on statistical learning model for repairing missing traffic data mainly include typically principal component analysis based methods, such as PPCA, BPCA, KPPCA, etc. Qu et al. [6] have proposed PPCA method to recover the missing data. Since then, PPCA has been widely used to fill missing data, which can improve imputation performance after properly considering the space-time dependence [7, 23, 36]. PPCA combines the main component analysis (PCA) and the maximum likelihood estimate (MLE). BPCA is slightly modified on the basis of PPCA, both PCA and BPCA being based on EM (expectation maximization) to interpolate missing data.

Li et al. [15] have compared the estimation performance of these early methods from the aspects of reconstruction error, statistical characteristics and running speed. The results show that the statistical learning method is more effective than the interpolation method and the traffic prediction method using the data of a single detector. Among the statistical learning methods, PPCA has the best performance and has strong robustness to weather change [6, 7]. However, the traffic data is considered as a matrix which limits the implicit multidimensional characteristics of traffic data, so that the accuracy of these methods needs to be further improved.

The traffic parameters (mainly speed, traffic volume, occupancy rate, etc.) usually have inherent space-time patterns and correlations, which are also in essence multidimensional. Most early recovery models do not min deeply the spatial-temporal correlation between traffic data with limited restoration ability. In recent years, many experts and scholars have begun to focus on the temporal and spatial patterns of traffic data, and repairing missing data can be realized by constructing the multidimensional features of traffic data into multidimensional matrices, such as tensor patterns with good recovery effect. Tensor-based model is also considered one of the most effective methods in repairing missing data. A tensor-based repair method was not involved in the earlier classification of methods. Therefore, it is necessary to reclassify the existing imputation methods for traffic flow.

3.2 New Classification of Imputation Methods for Missing Traffic Data

Different from the above classification, the imputation methods can also be divided into four categories according to the repairing mechanism.

Historical Data Based Imputation Methods: The historical trend method recovering the missing data according to the average value of the historical data is simple and feasible. The method can handle to continuous abnormal data. But, it can't reflect the change of the traffic in the real state, the historical data is smoothly processed, and the fluctuation characteristic of the data cannot be preserved, so that the processed data has partial characteristic distortion. And the abnormal value can be recovered by establishing the traffic historical database. The method can reflect the trend of the data, but can't characterize the microcosmic fluctuation characteristics of traffic data.

Time Series Based Methods: Traffic data is time series data in essence. The traditional methods repairing missing data mostly use multiple data prediction methods based on time series. Various repair methods based on time series can be also used to repair traffic data. These methods based on time series repair missing data by the average values of adjacent periods. And the missing values can also be supplemented by traffic parameter prediction methods, such as statistical learning methods, genetic optimization algorithms, classification and feature selection algorithms, maximum entropy methods, Kalman filter prediction methods, neural network methods and other algorithms. These methods are based on a complete sequence of road traffic data. The problem of the mass loss road traffic data can be solved by the non-parametric model method, the principal component analysis method, the maximum expectation method, the dynamic image model method and the state space neural network method. Time series data can only cover a small amount of space-time information, so these methods can only deal with a small amount of loss data.

Spatial Correlation Based Methods: The imputation methods based on spatial correlation of traffic data reflect the actual traffic spatial state, but that is difficult to show the non-uniform time-varying features of road network traffic. The method based on Grey residual GM(1, N) model is a typical method based on spatial correlation, which improves the accuracy to a certain extent. This method can improve the recovery mean based on historical data.

Spatiotemporal Correlation Based Methods: When modeling traffic data, it is very important to consider strong space-time correlation model. Recently, a tensor-based approach has been used to deal with missing data [22, 37]. These methods model the interaction between multiple traffic variables as multidimensional array (tensor) allowing the combination of multiple correlation between different variables. Tensor decomposition is a standard technique to capture multidimensional structure of data. There are two main tensor based methods in the previous study. The first is the method based on low-rank tensor without decomposition structures [21], avoiding the non-convex optimization problem. Another method is tensor decomposition, the high-order extension of the singular value decomposition (SVD) algorithm [38].

It is found that in the four kinds of methods, the method based on temporal and spatial correlation has the best effect and the highest accuracy. Therefore, the fourth chapter focuses on the imputation methods based on temporal and spatial correlation.

4 Imputation Methods Based on Space-Time Correlation

There are many methods to repair missing data based on spatial-temporal correlation. The mainstream methods are mainly divided into the following three categories.

4.1 Regression-Based Analysis Methods

Regression analysis can be used to study spatial-temporal correlation of data. The core work is to construct the spatial-temporal relationship between missing value and observed value. The early research is based on time series, mainly including ARIMA model, factor regression and other models. These models have some defects, such as poor extensibility, insufficient consideration of space-time correlation and low accuracy, and so on.

4.2 Matrix-Based Repair Methods

Therefore, in order to overcome these difficulties, many later studies try to couple the temporal and spatial information of traffic data, construct a matrix model to estimate missing value. The traffic time series is reconstructed by interpolating missing data, and a multi-dimensional matrix is established to describe the traffic flow. These matrix-based methods mainly include PPCA method, spatial-temporal multi-view learning algorithm, Bayesian non-parametric multi-output Gaussian model, etc. In many existing studies, matrix decomposition method is better than regression analysis in repairing missing data [22]. Although the matrix-based method has better performance when the missing rate is low, there are still many challenges. The traffic data can be structured into matrix model, and the spatial and temporal information of traffic data can't be fully utilized at the same time. Consequently, the matrix-based method can't deal with the missing data when the loss rate is large, especially in the extreme case of one or more days of data loss.

4.3 Imputation Methods Based on Tensor

In order to overcome the shortcomings of the matrix, a multi-dimensional matrix that is the tensor model, is proposed to model the traffic data, reserving the multi-way property of the traffic data, extracting potential factors in each tensor pattern, combining and utilizing a multi-mode correlation. A series of studies on the multi-dimensional data, the tensor method is proved to be a good analytical tool for processing multidimensional data. In recent years, the tensor-based method has been introduced into the repairing missing traffic data [1, 2, 25], with good performance. The dynamic tensor is constructed by the traffic time series. The periodicity, trend and temporal and spatial correlation of traffic data are excavated from multiple time granularity, and imputation missing data by tensor decomposition. The Bayesian probability matrix factorial decomposition model is extended to high-order tensor and applied to loss data, which can capture the global structure by high-order decomposition.

Background of Tensor-Based Methods: Tensor decomposition was initially proposed by Hitchcock [39], but only the expression of tensor was put forward, its application was not clarified until 1966, Tucker [40] applied tensor method to analyze data, and then proposed tucker decomposition, which is one of the most common tensor decomposition methods up to now. This method decomposes the given tensor into kernel tensor and factor matrix in a sequence. Carroll and Chang [41] proposed another tensor decomposition method, CP(CANDECOMP/PARAFAC) decomposition, which is now an important decomposition method. In 2009, in the research of Kolda and Bader [42], two decomposition methods were proposed which were both higher-order generalizations of the singular value decomposition (SVD). 2013, Tan [1] introduced the tensor model of matrix extension to model the traffic data for the first time. Tensor decomposition is a multi-linear structure [43], which inherits the advantages of interpolation method based on matrix pattern and the multidimensional internal correlation of traffic data, and the purpose of tensor decomposition is to obtain potential multivariate linear factors from partial data to estimate the missing value [44, 45].

Many studies have shown that tensor decomposition method can effectively recover missing data [7, 22–24, 39, 46]. Ran et al. [23] employed HaLRTC. Liu et al. [21] proposed a low rank tensor calculation algorithm to estimate the missing traffic data. Tensor decomposition model is used to repair missing traffic speed data [22, 25]. Goulart et al. [24] proposed a tucker decomposition method for the core tensor. Most of these studies on tensor computation rely on trace norm minimization to find the low rank approximate of the original incomplete tensor [21, 23]. But these methods only calculate a single point data, it is prone to occur over-fitting the phenomenon when the missing rate is large. Therefore, it is very difficult to deal with sparse tensors and capture global information [38, 47]. To solve this problem, Bayesian inference methods such as MCMC and variation inference methods are used to design and use tensor decomposition to overcome sparse tensor and extreme cases, ultimately obtaining better results [45, 48, 49].

In previous studies, a large number of experiments focused on repairing random missing data [1, 22, 23, 47], the others focused on prediction problem [50]. However, structured missing [44] is not analyzed in full detail. In the research of Chen et al. [25], these problems are well solved. A three-step framework based on tucker decomposition is proposed to complete the recovery task by discovering space-time patterns and underlying structures from incomplete data. The truncated singular value decomposition (SVD) is introduced to capture main latent features along each dimension. When the loss rate is between 20% and 80%, the method can also successfully repair the data. The tensor describes a richer algebra structure, which can encode more information [38]. Therefore, it is very important to structure the original traffic data with a tensor.

The Basic Theory of Tensor: Tensor used in modeling traffic flow data can capture the correlation between multiple modes of data, such as day, week, time and space composed a tensor. Most tensors have higher-order properties, however, low rank tensor is easier to repair missing data and complete tensor decomposition. Many researchers use small Tucker rank or n-rank to recover tensors [51]. The basic theory is as follows. The n-mode Tucker rank of a N-dimensional tensor is $\mathrm{A} \in R^{I_1 \times I_2 \times \cdots I_N}$, denoted by r_n, is the rank of the mode-n unfolding matrix $A_{(n)}$.

$$r_n = rank_n(\mathrm{A}) = rank(A_{(n)}) \tag{1}$$

Here, the tensor $\mathrm{A} \in R^{I_1 \times I_2 \times \cdots I_N}$ is defined as $A_{(n)}$. The tensor element $(i_1 i_2, \ldots, i_N)$ is mapped to the matrix element (i_n, j), where

$$j = 1 + \sum_{\substack{k=1 \\ k \neq n}}^{N} (i_k - 1) J_k,\ J_k = \prod_{\substack{m=1 \\ m \neq n}}^{k-1} I_m, A_{(n)} \in R^{I_n \times J},\ J = \prod\nolimits_{\substack{k=1 \\ k \neq n}}^{N} I_k \tag{2}$$

The n-mode product of a tensor $\chi \in R^{I_1 \times I_2 \times \cdots \times I_N}$, a matrix $U \in R^{J \times I_n}$ can be represented as $\chi \times {}_n U$, and dimensions $I_1 \times I_2 \times I_{n-1} \times J \times I_{n+1} \times \cdots \times I_N$. Then $y = \chi \times {}_n U$ equals to

$$Y_{(n)} = UX_{(n)} \,. \tag{3}$$

Theoretically, the tensor approach based on low rank approximation can be transformed into the optimization problem.

$$\min_{X \in A} \sum_{n=1}^{N} rank(\chi_{(n)}), s.t. \, \chi_{\Omega} = A_{\Omega} \tag{4}$$

Here, the elements of A in the set Ω are known, the rest is lost. The problem is difficult to solve which is non-convex. In some literatures, the nuclear norm $|| \, ||_*$ is introduced to solve this problem [52, 53].

$$||A||_* = \sum_{i=1}^{N} \alpha_i \, ||A_{(i)}||_* \tag{5}$$

Where, α_i is the weighted parameter, the minimization problem can be simplified as:

$$\min_{X \in A} ||\chi||_*, s.t. \, \chi_{\Omega} = A_{\Omega} \tag{6}$$

As shown above, the previous non-convex problem is transformed into a convex problem. Many tensor methods based on opportunity norm have been proposed [21, 54]. For detailed basic and related applications of the tensor imputation method, the reader can refer to numerous literatures, such as literatures [1, 2, 23–25].

Shortcomings and Potential Research of Tensor-Based Imputation Methods: To sum up, the repair of missing data is a hot topic, especially for the enhancement of traffic data quality, but which is still a challenging and difficult problem. Firstly, what is the hidden mode of traffic data, how to mine these hidden patterns has not been well solved. Secondly, There are still some difficult problems, such as how to use these hidden patterns to better help recover the missing data, how to improve the interpretability of the model. Although tensor decomposition technology has been proved to be effective in previous studies, which have not been described and tested in detail. Finally, it is necessary to obtain the global structure of traffic data by decomposing high-order tensor, however reducing dimensions of high-dimensional data is a challenging problem.

The Bayesian probability matrix factorial decomposition model is extended to high-order tensor and applied to calculating space-time traffic data. Third-order tensor structure is superior to matrix and fourth-order tensor in maintaining dataset information. The potential research direction is to integrate physical prediction models (such as automatic regression (AR), automatic regression moving average (ARMA)) and automatic regression comprehensive moving average (ARIMA) into the time dimension of tensor observation in the future. In the future, we can further utilize the multi-dimensional nature of traffic data, discover spatiotemporal traffic patterns, improve data quality, and design multi-source data fusion schemes, especially considering the fusion of different traffic parameters such as traffic volume and speed into their multivariate relationships. How to select the appropriate rank in different environments in the future is a question worth considering [2].

5 Evaluate the Repairing Performance Criteria of Imputation Methods

The average absolute error deviation (MAED) and root mean square error (RMSE) are usually used to evaluate the interpolation performance of imputation methods. In addition, other two indicators that are the normalized root mean square error (NRMSE) and the normalized mean absolute error (NMAE) can also be used for performance evaluation of repair methods [6, 25, 30].

5.1 MAED

$$MADE = \frac{\sum_{m=1}^{M} |err_m - \overline{err}|}{M}, \; err_m = \left|t_{org}^m - t_{est}^m\right|, \; \overline{err} = \frac{\sum_{m=1}^{M} \left|t_{org}^m - t_{est}^m\right|}{M} \tag{7}$$

Where, t_{org}^m is the real value of the missing entry, t_{est}^m is the estimated values of the missing entry, M denotes the amount of testing entries that were used.

5.2 RMSE

The root mean square error (RMSE) reflects the average performance of interpolation method, the smaller the RMSE, the better the interpolation performance.

$$RMSE = \sqrt{\frac{\sum_{m=1}^{M} \left(t_{org}^m - t_{est}^m\right)^2}{M}} \tag{8}$$

5.3 NRMSE

The normalized root mean square error (NRMSE):

$$NRMSE = \sqrt{\frac{1}{M}\sum_{m=1}^{M} \left(t_{org}^m - t_{est}^m\right)^2} \Big/ \sum_{m=1}^{M} t_{org}^m \tag{9}$$

5.4 NMAE

The normalized mean absolute error (NMAE):

$$NMAE = \sum_{m=1}^{M} | t_{org}^{m} - t_{est}^{m} | \Big/ \sum_{m}^{M} t_{org}^{m} \quad (10)$$

Among the above four evaluation indicators, RMSE is the most common indicator used to assess the accuracy of repair methods for missing traffic data [1, 6, 7, 23–25].

6 Conclusion and Future Research

In many applications of traffic data, missing or incomplete data is a common defect. The traffic data may contain unknown characteristics for different reasons, such as the sensor fault produces distorted, unmeasured values, the data is obscured by noise, and no response in the investigation. Processing missing data has become the basic requirement for follow-up application of traffic data. Inappropriate processing to missing data may lead to large errors or seriously affect the accuracy of subsequent research. This work reviews the most important means of filling missing traffic data, especially considering the method based on tensor. Usually, the missing traffic data is caused by different reasons, and the mechanism of each imputation method is also different, the accuracy of various methods is slightly different. Therefore this paper combs the methods of acquisition traffic data and the causes of missing data, and then classifies and arranges the repair methods according to different types of missing data. Various repair methods are systematically analyzed and compared. Early methods can be divided into four different types, namely, based on historical data method, based on time series repair method, based on spatial association repair method, based on temporal and spatial correlation repair method. In most cases, these methods can repair the missing data, and the repair method based on temporal and spatial correlation has the best repair effect, and the tensor based repair method has the best accuracy. This paper provides some accuracy evaluation based on repair methods, and enumerates the four most commonly used indexes to evaluate which interpolation methods of missing data have the highest accuracy and can restore the complete data set to the highest extent. Obviously, the research direction can still be expanded. Although there are some difficulties in the decomposition of higher-order tensors, the accuracy of imputation methods needs to be further improved, and the widely used tensor-based repair methods have not yet been completely solved, hoping to provide some useful references for researchers to discuss and choose methods.

References

1. Tan, H., Feng, G., Feng, J., et al.: A tensor-based method for missing traffic data completion. Transp. Res. Part C Emerg. Technol. **28**, 15–27 (2013)
2. Chen, X., He, Z., Sun, L.: A Bayesian tensor decomposition approach for spatiotemporal traffic data imputation. Transp. Res. Part C Emerg. Technol. **98**, 73–84 (2019). https://doi.org/10.1016/j.trc.2018.11.003
3. Zhang, J., Wang, F.Y., Wang, K., et al.: Data-driven intelligent transportation systems: a survey. IEEE Trans. Intell. Transp. Syst. **12**(4), 1624–1639 (2011). https://doi.org/10.1109/TITS.2011.2158001
4. Chen, C., Wang, Y., Li, L., Hu, J., Zhang, Z.: The retrieval of intra-day trend and its influence on traffic prediction. Transp. Res. Part C Emerg. Technol. **22**, 103–118 (2012). https://doi.org/10.1016/j.trc.2011.12.006
5. Al-Deek, H.M., Venkata, C., Chandra, S.R.: New algorithms for filtering and imputation of real-time and archived dual-loop detector data in I-4 data warehouse. Transp. Res. Rec. J. Transp. Res. Board **1867**, 116–126 (2004). https://doi.org/10.3141/1867-14
6. Qu, L., Li, L., Zhang, Y., Hu, J.: PPCA-based missing data imputation for traffic flow volume: a systematical approach. IEEE Trans. Intell. Transp. Syst. **10**(3), 512–522 (2009). https://doi.org/10.1109/TITS.2009.2026312
7. Li, L., Li, Y., Li, Z.: Efficient missing data imputing for traffic flow by considering temporal and spatial dependence. Transp. Res. Part C Emerg. Technol. **34**(9), 108–120 (2013). https://doi.org/10.1016/j.trc.2013.05.008
8. Vlahogianni, E.I., Karlaftis, M.G., Golias, J.C.: Short-term traffic forecasting: where we are and where we're going. Transp. Res. Part C Emerg. Technol. **43**, 3–19 (2014)
9. Schafer, J.L.: Analysis of Incomplete Multivariate Data. CRC Press, Boca Raton (1997)
10. Buuren, S.V.: Flexible Imputation of Missing Data. CRC Press, Boca Raton (2012)
11. Arteaga, F., Ferrer, A.: Dealing with missing data in MSPC: several methods, different interpretations, some examples. J. Chemom. **16**(8–10), 408–418 (2002)
12. Kondrashov, D., Ghil, M.: Spatio-temporal filling of missing points in geophysical data sets. Nonlinear Process. Geophys. **13**(2), 151–159 (2006)
13. Sainani, K.L.: Dealing with missing data. PM&R **7**(9), 990–994 (2015)
14. García-Laencina, P.J., et al.: Pattern classification with missing data: a review. Neural Comput. Appl. **19**(2), 263–282 (2010). https://doi.org/10.1007/s00521-009-0295-6
15. Li, L., Li, Y., Li, Z.: Missing traffic data: comparison of imputation methods. IET Intell. Transp. Syst. **8**(1), 51–57 (2014). https://doi.org/10.1049/iet-its.2013.0052
16. Tak, S., Woo, S., Yeo, H.: Data-driven imputation method for traffic data in sectional units of road links. IEEE Trans. Intell. Transp. Syst. **17**(6), 1762–1771 (2016). https://doi.org/10.1109/TITS.2016.2530312
17. Sun, B., Ma, L., et al.: An improved k-nearest neighbours method for traffic time series imputation. In: 2017 Chinese Automation Congress (CAC), pp. 7346–7351. IEEE (2017)
18. Zefreh, M.M., Torok, A.: Single loop detector data validation and imputation of missing data. Measurement **116**, 193–198 (2018). https://doi.org/10.1016/j.measurement.2017.10.066
19. Zou, H., Yue, Y., Li, Q., Yeh, A.G.O.: An improved distance metric for the interpolation of link-based traffic data using kriging: a case study of a large-scale urban road network. Int. J. Geogr. Inf. Sci. **26**, 667–689 (2012)
20. Shamo, B., Asa, E., Membah, J.: Linear spatial interpolation and analysis of annual average daily traffic data. J. Comput. Civil Eng. **29**, 04014022 (2015)

21. Liu, J., Musialski, P., Wonka, P., Ye, J.: Tensor completion for estimating missing values in visual data. EEE Trans. Pattern Anal. Mach. Intell. **35**(1), 208–220 (2013)
22. Asif, M.T., Mitrovic, N., Dauwels, J., Jaillet, P.: Matrix and tensor based methods for missing data estimation in large traffic networks. IEEE Trans. Intell. Transp. Syst. **17**(7), 1816–1825 (2016). https://doi.org/10.1109/TITS.2015.2507259
23. Ran, B., Tan, H., Wu, Y., Jin, P.J.: Tensor based missing traffic data completion with spatial-temporal correlation. Phys. Stat. Mech. Appl. **446**, 54–63 (2016)
24. Goulart, J.H.M., Kibangou, A.Y., Favier, G.: Traffic data imputation via tensor completion based on soft thresholding of Tucker core. Transp. Res. Part C Emerg. Technol. **85**, 348–362 (2017). https://doi.org/10.1016/j.trc.2017.09.011
25. Chen, X., He, Z., Wang, J.: Spatial-temporal traffic speed patterns discovery and incomplete data recovery via SVD-combined tensor decomposition. Transp. Res. Part C Emerg. Technol. **86**, 59–77 (2018). https://doi.org/10.1016/j.trc.2017.10.023
26. Payne, H.J., Helfenbein, E.D., Knobel, H.C.: Development and testing of incident detection algorithms, volume 2: research methodology and detailed results. Federal Highway Administration, Washington, D.C. (1976)
27. Jacobson, L.N., Nihan, N.L., Bender, J.D.: Detecting erroneous loop detector data in a freeway traffic management system. Transp. Res. Rec. (1287), 151–166 (1990)
28. Rubin, D.B.: Inference and missing data. Biometrika **63**, 581–592 (1976)
29. Tang, J., Zhang, G., Wang, Y., Wang, H., Liu, F.: A hybrid approach to integrate fuzzy C-means based imputation method with genetic algorithm for missing traffic volume data estimation. Transp. Res. Part C Emerg. Technol. **51**, 29–40 (2015)
30. Duan, Y., Lv, Y., Liu, Y.L., Wang, F.Y.: An efficient realization of deep learning for traffic data imputation. Transp. Res. Part C Emerg. Technol. **72**, 168–181 (2016)
31. Pigott, T.D.: A review of methods for missing data. Educ. Res. Eval. **7**(4), 353–383 (2001). https://doi.org/10.1076/edre.7.4.353.8937
32. Yin, W., Murray-Tuite, P., Rakha, H.: Imputing erroneous data of single-station loop detectors for nonincident conditions: comparison between temporal and spatial methods. J. Intell. Transp. Syst. **16**(3), 159–176 (2012)
33. Xu, J.R., Li, X.Y., Shi, H.J.: Short-term traffic flow forecasting model under missing data. J. Comput. Appl. **30**, 1117–1120 (2010)
34. Lee, S., Fambro, D.B.: Application of subset autoregressive integrated moving average model for short-term freeway traffic volume forecasting. Transp. Res. Rec. J. Transp. Res. Board **1678**, 179–188 (1999)
35. Castro-Neto, M., Jeong, Y.S., Jeong, M.K., et al.: Online-SVR for short-term traffic flow prediction under typical and atypical traffic conditions. Expert Syst. Appl. **36**, 6164–6173 (2009). https://doi.org/10.1016/j.eswa.2008.07.069
36. Chiou, J.M., Zhang, Y.C., Chen, W.H., et al.: A functional data approach to missing value imputation and outlier detection for traffic flow data. Transp. B Transp. Dyn. **2**(2), 106–129 (2014). https://doi.org/10.1080/21680566.2014.892847
37. Tan, H., Feng, J., Chen, Z., et al.: Low multilinear rank approximation of tensors and application in missing traffic data. Adv. Mech. Eng (2014). https://doi.org/10.1155/2014/157597
38. Anandkumar, A., Ge, R., Hsu, D., Kakade, S.M., Telgarsky, M.: Tensor decompositions for learning latent variable models. J. Mach. Learn. Res. **15**, 2773–2832 (2014)
39. Hitchcock, F.L.: The expression of a tensor or a polyadic as a sum of products. J. Math. Phys. **6**, 164–189 (1927). https://doi.org/10.1002/sapm192761164
40. Tucker, L.: Some mathematical notes on three-mode factor analysis. Psychometrika **31**(3), 279–311 (1966)

41. Carroll, J.D., Chang, J.J.: Analysis of individual differences in multidimensional scaling via an N-way generalization of "Eckart-Young" decomposition. Psychometrika **35**, 283–319 (1970)
42. Kolda, T.G., Bader, B.W.: Tensor decompositions and applications. SIAM Rev. **51**(3), 455–500 (2009). https://doi.org/10.1137/07070111X
43. Schifanella, C., Candan, K.S., Sapino, M.L.: Multiresolution tensor decompositions with mode hierarchies. ACM Trans. Knowl. Discov. Data **8**(2), 10 (2014)
44. Acar, E., Dunlavy, D.M., Kolda, T.G., Mørup, M.: Scalable tensor factorizations for incomplete data. Chemom. Intell. Lab. Syst. **106**(1), 41–56 (2011)
45. Zhao, Q., Zhang, L., Cichocki, A.: Bayesian CP factorization of incomplete tensors with automatic rank determination. IEEE Trans. Pattern Anal. Mach. Intell. **37**(9), 1751–1763 (2015). https://doi.org/10.1109/TPAMI.2015.2392756
46. Wang, Y., Zheng, Y., Xue, Y.: Travel time estimation of a path using sparse trajectories. In: Proceedings of the 20th ACM SIGKDD International Conference on Knowledge Discovery and Data Mining, KDD 2014, pp. 374–383 ACM (2014)
47. Salakhutdinov, R., Mnih, A.: Bayesian probabilistic matrix factorization using Markov chain Monte Carlo. In: Proceedings of the 25th International Conference on Machine Learning (ICML) (2008). https://doi.org/10.1145/1390156.1390267
48. Xiong, L., Chen, X., Huang, T.K., Schneider, J., Carbonell, J.G.: Temporal collaborative filtering with Bayesian probabilistic tensor factorization. In: SIAM International Conference on Data Mining, pp. 211–222 (2010). https://doi.org/10.1137/1.9781611972801.19
49. Rai, P., Wang, Y., Guo, S., Chen, G., Dunson, D., Carin, L.: Scalable Bayesian low-rank decomposition of incomplete multiway tensors. In: Proceedings of the 31st International Conference on Machine Learning (ICML), vol. 32, pp. 1800–1808 (2014)
50. Tan, H., Wu, Y., Shen, B., Jin, P.J., Ran, B.: Short-term traffic prediction based on dynamic tensor completion. IEEE Trans. Intell. Transp. Syst. **17**(8), 2123–2133 (2016)
51. De Lathauwer, L., De Moor, B., Vandewalle, J.: A multilinear singular value decomposition. SIAM J. Matrix Anal. Appl. **21**(4), 1253–1278 (2000)
52. Candès, E.J., Recht, B.: Exact matrix completion via convex optimization. Found. Comput. Math. **9**(6), 717–772 (2009). https://doi.org/10.1007/s10208-009-9045-5
53. Cai, J.F., Candès, E.J., Shen, Z.: A singular value thresholding algorithm for matrix completion. SIAM J. Optim. **20**(4), 1956–1982 (2010). https://doi.org/10.1137/080738970
54. Gandy, S., Recht, B., Yamada, I.: Tensor completion and low-n-rank tensor recovery via convex optimization. Inverse Probl. **27**(2), 1–20 (2011). https://doi.org/10.1088/0266-5611/27/2/025010

Big Data and Cloud Computing

Mining and Analysis Based on Big Data in Public Transportation

Yinxin Bao[1], Chengyu Zhang[1], and Quan Shi[1,2(✉)]

[1] School of Information Science and Technology, Nantong University, Nantong, China
sq@ntu.edu.cn

[2] School of Traffic, Nantong University, Nantong, China

Abstract. City bus is an important part of the city transportation system. Reasonable and healthy bus line can greatly improve the city's economy and environment. This paper is based on mass data generated by the public transportation system of Nantong, such as GPS data of buses and passengers' swiping card data. Through the cleaning, matching, calculation and mining of the original bus data, this paper can provide decision-making basis for bus companies in terms of transfer rate, non-straight line coefficient, scheduling conflict rate, station accessibility, station coverage and passenger number. The LSTM model is used to predict the arrival time of the bus, and the accuracy of the model is improved by influencing factors such as weather change and direction.

Keywords: Big data in public transportation · Bus operation indicator · LSTM · Data visualization

1 Introduction

Urban public transport is an important part of urban public facilities. A mature and reasonable public transport network directly affects the economic development and construction of the city. In the context of the era of big data and Internet of Vehicles, expanding scale of advanced intelligent and informational public transport system construction, along with the breadth and depth of data collection has increased dramatically. The intelligent bus subsystem has accumulated massive traffic data for the development of smart traffic [1]. The paper selects urban buses as the research object for in-depth analysis and research. According to the data generated for each submodule of the intelligent public transport system, analysis of bus operation time, number of passengers, regional bus accessibility, site accessibility, travel demand and other operation status and characteristics. Provide decision-making basis for bus operation analysis.

K. Li et al. (Eds.): ISICA 2019, CCIS 1205, pp. 681–688, 2020.
https://doi.org/10.1007/978-981-15-5577-0_54

2 Overview of Related Technologies

2.1 Development Technology

The data processing techniques used in this paper mainly include: Hadoop, HDFS, HBase, MapReduce, Spark.

Hadoop is a distributed system infrastructure developed by the Apache Foundation [2]. It is suitable for data storage and computation on multi-computer cluster. The two most important parts are MapReduce Distributed Computing System and HDFS Distributed File System [3]. Hadoop provides a secure and efficient access service and a secure and reliable large data processing engine for massive data processing.

HBase mainly solves the problem of high-speed storage of large data and reading distributed database.

HDFS is the basis of data storage management in Hadoop system. It is a high fault-tolerant storage system. It can not only detect and deal with data loss caused by hardware failures, but also store and share data of each computing node across multiple computers.

MapReduce is a parallel computing model between a large number of data sets, which is mainly used for large-scale data computing. MapReduce mainly consists of two parts: Map and Reduce. 'Map' is mapping data node, and 'Reduce' means simplifying data.

Spark is a framework for data reading and data computing through computer memory. In terms of speed, Spark uses and extends MapReduce computing model to achieve a universal and efficient computing platform [4]. Spark supports more computing models, including flow processing and interactive queries.

2.2 Development Tool

The development tools used in this paper mainly include: Flask back-end framework, Vue front-end framework and TensorFlow framework.

Flask is a back-end framework written in Python language. It uses Werkzeug's WSGI toolbox and Jinja2 as its template engine. Vue is a progressive framework for building user interfaces [5]. It is a layer-by-layer application from the bottom up. The core of Vue lies in the layer, which has a wide range of third-party library resources and project integration.

TensorFlow is an open source software library that uses data flow diagrams for numerical computation [6]. TensorFlow is an open source library developed by the Google team. It can be built on multiple platforms because of its flexible and expansive architecture through the calculation of deep neural network.

3 Analysis of Bus Operation Characteristics

3.1 Raw Data Manage

The data used in this paper are mainly provided by Nantong Public Transport Company. The original data are divided into two parts: dynamic data and static data, as shown in Table 1. The dynamic data are the resident's bus card swipe data, bus dispatch data, bus passenger flow counter data, etc. The static data are mainly the location data of bus stations, bus line data and so on.

Table 1. Types of original bus data.

Data type	Data	Data description
Static data	Bus station data	Site coordinates, Site names, Site numbers
	Bus line data	Line number, Starting station, Terminal station, Passing station
Dynamic data	Bus IC card	IC card number, IC card type, Line number, Transaction time
	Bus dispatch data	Departure time, Line number
	Bus location data	GPS Data

Static Data Processing. The static data used in this paper are mainly bus station data and bus route network data. Bus stop data mainly includes the site coordinates, site number, site name and so on. Bus route network data mainly includes starting station information, terminal information and route number. This paper integrates two kinds of static data and integrates the GPS data of the station into the bus line data, which can display the road network information of the bus system visually on the map [7].

Dynamic Data Processing. The dynamic data used in this paper are mainly divided into IC card data and bus GPS data. The original data mainly come from bus card swipe data provided by bus companies and GPS data during bus driving.

The IC card data and bus GPS data are processed respectively.

1. Data cleaning. Clean the bus IC card data and bus GPS data to remove invalid or incorrect data from IC card data and GPS data.
2. Data preprocessing. The time variables in the data are cut and spliced into time stamps, and the GPS data are converted into coordinates under the GCJ-09 standard.
3. Data matching and integration. The collected coordinate data are mapped and the GPS data deviating from the road are eliminated. Some data of original IC card swipe data contain a large number of data fields, some of which are not analyzed and studied in this paper. These redundant data need to be integrated and deleted in order to reduce the waste of time and memory space in subsequent data processing and calculation.

3.2 Bus Operation Indicators

The bus operation indicators analyzed in this paper mainly include: bus transfer rate, non-linear coefficient, dispatch conflict, site accessibility, site coverage and IC card data statistics [8].

As one of the important indicators to evaluate the quality of public transport service, the transfer rate of public transport needs to be analyzed. Transfer data of Nantong passengers are shown in Fig. 1.

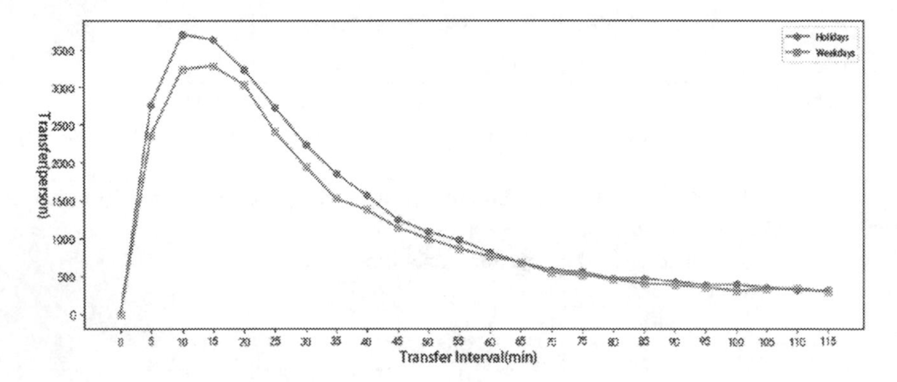

Fig. 1. Public switching multiplier data

Non-linear coefficient is an important index in the layout planning of public transport network [9]. The larger the non-linear coefficient is, the longer the detour distance of the line is, the lower the comfort of the residents. Excessive non-linear coefficient will lead to long ride time and uneven local passenger flow, while too low non-linear coefficient will lead to inconvenience for residents to transfer. The non-linear coefficient of Nantong's lines is shown in Fig. 2.

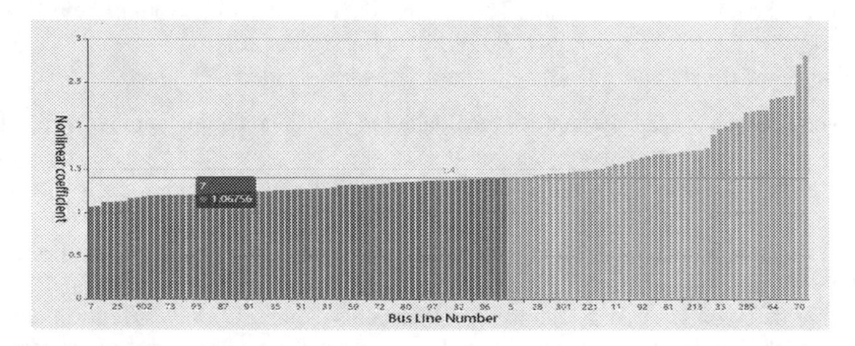

Fig. 2. Non-linear coefficients

Bus scheduling relies on fixed time intervals for departure. Because of road condition changes, weather factors and other factors, two buses on the same line will pass through the same station one after another. When two or more buses pass the same station within five minutes, dispatching conflicts are considered to occur, as shown in Fig. 3.

Fig. 3. Bus dispatch conflict

As an important part of analyzing the service quality standard of urban public transport, the accessibility of bus stops affects the travel efficiency of residents, and is also an important basis for residents to choose the way of travel.

Make statistics of different types of bus card swiping data, that is, analyze the total swiping data of one day, count the number of different time periods and different types of CARDS, and understand the characteristics of different groups.

Bus station coverage rate is an important index to evaluate the service quality of public transportation system in this region. The coverage of bus stops directly affects the convenience of residents to travel. According to calculation, the coverage rate of 300 meters service area of Nantong bus station is about 64.5%, which is 50% higher than the national standard in GB T 22484-2016.

4 Bus Arrival Prediction Model

4.1 Model Selection

The prediction model used in this paper is LSTM, which is a variant of RNN. Compared with RNN, LSTM solves the problem of long-term dependence of RNN [10]. RNN is applied to current tasks by connecting the factors that influence the previous event information. However, when predicting data sets with large time intervals, the gradient will disappear due to the increase of dependence factors. LSTM solves the problem of long-term dependence by selective forgetting and long-term memory of processed information.

4.2 Data Preprocessing

In order to improve the computational efficiency and fitting effect of the algorithm, it is necessary to remove variables that have no or minimal impact on the prediction results before data prediction by LSTM model. This paper uses Lasso algorithm to judge the weight of all variables in the original data. The calculation method is shown in formula (1).

$$\check{\beta}^{lasso} = min\left\{\frac{1}{N}\sum\nolimits_{i=1}^{N}\left(y_i - \beta_0 - x_i^T\beta\right)^2\right\} \tag{1}$$

In formula (1), x is the training data matrix, β is the regression coefficient, and y is the training label. The formation of training data sets is shown in Table 2.

Table 2. Training data table

Attribute	Explain
UPDAWN	Up and down
WEATHER	Weather condition
STARTTIME	Departure time
DISTANCE	Site distance
WEEKDAY	Date
STOP	Site number
BUSNO	Vehicle number
STOPTIME	Arrival time

4.3 LSTM Model

LSTM is a chain-structured cyclic neural network model, which predicts the arrival time of the bus by calculating the forgetting gate, input gate and output gate. Its structure is shown in Fig. 4.

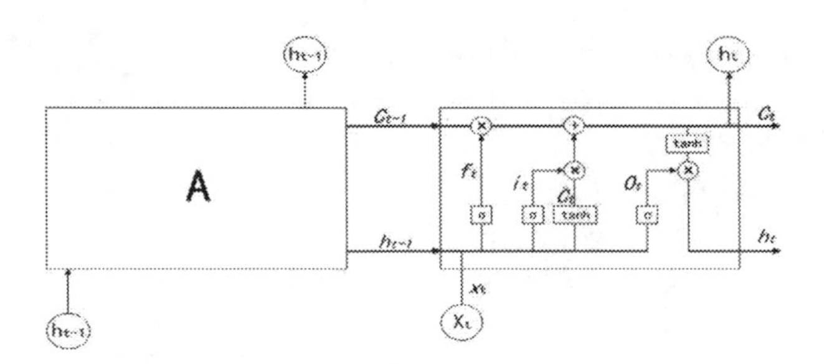

Fig. 4. LSTM structure

After data input, the first step is through the forgetting gate, which determines the current cell data input and the information discarded in the last cell state. The forgetting gate reads h_{t-1} and x_{t-1} and outputs a value between 0 and 1 through formula (2) and updates it to the state variables of C_{t-1} cells.

$$f_t = \sigma\left(W_f \cdot [h_{t-1}, x_t] + b_f\right) \tag{2}$$

In formula (2), σ denotes the activation function Sigmoid. The definition of Sigmoid function is shown in formula (2). W_f represents the weight of forgetting gate and b_f represents the bias matrix of forgetting gate.

$$\sigma(x) = \frac{1}{1 + e^{-x}} \tag{3}$$

In formula (3), x represents the input variable, and the value of the σ function is between 0 and 1.

The updated information is determined by formula (4) and formula (5). Through the Sigmoid function which data is worth updating, through the tanh function to create a new candidate value vector.

$$i_t = \sigma(W_i \cdot [h_{t-1}, x_t] + b_i) \tag{4}$$

$$\check{C} = tanh(W_C \cdot [h_{t-1}, x_t] + b_C) \tag{5}$$

In formula (4) and formula (5), Wi represents the input gate weight, bi refers to the input gate bias matrix, WC refers to the weight matrix of tanh function, b_C refers to the bias matrix of tanh function, and Č is the candidate cell state.

Formula (6) is used to determine the cell status of the current cell.

$$C_t = f_t \cdot C_{t-1} + i_t \cdot \check{C}_t \tag{6}$$

The output information is determined by formula (7) and formula (8).

$$O_t = \sigma(W_o \cdot [h_{t-1}, x_t] + b_o) \tag{7}$$

$$h_t = O_t \cdot tanh(C_t) \tag{8}$$

4.4 Fitting Results

The fitting rate of LSTM model is about 93.5921%.

5 Summary

This paper implements a data mining and analysis system based on urban bus data, which is mainly divided into four parts: data storage, data analysis and calculation, data prediction and data visualization. The data storage in the system is distributed through Spark framework. Data processing is mainly to sort out the original data and calculate the bus operation indexes such as the bus line's non-linear coefficient, passenger transfer rate, bus station coverage rate, etc. Data prediction is mainly based on LSTM model to predict bus arrival time. Data visualization is achieved by using a front-end framework (Vue) and a back-end framework (Flask).

Acknowledgments. The National Natural Science Foundation of China (61771265), The Science and technology Project of Nantong (CP12017001).

References

1. Zhou, H.: Research on traffic congestion management based on big data rule mining [1]. Stat. Inf. BBS **32**(05), 96–101 (2017)
2. Feng, X., Wang, W.: Research on Hadoop and Spark application scenarios [1]. Comput. Appl. Res. **35**(09), 2561–2566 (2018)
3. Hao, Y.: Design of CDH-based data visualization platform [2]. Chengdu University of Technology (2018)
4. Liu, P.: Design and implementation of data management platform based on Spark [2]. Zhejiang University (2016)
5. Wang, Z.: Design and implementation of development platform based on Vue.js [2]. Guangdong University of Technology (2018)
6. Wang, Q., Liang, J.: Research on traffic sign recognition based on TensorFlow [1]. Value Eng. **38**(27), 204–206 (2019)
7. Li, M.: Public transportation intelligent scheduling and location publishing system based on GPS [1]. Comput. Prod. Circ. **10**, 127–128 (2018)
8. Chen, J., Lv, Y., Cui, M.: Judgment method of passenger disembarkation station with IC card based on travel mode [1]. J. Xi'an Univ. Arch. Technol. (Nat. Sci. Ed.) **50**(01), 23–29 (2018)
9. Zhao, S., Lu, Y., Hao, L.: Intercity bus route planning based on node importance in urban agglomeration [1]. Highw. Mot. Transp. (05), 24–26+30 (2018)
10. Yang, Q., Wang, C.: Prediction of global stock index based on deep learning LSTM neural network [1]. Stat. Res. **36**(03), 65–77 (2019)

The Study on Low Laser Damage Technology of SE Solar Cell

Shuaidi Song, Chunhua Sheng, Rui Cao, Tan Song, and Qiang Wang(✉)

The College of Information Science and Technology, Nantong University, Nantong 226019, China
wang_q@ntu.edu.cn

Abstract. The combination of the passivated emitter and rear cells (PERC) technology and the laser selective emitter (SE) preparation process was one of the effective methods to improve the efficiency of the PERC cells developed in recent years. However, in the SE preparation process, the phosphorosilicate glass (PSG) layer was prone to ablation, which led to the degradation of solar cell performance. By studying the effects of laser wavelength, laser power, pulse frequency and scanning speed on the surface morphology, surface sheet resistance and solar cells performance of silicon wafers, the preparation method of high efficiency and low damage SE solar cells was explored. The experimental results show that the optimun cell performance can be obtained under the condition of laser wavelength of 1064 nm, laser power of 2 W and pulse frequency of 30 Hz. The conversion efficiency of SE solar cells is up to 22.01%, which is about 0.51% higher than that of the solar cells prepared by traditional process.

Keywords: Selective emitter (SE) · Laser · Selective doping · PSG · Sheet resistance

1 Introduction

In recent years, PERC have widely used in the photovoltaic (PV) market due to higher efficiencies. The preparation of laser SE-doping process has become the standard technology of PERC. The selective doping of the solar cell can realize the light-doped in the non-electrode region and heavy doping in the electrode region. This structure can not only form good ohmic contact, reduce the series resistance of cells, improve the fill factor and short-wave response, but also reduce the recombination of minority carriers on the surface and increase the open circuit voltage [1–3]. The lower surface impurity concentration can further reduce the surface state density in the non-electrode region, in order to achieve better surface passivation effect and reduce the surface recombination of photogenerated minority carriers [4, 5].

The SE-doping process was usually performed after the diffusion process of the single crystalline silicon (SC-Si). Since the laser has a short action time and high energy on the silicon surface, it can effectively achieve heavy doping of the gate region without causing secondary distribution of impurities in other regions. In order to make better use of laser energy and reduce the influence on other regions, the laser selective

K. Li et al. (Eds.): ISICA 2019, CCIS 1205, pp. 689–698, 2020.
https://doi.org/10.1007/978-981-15-5577-0_55

doping process usually used the expensive nanosecond pulsed laser [6–8] as laser source to scan PSG layer after diffusion, and drove the impurities into the wafer. Generally, the laser selective doping process can increase conversion efficiency of solar cells by about 0.3–0.4% [6–8].

Although the laser selective doping process can effectively improve the conversion efficiency of solar cells, high-energy lasers can damage the PSG layer during the process (so-called laser ablation). In the subsequent alkaline polishing process, the protective effect of the damaged PSG layer on the silicon surface was reduced, thereby causing over-corrosion of the electrode region and reducing the conversion efficiency. In order to avoid the corrosion of the gate region caused by the damage of the PSG layer, the high-efficiency immersion alkaline polishing process has to be changed into chain alkaline polishing process, which seriously degraded the production efficiency and increased the manufacturing cost. Therefore, it was necessary to study a low PSG layer damage SE process to improve conversion efficiency and reduce production cost. In this paper, the process parameters such as laser wavelength, laser power, pulse frequency and scanning speed were studied, which was useful to explore a high-efficiency, low-damage and low-cost laser SE solar cell process.

2 Experiments

SC-Si wafers (156.75 mm × 156.75 mm, p-type, [100], resistivity of 0.5–1.5 $\Omega \cdot$ cm) were provided by Jiangsu Econess-Energy Co., Ltd. The CTGD-F-10/20 fiber laser produced by Wuhan Chutian Laser was used to study the effects of different laser wavelength, laser power, pulse frequency and scanning speed on laser-doped SE solar cells. The laser beam diameter of the laser was 120 μm, the laser wavelength was 355 nm, 532 nm and 1064 nm respectively; the laser power was 2, 3, 4 and 6 W respectively; the pulse frequency was 20, 30, 40, 50 and 60 Hz respectively; Scanning speed was 300, 400, and 500 mm/s respectively.

Figure 1 showed the PERC SC-Si cell structural mapping. First, the SC-Si wafers was cleaned to remove the surface damage layer and impurities adhering to the surface, then alkaline texture. Then, the phosphorus diffusion process was carried out to prepare the SC-Si wafer with the sheet resistance of 69 Ω/□. The impurities in the PSG layer diffused into wafers to form heavily doped regions through laser selective doping process. The wet back polish edge to realize high sheet resistance on the front. After that, the front passivation, backside lamination passivation, front anti-reflection film, and laser grooving were prepared. Finally, the positive electrode and the back electrode were prepared by screen printing and sintered to make solar cells.

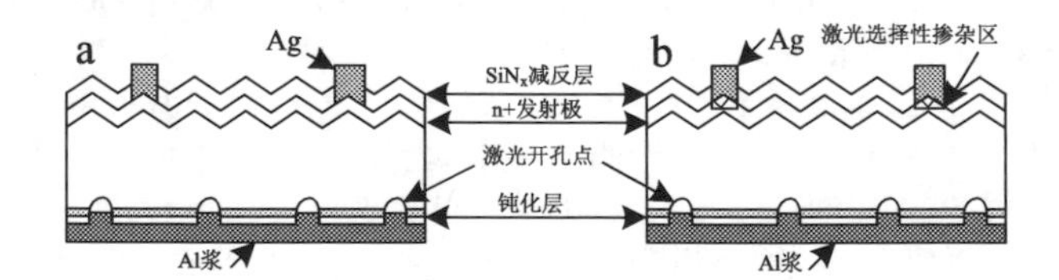

Fig. 1. The structure of PERC SC-Si solar cells

Laser doping experiments were carried out at different positions on the same silicon wafer to avoid divergence. The silicon wafer was divided into 100 regions (the size of each point is 15 mm × 15 mm), Hitachi s-4800 scanning electron microscope (SEM) was used to observe the surface morphology of the cell, and RST-9 four point probe tester was used to test the sheet resistance.

Through the analysis of the experimental results, the conventional method and the silicon solar cells obtained under the optimal conditions were prepared in Jiangsu Econess-Energy Co., Ltd., and the electrical properties were tested.

3 Experimental Results and Analysis

3.1 The Effect of Laser Wavelength on the Surface of the Cell

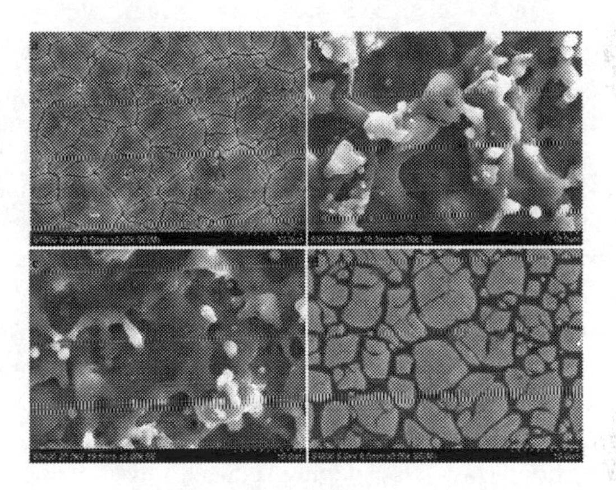

Fig. 2. SEM images of different laser wavelengths (a) substrate (b) 355 nm (c) 532 nm (d) 1064 nm

Figure 2 showed the influence of different laser wavelengths on the wafer surface under the conditions of laser power 2 W, pulse frequency 60 Hz, scanning speed 500 mm/s. Under this process condition, the single-pulse laser energy was low, and it was easy to compare the effect of different lasers wavelengths on the surface morphology of silicon. It can be seen from Fig. 2 that the damage degree of wafer surface gradually decreases with the increase of laser wavelength. When the laser wavelength was 1064 nm, the wafer surface was almost undamage, which was because the long-wave laser acted on the silicon wafer mainly by thermal effect, while the short-wave laser can be absorbed by silicon atoms, which easily broke the lattice structure and made the surface present molten state. which was not conducive to the formation of good ohmic contact of cells, reducing the conversion efficiency. Therefore, the selective doping process of the silicon wafer should use 1064 nm laser wavelength.

3.2 The Effect of Laser Frequency on the Sheet Resistance

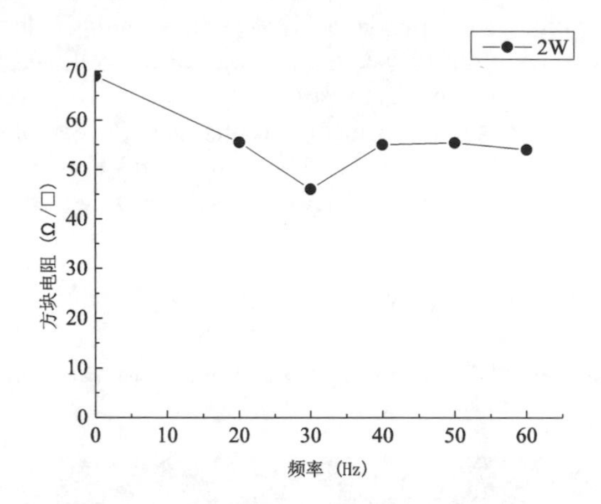

Fig. 3. Sheet resistance of SE solar cell at different laser frequencies

Figure 3 showed the sheet resistance of SE cells with different pulse frequencies under the conditions of laser wavelength 1064 nm, laser power 2 W, and scanning speed 500 mm/s. It was known from Fig. 3 that the surface sheet resistance decreased and then increased with the increase of the pulse frequency, and the sheet resistance was the lowest when the pulse frequency was 30 Hz. The main factor determining the surface sheet resistance on the wafer surface was the impurity concentration. Two dominant factors of doping concentration: the impurities concentration in the external doped silicon and the impurities concentration escaped from the surface during the doping process. The expression was: Net doping concentration = doping concentration-escaping impurity concentration. The impurity concentration retained on the silicon surface determined sheet resistance.

Generally, when the higher laser energy, the higher the energy obtained by the PSG layer during laser doping, and the higher the concentration of impurities incorporated into the silicon from the PSG layer. If the PSG layer was not damaged, the surface sheet resistance will decrease with the increase of laser energy. Conversely, if the PSG layer was ablated by laser, the dopant will escape during the doping process, thus reducing the surface doping concentration.

For the pulsed laser, the lower pulse frequency, the greater the energy of the single pulse. The energy of the single pulse (P_s) can be defined as

$$p_s = \frac{P}{f} \tag{1}$$

Where P is the laser power and f is the pulse frequency. When the laser power was 2 W, the sheet resistance should be reduced as the pulse frequency decreased, but it can be seen from Fig. 3 that the sheet resistance corresponding to the 30 Hz was the lowest. This was due to the PSG layer with the function of protecting dopant and preventing dopant from escaping. The 20 Hz pulse frequency has ablation on the PSG layer, which destroyed the PSG layer and impurities escaped, thus reducing the doping concentration on the surface and improving the sheet resistance. As shown in the SEM image of Fig. 4, when the laser pulse frequency was greater than 30 Hz, the PSG layers were not damaged. Therefore, when the pulse frequency was 30 Hz, the wafer sheet resistance decreased deeply, about 34.8%. When the pulse frequency was further increased, the laser energy was lower, the impurity mass driven into the silicon wafer was less, and the sheet resistance on the silicon wafer surface was relatively closer.

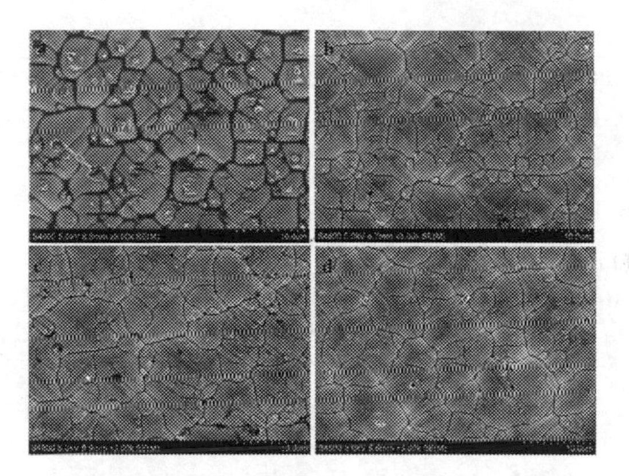

Fig. 4. Sheet resistance of cell at different frequencies (a) 2 W 20 Hz (b) 2 W 30 Hz (c) 2 W 40 Hz (d) 2 W 50 Hz

3.3 The Effect of Laser Power on the Sheet Resistance

Figure 5 showed the sheet resistance on the surface at the pulse frequency 30 Hz and laser power 2, 3, 4, and 6 W, respectively. It can be seen that the sheet resistance on the

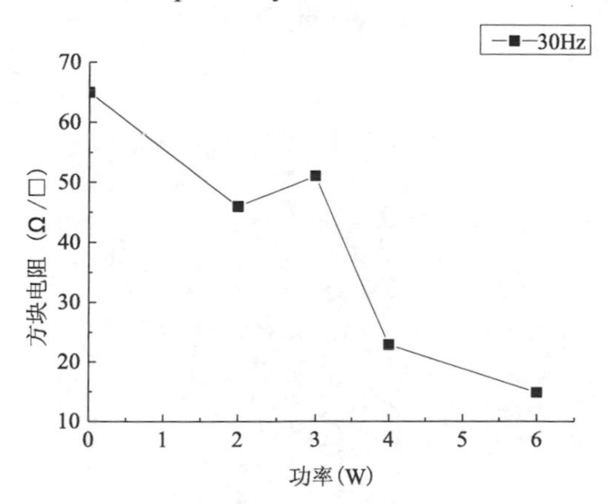

Fig. 5. Sheet resistance of SE battery at different laser powers and frequencies

surface decreased, then increased and then decreased, showing a general downward trend, with the increase of laser power.

As the laser power increased, the impurities obtained energy increasingly in the PSG layer, so the surface resistance of the silicon wafer generally showed the downward trend, and the laser melting aperture also raised. The laser melting aperture (*d*) can be expressed by the following formula [9]:

$$d = 2\left[\frac{3E}{\pi(L_{\mathrm{B}} + 2L_{\mathrm{M}})}\right]^{\frac{1}{3}} \tag{2}$$

Where E is the laser energy, L_B is the specific energy of material vaporization, L_M is the specific energy of melting heat of the material. For the same material, L_B and L_M were fixed, so the aperture d was proportional to E, which means that when the pulse frequency was constant, as the laser power increased, the single laser pulse energy and the melting aperture increase. This inevitably led to the volatilization of impurities during the laser doping process. When the laser power was 3 W, since the PSG layer was ablated, the impurities diffused into the silicon wafer escape during the laser process, resulting in the decrease of the net doping concentration and the increase of the surface resistance of the silicon wafer.

When the laser power continued to increase, although the ablation damage of the PSG layer on the silicon wafer surface was serious, more impurities were driven into the silicon wafer, the driving depth increased, the net impurity concentration increased, and the surface resistance of the silicon wafer decreased. Figure 6 showed the SEM image of the surface morphology of silicon wafer under different laser power when the pulse frequency was 30 Hz. It can be seen from Fig. 6 that the surface damage of silicon wafer increased gradually with the increase of laser power.

In summary, the laser process with laser wavelength 1064 nm, laser power 2 W and pulse frequency 30 Hz will not damage the PSG layer, so as to ensure that the selective doping area will not be corroded in the post alkaline polishing process.

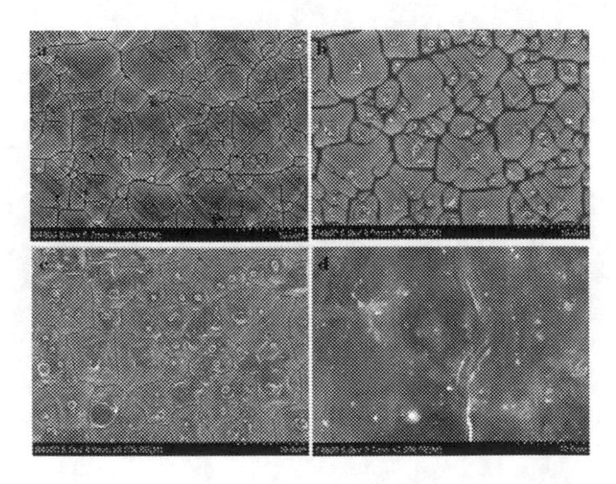

Fig. 6. SEM images of different laser powers (a) 2 W 30 Hz (b) 3 W 30 Hz (c) 4 W 30 Hz (d) 6 W 30 Hz

3.4 Effect of Impurity Volatilization on the Surface Concentration and Impurity Concentration Distribution of Silicon Wafer During Laser Doping

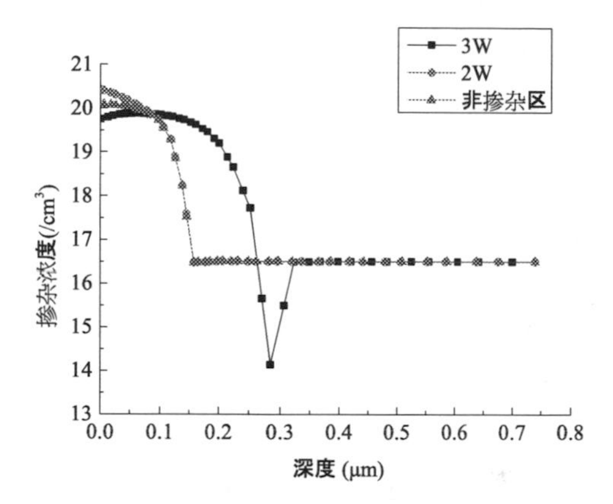

Fig. 7. SE cell impurity concentration profile Images of doping region or unselective doping region

In order to further study the influence of the surface sheet resistance on the doping concentration and the doping concentration distribution in the silicon wafer, the simulation of the doping concentration distribution in the silicon wafer after the laser action was carried out by using the SILVACO software. The simulation results were shown in Fig. 7. In the figure, *C* is the doping concentration and *H* is the depth. It can be seen from Fig. 6 that when the pulse frequency was 30 Hz and the laser power was greater than 3 W, the surface of the silicon wafer was damaged. Therefore, when the laser power was 2 W, 3 W, the surface sheet resistance value 46 Ω/□ and 51.1 Ω/□ were selected as the sheet resistance values of silicon surface (depth 0), and the block resistance value was 69 Ω/□.

It can be seen from Fig. 7 that under 2 W laser power, the highest doping concentration was located on the surface, because PSG layer was not damaged, single pulse energy was low, and the impurity mass driven into silicon wafer decreases with the increase of depth. However, under the effect of 3 W laser power, although the laser energy increased, the PSG layer was destroyed, impurities were easy to escape from the surface, the surface doping concentration decreases, the highest value of silicon doping concentration was below the surface, and the surface block resistance increased.

3.5 The Effect of Laser Scanning Speed on the Sheet Resistance

In order to be compatible with the production efficiency of solar cells, the influence of laser scanning speed (*V*) on the sheet resistance on the surface of silicon wafer was further studied. Figure 8 shows the relationship between the sheet resistance on the surface of solar cells and laser scanning speed after selective doping with laser frequency 30 Hz, laser power 2 W and scanning speed 300, 400 and 500 mm/s, respectively.

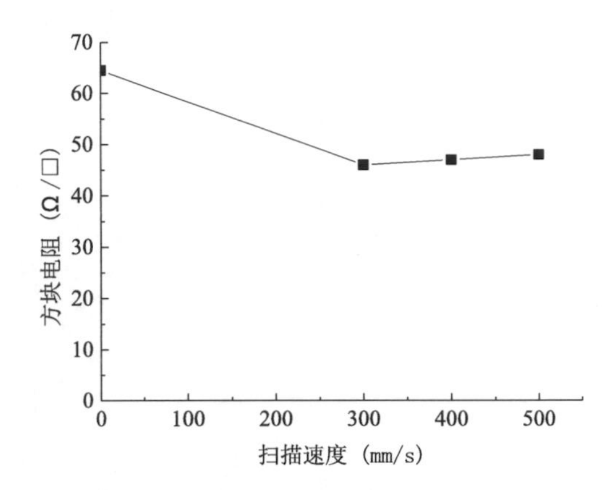

Fig. 8. Sheet resistance of SE solar cell at different laser scanning speeds

It can be seen from Fig. 8 that as the increase of scanning speed, the sheet resistance value of the solar cells showed a slight upward trend. It was due to the increase of interval between the two laser pulses and the thermal effect superposition effect was poor, so the sheet resistance on the surface increased.

3.6 Solar Cell Electrical Performance

Table 1 showed the comparison of various electrical performance parameters of SE solar cell and non-SE solar cell under the conditions of laser wavelength 1064 nm, laser power 2, 3, 4 and 6 W, pulse frequency 30 and 40 Hz, scanning speed 500 mm/s, respectively. It can be seen from Table 1 that the damage of PSG layer in SE process will affect the conversion efficiency, and the solar cell without damage of PSG layer has the highest conversion efficiency. Under the optimal SE process, the average photoelectric conversion efficiency of SE solar cell was 0.51% higher than that of non-SE solar cell, which showed that SE process can effectively improve the conversion efficiency. Because the PSG layer was not damaged, the selective doping electrode area was well protected, and the compatibility with the existing solar cell preparation process was good.

Table 1. Comparisons of electrical performance parameters of SE solar cells before and after laser doping

	V_{oc}/mV	I_{sc}/A	FF/%	E_{ta}/%
PERC	668	9.67	81.36	21.50
PERC+SE (2 W, 30 Hz)	681	9.72	81.30	22.01
PERC+SE (2 W, 40 Hz)	668	9.67	81.47	21.53
PERC+SE (3 W, 30 Hz)	676	9.70	81.24	21.81
PERC+SE (4 W, 30 Hz)	677	9.68	80.96	21.71
PERC+SE (6 W, 30 Hz)	676	9.67	80.87	21.64

Fig. 9. PERC-SE solar cells image

Figure 9 was the diagram of PERC-SE solar cell prepared under the conditions of the laser power 2 W, pulse frequency 30 Hz, and scanning rate 500 mm/s. The solar cell area was 24,432 mm^2.

4 Conclusion

In this paper, the 1064 nm laser was used to study the low damage laser selective doping process of SC-Si solar cells. The effects of laser on the ablation of PSG layer and the surface resistance of grid area were studied under the conditions of pulse frequency 20–60 Hz, laser power 2–6 W and scanning speed 300–500 mm/s. The results showed that when the pulse frequency was 30 Hz and the laser power was 2 W, there was no damage on the surface of the silicon wafer. At the same time, the sheet resistance on the surface of the silicon wafer was reduced by about 35%, achieving the purpose of SE selective doping. The laser scanning speed has little effect on the sheet resistance of silicon wafer. The conversion efficiency of SE solar cell was improved by 0.51% by preparing SE silicon wafer into solar cells.

Acknowledgements. This work was supported by Major Natural Science Research Project of Jiangsu Higher Education Institutions of China (19KJ320004), and Jiangsu Province Postgraduates Research and practice innovation program of China (KYCX19_2059).

References

1. Binetti, S., Le Donne, A., Rolfi, A.: Picosecond laser texturization of mc-silicon for photovoltaics: a comparison between 1064 nm, 532 nm and 355 nm radiation wavelengths. Appl. Surf. Sci. **371**, 196–202 (2016)
2. Chen, N., Shen, S., Du, G.: Enhanced photovoltaic properties for rear passivated crystalline silicon solar cells by fabricating boron doped local back surface field. J. Wuhan Univ. Technol.-Mater. Sci. Ed. **32**(6), 1323–1328 (2017). https://doi.org/10.1007/s11595-017-1748-x
3. Shanmugam, V., Khanna, A., Basu, P.K., Aberle, A.G., Mueller, T., Wong, J.: Impact of the phosphorus emitter doping profile on metal contact recombination of silicon wafer solar cells. Solar Energy Mater. Solar Cells **147**, 171–176 (2016)
4. Cho, E., Ok, Y.W., Dahal, L.D., Das, A., Upadhyaya, V., Rohatgi, A.: Comparison of $POCl_3$ diffusion and phosphorus ion-implantation induced gettering in crystalline Si solar cells. Solar Energy Mater. Solar Cells **157**, 245–249 (2016)
5. Lin, D., et al.: Incorporation of deep laser doping to form the rear localized back surface field in high efficiency solar cells. Solar Energy Mater. Solar Cells **130**, 83–90 (2014)
6. 庞恒强. 激光掺杂制作选择性发射极晶体硅 PERC 电池的工艺研究. 材料导报, **32**(S2), 21–29 (2018)
7. 马红娜,赵学玲,张红妹.激光掺杂制作选择性发射极电池扩散工艺的研究.光电子技术, **36**(04), 270–273+282 (2016)
8. 王雪,豆维江,秦应雄,巨小宝.多晶硅太阳电池激光掺杂选择性发射极. 光子学报, **43**(06), 48–52 (2014)
9. 陈家璧，彭润玲. 激光原理及应用：第 2 版. 北京：电子工业出版社, 171–174 (2008)

Information Security Risk and Protective Measures of Computer Network in Big Data Age

Lei Deng[1(✉)], Haiping Li[1], and Fanchun Li[2]

[1] College of Information Engineering, Jiangxi College of Applied Technology, Ganzhou 341000, Jiangxi, China
qjdenglei@163.com, pingxie2010@163.com
[2] College of Social Management, Jiangxi College of Applied Technology, Ganzhou 341000, Jiangxi, China
lifanchun1209@163.com

Abstract. Along with the rapid development of Internet information technology, the field of network information technology has become more and more widely applied, and it is followed by the "big data era". Under the development of big data era, enterprises cannot leave the Internet and information technology applications on the road of sustainable development, but the competition relationship between enterprises is increasingly prominent, and the development prospect of enterprises is not smooth, which is a development environment where opportunities and challenges coexist. In development, enterprises usually use Internet computer technology for information transmission and data preservation, so there must be secrets in the enterprise website that affect the continuous development of the enterprise, once stolen, Will bring fatal impact to the enterprise. Therefore, the network information security in big data's era needs to be guaranteed. This paper mainly studies how to use effective security risk monitoring and protection measures to remove the hidden dangers of network information and ensure network security management based on big data background.

Keywords: Big data age · Computer information security · Risks · Protection · Measures

1 Big Data and Computer Network Security

1.1 The Connotation of Big Data

So far, no unified definition of big data has been specified, but generally speaking, the connotation of big data can be analyzed from the following three perspectives:

(1) From the point of view of object, big data is a kind of networked data combination. Big data technology is much more capable than typical database software management, collection, storage and analysis, but it needs to be attached great importance to big data is not simply worthless data stacking, on the contrary, even if the ability of big data or the content that can be stored is very large, it does not

K. Li et al. (Eds.): ISICA 2019, CCIS 1205, pp. 699–708, 2020.
https://doi.org/10.1007/978-981-15-5577-0_56

necessarily bring too much value to a certain industry. Generally speaking, the amount of information stored in the database under big data technology is relatively “novel”. Only when the relevant information is excavated in the huge big data, can it be used as valuable data.

(2) Based on a technical perspective, Big data does not exist independently, nor does he specifically point to a kind of technology or database, but in all kinds of databases, through some kind of information search method, carries on the valuable information integration or the technical way to obtain the technology. Compared with “massive data” and “large-scale data”, big data has different conceptual characteristics. The biggest difference is that big data is a concept that contains behavior, that is, big data processes data objects as a technology when searching for data, which is a kind of ability, which can be said to be a verb, which is not the same as the special noun “massive data”. In order to be able to complete the big effectively. The ability behavior of data technology, in order to search valuable data information in the vast database, it is necessary to activate big data technology and use it in a multi-disciplinary, multifunctional, multi-technology and flexible way, so that it can effectively retrieve the information. In addition, data mining, distributed processing and so on all need big data technology to complete. Thus it can be seen that big data has a powerful information integration function and is a tool for mining valuable information in the Internet environment.

(3) Based on the application perspective, big data technology can be applied in many fields, its application mode is according to big data’s work characteristics, carries on the valuable information acquisition, carries on the collection to the specific data way. The reason why big data has been widely used in various industries and fields has become an indispensable technical tool for the development of enterprises. The reason lies in the fact that big data’s application characteristics lie in “integration”, which has the characteristics of “one-to-one” work and is closely related to the specific work content.

1.2 Meaning of Computer Network Security

From the concept point of view, the computer network is a kind of security measure to keep the data confidential and preserved in the Internet environment. In the daily work, people often need to use the computer for information transmission or storage. In the process of information transmission, it is necessary to carry on the security protection to the information, to ensure that the confidential information will not be stolen by the illegal elements, and to protect the interests of the enterprise. This is the Internet environment computer network security. The hidden dangers of computer network security are as follows: when carrying out data information transmission, it is inevitable that it is necessary to transmit information through a variety of information transmission channels and media. The more channels through which information is transmitted, the more channels it means to keep information confidential. The weaker the sex, the easier it is to encounter network system vulnerabilities and human damage, and the stronger the probability that data information will be stolen, resulting in hidden dangers in network security. There are three main reasons for the hidden danger of computer information security: first, because the economy is gradually moving towards the trend of globalization, the area of network information dissemination in this environment is very extensive, which is easy to cause network security accidents; Second, the network is a virtual space, sending information on the virtual space has a certain degree of freedom. As long as people have a mobile society, they can express their own views or opinions on the network, and can receive a large amount of information every minute. People can unscrupulously publish any views on the network, but if there is no special supervisor of network ideas, it may bring hidden dangers to network security. Third, the openness of the network is very large, and the information in the network environment is vulnerable to virus intrusion, resulting in security risks.

2 The Risk of Computer Network in the Era of Big Data

2.1 Vulnerabilities in the Computer System Itself

Vulnerability is a kind of defect of computer operation, which generally exists in software, hardware, computer system and protocol. The vulnerability gives computer attackers the opportunity to destroy the system or steal information according to the direction of vulnerability in the user's computer without authorization. About the vulnerabilities of the system itself, the Windows system and the Solaris system have system vulnerabilities to a certain extent; for example, there may be some logic errors in the Intel Pentium chip, and there are also some programming errors in the earlier versions of Send mail, such as if the Ftp service is carried out. When the service is set up, the configuration error in the Unix system will also lead to the security hidden danger of the system, so that the wrongdoers can take advantage of it organically, which will affect the security of the system and the confidentiality of the information. All of these belong to the security vulnerabilities of the system itself. In addition to the first time, people use computers, not to apply the functions of the computer itself. The computer after the manufacturer leaves the factory is a bare machine, and there is no software, which means that there is no function. The computer used by people in daily work is after downloading and installing the necessary software before they can work. The software download and installation process is a possible security vulnerability Links, especially the software downloaded on the website, is not secure, the virus has a great impact on the computer, may cause file loss.

2.2 Hidden Dangers of Network Information Security in Big Data Environment

Mass information is synonymous with big data era. In big data era, the function of computer in Internet environment is very powerful, which can collect and spread massive information, which facilitates people's daily work and study in some respects, but it is precisely this environment with information search that brings people information security risk and becomes a threat for people to store files with computer. In detail, these potential security risks affect the security attributes of information storage and dissemination, once there are security risks, then leakage and damage is very common. In addition, the hidden danger of network information security is also reflected in the means of virus invasion. Viruses have a certain impact on the integrity and security of files. If, in the development of enterprises, file depositors put files on the computer, but the simultaneous virus invades the computer, then the important files on the computer may be lost and the contents of the files are no longer complete, which has a great impact on enterprises or file managers, and lightly affects the work content of small departments. The important thing is to delay the whole development progress of the enterprise, even more, the confidential documents of the enterprise are obtained by the peers, and their own labor achievements become the handheld of others, which affects the overall development of the enterprise.

2.3 Safety Risks Caused by Artificial Improper Operation

The hidden dangers of computer information caused by improper operation can be divided into two situations: one is that the user accidentally carries out some operations, which unwittingly leads to the occurrence of hidden danger accidents; the other is that hackers deliberately operate the user's computer through illegal technology in an attempt to destroy the security of the computer. Although the application scope of computer network has become very extensive in recent years, there are also special network supervision departments, but in practical operation, because people do not have a comprehensive understanding of computer, because of the lack of level of operation mode, there are operation errors, which makes the problem of network information security appear frequently. And it's because of the right plan. The operation of the calculation is not skillful, and the user will not invest in the computer network information security, so the result is that the user causes the computer network security by using the improper operation mode, and divulges his personal privacy to the hacker. In this case, the lawbreakers will take bad means to resell the user's personal information and cause another kind of computer user injury. In the era of big data, the biggest threat to the application of computer network information data is that hackers destroy the security of user information, carry out information acquisition behavior, and cause the loss of users' interests. Based on this, the user's awareness of computer security should be strengthened. to prevent non-sending molecules from obtaining information.

2.4 The Hidden Danger of Network Information Security is Caused by the Means of Hackers Attack

The use of computer in big data environment is necessary to browse the information on the website and download the relevant information, but it is in the process of information download that hackers can take advantage of it. Hackers usually use computer system attacks to carry out user computer data, information or system damage, hackers attack means are more covert, computer knowledge is not deep in general can not find the offensive means of hackers in time, unless users have a certain understanding of hackers' attack techniques, which leads to greater damage rate of computer systems. Nowadays, through the application of computer, people can publish network information at any time, and the network information can be presented at any time. In this environment, the application software which uses computer to obtain information can not directly judge whether there is a virus in the information of the visitor, and the hidden attack behavior of hackers can not be detected in time. Therefore, it is easy for hackers to steal information by illegal means, which brings a great negative impact on people's life or work, and is a threat and hidden danger to the development of computer network security.

2.5 Lack of Sound Network Security Management

Although the network is a virtual world, the dissemination of information in the network environment also has a certain impact on people's ideological concept. The network environment needs to be maintained. Only by constructing perfect network security management measures can we ensure the progress of the network world order. Otherwise, bad ideas and bad behavior exposed to the network environment will cause the network to become a negative tool affecting people's normal life and a harmful tool. Therefore, in order to use Internet computer information technology in an orderly manner, it is necessary to strengthen the management of network security. However, in real life, the lack of perfect network security management is not only manifested in the lack of managers who work normally on the network of special readers. It is also manifested in the lack of relevant management system, so that external factors may become the destruction of the working environment of users' computers. Under the era of big data, the computer network environment lacks the perfect network security management strategy analysis: first of all, it is manifested in the daily computer maintenance link of the computer user, the maintenance measure is not in place or the management mode is incorrect is one of the reasons that causes the computer network security to be destroyed; Secondly, the administrator who specializes in the management of network information security does not pay more attention to management and management vigilance, and is perfunctory in the work of information security management, so that the security of computer network information can not be guaranteed. Then it leads to the hidden danger of computer information security. In addition, when most colleges and universities or training institutions carry out personnel information management, due to the lack of supervision, the information of students or staff is stolen by others and transferred to a third party in the form of trading, forming illegal commercial means, which leads to the invasion of the privacy of students or staff.

3 Analysis of Protective Measures for Information Security Risk of Computer Network

3.1 Rational Application of Network Monitoring Systems and Firewalls

Network monitoring system, also known as network information monitoring system, is a kind of security monitoring method which is planned in advance. Under the environment of big data, the business network information is diversified. Through the network monitoring system, the hidden danger problem of network security can be taken to prevent the hidden danger of network security before it occurs. First of all, it is necessary to build a set of scientific and perfect network information security management system, through the use of scientific and reasonable network security technology to manage and protect the security of computer network information; secondly, in order to ensure that computer network information is not attacked by new computer viruses or malware, it is necessary to use the function of firewall and computer. Network security monitoring measures combine with each other to prevent the data or information in the process of transmission from being intercepted or copied by other illegal means; in addition, there are many and miscellaneous personnel in the use of network data in public places, so it is necessary to strengthen the application of firewall technology in network sharing, and the installation of protective wall technology in public places is beneficial to improve the security of network information transmission. Generally speaking, firewall technology can be divided into two management directions: internal management and external management, compared with external management, in most cases, the role of internal management is stronger than external management, that is to say, the security of internal management is in many cases better than that of external management. The reason is obvious. Therefore, the more important data information can be stored in the internal management. In view of the hidden dangers in the operation of computer systems, firewalls can also play a role in clearing hidden dangers. In the era of big data, the fields involved in computer technology are far from what people are waiting to see at present, but from another perspective, the stronger the application ability of computer technology, the wider the scope of application, which means that the computer may face more kinds of viruses. It is necessary to reasonably apply network monitoring system and firewall technology to cut off virus spread channels and bring application security to computer users.

3.2 Measures to Strengthen Network Information Management

The management of network information plays an important role in the use of computers. The main reason for the security hazard of most computer network information is that the network information manager does not pay more attention to the management of network information security. Therefore, the author believes that it is necessary to manage the information security of computer network through the following ways: firstly, whether it is a group computer or a personal computer, it is necessary to raise the awareness of the measures for information security management and protection in the process of use. Personal use of computer, preliminary use of the need to know the computer. The hidden danger factors of network information, the security characteristics of information use, and the secure storage or management of a large number of information in the network environment. Based on the background of big data, the frequency of learning work using computer is very large, and the probability of virus intrusion is also large. Therefore, it is necessary to carry out network information security management through scientific and effective security hidden danger protection measures. Network information security managers play a very important role in this link, and staff need to be aware of this. At the same time, in the internal operation and project development, enterprises need to manage the computer network information safely and recruit special network information. The person in charge of security adopts a scientific and effective dynamic management order and system to find the changes in the operation of network information security in the development of enterprises in time, so as to prevent hackers from taking advantage of the opportunity to enter. In the project of network information security management, the person in charge needs to strengthen the security theory research and protection measure detection of network information security, and construct a computer technology platform with invalidity. In addition, the most important thing to pay attention to is that when the enterprise staff carries out the operation of computer data or information, they need to improve the awareness of security protection. Based on the sufficient knowledge of the computer operation specification, the staff can be exposed to the more confidential plan. Computer file operation, so as not to cause uncontrollable consequences.

3.3 Installation of Computer Antivirus Software

Antivirus software, also known as antivirus software and antivirus software, was invented to kill data or information that has security risks to computers, such as computer viruses, malware, Trojan horses, and so on. Nowadays, whether it is for business operations or personal applications, when purchasing bare metal, businesses will recommend antivirus software suitable for this model to users. The types of antivirus software are diverse, and their functions and working methods are not the same. If some antivirus software only needs to segment some storage from memory, it can judge whether there is a single antivirus software by comparing all its virus libraries with the data involved in the operation of the computer. If there is a consistent virus signature, it is judged that there is a virus in the data. Some antivirus software is executed according to the program submitted by the user or virtual execution system, and makes a correct judgment according to the actual results or reflected behavior. Under the background of big data, computer development technology is spreading, and the area covering people's daily life has been expanded, but there are also programs that can threaten the normal use of computer network information technology, that is, virus. The virus poses a great threat to the computer information management system, and may also cause damage to the privacy and security of the information. According to our country, The research results of computer security performance and virus development that affect computer security show that there are many forms of virus in the process of using computer. The invasion of virus can directly destroy the data in computer system, and the virus is contagious and can spread according to a certain line. After a long period of virus research, the Ministry of virus Management and Security Protection of China has launched the technology of "cloud security". Cloud security technology can connect the virus killing software with the client, check and kill the malicious virus data or software that appear in the operation of the computer, and restore the network information security environment of the computer.

3.4 Strengthen the Awareness and Strategy of Network Hackers

At present, with the gradual expansion of the scope of computer technology coverage, people are more and more unable to leave the use of computer information technology in the work. In the network environment, what can not be completely avoided is the hacker behavior hidden in the dark. The characteristic of hackers is that they are very familiar with the design of the virus and know the channels and paths of virus transmission. Therefore, there are some hackers who are subjected to commercial transactions by competitors in the industry and take illegal acts. By stealing information from the relevant enterprises and getting rich compensation, the enterprises that employ hackers know all the inside information of their competitors very well, and thus have a more competitive advantage in the market. In order to be able to strengthen the network environment. The author believes that the professional management department which needs to manage the network information environment integrates and detects a large number of data, monitors the information security of the network environment in real time, and carries on the effective security protection according to the existing data strategy. At the same time, in order to strengthen the technology of identity tracking of network hackers, once it is found that hackers are not issued, it is necessary to impose legal sanctions on hackers. In practical application, the firewall technology design of computer information security in network environment can be strengthened by isolating the computer system, so as to reduce the aggression of hackers to computer technology. In addition to the above, digital technology can also be strengthened. The application strictly controls the access authority of computer information and constructs the network security environment with identity characteristics, that is, the unauthenticated user information can not access or modify the important information. In this way, the use of computer information technology will be more secure, but also can protect the security of the network.

3.5 Use of Data Encryption Technology to Ensure Data Security

In the era of big data, the risks and loopholes of computer network information security did not come out of thin air, but after the design of virus programmers, waiting for the computer users to "throw into the network", such as installing software on unknown websites and downloading files with viruses, can be called the reasons for the hidden dangers of computer systems or data. In order to strengthen the security performance of data and information, it is necessary to encrypt the data of computer network safely to ensure the security of information transmission and the stability of storage. Data encryption can be said to be a long history of technology, refers to the encryption algorithm and encryption key to convert clear text into ciphertext, while decryption Ciphertext is restored to clear text through decryption algorithm and decryption key. Cryptography is the core technology of data encryption. Therefore, in order to ensure the security of data transmission through data encryption measures, we need to have full knowledge of cryptography. It should be noted that data encryption technology needs to cooperate with firewall technology in order to play a tenacious role in risk resistance. Data encryption technology is a reliable technology. The purpose of data transmission encryption technology is to encrypt the data stream in transmission, usually there are two kinds of encryption: line encryption and end-to-end encryption. Line encryption focuses on the online road, regardless of the source and destination, and is the transmission of confidential information through each line. Different encryption keys are used to provide security. End-to-end encryption refers to the automatic encryption of information by the sender and packet encapsulation by TCP/IP, and then passes through the Internet as unreadable and unrecognized data.

When the information reaches its destination, it will be automatically reorganized, decrypted, and become readable data. Because of the ciphertext setting in the computer network information, some illegal elements can not crack the information, which can effectively ensure the security of information transmission and storage.

4 Conclusion

In a word, with the rapid development of computer technology, computers are used in various fields, including life, work and learning. In such an environment, people can no longer work normally when they leave the computer. In the future development, the application field of computer will only show an unabated development trend. It is such an environment in which people can only carry out daily work through computer information technology, so many important files will be stored in the computer used at work. However, due to the spread and destruction of the virus, the security of computer information can not be fully guaranteed. Based on this, how to solve the problem of viruses and hackers. The intrusion problem has become the focus of attention of industry researchers. In the era of big data, it is inevitable to store or transmit data through computer, and these two behaviors are prone to virus infection to a great extent. The author thinks that the privacy information management can be carried out by applying for a special account, and the firewall can be set up to avoid the loophole of computer system, data or information to the greatest extent. Through the installation of antivirus software, increase the awareness of virus protection, strengthen the network information security monitoring of professional departments, set up firewalls and other protective measures to ensure the security of network information transmission, once the security hidden dangers are found, it is necessary to Its in-depth research and analysis to ensure the security and stability of users using the computer.

Acknowledgement. This work was jointly supported by the Key Research and Development Project of Gan-zhou, the name is “Research and Application of Key Technologies of License Plate Recognition and Parking Space Guidance in Intelligent Parking Lot”, the Education Department of Jiangxi Province of China Science and Technology research projects with the Grant No. GJJ181265.

References

1. Wang, D., Ju, J.: Research on computer network information security and protection strategy in big data era. Wirel. Interconnect. Technol. (24), 40/41 (2015)
2. Feng, S.: A brief analysis of computer network security and preventive measures in big data’s era. Econ. Manag. Full Text (7), 00149 (2016)
3. Yang, Y., Zhang, Y.: Research on network Information Security under the background of big data. Autom. Instrum. (10), 149/150 (2016)
4. Long, Z., Lihao, W., Zhiheng, L., et al.: Research on computer network information security protection strategy and evaluation algorithm. Mod. Electron. Technol. **38**(454(23)), 89–94 (2015)

5. Li, Q., Ren, T., Xiaohu, W., et al.: Research on computer network information management and security protection countermeasures. Autom. Instrum. (9), 1 (2017)
6. Wen, A., Zhang, T., Liu, J.: Discussion on information security technology and its development trend based on computer network. Coal Technol. **31**(5), 248 (2012)
7. Wang, X., Hu, B.: Effective application of virtual private network technology in computer network information security. J. Xichang Univ. (Nat. Sci. Ed.) **30**(4), 36 (2016)
8. Xie, J.: Research on application value of virtual private network technology in computer network information security. Inf. Constr. (4), 6364 (2016)
9. Yuan, X.: Analysis on the threat and countermeasures of computer network information security under the New situation. Sci. Technol. Econ. Mark. (5), 104/105 (2016)
10. Deng, C., Tang, Y., Xue, Y., et al.: Construction of computer Network Information Security Management system. Inf. Comput. (Theor. Ed.), (5), 185/186 (2016)
11. Huang, C.: Analysis of computer network security strategy. Sci. Technol. Inf. (Acad. Res.) (16), 248/246 (2007)
12. Xin, F., Liu, K.: Research on computer network information security and preventive measures. J. Jiamusi Vocat. Coll. (12), 442 (2013)
13. Tian, Y., Shi, Q.: On the computer network security and preventive measures in big data's era. Comput. Program. Ski. Maint. (10), 90 (2016)
14. Lai, D., Chen, H.: Computer network information security and protective measures under the background of big data. China New Commun. **20**(20), 146 (2018)
15. Huang, J.: Discussion on the problems and countermeasures of computer network security in big data's era. Intelligence (29), 23 (2018)
16. Yang, L.: Research on common loopholes and preventive measures of network security in big data era. Think Tank Era (39), 138143 (2018)

Anti-lost Intelligent Tracker Based on NB-IoT Technology for the Elderly

Juanjuan Tao(✉), Shucheng Xie, Jinwei Jiang, and Wenbin Wei

College of Computer Science, Guangdong University of Science and Technology, Dongguan, Guangdong, China
taojuan157@126.com

Abstract. In order to promote the development of wearable device informatization and solve the problem of frequent lost of the elderly, we attempt to research an anti-lost intelligent tracker based on Narrow Band Internet of Things (NB-IoT) technology. It covers functions such as remote acquisition and data storage of multi-sensor node information from sensors. The system collects environmental information and health information in real time and transmits them remotely to the IoT cloud platform through NB-IoT technology. The real-time position of the elderly can be taken on the map in APP. The experimental results show that the APP can indicate the specific location of the target on the map in real time and the environmental information in the meanwhile. Through monitoring the health information, the health status of individuals in the environment can be timely known to ensure safety. This paper develops an anti-lost intelligent tracker which uses NB-IoT technology to replace WiFi and Bluetooth. We use NB-IoT technology to achieve data aggregation and long-distance transmission to IoT telecom cloud platform. Compared with other devices on the market, the proposed system has the characteristics of low-consumption, power management and endurable continuous positioning.

Keywords: Wearable device · IoT cloud platform · NB-IoT · Intelligent tracker · Low power consumption

1 Introduction

The elderly are always the special focus of the society. In particular, the "demented old people", who are weak in judgment and mobility, are easily lost in workplaces where people are mixed. Furthermore, the health problems of the elderly are also increasingly prominent problems. The health management of the elderly should be gradually paid close attention to. In order to meet the needs of the health and safety of the elderly, to better protect the elderly from being unsafe, to achieve timely records of location in the map and effective control and management of disease occurrence or development, we put forward the feasibility of using wearable smart devices to help build healthy living habits. The idea is to shift from medical treatment to prevention, therefore the topic has practical significance.

According to the 2015 White Paper regarding the Lost of Chinese Elderly, the number of the annual lost of Chinese elderly people reached up to 500,000. In February

K. Li et al. (Eds.): ISICA 2019, CCIS 1205, pp. 709–723, 2020.
https://doi.org/10.1007/978-981-15-5577-0_57

2018, Today's Headlines published the annual public welfare tracing data: one third of the lost are elderly people. There are several reasons behind that, such as lost, mental illness and dementia. Since the beginning of the 21st century, the aging population of China has accelerated its development. According to the National Bureau of Statistics, by the end of 2018, China's population of over 60-year-old had reached 249.49 million, it is the first time it exceeded the population of 0–15-year-old, covering 17.9% of the total population. The National Committee on Ageing pointed out that China's elderly population over 60 years old will double to 400 million by 2033, and will reach one third of the national population by 2050. Accompanied by the wave of aging is the problem of the elderly who have been reported frequently. It is clear that the lost of the elderly has become a dominant problem. Therefore, it is urgent!

At present, there are some products with similar functions at home and abroad. Most of the mainstream products are wearable devices with certain computing capability. The physical condition and location information of the elderly are shown on the terminal by connecting mobile phones and other devices [1, 2]. In fact, some anti-lost devices that have shown on the market are mostly by way of Bluetooth, Wi-Fi, GPRS and other technology for communication and positioning, however, there are still some drawbacks in the process of use: for one thing, the effective transmission distance is limited when encountering underground garages and remote mountainous areas, hence the device cannot work; for another, the power consumption is large, the standby time is too short, as a result, the battery must be charged frequently or the battery must be replaced in time [3].

Hence, it is necessary to investigate and design an anti-lost intelligent tracker with accurate positioning, high safety factor and high endurance.

With the rapid development of Internet of Things technology, low-power wide-area networks for device access have emerged in recent years, such as the Narrow Band Internet of Things (NB-IoT), which is a kind of cellular network of characterizing in low-cost, wide-coverage, and low-power consumption [4]. In order to overcome the shortcomings like high power consumption, high cost, limited transmission distance, and solve the problems in terms of latency, frequency and rate, and the sensitive problem to IoT services in process of environmental awareness monitoring and human physiological indicators [5], we attempt to develop a monitoring system mainly using this technology. The anti-lost intelligent tracker for the elderly with NB-IoT technology comprehensively adopts STM32L476RGT6 and sensor terminals to collect sensing information in real time, by way of the NB module to achieve no-gateway uploaded data. The cloud platform receives the data and saves it concurrently and sends the data to the remote monitoring applicant system that makes the HTTP request.

2 System Architecture

The system is designed to cover functions including data acquisition, remote transmission, storage management, remote monitoring, and hazard alarm, which includes the collection and management of environmental information, positioning information and human physiological state information. The system is consisted of a four-layer structure: a sensing layer, a transport layer, a platform layer, and an application layer. The system architecture diagram is shown in Fig. 1.

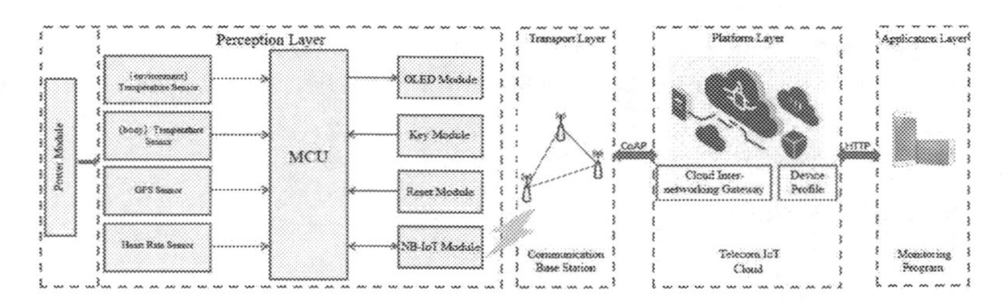

Fig. 1. Chart of smart tracking system for the elderly

The first layer is the sensing layer, which mainly includes a power module, a sensor module (temperature and humidity sensor, body temperature sensor, GPS sensor, heart rate sensor, etc.), STM32L476RGT6 (Microcontroller Unit, MCU), OLED display module, button module, reset module, and the NB module in charge of accessing the cellular network. The sensing layer realizes node data acquisition, and connects the physical entity to the transport layer through the NB communication module. The second layer is the transport layer, which mainly includes the core network and the communication base station. After the sensing layer collects the environment-aware information and the human health data, data streams will be sent to the nearby communication base station through the core network of the transport layer to establish a user pane bearer for the NB-IoT terminal, passing the uplink and downlink business data. The third layer is the platform layer, which is responsible for converging related data obtained by the access network, and forwarding it to the monitoring system for specific processing according to different sensing data types, also opening interfaces to the system to facilitate data acquisition. The fourth layer is the application layer. On the platform basis, the tracking data monitoring system is deployed on the IoT telecom cloud platform. The query interface can be used to monitor the platform data, and also can command issues to the underlying control module to achieve an overrun alarm.

The division of working at each layer is clear, and the two functions are mainly realized: first, the data reported from the sensing layer to the application layer realizes the remote monitoring of the sensing information and the health data; secend, commanding issues from the application layer to the sensing layer achieves the intelligent control of the device. The system flow chart is shown in Fig. 2.

2.1 Data Uplink

The temperature and humidity sensor device node collects relevant data and encodes according to the data frame format, and sends the encoded data to the NB-IoT module through the MCU serial ports in the form of an AT command. After receiving the AT command, the NB-IoT chip automatically encapsulates the payload into the CoAP protocol information and sends it to the previously configured cloud platform. Then the platform automatically analyses the CoAP protocol packet [6–8], then finds the matching coder-decoder plugin which is in accord with the device profile file. It means that platform analyses payload to JSON data matching the service described in the device profile file, and saves it on the platform. Thus, the tracker system obtains the data on the platform via the query interface, and remotely monitors the data changes. In each monitoring cycle, platform sends data to the designated server through a POST message.

2.2 Command Downlink

The monitoring system can issue control commands to the platform. If the platform detects that the device is online, the command is immediately issued. If the platform determines that the device is offline, then the command is cached in the database of the platform. So we learn The NB device reporting data at a certain time. After receiving the data, the platform will retrieve whether the corresponding device has a valid unsent command in the database. If so, the command is issued. The issued commands are encoded by way of the coder-decoder plug-in (JSON to hexadecimal stream) and sent to the device. The device receives the command, executes the overrun alarm control command, and returns the result.

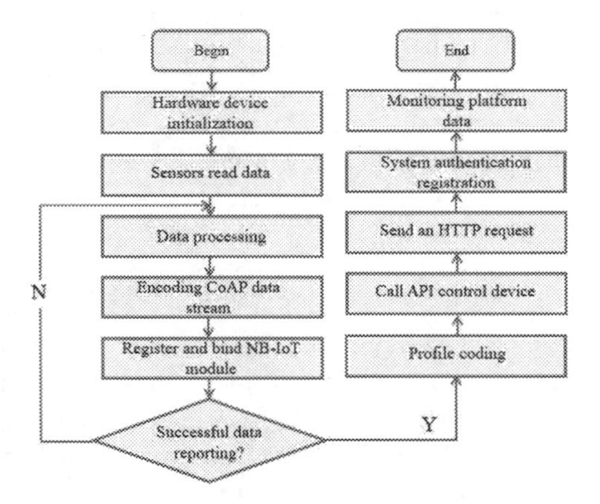

Fig. 2. Flow chart of perception layer to cloud platform

3 System Design

3.1 Hardware Device and Software Environment

The software and hardware equipment selection and environment required for the system designed in this paper are shown in Table 1. The MCU selects the STM32L476RGT6 chip, and the NB-IoT module uses the BC95-B5 wireless communication module. The sensors are responsible for environmental sensing parameters such as temperature, humidity, latitude and longitude, health information including body temperature and heart rate. Module units send the collected data through the NB-IoT module using CoAP protocol, and receives control commands from the platform.

Table 1. Hardware and software equipment detail list

Equipment name	Equipment type and develop environment
MCU microprocessor	STM32L476RGT6
Power	5 V, 3000 mA·h rechargeable power supply
Display module	OLED
(environment) Temperature and humidity sensor	DHT11
(body) Temperature sensor	MLX90614ESF
GPS sensor	AIR530
Heart rate sensor	MAX30102
NB module	BC95-B5 module
Emulator	SEGGER JLINK Emulator
SIM card	China Telecom 4G NB-IoT card
IoT platform	OceanConnect
Serial debugging	ComAssistant
Application layer interface docking	tomcat+JKS key library
Android develop environment	Android Studio 3.3.0.0, Gradle 4.10.1, targetSdkVersion 28
Database	Litepal+SQLite

3.2 Perception Layer Design

The design of the perception layer is mainly divided into five components: main control board design, communication module circuit design, data frame format design, algorithm module design and embedded development. The NB-IoT main control board and communication module circuit are designed as the core of the sensing layer design. The main control board is responsible for collecting information of health and positioning, and the function of the communication module is to interact with the data of the NB-IoT base station.

3.2.1 Main Control Board Design

The main control board design block diagram is shown in Fig. 3. It is obvious that temperature and humidity sensors, body temperature sensors, GPS sensors, and heart rate sensors are used to collect temperature and humidity, body temperature, latitude and longitude, and human heart rate data respectively. The 32-bit processor (MCU) is the core of the entire main control board. It is responsible for data acquisition, processing, analysis, and communication with the communication module. The STM32L476RGT6 is the main control chip. The chip is mainly used to collect and process data, and store the processed data in the internal storage module. The setting of system operating parameters can be done by AT commands. For instance, AT +NMSTATUS (Message Registration Status B657SP1 or later) determines the connection between the current module and the platform.

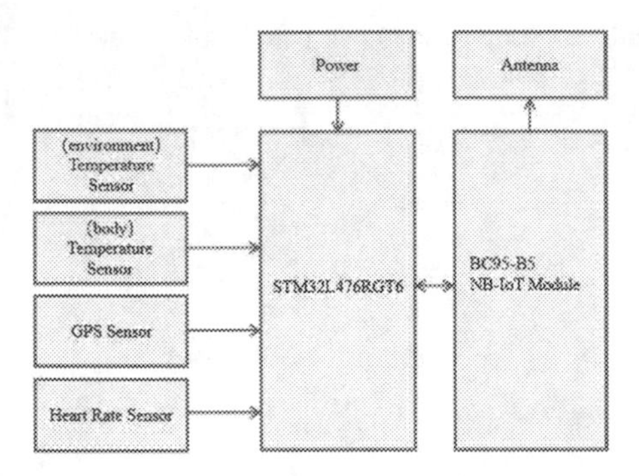

Fig. 3. Design block diagram of main control board

After the MCU sends data to the NB-IoT module, the NB-IoT module creates a transmission channel between the base station and the base station. The uplink data-sending uses binary phase shift keying (BPSK) or QPSK modem and uses single carrier frequency division multiple access (SC-FDMA) technology. Orthogonal Frequency Division Multiple Access technology (OFDMA) is used in the downlink, while the sub carrier interval is maintained at 15 kHz [9]. In addition, it can also monitor commands from the base station, and it can respond to the command in time when it receives the command request signal. The system uses the internal timer to collect data signals, processing them through the ADC unit packaged in the sensors, and encodes the processed data in conformity with the data frame design format, and then stores it in the internal memory. The body temperature sensor MLX90614ESF and the heart rate sensor MAX30102 collect the values and convert them into digital signals. The serial signals are sent to the NB-IoT module via the serial port in the form of AT+NMGS commands. After receiving the AT commands, the module automatically encapsulates the payload into a CoAP protocol message and sends it to the ready bound platform OceanConnect via the telecommunication base station.

3.2.2 Communication Module Circuit Design

The NB module is used to receive and forward collected data. The module is programmed by Keil uVision4 to realize the control of the NB module, and the monitored data is sent out through the antenna by the module, and it is judged whether the SIM card information is completely read or not, and then is matched to the corresponding state.

This design uses the BC95-B5 wireless communication module to report monitoring data and receive over-limit alarm commands. BC95-B5 is a wireless communication module based on NB-IoT technology. It uses NB-IoT communication protocol (3GPP Rel.13) to establish connection with network operator's equipment, mainly working in the licensed band of 850 MHz [10–12]. Since the current minimum deep-sleep is up to 5 μA, power consumption is largely saved. The SIM card stores user's information, encryption key, available traffic and time information, etc., to meet the management needs of intelligent hardware and smart wearable devices for device networking.

Combined with the above various types of sensors, other electronic components, BC95-B5 radio frequency chip, and radio frequency antenna together constitute a terminal data acquisition node of the wearable device. The terminal is shown in Fig. 4.

Fig. 4. Physical drawing of terminal hardware

3.2.3 Data Frame Format Design

On the purpose of improving the accuracy of data matching and convenience for hardware and software devices to capture and resolve data, the frame structure is adopted to reduce the probability of data packets being transmitted during transmission. According to the application architecture, the terminal device is distinguished by the method of using the International Mobile Subscriber Identification Number (IMSI) in the communication, and the device terminal has a unique IP address and an IMSI number, and the two are added to the transmitted data packet. Factors are used to ensure the delivery of data packets and timely response. The frame format content includes four parts: frame header, frame length, the control word and data.

The frame header portion occupies 2 Bytes. As the starting character of each frame data, the fixed frame header content is 0x7E 0x7E; the frame length portion occupies 2 Bytes, indicating the data length of the frame and including only the length of the control word part and the data part; the control word part occupies 1 Byte, which is divided into 4 parts—operation type, data type, data item and reserved bit; the number of bytes occupied by the data part is not fixed, depending on the actual length of data transmission, this part stores the data content that is actually sent. The control word part occupies 1 Byte, and contains 8 bits in total. The data type value of Bit7 and Bit6 is "00", indicating that the data is reported, "01" is the command, and the data or command related to the temperature and humidity module is transmitted by Bit5 and Bit4; Bit3 and Bit2 transmit data or commands related to the latitude and longitude acquisition module; Bit1 and Bit0 transmit data or commands related to the body temperature and heart rate module. When the operation type is "00", it means the corresponding ambient temperature measurement value, environmental humidity measurement value, human body temperature measurement value and heart rate measurement value, and the remaining values are reserved; when the operation type is

"01", the data item value "00" means to open the measurement, "01" means to close the measurement, "10" means to obtain the measured value, and the rest of the values are reserved; when the operation type is the command response, the data item values are reserved.

3.2.4 Algorithm Module Design

In order to improve the accuracy of heart rate monitoring, a new photoelectric volumetric algorithm was developed for a accurate analysis of user's heart rate by using the improved photoplethysmography algorithm in the heart rate monitoring module.

During the blood circulation process, when the heart contracts and ejects blood, and the blood passes through the arterial system into the microvasculature such as the arterioles, capillaries, and venules in the perivascular vessels of the human body, the blood volume of the microvessels is the largest at this time; when the heart is diastolic, the ejection is stopped. When the blood returns to the heart through the venous system, the blood volume of the microvessels is minimal at this time. It can be seen that the blood volume of the microvessels is a pulsating change with the heart beat, and this pulsatile change is called a volume pulse wave. Such volumetric pulse wave blood can generally be obtained by photoplethysmography (PPG) [13–15].

PPG measures the pulse. The basic principle is to measure the pulse and oximetry by using the human tissue to cause different light transmittance when the blood vessel beats [16]. When the volume of the microvessels in the light-transmitting region changes, the amount of light absorbed by the blood will change, while the absorption of light by other tissues such as skin, muscle, bone and part of the venular blood is constant. Light-emitting tubes generally employ light-emitting diodes of a specific wavelength selective for oxyhemoglobin (HbO_2) and hemoglobin (Hb) in arterial blood.

The electrical signal changing period of the photoelectric transducer is the pulse rate. The heart rate monitoring algorithm was designed according to the definition of blood oxygen saturation. The heart rate was expressed by *HeartRate*, and the volume of oxygenated hemoglobin and hemoglobin in the blood was represented by C_{HbO_2} and C_{Hb}, respectively. The procedures as follows are useful in analyzing how to reach an outcome.

Using the frequency of the photoplethysmographic pulse wave, that is, the pulse rate, approximately represents the heart rate. From the formula, the pulse rate is the quotient of the oxyhemoglobin volume and the sum of oxyhemoglobin and hemoglobin, and then corrected by subtracting the calibration parameters:

$$HeartRate = \frac{C_{HbO_2}}{C_{HbO_2} + C_{Hb}} \times 100\% - \zeta \; where \; databuf[x] > V_H$$

When the lower boundary of the narrow band is touched, the parameters are corrected to the measurement range:

$$HeartRate = \frac{C_{HbO_2}}{C_{HbO_2} + C_{Hb}} \times 100\% - \zeta \; where \; databuf[x] < V_L$$

Find the arithmetic figures for each of the 5 sets of data buffers to get the result:

$$HeartRateAvg = \frac{\sum_{x=1}^{5} HeartRate}{5}$$

The hardware program reads the FIFO buffer queue of the MAX30102 through the hardware I2C program, takes out the converted light intensity value, and stabilizes the data fluctuation in a frequency band via the algorithm processing of steps (1) and (2). Given a harmonic parameter ζ, by repeatedly debugging, ζ is taken as 0.23 eventually. When the sampled value is higher than the upper limit of the threshold, step (1) is performed; and when the sampled value is lower than the lower limit of the threshold, step (2) is performed. Finally, the five sample values in the group are averaged and assigned to the final value of *HeartRateAvg*.

Compared with the traditional photoelectric volume method, the sampling frequency band is moved down to the ideal range by the harmonic parameters, and the invalid data points are ignored. In this way, the sampling of the photoplethysmographic pulse wave signal can be neglected during the tracing because of the program simulation estimation and the sharp point on the image generated by the hardware basis error. And the parameter calibration is used to stabilize it in the slower narrow band, so the parameter reconciles the sampled value. Neglecting invalid data points generated by bumps and sharp points can improve the effectiveness and reliability of the algorithm for controlling data.

3.2.5 Embedded Development

The ultimate goal of the perception layer design is to complete the uplink of data. Data uplink business is the most striking one in the IoT business.

This paper uses the Keil integrated development environment to program and connects it to the STM32 main chip via the SEGGER J-Link emulator. After compilation, download and simulation, then we program codes to the MCU, and complete the development. The specific functions include data acquisition, signal conversion, data transmission, and receiving commands.

3.3 Transport Layer Design

In this section, we describe main principle of transport layer. The environment information and the human health data are transmitted by the NB module and transmitted to the nearby communication base station via the core network of the transmission layer. The transport layer includes the core network and communication base stations. The core network mainly performs the function of interacting with the non-access layer, and forwards the data stream monitored by the ambient temperature and humidity, latitude and longitude to the IoT platform OceanConnect for processing.

The communication base station is set up by an operator. The NB-IoT radio access network is composed of one or more base stations (eNBs), which are responsible for management of radio resources, including radio bearer control, radio access control, connection mobility control, and uplink and downlink resource dynamic allocation scheduling and encryption of user's data streams, IP header compression and other functions [17] (Fig. 5).

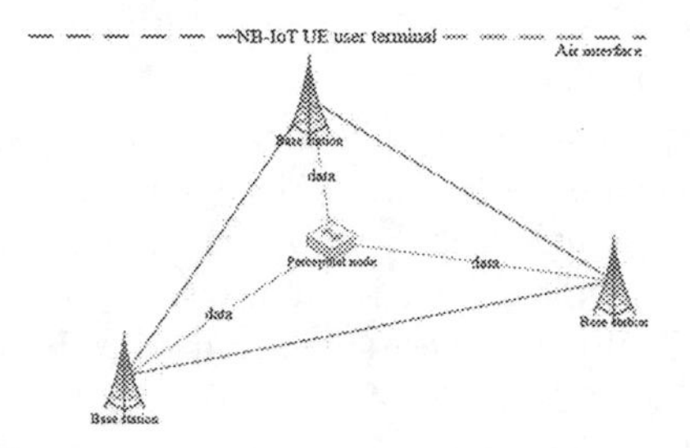

Fig. 5. NB-IoT network typology

3.4 Platform Layer Design

In this section, it is developed to the modeling and processing of platform design. The NB module is bound to the OceanConnect cloud platform, and the deployment application system is created. The data stream transmitted from the RF module is stored in the newly created data stream, and the profile file and the coder-decoder plug-in are written to implement unified management of the NB-IoT device [18]. Besides, the interface to the deployed monitoring system is open for remote monitoring of platform data.

If the platform receives the reported data, it is decoded into matching JSON data according to the coder-decoder plug-in and stored for monitoring remote monitoring of the system; if the platform receives the command issued by the monitoring system, it is encoded into hexadecimal streams by the coder-decoder plug-in. The binary stream is sent to the device to implement intelligent control of the device. The platform layer design flow chart is shown in Fig. 6.

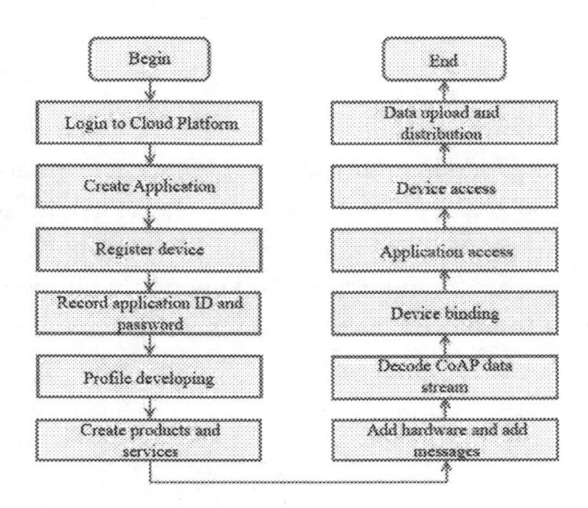

Fig. 6. Flow chart of cloud platform design

This paper uses the SP Portal front-end interface provided by the IoT platform to implement basic application management, device management, data viewing, and command viewing for tracker monitoring. We ought to login to the SP Portal (the account name and password will be sent along with the platform resources) to create an "app". "Application" is a mapping of the monitoring system on the platform. After the application is created, the platform will return "appid" and "secret". These two values are used to monitor the unique binding of the system and the NB devices.

3.4.1 Profile Design

In order to ensure the uniqueness of the data monitored by different monitoring applications, and to accurately save data such as ambient temperature to the cloud platform. It is indispensable to establish a mapping relationship between different sensing nodes and the cloud platform. So this paper designs the mapping file Profile.

The content of the file is mainly composed of "Device Capability" and "Service". The first one contains "manufacturerId", "manufacturerName", "deviceType", "model", "protocolType", etc. These fields need to fit into instructions of the NB modules. The second part is the device service information (Service), which defines the data fields of the environment parameters and health information, including temperature (entemperature), humidity (humidity), body temperature (botemperature). The heart rate, the longitude, and the latitude require the uplink data reported by the device and the downlink data sent by the server to the device.

3.4.2 Coder-Decoder Plugin Design

After the data monitored by the sensing node uplink to the platform through the core network of the transport layer via the NB module, then the data needs to be encoded. Similarly, the platform needs to decode the command after receiving the control command issued by the system. Therefore, a coder-decoder plug-in is designed on the platform to realize the encoding of the data and the decoding of the command.

In view that the device are usually sensitive to high power consumption requirements, the application layer data is generally in hexadecimal format. When the application layer data is protocol resolved on the IoT platform, it will be converted into a unified JSON format to facilitate monitoring of system usage. To implement the conversion of hexadecimal messages and JSON-formatted messages, the IoT platform requires a coder-decoder plugin provided by the device vendor. Moreover, one NB module corresponds to one coder-decoder plugin.

3.5 Application Layer Design

The application layer is the local client mobile application, which mainly implements two major functions: one can subscribe to the data change message in advance, or obtain the environmental data and health information sent by the NB monitoring device to the IoT platform through the data query interface. The other one is that IoT platform sends commands to control the operation of the underlying module. The monitoring system sends commands to the sensing layer NB device, and the device responds according to the command.

The application server communicates with the platform through the HTTP protocol, and controls the device by calling the open API of the platform, and the platform pushes the data reported by the device to the application server. What's more, the platform supports protocol analysis of device data and conversion to standard JSON formatted data.

The local mobile application receives the data on the platform through the data query interface file "QueryDeviceHistoryData.java" to perform remote monitoring of the usage group, and saves it to database, and designs the GUI page for display.

After above steps, APP is used to show the human health and environmental conditions obtained by the sensor. It also imports the Baidu LBS service, which can indicate the location of the elderly and record its historical walking trajectory.

4 Results

In order to verify the timely monitoring and precise control of each distribution parameter during normal operation of the system, we take a test in different weather (cloudy, sunny, rainy). Data is monitored at different locations at different time (9:00, 10:00, 16:00, 17:00), collected separately and recorded. The data monitored on the platform are shown in Fig. 7.

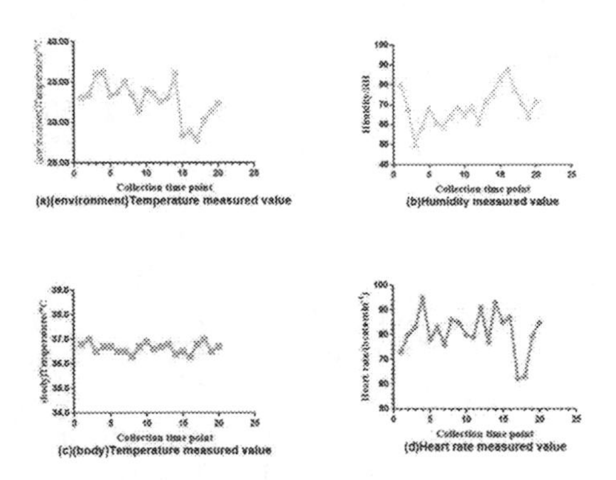

Fig. 7. Curve of measured values from cloud platform

The sensor node collects data and sends them to the cloud platform. In the historical data of the cloud platform, the service and data details can be viewed. The application APP subscribes to the cloud platform data, so the data are captured from the cloud platform and parsed, and shown on the APP UI interface. As is shown in Fig. 8, the subscribed data are shown on the map with the Baidu map LBS service and by calling the construction of the customized layer according to the position measured by the sensor, thereby revealing the historical trajectory. The real-time positioning test chart is shown in Fig. 9.

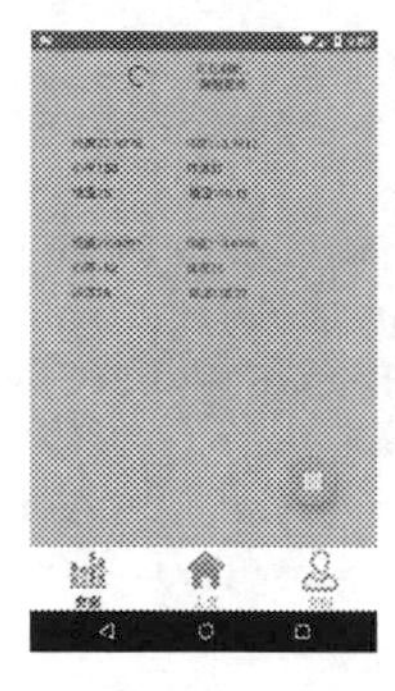

Fig. 8. Data shown on APP interface

We set up the upload period of the sensor node to 30 min. As is shown in Fig. 7, 8 and 9, the experimental results show that platform can indicate the monitoring data of each sensor accurately and promptly. The system runs stably under real-time monitoring on site, and realizes precise monitoring and control of environmental factors and health information, as well as network communication functions of wireless sensors to meet the requirements.

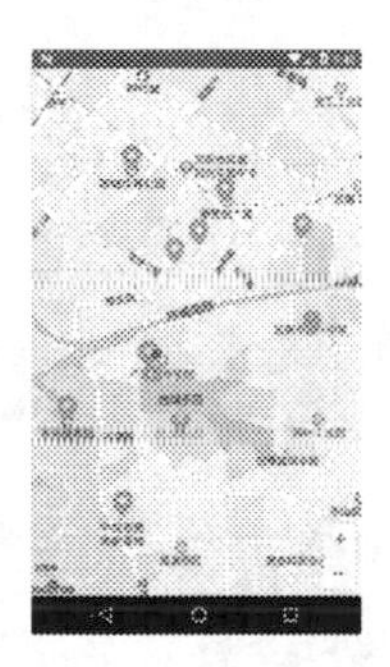

Fig. 9. Indicate location of individuals in the map

5 Conclusion

In order to promote the development of information technology for wearable devices and solve the problem of frequent lost of the elderly, this paper develops an anti-lost intelligent tracker based on NB-IoT technology, which realizes remote data acquisition and storage of multi-sensor node information such as ambient temperature and humidity, body temperature and the heart rate. The experiment is completed by a local data acquisition record over a period of time. Conclusions are as follows:

1) The monitoring system uses NB-IoT technology to deploy the network structure and establishes contact with the base station through the design of the plug-in and profile. The sensor node is used to get environmental parameters. Afterwards, the

NB module reports to the cloud platform via the core network, and the monitoring system calls the query interface to realize the individual's online remote monitoring.

2) This paper outlines the functions regarding how to make the elderly safe and how to track activity-information about the elderly. The monitoring application can show the specific location of the individual target on the map in real time, and the environmental information of the target individual's surroundings at this time. By monitoring the health information, the physical state of individual could be known. Thus, the action is to ensure safety. The GPS positioning accuracy is about 5–50 m, and the location based service is about 50–1000 m.
3) The system can be in the standby mode silently for 24 h. It is excitation mode when sending data. After stopping sending data, it will automatically enter the sleep mode to achieve intelligent sleep. Continuous positioning can last for 12–15 days and it has a high battery endurance.

It should be noted that this study reveals that the NB-IoT technology is stable and reliable, data acquisition is convenient. It is evident that commanding response is timely, which can provide technical support and reference for the technology in a wider field such as medical prevention.

References

1. Kroll, R.R., et al.: Use of wearable devices for post-discharge monitoring of ICU patients: a feasibility study. J. Intensive Care **5**(1), 64 (2017)
2. Liu, D., Chen, J., Deng, Q., Konate, A., Tian, Z.: Secure pairing with wearable devices by using ambient sound and light. Wuhan Univ. J. Nat. Sci. **22**(4), 329–336 (2017)
3. Ge, X.: Design of Handheld Positioning Tracker Based on GPS/GSM. In: Proceedings of 2017 IEEE 3rd Information Technology and Mechatronics Engineering Conference (ITOEC 2017), p. 4. IEEE Beijing Section, Global Union Academy of Science and Technology, Chongqing Global Union Academy of Science and Technology, Chongqing Geeks Education Technology Co., Ltd: IEEE BEIJING SECTION (2017)
4. Haridas, A., Rao, V.S., Prasad, R.V., Sarkar, C.: Opportunities and challenges in using energy-harvesting for NB-IoT. ACM SIGBED Rev. **15**(5), 7–13 (2018)
5. Salva-Garcia, P., Alcaraz-Calero, J.M., Wang, Q., Bernabe, J.B., Skarmeta, A., Nicopolitidis, P.: 5G NB-IoT: efficient network traffic filtering for multitenant IoT cellular networks. Secur. Commun. Netw. **2018** (2018)
6. Anna, L., Antti, R., Juha, S.: Impact of CoAP and MQTT on NB-IoT system performance. Sensors (Basel, Switzerland) **19**(1), 7 (2018)
7. Mukherjee, S., Biswas, G.P.: Networking for IoT and applications using existing communication technology. Egypt. Inf. J. **19**, 107–127 (2017)
8. Kaku, E., Lomotey, R.K., Deters, R.: Using provenance and CoAP to track requests/responses in IoT. Proc. Comput. Sci. **94**, 144–151 (2016)
9. Zhang, G., Yao, C., Li, X.: Research on joint planning method of NB-IoT and LTE. Proc. Comput. Sci. **131**, 985–991 (2018)
10. Ratasuk, R., Mangalvedhe, N., Zhang, Y., Robert, M.: Overview of narrowband IoT in LTE Rel-13. In: Standards for Communications and Networking (CSCN), pp. 1–7, October 2016
11. Zhang, H., Li, J., Bo, W., et al.: Connecting intelligent things in smart hospitals using NB-IoT. IEEE Internet Things J. **62**(5), 1550–1560 (2018)

12. Wang, Z.: NB-IoT technology based on cellular network. In: Proceedings of 2017 5th International Conference on Mechatronics, Materials, Chemistry and Computer Engineering (ICMMCCE 2017), p. 5. Research Institute of Management Science and Industrial Engineering: Computer Science and Electronic Technology International Society (2017)
13. Tine, P., Christophe, M., Ruth, V.H., Frederik, V., Pieter, V., Bert, V.: Mobile phone-based use of the photoplethysmography technique to detect atrial fibrillation in primary care: diagnostic accuracy study of the FibriCheck app. JMIR mHealth uHealth **7**(3), e12284 (2019)
14. JaeWook, S., Jaegeol, C.: Noise-robust heart rate estimation algorithm from photoplethysmography signal with low computational complexity. J. Healthc. Eng. **2019** (2019)
15. Gu, J., et al.: Algorithm for evaluating tissue circulation based on spectral changes in wearable photoplethysmography device. Sens. Bio-Sens. Res. **22**, 100257 (2019)
16. Fan, Y.-Y., et al.: Diagnostic performance of a smart device with photoplethysmography technology for atrial fibrillation detection: pilot study (pre-mAFA II registry). JMIR mHealth and uHealth **7**(3), e11437 (2019)
17. Anonymous. Cryptography Research, Inc.; Cryptography Research and IP Cores Announce Agreement for Differential Power Analysis Countermeasures Patents. Computers, Networks & Communications (2010)
18. Kuo, F.C.: Mob. Netw. Appl. **24**, 853 (2019). https://doi.org/10.1007/s11036-018-1092-1

Performance Optimization of Cloud Application at Software Architecture Level

Xin Du[1,2], Youcong Ni[1,2(✉)], Peng Ye[3], Xin Wang[1], and Ruliang Xiao[1]

[1] College of Mathematics and Informatics, Fujian Normal University, Fuzhou 350117, China
xindu79@126.com

[2] Fujian Provincial Engineering Technology Research Center for Public Service, Fuzhou 350117, China

[3] College of Mathematics and Computer, Wuhan Textile University, Wuhan 430200, China

Abstract. Due to the limited search space in the existing performance optimization approaches at software architectures of cloud applications (SAoCA) level, it is difficult for these methods to obtain the cloud resource usage scheme with optimal cost-performance ratio. Aiming at this problem, this paper firstly defines a performance optimization model called CAPOM that can enlarge the search space effectively. Secondly, an efficient differential evolutionary optimization algorithm named MODE4CA is proposed to solve the CAPOM model by defining evolutionary operators with strategy pool and repair mechanism. Further, a method for optimizing performance at SAoCA level, called POM4CA is derived. Finally, two problem instances with different sizes are taken to conduct the experiments for comparing POM4CA with the current representative method under the light and heavy workload. The experimental results show that POM4CA method can obtain better response time and spend less cost of cloud resources.

Keywords: Cloud application · Software architecture · Performance optimization · Differential evolution

1 Introduction

Performance is not only an important quality attribute of software systems, but also a vital factor to determine success or failure of a system. Cloud platforms [1] including cloud infrastructure platform and cloud service platform, provide scalable resource sharing and management services to cloud applications through virtualization technologies. This can strongly ensure the performance of cloud applications. Performance optimization in the design phase of cloud application development can predict the performance of system earlier to support the construction of cloud resource scheme with high cost-performance ratio.

As an important software artifact, the software architecture of cloud application (SAoCA) [2] not only describes the structural relationships and behavior interactions

K. Li et al. (Eds.): ISICA 2019, CCIS 1205, pp. 724–738, 2020.
https://doi.org/10.1007/978-981-15-5577-0_58

among software components, but also gives deployment schemes of these software components on the virtual machine pool. Compared with the traditional SA [3], SAoCA can model the virtual machine resources used by software components with variable workload at any time by resource deployment scheme. But it also brings greater challenges to performance optimization. With the increase of the scale and complexity of cloud application system, a growing number of virtual machine resource pools are used to deploy the software components. And the types of hardware configuration of virtual machines that are provided for resource pools are also increasing. In addition, the number of virtual machines in each resource pool should dynamically be adjusted according to the time-varying workload, which will lead to the increase of the performance optimization space at SAoCA level. For example, a cloud application uses two resource pools, each of which has 100 different types of available virtual machines. Given that the observation interval of virtual machine resources used by the cloud application is set as one day, the resources are allocated every hour according to the workload and 5000 virtual machines can be used in each resource pool every hour at most. Then the number of usage schemes for cloud resource within one day will also exceed $5.96 * 10^{136}$. How to find the usage schemes of cloud resources with optimal cost-performance ratio in such a huge space has become a key problem in cloud application development.

To solve this problem, some metaheuristic-based performance optimization methods at SAoCA level have emerged based on the existing achievements on performance modeling and evaluation [2, 4–13]. Zheng et al. [14] obtained better response time of cloud application by using ant colony algorithm to optimize resource allocation. To reduce the cost of cloud resources, Pandey [15] applied particle swarm optimization algorithm to optimize cloud resource scheduling. Ardagna et al. [16, 17] proposed a two-stage optimization method to improve search efficiency and quality by combing mixed integer linear programming [18] and Tabu algorithm [19]. At present, most of the existing metaheuristic-based methods abstract the performance optimization problem that satisfying the performance requirements and minimizing the usage cost of cloud resources at SAoCA level into a single objective optimization problem. It leads to the limited search space so that the usage scheme of cloud resource with optimal cost-performance ratio is hard to be obtained. To address the above limitations, we propose a new performance optimization approach at SAoCA level, named POM4CA. The main contributions of this paper are as follows:

(1) A multi-objective performance optimization model at SAoCA level named CAPOM is defined. This model can help to increase the search space and provide support for searching the usage scheme of cloud resource with optimal cost-performance ratio.
(2) A differential multi-objective performance optimization algorithm at SAoCA level named MODE4CA is proposed. This algorithm can efficiently solve CAPOM model by defining the genetic operators with strategy pool and repair mechanism.
(3) Compared with Ardagna method [16, 17] under two cases with different size and different workloads. The results show that the POM4CA method can obtain better response time and spend less cost of cloud resources.

The rest of this paper is organized as follows. Section 2 gives the definition of CAPOM model. Section 3 proposes our MODE4CA algorithm. Section 4 presents our experimental results and analysis. Section 5 presents our conclusions and discusses opportunities for future work.

2 The Multi-objective Performance Optimization Model at SAoCA Level (CAPOM)

This section presents CAPOM, which gives the formal description of relationship among the type and number of virtual machines, the cost of cloud resource usage and response time of cloud application. Notations used by CAPOM is presented in Table 1. Some related definitions are given as follows.

Table 1. The symbols used in CAPOM

Symbol	Explanation
T	The observation period of performance evaluation for a cloud application
t	An observation interval in T
t_j	The jth observation interval in T
n	The number of observation intervals in $T(n = T/t)$
I	The total number of resource pools
V	A set of all the types of available virtual machine. The total number is $\mid$ V $\mid$ and each type is numbered from 1 to $\mid$ V $\mid$
$V_{t_j,i}$	The types of virtual machines for resource pool i in observation interval t_j
$Z_{t_j,i}$	The number of virtual machines in resource pool i in observation interval t_j. The upper bound is expressed as $Z^U_{t_j,i}$
SAoCA_0	Initial SAoCA
$SAoCA_{t_j}$	The SAoCA in observation interval t_j
$rest(SAoCA_{t_j})$	Return the response time of SAoCA in observation interval t_j
$cost(SAoCA_{t_j})$	Return the cost of cloud resource related to SAoCA in observation interval t_j
R_0	The upper bound of response of SAoCA

Definition 1 (Search Space Ω). The Search Space Ω is the set of candidate solutions $\widehat{\chi}$ to solve performance optimization problem at SAoCA level. The definition of $\widehat{\chi}$ is shown in Eq. (1).

$$\hat{\chi} = <<(V_{t_1,1}, Z_{t_1,1}), \cdots, (V_{t_1,i}, Z_{t_1,i}) \cdots, (V_{t_1,I}, Z_{t_1,I}) > , \cdots <V_{t_j,1}, Z_{t_j,1}), \cdots, (V_{t_j,i}, Z_{t_j,i}) \cdots, (V_{t_j,I}, Z_{t_j,I}) > > , 1 \leq j \leq \mathrm{n}, V_{t_j,i} \in \mathrm{V}\,(1 \leq \mathrm{i} \leq \mathrm{I}), 1 \leq Z_{t_j,i} \leq Z^U_{t_j,i} \quad (1)$$

Definition 2 (The usage scheme of cloud resource in observation interval t_j). $\hat{\chi}_{t_j}$ is defined as the usage schemes of cloud resource of candidate solution $\hat{\chi}$ in observation interval t_j. Here $\hat{\chi}_{t_j} = <(V_{t_j,1}, Z_{t_j,1}), \cdots, (V_{t_j,i}, Z_{t_j,i}) \cdots, (V_{t_j,I}, Z_{t_j,I}) >$.

Definition 3 (Solution). The solution $\chi \in \Omega$ of performance optimization problem at SAoCA level should satisfy the constraint defined in Eq. (2).

$$rest_{\chi}^{t_j} = rest(genSAoCA(\hat{\chi}_{t_j}, \mathrm{SAoCA_0})),\ rest_{\chi}^{t_j} \leq R_0,\ 1 \leq j \leq \mathrm{n} \tag{2}$$

In Eq. (2), $rest_{\chi}^{t_j}$ represents the upper bound of the response time of cloud application in observation interval t_j. Function $genSAoCA(\hat{\chi}_{t_j}, \mathrm{SAoCA_0})$ returns the SAoCA_{t_j} by taking $\mathrm{SAoCA_0}$ and $\hat{\chi}_{t_j}$ as inputs.

Definition 4 (Objective Function). Functions $f_{cost}(\chi)$ and $f_{rest}(\chi)$ are two objective functions on the cost of cloud resource usage and the response time of cloud application at SAoCA level. They are defined in Eq. (3) and Eq. (4). The smaller values of $f_{cost}(\chi)$ and $f_{rest}(\chi)$ mean better.

$$f_{cost}(\chi) = \sum\nolimits_{1 \leq j \leq \mathrm{n}} cost(genSAoCA(\hat{\chi}_{t_j}, \mathrm{SAoCA_0})) \tag{3}$$

$$f_{rest}(\chi) = \sum_{1 \leq j \leq n} ((-1) * \gamma(\chi, t_j) * (sd_{Drest}(\chi, t_j))) \tag{4}$$

$$sd_{Drest}(\chi, t_j) = \begin{cases} 1 - \left(\frac{rest_{\chi}^{t_j}}{\mathrm{R_0}}\right) & \mathrm{rest}_{\chi}^{t_j} < \mathrm{R_0} \\ \left(\frac{\mathrm{R_0}}{rest_{\chi}^{t_j}}\right) - 1 & \text{otherwise} \end{cases} \tag{5}$$

$$\gamma(\chi, t_j) = \begin{cases} 1 & \text{if} (sd_{Drest}(\chi, t_j) \leq 0 \vee (j = 1)) \\ \gamma(\chi, t_{j-1}) + 1 & \text{otherwise} \end{cases} \tag{6}$$

In Eq. (5) $sd_{Drest}(\chi, t_j)$ has two situations. When $rest_{\chi}^{t_j}$ satisfies the requirement, the value of $sd_{Drest}(\chi, t_j)$ is greater than 0. It indicates that the bigger values mean higher degree of demand satisfaction. Otherwise, the value of $sd_{Drest}(\chi, t_j)$ is less than 0. It indicates that the smaller values mean higher degree on demand dissatisfaction. $\gamma(\chi, t_j)$ represents the cumulative weighting of the degree of demand satisfaction for $rest_{\chi}^{t_j}$. The initial value of $\gamma(\chi, t_j)$ is set as 1.

Definition 5 (The set of pareto optimal solutions). The set of pareto optimal solutions Opt^* is defined in Eq. (7). The dominance relation is defined in Eq. (8).

$$Opt^* = \{\chi^* \mid \chi^* \in Opt \wedge rest_{\chi^*}^{t_j} \leq R_0 \wedge 1 \leq j \leq \mathrm{n})\},\ Opt = \{\hat{\chi} \mid \nexists \hat{\chi}' \in \Omega \wedge (\hat{\chi}' \prec \hat{\chi})\} \tag{7}$$

$$\hat{\chi}' \prec \hat{\chi} \triangleq (f_{cost}(\hat{\chi}') \leq f_{cost}(\hat{\chi}) \wedge f_{rest}(\hat{\chi}') < f_{rest}(\hat{\chi})) \vee (f_{cost}(\hat{\chi}') < f_{cost}(\chi) \wedge f_{rest}(\hat{\chi}') \leq f_{rest}(\hat{\chi})) \tag{8}$$

Definition 6 (The performance optimization at SAoCA level). The performance optimization at SAoCA level can be abstracted as two-objective optimization problem defined in Eq. (9).

$$\text{Minimize}\, F(\chi) = (f_{cost}(\chi), f_{rest}(\chi)) \tag{9}$$

3 Multi-objective Differential Evolution Algorithm for Performance Optimization at SAoCA Level (MODE4CA)

To solve the CAPOM, a multi-objective differential evolution algorithm for performance optimization named MODE4CA is proposed based on MODE algorithm [20].

3.1 Individual Encoding

The encoding of individual X is randomly generated in the interval t_j according to Eq. (10) and Eq. (11), which is shown in Fig. 1. The odd bit and even bit in an individual X indicate the type and the number of virtual machines, respectively. And the length of individual X is n * I * 2. *ceil* is *a* ceiling function.

$$V_{t_j,i} = ceil(1 + rand(0,1) * (|\mathrm{V}| - 1)) \tag{10}$$

$$Z_{t_j,i} = ceil(1 + rand(0,1) * (Z^{U}_{t_j,i} - 1)) \tag{11}$$

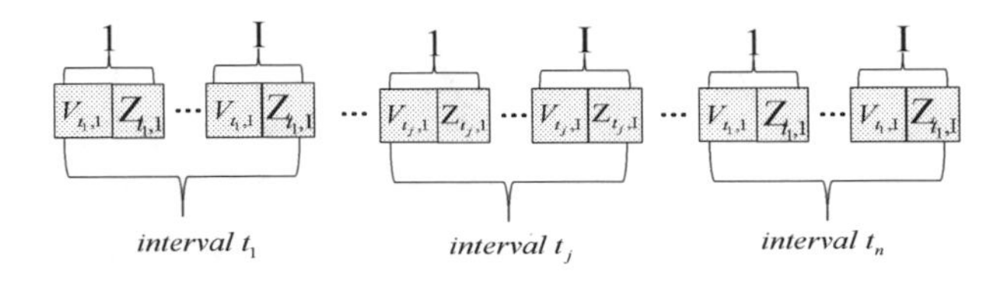

Fig. 1. Individual encoding

3.2 Mutation Operator

The mutation operator adopts single point mutation based on mutation strategy pool and repair mechanism. It includes three calculation steps: generating mutation position, mutation according to selected strategy, check and repair.

Firstly, an integer in [1, $n * I * 2$] is randomly generated as a mutation position of mutant vector X. If this integer is odd, the type of virtual machines will be mutated. Otherwise, the number of virtual machines will be mutated.

Secondly, a strategy is selected from mutation strategy pool [21] shown in Table 2. A new mutant vector X' will be generated by applying the selected strategy to mutant vector X. In Table 2, four kinds of strategies of DE/best/1, DE/best/2, DE/rand/2 and DE/rand/1 are selected to prevent population from getting into local optimum and accelerate convergence rate. In addition, F is the scaling factor.

Table 2. Mutation strategy used in this paper

No.	Differential strategy	Mutation strategy
1	DE/best/2	$X' = X_b + F \cdot (X_{r1} - X_{r2}) + F \cdot (X_{r3} - X_{r4})$
2	DE/best/1	$X' = X_b + F \cdot (X_{r2} - X_{r3})$
3	DE/rand/2	$X' = X_{r1} + F \cdot (X_{r2} - X_{r3}) + F \cdot (X_{r4} - X_{r5})$
4	DE/rand/1	$X' = X_{r1} + F \cdot (X_{r2} - X_{r3})$

Thirdly, if the number of virtual machine type is out of range, X' is repaired according to Eq. (12). Otherwise, it is repaired according to Eq. (13).

$$V'_{t_j,i} = \begin{cases} \text{ceil}\left(\frac{V_{t_j,i}+1}{2}\right), & \text{if } K'_{t_j,i} < 1 \\ \text{ceil}\left(\frac{V_{t_j,i}+|V|}{2}\right), & \text{if } K'_{t_j,i} > |V| \end{cases} \tag{12}$$

$$Z'_{t_j,i} = \begin{cases} \text{ceil}\left(\frac{Z_{t_j,i}+1}{2}\right), & \text{if } Z'_{t_j,i} < 1 \\ \text{ceil}\left(\frac{Z_{t_j,i}+Z^U_{t_j,i}}{2}\right), & \text{if } Z'_{t_j,i} > Z^U_{t_j,i} \end{cases} \tag{13}$$

3.3 Crossover Operator

The crossover strategy pool [21] is composed of binomial distribution strategy and exponential distribution strategy. Firstly, a crossover strategy is randomly selected from crossover strategy pool and a crossover position is decided randomly. Then the trial vector U is generated by crossing the target vector X and mutant vector X'. The process of crossover is similar to that of mutation above mentioned.

3.4 Selection Operator

MODE4CA adopts greedy strategy described in Eq. (14) to select individuals for the next population. In Eq. (14), $\tilde{P}(g)$ represents the gth temporary population. It is set as empty before the evolution of gth population.

$$\tilde{P}(g) = \begin{cases} \tilde{P}(g) \cup X, \text{ if } X \prec U \\ \tilde{P}(g) \cup U, \text{ else} \end{cases} \tag{14}$$

3.5 The MODE4CA Algorithm

The MODE4CA algorithm is shown in Algorithm 1. The truncation method in row 12 for $\tilde{P}(g)$ is same with that in Ref. [20]. They all use non-dominance rank and crowding distance to sort individuals, and then select NP individuals.

Algorithm 1. The MODE4CA algorithm

Input:	Population size *NP*, mutation rate p_m, crossover rate p_c, evolutionary generation t
Output:	The set of pareto optimal solutions Opt^*
1	Evolutionary generation $t \leftarrow 0$
2	Generate initial population $P(t)$ randomly whose size is *NP*
3	Evaluate the fitness value of each individual *X* in $P(t)$
4	While stopping criterion not met do
5	Empty the temporary population $\tilde{P}(t)$
6	for ($i := 1$ *to* NP) do
7	Generate mutation vector X' based on mutation rate p_m
8	Generate trial vector *U* by executing crossover on X' and target vector *X* based on crossover rate p_c
9	Generate temporary population $\tilde{P}(t)$ by selecting individual from *X* and *U*
10	End for
11	If $\|\tilde{P}(t)\| > NP$ do
12	Generate new population whose size is *NP* by truncating $\tilde{P}(t)$
13	End if
14	$P(t+1) \leftarrow \tilde{P}(t)$, and permutate the individual of $P(t+1)$ randomly, $t \leftarrow t+1$
15	End while
16	Solve Opt^* based on definition 5
17	Out Opt^*

4 Experiment

This section presents our experimental study. Specifically, Sect. 4.1 introduces problem instances that we have used. Section 4.2 illustrates the statistical test methods for our analysis. Section 4.3 presents the experimental setup. Section 4.4 gives the research questions we set out to answer and the assessment metrics. Section 4.5 analyzes the experimental results.

4.1 Problem Instances

As shown in Table 3, two problem instances of Meeting in the Cloud(MIC) and Open For Business (OfBiz) with heavy and light workloads are selected from paper [16]. MIC is a social network application with small scale, while OfBiz is a business automation system with large scale.

Table 3. The size of MIC and OfBiz

	MIC	OfBiz
Quantity of components	9	12
Connections	9	9
The number of resource pools	2	2

4.2 Statistical Test

In this paper, POM4CA and Ardagna run 30 times for each problem instance independently. And the nonparametric Wilcoxon rank-sum test [22] is used to check for statistical significance. Wilcoxon test is a safe test to apply since it raises the bar for significance by making no assumptions about underlying data distributions. We set the level of significance α at 0.05.

4.3 Experimental Setup

In the experimental environment of Intel(R) Core (TM) 2, 2.66 GHz processor, 2G memory and ubuntu 16.04.1 operating system, we install and run the experiments. To compare fairly, the parameter settings of cloud resource and compared algorithms are set to same. Table 4 presents the parameter settings of cloud resource in POM4CA and Ardagna [16]. Table 5 gives the parameter settings of POM4CA and Ardagna.

Table 4. Parameter settings of cloud resource in POM4CA and Ardagna

No.	Parameter	Value
1	$\|V\|$	130
2	$Z^{U}_{t_j,i}$	5000
3	n	24

Table 5. Parameter settings of POM4CA and Ardagna

Parameter	Value
Population size	30
Mutation rate	0.6
Crossover rate	0.8
Maximum generations	50
Scaling factor F	[0.5, 1]

4.4 Research Questions and Assessment Metrics

1. **RQ1 (usefulness)**: whether POM4CA can reduce the cost of cloud resources compared with Ardagna while it satisfies the requirement of response time?

Ardagna [16] is a kind of method that can obtain better solution among the methods of performance optimization at SAoCA level. To answer RQ1, usefulness of POM4CA can be validated.

For the quantitative analysis of RQ1, $\mathrm{DR}_{\mathrm{cost}}$ and cost% is designed as two metrics and shown in Eq. (15) and Eq. (16), respectively. In Eq. (15), C_a is the cost of cloud resource usage related to the solution obtained by Ardagna. Opt^* is the optimal Pareto solution set obtained by POM4CA. η_{cost} is a subset of Opt^* and composed of the solutions whose corresponding cost of cloud resource usage is less than C_a. $\mathrm{DR}_{\mathrm{cost}}$ indicates that the proportion of solutions in Pareto solution set obtained by POM4CA that is prior to that of Ardagna with respect to cost of cloud resource usage. In Eq. (16), cost% indicates that the average reduction percentage of cost of cloud resource usage is obtained by POM4CA against Ardagna. The bigger value of cost% means greater reduction percentage of cost.

$$\eta_{cost} = \{\chi | \chi \in Opt^* \wedge \mathrm{f}_{\mathrm{cost}}(\chi) \le C_a\} \quad \mathrm{DR}_{\mathrm{cost}} = \frac{|\eta_{cost}|}{|Opt^*|} \tag{15}$$

$$\mathrm{cost\%} = \sum_{\chi \in \eta_{cost}} \frac{C_a - \mathrm{f}_{\mathrm{cost}}(\chi)}{|\eta_{cost}| * \mathrm{f}_{\mathrm{cost}}(\chi)} \tag{16}$$

2. **RQ2 (effectiveness):** how effective are the proposed POM4CA compared to Ardagna in terms of the response time when POM4CA satisfies the requirement of response time and spends less cost of cloud resource? The effectiveness of POM4CA can be validated by answering RQ2,

To answer RQ2 quantitatively, two metrics of DR_{rest} and rest% are proposed and shown in Eq. (17) and Eq. (18), respectively. In Eq. (17),$\mathrm{R}_{\chi_a}^{t_j}$ are the response time in observation interval t_j based on the solution χ_a obtained by Ardagna. $\mathrm{rest}_{X}^{t_j}$ represents the response time in observation interval t_j based on the solution X obtained by POM4CA. T_X is the set of observation intervals in which the response time corresponding to solution X is better than the solution χ_a. η_{rest} is the set of solutions whose response time is better than $\mathrm{R}_{\chi_a}^{t_j}$ in more than half of observation intervals of T_X. And η_{rest} is a subset of η_{cost}. $\mathrm{DR}_{\mathrm{rest}}$ represents the proportion of the solutions of η_{rest} in η_{cost}. In Eq. (19), rest% indicates that the average reduction percentage of response time is obtained by POM4CA against Ardagna. The bigger value of rest% means greater reduction percentage of response time of system.

$$T_X = \{t_j | 1 \leq j \leq n \wedge \mathrm{rest}_X^{t_j} < \mathrm{R}_{\chi_a}^{t_j}\}, \eta_{rest} = \{X | X \in \eta_{cost} \wedge \bigwedge |\mathrm{T_X}| > \frac{\mathrm{n}}{2}\},$$
$$\mathrm{DR_{rest}} = \frac{|\eta_{rest}|}{|\eta_{cost}|} \tag{17}$$

$$N_t = \sum\nolimits_{X \in \eta_{rest}} |\mathrm{T_X}|, \ \mathrm{rest\%} = \frac{\sum\nolimits_{X \in \eta_{rest}} \sum\nolimits_{t_j \in \mathrm{T_X}} ((\mathrm{R}_{\chi_a}^{t_j} - \mathrm{rest}_X^{t_j}))/\mathrm{rest}_X^{t_j})}{N_t} \tag{18}$$

4.5 Experimental Results and Analysis

(1) Result for RQ1 (Usefulness)

It can be seen from the p-value of Table 6 that POM4CA is significantly better than Ardagna with respect to $\mathrm{DR_{cost}}$ under the premise of ensuring performance requirements. What's more, in Fig. 2, $\mathrm{DR_{cost}}$ is more than 55% on average under light workload and more than 50% on average under heavy workload in MiC case. In OfBiz case, $\mathrm{DR_{cost}}$ reaches 60% on average under light workload and about 50% on average under heavy workload.

Table 6. The p-value of the rank-sum test on the $\mathrm{DR_{cost}}$ when POM4CA compares to Ardagna in two cases of MiC and OfBiz under heavy and light workload

	MiC		OfBiz	
	Light workload	Heavy workload	Light workload	Heavy workload
p-value	6.3403e-05	6.3864e-05	6.3403e-05	6.3864e-05

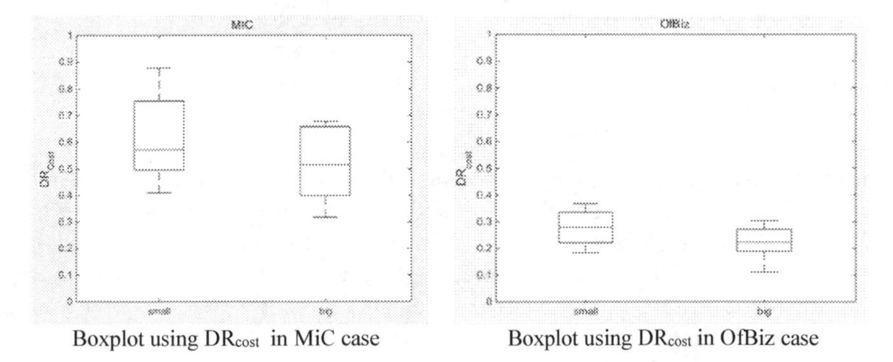

Boxplot using $\mathrm{DR_{cost}}$ in MiC case   Boxplot using $\mathrm{DR_{cost}}$ in OfBiz case

Fig. 2. Boxplot using the $\mathrm{DR_{cost}}$ when POM4CA compares to Ardagna in two cases

Table 7. The cost reduction of POM4CA compared to Ardagna in case of light and heavy load in MiC case and OfBiz case

	MiC		OfBiz	
	Light workload	Heavy workload	Light workload	Heavy workload
$cost_{best}\%$	93.1%	82.5%	37.4%	33.8%
$cost_{avg}\%$	61.5%	50.9%	30.1%	21.1%
$cost_{worst}\%$	18.7%	13.2%	25.0%	17.6%

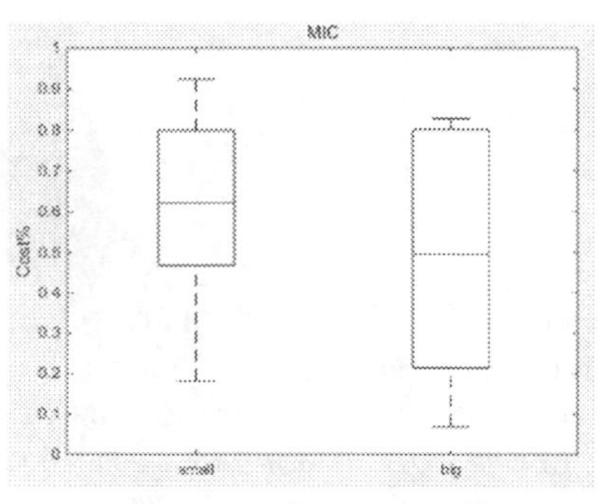

Boxplot using the cost% in MiC

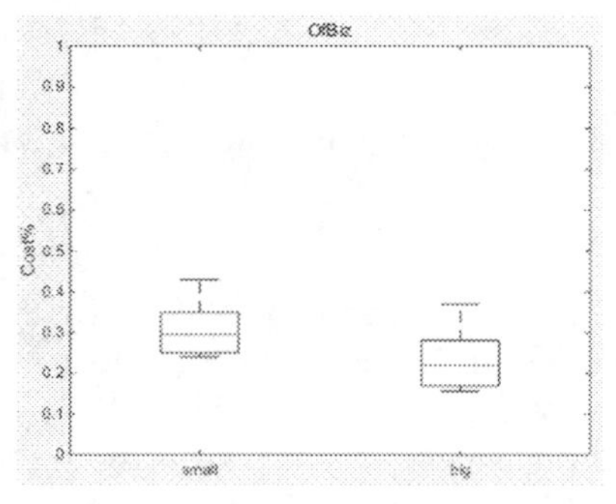

Boxplot using the cost% in OfBiz

Fig. 3. Boxplot using the cost% when POM4CA compares to Ardagna in MiC and OfBiz

In Table 7, $cost_{best}\%$, $cost_{avg}\%$, and $cost_{worst}\%$ are the optimal, average and worst values of average reduction cost of cloud resource usage. In MiC case, $cost_{best}\%$, $cost_{avg}\%$, and $cost_{worst}\%$ are 93.1%, 61.5% and 18.7% under light workload. They are 82.5%, 50.9% and 13.2% under heavy workload, respectively. In OfBiz case, $cost_{best}\%$, $costavg\%$, and $cost_{worst}\%$ are 37.4%, 30.1%, and 25.0% under light workload. They are 33.8%, 21.1% and 17.6% under heavy workload. Figure 3 shows this conclusion more intuitively.

In addition, the reduction amplitude of cloud resource usage cost in MiC is greater than that of OfBiz case. The reason is that it is easier to search for better solutions for the case with smaller size under the same search conditions.

(2) Result for RQ2 (effectiveness)

From the p-value of Table 8, we can see that POM4CA is significantly better than Ardagna in terms of DR_{rest} when the set of solutions obtained by POM4CA meets the performance requirements and has less cost of cloud resource usage. In addition, in Fig. 4, DR_{rest} is over 80% on average under light workload and over 60% on average under heavy load in MiC case. In OfBiz case, DR_{rest} is more than 15% on average under light workload and more than 15% on average under heavy workload.

Table 8. The p-value of the rank-sum test on the DR_{rest} when POM4CA compares to Ardagna methods in two cases of MiC and OfBiz under heavy and light workload

	MiC		OfBiz	
	Light workload	Heavy workload	Light workload	Heavy workload
p-value	6.3403e-05	6.3864e-05	6.3864e-05	6.2944e-05

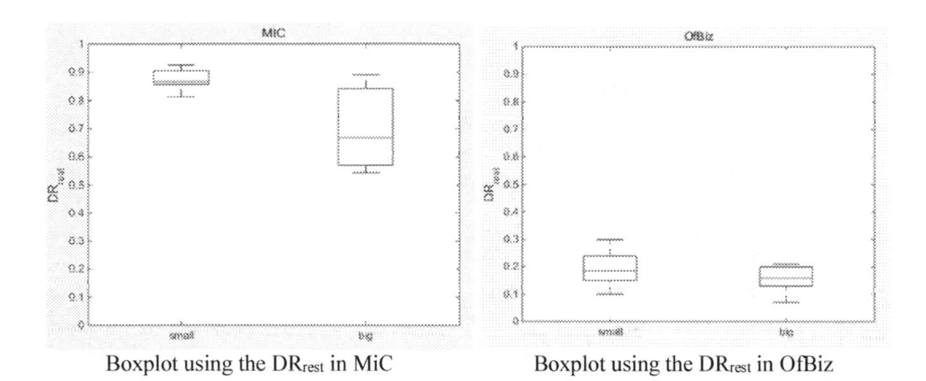

Boxplot using the DR_{rest} in MiC Boxplot using the DR_{rest} in OfBiz

Fig. 4. Boxplot using the DR_{rest} when POM4CA compares to Ardagna in MiC and OfBiz

In Table 9, $rest_{best}\%$, $rcst_{avg}\%$, and $rest_{worst}\%$ are the optimal, average and worst values of average reduction degree of response time. In MiC case, $rest_{best}\%$, $rest_{avg}\%$, and $rest_{worst}\%$ are 76.0%, 55.1% and 40.5% under light workload. They are 76.3%, 65.2% and 49.9% under heavy workload respectively. In OfBiz case, $rest_{best}\%$, $rest_{avg}\%$, and $rest_{worst}\%$ are 22.6%, 21.3% and 18.2% under light workload. They are 20.7%, 17.9% and 15.6% under heavy workload respectively. Figure 5 shows this conclusion more intuitively.

Table 9. The reduction degree of response time in POM4CA against Ardagna

	MiC		OfBiz	
	Light workload	Heavy workload	Light workload	Heavy workload
$rest_{best}\%$	76.0%	76.3%	22.6%	20.7%
$rest_{avg}\%$	55.1%	65.2%	21.3%	17.9%
$rest_{worst}\%$	40.5%	49.9%	18.2%	15.6%

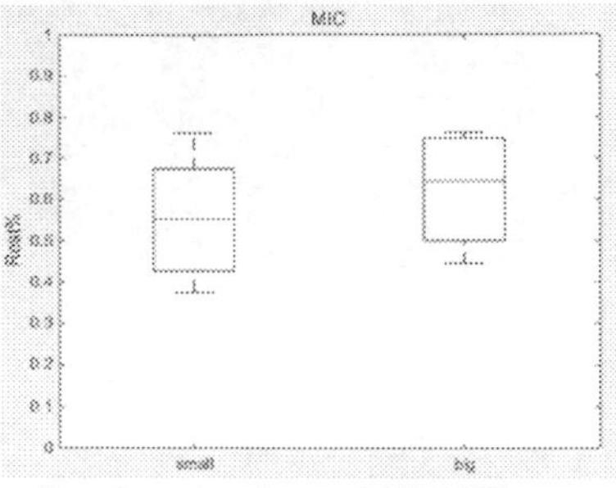

Boxplot using the rest% in MiC case

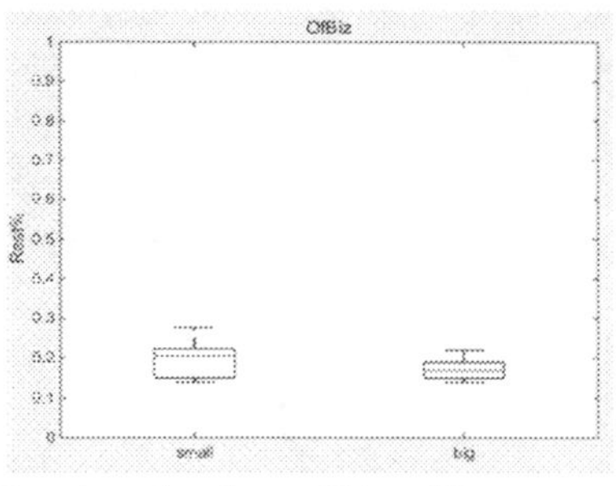

Boxplot using the rest% in OfBiz case

Fig. 5. Boxplot using the rest% when POM4CA compares to Ardagna in MiC and OfBiz

5 Conclusion and Future Work

This paper proposes a performance optimization method POM4CA for cloud application at SAoCA level. Firstly, a performance multi-objective optimization model CAPOM is defined to increase the search space to find high cost-performance solutions. Then a differential multi-objective evolutionary algorithm named MODE4CA is designed. MODE4CA efficiently solves the CAPOM model by designing mutation operators and crossover operators with strategy pool and repair mechanism. The experimental results show that the POM4CA can achieve better response times and lower cloud resource usage costs than those of Ardagna.

In the future, we will introduce surrogate model to effectively reduce the high computational cost of performance evaluation during the evolutionary process.

Acknowledgments. This work is supported by the Royal Society International Exchanges (IE151226), the Natural Science Foundation of Fujian Province (No. 2017J01498), Talent support program of High school in the new century of Fujian Province(Year 2017), the project of Fujian Normal University(The division of Fujian Province [2016] 14), the Natural Science Foundation of Hubei Province (No. 2018CFB689).

References

1. Cusumano, M.: Cloud computing and SaaS as new computing platforms. Commun. ACM **53**(4), 27–29 (2010)
2. Franceschelli, D., Ardagna, D., Ciavotta, M., Di Nitto, E.: SPACE4CLOUD: a tool for system performance and costevaluation of cloud systems. In Proceedings of the 2013 International Workshop on Multi-cloud Applications and Federated Clouds, New York, NY, USA, pp. 27–34 (2013)
3. Taylor, R.N., Medvidovic, N., Dashofy, E.M.: Software Architecture: Foundations, Theory, and Practice. Wiley Publishing, Chichester (2009)

4. Guillén, J., Miranda, J., Murillo, J.M., Canal, C.: A UML profile for modeling multicloud applications. In: Lau, K.-K., Lamersdorf, W., Pimentel, E. (eds.) ESOCC 2013. LNCS, vol. 8135, pp. 180–187. Springer, Heidelberg (2013). https://doi.org/10.1007/978-3-642-40651-5_15
5. Bacigalupo, D.A., et al.: Managing dynamic enterprise and urgent workloads on clouds using layered queuing and historical performance models. Simul. Model. Pract. Theory **19** (6), 1479–1495 (2011)
6. Ardagna, D., Casale, G., Ciavotta, M., Pérez, J.F., Wang, W.: Quality-of-service in cloud computing: modeling techniques and their applications. J. Internet Serv. Appl. **5**(1), 1–17 (2014). https://doi.org/10.1186/s13174-014-0011-3
7. Núñez, A., Vázquez-Poletti, J.L., Caminero, A.C., Castañé, G.G., Carretero, J., Llorente, I. M.: iCanCloud: a flexible and scalable cloud infrastructure simulator. J. Grid Comput. **10**(1), 185–209 (2012). https://doi.org/10.1007/s10723-012-9208-5
8. Chauhan, M.A., Babar, M.A., Sheng, Q.Z.: A reference architecture for a cloud-based tools as a service workspace. In: 2015 IEEE International Conference on Services Computing (SCC), pp. 475–482 (2015)
9. Alhamazani, K., Ranjan, R., Mitra, K., Rabhi, F., Jayaraman, P.P., Khan, S.U., Guabtni, A., Bhatnagar, V.: An overview of the commercial cloud monitoring tools: research dimensions, design issues, and state-of-the-art. Computing **97**(4), 357–377 (2014). https://doi.org/10.1007/s00607-014-0398-5
10. Merkel, D.: Docker: lightweight linux containers for consistent development and deployment. Linux J. **2014**(239), 2 (2014)
11. Kumar, R., Jain, K., Maharwal, H., Jain, N., Dadhich, A.: Apache CloudStack: Open Source Infrastructure as a Service Cloud Computing Platform. ResearchGate, Berlin (2014)
12. Sefraoui, O., Aissaoui, M., Eleuldj, M.: OpenStack: toward an open-source solution for cloud computing. IJCA **55**(3), 38–42 (2012)
13. Critical Evaluation on jClouds and Cloudify Abstract APIs against EC2, Azure and HP-Cloud - IEEE Conference Publication. https://ieeexplore.ieee.org/document/690318. Accessed 27 Oct 2019
14. Hua, X., Zheng, J., Hu, W.: Ant colony optimization algorithm for computing resource allocation based on cloud computing environment. J. East China Normal Univ. (Nat. Sci.) **1** (1), 127–134 (2010)
15. Pandey, S., Wu, L., Guru, S.M., Buyya, R.: A particle swarm optimization-based heuristic for scheduling workflow applications in cloud computing environments. In: 2010 24th IEEE International Conference on Advanced Information Networking and Applications (AINA), pp. 400–407 (2010)
16. Ardagna, D., Gibilisco, G.P., Ciavotta, M., Lavrentev, A.: A Multi-model optimization framework for the model driven design of cloud applications. In: Le Goues, C., Yoo, S. (eds.) SSBSE 2014. LNCS, vol. 8636, pp. 61–76. Springer, Cham (2014). https://doi.org/10.1007/978-3-319-09940-8_5
17. Ciavotta, M., Ardagna, D., Gibilisco, G.P.: A mixed integer linear programming optimization approach for multi-cloud capacity allocation. J. Syst. Softw. **123**, 64–78 (2017)
18. Kondili, E., Pantelides, C.C., Sargent, R.W.H.: A general algorithm for short-term scheduling of batch operations—I. MILP formulation. Comput. Chem. Eng. **17**(2), 211–227 (1993)

19. Glover, F.: Tabu Search—Part I. ORSA J. Comput. **1**(3), 190–206 (1989)
20. Zhang, J., Sanderson, A.C.: JADE: adaptive differential evolution with optional external archive. IEEE Trans. Evol. Comput. **13**(5), 945–958 (2009)
21. Mendes, R., Mohais, A.S.: DynDE: a differential evolution for dynamic optimization problems. In: 2005 IEEE Congress on Evolutionary Computation, Edinburgh, Scotland, UK, vol. 3, pp. 2808–2815 (2005)
22. Arcuri, A., Briand, L.: A practical guide for using statistical tests to assess randomized algorithms in software engineering. In: Proceedings of the 33rd International Conference on Software Engineering, Waikiki, Honolulu, HI, USA, pp. 1–10 (2011)

Mining and Analysis of Big Data Based on New Energy Public Transit

Yinxin Bao[1] and Quan Shi[1,2](✉)

[1] School of Information Science and Technology, Nantong University, Nantong, China
sq@ntu.edu.cn
[2] School of Traffic, Nantong University, Nantong, China

Abstract. With the rapid development of new energy vehicles, urban public transport vehicles are gradually dominated by new energy public vehicles. This article collects and analyzes various data in the system of new energy bus, and mines the basic information needed for the Intelligent bus system. It also analyzes and integrates related data, explores the operation characteristics and laws of new energy buses, and establishes a new energy bus data mining and analysis system to provide important decision-making basis for the operation and management of new energy buses. This paper combines big data processing technology with visualization technology to realize the new energy public transportation data visualization system. This system can provide bus operators with the basis for optimization of public transportation network, which greatly improves the management level of public transportation enterprises and the operating efficiency of public transportation vehicles.

Keywords: Big data of new energy bus · Operational analysis · Data mining · Data visualization

1 Introduction

Nowadays, sustainable development has been strongly advocated by the government, urban public transport vehicles are gradually transitioning from fuel vehicles to new energy vehicles. The advantages of new energy public vehicles are low pollution, high utilization rate, low noise, convenient maintenance and high safety performance, which is the trend of public transportation development in the future [1]. New energy buses have abundant data resources. As long as relevant data are used and data collection and analysis are done well, the basic information needed by the intelligent bus system can be mined. By analyzing the data information from different sources and integrating them organically, the characteristic values and rules related to bus operation can be found out, which can provide important decision-making basis for the operation and management of new energy bus system.

K. Li et al. (Eds.): ISICA 2019, CCIS 1205, pp. 739–746, 2020.
https://doi.org/10.1007/978-981-15-5577-0_59

2 Data Acquisition and Data Preprocessing

2.1 New Energy Bus Data Collection

At present, data collection technologies commonly used on buses include: bus card swiping device, bus track positioning device, passenger flow counter, etc. Data collection methods are as follows.

1) **New energy bus credit card data collection method**
   At present, the one-vote system is more commonly used in China. The passengers complete the card swiping through the handheld IC card on the bus, and each time the card is swiped, the passenger swipe data is generated at the card swiping machine terminal [2, 3].
2) **New energy bus trajectory data collection method**
   By carrying a positioning system, the new energy bus can send real-time information such as vehicle position coordinates, speed, running state and running direction to the dispatching center [4].
3) **New energy bus passenger flow data collection method**
   The new energy buses put into operation in Nantong are equipped with passenger flow counters, which can collect the arrival time of the bus and the data of the upper and lower passenger flow at the station [5].

2.2 New Energy Bus Data Preprocessing

This section mainly preprocesses the data of passenger swipe card data, bus track data, passenger flow counter data, station and line data of new energy bus.

The spark SQL in spark is mainly used for data preprocessing of passenger swiping card data to eliminate the error location in the original data. The data preprocessing of the bus location data is mainly based on the location of the bus, considering that the bus runs in the urban area of Nantong City, so the data out of this range are eliminated [6]. Because the passengers on the bus will walk back and forth, resulting in repeated counting of the passenger flow counter, considering that the standard number of passengers in Nantong city is 35 to 43, so the data of the number of passengers exceeding the standard number is eliminated. Analyze and process station, line number, up and down lines, station sequence and positioning data. Through data analysis, this kind of data mainly exists the situation of latitude data missing.

3 New Energy Bus Data Analysis and Mining

3.1 Data Analysis

1) **Analysis of passenger flow data**

Nantong City passenger flow analysis is mainly aimed at the analysis of bus card data and passenger flow counter data. Firstly, the bus card data is analyzed according to the line, operation time and bus card data statistics. The card swiping data of the public transport card is counted according to the card swiping personnel and the public transport operation line, among which the card swiping personnel are divided into three types: student card, senior card and other cards. Because of the prepossessing of the original card swiping data, the original card swiping data is divided into IC card, M1 card and citizen card. As shown in Fig. 1, spark SQL is used to analyze the processed data, and python calls Matplotlib library to display the card reading data [7, 8].

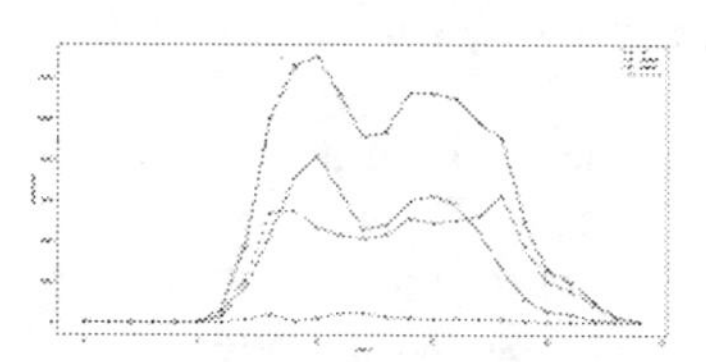

Fig. 1. Swipe statistics

2) **Trajectory data analysis**

Due to the complexity of the road conditions, the bus arrival time will deviate from the original planned arrival time. When this happens, it will lead to the aggregation of buses departing at different times in the line and reduce the efficiency of the bus operation line [9]. In this paper, the GPS track data of new energy bus is used to analyze the phenomenon of bus aggregation, and a reasonable model is established. The aggregation calculation of new energy bus in Nantong is carried out. Figure 2 shows the traffic flow on a line in Nantong on October 1, 2018. The more dense the number of points, the more bus aggregation occurs.

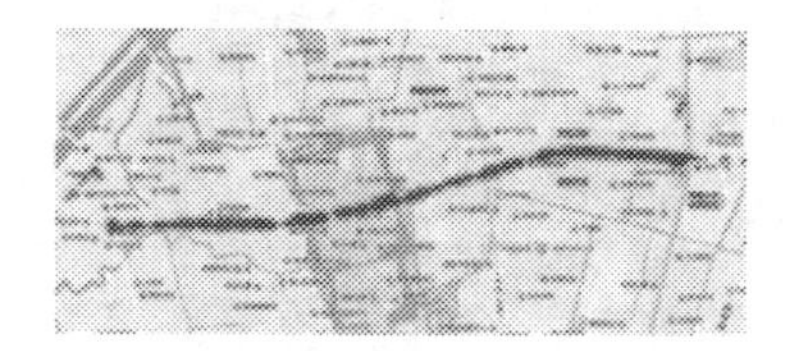

Fig. 2. A line conflict situation on October 1, 2018

3) **Line network analysis**

In order to better plan the operation scheme of Nantong new energy bus, this paper makes a comprehensive evaluation of the existing network and routes of Nantong city, and selects three indexes of site coverage rate, site density and non-linear coefficient to analyze.

Bus station coverage rate is an important basis to judge whether the allocation of regional public transport resources is reasonable. In general, site coverage is defined as the coverage radius of a bus stop within a specified area of 300 m or 500 m. The coverage area can be obtained by calculating the total coverage area of the site divided by the total area of the specified area. Nantong bus station information table data has 1660, this paper calculates the site coverage of each site. The result shows that the service coverage of 300 m in Nantong bus station is about 64.5%, which is much higher than 50% of the national public tram passenger service standard.

The site density is the ratio of the total number of stations parked by the bus to the total area of the city. This feature reflects the distribution of the bus station under the current area and its adaptation. Using Nantong urban station site data, statistics were made on all bus stations in Nantong City, and the density of Nantong bus stations reached 2.295.

The nonlinear coefficient is the ratio of the actual distance of a line to the first and last distance of a line. The larger the nonlinear coefficient, the larger the service area of the line. The smaller the coefficient is, the smaller the service area of the line is. If the service area is too small, it can be solved by adding bus lines.

3.2 Data Mining

Accurate prediction of bus arrival time can not only significantly improve the operational efficiency and service level of public transportation, but also significantly promote the development of public transportation. This chapter will study the arrival time prediction of Nantong New Energy Bus and explore how to establish a prediction model for the arrival time of new energy buses.

1) **Model establishment**

Based on the study of RNN recurrent neural network, this paper establishes the GRU neural network model to complete the prediction of bus arrival time. Different from the traditional LSTM, the GRU model only has two gates, which do not control and retain internal memory. Therefore, the model is simpler and more efficient than LSTM. Figure 3 shows the GRU network model diagram used in this paper [10].

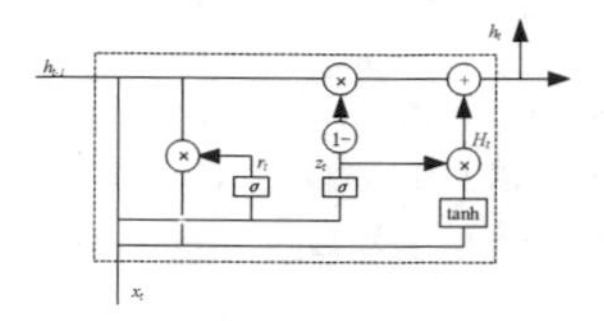

Fig. 3. GRU network model diagram

2) **Model calculation**
Data of different categories need to be processed by different standardized methods. For fixed types of data defined by category labels such as car number, week, and weather, zero mean method can be used for standardized processing. The formula is shown in Eq. (1).

$$m^* = \frac{(x - \mu)}{\sigma} \tag{1}$$

In Eq. (1), m^* represents the normalized data, x represents the pre-processed data, μ represents the sample mean with characteristics, and σ represents the standard deviation of the sample.

In this project, Lasso algorithm is mainly used to filter the original eigenvalues of new energy bus, as shown in Eq. (2).

$$\widehat{L}_{Lasso} = \text{argmin}_\beta \sum\nolimits_{i=1}^{n} (m_i - L_0 - \sum\nolimits_{j=1}^{p} x_{ij}\beta_j)^2 \tag{2}$$

In Eq. (2), x_{ij} is the j-th variable of the i-th group of eigenvalues. Let row vector L be the regression coefficient in the eigenvalue and m be the training label.

In Fig. 3, this paper analyzes the network element of the layer. When the time step is step t, the update gate calculates the data through the Eq. (3).

$$z_t = \sigma\left(W^{(z)}x_t + U^{(z)}h_{t-1}\right) \tag{3}$$

In Eq. (3), x_t represents the t-th component of the input value x, $W^{(z)}$ represents the weight matrix, h_{t-1} holds the information of the previous time step, and linearly transforms through the weight matrix $U^{(z)}$.

The purpose of the update gate is to add it and convert the result through the Sigmoid function to compress the activation result between 0 and 1. The update gate determines how much historical data is passed to the future, reducing the risk of gradients disappearing. The reset gate determines the forgetting process for the data and can be expressed by Eq. (4).

$$r_t = \sigma\left(W^{(r)}x_t + U^{(r)}h_{t-1}\right) \tag{4}$$

3) **Model training**

After model training, Fig. 4 shows the Loss function curve of GRU network model under Tensorboard environment. It can be seen that the model rapidly converges during training, and the Loss value function tends to be stable after the 10th cycle.

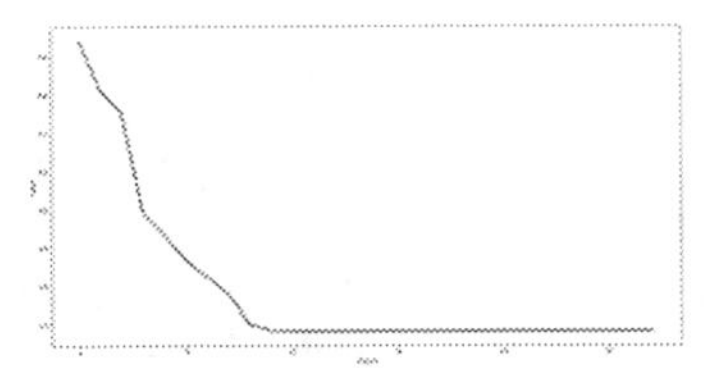

Fig. 4. Loss function curve

4) **Model fitting**

After the model training is completed, this paper compares four days of bus operation data with the predicted results. Figure 5 shows the fitting curve of GRU model prediction and real data. As shown in the figure, the fitting effect is relatively ideal. By using the linear regression fitting index r-square as the model evaluation index, the fitting degree of GRU neural network is 94.6138%.

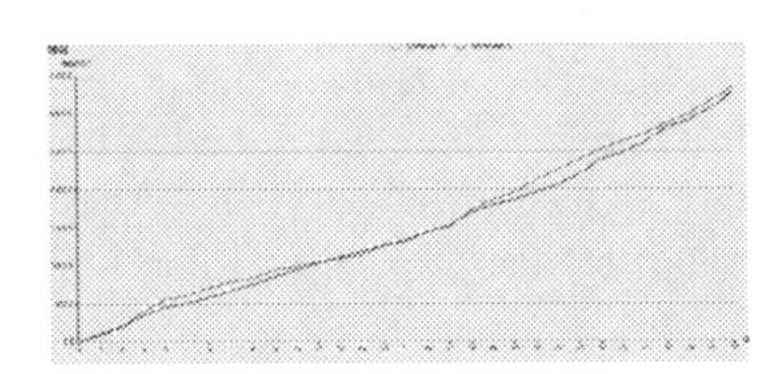

Fig. 5. GRU model prediction and real data fitting curve

4 Realization of New Energy Public Transport Visualization Platform

The core functions of the system are divided into the following parts: bus line visualization module, bus station visualization module, bus passenger flow visualization module and bus arrival station prediction module.

In the bus line visualization module, the line query function is designed, as shown in Fig. 6(a), to provide line query service for users. When the user wants to access a place, he can get whether the current line is accessible. The bus station visualization module realizes the visual query function of Nantong bus station. This module shows through the thermal map of the website that the dark sites are more densely distributed on the map, whereas the light sites are more loosely distributed on the map. Passenger flow visualization module is a visual analysis of bus passenger flow status. The system can divide passenger card data into elderly card, student card and ordinary card according to the type. This module can show the bus swiping card and the number of passengers in Nantong city. The bus arrival prediction module shown in Fig. 6(b) details the bus arrival prediction model based on GRU neural network. This system can accurately predict the bus arrival time.

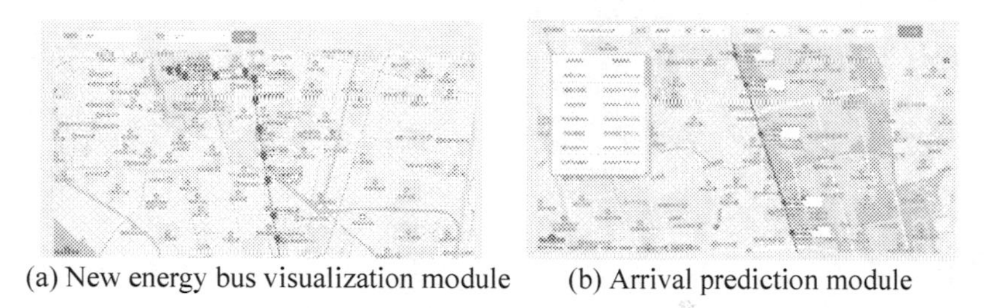

(a) New energy bus visualization module (b) Arrival prediction module

Fig. 6. Visual module

5 Conclusion

This paper aims to analyze the problems and deficiencies existing in the current public transportation system through the visual analysis and data mining of new energy public transportation data. On the basis of studying the theoretical basis of big data processing technology, data mining, analysis and processing, and data visualization, this paper analyzes the operation and route indicators of new energy buses by combining bus passenger swipe card data, bus network data, and bus positioning data. The visualization of new energy bus data mining system is realized to provide decision support for bus operators.

Acknowledgments. The National Natural Science Foundation of China (61771265), The Science and technology Project of Nantong (CP12017001).

References

1. Xingqiang, C., Ke, T., Yuechuan, C., et al.: Research on big data decision support platform for new energy public transport vehicles. Urban Public Transp. **12**(03), 25–29 (2017)
2. Qunyong, Z., Keyun, S., Zhijie, Z.: Time and space analysis of bus passenger flow based on massive ic card data. J. Guizhou Univ. (Nat. Sci.) **35**(06), 93–98+105 (2018)
3. Ting, Y., Guilian, F., Xinghui, Ma.: Bus passenger flow calculation based on bus IC card information. Traffic Eng. **18**(06), 51–56 (2018)
4. Andeng, Y.: Bus od calculation based on bus GPS and IC card data. Harbin Institute of Technology (2017)

5. Zhouquan, W.: Research on temporal and spatial distribution of bus passenger flow based on IC card data and GPS data. Southwest Jiaotong University (2016)
6. Andrienko, N., Andrienko, G., Rinzivillo, S.: Leveraging spatial abstraction in traffic analysis and forecasting with visual analytics. Inf. Syst. **57**(02), 172–194 (2016)
7. Zhu, J.: Research on data mining of electric power system based on Hadoop cloud computing platform. Int. J. Comput. Appl. **41**(4), 22–24 (2019)
8. Chaolong, J., Hanning, W., Lili, W.: Research on visualization of multi-dimensional real-time traffic data stream based on cloud computing. Procedia Eng. **137**(8), 24–26 (2016)
9. Perveen, S., Kamruzzaman, M., Yigitcanlar, T.: What to assess to model the transport impacts of urban growth? A Delphi approach to examine the space–time suitability of transport indicators. Int. J. Sustain. Transp. **13**(8), 31–34 (2019)
10. Ke, K., Hongbin, S., Chengkang, Z., Brown, C.: Short-term electrical load forecasting method based on stacked auto-encoding and GRU neural network. Evol. Intell. **12**(3), 2445 (2019)

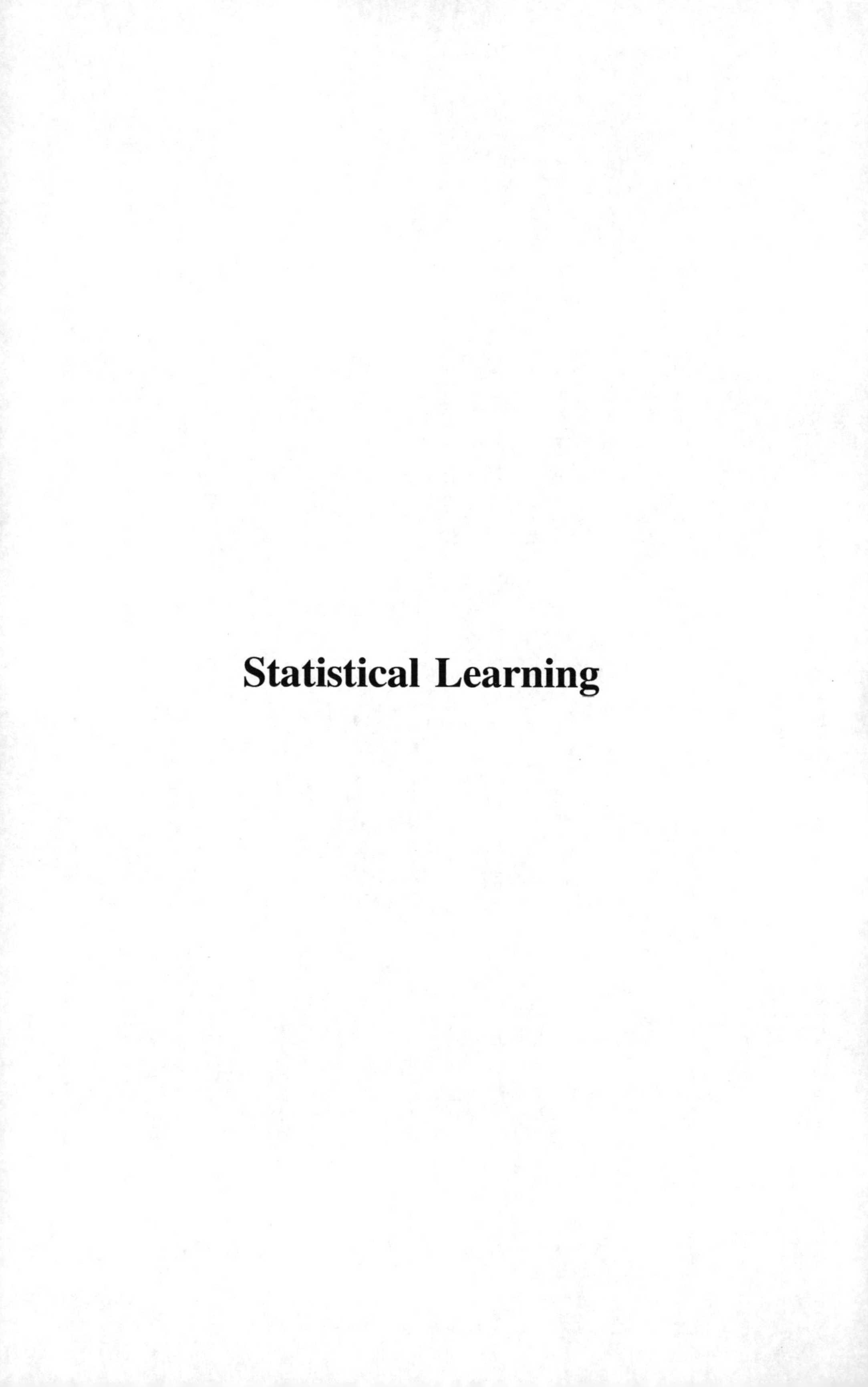

Statistical Learning

The Network Design of License Plate Recognition Based on the Convolutional Neural Network

Xingzhen Tao[1(✉)], Fahui Gu[2], and Shumin Xie[1]

[1] College of Information Engineering, Jiangxi College of Applied Technology, Ganzhou, China
348627805@qq.com

[2] Office of Academic Studies, Guangdong Vocational and Technical College, Guangzhou 341000, Jiangxi, China

Abstract. Due to the generation of data sets and the rapid improvement of GPU computing performance, in-depth learning has undergone qualitative changes and development in the past decade. Various excellent convolutional neural network models have been implemented and verified, which accelerates the application of the convolutional neural network in various fields. A license plate recognition based on the convolutional neural network is pro-posed for the application of large underground parking lots. An end-to-end identification network framework without segmentation characters is designed. At the same time, sequence information is added to the convolutional neural network for improving the license plate recognition rate. Compared with the existing step-by-step license plate detection and recognition method, the joint solution of a single network can avoid the error accumulation in the intermediate process. At the same time, it can improve the accuracy rate, save the recognition time, accelerate the vehicle entering and leaving time and avoid traffic congestion.

Keywords: Convolutional neural network · License plate recognition · End-to-end identification network

1 Introduction

The phenomenon of "difficult parking and disorderly parking" on the ground is getting worse and worse, which has seriously affected the urban environment and traffic order. It is undoubtedly the most effective way to solve this problem by vigorously promoting the construction of urban underground parking lots through digging tapping the potential, making a renovation and starting a new construction. In the traditional underground parking lot, there are vehicle railings at the entrance and exit. When entering the parking lot, vehicles need to be intercepted manually for registration or card taking registration. When leaving the parking lot, the vehicles can be removed only after completing the manual payment, which is inefficient. In recent years, the intelligent recognition system is widely used in underground parking, but the parking lot gate is usually installed at the underground exit, which has the problems of dim

K. Li et al. (Eds.): ISICA 2019, CCIS 1205, pp. 749–758, 2020.
https://doi.org/10.1007/978-981-15-5577-0_60

light, large slope, narrow access road and so on. Therefore, the correct recognition rate and time of vehicle license plates are very important.

2 Related Work

The intelligent license plate recognition system is an indispensable part of intelligent underground parking lots [1], which is of great significance to realize automatic management of parking lots. The accurate recognition rate and recognition time of license plate are important indicators to consider. The main problems of license plate recognition technology include license plate location and character recognition [2]. The steps of a common license plate recognition method are shown in Fig. 1.

Fig. 1. The steps of license plate recognition.

2.1 License Plate Location

For license plate location, the currently used techniques are divided into two categories [3]. The first category is based on feature extraction methods and can be divided into gray feature based methods and color feature based methods. The second category is based on machine learning methods. It can be divided into a method based on pattern recognition and a method based on deep learning. The document [4] uses the unique color of the license plate to locate the license plate. However, when the light is weak, the resolution is low and the license plate picture is unclear, so the recognition rate of the license plate will be affected. The gray scale feature of license plate area contains enough information for license plate location, but the license plate location method based on gray scale feature requires a lot of calculation time. The method of license plate location based on edge and color assistance proposed by Abolghasemi et al. increases the comparison of similar regions of license plates, thus avoiding erroneous location of license plate regions, but this method is especially affected by illumination.

In recent years, due to the development of the big data and computer capability, and the rapid development of deep learning [5–7], license plate detection based on depth learning has been applied in intelligent license plate recognition system and achieved good results. Mainstream target recognition frameworks include Faster R-CNN [8, 9]. The document designed a license plate detector based on YOLO [10] and achieved good recognition accuracy. The document [11] reduces the difficulty of image recognition by preprocessing the image, and then locates the license plate through CNN network. These methods regard the license plate as an approximate rectangle. Although it has high accuracy for license plate recognition in complex environment, there is a large amount of redundancy of background information for the recognition results of tilted license plates, which will cause great interference for subsequent character cutting and recognition.

2.2 Character Segmentation

The domestic standard license plate contains 7 characters, the first character is Chinese characters, the second is uppercase English characters, and the last 5 characters are a mixture of numbers and letters. It is a key problem to How to accurately and effectively identify the 7 characters segmented by vertical projection is a key problem. Currently, the commonly used license plate character segmentation technologies include projection-based license plate character segmentation method and stroke-based license plate character segmentation method [12]. The traditional template matching method has a simple process and is greatly disturbed by additional factors, such as uneven illumination or shape change of the obtained image, so its recognition rate and robustness are not high. The projection-based license plate character segmentation method is simple and effective. Once the license plate area is determined, the characters can be segmented according to the length and height of the characters; however, such methods have poor adaptability, low resolution, and very poor detection effect for license plate images with poor light. The license plate character segmentation method based on stroke construction has a good adaptability to noise, dust, angle and other conditions [13]. However, this kind of algorithm is very complex, and its running efficiency is low, so it cannot reach the level of real-time application. The license plate character detection is a key part of license plate recognition system. The main task is to separate the characters in the license plate one by one for subsequent license plate character recognition. The effect of license plate character detection directly determines the effect of license plate character recognition. However, at the entrance and exit of the underground parking lot, it is a difficult problem to accurately segment each character in the license plate picture because the license plate is affected by factors such as illumination, angle change of the license plate, dust cover on the license plate, etc.

2.3 The License Plate Recognition

After the license plate characters are segmented, the license plate characters need to be identified. The method based on template matching is widely used in the license plate character recognition technology. However, this method is susceptible to factors such as the segmentation results of license plate characters, the noise of license plate region, and whether the license plate region is clear or not. The support vector machine (SVM) classifier is also used for the license plate character recognition [14]. The SVM-based license plate character recognition method needs to specify the extracted features, and the selection of different features will directly affect the effect of the license plate character recognition. In addition, there are also methods based on BP neural network [15]. The license plate character recognition method based on BP neural network has high learning ability, but it has the defects of complex structure and slow convergence. In recent years, methods based on depth learning have also been applied to the license plate character recognition, but they are all based on license plate character segmentation, which requires relatively long recognition time. The accuracy of the license plate character recognition depends on the effect of license plate character segmentation. For all the license plate recognition methods which are based on license plate character segmentation, the vast majority of experiments are carried out in an ideal environment.

When applied to underground parking lots, the accuracy of the license plate recognition will be greatly reduced due to weak illumination, slope inclination and other conditions.

2.4 The End-to-End Identification

With the development of the in-depth neural network and the improvement of data sets, the direct whole recognition method of license plate without character segmentation has been continuously proposed. Li et al. considered the license plate recognition as a sequence labeling problem and the convolutional neural network (CNNs) is used to extract a series of feature vectors from the license plate boundary frame in the way of sliding window [16]. Then RNN with CTC loss is used to label sequence data, and finally the recognition result without dividing characters is obtained. The recognition accuracy of the model is high, but it is time consuming.

3 System Network Model

Previous research works on the license plate detection and recognition usually divides the detection and recognition of license plates into two sub-tasks, which are solved by different methods. Before the completion of identification, the license plate should be detected firstly so as to find the position of the license plate on the map, and latter to divide the license plate character [17]. actually, the two sub-tasks of the license plate detection and recognition are highly correlated in nature [18]. The accurate license plate detection helps to improve the recognition accuracy, and accurate recognition results can be used to eliminate false alarms during the detection process. Therefore, this paper proposes a unified neural network framework, which unifies the license plate detection and recognition tasks into one framework. Through a deep neural network, the two sub-tasks of license plate location and recognition can be completed simultaneously, which will improve the license plate recognition efficiency. The specific network structure is shown in the Fig. 2.

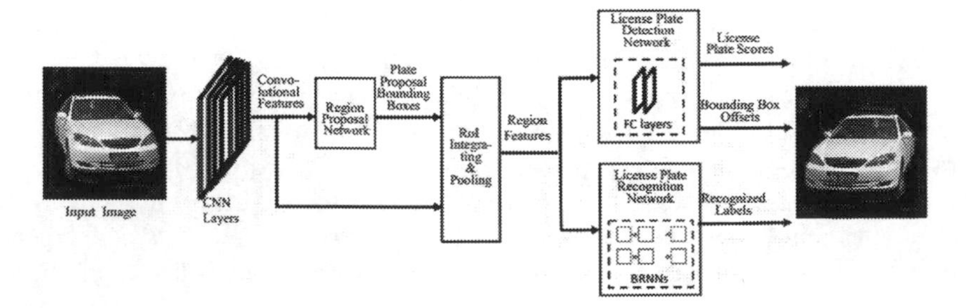

Fig. 2. The specific network structure.

3.1 Model Network Structures

As shown in Fig. 2 that the overall structure of our model consists of several convolutional layers, a region proposal network for license plate proposals generation, proposal integrating and pooling layer, multi-layer perceptrons for plate detection and bounding box regression, and RNNs for plate recognition. Given an input RGB image with a single forward evalution, the net-work outputs scores of predicted bounding boxes being license plates, bounding box offsets with a scale-invariant translation and log-space height/width shift relative to a proposal, as well as the recognized license plate labels at the same time. The extracted region features are used by both detection and recognition, which not only shares computation, but also reduces model size.

The license plate detection and recognition algorithm based on depth neural networks proposed in this paper can complete the task of license plate detection and recognition at the same time. In the network structure, the convolution features of the shallow layer and the convolution features of the deep layer are simultaneously utilized, thus improving the expressiveness of the convolutional neural network and further improving the recognition accuracy of the model. The model in this paper can be trained end-to-end and the license plate detection and recognition share a set of shallow convolution feature extraction networks. Therefore, the model can greatly reduce the amount of computation in the previous convolution feature extraction process, improve the recognition efficiency and speed up the vehicle entering and leaving time.

1) CNN Layers: shallow feature extraction networks. The residual network in ResNet is used to reduce the calculation and parameter amount, and the residual learning structure can be realized through forward neural network +shortcut connection, as shown in the structure diagram. Moreover, the shortcut connection is equivalent to simply performing equivalent mapping, which will not generate additional parameters or increase computational complexity. Moreover, the entire network can still undergo end-to-end back propagation trainings.

2) Candidate box generation networks: The RPN network layer is proposed in the Faster R-CNN network model designed by Ren et al. This is a full volume network, which inputs an image of any size and outputs a set of candidate boxes [19]. The generated candidate frames are sent to the training network of Fast R-CNN for target detection and accurate positioning. After inputting a feature map, The RNP uses a sliding window of n * n (where n is 3, i.e. a sliding window of 3 * 3). Each sliding window is mapped to a low-dimensional feature (ZF network is used in this paper, and the feature is 256-d). This feature is input into two full connection layers, one is used for border regression and the other is used for classification. The position of each sliding window is simultaneously predicted for multiple areas. The maximum number recommended for each region is k (where k is 9), so the border regression layer has 4k outputs and the classification layer has 2k outputs. The value output by the classification layer cls indicates that the candidate frame is the foreground or the background, while the border regression layer reg outputs the coordinates and offsets of the candidate frame. The RPN network structure is shown in the Fig. 3.

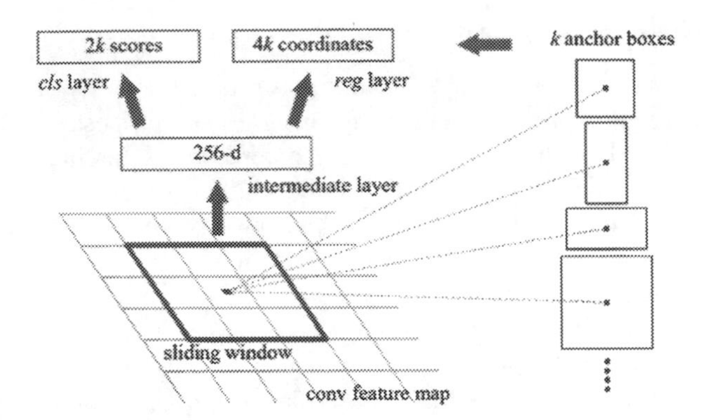

Fig. 3. The RPN network structure.

According to the size and aspect ratio of the China's license plate, three different scales (128, 192, 256) and two different aspect ratios (3, 2) are designed, corresponding to each position on the convolution map. The k (k = 6) anchor of different scales and aspect ratios generated in advance can be obtained through the RPN network. The license plate classification layer outputs 2K scores, corresponding to the probability of each anchor frame at each position being a license plate. Reg layer outputs 4K values, which respectively correspond to the offset coefficient of the nearest calibration frame at each anchor frame. Given an anchor frame with a center of (xa, ya) and a length and width of wa and ha respectively, the four values output by the regression layer are (tx, ty, tw, th), where tx, ty is the scaling factor and tw, th is the logarithmic value of the offset at the anchor frame center point. The new position calculation formula of anchor frame after passing through regression layer is as follows:

$$\mathrm{x} = x_a + t_x w_a \quad \mathrm{y} = y_a + t_y h_a$$

$$\mathrm{w} = w_a \exp(t_w) \quad \mathrm{h} = h_a \exp(t_h)$$

Where x and y are the coordinates of the center point of the anchor frame after the regression, and w and h are its width and height. Both the cls layer and reg layer will receive their own loss function and give the value of the loss function. At the same time, they will give back-propagation data according to the result of derivation. The loss function is calculated as follows:

$$\mathrm{L}(\{p_i\}, \{\mathrm{t}_i\}) = \frac{1}{N_{cls}} \sum\nolimits_i L_{cls}(p_i, p_i^*) + \lambda \frac{1}{N_{reg}} \sum\nolimits_i p_i^* L_{reg}(t_i, t_i^*)$$

$$t_x = (x - x_a)/w_a \quad t_y = (y - y_a)/h_a$$

$$t_w = \log(w/w_a) \quad t_h = \log(h/h_a)$$

$$t_x^* = (x^* - x_a)/w_a \quad t_y^* = (y^* - y_a)/h_a$$

$$t_w^* = \log(w^*/w_a) \quad t_h^* = \log(h^*/h_a)$$

3) The region normalization: The RoI Pooling operation is performed for each region candidate frame on the feature layer to obtain a fixed size feature representation. The RoI pooling layer can significantly accelerate training and testing, improve detection accuracy, and allow end-to-end trainings of the target detection system. The specific operation of ROI pooling is as follow:

a) According to the input image, the ROI is mapped to the corresponding position of the feature map;
b) Divide the mapped area into sections of the same size (the number of sections is the same as the dimension of the output);
c) Perform a max pooling operation on each section.

4) The License Plate Detection Network: The goal of the license plate detection network is to determine whether the candidate Region of Interests is a license plate and refine the coordinates of the candidate to complete the location of license plates.

5) The License Plate Recognition Network: The target of license plate recognition network is to recognize the characters of the Region of Interests based on the extracted features. In order to avoid the challenges brought by the license plate segmentation, the license plate recognition problem is regarded as a sequence labeling problem. A bidirectional RNN network based on BLSTM + CTC is proposed in the system model, as shown in the figure. The convolution layer proposes a feature sequence from the input picture and inputs it to the bidirectional regression neural network layer. in the feature sequence x = x1, x2, … xT, the regression layer predicts the distribution yi of each frame label xi. Finally, the preliminary prediction is converted into a tag sequence through a transcription layer. In order to solve the problem that the length of the character sequence directly recognized by CNN is much smaller than the input feature frame sequence, the CTC loss function is added at the end of the network. A blank is added to the label symbol set, and then the label is marked with RNN. Finally blank symbols and predicted repeated symbols are eliminated (Fig. 4).

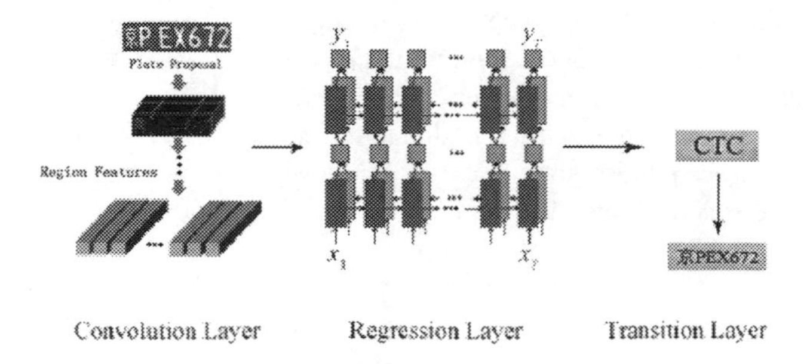

Fig. 4. The BLSTM + CTC system model.

4 The Experiments and Conclusions

4.1 The Database

The experimental data in this paper are provided by OpenITS, it is provided by Sun Yat-Sen University. The open dataset VRID for vehicle re-identification contains 10,000 images, which are captured by 326 surveillance cameras within 14 days. The resolutions of images are distributed from 400×424 to 990×1134. VRID collects 1000 vehicle IDs (vehicle identities) of top 10 common vehicle models (Table 1) to reconstruct the interference with the same vehicle model in the real world. The vehicle IDs belong to the same model have very similar appearance and their differences appears in the area of the logo and accessories. Besides, each vehicle IDs contains 10 images which are in various illuminations, poses and weather condition.

Table 1. The 10 vehicle models in the dataset

Vehicle model	Vehicle IDs	Total images
Audi_A4	100	1000
Honda_Accord	100	1000
Buick_Lacrosse	100	1000
Volkswagen_Magotan	100	1000
Toyota_Corolla_I	100	1000
Toyota_Corolla_II	100	1000
Toyota_Camry	100	1000
Ford_Focus	100	1000
Nissan_Tiida	100	1000
Nissan_Sylphy	100	1000

4.2 The Experimental Result

The experimental running environment is Win10 64 bit operation system, Python language is used for program editing, and the end-to-end license plate recognition neural network is constructed. Firstly, input data to train the neural network, get the classification model of convolution neural network according to the output results, then input the test data set to test the recognition results, and finally output the correct rate and recognition time of network recognition. The recognition accuracy rate of end-to-end license plate recognition neural network proposed in this paper is 95.4%, and the time is 200 ms. Compared with the document [14], the recognition accuracy is improved and the recognition time is greatly reduced (Table 2), which is conducive to speeding up the speed of vehicles entering and leaving the underground parking lot.

Table 2. The results of two algorithms

Method	Performance (%)	Speed (per image) (ms)
Ours (End-to-End)	95.4	300
Document [14] (SVM)	94.6	643

4.3 Conclusions

The system model proposed in this paper can be used for end-to-end training and complete the license plate detection and recognition tasks simultaneously in one time. The entire framework does not require artificially extracted features, all of which are obtained by deep neural networks through learning, and can be trained end-to-end, greatly shortening the training time of the model. The license plate detection and license plate recognition tasks are combined together, and the problem of license plate character segmentation is avoided at the same time. Compared with the two separate sub-tasks, the method is more efficient, and the character recognition is no longer affected by the character segmentation results.

Acknowledgement. This work was supported by the Key Research and Development Project of Ganzhou,the name is “Research and Application of Key Technologies of License Plate Recognition and Parking Space Guidance in Intelligent Parking Lot”.

References

1. Mohandes, M., et al.: Preference-based smart parking system in a university campus. IET Intell. Transp. Syst. **13**(2), 417–423 (2019)
2. Saha, S.: A Review on Automatic License Plate Recognition System (2019)
3. Bei, C., Cao, W., He, Y.: An automated vision system for vehicle license plate localization, segmentation in real life scene. Int. J. Pattern Recogn. Artif. Intell. **32**(7), 1855010 (2017)
4. Wang, C.Q., et al.: The research on locating license plate of vehicle based on color conversion and projection. In: Advanced Materials Research, vol. 655–657, pp. 769–772 (2013)
5. Namysl, M., Konya, I.: Efficient, lexicon-free OCR using deep learning (2019)
6. Yao, Z., Yi, W.: Bionic vision system and its application in license plate recognition. Nat. Comput. **3**, 1–11 (2019)
7. Li, H., et al.: Reading car license plates using deep neural networks. Image Vis. Comput. **72**, 14–23 (2018). S0262885618300155
8. Ren, S., et al.: Faster R-CNN: towards real-time object detection with region proposal networks. IEEE Trans. Pattern Anal. Mach. Intell. **39**(6), 1137–1149 (2017)
9. Jacobs, I.S., Bean, C.P.: Fine particles, thin films and exchange anisotropy. In: Rado, G.T., Suhl, H. (eds.) Magnetism, vol. III, pp. 271–350. Academic, New York (1963)
10. Redmon, J., et al.: You only look once: unified, real-time object detection (2015)
11. Laroca, R., et al.: A robust real-time automatic license plate recognition based on the YOLO detector, pp. 1–10 (2018)
12. Lu, Y., Shridhar, M.: Character segmentation in handwritten words—an overview. Pattern Recogn. **1**(1), 77–96 (1996)

13. Khamdamov, R.Kh., Rakhmanov, H.E.: A character segmentation algorithm for vehicle license plates. Cybern. Syst. Anal. **55**(4), 649–654 (2019). https://doi.org/10.1007/s10559-019-00173-0
14. Khan, M.A., et al.: License number plate recognition system using entropy-based features selection approach with SVM. IET Image Process. **12**(2), 200–209 (2018)
15. Liu, W.B., Wang,T.: The character recognition of vehicle's license plate Based on BP neural networks. In: Applied Mechanics and Materials, vol. 513–517, 3805–3808 (2014)
16. Hui, L., Peng, W., Shen., C.: Toward end-to-end car license plate detection and recognition with deep neural networks. IEEE Trans. Intell. Transp. Syst. **20**(3), 1126–1136 (2019)
17. Yang, X., Wang, X.: Recognizing License Plates in Real-Time (2019)
18. Chen, L.W., Ho, Y.F.: Centimeter-grade metropolitan positioning for lane-level intelligent transportation systems based on the internet of vehicles. IEEE Trans. Ind. Inf. **15**(3), 1 (2019)
19. Nabati, R., Qi, H.: RRPN: Radar Region Proposal Network for Object Detection in Autonomous Vehicles (2019)

Research on Intelligent Algorithm for Image Quality Evaluation Based on Image Distortion Type and Convolutional Neural Network

Lei Deng[1(✉)], Fahui Gu[2], and Shumin Xie[1]

[1] College of Information Engineering, Jiangxi College of Applied Technology, Ganzhou 341000, Jiangxi, China
qjdenglei@163.com, 459955101@qq.com
[2] Office of Educational Administration, Guangdong Vocational and Technical College, Foshan 528041, Guangdong, China
gufahui@139.com

Abstract. In order to detect different images under different distortion conditions, this paper classifies image distortion types and proposes video image distortion detection and classification based on convolutional neural networks. By segmenting the input video image to obtain a small block image, and then using the active learning feature of the convolutional neural network, the positive and negative cases are equalized, the adaptive learning rate is slowed down and the local minimum problem is solved. The type of distortion is mainly predicted by the SoftMax classifier image, and then the video image prediction type obtained by the majority voting rule is used. The objective quality evaluation algorithm based on image distortion type and convolutional neural network is analyzed. Using LIVE (simulation standard image library) and actual monitoring video library, the final accurate results of the two performance tests are not much different. The overall classification accuracy rate is significantly higher than other algorithms. After the positive and negative case equalization and adaptive learning rate are introduced, it is found that the CNN classification accuracy can be significantly improved. The final test results also confirm that this method can actively learn image quality features, and improve the accuracy of video image classification detection. It is applicable to image quality detection of all video image distortion conditions, and has strong practicability and robustness.

Keywords: Image distortion type · Convolutional neural network · Image quality · Intelligent algorithm

With the widespread adoption of video surveillance systems and the rapid growth of the size of network video users, video images are increasingly used to represent and communicate information. However, the quality of video images may introduce various distortion problems due to equipment damage, camera shake or environmental factors, such as excessive or too dark anomalies caused by improper exposure setting. The introduction of these distortions will hinder people. Obtaining information from video images, especially in video surveillance systems, can cause serious distortions in monitoring. Therefore, in order to monitor and optimize the video image acquisition,

K. Li et al. (Eds.): ISICA 2019, CCIS 1205, pp. 759–771, 2020.
https://doi.org/10.1007/978-981-15-5577-0_61

transmission and processing, it is important to study the distortion detection of video images and classify them for timely processing. Video image distortion detection can be divided into subjective detection and objective detection. Subjective detection is the most reliable method, but it is susceptible to objective conditions and subjective emotions, resulting in unsatisfactory and time-consuming test results. Therefore, objective detection methods are more favored in practical applications. Currently, there are some objective detection methods designed for different types of distortion. In the past, the Sobel operator based edge energy sharpness detection algorithm, image analysis based color shift detection method and adaptive noise evaluation algorithm based on noise point neighborhood pixel gray value were used to blur and color the video image respectively. And noise is detected. Some researchers have proposed some image processing-based algorithms to detect the blur, color cast, stripes, snowflakes, and wrong screens of the images in the surveillance video. Wang Zhengyou et al. used a spread function (LSF) to obtain a point spread function (PSF) to detect defocused images. The above algorithms can only detect specific image distortion problems. When the distortion type of the video image changes, a new algorithm needs to be selected. Therefore, it is necessary to study a detection algorithm commonly used for all types of video image distortion types.

The above algorithms can only detect specific image distortion problems. When the distortion type of the video image changes, a new algorithm needs to be selected. Therefore, it is necessary to study a detection algorithm commonly used for all types of video image distortion types. Here we briefly review the existing related methods and analyze them. Fang et al. proposed a quality evaluation method for contrast distortion without reference, which is based on some statistical characteristics of natural images, including mean, variance, skewness, peak and Information entropy. Although these two methods can predict the quality of the image, there are certain deficiencies. For example, the characteristics of the two methods are artificially designed, and the number of features is also large, so that the calculation time complexity is high. It is easy to cause problems such as over-fitting. Conversely, in this paper, our proposed method is based on convolutional neural networks. By reasonably designing convolutional neural networks and training networks, we can automatically learn quality-related features, thus achieving higher prediction performance, and the network. The model can be directly applied after one training, and the calculation time complexity is greatly reduced, designed to provide reference for similar research.

1 Convolutional Neural Network Contrast Distortion Image Quality Evaluation Method

1.1 Positive and Negative Example Equalization

The positive and negative case equalization method can artificially increase the data volume of the sample, avoiding the decrease of the classification accuracy rate due to the uneven distribution of the sample, and is the simplest and most commonly used method for preventing over-fitting. The actual surveillance video library used in the experiment contains 1640 normal video, 455 fuzzy abnormal video, 122 color cast abnormal video, 23 over bright abnormal video and 40 over dark abnormal video. Since the distribution of different types of samples is not uniform, the positive and negative equilibrium of the monitoring video library is realized by using two affine transformations of Ar (rotation) and As (size scaling). It mainly includes the following processes:

(1) Calculate the ratio of the positive sample number Np and the negative sample number Nn:

$$\rho = \frac{N_\rho}{N_n} \tag{1}$$

When the value of p is within a reasonable interval, the sample distribution can be considered to be more uniform, and the classification performance of the algorithm will not be affected. According to the experimental results, this reasonable interval is set: $\tilde{\rho} = [\frac{\sqrt{2}}{2}, \sqrt{2}]$.

(2) At the $\rho \in \tilde{\rho}$ time, the operation is stopped; otherwise, the sample is subjected to positive and negative case equalization. K denotes a meta-sample, and K' denotes a new sample after K undergoes affine transformation:

$$K' = A_r A_s K \tag{2}$$

$A_r = \begin{bmatrix} \cos\theta, \sin\theta \\ -\sin\theta, \cos\theta \end{bmatrix} (\theta \in [0, 2\pi]), A_S = \begin{bmatrix} s, 0 \\ 0, s \end{bmatrix} (s \in [\frac{\sqrt{2}}{2}, \sqrt{2}])$, θ And s indicates the rotation angle and scaling.

(3) θ And s Repeat step (2) with random values until the end $\rho \in \tilde{\rho}$.

1.2 Local Contrast Normalization

Local contrast normalization can avoid supersaturation of neurons, enhance the generalization of the network, effectively eliminate the influence of brightness and contrast variance on the network, and greatly reduce the dependence between adjacent factors. Before the network training, this paper firstly compares the extracted image blocks with local contrast, and sets the brightness value of the image (i, j) to I(i, j), and the localized normalized brightness value is I' (i, j), the normalization method can be expressed as:

$$I'(i,j) = \frac{I(i,j) - \mu}{\sigma + C} \tag{3}$$

In the formula: $i \in \{1, 2, \ldots, M\}$, $j \in \{1, 2, \ldots, N\}$, M and N are the dimensions of the image block respectively, C is a constant value of 1, and the denominator is avoided to be 0, μ, σ which is the mean and standard deviation of the image block pixel values, respectively.

1.3 LRel Function

Previous literature studies have suggested that cortical neurons are rarely in maximum activation and that the activation function can be modeled using the ReL function. A neuron with a nonlinear characteristic of the ReL function is called a ReLU. The use of ReLU in deep neural networks makes it easier for networks to obtain sparse representations of data, achieve convergence faster, and reduce training time. However, the ReLU function has a potential disadvantage: Since the gradient-based optimization algorithm does not adjust the unactivated neuron weights, when a neuron is inactive, its gradient value is 0, resulting in this nerve. Yuan will never be activated. Therefore, there are previous literatures that propose LReL functions based on the ReL function, as shown in the following formula:

$$h^{(i)} = \begin{cases} w^{(i)T}x, & w^{(i)T}x > 0 \\ a \times w^{(i)T}x, & w^{(i)T}x \leq 0 \end{cases} \tag{4}$$

Where: a is a small non-zero constant, which makes the LReL function still obtain a smaller non-zero gradient value when the neuron is inactive, which enhances the robustness of the optimization algorithm. When a is 0, the ReLU function and the LReL function are equivalent. In this study, the LReL function is used as the activation function of the convolution layer, taking a = 0.01.

1.4 Training Network

The weights $\theta = [W1, W2, \ldots, Wl]$ in the convolutional neural network can be learned by minimizing the cost function. The mathematical expression of the cost function is:

$$J(\theta) = -\frac{1}{n}\sum_{i=1}^{N}[\sum_{j=1}^{K} 1\{y^{(i)} = j\} \log \frac{e^{\theta_j^T x^{(i)}}}{\sum_{l=1}^{K} e^{\theta_j^T x^{(i)}}}] \tag{5}$$

Where: the training sample size is N; the total number of sample categories is K, $1\{\cdot\}$ represents the indicative function, 1 is the true proposition value, and 0 is the false proposition value. $x^{(i)}$ and $y^{(i)}$ the input image block and its label. In the network training process, a random gradient descent algorithm and a backpropagation algorithm are used to minimize the cost function and update the weight. The momentum term is introduced to speed up the learning. The initial value of the momentum term $P_s = 0.9$. After 10 training, the linear decrease $P_s = 0.5$ remains unchanged. The ε_0 learning rate is initialized to 0.1. According to the adaptive change of the loss function value, the learning rate is reduced by half when the cost function value enters the lag phase. The adaptive learning rate formula is expressed as:

$$\varepsilon^t = \begin{cases} 0.5 \times \varepsilon^{t-1}, & L^t \geq L^{t-1} \\ \varepsilon^{t-1}, & L^t < L^{t-1} \end{cases} \tag{6}$$

The training method is batch processing, and the batch processing amount is 100 weight update:

$$W^t = W^{t-1} + \Delta W^t \tag{7}$$

The last fully connected layer of the convolutional neural network uses the dropout method, which randomly sets the output value of the hidden unit to 0 with a probability of 0.5 when training the network, and hides the output of the unit when predicting the category of the test sample with the trained network. The value is halved. Using the dropout method can reduce the complex adaptability between neurons, forcing neurons to learn more robust features and prevent over-fitting of neural networks. At the same time, with different dropout masks, dropout can achieve a model averaging in deep neural network structures. After the network training is completed, according to the majority voting rule, the prediction class of the input image is determined by the majority of the image blocks corresponding to the input image.

2 Image Quality Objective Evaluation Algorithm Based on Image Distortion Type and Convolutional Neural Network

2.1 Convolutional Neural Network Design

In recent years, convolutional neural networks have achieved great success in dealing with computer vision problems, greatly improving the performance of prediction. In image quality evaluation, quality evaluation algorithms based on convolutional neural networks have also been proposed to verify convolution. Neural networks are used for the effectiveness of quality assessment questions. The convolutional neural network designed in this paper, the network input is an image, including three convolutional pooling layer units and three fully connected layers, respectively as 1 ~ 3 convolutional layer, 1 ~ 3 pooling layer, 1 ~ 3 Fully connected layer, the network output is a fraction, indicating the quality of the image. In the network, we use the ReLU function as the activation function of the neuron, and the formula is as follows:

$$C_{j+1} = \max(0, W_j \otimes C_j + B_j) \tag{8}$$

In the formula: extract the feature graph representation in the j-layer network C_{j+1}, the j-th layer network weight representation W_j; the offset parameter representation B_j, the convolution operation representation $\otimes$; the maximum function represents max, and the pooling method we select the maximum pooling, i.e.:

$$C_{j+1} = \max_R C_j \tag{9}$$

Where: a pooled area of the feature map represents R; the maximum value in the valued area represents: C_{j+1}.

The specific configuration information of each layer of the network (see Table 1), as can be seen from the table, the input is a gray image block with a size of 32 × 32, and the output is the output of the last fully connected layer and the fully connected layer., a value that represents the quality of the image.

Table 1. Specific configuration information of each layer of the convolutional neural network.

Layer name	Filling	Filter size	Number of filters	Stride	Output size
Input	0				32 × 32 × 1
Convolution layer 1	0	5 × 5	48	1	28 × 28 × 48
Pooling layer 1	0	2 × 2		2	14 × 14 × 48
Convolution layer 2	0	5 × 5	64	1	10 × 10 × 64
Pooling layer 2	0	2 × 2		2	5 × 5 × 64
Convolution layer 3	0	4 × 4	80	1	2 × 2 × 80
Pooling layer 3	0	2 × 2		2	1 × 1 × 80
Fully connected layer 1	0	1 × 1	256	1	1 × 1 × 256
Fully connected layer 2	0	1 × 1	128	1	1 × 1 × 128
Fully connected layer 3	0	1 × 1	1	1	1 × 1 × 1

2.2 Convolutional Neural Network Training

2.2.1 Preparing Training Samples

After the convolutional neural network structure is defined, we train the network to have the ability to evaluate quality. We first need to prepare the training samples and the quality labels corresponding to the samples. We use the existing contrast distortion image library CID2013 to train us. The network model covers a total of 400 contrast-distortion images in CID2013. Each image corresponds to a subjective quality score of 768 × 512 resolution. By dividing each image into 32 × 32 blocks, 384 image blocks are finally obtained. The 400 frames can obtain 153,600 image blocks, and the SSIM value between each of the distorted image blocks and its original image block is calculated, and the quality of the distorted image block is marked, so that the convolutional neural network can be trained.

2.2.2 Network Training

After preparing the training data, we import the training data into the convolutional neural network to train the network. First, use the norm to define the loss function of the network training:

$$J = \sqrt{\|f(x, W) - y\|^2} \tag{10}$$

Where: x is the input training image block, W represents the weight of the network, x quality mark represents y, which is the SSIM value, f(x, W) represents the input x, and the convolutional neural network output value is minimized J, Train the entire convolutional neural network. Here, the gradient descent (SGD) backpropagation algorithm is mainly used, and the weight update process formula is as follows:

$$\Delta_{i+1} = m \times \Delta_i - \eta \times \frac{\partial L}{W_i^j} \quad (11)$$

$$W_{i+1}^j = W_i^j + \Delta_{i+1} - \lambda \times \eta \times W_i^j \quad (12)$$

2.2.3 Image Quality Prediction

After the convolutional neural network is trained, the network is used to evaluate the quality of the new contrast-distorted image. Since the training uses a 32 × 32 image block size, the new image is divided into non-overlapping 32 × 32 image blocks, and then each image block is input into the network, and the network will give the quality score of the image block., denoted as qi, which represents the output quality of the i-th image block. Assuming that the image has a total of N image blocks, then we use the mass mean of all image blocks to define the quality of the entire image, indicating:

$$Q = \frac{1}{N}\sum_{i}^{N} q_i \quad (13)$$

3 Test Results and Analysis

3.1 Experimental Data

In the research of image quality evaluation, three statistical indicators are usually used to evaluate the prediction performance of the proposed algorithm, which are Spearman rank correlation coefficient (SROCC), Kendall rank correlation coefficient (KROCC), and Pearson linear correlation coefficient (PLCC). SROCC, KROCC and PLCC are the consistency of the quality score given by the calculation algorithm and the subjective quality score given by the tester. The closer the consistency is to 1, the higher the performance of the algorithm.

The actual monitoring video library comes from the security city's safe city project, which includes 1640 normal video, 455 fuzzy abnormal video, 122 color cast abnormal video, 23 over-bright abnormal video, and 40 dark abnormal video. Excessive and too dark anomalies in the video library are caused by improper exposure settings, and subjective detection is easier to discriminate. In the color cast video, there are many kinds of color cast problems such as reddish "yellow", "green" and "blue", which are caused by the error between the image captured by the camera and the real color of the object. The cause of the blurred video is the most complicated, by motion. Fuzzy "defocus blur" gray layer occlusion and other factors are generated. In the experiment, the video image is generated from the actual monitoring video library video, there are different resolutions, and the positive and negative samples are equalized by one analysis above. Obtain a more evenly distributed experimental sample (see Fig. 1) as part of the video image of the actual surveillance video library.

Fig. 1. Actual surveillance video library video image. Note: Normal video, over-bright video, over-dark video, partial color video, blurred video.

3.2 Evaluation Method

According to the traditional method of image classification, this paper calculates the classification error between the predicted category and the real category of the image, and uses the confusion matrix to obtain the classification accuracy of various images and all images, in order to analyze the experimental results. In this paper, the experimental samples are randomly divided into three subsets, of which 60% are training sets, 20% are validation sets, and the remaining 20% are test sets. Using cross-validation training, the training set, verification set, and test set are absolutely isolated from the image content.

3.3 Simulation Standard Image Library Experimental Results Analysis

In order to more intuitively describe which types are easily confused when sorting, the Simulation Standard Image Library (LIVE) test focuses on the confusion matrix between classifications of various types. The confusion matrix is listed as the algorithm classification prediction type, the behavior is true distortion type, and the value in the matrix represents the confusion probability between the distortion types on the corresponding row and column (see Table 2). Compare the method and BIQI, DIIVINE, and BRISQUE to the various types of distorted images in the simulation standard image library test and the overall classification accuracy. The classification accuracy of this method is better than that of the other three methods. The experimental results show that the proposed method can overcome the influence of image content on image distortion detection, and actively learn image quality related features, and has good recognition performance for various types of distorted images.

Table 2. Classification accuracy of different algorithms for LIVE library test sets.

Method	Image type					
	JP2K	JPEG	WN	GBLUR	FF	ALL
BIQI	86.38	88.49	92.61	68.77	66.68	80.82
DIIVINE	80.01	80.11	100.00	90.01	73.45	83.76
BRISQUE	82.91	88.90	100.00	96.77	83.36	88.61
This article	88.15	90.93	100.00	97.38	84.02	92.23

3.4 Analysis of Experimental Results of Actual Monitoring Video Library

In this paper, the confusion matrix of classification prediction between various types of images in the actual monitoring video library test set is compared with the above. The method of introducing positive and negative example equalization method and adaptive learning rate before and after this method, various video images and overall classification Accuracy (see Table 3). It can be found that the method has good subjective consistency for the classification and detection of video images, and the overall classification accuracy rate reaches 92.88%. After introducing the positive and negative case equalization method and the adaptive learning rate, the network performance is greatly improved. The experimental results show that the proposed method can achieve better classification and detection results for the actual surveillance video images, and has strong practicability.

Table 3. Actual monitoring video library test set classification accuracy rate.

Method	Image type					
	Normal	Over bright	Too dark	Color cast	Blurry	ALL
This paper has no positive and negative case equalization	86.68	75.03	80.05	65.89	86.69	78.69
There is no adaptive learning rate in this paper	92.00	95.46	94.48	93.46	75.08	90.45
This article	96.18	96.45	95.68	94.15	87.59	92.89

3.5 Parameter Comparison Experiment Results Analysis

Through the control variable method, the experiment uses the actual monitoring video library to detect the influence of the size of the image block and convolution kernel on the classification performance of the proposed method. The image block mainly includes two aspects of the size of the image block and the number of image blocks.

(1) Image block size. The experiment compares the effect of different size image blocks on the classification accuracy. The results show that if there is no overlapping sampling, the number of image blocks will change with the size of the image block. The larger the size of the image block, the more the number Less, does not meet the principle of control variable law. Therefore, in the experiment, the overlapping sampling method is adopted, and the sampling step size is fixed to 32, and the influence of the edge of the image is removed, and the size of the image block is changed, and the number thereof remains substantially unchanged. When the size of the block is 56×56, the classification accuracy is up to 92.88%.
(2) The step size of the sampling. In order to observe the influence of different numbers of image blocks on the network performance, the size of the image block is fixed to 56×56 in the experiment, and different number of image blocks are obtained by changing the sampling step size. The overall classification accuracy rate shows that the sampling step size is 32. When the sampling step size is 32, the overall classification accuracy is the highest.
(3) Convolution kernel size. The effect of the size of the convolution kernel on the performance of the algorithm was tested without changing the other structures of the algorithm. The experimental results show that under different convolution kernel conditions, the overall classification accuracy of the algorithm is higher, and the classification accuracy is smaller with the size of the convolution kernel. The experimental results show that the size of the convolution kernel has little effect on the performance of the algorithm.

Although FTQM only refers to part of the original information, we can speculate that this part of the information can be used to evaluate the image quality of contrast distortion. On the contrary, the full reference method can refer to all the information, which contains a lot of redundant information. The information is ineffective for contrast distortion, so the overall evaluation performance is degraded. Observing the non-reference evaluation method, the performance of all non-reference methods is reduced. The proposed method exceeds the FANG method, and the accuracy is improved to close to 0.5, achieving excellent prediction performance. The three methods of NIQE, IL-NIQE and BQMS have poor prediction performance, and the results are all below 0.3. The final analysis found that the full reference and semi-reference methods have achieved high prediction accuracy, and the indicators show that the prediction results exceed 0.9. Compared with the non-reference method, we see that the FANG method has the lowest performance and IL-NIQE can achieve better prediction. performance. The accuracy of NIQE and BQMS prediction is general. The proposed method achieves the best prediction performance with an accuracy of about 0.6.

4 Conclusion

The key to video image quality detection and classification and image quality evaluation is the quality of feature extraction. Some two-step frameless reference image quality evaluation algorithms, such as BIQI, DIIVINE, BRISQUE, etc., must first identify the distorted image and then evaluate the image quality. Based on the statistical characteristics of distorted images, BIQI uses the image distortion type obtained by SVM classifier. DIIVINE can use wavelet transform to decompose the distorted image, then use Gaussian scale hybrid model to fit the wavelet coefficients, and use the fitting parameters as feature vectors to identify the distorted image. BRISQUE extracts the spatial statistical features based on the normalized luminance coefficient distribution, and uses the support vector machine to obtain the image distortion category. Some researchers use natural images and their corresponding normalized luminance images and local standard deviation images as input, use autocorrelation mutual information to describe the correlation of adjacent pixels of input images, and combine multi-directional and multi-scale analysis to extract mutual information features. Achieve classification of distorted images. In addition, there are some methods for extracting distorted image features, such as extracting the mean, variance, gradient, entropy and other perceptual features of the image. This paper introduces an image quality evaluation method specifically for contrast distortion. In this method, we design a convolutional neural network including a three-layer convolution layer "three-layer pooling layer and three-layer full-connection layer. The network allows it to automatically extract features related to contrast distortion to predict image quality. Finally, we verify the effectiveness of the proposed method and surpass the same quality evaluation method. However, through the experimental results, we can find that the proposed method still has room for improvement in prediction performance, so in the future work, we will further study the better evaluation method of contrast distortion and continue to improve the accuracy of quality prediction. In summary, the convolutional neural network (SNN) algorithm is used to detect and identify the four types of distortion, such as excessive anomaly, over-dark anomaly, color cast and blur, which appear in the actual surveillance video image. For the over-fitting and local minimum problems, the positive and negative example equalization method and adaptive learning rate are introduced to improve the learning ability of the network, which has better convergence and robustness.

Acknowledgement. This work was jointly supported by the Key Research and Development Project of Gan-zhou, the name is "Research and Application of Key Technologies of License Plate Recognition and Parking Space Guidance in Intelligent Parking Lot", the Education Department of Jiangxi Province of China Science and Technology research projects with the Grant No. GJJ181265.

References

1. Fan, C., Zhang, Y., Feng, L., et al.: No reference image quality assessment based on multi-expert convolutional neural networks. IEEE Access **6**, 8934–8943 (2018)
2. Wang, H., Fu, J., Lin, W., et al.: Image quality assessment based on local linear information and distortion-specific compensation. IEEE Trans. Image Process. **26**(2), 915–926 (2017)
3. Eerola, T., Lensu, L., Kälviäinen, H., et al.: Study of no-reference image quality assessment algorithms on printed images. J. Electron. Imaging **23**(6), 061106 (2014)
4. Visual quality assessment: recent developments, coding applications and future trends. APSIPA Trans. Signal Inf. Process. **2**, e4 (2013)
5. Ma, L.J., Zhao, C.H.: An effective image fusion method based on nonsubsampled contourlet transform and pulse coupled neural network. Adv. Mater. Res. **756–759**, 3542–3548 (2013)
6. Wang, Z., Ma, Y., Cheng, F., et al.: Review of pulse-coupled neural networks. Image Vis. Comput. **28**(1), 5–13 (2010)
7. Zhang, L., Zhang, L., Bovik, A.C.: A feature-enriched completely blind image quality evaluator. IEEE Trans. Image Process. **24**(8), 2579–2591 (2015)
8. Chetouani, A., Beghdadi, A., Deriche, M.: A hybrid system for distortion classification and image quality evaluation. Signal Process. Image Commun. **27**(9), 948–960 (2012)
9. Yong, C., Qiang, F., Feng, S.: Sparse image fidelity evaluation based on wavelet analysis. J. Electron. Inf. Technol. **37**, 2055–2061 (2015)
10. Wang, H.: A new algorithm for integrated image quality measurement based on wavelet transform and human visual system. Proc. SPIE Int. Soc. Opt. Eng. **6034**, 60341K-1–60341K-7 (2006)
11. Chang, C.C., Lin, M.H., Hu, Y.C.: A fast and secure image hiding scheme based on lsb substitution. Int. J. Pattern Recogn. Artif. Intell. **16**(04), 399–416 (2002)
12. Singh, P., Chandler, D.M.: F-MAD: a feature-based extension of the most apparent distortion algorithm for image quality assessment. Proc. SPIE Int. Soc. Opt. Eng. **8653**, 86530I (2013)
13. Mou, X., Imai, F.H., Xiao, F., et al.: SPIE Proceedings [SPIE IS&T/SPIE electronic imaging - San Francisco airport, California, USA (Sunday 23 January 2011)] digital photography VII - Image quality assessment based on edge. Digit. Photogr. VII **7876**, 78760N (2011)
14. Zhu, L.: Image quality evaluation method based on structural similarity. In: Proceedings of SPIE - The International Society for Optical Engineering, vol. 6790, pp. 67905L-67905L-10 (2007)
15. Hassen, R., Wang, Z., Salama, M.M.A.: Image sharpness assessment based on local phase coherence. IEEE Trans. Image Process. **22**(7), 2798–2810 (2013)
16. Dony, R.D., Coblentz, C.L., Nabmias, C., et al.: Compression of digital chest radiographs with a mixture of principal components neural network: evaluation of performance. RadioGraphics **6**(6), 1481–1488 (1996)
17. Oliveira, S.A.F., Alves, S.S.A., Gomes, J.P.P., et al.: A bi-directional evaluation-based approach for image retargeting quality assessment. Comput. Vis. Image Underst. **168**, 172–181 (2017). S1077314217302035
18. Dendi, S.V.R., Dev, C., Kothari, N., et al.: Generating image distortion maps using convolutional autoencoders with application to no reference image quality assessment. IEEE Signal Process. Lett. **26**, 1 (2018)
19. Bin, J., Jiachen, Y., Zhihan, L., et al.: Wearable vision assistance system based on binocular sensors for visually impaired users. IEEE Internet Things J. **6**, 1 (2018)
20. Hui, C., Chaofeng, L.I.: Stereoscopic color image quality assessment via deep convolutional neural network. J. Front. Comput. Sci. Technol. **12**, 1315–1322 (2018)

21. Rehman, A.U, Rahim, R., Nadeem, M.S., et al.: End-to-end Trained CNN Encode-Decoder Networks for Image Steganography (2017)
22. Ding, Y., Deng, R., Xie, X., et al.: No-reference stereoscopic image quality assessment using convolutional neural network for adaptive feature extraction. IEEE Access **6**, 37595–37603 (2018)
23. Long, M., Ouyang, C., Liu, H., et al.: Image recognition of Camellia oleifera diseases based on convolutional neural network and transfer learning. Nongye Gongcheng Xuebao/Trans. Chin. Soc. Agric. Eng. **34**(18), 194–201 (2018)
24. Li, Y., Liu, D., Li, H., et al.: Convolutional neural network-based block up-sampling for intra frame coding. IEEE Trans. Circuits Syst. Video Technol. **28**(9), 2316–2330 (2017)
25. Jia, C., Wang, S., Zhang, X., et al.: Content-aware convolutional neural network for in-loop filtering in high efficiency video coding. IEEE Trans. Image Process. **28**, 3343–3356 (2019)
26. Ahn, N., Kang, B., Sohn, K.A.: Image distortion detection using convolutional neural network (2018)
27. Jongyoo, K., Anh-Duc, N., Sanghoon, L.: Deep CNN-based blind image quality predictor. IEEE Trans. Neural Netw. Learn. Syst. **30**, 1–14 (2018)
28. Mngenge N A . An adaptive quality-based fingerprints matching using feature level 2 (minutiae) and extended features (pores) (2013). Nelwamondo F.v.prof
29. Miao, Z., Xu, H., Chen, Y., et al.: An Intelligent computational algorithm based on neural network for spatial data mining in adaptability evaluation. Proc. SPIE Int. Soc. Opt. Eng. **7146**, 71461 (2009)
30. Yuan, C.H., Zhang, M., Gao, S.W., et al.: Research on method of state evaluation and fault analysis of dry-type power transformer based on self-organizing neural network. Appl. Mech. Mater. **303–306**, 562–566 (2013)

Artificial Bee Colony Algorithm Based on New Search Strategy

Minyang Xu[1,2], Hui Wang[1,2](✉), Songyi Xiao[1,2], and Wenjun Wang[3]

[1] Jiangxi Province Key Laboratory of Water Information Cooperative Sensing and Intelligent Processing, Nanchang Institute of Technology, Nanchang 330099, China
huiwang@whu.edu.cn

[2] School of Information Engineering, Nanchang Institute of Technology, Nanchang 330099, China

[3] School of Business Administration, Nanchang Institute of Technology, Nanchang 330099, China

Abstract. Artificial bee colony (ABC) is a popular swam intelligence algorithm. In ABC, a food source is considered as a feasible solution. Bees flying in the sky and searching food sources is converted into an optimization process. In contrast with other swarm intelligence algorithms, ABC has fewer parameters and stronger search ability. Though ABC excels at exploration, it does not perform well in exploitation. This paper proposes an improved ABC algorithm based on a new search strategy (called NSSABC), in which some current global best solutions are preserved and they are used to guide the search. Experiment was performed on some classical problems and results shows the proposed strategy greatly improves the optimization ability of ABC.

Keywords: Artificial bee colony algorithm · External archive · Global best solution · New search method

1 Introduction

Swarm intelligence algorithms have some advantages in handling complex and difficult optimization problems [1–3]. So, it has received more and more attention in optimization community. At present, the main swarm intelligence algorithms include ACO, PSO and ABC. Compared with PSO and ACO, ABC has fewer parameters. Some other studies proved that ABC has certain advantages [3–8].

However, ABC also has certain disadvantages. Some studies showed that the randomness of the neighborhood selection method is too strong [4–8]. So, ABC exhibits good exploration and poor exploitation. To conquer this shortcoming, there were some improved methods for search strategy design [9–17]. For example, an enhanced ABC based on global best solution was proposed in [9]. In [10], Wang et al. utilized an external archive to guide ABC. In [11], different guidance information was used to improve the solution. Banharnsakun et al. [12] used a best-so-far ABC to effectively solve the JSSP problem. In [13], ABC was parallelized [13]. Tuba et al. [14]

K. Li et al. (Eds.): ISICA 2019, CCIS 1205, pp. 772–779, 2020.
https://doi.org/10.1007/978-981-15-5577-0_62

introduced adaptive steering adjustments in ABC. In [15], inertia weights and acceleration factors were used to strengthen the search equation. Bi and Wang [16] designed a fast mutation-based ABC. In [17], DE was utilized to strengthen the search equation. It can be seen from the above literature that most of improvements on ABC aim to improve their search equations.

To further strengthen the performance, an improved ABC based on external archive is presented. The new method is motivated by GABC [9] and JADE [18]. In NSSABC, we use an external archive [10] to preserve the global best solution in the search equation of GABC. Those best solutions contained in the external archive are updated as the number of iterations increases. Simulation test experiments show NSSABC algorithm is superior to the standard ABC, GABC and IABC on most test functions.

The remainder of this paper is organized as follows. The descriptions of ABC are briefly given in Sect. 2. The proposed NSSABC is described in Sect. 3. Simulation experiments are conducted in Sect. 4. Finally, this work is concluded in Sect. 5.

2 Artificial Bee Colony

In ABC, there are four main processes: initialization process, employ bee process, onlooker bee process, and scout bee process. At the initialization stage, ABC randomly generates *SN* solutions (food sources) by the following formula.

$$x_{i,j} = x_{j,\min} + rand(0,1) \cdot (x_{j,\max} - x_{j,\min}) \tag{1}$$

where $i = 1, 2 \ldots, SN$, $j = 1, 2 \ldots, D$, SN is the population size, D is the dimension size, $rand(0, 1)$ randomly generated in [0, 1] and $[x_{j,min}, x_{j,max}]$ is the constraint of search range.

In the process of employing bee, they have their own food sources. Each employ bee finds other food sources around its own food source by the following formula.

$$v_{i,j} = x_{i,j} + \phi_{i,j}(x_{i,j} - x_{k,j}) \tag{2}$$

where $i = 1, 2 \ldots, SN$, v_i is a new solution, x_k is a randomly chosen solution ($k \neq i$), and $\phi_{i,j} \in [-1, 1]$ is a random weight.

In the process of onlooker bee, they select the food source according to the selection formula as follows.

$$p_i = \frac{f_i}{\sum_{i=1}^{SN} f_i} \tag{3}$$

where p_i is the selection probability, and f_i is the fitness value. It is apparently that the probability and the fitness value are positively correlated. It means that a better solution has a larger selection probability.

In the process of scout bee, if the food source could not be further updated within some iterations, it is supposed to be discarded. Assuming that the discarded solution is x_i, the scout bee randomly produces a new one by formula (1) instead of x_i.

3 ABC Based on New Search Strategy

Global exploration capabilities and local exploitation capabilities are two very important aspects for assessing the optimization ability of a swarm intelligence algorithm. So, balancing the two capabilities has influences on the performance of ABC.

For example, in GABC, the global best solution is used as a directive solution. Compared with the standard ABC, GABC further enhances the exploitation capabilities. However, the gbest may cause the entire swarm to a wrong search direction. It is pointed out in [7] that the search formula in GABC may be in the opposite direction, which easily leads to the oscillation during the search.

In [10], an external archive in IABC was used to lead the search, in which a concept of external archive makes full use of the global optimal solution. However, IABC causes all solutions converging toward the global best solution at different stages. It may result in the loss of swarm diversity.

To solve the above problems, a new ABC based on external archive is deigned, which combines IABC and GABC. In NSSABC, we add the external archive of IABC to the GABC search strategy to replace the guidance of the gbest in GABC. The external archive focuses on preserving different global best solutions. With the growth of iterations, the external archive is updated. Compared with GABC, NSSABC uses some good solutions (not the best one) to lead the search. It may correct the wrong search direction and avoid local minima. Unlike IABC, NSSABC does not use the current gbest to lead the search. It reduces the degree of convergence of all solutions to the gbest, and the swarm diversity increases. In summary, NSSABC can maintain strong exploration and exploitation capabilities.

$$v_{i,j} = x_{i,j} + \phi_{i,j}(x_{i,j} - x_{k,j}) + \psi_{i,j}\left(x_{Arc,j} - x_{i,j}\right) \quad (4)$$

where x_{Arc} is randomly chosen from the external archive, and x_k is randomly selected from the swarm (k $\neq$ i). $\psi_{i,j} \in [0, 1.5]$ and $\phi_{i,j} \in [-1, 1]$ are two random numbers.

Initially, an external archive A is created, and m is used as the threshold size of the archive A. After each generation, the archive A is updated. If the gbest is improved, then gbest will be preserved in A and the size increases 1 [10]. If the current archive is greater than m, we randomly delete a solution in the archive A.

Compared with the standard ABC, NSSABC only modifies the search strategy. Therefore, they have the same time complexity.

Table 1. Benchmark functions

Functions	Search range	Global optimum
Sphere (f_1)	[−100,100]	0
Schwefel 2.22 (f_2)	[−10,10]	0
Schwefel 1.2 (f_3)	[−100,100]	0
Schwefel 2.21 (f_4)	[−100,100]	0
Rosenbrock (f_5)	[−30,30]	0
Step (f_6)	[−100,100]	0
Quartic with noise (f_7)	[−1.28,1.28]	0
Schwefel 2.26 (f_8)	[−500,500]	−12569.5
Rastrigin (f_9)	[−5.12,5.12]	0
Ackley (f_{10})	[−32,32]	0
Griewank (f_{11})	[−600,600]	0
Penalized (f_{12})	[−50,50]	0

4 Experimental Study

To assess the optimization ability of NSSABC, some famous problems are applied in the simulation tests. Table 1 presents the brief descriptions of these benchmark functions.

In the experiment, the SN and the control parameter limit are set to 100. The threshold m is set to 5. For D = 30 and D = 100, the MAX FEs is equal to 1.5E+05 and 5.0E+05, respectively.

Table 2 gives the comparison results of four ABC algorithms when D = 30. From the results, NSSABC surpasses other three ABCs on all test cases except for f_3. NSSABC finds the global optimal solution on the three test functions f_6, f_9, and f_{11}. For f_1 and f_2, NSSABC greatly improves the quality of solutions comparing with IABC and GABC. However, GABC outperforms NSSABC on f_3. NSSABC, GABC and IABC achieve the same result on both f_6 and f_9. Overall, NSSABC provides better solutions than GABC, ABC and IABC.

Table 3 displays the results of four ABCs when D = 100. Like D = 30, these four algorithms do not achieve good results on f_3, f_4, and f_5. When D = 100, ABC falls into local minima, but GABC, IABC and NSSABC can still find the global optimal solution. In general, for D = 100, the test results of these four ABC algorithms are worse than those of D = 30. For f_9 and f_{11}, NSSABC can still find the global optimal solution, but IABC and GABC fail to do.

Table 2. Results achieved by ABC, IABC, GABC and NSSABC when D = 30

Fun	ABC		IABC		GABC		NSSABC	
	Mean	Std	Mean	Std	Mean	Std	Mean	Std
f_1	1.14E−15	3.58E−16	1.67E−35	6.29E−36	4.52E−16	2.79E−16	**1.56E−46**	**4.67E−46**
f_2	1.49E−10	2.34E−10	3.09E−19	3.84E−19	1.43E−15	3.56E−15	**1.53E−25**	**3.28E−25**
f_3	1.05E+04	3.37E+03	5.54E+03	2.71E+03	**4.26E+03**	**2.17E+03**	4.45E+03	6.59E+03
f_4	4.07E+01	1.72E+01	1.06E+01	4.26E+00	1.16E+01	6.32E+00	**7.86E+00**	**1.28E+00**
f_5	1.28E+00	1.05E+00	2.36E−01	3.94E−01	2.30E−01	3.72E−01	**1.03E−01**	**2.42E−01**
f_6	**0.00E+00**	**0.00E+00**	**0.00E+00**	**0.00E+00**	**0.00E+00**	**0.00E+00**	**0.00E+00**	**0.00E+00**
f_7	1.54E−01	2.93E−01	4.23E−02	3.02E−02	5.63E−02	3.66E−02	**1.67E−02**	**8.78E−03**
f_8	−12490.5	5.87E+01	**−12569.5**	1.31E−10	**−12569.5**	3.25E−10	**−12569.5**	**2.84E−10**
f_9	7.11E−15	2.28E−15	**0.00E+00**	**0.00E+00**	**0.00E+00**	**0.00E+00**	**0.00E+00**	**0.00E+00**
f_{10}	1.60E−09	4.32E−09	3.61E−14	**1.76E−14**	3.97E−14	2.83E−14	**2.19E−14**	1.87E−14
f_{11}	1.04E−13	3.56E−13	**0.00E+00**	**0.00E+00**	1.12E−16	2.53E−16	**0.00E+00**	**0.00E+00**
f_{12}	5.46E−16	3.46E−16	3.02E−17	**0.00E+00**	4.03E−16	2.39E−16	**3.01E−17**	**0.00E+00**

Table 3. Results achieved by ABC, IABC, GABC and NSSABC when D = 100

Fun	ABC		IABC		GABC		NSSABC	
	Mean	Std	Mean	Std	Mean	Std	Mean	Std
f_1	7.42E−15	5.89E−15	3.23E−33	1.45E−34	3.37E−15	7.52E−16	**1.77E−44**	**8.64E−45**
f_2	1.09E−09	4.56E−09	4.82E−18	3.53E−18	6.54E−15	2.86E−15	**1.46E−23**	**1.19E−23**
f_3	1.13E+05	2.62E+04	9.76E+04	2.81E+04	9.28E+04	2.71E+04	**8.60E+04**	**2.23E+04**
f_4	8.91E+01	4.37E+01	8.29E+01	**1.28E+01**	8.37E+01	3.68E+01	**7.72E+01**	**2.58E+01**
f_5	3.46E+00	4.29E+00	2.97E+00	2.72E+00	2.08E+01	3.46E+00	**6.61E−01**	**5.64E−01**
f_6	1.58E+00	1.68E+00	**0.00E+00**	**0.00E+00**	**0.00E+00**	**0.00E+00**	**0.00E+00**	**0.00E+00**
f_7	1.96E+00	2.57E+00	7.45E−01	2.27E−01	9.70E−01	7.32E−01	**3.93E−01**	**1.25E−01**
f_8	−40947.5	7.34E+02	−41898.3	3.21E−10	−41898.3	5.68E−10	**−41898.2**	**2.58E−10**
f_9	1.83E−11	2.27E−11	1.42E−14	2.63E−14	1.95E−14	3.53E−14	**0.00E+00**	**1.03E−14**
f_{10}	3.54E−09	7.28E−10	1.50E−13	4.87E−13	1.78E−13	5.39E−13	**1.35E−13**	**3.94E−13**
f_{11}	1.12E−14	9.52E−15	7.78E−16	5.24E−16	1.44E−15	3.42E−15	**0.00E+00**	**1.16E−15**
f_{12}	4.96E−15	3.29E−15	9.05E−18	0.00E+00	2.99E−15	4.37E−15	**9.04E−18**	**2.89E−15**

Figure 1 shows a comparison of the convergence performance of NSSABC and other ABCs. It is obvious that NSSABC is much faster than the other three algorithms except for f_5, f_6, and f_9. The main reason is that the new search strategy based on GABC and IABC helps speed up the convergence on most of the test functions.

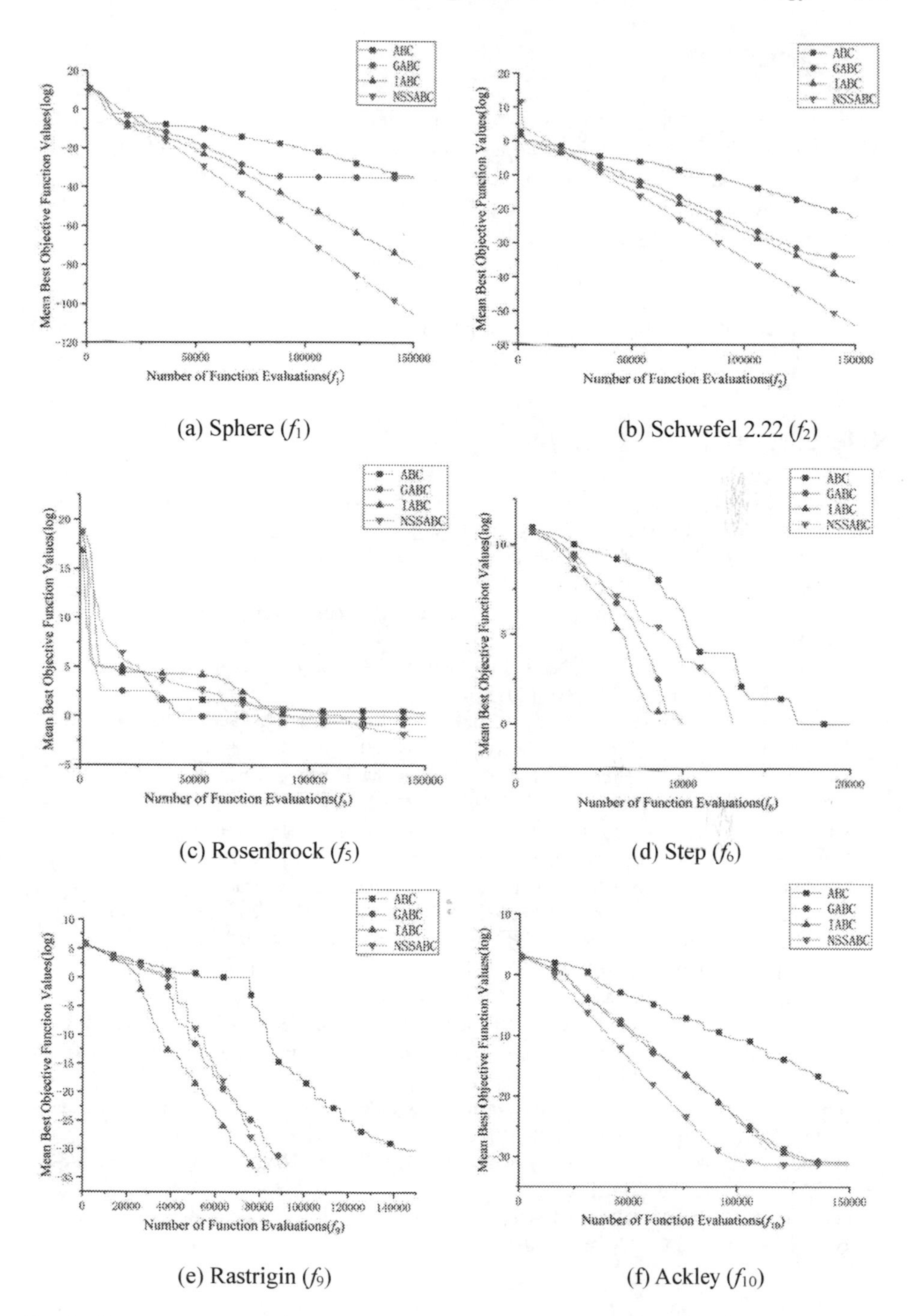

(a) Sphere (f_1)

(b) Schwefel 2.22 (f_2)

(c) Rosenbrock (f_5)

(d) Step (f_6)

(e) Rastrigin (f_9)

(f) Ackley (f_{10})

Fig. 1. The convergence characteristics of ABC, GABC, IABC and NSSABC on some functions for D = 30.

5 Conclusions

In this paper, we propose an improved ABC algorithm called NSSABC, which uses the information of the global best solution to guide the search. An external archive aims to preserve the updated global best solutions. It avoids the oscillating search to some extent. NSSABC can maintain good exploration and exploitation capacities. Experimental results show NSSABC algorithm is superior to the ABC, IABC, and GABC algorithms.

Acknowledgement. This work was supported by the National Natural Science Foundation of China (No. 61663028).

References

1. Wang, M.: Research on micro grid optimal scheduling based on particle swarm optimization. J. Ningxia Normal Univ. **40**(10), 85–90 (2019)
2. Shi, J., Liu, Z., Pan, S.: Path planning improvement strategy of ant colony algorithm. Fire Command Control **44**(10), 153–162 (2019)
3. Zhang, M.X., Ma, X., Duan, Y.M.: Improved artificial bee colony. J. Xidian Univ. **42**(2), 65–70 (2015)
4. Du, Z.X., Liu, G.Z.: Artificial bee colony with global and unbiased search strategy. Acta Electronica Sinica **46**(2), 308–314 (2018)
5. Lin, Q., Zhu, M., Li, G.: A novel artificial bee colony with local and global information interaction. Appl. Soft Comput. J. **62**(3), 702–735 (2018)
6. Gao, W.F., Liu, S.Y., Huang, L.L.: A novel artificial bee colony algorithm with new probability model. Soft. Comput. **22**(7), 2217–2243 (2018)
7. Du, Z.X., Liu, G.Z., Zhao, X.H.: Integrated learning artificial bee colony algorithm. J. Xi'an Univ. Electr. Sci. Technol. **46**(2), 124–131 (2019)
8. Kucukkoc, I., Buyukozkan, K., Satoglu, S.I., Zhang, D.Z.: A mathematical model and artificial bee colony algorithm for the lexicographic bottleneck mixed-model assembly line balancing problem. J. Intell. Manufact. **30**(8), 2913–2925 (2019)
9. Zhu, G., Kwong, S.: Gbest-guided artificial bee colony algorithm for numerical function optimization. Appl. Math. Comput. **217**(7), 3166–3173 (2010)
10. Wang, H., Wu, Z.J., Zhou, X.Y., Rahnamayan, S.: Accelerating artificial bee colony algorithm by using an external archive. In: Proceedings of IEEE Congress on Evolutionary Computation, pp. 517–521 (2013)
11. Yurtkuran, A., Emel, E.: An adaptive artificial bee colony algorithm for global optimization. Appl. Math. Comput. **271**(10), 1004–1023 (2015)
12. Banharnsakun, A., Sirinaovakul, B., Achalakul, T.: Job shop scheduling with the best -so-far ABC. Eng. Appl. Artif. Intell. **25**(2), 583–593 (2011)
13. Subotic, M., Tuba, M., Stanarevic, N.: Different approaches in parallel of the artificial bee colony algorithm. Int. J. Math. Models Meth. Appl. Sci. **5**(4), 755–762 (2011)

14. Tuba, M., Bacaninand, N., Stanarevic, N.: Guided artificial bee colony algorithm. In: Proceedings of the 5th European Conference on European Computing Conference (ECC 2011), pp. 398–403 (2011)
15. Li, G., Niu, P., Xiao, X.: Development and investigation of efficient artificial bee colony algorithm for numerical function optimization. Appl. Soft Comput. J. **12**(1), 320–332 (2011)
16. Bi, X., Wang, Y.: An improved artificial bee colony algorithm. In: Proceedings of the 3rd International Conference on Computer Research and Development (ICCRD), pp. 174–177 (2012)
17. Gao, W.F., Liu, S.Y.: A modified artificial bee colony algorithm. Comput. Oper. Res. **39**(3), 687–697 (2012)
18. Zhang, J., Sanderson, A.C.: JADE: adaptive differential evolution with optional external archive. IEEE Trans. Evol. Comput. **13**(5), 945–958 (2009)

Regression Network for Real-Time Pedestrian Detection

Wanjuan Song[1,2(✉)] and Wenyong Dong[1,3(✉)]

[1] Computer School, Wuhan University, Wuhan, China
{key_swj, dwy}@whu.edu.cn
[2] Hubei University of Education, Wuhan, China
[3] School of Software, Nanyang Institute of Technology, Nanyang, Henan, China

Abstract. This paper presents a real-time robust pedestrian detection algorithm based on Convolutional Neural Network (CNN). With the success of CNN in image recognition, CNN has also been used in pedestrian detection, but these methods are still difficult to use the case of real-time application. The proposed method makes full use of the powerful feature representation and faster recognition ability of CNN to meet the real-time request for pedestrian detection. To do so, we regard the pedestrian detection as a regression problem, and a CNN model is employed to solve it. Since a new loss function is introduced, the training can be completed end-to-end, and the trained CNN can directly map an image to the location and confidence of pedestrian bounding box without feature abstraction. We verify the proposed method on the common pedestrian detection dataset of Caltech, and experimental results show that it owns lower miss rate and higher detection frame rate compared with some popular methods.

Keywords: Machine vision · Pedestrian detection · Convolutional neural network · Model update

1 Introduction

Pedestrian detection has a wide range of application scenarios, such as video surveillance, automatic driving, etc. [1]. Real-time robust pedestrian detection method is the key to these applications. The goal of pedestrian detection algorithm is to locate pedestrians correctly in real time in images, but this goal is difficult to achieve, because the algorithm needs to get a trade-off between real-time and accuracy [2]. In addition, pedestrian congestion, mutual occlusion and various changes of pedestrian posture pose great challenges to pedestrian detection algorithms.

General pedestrian detection framework [3, 13] includes pedestrian candidate box generation, feature extraction and pedestrian candidate box verification. Classical pedestrian detection usually uses sliding window method to generate pedestrian candidate boxes, choosing HOG [4], SIFT [5], ACF [6] as features, and using SVM, Adaboost and other classifiers to classify the generated pedestrian candidate boxes based on the extracted features, in which the pedestrian candidate boxes divided into positive samples are the final pedestrian detection Result. In recent years, with the wide success of Convolutional Neural Network (CNN) in image recognition [14], CNN has

K. Li et al. (Eds.): ISICA 2019, CCIS 1205, pp. 780–787, 2020.
https://doi.org/10.1007/978-981-15-5577-0_63

also been used in pedestrian detection. For example, in document [7], a cascade CNN is proposed to classify candidate frames. Then, the influence of network structure of CNN on pedestrian detection performance is analyzed in document [8]. With the great success in the area of target detection in Regional Proposal Network (RPN) [9], a new pedestrian detection method is proposed based on this document [10]: generating pedestrian candidate boxes based on RPN and classifying these pedestrian candidate boxes with Boosted Forest; at the same time, the paper [11] aims at the pedestrian candidate boxes. Pedestrian candidate boxes of different scales train multiple detection networks, and combine the output of multiple detection networks to get the final pedestrian detection results. However, these pedestrian detection algorithms often need to go through complex processing steps, which are difficult to achieve real-time application, and it is difficult to combine multiple steps to optimize.

A new real-time pedestrian detection method based on CNN is proposed in this paper. This method constructs pedestrian detection as a regression problem. Each image needs only one forward propagation of the network, and directly outputs the position and confidence of the pedestrian bounding frame. Therefore, it can meet the needs of real-time applications, and the network can train end-to-end. It avoids the problem that complex steps are difficult to optimize jointly. In addition, a multi-scale feature representation method is proposed to further improve the performance of pedestrian detection.

2 Pedestrian Detection Based on Regression Network

Recently, Joseph Redmon et al. [12] proposed a general target detection algorithm based on regression convolution network, Yolo universal target detection field has achieved the same detection performance as classified convolution network. Inspired by this, this paper constructs a convolution neural network structure and designs a loss function based on square loss. So that the training can be completed end-to-end, the test can directly output the location and confidence of pedestrian bounding box with only one forward propagation of the network.

As shown in Fig. 1, the algorithm can directly output the target detection result in the image through only two steps of feature coding and feature decoding. The feature coding module constructs the depth feature representation of the image through multi-layer convolution layer. Its advantage lies in making full use of the strong feature representation ability of convolution neural network and feature solution. The code module is constructed by a multi-layer full connection layer, and the last layer is the prediction layer: the position of the pedestrian candidate box and the corresponding confidence are regressed, and the output of the detection box whose confidence is greater than a certain threshold is taken as the final pedestrian detection result. In order to accomplish this task, the following will focus on the training process of the network and the design of loss function.

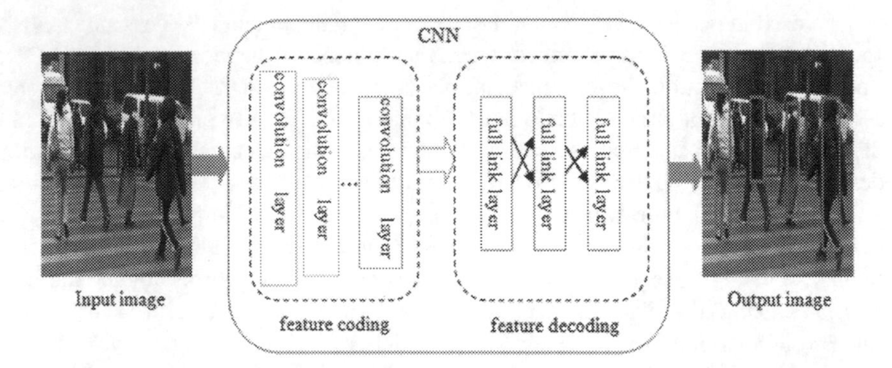

Fig. 1. Framework of our proposed method

Firstly, an image is divided into $S \times S$ grids, each of which is responsible for predicting the location and confidence of B pedestrian candidate frames. For this reason, our loss function consists of two parts: location loss and confidence loss. The definition of location loss L_{loc} is as Formula 1.

$$L_{loc} = \sum_{i=1}^{S^2} \sum_{j=1}^{B} l_{ij}^{obj} (x_i - g_i)^2 \tag{1}$$

Where l_{ij}^{obj} is an indicator function, if the jth pedestrian candidate box in grid i has the largest bounding box with the pedestrian bounding box labeled in the training set, then $l_{ij}^{obj} = 1$, otherwise $l_{ij}^{obj} = 0$; x_i and g_i represent the parameters of the predicted pedestrian bounding box and the pedestrian bounding box labeled in the training set, respectively (including the coordinates and width of the central position). Confidence loss L_{loc} is defined as follows:

$$L_{con} = \sum_{i=1}^{S^2} \sum_{j=1}^{B} l_{ij}^{obj} (c_i - \widehat{c}_i)^2 + \lambda_{noobj} \sum_{i=1}^{S^2} \sum_{j=1}^{B} l_{ij}^{noobj} (c_i - \widehat{c}_i)^2 \tag{2}$$

c_i represents the confidence of the predicted pedestrian bounding frame, $\widehat{c}_i$ represents the confidence of the marked pedestrian bounding frame, $\widehat{c}_i = 1$ represents the existence of pedestrian targets, and $\widehat{c}_i = 0$ represents the absence of pedestrian targets. Because there are no pedestrian targets in a large number of grids, the confidence of pedestrian candidate boxes predicted by these grids should be close to 0, which will make the network difficult to converge. Therefore, this paper sets a weight parameter λ_{noobj} for grids without pedestrian targets. Where λ_{noobj} is an indicator function, if the jth pedestrian candidate box in grid i does not contain a pedestrian target, then $l_{ij}^{noobj} = 1$, otherwise $l_{ij}^{noobj} = 0$.

In combination (1) and (2), the ultimate objective $L_{overall}$ of our optimization is to locate the weights of loss and confidence loss:

$$L_{overall} = \lambda_{loc}L_{loc} + \lambda_{con}L_{con} \tag{3}$$

Where, λ_{loc} and λ_{con} represent the weight of location loss and confidence loss respectively. The training process uses the stochastic gradient descent algorithm to train the network. Details of the training can be found in the experimental part.

3 Experiments and Results Analysis

3.1 Test Data and Evaluation Criteria

In this paper, the most popular and challenging Caltech pedestrian detection data set [3] is selected to test the algorithm, and the current mainstream pedestrian detection algorithms are compared and tested. Caltech pedestrian detection data set consists of 11 subsets (the first six subsets as training sets and the last five as test sets). Each subset contains 6–13 videos of about 1 min in length. It contains a total of about 2500 frames of images and 350 pedestrian bounding frames that have been labeled. For the sake of fairness, referring to the general settings in reference [10, 11], the training set of all algorithms contains 42782 images and the test set contains 4024 images.

Log-average miss rate (L-AMR) is used as a general evaluation standard. L-AMR is calculated by taking 10 logarithms (x-axis range is 10–2 to 100) according to the number of false positives per image (FPPI), and the corresponding average miss detection rate is used as a comparison result. Caltech Pedestrian Detection Data Set has different evaluation settings depending on the pedestrian height and the degree of occlusion. Details are shown in Table 1.

Table 1. Evaluation setting on caltech dataset

Evaluation setting	Pedestrian height	Whether to occlude
All	>20 pixels	Unshielded, Partial, Severe occlusion
Reasonable	>50 pixels	Unshielded, Partial occlusion
Far	<30 pixels	\
Medium	>0 pixels	\
Near	>80 pixels	\
Occ. 1	\	Unshielded
Occ. 2	\	Partial occlusion
Occ. 3	\	Severe occlusion

3.2 Experimental Design

In order to make the network training converge faster, set 0.5 in formula (2), 2 in formula (3) and 1 in formula (3). In the network structure, the network weight of the first three layers of convolution layer comes from the pre-trained VGG-f model, and all layers are randomly initialized. ReLU layer (leaky parameter is 0.1) is added after each layer of convolution layer.

In the whole training process, 128 randomly disrupted image samples were trained for 50 generations. The learning rate of the first 30 generations was 0.001, and that of the second 20 generations was 0.0001. For the sake of fairness, referring to the general settings in reference [10, 11], the training set of all algorithms contains 42782 images and the test set contains 4024 images. In order to verify the effectiveness of the proposed algorithm, several new methods using convolutional neural network are selected to carry out comparative experiments: DeepCascade [7], MS-CNN [11] and RPN+BF [10]. At the same time, in order to verify the effectiveness of the proposed multi-scale feature representation, two versions of the algorithm are implemented. The first one removes the multi-scale feature representation in Fig. 2 and the second one uses the multi-scale feature representation in Fig. 2, which is named MS.

3.3 Result Analysis

Quantitative Analysis
Table 2 shows the overall performance of all evaluation algorithms under different evaluation settings. Roughening represents the optimal performance and italics represents the sub-optimal performance. From Table 2, it can be found that the proposed algorithm achieves optimal or sub-optimal performance under different evaluation settings. When pedestrian targets are more conventional (the Reasonable column in Table 2), the proposed algorithm can achieve better detection results than other algorithms by using multi-scale feature representation. It is worth noting that the row-based algorithm has better detection performance than other algorithms. When the human target is far away from the camera and the target scale is small (Far column in Table 2), the detection performance of the algorithm in this paper is far beyond that of all other contrast algorithms. Moreover, the detection performance of the algorithm in different scales of the target (Medium and Near in Table 2) is excellent, which fully verifies the proposed method. The validity of network structure represented by multi-scale features. In addition, we can see from Table 2 that our algorithm still performs well under different degrees of occlusion. In order to show the performance comparison of different algorithms under different false alarm situations more comprehensively, the L-AMR curve set by 'Reasonable' is given in Fig. 2. It can be found from the graph that MS, a pedestrian detection algorithm based on multi-scale feature representation, achieves the best performance under different FPPI condition.

Qualitative Analysis
In order to further verify the effectiveness of the proposed multi-scale feature representation network structure, Fig. 3 shows some challenging pedestrian detection results of the proposed algorithm SS without multi-scale feature representation and MS with multi-scale feature representation)on the test set. Pedestrians in Figs. 3(a) and (b) are shown. The target is far away from the camera and has a small scale. Proposed-SS misses detection. In figure (b), there is no output of pedestrian detection results. MS can locate the target well, but in figure (b), it is still impossible to separate two pedestrians. Only one pedestrian surrounds the two pedestrian targets. In Fig. 3(c) and (d), the pedestrian target

is close to the camera and has medium scale, but the mutual occlusion between pedestrians and pedestrians is more serious. Proposed-SS takes the two occluded pedestrian targets as the image. A test result output loses the ability to discriminate different pedestrian targets, while MS can distinguish the different pedestrian targets which are occluded from each other very well. Four and three pedestrian test results are output in Fig. 3(c) and (d), respectively, which fully verifies that the network structure represented by multi-scale features in this paper can be extremely effective. The recall rate of the algorithm is greatly improved.

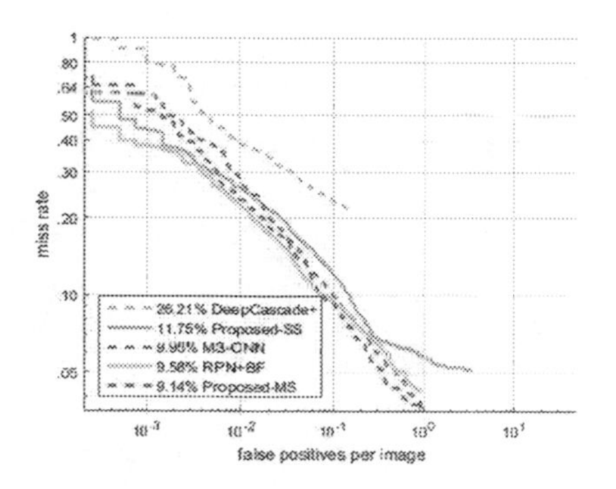

Fig. 2. L-AMR curve on reasonable setting

Table 2. Performance for different methods on various settings (miss rate%)

Algorithm	Reasonable	All	Far	Medium	Near	Occ.1	Occ. 2	Occ. 3
DeepCascade	27.2	69.8	100	62.7	9.9	20.5	48.6	75.2
MS-CNN	10.9	61.9	98.2	49.9	2.7	8.5	19.4	58.3
RPN+BF	9.8	65.7	100	54.5	2.4	7.9	23.5	69.5
SS	12.7	64.6	97.0	56.4	4.0	9.3	23.2	68.2
MS	9.6	62.2	78.2	46.2	2.6	3.5	17.1	51.5

Fig. 3. Some pedestrian detection results on test set

Velocity Analysis

The network model in this paper is based on the Caffe toolbox to carry out algorithm experiments. The experimental platforms for all algorithms testing are Intel Core i7 CPU with 3.4 GHz main frequency and NVIDIA GTX980 (4 GH) GPU. Table 3 gives the average processing speed of the algorithm on CPU and GPU. It can be seen from Table 3 that Proposed-M uses more complex network structure. The processing speed of S is slightly lower than Proposed-SS, but it has great advantages in detection performance. Moreover, the detection speed of 50.6 FPS on GPU can meet the requirements of real-time applications.

Table 3. Average detection speed of our methods

Algorithm	CPU (Intel Core i7, 3.4 GHz)	GPU (NVIDIA GTX980, 4G)
SS	26.5	72.0
MS	18.2	50.7

4 Conclusion

Aiming at the problem that existing pedestrian detection methods based on depth feature representation are difficult to complete real-time detection, a real-time robust pedestrian detection algorithm based on multi-scale feature representation is proposed. In this method, pedestrian detection is constructed as a regression problem, so that each image detection only needs to perform a network forward propagation, so it can fully meet the needs of real-time detection. In this paper, a multi-scale feature representation network structure is designed to make the detection performance of pedestrian targets with different scales have better robustness. Finally, the algorithm is tested in Caltech pedestrian detection data set [3]. The experimental results show that the proposed algorithm achieves better detection performance than the existing pedestrian detection methods based on depth feature representation.

Acknowledgements. This research work is supported by the National Natural Science Foundation of China under Grant No. 61672024, 61170305 and 60873114, and National Key R\&D Program of China (No. 2018YFB0904200 and 2018YFB2100500)

References

1. Huang, K., Chen, X., Kang, Y., Tan, N.: Review of intelligent video surveillance. J. Comput
2. Wang, X.: Intelligent multi-camera video surveillance: a review. Pattern Recognit. Lett. **34** (1), 3–19 (2013)
3. Dollár, P., Wojek, C., Schiele, B., Perona, P.: Pedestrian detection: an evaluation of the state of the art. IEEE Trans. Pattern Anal. Mach. Intell. **34**, 212–226 (2012)
4. Dalal, N., Triggs, B.: Histograms of oriented gradients for human detection. In: IEEE International Conference on Computer Vision and Pattern Recognition, pp. 886–893 (2005)

5. Lowe, D.G.: Distinctive image features from scale-invariant keypoints. Int. J. Comput. Vis. **60**(2), 91–110 (2004)
6. Ren, S., He, K., Girshick, R., et al.: Faster R-CNN: towards real-time object detection with region proposal networks. In: Advances in Neural Information Processing Systems, pp. 91–99 (2015)
7. Dollár, P., Appel, R., Belongie, S., Perona, P.: Fast feature pyramids for object detection. IEEE Trans. Pattern Anal. Mach. Intell. **36**(8), 1532–1545 (2014)
8. Angelova, A., Krizhevsky, A.: Real-time pedestrian detection with deep network cascades. In: European Conference on Computer Vision, pp. 127–141 (2015)
9. Hosang, J.H., Omran, M., Benenson, R., Schiele, B.: Taking a deeper look at pedestrians. In: IEEE International Conference on Computer Vision, pp. 1305–1312 (2015)
10. Girshick, R.: Fast R-CNN. In: IEEE International Conference on Computer Vision, pp. 2544–2550 (2015)
11. Zhang, L., Lin, L., Liang, X., He, K.: Is faster R-CNN doing well for pedestrian detection. In: European Conference on Computer Vision, pp. 702–715 (2016)
12. Li, J., Liang, X., Shen, S., Xu, T., Yan, S.: Scale-aware fast R-CNN for pedestrian detection. In: European Conference on Computer Vision, pp. 83–596 (2016)
13. Li, Z., Chen, Z., Wu, Q.M.J., et al.: Real-time pedestrian detection with deep supervision in the wild. SIViP **13**(8), 1–9 (2019)
14. Lu, P., Lu, K., Wang, W., Zhang, J., Chen, P., Wang, B.: Real-time pedestrian detection in monitoring scene based on head model. In: Huang, D.-S., Jo, K.-H., Huang, Z.-K. (eds.) ICIC 2019. LNCS, vol. 11644, pp. 558–568. Springer, Cham (2019). https://doi.org/10.1007/978-3-030-26969-2_53

Dynamic Gesture Recognition Based on HMM-DTW Model Using Leap Motion

Geng Tu[1,2], Qingyuan Li[3], and Dazhi Jiang[1,2](✉)

[1] Department of Computer Science, Shantou University, Shantou 515063, China
{19gtu, dzjiang}@stu.edu.cn
[2] Key Laboratory of Intelligent Manufacturing Technology, Ministry of Education, Shantou University, Shantou 100091, China
[3] Electrical Engineering and the Automation, China University of Mining and Technology, Beijing, China
864892368@qq.com

Abstract. In recent years, gesture recognition that has the extensive application future is gradually the hotspot of the human-computer interaction. With the emergence of depth sensors such as Leap Motion, more depth information can be obtained, which provides a reliable data guarantee for recognizing complex gestures and better application scenarios. In this paper, five common dynamic gestures are defined firstly. Then, Leap Motion sensor is used to acquire the feature data of the hand. After comparing the recognition effects of HMM and DTW algorithms, a hybrid HMM-DTW model is proposed for dynamic gesture recognition. According to Kappa coefficient, it can be noticed that the model is substantial.

Keywords: Leap Motion · HMM-DTW · Dynamic gesture recognition

1 Introduction

Human-computer interaction has increasingly become an important part of people's daily life. Especially in recent years, with the rapid development of computer technology, the research of human-computer interaction technology in line with human communication habits has become very active, and gesture recognition research is conforming to this trend.

As a common way of communication, gesture not only has the characteristics of simplicity and image, but also contains abundant information. Therefore, gesture recognition can be used not only in the field of human-computer interaction, but also in other fields, such as medicine [1], military robotics [2], virtual experiment [3] and so on.

Because of the diversity, ambiguity, differences and technical limitations of gesture itself, it is unrealistic to solve the problem of recognition of general gesture as a whole in the foreseeable future. In order to find a breakthrough, it is necessary to study gesture classification and gesture usage in interpersonal communication, so as to determine the rationality. Identifying the scope and establishing interaction models that conform to human behavior habits could provide the technical guidance.

K. Li et al. (Eds.): ISICA 2019, CCIS 1205, pp. 788–798, 2020.
https://doi.org/10.1007/978-981-15-5577-0_64

Five kinds of common dynamic gestures are defined in this paper, and the reasonable recognition range is determined. Furthermore, two feature extraction algorithms are proposed for Hidden Markov Model (HMM) and Dynamic Time Warping (DTW) gesture recognition respectively. After comparing the recognition effects of HMM algorithm and DTW algorithm, a hybrid HMM-DTW model is proposed for gesture recognition.

2 Related Work

In today's world, with the continuous development of science and technology, gesture recognition technology has also made a qualitative leap. Generally speaking, there are two ways to realize computer gesture recognition technology.

Using special hardware devices, such as data gloves, for feeding back the information of each joint. The main research could be summarized as the follows. Grimes [4] designed the first data input glove for understanding sign language, which uses sensors installed on gloves to measure hand movement. Data gloves developed by Zimmerman use thin flexible plastic tubes and light sources with detectors to record joint angles [5]. Subsequently, Stanford University developed network gloves and commercialized them by Virtual Technologies Inc. The company is equipped with 18 or 22 piezoresistive sensors [6]. DT data gloves commercialized by the fifth-dimension technology use fiber optic flexor sensor [7]. The recently developed Strin gloves are commercialized by Teiken Ltd., which uses 24 induction coders to record the angle of the finger joint [8, 9]. Liang and other researchers from Taiwan University have used VPL gloves to recognize 250 gestures in their textbooks. Although these glove devices can solve the problem of gesture recognition, they are based on optical principles and environmental lighting conditions. In addition, gloves with flexible sensors can measure the overall bending of the fingers, not the angle at the joints. This will lead to poor performance of data gloves, expensive glove equipment and unfriendly interaction with users.

The other is to use computer vision technology for gesture recognition, such as collecting gesture images or videos with one or more cameras, then analyzing and modeling the feature information, and finally recognizing gestures. Starner and Entland [10] are one of the pioneers of sign language recognition. Based on the camera, the Hidden Markov Model (HMM) is used for real-time gesture recognition of American Sign Language (ASL). Subsequently, Sarkaleh et al. proposed a British Sign Language Recognition System, which describes the shape of the hand through a histogram of the directional gradient (hog) descriptor. [11] A Persian sign language recognition system based on discrete wavelet transform for gesture image feature extraction and neural network recognition was developed. [12] It was the first group to report recognition of Indian Sign Language (ISL) using visual-based systems and Euclidean Distance Measurement Classifier. Kishore [13] and others proposed that the histogram of edge frequency (HOEF) feature is more efficient than hog feature in ISL recognition. In addition, because gesture systems are mostly color-based trackers, the types and backgrounds of clothing are limited. Paquin [14, 15] proposed a skin color segmentation method to track users' hands and heads. With the emergence of deep information acquisition devices such as Microsoft Kinect and Leap Motion, the research of gesture recognition is more convenient. For example, Yi Li [16] uses Kinect to recognize nine predefined gestures. Matthew [17] uses Kinect to obtain depth information and RGB feature information, and elaborates a mainstream gesture recognition technology in the contemporary era. However, Kinect is developed for human motion tracking. In the research of Kim et al., it has been proved that the accuracy of hand motion tracking using Kinect is much lower than LeapMotion [18], which is specially designed for hand motion tracking. Leap Motion, developed by Leap, is a new non-contact hand tracking sensor. It has high tracking accuracy and provides a large number of software interfaces for gesture recognition. Zubrycki et al. controlled a three-finger gripper through a Leap Motion. Guerrerorinco et al. developed the control system of the manipulator by Leap Motion. In this paper, Leap Motion sensor is used to collect real-time feature information and trajectory information of the hand. Through continuous experiments, different solutions for static and dynamic gestures are finally obtained.

3 The Definition of Dynamic Gestures

In the long social practice, gestures have accumulated and produced rich connotations. In this paper, the gestures that can express their meanings without relying on coherent movements and only relying on individual gesture forms are defined as static gestures, such as stones, scissors, cloth, etc. It defines the gestures that need to act coherently to express their meanings as dynamic gestures, such as “love” in sign language.

There are many kinds of gestures, and their meanings are also different in different rules. Therefore, in this paper, five different dynamic gestures are defined in Fig. 1.

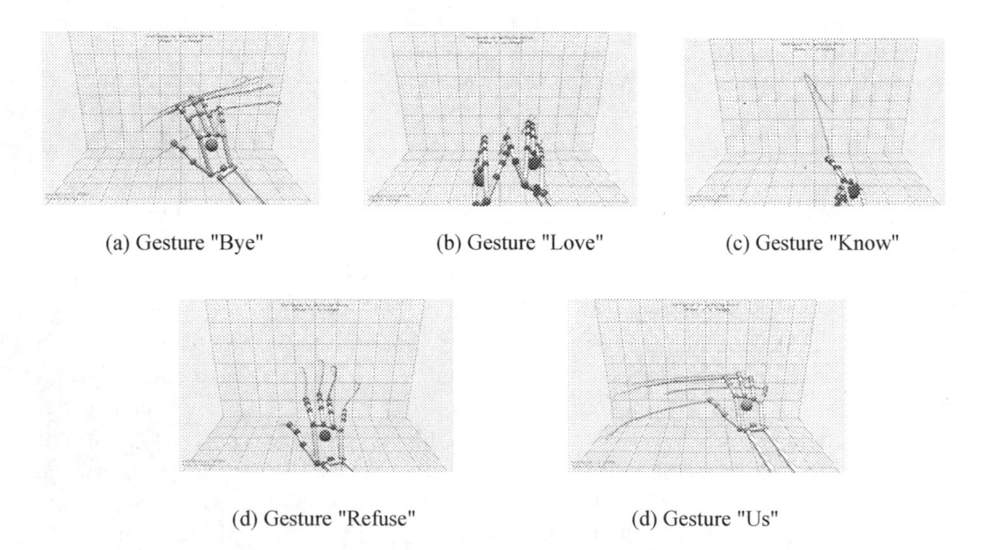

(a) Gesture "Bye" (b) Gesture "Love" (c) Gesture "Know"

(d) Gesture "Refuse" (d) Gesture "Us"

Fig. 1. Five different dynamic gestures are defined.

4 The Feature Extraction of Gestures

There are many features of three-dimensional trajectory, such as feature extraction technology based on spatial position, direction angle, angular velocity, tangential velocity, etc. However, two feature extraction algorithms are proposed in this paper. One is to extract three-dimensional motion trajectory and calculate the weighted sum of squares of three-dimensional data as the feature data of DTW model.

$$D[i] = \sqrt{[x[i] - x_c]^2 + [y[i] - y_c]^2 + [z[i] - z_c]^2}, i = 1, \ldots, N \quad (1)$$

$$x_c = \frac{1}{N}\sum_{i=1}^{N} x[i] \quad y_c = \frac{1}{N}\sum_{i=1}^{N} y[i] \quad z_c = \frac{1}{N}\sum_{i=1}^{N} z[i]$$

Among them, $x[i]$, $y[i]$, $z[i]$ represents coordinate points on X, Y and Z axes respectively.

In addition, it is pointed out in [19] that the feature based on direction angle is the best for recognition. Therefore, the three-dimensional trajectory is projected onto the XOY plane, and then the direction angle between the two points is calculated. Here, assuming that $P_1(x_1, y_1)$ and $P_2(x_2, y_2)$ are two points on the XOY plane, the process of calculating the direction angle is as follows.

$$\alpha(P_1, P_2) = \begin{cases} \arctan(\frac{y_2 - y_1}{x_2 - x_1}) + \pi, (x_2 - x_1 < 0 \cap y_2 - y_1 > 0) \\ \arctan(\frac{y_2 - y_1}{x_2 - x_1}) + 2\pi, (x_2 - x_1 > 0 \cap y_2 - y_1 < 0) \\ \arctan(\frac{y_2 - y_1}{x_2 - x_1}), otherwise \end{cases} \quad (2)$$

Therefore, the following HMM feature data can be extracted by quantizing the orientation angle with chain code in 16 directions:

$$P_m(x_m, y_m) = \frac{1}{T}(\sum_{i=1}^{T} x_i, \sum_{i=1}^{T} y_i) \quad (3)$$

Among them, the coordinates of P_m are the average values of the coordinates of X and Y axes, and T is the length of the trajectory sequence.

5 Gesture Recognition Algorithm

In gesture recognition algorithm, this paper first describes the basic principles of HMM and DTW algorithm, and then combines HMM and DTW to propose a hybrid HMM-DTW model for gesture recognition.

5.1 HMM Model

In view of the simplicity of feature sequence, in order to avoid frequent local optimum solutions in model training, we stipulate that each state in HMM model can only be transferred to itself or to the next unique state. In addition, it is found that the selection of hidden state number *N* in HMM model can greatly improve the recognition effect of the model. Generally speaking, the hierarchical Dirichlet process hidden Markov model (HDP-HMM), which is based on Dirichlet process, is a common method to determine the number of hidden states. However, HDP-HMM model is too complex and inconsistent with the real physical situation of the original problem. It is impossible for all phenomena to be a Markov process and not necessarily know the observation probability of each state, and human behavior. It is not necessarily a finite-state mixture of Gauss. Therefore, in this paper, we use Bayesian Information Criterion-based violence search algorithm to calculate the hidden state *N*. The results are shown in Fig. 2 below. From left to right, from top to bottom, the logarithmic relationship between the implicit state number of the dynamic gesture "Bye", "Know", "Love", "Refuse" and "Us" HMM model and the growth rate of recognition rate is respectively analyzed. The last one is a summary of all the previous results.

Therefore, there are two methods to select the hidden layer: one is to select the highest recognition rate to determine the number of hidden states as shown in Table 1, the other is to determine the number of hidden states according to the growth rate as

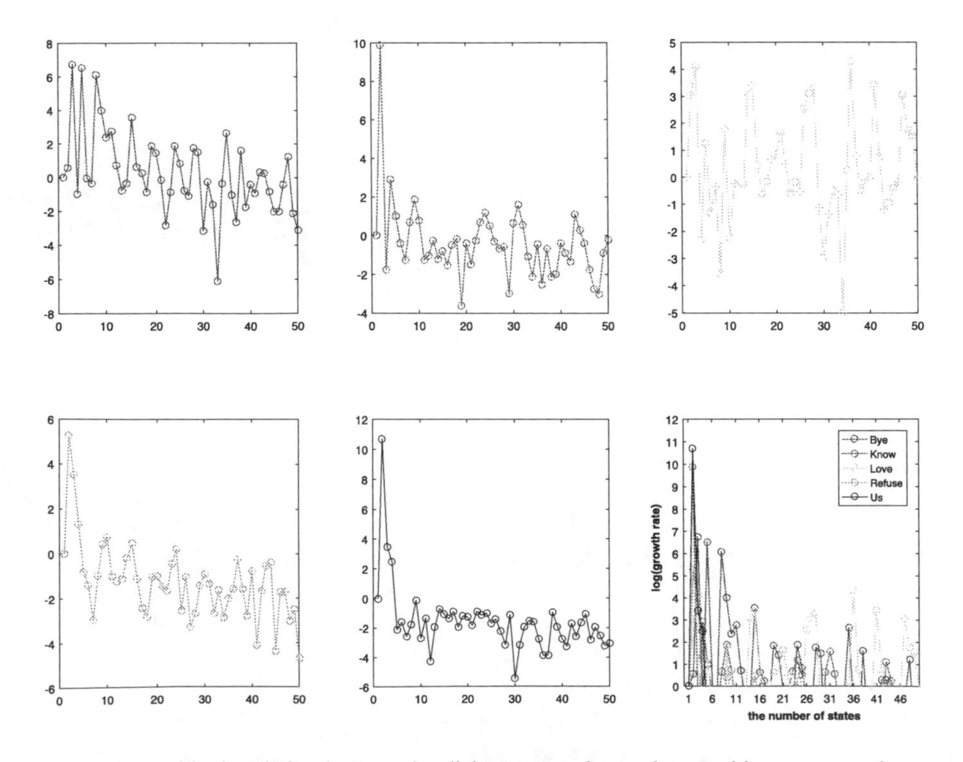

Fig. 2. Logarithmic relation between implicit state number and recognition rate growth rate

shown in Table 2, that is, to determine the number of hidden states by weighing the peak growth rate of the number of states. Finally, according to the experimental results of Fig. 3 and Fig. 4, the number of implicit states shown in Table 2 is selected.

Table 1. The number of implicit states of each gesture under method 1

	Bye	Know	Love	Refuse	Us
States	22	19	24	20	23

Table 2. The number of implicit states of each gesture under method 2

	Bye	Know	Love	Refuse	Us
States	8	5	13	4	4

In addition, the HMM model also has the following parameters A (state transition matrix), B (confusion matrix), π (initial state probability), where A denotes the probability of the implicit state from i to j, B denotes the probability of the implicit state from i to j, and π denotes the probability distribution of the initial state.

State transition matrix A:

$$A = \begin{bmatrix} 1-\delta & \delta & \ldots & \ldots & 0 \\ 0 & 1-\delta & \delta & \ldots & 0 \\ 0 & \ldots & \ldots & \ldots & \ldots \\ 0 & \ldots & \ldots & 1-\delta & \delta \end{bmatrix}$$

Among them, M is the sample size and the length of group I samples. confusion matrix B is defined as follows.

$$B = \begin{bmatrix} 1/L & 1/L & \ldots & 1/L \\ \ldots & \ldots & \ldots & \ldots \\ 1/L & 1/L & \ldots & 1/L \end{bmatrix}$$

Matrix B is a $N * L$ dimension matrix. In this paper, the 16-direction chain code is used to discretize the feature data, so L is 16. For the initial state π, Because we use the HMM model, so the $\pi = \begin{bmatrix} 1 & 0 & \ldots & 0 \end{bmatrix}^T$.

5.2 DTW Algorithm

Dynamic Time Warping (DTW) is an algorithm to compare the similarity of two time series with different lengths. It is also widely used in information retrieval, gesture recognition, speech recognition and other issues.

Suppose there are two characteristic time series:

$$A = \{a_1, a_2, \ldots, a_i, \ldots a_n\} \text{ and } B = \{b_1, b_2, \ldots, b_j, \ldots b_m\}$$

Among them, the template sequence is A, and the values of B, n and m are not necessarily the same, but the dimensions of a_i and b_j must be the same.

When the sequence length of A and B is different: Firstly, the time dimension is normalized, and then template matching and recognition are performed after aligning the sequence length of A and B. In the whole learning process of the algorithm, DTW algorithm does not need to change the time dimension of the matching sequence. The whole process is to adjust the template sequence dynamically while matching. The concrete method is to find a time distortion function under the condition of minimum distortion, and then map the time dimension data of the sequence to the time dimension of the template sequence, that is, to find the optimal path C, and the total weighted distance of any matching point on the path is the smallest.

$$C = \{c_1, c_2, \ldots, c_N\}$$

Among them, the path length is N, $c_k = (i_k, j_k)$ denotes: K matching point, which is composed of feature data of sequence A and B.

5.3 HMM-DTW Algorithm

In this paper, when using HMM algorithm to recognize gestures "Refuse" and "Known", it is found that the three-dimensional gesture trajectory is projected into two-dimensional space. The trajectories of the two gestures are very similar. Therefore, it is difficult to effectively distinguish gestures using a single HMM algorithm. Although DTW algorithm is not as effective as HMM algorithm on the whole, it can make good use of depth information to distinguish gestures. Therefore, this paper uses DTW algorithm to learn the feature information of gestures twice, that is, HMM-DTW model. The basic method is to input the depth information of dynamic gesture "Know" and "Refuse" into DTW algorithm, and then output the result after making comprehensive decision with HMM model.

6 Experimental Result

The confusion matrix derived from the HMM model is shown in the following Fig. 3 and Fig. 4. As can be seen from these two figures, the implicit state number of HMM model is determined by weighing the size of the state number and the peak growth rate shown in Fig. 4 under the second way, which is better than the overall recognition under the first way. However, the recognition effect of gesture "Know" is not ideal. The main reason is that when HMM algorithm projects the three-dimensional trajectories of two gestures into two-dimensional space when recognizing gestures "Refuse" and "Know", the trajectories of the two gestures are very similar, and the HMM model is difficult to distinguish effectively. Therefore, in the follow-up work, the HMM-DTW hybrid model is proposed to distinguish gestures more effectively.

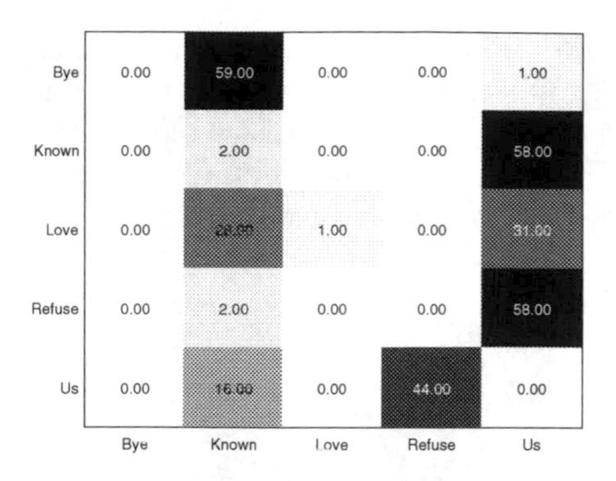

Fig. 3. Confusion matrix under method 1

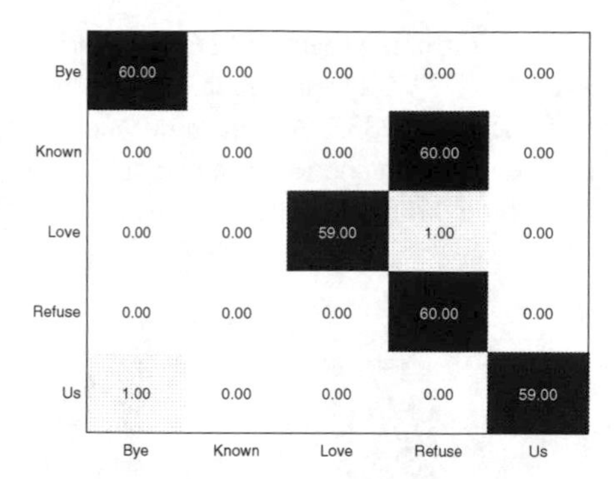

Fig. 4. Confusion matrix under method 2

The learning results and the best path of DTW algorithm of gesture "Refuse" and "Know" are shown in the following figures.

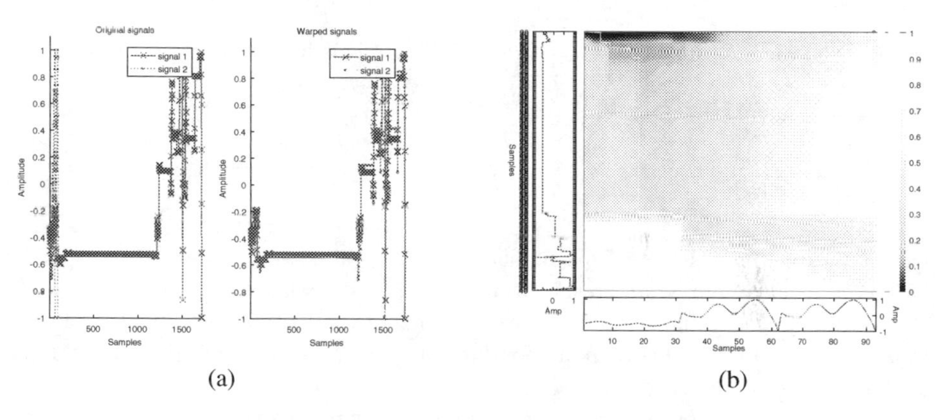

(a) (b)

Fig. 5. (a) Result of gesture "Know". (b) Path of gesture "Know"

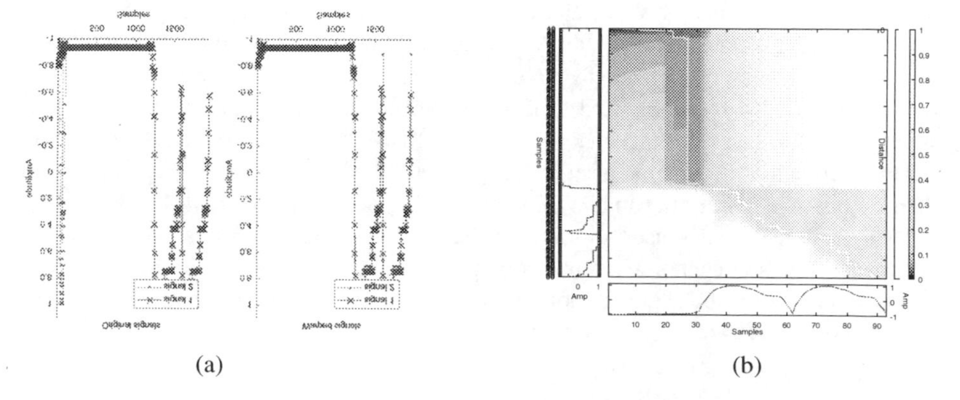

(a) (b)

Fig. 6. (a) Result of gesture "Refuse". (b) Path of gesture "Refuse"

The confusion matrix of DTW algorithm is shown in Fig. 7 and the confusion matrix of HMM-DTW hybrid model is shown in Fig. 8. According to the Fig. 8, the kappa coefficient is 0.9331. According to the first type of analysis criteria of kappa coefficient, the combination model is substantial. Therefore, for the recognition of dynamic gestures, the combination model based on HMM-DTW can solve the problem well.

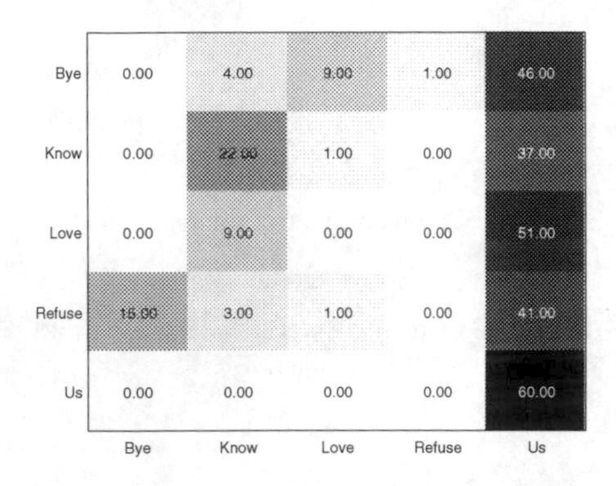

Fig. 7. Obfuscation matrix of DTW algorithm

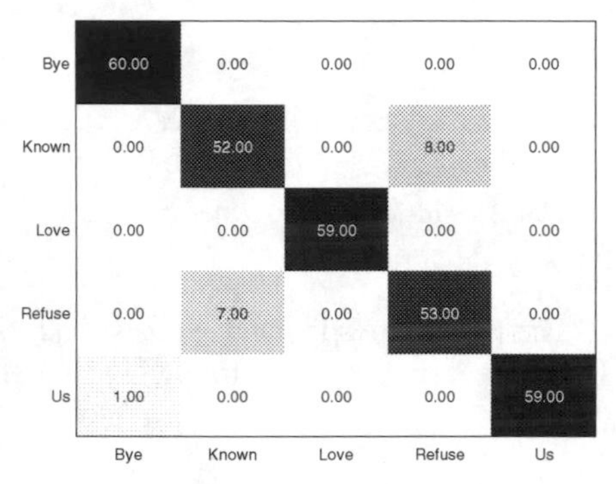

Fig. 8. Obfuscation matrix of reconstructed model

7 Conclusion

Gesture recognition is one of the human-computer interaction modes. This paper is based on the existing research results to further study the related algorithms of gesture recognition. Its main innovation is to propose a feature extraction algorithm for gesture trajectory and an HMM-DTW model for dynamic gesture recognition. From the Kappa coefficient, the model is substantial.

In feature extraction, the three-dimensional trajectory is extracted firstly, and the weighted sum of squares of the three-dimensional data is calculated as the feature data. Secondly, the three-dimensional trajectory is projected to the XOY plane in the right-hand regular coordinate system, and the direction angle between each point and the center point is extracted by using the knowledge of direction angle, then the continuous feature sequence is discretized by 16-direction chain code. In the aspect of gesture recognition, HMM model is used to learn gesture features first, and then DTW algorithm is used to learn gesture features twice, that is, HMM-DTW model. It inputs the depth information of dynamic gesture "Know" and "Refuse" into DTW algorithm, and then outputs the result after making comprehensive decision with HMM model.

The gesture samples used in this paper are all individual gestures. The proposed HMM-DTW model cannot recognize continuous gestures effectively. In the future, the work will focus on the segmentation and extraction of continuous gestures and other related work to realize the recognition of continuous gestures.

Acknowledgements. The authors would like to thank anonymous reviewers for their very detailed and helpful review. This work was supported by National Natural Science Foundation of China (61902232, 61902231), Natural Science Foundation of Guangdong Province (2019A-1515010943), Key Project of Basic and Applied Basic Research of Colleges and Universities in Guangdong Province (Natural Science) (2018KZDXM035), The Basic and Applied Basic Research of Colleges and Universities in Guangdong Province (Special Projects in Artificial Intelligence) (2019KZDZX1030).

References

1. Fujii, K., Gras, G., Salerno, A., Yang, G.Z.: Gaze gesture based human robot interaction for laparoscopic surgery. Med. Image Anal. **44**, 196 (2017)
2. Yamauchi, B.M.: PackBot: a versatile platform for military robotics. Proc. SPIE Int. Soc. Opt. Eng. **5422**, 228–237 (2004)
3. Kim, Y., Cho, S., Fong, S., Park, Y.W., Cho, K.: Design of hand gesture interaction framework on clouds for multiple users. J. Supercomput. **73**(7), 2851–2866 (2016). https://doi.org/10.1007/s11227-016-1722-y
4. Grimes, G.J.: Digital data entry glove interface device. US Patent 4,414,537, 8 November 1983
5. Zimmerman, T.G.: Optical flex sensor. US Patent 4,542,291, 17 September 1985
6. Mulder, A., Fels, S.: Sound sculpting: manipulating sound through virtual sculpting. In: Proceedings of the 1998 Western Computer Graphics Symposium, pp. 15–23 (1998)
7. Kakoty, N.M., Sharma, M.D.: Recognition of sign language alphabets and numbers based on hand kinematics using a data glove. Procedia Comput. Sci. **133**, 55–62 (2018)
8. Kuroda, T., Tabata, Y., Goto, A., et al.: Consumer price data-glove for sign language recognition. In: Proceedings of 5th International Conference on Disability, Virtual Reality Associated Technologies, Oxford, UK, pp. 253–258 (2004)
9. Dipietro, L., Sabatini, A.M., Dario, P.: A survey of glove-based systems and their applications. IEEE Trans. Syst. Man Cybern. C (Appl. Rev.) **38**(4), 461–482 (2008)
10. Starner, T., Pentland, A.: Real-time American Sign Language Recognition from Video Using Hidden Markov Models. Motion-Based Recognition, pp. 227–243. Springer, Dordrecht (1997)
11. Sarkaleh, A.K., Poorahangaryan, F., Zanj, B., et al.: A neural network based system for persian sign language recognition. In: 2009 IEEE International Conference on Signal and Image Processing Applications, pp. 145–149. IEEE (2009)
12. Nandy, A., Mondal, S., Prasad, J.S., et al.: Recognizing & interpreting indian sign language gesture for human robot interaction. In: 2010 International Conference on Computer and Communication Technology (ICCCT), pp. 712–717. IEEE (2010)
13. Lilha, H., Shivmurthy, D.: Evaluation of features for automated transcription of dual-handed sign language alphabets. In: 2011 International Conference on Image Information Processing, pp. 1–5. IEEE (2011)
14. Paquin, V., Cohen, P.: A vision-based gestural guidance interface for mobile robotic platforms. In: Sebe, N., Lew, M., Huang, Thomas S. (eds.) CVHCI 2004. LNCS, vol. 3058, pp. 39–47. Springer, Heidelberg (2004). https://doi.org/10.1007/978-3-540-24837-8_5
15. Chaudhary, A.: Robust Hand Gesture Recognition for Robotic Hand Control. Springer, Singapore (2018). https://doi.org/10.1007/978-981-10-4798-5

16. Li, Y.: Hand gesture recognition using kinect. In: 2012 IEEE International Conference on Computer Science and Automation Engineering, pp. 196–199. IEEE (2012)
17. Tang, M.: Recognizing hand gestures with microsoft's kinect. Department of Electrical Engineering of Stanford University:[sn], Palo Alto (2011)
18. Kim, Y., Kim, P.C.W., Selle, R., et al.: Experimental evaluation of contact-less hand tracking systems for tele-operation of surgical tasks. In: 2014 IEEE International Conference on Robotics and Automation (ICRA), pp. 3502–3509. IEEE (2014)
19. Yoon, H.S., Soh, J., Bae, Y.J., et al.: Hand gesture recognition using combined features of location, angle and velocity. Pattern Recogn. **34**(7), 1491–1501 (2001)

Learning Target Selection in Creating Negatively Correlated Neural Networks

Yong Liu(✉)

School of Computer Science and Engineering, The University of Aizu, Aizu-Wakamatsu, Fukushima 965-8580, Japan
yliu@u-aizu.ac.jp

Abstract. In an ensemble system with a number of learning modules, it is important for these modules to be cooperative in solving a given task. One measurement on their cooperation is to ensure that these learning modules should be negatively correlated while the expected mean squared errors of the ensemble with these negatively correlated modules should be as small as possible. This paper summarizes three different approaches to how to set target values on the given data so that negative correlation learning could be more effective in creating negatively correlated neural networks. Rather than fixing the learning targets in the whole learning process, it would be better to adaptively select different learning targets among individual networks in the ensemble system so that all the individual networks could learn to be cooperative through the interactive learning. The first two approaches specified the target values based on either the ensemble's behavior or the individual module's behavior. The third approach introduced the different targets between two randomly sampled data sets.

1 Introduction

Currently deep learning has attracted most of attentions due to its many successful applications [1–3]. Meanwhile, researchers have started to realize the importance of deep network architecture design in the applications. In practice, deep network architectures have often been set by the experts by either testing some candidate architectures, or modifying the existing useful architecture [4,5]. Besides the difficulty in designing the deep network architectures, understanding of the functions of the layers or the nodes in a deep network has become a big challenge after the deep network has grown too big [6].

As one of final goals for deep networks is to implement brain-like learning system, it would be helpful to see how the human brain architecture. Rather than a monolithic system such as a deep network, the brain is known as a modular architecture consisting of some specialized modules for vision, language, and others. If a learning system could also be organized into a set of specialized modules, the functions of these modules might provide the insight into the understanding of the learning system. Certainly these modules should be specialized and cooperative as well. One way to create such specialized modules is

K. Li et al. (Eds.): ISICA 2019, CCIS 1205, pp. 799–807, 2020.
https://doi.org/10.1007/978-981-15-5577-0_65

to let them learn the different tasks or the divided subtasks. It requires that the task decomposition is available. Besides, it would be uncertain whether these specialized modules would be cooperative. The cooperation among them should be ensured in their learning process except for the task decomposition. The other way is to let a set of modules learn how to divide the learning tasks automatically through learning so that cooperation could be built up in the same learning process.

Negative correlation learning is an ensemble learning method that trains a set of modules interactively and cooperatively on the same train data [7–10]. The learned modules could be much negatively correlated in solving some function regression tasks. It is because no restrictions were made on the output ranges for such regression tasks. However, for the classification problems, the output range is often limited between 0 and 1 in order to let the output represent the certain probabilities of falling into one class. When the output values are limited, negative correlation among the individual modules would be hard to be achieved especially if negative correlation learning would be proceeded for long.

This paper reviewed and compared three different approaches to selecting learning targets in negative correlation learning. The purpose is to see how much the individual networks should be negatively correlated in an ensemble. Although the theoretical results suggest that negative correlation in an ensemble would help, experimental results had shown that the high negative correlation had led to overfitting on some data sets. Through the comparisons among three different approaches, it is to examine how to set up learning functions properly in order to achieve the robust performance. Rather than fixing the learning targets in the whole learning process, it would be better to adaptively select different learning targets among the individual networks in the ensemble system so that all the individual networks could learn to be copperative through the interactive learning. The first two approaches specified the target values based on either the ensemble's behavior or the individual module's behavior. The third approach introduced the different targets between two randomly sampled data sets.

The next three sections explained the implementations of the aforementioned three approaches, and showed the comparison results. Section 2 describes how different targets were set in negative correlation learning. Section 3 discusses how the different targets changed both the behaviors and the performance of the learned ensembles and individual modules. Finally, Sect. 4 provides some suggestions in setting learning targets in ensemble learning based on the comparison results.

2 Three Adaptive Learning Error Functions in Negative Correlation Learning

An ensemble system combines a set of M individual learning modules in solving the given tasks. There are several combination methods for defining the output of an ensemble from these individual modules, such the weighted linear or nonlinear combinations, and the voting method. In the following discussions, a simple average of all the individual neural networks is chosen as the output $F(\mathbf{x}(n))$ for the neural network ensemble on the n-th data $(\mathbf{x}(n), y(n))$, $n = 1, \cdots, N$ in a given training set:

$$F(\mathbf{x}(n)) = \frac{1}{M}\Sigma_{i=1}^{M} F_i(\mathbf{x}(n)) \tag{1}$$

where $F_i(\mathbf{x}(n))$ is the output of the i-th neural network on $\mathbf{x}(n)$.

Negative correlation learning (NCL) trains all the individual networks to learn the data together, and be different to the other networks through the interactive learning [7]. In NCL, the learning function is normally defined as:

$$E_i(\mathbf{x}(n)) = \frac{1}{2}\left[(F_i(\mathbf{x}(n)) - y(n))^2 - \lambda(F_i(\mathbf{x}(n)) - F(\mathbf{x}(n)))^2\right] \tag{2}$$

where λ is a learning parameter in [0,1] although it could be out of this range. By minimizing $E_i(\mathbf{x}(n))$ on $(\mathbf{x}(n), y(n))$, each network would be trained to learn the object target $y(n)$, and pushed away from the ensemble output.

Gradient based learning could be applied to train all the individual networks on minimizing the error function in Eq. (2) simultaneously and interactively. The partial derivative of $E_i(\mathbf{x}(n))$ can be obtained from Eq. (2):

$$\frac{\partial E_i(\mathbf{x}(n))}{\partial F_i(\mathbf{x}(n))} = (1 - \lambda)(F_i(\mathbf{x}(n)) - y(n)) + \lambda(F(\mathbf{x}(n)) - y(n)) \tag{3}$$

At $\lambda = 1$, a simplified derivative of E_i would become

$$\frac{\partial E_i(\mathbf{x}(n))}{\partial F_i(\mathbf{x}(n))} = F(\mathbf{x}(n)) - y(n) \tag{4}$$

This simplified derivative would let all the individual networks apply the same error signals in Eq. (4) regardless which network would be trained. It would be more effective in creating the negatively correlated networks if the sign of $(F(\mathbf{x}(n)) - y(n))$ would be different to that of $(F_i(\mathbf{x}(n)) - y(n))$. In solving the function regression tasks, it would be easy to have the different signs between them. It is because that both the ensemble output and the individual output could be much larger or smaller than the object targets. However, in dealing with the classification problems, the values of $F_i(\mathbf{x}(n))$ is often limited between 0 and 1 so that they could be regarded as the certain probabilities of falling in one class. In this case, $(F(\mathbf{x}(n)) - y(n))$ and $(F_i(\mathbf{x}(n)) - y(n))$ would always have the same sign if the object target $y(n)$ would be either 0 or 1 in the case of two-class problems.

The first two learning error selections (LES) had been proposed, including LES based the ensemble (LESE) and LES based the individual (LESI). In LESI [8], the individual network in the ensemble could adaptively select three different error functions in learning at the different conditions. If the data were wrongly classified by the ensemble, all the individual networks would use the error function defined in Eq. (4). If the data were correctly classified by the ensemble, the i-th model would either choose the learning signal in Eq. (5):

$$\frac{\partial E_i(\mathbf{x}(n))}{\partial F_i(\mathbf{x}(n))} = F(\mathbf{x}(n)) - |y(n) - \beta| \tag{5}$$

or the learning signal in Eq. (6):

$$\frac{\partial E_i(\mathbf{x}(n))}{\partial F_i(\mathbf{x}(n))} = F(\mathbf{x}(n)) - |y(n) - \gamma| \tag{6}$$

In the case of $y(n)$ with 0 or 1 for the two-class classification problem, β is a small value between 0 and 0.5 while γ is between 0.5 and 1. Although the selection between Eq. (5) and Eq. (6) could be done randomly, it had actually been made based on the following measurement. If the i-th model output were closer to $y(n)$ than the ensemble output, the learning signal in Eq. (5) would be applied. Otherwise, the learning signal in Eq. (6) would be used. The idea of LESI is to let each individual network adaptively decide the targets on learning each data point based on how well it had learned the data so far.

Different to LESI, LESE would define the targets for the given data based on how well an ensemble had learned the data [9]. For the two-class classification problem with the targets 0 and 1, the n-th data $(\mathbf{x}(n), y(n))$ would be correctly classified if the absolute difference $|F(\mathbf{x}(n)) - y(n)|$ were less than a positive T between 0 and 0.5. Therefore, T could tell how well the ensemble might have learned the data. The smaller T would be, the more the data might have been learned. In LESE, if $|F(\mathbf{x}(n)) - y(n)|$ were larger than T, all the individual network would still apply the learning signal in Eq. (4). Otherwise, the following error function would be defined for the i-th network:

$$E_i(\mathbf{x}(n)) = (F_i(\mathbf{x}(n)) - 0.5)(F(\mathbf{x}(n)) - 0.5) \tag{7}$$

Therefore, with the bigger T, more individual networks would be expected by applying Eq. (7) in LESE.

For both LESI and LESE, all the given training data had been used in the learning process. Data sampling had also tested in NCL. LES by data sampling (LESD) was implemented by randomly dividing the given data into two subsets for each individual network [10]. A vector of learning control $\mathbf{c_i}$ was randomly generated for the i-th network, where $c_i(n)$ would be 1 in a probability of α or 0 otherwise. Based on $\mathbf{c_i}$, the same training data was actually randomly divided into two subsets for training the i-th network. In one subset, each data was labeled by 1. In the other subset, each data was marked by 0. Because of random generation, the number of data in the two subsets could be different. Meanwhile, different individual networks would also have the two different subsets. For each individual network, it would apply the learning signal in Eq. (4) on the subset with label 1. On the other subset with label 0, the learning signal in Eq. (7) would be used. α could be a positive value between 0 and 1 in LESD. For the smaller α, the less data would be used in learning object targets where the more data would be used to control the difference among the individual networks.

3 Comparisons Among Different Learning Targets

The data set used for comparing three different approaches is the Australian credit card assessment data set from the UCI machine learning bench-mark repository. This data set consists of total 690 examples with 14 attributes. Error rates were measured to show the performance while the overlapping rates between a pair of two individual networks were counted to show the similarities. The overlapping rate between a pair of two individual networks reflect percentage of the number of the same classifications made by the two networks. All the results were averaged from 100 runs through 10 times of 10-fold cross-validation on the card data.

3.1 Results of LESI

$\beta = 0.1$ and $\gamma = 0.6$ were used in LESI. When the number of individual networks was increased from 50 to 200 in the ensembles, the error rates of these ensembles were very close each other in both LESI and the original NCL shown in Tables 1 and 2. Although individual networks sometimes might be stopped in learning the object targets in LESI, the ensembles by LESI learned slightly worse than those by NCL on the training data. The ensembles by LESI achieved rather better results on the testing sets.

Table 1. Ensemble performance by LESI on the average error rates from 100-runs on the card data. Three ensemble architectures were tested with 50, 100, and 200 individual networks respectively. Each network has one hidden layer with 5 hidden nodes.

Structures	250		500		1000		2000	
	Training	Testing	Training	Testing	Training	Testing	Training	Testing
Ens-50-5	0.0838	0.1352	0.0698	0.1329	0.0619	0.1310	0.0570	0.1317
Ens-100-5	0.0824	0.1348	0.0702	0.1341	0.0625	0.1328	0.0577	0.1309
Ens-200-5	0.0821	0.1349	0.0699	0.1322	0.0630	0.1320	0.0577	0.1330

3.2 Results of LESE

Five values of T from 0 to 0.4 with step 0.1 were tested in LESE. At $T = 0$, LESE became the original NCL. As shown in Table 3, the larger T was in LESE, the lower the error rates the ensembles by LESE obtained. When T was increased from 0 to 0.3, the training error rates of the ensembles by LESE decreased slightly. At $T = 0.4$, the training error rates of the ensembles dropped quickly. However, it suggested that overfitting appeared in LESE at $T = 0.4$ on the testing set. Compared to the lower error rates by the ensembles, the individual networks rather had the much higher error rates listed in Table 4. The error rates of the individual networks could be near to a pure random guess with the error rate at 0.5. Difference among the individual networks certainly contributed

Table 2. Ensemble performance by NCL on the average error rates from 100-runs on the card data. Three ensemble architectures were tested with 50, 100, and 200 individual networks respectively. Each network has one hidden layer with 5 hidden nodes.

Structures	250		500		1000		2000	
	Training	Testing	Training	Testing	Training	Testing	Training	Testing
Ens-50-5	0.0702	0.1355	0.0597	0.1359	0.0525	0.1352	0.0481	0.1359
Ens-100-5	0.0696	0.1370	0.0588	0.1354	0.0522	0.1358	0.0476	0.1348
Ens-200-5	0.0699	0.1358	0.0592	0.1375	0.0521	0.1364	0.0477	0.1362

cooperations. However, too much difference dragged the performance down when the difference between each pair of two individual networks reached nearly 50% in Table 5.

Table 3. Ensemble performance by LESE on the average error rates from 100-runs on the card data. There are 50 individual networks in the ensemble. Each network has one hidden layer with 5 hidden nodes.

T	100		500		1000		2000	
	Training	Testing	Training	Testing	Training	Testing	Training	Testing
0	0.0924	0.1377	0.0597	0.1359	0.0525	0.1352	0.0481	0.1359
0.1	0.0933	0.1365	0.0564	0.1330	0.0508	0.1358	0.0460	0.1357
0.2	0.0944	0.1352	0.0552	0.1352	0.0479	0.1329	0.0426	0.1339
0.3	0.0993	0.1388	0.0546	0.1345	0.0406	0.1364	0.0322	0.1374
0.4	0.1024	0.1420	0.0483	0.1381	0.0273	0.1436	0.0145	0.1480

3.3 Results of LESD

Only the results of LESD with $\alpha = 0.5$ were compared where the results of LESD with other values of α were not included here. Compared to the original NCL in Tables 6, and 7, not only could LESD prevent the ensembles from underfitting by learning the object targets on one subset, but also avoid overfitting by learning difference on the other subset. Besides the difference at the level of the ensembles by NCL and LESD, the difference among the individuals was much bigger. It could be seen in Tables 8, and 9 that individuals by LESD had nearly 4-time high error rates as those by NCL.

Table 4. Individual performance by LESE on the average error rates from 100-runs on the card data. There are 50 individual networks in the ensemble. Each network has one hidden layer with 5 hidden nodes.

T	100		500		1000		2000	
	Training	Testing	Training	Testing	Training	Testing	Training	Testing
0	0.1431	0.1774	0.1106	0.1796	0.1033	0.1822	0.0978	0.1848
0.1	0.1841	0.2150	0.1672	0.2268	0.1600	0.2279	0.1551	0.2290
0.2	0.2509	0.2739	0.2414	0.2895	0.2359	0.2909	0.2318	0.2926
0.3	0.3328	0.3472	0.3237	0.3588	0.3179	0.3610	0.3136	0.3624
0.4	0.4228	0.4307	0.4109	0.4322	0.4050	0.4336	0.4012	0.4349

Table 5. Average overlapping rates among individual neural networks by LESE from 100-runs on the card data. There are 50 individual networks in the ensemble. Each network has one hidden layer with 5 hidden nodes.

T	100		500		1000		2000	
	Training	Testing	Training	Testing	Training	Testing	Training	Testing
0	0.8450	0.8377	0.8655	0.8345	0.8706	0.8294	0.8743	0.8252
0.1	0.7714	0.7649	0.7633	0.7419	0.7678	0.7411	0.7704	0.7395
0.2	0.6683	0.6650	0.6545	0.6413	0.6561	0.6388	0.6573	0.6375
0.3	0.5726	0.5714	0.5660	0.5595	0.5670	0.5574	0.5680	0.5562
0.4	0.5090	0.5087	0.5098	0.5077	0.5107	0.5074	0.5113	0.5074

Table 6. Ensemble performance by NCL with $\lambda = 1$ on the average error rates from 100-runs on the card data. There are 100 individual one-hidden-layer networks in the ensemble. Four different numbers of hidden nodes were tested, including 2, 5, 10, and 20.

epoch	250		500		1000		2000	
nodes	Training	Testing	Training	Testing	Training	Testing	Training	Testing
2	0.081	0.138	0.073	0.140	0.069	0.140	0.067	0.140
5	0.068	0.139	0.058	0.137	0.051	0.138	0.047	0.136
10	0.065	0.139	0.052	0.138	0.041	0.138	0.030	0.137
20	0.068	0.139	0.057	0.137	0.044	0.137	0.028	9.141

Table 7. Ensemble performance by LESD on the average error rates from 100-runs on the card data. There are 100 individual one-hidden-layer networks in the ensemble. Four different numbers of hidden nodes were tested, including 2, 5, 10, and 20.

epoch	250		500		1000		2000	
nodes	Training	Testing	Training	Testing	Training	Testing	Training	Testing
2	0.089	0.138	0.083	0.135	0.080	0.137	0.078	0.137
5	0.083	0.138	0.073	0.137	0.068	0.136	0.064	0.135
10	0.082	0.137	0.071	0.137	0.063	0.136	0.057	0.135
20	0.085	0.136	0.073	0.136	0.065	0.137	0.056	0.137

Table 8. Individual network performance by NCL with $\lambda = 1$ on the average error rates from 100-runs on the card data. There are 100 individual one-hidden-layer networks in the ensemble. Four different numbers of hidden nodes were tested, including 2, 5, 10, and 20.

epoch	250		500		1000		2000	
nodes	Training	Testing	Training	Testing	Training	Testing	Training	Testing
2	0.147	0.185	0.141	0.186	0.138	0.186	0.137	0.187
5	0.126	0.179	0.113	0.180	0.106	0.182	0.101	0.184
10	0.121	0.178	0.105	0.179	0.091	0.181	0.078	0.185
20	0.128	0.180	0.111	0.178	0.095	0.180	0.077	0.183

Table 9. Individual network performance by LESD on the average error rates from 100-runs on the card data. There are 100 individual one-hidden-layer networks in the ensemble. Four different numbers of hidden nodes were tested, including 2, 5, 10, and 20.

epoch	250		500		1000		2000	
nodes	Training	Testing	Training	Testing	Training	Testing	Training	Testing
2	0.338	0.354	0.339	0.355	0.339	0.356	0.340	0.357
5	0.339	0.355	0.340	0.358	0.342	0.360	0.343	0.361
10	0.339	0.355	0.341	0.358	0.344	0.360	0.347	0.363
20	0.340	0.356	0.341	0.358	0.344	0.361	0.347	0.364

4 Conclusions

From the comparisons among the different learning functions, it is shown that the learning functions should be adaptively adjusted in order to build up the cooperation among the individual networks in an ensemble. Individual networks could decide their own learning functions based on either their own behaviors or the ensemble's behaviors. If the data set were small, it would be better for all the individual networks to learn the whole data set. In contrast, if the data set were big, it might be better for each individual network to learn a small portion of the set. Basically each individual network would be required to just be a weak learner better than random guess. Therefore, learning for each individual network would not be hard for the big data problems. The more important thing for ensemble learning is to make all of individual modules cooperate each other in the ensemble.

References

1. Peng, H., Deng, C., Wu, Z., Liu, Y. (eds.): ISICA 2018. CCIS, vol. 986. Springer, Singapore (2019). https://doi.org/10.1007/978-981-13-6473-0
2. Liu, Y., Wang, L., Zhao, L., Yu, Z. (eds.): ICNC-FSKD 2019: Volume 1. AISC, vol. 1074. Springer, Cham (2020). https://doi.org/10.1007/978-3-030-32456-8
3. Liu, Y., Wang, L., Zhao, L., Yu, Z. (eds.): ICNC-FSKD 2019: Volume 2. AISC, vol. 1075. Springer, Cham (2020). https://doi.org/10.1007/978-3-030-32591-6
4. Li, K., Li, W., Chen, Z., Liu, Y. (eds.): ISICA 2017. CCIS, vol. 873. Springer, Singapore (2018). https://doi.org/10.1007/978-981-13-1648-7
5. Li, K., Li, W., Chen, Z., Liu, Y. (eds.): ISICA 2017. CCIS, vol. 874. Springer, Singapore (2018). https://doi.org/10.1007/978-981-13-1651-7
6. Bengio, Y., Courville, A., Vincent, P.: Representation learning: a review and new perspectives. IEEE Trans. Pattern Anal. Mach. Intell. (PAMI) **35**(8), 1798–1828 (2013)
7. Liu, Y., Yao, X.: Simultaneous training of negatively correlated neural networks in an ensemble. IEEE Trans. Syst. Man Cybern. Part B: Cybern. **29**(6), 716–725 (1999)
8. Liu, Y.: Learning targets for building cooperation awareness in ensemble learning. In: the 9th IEEE International Conference on Awareness Science and Technology, 4 p. IEEE (2018)
9. Liu, Y.: Awareness learning for balancing performance and diversity in neural network ensembles. In: Liu, Y., Wang, L., Zhao, L., Yu, Z. (eds.) ICNC-FSKD 2019. AISC, vol. 1074, pp. 113–120. Springer, Cham (2020). https://doi.org/10.1007/978-3-030-32456-8_12
10. Liu, Y.: Computational awareness for learning neural network ensembles. In: The 2017 IEEE International Conference on Information and Automation (2017)

Author Index

Printed in the United States
By Bookmasters